# Numerical Analysis
A Comprehensive Introduction

# Numerical Analysis
## A Comprehensive Introduction

**H. R. Schwarz**
*University of Zürich*
*Switzerland*

with a contribution by

**J. Waldvogel**
*Swiss Federal Institute of Technology,*
*Zürich*

JOHN WILEY & SONS
Chichester · New York · Brisbane · Toronto · Singapore

Copyright © 1989 by John Wiley & Sons Ltd.

First published as *Numerische Mathematik*, 2 Aufl.,
by H. R. Schwarz, © 1988. This edition is published
by permission of Verlag B. G. Teubner, Stuttgart,
and is the sole authorized English translation of the
original German edition.

All rights reserved.

No part of this book may be reproduced by any means,
or transmitted, or translated into a machine language
without the written permission of the publisher.

*Library of Congress Cataloging-in-Publication Data:*
Schwarz, Hans Rudolf
  [Numerische Mathematik. English]
  Numerical analysis: a comprehensive introduction/H.R. Schwarz:
with a contribution by J. Waldvogel.
     p. cm.
  Translation of: Numerische Mathematik.
  Bibliography: p.
  Includes index.
  ISBN 0 471 92064 9
  1. Numerical analysis.  I. Waldvogel, Jörg.  II. Title.
QA297.S38513 1989                                    89–5538
519.4—dc19                                              CIP

*British Library Cataloguing in Publication Data:*
Schwarz, H.R.
  Numerical analysis.
  1. Numerical analysis
  I. Title
  519
  ISBN 0 471 92064 9

Typeset by Thomson Press (India) Ltd, New Delhi, India.
Printed and Bound in Great Britain at the Anchor Press, Tiptree Essex.

# Preface

This book arose from the explicit wish of my esteemed teacher, Professor Dr E. Stiefel, who entrusted me with the legacy of rewriting his comprehensive textbook (Stiefel, 1976) and adapting it to modern developments and requirements. Professor Stiefel's greatest concern was always to give a clear and detailed presentation. I have attempted to follow his personal philosophy in writing this introductory textbook, by presenting the basic methods of numerical analysis in detail.

The book resulted from four terms of lecture courses given at the University of Zürich each consisting of four hours per week. The intention was to enable the students to solve problems successfully in applied mathematics using numerical methods, or at least to provide them with a foundation to enable them to study more advanced and specialized publications. The book is addressed to mathematicians, physicists, engineers, students of information studies and of natural sciences. It assumes the basic mathematical knowledge taught in undergraduate courses at universities or colleges of technology.

The topics represented are strongly algorithmic, to take into account the fact that personal computers are now in widespread use. The numerical methods are introduced by establishing the necessary theoretical background. The algorithmic description is given in a form which can easily be translated into the preferred programming language of the reader, thus making the computer implementation of a method straightforward. The user is thus offered the opportunity to study the performance of the methods by completing the algorithms with appropriate input and output statements under the necessary conditions. Additional problems can be found, for instance, in Collatz and Albrecht (1971) and Hainer (1983).

My colleague Professor Dr J. Waldvogel at the ETH Zürich, who established the special results in the field of numerical quadrature was kind enough to contribute Sections 8.1 and 8.2 with the corresponding problems. I thank him for the valuable contribution. I also want to thank my assistants Dipl.-Math. W. Businger and Dr H. P. Märchy for their critical review of the manuscript and for their stimulating suggestions which improved the presentation of the book. Finally I want to thank the publishers B. G. Teubner for publishing the book and for their courteous cooperation.

Zürich, Summer 1985                                             H. R. Schwarz

# Preface to the English Edition

The English text essentially corresponds to the second edition of the original book, in which the known printing errors have been eliminated and the formulation has been improved where necessary. The English manuscript was prepared by myself, but I am indebted to Dipl.-Math. W. Businger for his valuable help in correcting the text and suggesting better wording. I also want to thank Mrs U. Henauer for typing and preparing the manuscript. I want to express my thanks to Mr J. M. Taylor and Mr J. F. Blowey of the University of Sussex for improving the readability of the text. Finally I want to thank John Wiley and Sons Limited for considering this book as part of their series of publications.

Zürich, Spring 1988                                                 H. R. Schwarz

# Contents

**Preface** . . . . . . . . . . . . . . . . . . . . . . . . . . v

**Preface to the English Edition** . . . . . . . . . . . . . vii

**1 Systems of Linear Equations** . . . . . . . . . . . . . . 1
   1.1 Gaussian algorithm . . . . . . . . . . . . . . . . . . . . 1
      1.1.1 The fundamental process . . . . . . . . . . . . . 1
      1.1.2 Pivotal strategies. . . . . . . . . . . . . . . . . . . 9
      1.1.3 Supplements . . . . . . . . . . . . . . . . . . . . 16
   1.2 Accuracy, error estimates . . . . . . . . . . . . . . . . 20
      1.2.1 Norms . . . . . . . . . . . . . . . . . . . . . . . 20
      1.2.2 Error estimates, condition . . . . . . . . . . . . . 25
   1.3 Systems with special properties . . . . . . . . . . . . . 29
      1.3.1 Symmetric, positive definite systems . . . . . . . . 30
      1.3.2 Banded systems . . . . . . . . . . . . . . . . . . 35
      1.3.3 Tridiagonal systems. . . . . . . . . . . . . . . . . 37
   1.4 Exchange step and inversion of matrices . . . . . . . . . 41
      1.4.1 Linear functions, exchange . . . . . . . . . . . . 42
      1.4.2 Matrix inversion . . . . . . . . . . . . . . . . . . 44
   1.5 Exercises . . . . . . . . . . . . . . . . . . . . . . . . . 46

**2 Linear Programming** . . . . . . . . . . . . . . . . . . . 49
   2.1 Introductory examples, graphical solution . . . . . . . . 49
   2.2 The simplex algorithm . . . . . . . . . . . . . . . . . . 54
   2.3 Supplements to the simplex algorithm. . . . . . . . . . . 63
      2.3.1 Degeneracy. . . . . . . . . . . . . . . . . . . . . 63
      2.3.2 Nonunique solution. . . . . . . . . . . . . . . . . 66
      2.3.3 Unbounded objective function . . . . . . . . . . . 67
   2.4 General linear programs . . . . . . . . . . . . . . . . . 68
      2.4.1 Treatment of free variables . . . . . . . . . . . . 68
      2.4.2 Coordinate shift method . . . . . . . . . . . . . . 69

|     |       |                                                              |     |
| --- | ----- | ------------------------------------------------------------ | --- |
|     | 2.4.3 | The two-phase method                                         | 73  |
| 2.5 |       | Discrete Chebyshev approximation.                            | 77  |
| 2.6 |       | Exercises.                                                   | 83  |

# 3 Interpolation . . . . . . . . . . . . . . . . . . . 85
- 3.1 Existence and uniqueness of polynomial interpolation . . . . . 85
- 3.2 Lagrange interpolation . . . . . . . . . . . . . . . . . 86
  - 3.2.1 Technique of computation . . . . . . . . . . . 86
  - 3.2.2 Applications . . . . . . . . . . . . . . . . 90
- 3.3 Error estimates. . . . . . . . . . . . . . . . . . . 96
- 3.4 Newton interpolation . . . . . . . . . . . . . . . . 100
- 3.5 Aitken–Neville interpolation . . . . . . . . . . . . . . 108
  - 3.5.1 Aitken's and Neville's algorithms . . . . . . . . . 109
  - 3.5.2 Extrapolation and the Romberg scheme . . . . . . . 111
  - 3.5.3 Inverse interpolation . . . . . . . . . . . . . 114
- 3.6 Rational interpolation . . . . . . . . . . . . . . . . 116
  - 3.6.1 Formulation of the problem, difficulties. . . . . . . . 116
  - 3.6.2 Special interpolation problem, Thiele's continued fraction . . 118
- 3.7 Spline interpolation . . . . . . . . . . . . . . . . . 125
  - 3.7.1 Characterization of the spline function . . . . . . . . 126
  - 3.7.2 Computation of the cubic spline function . . . . . . . 128
  - 3.7.3 General cubic spline functions . . . . . . . . . . 133
  - 3.7.4 Periodic cubic spline interpolation . . . . . . . . . 136
  - 3.7.5 Smooth two-dimensional curves . . . . . . . . . . 139
- 3.8 Exercises . . . . . . . . . . . . . . . . . . . . . 141

# 4 Approximation of Functions . . . . . . . . . . . . . . . 143
- 4.1 Fourier series . . . . . . . . . . . . . . . . . . . 143
- 4.2 Efficient evaluation of Fourier coefficients . . . . . . . . . 155
  - 4.2.1 Runge's algorithm . . . . . . . . . . . . . . 155
  - 4.2.2 The fast Fourier transform . . . . . . . . . . . 159
- 4.3 Orthogonal polynomials . . . . . . . . . . . . . . . 169
  - 4.3.1 The Chebyshev polynomials . . . . . . . . . . . 169
  - 4.3.2 Chebyshev interpolation . . . . . . . . . . . . 178
  - 4.3.3 Legendre polynomials . . . . . . . . . . . . . 182
- 4.4 Exercises . . . . . . . . . . . . . . . . . . . . . 188

# 5 Nonlinear Equations . . . . . . . . . . . . . . . . . 191
- 5.1 The Banach fixed point theorem. . . . . . . . . . . . . 191
- 5.2 Behaviour and order of convergence . . . . . . . . . . . 195
- 5.3 Equations in one unknown . . . . . . . . . . . . . . 203
  - 5.3.1 Bisection, regula falsi, secant method. . . . . . . . . 203
  - 5.3.2 Newton's method . . . . . . . . . . . . . . 209
  - 5.3.3 Interpolation methods. . . . . . . . . . . . . 213

| | | |
|---|---|---|
| 5.4 | Equations in several unknowns | 216 |
| | 5.4.1 Fixed point iteration and convergence | 216 |
| | 5.4.2 Newton's method | 218 |
| 5.5 | Zeros of polynomials. | 224 |
| 5.6 | Exercises | 235 |

## 6 Eigenvalue Problems — 238

| | | |
|---|---|---|
| 6.1 | The characteristic polynomial, difficulties. | 238 |
| 6.2 | Jacobi methods. | 241 |
| | 6.2.1 Elementary rotations | 242 |
| | 6.2.2 The classical Jacobi method | 244 |
| | 6.2.3 Cyclic Jacobi method | 250 |
| 6.3 | Transformation methods | 253 |
| | 6.3.1 Transformation into Hessenberg form | 254 |
| | 6.3.2 Transformation into tridiagonal form | 258 |
| | 6.3.3 Fast Givens transformation | 260 |
| | 6.3.4 Hyman's method. | 264 |
| 6.4 | QR algorithm | 269 |
| | 6.4.1 Fundamentals of the QR transformation | 270 |
| | 6.4.2 Practical implementation, real eigenvalues | 275 |
| | 6.4.3 QR double step, complex eigenvalues | 280 |
| | 6.4.4 QR algorithm for tridiagonal matrices | 286 |
| | 6.4.5 Computation of eigenvectors | 290 |
| 6.5 | Exercises | 291 |

## 7 Method of Least Squares — 294

| | | |
|---|---|---|
| 7.1 | Linear problems, normal equations. | 294 |
| 7.2 | Methods of orthogonal transformation | 299 |
| | 7.2.1 Givens transformation. | 300 |
| | 7.2.2 Special computational techniques | 306 |
| | 7.2.3 Householder transformation. | 309 |
| 7.3 | Singular value decomposition. | 315 |
| 7.4 | Nonlinear problems | 319 |
| | 7.4.1 The Gauss–Newton method. | 320 |
| | 7.4.2 Minimization methods. | 324 |
| 7.5 | Exercises | 328 |

## 8 Numerical Quadrature — 330

| | | |
|---|---|---|
| 8.1 | The trapezoidal method. | 331 |
| | 8.1.1 Problem and notation. | 331 |
| | 8.1.2 Definition of the trapezoidal method and improvements | 331 |
| | 8.1.3 The Euler–MacLaurin formula. | 335 |
| | 8.1.4 The Romberg procedure | 337 |
| | 8.1.5 Adaptive quadrature | 339 |

|     |     |
| --- | --- |
| 8.2 Transformation methods | 342 |
|     8.2.1 Periodic integrands | 343 |
|     8.2.2 Integrals over **R** | 345 |
|     8.2.3 Transformation methods | 347 |
| 8.3 Interpolation quadrature formulae | 350 |
|     8.3.1 Newton–Cotes quadrature formulae | 351 |
|     8.3.2 Spline quadrature formulae | 359 |
| 8.4 Gaussian quadrature formulae | 361 |
| 8.5 Exercises | 368 |

## 9 Ordinary Differential Equations — 371

| | |
| --- | --- |
| 9.1 Single step methods | 371 |
|     9.1.1 Euler and the Taylor series method | 371 |
|     9.1.2 Discretization errors, order of convergence | 375 |
|     9.1.3 Improved polygonal method, trapezoidal method, Heun's method | 379 |
|     9.1.4 Runge–Kutta methods | 384 |
|     9.1.5 Implicit Runge–Kutta methods | 392 |
|     9.1.6 Differential equations of higher order and systems | 395 |
| 9.2 Multistep methods | 398 |
|     9.2.1 Adams–Bashforth methods | 398 |
|     9.2.2 Adams–Moulton methods | 401 |
|     9.2.3 General linear multistep methods | 404 |
| 9.3 Stability | 413 |
|     9.3.1 Inherent instability | 413 |
|     9.3.2 Absolute stability | 415 |
|     9.3.3 Stiff differential equations | 422 |
| 9.4 Exercises | 427 |

## 10 Partial Differential Equations — 430

| | |
| --- | --- |
| 10.1 Elliptic boundary value problems, finite differences | 430 |
|     10.1.1 Formulation of the problem | 430 |
|     10.1.2 Discretization of the problem | 432 |
|     10.1.3 Grid points near the boundary, general boundary conditions | 439 |
|     10.1.4 Discretization errors | 451 |
|     10.1.5 Supplements | 464 |
| 10.2 Parabolic initial boundary value problems | 468 |
|     10.2.1 One-dimensional problems, explicit method | 468 |
|     10.2.2 One-dimensional problems, implicit method | 474 |
|     10.2.3 Diffusion equation with variable coefficients | 479 |
|     10.2.4 Two-dimensional problems | 482 |
| 10.3 Finite element method | 487 |
|     10.3.1 Fundamentals | 487 |
|     10.3.2 Principle of the finite element method | 490 |
|     10.3.3 Elementwise treatment | 492 |

*Contents* xiii

    10.3.4 Compilation and solution of the linear equations . . . . . 499
    10.3.5 Examples . . . . . . . . . . . . . . . . . . 500
  10.4 Exercises . . . . . . . . . . . . . . . . . . . . . 503

**References** . . . . . . . . . . . . . . . . . . . 507

**Index** . . . . . . . . . . . . . . . . . . . . . 513

# 1

# Systems of Linear Equations

The numerical solution of systems of linear equations plays an essential role in solving many problems of applied mathematics. In this chapter we show appropriate computational methods to solve the system of linear equations, taking computational errors into account. To begin with we develop a general algorithm to solve the system when the equations are linearly independent. Then we consider systems of equations with special properties, which frequently occur in applications and allow the algorithm to be simplified.

## 1.1 Gaussian Algorithm

### 1.1.1 The fundamental process

Given a system of $n$ linear equations

$$\sum_{k=1}^{n} a_{ik} x_k + b_i = 0 \quad (i = 1, 2, \ldots, n) \tag{1.1}$$

with $a_{ik}$ and $b_i$ real, $i, k = 1, 2, \ldots, n$, we seek the unknowns $x_k$. The system (1.1) can be abbreviated as

$$\mathbf{Ax} + \mathbf{b} = \mathbf{0} \tag{1.2}$$

where the $\mathbf{A}$ is an $n \times n$ matrix, $\mathbf{b}$ a constant vector and $\mathbf{x}$ the vector of the unknowns. In the following we assume that the coefficient matrix $\mathbf{A}$ is nonsingular, so that the existence and uniqueness of the solution, $\mathbf{x}$ of (1.2), is assured (Bunse and Bunse-Gerstner, 1984).

To simplify the notation, with a view to computer implementation, we use a schematic representation for the given equations (1.1), which for the case $n = 4$

has the following self-explanatory form (1.3)

| $x_1$ | $x_2$ | $x_3$ | $x_4$ | 1 |
|---|---|---|---|---|
| $a_{11}$ | $a_{12}$ | $a_{13}$ | $a_{14}$ | $b_1$ |
| $a_{21}$ | $a_{22}$ | $a_{23}$ | $a_{24}$ | $b_2$ |
| $a_{31}$ | $a_{32}$ | $a_{33}$ | $a_{34}$ | $b_3$ |
| $a_{41}$ | $a_{42}$ | $a_{43}$ | $a_{44}$ | $b_4$ |

$A_{2i} - A_{2i}\left(\dfrac{A_{ii}}{A_{ii}}\right) = 0$

(1.3)

In the Table (1.3) the coefficients $a_{ik}$ of the matrix **A** and the components of the vector **b** appear. The following three elementary operations may be applied to the scheme (1.3) without essentially changing the given system of equations (1.1):

(1) Permutation of rows;
(2) Multiplication of a complete row by a number $\neq 0$;
(3) Addition of a multiple of a row to a different one.

On the assumption $a_{11} \neq 0$, we subtract the multiple $(a_{i1}/a_{11})$ of the first row from the $i$th row for every $i \geq 2$. The scheme thus results in

| $x_1$ | $x_2$ | $x_3$ | $x_4$ | 1 |
|---|---|---|---|---|
| $a_{11}$ | $a_{12}$ | $a_{13}$ | $a_{14}$ | $b_1$ |
| 0 | $a_{22}^{(1)}$ | $a_{23}^{(1)}$ | $a_{24}^{(1)}$ | $b_2^{(1)}$ |
| 0 | $a_{32}^{(1)}$ | $a_{33}^{(1)}$ | $a_{34}^{(1)}$ | $b_3^{(1)}$ |
| 0 | $a_{42}^{(1)}$ | $a_{43}^{(1)}$ | $a_{44}^{(1)}$ | $b_4^{(1)}$ |

STEP 1 Divide $\dfrac{A_{i1}}{A_{11}}$

STEP 2

(1.4)

with the quotients

$$l_{i1} = a_{i1}/a_{11} \quad (i = 2, 3, \ldots, n) \tag{1.5}$$

and the elements in (1.4) are given by the formulae

$$a_{ik}^{(1)} = a_{ik} - l_{i1} a_{1k} \quad (i, k = 2, 3, \ldots, n) \tag{1.6}$$

$$b_i^{(1)} = b_i - l_{i1} b_1 \quad (i = 2, 3, \ldots, n). \tag{1.7}$$

The new scheme, (1.4), is equivalent to the system of equations (1.1). Only the first equation contains the unknown $x_1$, which can be expressed in terms of the other unknowns as follows

$$x_1 = -\left(\sum_{k=2}^{n} a_{1k} x_k + b_1\right) \bigg/ a_{11}. \tag{1.8}$$

Moreover, the scheme (1.4) contains, in general, a reduced system of $(n-1)$ equations for the unknowns $x_2, x_3, \ldots, x_n$. The transition from (1.3) to (1.4)

## 1.1 Gaussian Algorithm

corresponds to an *elimination step*, by which the solution of (1.1), with $n$ unknowns, is reduced to the solution of a smaller system with $(n-1)$ unknowns. We treat this reduced system in the same manner as the first step, whereby the first row in (1.4) remains unchanged. This row is, therefore, denoted as the first *final equation*, and $a_{11}$ is called the *pivotal element*.

With the additional assumption that $a_{22}^{(1)} \neq 0$, as the result of applying a second elimination step, we obtain from (1.4)

$$
\begin{array}{|cccc|c|}
\hline
x_1 & x_2 & x_3 & x_4 & 1 \\
\hline
a_{11} & a_{12} & a_{13} & a_{14} & b_1 \\
0 & a_{22}^{(1)} & a_{23}^{(1)} & a_{24}^{(1)} & b_2^{(1)} \\
0 & 0 & a_{33}^{(2)} & a_{34}^{(2)} & b_3^{(2)} \\
0 & 0 & a_{43}^{(2)} & a_{44}^{(2)} & b_4^{(2)} \\
\hline
\end{array}
\tag{1.9}
$$

With the auxiliary quantities

$$l_{i2} = a_{i2}^{(1)}/a_{22}^{(1)} \quad (i = 3, 4, \ldots, n) \tag{1.10}$$

the new elements in (1.9) are given by

$$a_{ik}^{(2)} = a_{ik}^{(1)} - l_{i2} a_{2k}^{(1)} \quad (i, k = 3, 4, \ldots, n) \tag{1.11}$$

$$b_i^{(2)} = b_i^{(1)} - l_{i2} b_2^{(1)} \quad (i = 3, 4, \ldots, n). \tag{1.12}$$

The scheme (1.9) contains the second final equation for $x_2$

$$x_2 = -\left(\sum_{k=3}^{n} a_{2k}^{(1)} x_k + b_2^{(1)}\right) \Big/ a_{22}^{(1)}. \tag{1.13}$$

By continuing the elimination process, we end up after $(n-1)$ steps with a scheme that contains only final equations. We define the quantities

$$a_{ik}^{(0)} := a_{ik} \quad (i, k = 1, 2, \ldots, n); \quad b_i^{(0)} := b_i \tag{1.14}$$

$$\left.\begin{array}{l} r_{ik} := a_{ik}^{(i-1)} \quad (k = i, i+1, \ldots, n) \\ c_i := b_i^{(i-1)} \end{array}\right\} \quad (i = 1, 2, \ldots, n) \tag{1.15}$$

Hence, the scheme of final equations reads

$$
\begin{array}{|cccc|c|}
\hline
x_1 & x_2 & x_3 & x_4 & 1 \\
\hline
r_{11} & r_{12} & r_{13} & r_{14} & c_1 \\
0 & r_{22} & r_{23} & r_{24} & c_2 \\
0 & 0 & r_{33} & r_{34} & c_3 \\
0 & 0 & 0 & r_{44} & c_4 \\
\hline
\end{array}
\tag{1.16}
$$

In general, the unknowns can be computed from (1.16) in the order $x_n$, $x_{n-1}, \ldots, x_2, x_1$ by means of the formula

$$x_i = -\left(\sum_{k=i+1}^{n} r_{ik} x_k + c_i\right) \Big/ r_{ii} \quad (i = n, n-1, \ldots, 2, 1) \tag{1.17}$$

The process described by equation (1.17) is called *back substitution*, because the final equations are used in the opposite order of their generation.

This computational procedure comprises the main elements of the *Gaussian algorithm*. It reduces the given system of linear equations $\mathbf{Ax} + \mathbf{b} = \mathbf{0}$ to a system $\mathbf{Rx} + \mathbf{c} = \mathbf{0}$ corresponding to (1.16), where $\mathbf{R}$ represents a *right triangular matrix* or *upper triangular matrix*

$$\mathbf{R} = \begin{pmatrix} r_{11} & r_{12} & r_{13} & \cdots & r_{1n} \\ 0 & r_{22} & r_{23} & \cdots & r_{2n} \\ 0 & 0 & r_{33} & \cdots & r_{3n} \\ \vdots & \vdots & \vdots & & \vdots \\ 0 & 0 & 0 & \cdots & r_{nn} \end{pmatrix} \tag{1.18}$$

where the diagonal elements $r_{ii}$ are not zero. The unknowns can immediately be derived from this final system.

**Example 1.1** Let us solve the system of equations

$$2x_1 + 3x_2 - 5x_3 + 10 = 0$$
$$4x_1 + 8x_2 - 3x_3 + 19 = 0$$
$$-6x_1 + x_2 + 4x_3 + 11 = 0$$

The Gaussian algorithm yields in two elimination steps

| $x_1$ | $x_2$ | $x_3$ | 1 |
|---|---|---|---|
| 2 | 3 | −5 | 10 |
| 4 | 8 | −3 | 19 |
| −6 | 1 | 4 | 11 |

| $x_1$ | $x_2$ | $x_3$ | 1 |
|---|---|---|---|
| 2 | 3 | −5 | 10 |
| 0 | 2 | 7 | −1 |
| 0 | 10 | −11 | 41 |

| $x_1$ | $x_2$ | $x_3$ | 1 |
|---|---|---|---|
| 2 | 3 | −5 | 10 |
| 0 | 2 | 7 | −1 |
| 0 | 0 | −46 | 46 |

The back substitution yields the unknowns $x_3 = 1$, $x_2 = -(7-1)/2 = -3$, $x_1 = -(-9-5+10)/2 = 2$.

When we implement the Gaussian algorithm on a computer, the numerical values of the successive schemes are stored in an array, so that at the end of the process the system of the final equations (1.16) is available. In practice it does not make much sense to actually fill the array below the diagonal with zeros. Rather, it is advisable to store the values of the quotients $l_{ik}$ at the appropriate places, as

## 1.1 Gaussian Algorithm

will be seen below. Thus instead of (1.16) we obtain the following scheme (1.19), in which a staircase line has been drawn between the $l$ and $r$ values in order to distinguish them clearly.

$$
\begin{array}{|cccc|c}
\hline
x_1 & x_2 & x_3 & x_4 & 1 \\
\hline
r_{11} & r_{12} & r_{13} & r_{14} & c_1 \\
l_{21} & r_{22} & r_{23} & r_{24} & c_2 \\
l_{31} & l_{32} & r_{33} & r_{34} & c_3 \\
l_{41} & l_{42} & l_{43} & r_{44} & c_4 \\
\hline
\end{array}
\tag{1.19}
$$

Now we aim to show the relationships that hold between the quantities in the scheme (1.19) and the coefficients given in (1.3). It is clear that we have to take into account the generation of the final values in (1.19). The value $r_{ik}$ in the $i$th row with $i \geqslant 2$ and $k \geqslant i$ is generated after $(i-1)$ elimination steps using the relationships (1.6), (1.11) and so on, and finally (1.15)

$$
\begin{aligned}
r_{ik} = a_{ik}^{(i-1)} &= a_{ik} - l_{i1}a_{1k}^{(0)} - l_{i2}a_{2k}^{(1)} - \cdots - l_{i,i-1}a_{i-1,k}^{(i-2)} \\
&= a_{ik} - l_{i1}r_{1k} - l_{i2}r_{2k} - \cdots - l_{i,i-1}r_{i-1,k} \quad (i \geqslant 2, k \geqslant i).
\end{aligned}
$$

From this we obtain the relationship

$$
a_{ik} = \sum_{j=1}^{i-1} l_{ij} r_{jk} + r_{ik}, \quad (k \geqslant i \geqslant 1), \tag{1.20}
$$

which also holds for $i=1$, since the sum is void. The value $l_{ik}$ in the $k$th column, with $k \geqslant 2$ and $i > k$, is obtained in the $k$th elimination step from $a_{ik}^{(k-1)}$ from

$$
\begin{aligned}
l_{ik} = a_{ik}^{(k-1)}/a_{kk}^{(k-1)} &= (a_{ik} - l_{i1}a_{1k}^{(0)} - l_{i2}a_{2k}^{(1)} - \cdots - l_{i,k-1}a_{k-1,k}^{(k-2)})/a_{kk}^{(k-1)} \\
&= (a_{ik} - l_{i1}r_{1k} - l_{i2}r_{2k} - \cdots - l_{i,k-1}r_{k-1,k})/r_{kk}.
\end{aligned}
$$

This equation is solved for $a_{ik}$ to yield the relation

$$
a_{ik} = \sum_{j=1}^{k} l_{ij} r_{jk} \quad (i > k \geqslant 1), \tag{1.21}
$$

which also holds for $k=1$ due to (1.5). The relationships (1.20) and (1.21) remind us of the rules of matrix multiplication. In addition to the right triangular matrix $\mathbf{R}$, (1.18), we define the *left triangular matrix* or *lower triangular matrix* $\mathbf{L}$ with ones on the diagonal

$$
\mathbf{L} = \begin{pmatrix}
1 & 0 & 0 & \cdots & 0 \\
l_{21} & 1 & 0 & \cdots & 0 \\
l_{31} & l_{32} & 1 & \cdots & 0 \\
\vdots & \vdots & \vdots & & \vdots \\
l_{n1} & l_{n2} & l_{n3} & \cdots & 1
\end{pmatrix}. \tag{1.22}
$$

With this definition the relationships (1.20) and (1.21) are equivalent to the matrix equation

$$\mathbf{A} = \mathbf{LR}. \tag{1.23}$$

*Theorem 1.1* *With the assumption that the pivotal elements $a_{11}, a_{22}^{(1)}, a_{33}^{(2)},\ldots$ are nonzero the Gaussian algorithm decomposes a nonsingular matrix $\mathbf{A}$ into the product of a lower triangular matrix $\mathbf{L}$ (1.22), and a right triangular matrix $\mathbf{R}$, (1.18).*

Finally the values $c_i$ with $i \geqslant 2$ are derived according to (1.7), (1.12) and (1.15) from

$$\begin{aligned} c_i &= b_i^{(i-1)} = b_i - l_{i1}b_1 - l_{i2}b_2^{(1)} - \cdots - l_{i,i-1}b_{i-1}^{(i-2)} \\ &= b_i - l_{i1}c_1 - l_{i2}c_2 - \cdots - l_{i,i-1}c_{i-1} \quad (i \geqslant 2). \end{aligned}$$

This gives the following relationships

$$b_i = \sum_{j=1}^{i-1} l_{ij}c_j + c_i \quad (i = 1, 2, \ldots, n), \tag{1.24}$$

which can be combined with the lower triangular matrix $\mathbf{L}$ as

$$\mathbf{Lc} - \mathbf{b} = \mathbf{0}. \tag{1.25}$$

Hence the vector $\mathbf{c}$ of the scheme (1.16) can be interpreted as the solution of the system (1.25). Since $\mathbf{L}$ is a left triangular matrix, the unknowns $c_i$ are computed in the order $c_1, c_2, \ldots, c_n$ by means of the process of *forward substitution* according to

$$c_i = b_i - \sum_{j=1}^{i-1} l_{ij}c_j \quad (i = 1, 2, \ldots, n). \tag{1.26}$$

As a result of this observation the solution $\mathbf{x}$ of a system of linear equations $\mathbf{Ax} + \mathbf{b} = \mathbf{0}$ can be computed by means of the Gaussian algorithm in three steps:

$$\begin{array}{ll} 1. \ \mathbf{A} = \mathbf{LR} & \text{(decomposition of } \mathbf{A}) \\ 2. \ \mathbf{Lc} - \mathbf{b} = \mathbf{0} & \text{(forward substitution } \rightarrow \mathbf{c}) \\ 3. \ \mathbf{Rx} + \mathbf{c} = \mathbf{0} & \text{(backward substitution } \rightarrow \mathbf{x}) \end{array} \tag{1.27}$$

The procedure (1.27) is a formal algorithm. It is based on the assumption that all pivotal elements $a_{11}, a_{22}^{(1)}, a_{33}^{(2)}, \ldots$ are nonzero. We now want to show, that this assumption can be fulfilled by appropriate permutations of rows, which, if necessary, are performed, before each elimination step. In this way the procedure becomes theoretically feasible.

*Theorem 1.2* *For any nonsingular matrix $\mathbf{A}$ there exists a row permutation prior to the kth elimination step of the Gaussian algorithm such that the kth diagonal element is nonzero.*

$DA \neq O$

## 1.1 Gaussian Algorithm

*Proof* From the assumed nonsingularity of the matrix **A**, it follows that the determinant $|\mathbf{A}|$ is nonzero. To prove the theorem we show connections between the row operations performed on the coefficients of the matrix **A** during the Gaussian algorithm, and elementary operations performed on the corresponding determinant.

Let us assume that $a_{11} = 0$. Then there exists at least one element $a_{i1} \neq 0$ in the first column, because otherwise the determinant of **A** would be zero, contrary to our assumption. The exchange of the $i$th row with the first one generates a nonzero pivotal element at the desired place. A permutation of rows changes the sign of a determinant. For later use we put $v_1 = 1$ if a permutation of rows is needed before the first elimination step, and $v_1 = 0$ if $a_{11} \neq 0$. It is well known that adding a multiple of the first row to subsequent rows does not change the value of the determinant. To simplify the notation, even after a possible permutation of rows, we denote the matrix elements by $a_{ik}$. In the case $n = 4$ we have the following relationships with the quantities defined by (1.6)

$$|\mathbf{A}| = (-1)^{v_1} \begin{vmatrix} a_{11} & a_{12} & a_{13} & a_{14} \\ a_{21} & a_{22} & a_{23} & a_{24} \\ a_{31} & a_{32} & a_{33} & a_{34} \\ a_{41} & a_{42} & a_{43} & a_{44} \end{vmatrix} = (-1)^{v_1} \begin{vmatrix} a_{11} & a_{12} & a_{13} & a_{14} \\ 0 & a_{22}^{(1)} & a_{23}^{(1)} & a_{24}^{(1)} \\ 0 & a_{32}^{(1)} & a_{33}^{(1)} & a_{34}^{(1)} \\ 0 & a_{42}^{(1)} & a_{43}^{(1)} & a_{44}^{(1)} \end{vmatrix}$$

$$= (-1)^{v_1} a_{11} \begin{vmatrix} a_{22}^{(1)} & a_{23}^{(1)} & a_{24}^{(1)} \\ a_{32}^{(1)} & a_{33}^{(1)} & a_{34}^{(1)} \\ a_{42}^{(1)} & a_{43}^{(1)} & a_{44}^{(1)} \end{vmatrix} \tag{1.28}$$

The considerations for the first elimination step similarly apply to the following reduced systems and their corresponding determinants. Hence, there exists at least one element $a_{j2}^{(1)} \neq 0$, because otherwise the determinant of order three in (1.28), and therefore $|\mathbf{A}|$, would vanish. If required, a permutation of the $j$th row with the second row generates a nonzero pivotal element at the place (2, 2) in the scheme.

Let us denote by $r_{kk}$, $k = 1, 2, \ldots, n - 1$, the pivotal element that is used in the $k$th elimination step that is in the diagonal even after a possible permutation of rows, and by $r_{nn}$ the pivot for the void $n$th step. An immediate consequence of the proof of theorem 1.2 is

**Theorem 1.3** *If there is a total of $V = \sum_{i=1}^{n-1} v_i$ permutations of rows in the course of the Gaussian algorithm, the determinant of **A** is given by*

$$|\mathbf{A}| = (-1)^V \prod_{k=1}^{n} r_{kk}. \tag{1.29}$$

The determinant $|\mathbf{A}|$ of the given system of equations is, apart from the sign, equal to the product of the $n$ pivotal elements of the Gaussian algorithm. It is easy to compute as a by-product of the solution of a system of linear equations. If no permutations of rows are required, this statement follows directly from the decomposition (1.23) with $|\mathbf{L}| = 1$ from

$$|\mathbf{A}| = |\mathbf{LR}| = |\mathbf{L}| \cdot |\mathbf{R}| = \prod_{k=1}^{n} r_{kk}.$$

The Gaussian algorithm is, an independent method of solving a system of linear equations as well as being a useful and efficient method for computing determinants. From (1.29) it follows that the product of the moduli of the pivotal elements of the Gaussian algorithm, for a given system of linear equations $\mathbf{Ax} + \mathbf{b} = \mathbf{0}$, is a fixed quantity, i.e. it is an invariant.

Theorem 1.2 assures us of the feasibility of the Gaussian algorithm in the case of a nonsingular matrix $\mathbf{A}$, if appropriate permutations of rows are carried out. The permutations of rows may be thought of as being done before beginning the computation, so that there is a nonzero pivotal element on the diagonal for each elimination step. These preliminary row operations can be described by means of an appropriate *permutation matrix* $\mathbf{P}$. This is a square matrix, which contains in each row and each column exactly one element equal to one, all other elements equal to zero and the determinant $|\mathbf{P}| = \pm 1$. As a consequence of Theorems 1.1 and 1.2, Theorem 1.4 follows.

*Theorem 1.4  For each nonsingular matrix $\mathbf{A}$ there exists a permutation matrix $\mathbf{P}$, such that $\mathbf{PA}$ is decomposable into the product of a left triangular matrix $\mathbf{L}$, (1.22), and a right triangular matrix $\mathbf{R}$, (1.18), in the following way*

$$\mathbf{PA} = \mathbf{LR}. \tag{1.30}$$

We will now modify the statement of algorithm (1.27) to take into account Theorem 1.4. The system of linear equations $\mathbf{Ax} + \mathbf{b} = \mathbf{0}$ is multiplied on the left by the permutation matrix $\mathbf{P}$ resulting in $\mathbf{P}(\mathbf{Ax} + \mathbf{b}) = \mathbf{PAx} + \mathbf{Pb} = \mathbf{0}$. The computation of the solution $\mathbf{x}$ by means of the Gaussian algorithm is modified as follows

$$\begin{array}{ll} 1. \ \mathbf{PA} = \mathbf{LR} & \text{(decomposition of } \mathbf{PA}) \\ 2. \ \mathbf{Lc} - \mathbf{Pb} = \mathbf{0} & \text{(forward substitution } \to \mathbf{c}) \\ 3. \ \mathbf{Rx} + \mathbf{c} = \mathbf{0} & \text{(backward substitution } \to \mathbf{x}) \end{array} \tag{1.31}$$

Hence, the permutations of the rows of the matrix $\mathbf{A}$ have to be applied to the constant vector $\mathbf{b}$ before the process of forward substitution.

Finally, we determine the amount of computational effort required for the solution of a system of $n$ linear equations in $n$ unknowns by means of the Gaussian algorithm. We only count multiplications and divisions as the essential operations and neglect additions. In the general $j$th elimination step of the

## 1.1 Gaussian Algorithm

decomposition $\mathbf{PA} = \mathbf{LR}$ we compute $(n-j)$ quotients $l_{ij}$, (1.5), and $(n-j)^2$ values $a_{ik}^{(j)}$, (1.6). Obviously $[(n-j)+(n-j)^2]$ essential operations are necessary, thus we obtain the following equation for the computation required for the decomposition

$$Z_{LR} = [(n-1)+(n-2)+\cdots+1] + [(n-1)^2+(n-2)^2+\cdots+1^2]$$
$$= \tfrac{1}{2}n(n-1) + \tfrac{1}{6}n(n-1)(2n-1) = \tfrac{1}{3}(n^3-n) \tag{1.32}$$

The computation of $c_i$ during the process of forward substitution needs, according to (1.26), $(i-1)$ multiplications. Hence the total number of multiplications for the forward substitution is

$$Z_F = [1+2+\cdots+(n-1)] = \tfrac{1}{2}n(n-1) = \tfrac{1}{2}(n^2-n) \tag{1.33}$$

Finally, the computation of $x_i$ according to (1.17) requires $(n-i)$ multiplications and a division. Therefore the number of essential operations for the backward substitution is given by

$$Z_B = (1+2+\cdots+n) = \tfrac{1}{2}n(n+1) = \tfrac{1}{2}(n^2+n) \tag{1.34}$$

Hence the computational effort required for the two, usually correlated, processes of the forward and backward substitution is

$$Z_{FB} = n^2 \tag{1.35}$$

essential operations. The complete Gaussian algorithm for solving $n$ linear equations requires

$$\boxed{Z_{\text{Gauss}} = \tfrac{1}{3}n^3 + n^2 - \tfrac{1}{3}n} \tag{1.36}$$

essential operations. So we can say that the amount of work, and hence the computing time, increases by a power of three with respect to the number of unknowns.

### 1.1.2 Pivotal strategies

Theorems 1.2 and 1.4 assure us of the existence of a nonzero pivotal element for successive elimination steps but they leave unanswered the question of how to choose a pivot. For numerical reasons the appropriate choice of the pivots is crucial in order to achieve a computed solution that is as accurate as possible. Moreover, every computer program requires a well defined rule on how to determine the pivotal element, the *pivotal strategy*.

The choice of successive pivotal elements along the diagonal is called the *diagonal strategy*. This simple strategy works without any permutations of rows, and it is to be expected that it is not always adequate for numerical reasons, although the pivotal elements are admissible theoretically.

**Example 1.2** We consider a system of two linear equations in two unknowns to illustrate the numerical consequences that happen when a pivotal element of small absolute size is used. We apply diagonal strategy and carry out the computation to an accuracy of five decimal places. This means that we simulate a computer with a five digit mantissa. All numbers generated can have at most five significant digits and all values have to be correctly rounded to five figures after each arithmetic operation. The first elimination step of the Gaussian algorithm yields

| $x_1$ | $x_2$ | 1 | $x_1$ | $x_2$ | 1 |
|---|---|---|---|---|---|
| 0.00035 | 1.2654 | −3.5267 | 0.00035 | 1.2654 | −3.5267 |
| 1.2547 | 1.3182 | −6.8541 | 0 | −4535.0 | 12636 |

With $l_{21} = 1.2547/0.00035 \doteq 3584.9$ we obtain the two values of the second row of the right-hand scheme to be $r_{22} = 1.3182 − 3584.9 \times 1.2654 \doteq 1.3182 − 4536.3 \doteq −4535.0$ and $c_2 = −6.8541 − 3584.9 \times (−3.5267) \doteq −6.8541 + 12643 \doteq 12636$. Back substitution yields $x_2 = −12636/(−4535.0) \doteq 2.7863$ and $x_1 = −(−3.5267 + 1.2654 \times 2.7863)/0.00035 \doteq −(−3.5267 + 3.5258)/0.00035 = 0.0009/0.00035 \doteq 2.5714$. While computing $x_1$, we observe a cancellation of leading digits. Substitution of the computed unknowns into the second given equation indicates the inconsistency. The exact value of $x_1$ to five significant digits is 2.5354, whereas the value obtained for $x_2$ is correct, just by chance.

The diagonal strategy is, however, feasible and even appropriate in a special case of a general system of linear equations.

**Definition 1.1** *A matrix* **A** *is called diagonally dominant, if in each row the absolute value of the diagonal element is larger than the sum of the absolute values of the remaining matrix elements of the same row, i.e.*

$$|a_{ii}| > \sum_{\substack{k=1 \\ k \neq i}}^{n} |a_{ik}| \quad (i = 1, 2, \ldots, n). \tag{1.37}$$

**Theorem 1.5** *Diagonal strategy is feasible for the solution of a system of linear equations whose coefficient matrix* **A** *is diagonally dominant.*

**Proof** From the hypothesis we have $|a_{11}| > \sum_{k=2}^{n} |a_{1k}| \geq 0$, and hence $a_{11} \neq 0$ is an admissible pivotal element for the first elimination step. We now show that the property of diagonal dominance is transferred to the reduced system of equations. After substitution of (1.5) into (1.6), the reduced elements are given by

$$a_{ik}^{(1)} = a_{ik} - \frac{a_{i1} a_{1k}}{a_{11}} \quad (i, k = 2, 3, \ldots, n). \tag{1.38}$$

## 1.1 Gaussian Algorithm

From this we derive the inequality for the diagonal elements

$$|a_{ii}^{(1)}| = \left|a_{ii} - \frac{a_{i1}a_{1i}}{a_{11}}\right| \geq |a_{ii}| - \left|\frac{a_{i1}a_{1i}}{a_{11}}\right| \quad (i=2,3,\ldots,n). \tag{1.39}$$

Thus the sum of absolute values of the off-diagonal elements of the $i$th row, $i = 2, 3, \ldots, n$, of the reduced system satisfies the following inequality, if we use the assumption (1.37) as well as the estimate (1.39) *Because of absolute value.*

$$\sum_{\substack{k=2 \\ k \neq i}}^{n} |a_{ik}^{(1)}| = \sum_{\substack{k=2 \\ k \neq i}}^{n} \left|a_{ik} - \frac{a_{i1}a_{1k}}{a_{11}}\right| \leq \sum_{\substack{k=2 \\ k \neq i}}^{n} |a_{ik}| + \left|\frac{a_{i1}}{a_{11}}\right| \sum_{\substack{k=2 \\ k \neq i}}^{n} |a_{1k}|$$

$$= \sum_{\substack{k=1 \\ k \neq i}}^{n} |a_{ik}| - |a_{i1}| + \left|\frac{a_{i1}}{a_{11}}\right| \left(\sum_{k=2}^{n} |a_{1k}| - |a_{1i}|\right)$$

$$< |a_{ii}| - |a_{i1}| + \left|\frac{a_{i1}}{a_{11}}\right|(|a_{11}| - |a_{1i}|) = |a_{ii}| - \left|\frac{a_{i1}a_{1i}}{a_{11}}\right| \leq |a_{ii}^{(1)}|.$$

Therefore $|a_{22}^{(1)}| > \sum_{k=3}^{n} |a_{2k}^{(1)}| \geq 0$ holds, $a_{22}^{(1)}$ is a nonzero pivotal element and diagonal strategy is feasible.

Since the pivotal element has to be nonzero, a natural rule is to choose the largest absolute value among the possible elements. We call this rule the maximal column pivoting strategy. Before the $k$th elimination step the index $p$ is determined such that

$$\max_{i \geq k} |a_{ik}^{(k-1)}| = |a_{pk}^{(k-1)}|. \tag{1.40}$$

If $p \neq k$, the $p$th row has to be exchanged with the $k$th row. Following this strategy the quotients $l_{ik} = a_{ik}^{(k-1)}/a_{kk}^{(k-1)}$ $(i > k)$ are bounded in absolute value by one. Hence, the factors, by which the current $k$th row has to be multiplied, are in absolute value less than or equal to one. This may have an advantageous effect on the propagation of rounding errors.

*Example 1.3* We treat the system of equations of example 1.2 by using the maximal column pivoting strategy. Hence, a permutation of the two rows has to be done before the first step. A computation with five digits yields the schemes

| $x_1$ | $x_2$ | 1 | $x_1$ | $x_2$ | 1 |
|---|---|---|---|---|---|
| 1.2547 | 1.3182 | −6.8541 | 1.2547 | 1.3182 | −6.8541 |
| 0.000 35 | 1.2654 | −3.5267 | 0 | 1.2650 | −3.5248 |

The quotient $l_{21} \doteq 0.000\,278\,95$ is very small and causes only minor changes in the elements of the reduced system. Backward substitution yields $x_2 \doteq 2.7864$ and

$x_1 = -(-6.8541 + 1.3182 \times 2.7864)/1.2547 \doteq -(-6.8541 + 3.6730)/1.2547 = 3.1811/1.2547 \doteq 2.5353$. The two values differ by just one unit in the last decimal place from the exact rounded values. The situation has indeed been improved by the maximal column pivoting strategy.

*Example* 1.4  In order to show up a weakness of the maximal column pivoting strategy we solve the following system of equations again to an accuracy of five digits. To avoid confusing permutations of rows, the equations are arranged in such a way that maximal column pivoting strategy becomes a diagonal strategy. In the second and third schemes we write the $l$ values instead of zeros

| $x_1$ | $x_2$ | $x_3$ | 1 |
|---|---|---|---|
| 2.1 | 2512 | −2516 | −6.5 |
| −1.3 | 8.8 | −7.6 | 5.3 |
| 0.9 | −6.2 | 4.6 | −2.9 |

| $x_1$ | $x_2$ | $x_3$ | 1 |
|---|---|---|---|
| 2.1 | 2512 | −2516 | −6.5 |
| −0.619 05 | 1563.9 | −1565.1 | 1.276 2 |
| 0.428 57 | −1082.8 | 1082.9 | −0.114 30 |

| $x_1$ | $x_2$ | $x_3$ | 1 |
|---|---|---|---|
| 2.1 | 2512 | −2516 | −6.5 |
| −0.619 05 | 1563.9 | −1565.1 | 1.276 2 |
| 0.428 57 | −0.692 37 | −0.700 00 | 0.769 30 |

Thus we obtain the solutions $x_3 \doteq 1.0990$, $x_2 = 1.0990$, $x_1 \doteq 5.1905$, whereas the exact values are $x_3 = x_2 = 1$ and $x_1 = 5$. The differences can be explained by the fact that the first elimination step produces coefficients $a_{ik}^{(1)}$ of the reduced system that are large in absolute value. Hence some loss of information takes place due to rounding. Moreover, we observe a dramatic cancellation of digits while computing $a_{33}^{(2)}$ in the second step. The reason for the bad result is that the pivotal element of the first elimination step is small in comparison with the maximum of the absolute values of the other matrix elements of the first row. According to formula (1.38) the elements $a_{i1}$ and $a_{1k}$ play a ~~somewhat~~ symmetric role, and this should be taken into account.

A simple measure to improve the situation consists of *scaling* the given

## 1.1 Gaussian Algorithm

equations such that the new coefficients $\tilde{a}_{ik}$ satisfy

$$\sum_{k=1}^{n} |\tilde{a}_{ik}| = 1 \quad (i = 1, 2, \ldots, n). \tag{1.41}$$

This scaling, with the effect of $|\tilde{a}_{ik}| \leq 1$, $(i, k = 1, 2, \ldots, n)$ influences the choice of the pivotal elements according to the maximal column pivoting strategy in an advantageous way.

*Example* 1.5 After scaling, the equations of Example 1.4 read, using an accuracy of five significant digits,

| $x_1$ | $x_2$ | $x_3$ | 1 |
|---|---|---|---|
| 0.000 417 49 | 0.499 39 | −0.500 19 | −0.001 2922 |
| −0.073 446 | 0.497 18 | −0.429 38 | 0.299 44 |
| 0.076 923 | −0.529 91 | 0.393 16 | −0.247 86 |

The maximal column pivoting strategy yields $a_{31}$ as the pivotal element. An appropriate row exchange takes place, and after the first elimination step we obtain

| $x_1$ | $x_2$ | $x_3$ | 1 |
|---|---|---|---|
| 0.076 923 | −0.529 91 | 0.393 16 | −0.247 86 |
| −0.954 80 | −0.008 7800 | −0.053 990 | 0.062 780 |
| 0.005 4274 | 0.502 27 | −0.502 32 | 0.000 053 000 |

The maximal column pivoting strategy requires another row exchange. The elimination step finally yields

| $x_1$ | $x_2$ | $x_3$ | 1 |
|---|---|---|---|
| 0.076 923 | −0.529 91 | 0.393 16 | −0.247 86 |
| 0.005 4274 | 0.502 27 | −0.502 32 | 0.000 053 000 |
| −0.954 80 | −0.017 481 | −0.062 771 | 0.062 781 |

Back substitution results in the values $x_3 \doteq 1.0002$, $x_2 \doteq 1.0002$ and $x_1 \doteq 5.0003$, which is quite a good approximation of the exact solution.

The scaling of the equations according to (1.41) is not carried over to the equations of the reduced systems, as can be seen from Example 1.5. Hence, the advantageous influence of the maximal column pivoting strategy on the choice of the pivot for the first step may be lost for the later elimination steps. Therefore, the reduced systems should be scaled again. However, we do not proceed in such a way, since the amount of work would be doubled and each scaling would produce additional rounding errors. In order to retain the idea, we do not perform an explicit scaling but use it implicitly to determine an appropriate pivot. We apply the maximal column pivoting strategy to the systems which are only thought to be scaled. This means that we choose the pivot such that it has the largest in absolute value of the candidates with respect to the sum of absolute values of the elements of the corresponding row. This *relative maximal column pivoting strategy* chooses the index $p$ prior to the $k$-th elimination step such that

$$\max_{k \leq i \leq n} \left( \frac{|a_{ik}^{(k-1)}|}{\sum_{j=k}^{n} |a_{ij}^{(k-1)}|} \right) = \frac{|a_{pk}^{(k-1)}|}{\sum_{j=k}^{n} |a_{pj}^{(k-1)}|}. \tag{1.42}$$

If $p \neq k$, then the $p$th and $k$th rows are exchanged. It is obvious that the quotients $l_{ik}(i > k)$ are no longer bounded in absolute value by one.

*Example 1.6* The system of equations of Example 1.4 is now solved by applying the relative maximal column pivoting strategy and using an accuracy of five significant digits. To clarify the steps, the sums of absolute values of the matrix elements $s_i = \sum |a_{ij}^{(k-1)}|$ and the relevant quotients $q_i = |a_{ik}^{(k-1)}|/s_i$ for the choice of the pivot are given to the right of the first two schemes. In the first step it is clear that the element $a_{31}$ is taken as the pivot, as was the case in Example 1.5. This is now the smallest in absolute value among the elements of the first column. In the second step another row exchange is necessary.

| $x_1$ | $x_2$ | $x_3$ | 1 | $s_i$ | $q_i$ |
|---|---|---|---|---|---|
| 2.1 | 2512 | −2516 | −6.5 | 5030.1 | 0.000 417 49 |
| −1.3 | 8.8 | −7.6 | 5.3 | 17.7 | 0.073 446 |
| 0.9 | −6.2 | 4.6 | −2.9 | 11.7 | 0.076 923 |

| $x_1$ | $x_2$ | $x_3$ | 1 | $s_i$ | $q_i$ |
|---|---|---|---|---|---|
| 0.9 | −6.2 | 4.6 | −2.9 | — | — |
| −1.4444 | −0.155 30 | −0.955 80 | 1.1112 | 1.1111 | 0.139 77 |
| 2.3333 | 2526.5 | −2526.7 | 0.266 60 | 5053.2 | 0.499 98 |

| $x_1$ | $x_2$ | $x_3$ | 1 |
|---|---|---|---|
| 0.9 | −6.2 | 4.6 | −2.9 |
| 2.3333 | 2526.5 | −2526.7 | 0.266 60 |
| −1.4444 | −0.000 061 468 | −1.1111 | 1.1112 |

## 1.1 Gaussian Algorithm

The unknowns are computed to be $x_3 \doteq 1.0001$, $x_2 \doteq 1.0001$, $x_1 \doteq 5.0001$. The determinant of **A** is given, according to (1.29), by $|\mathbf{A}| = (-1)^2 \times 0.9 \times 2526.5 \times (-1.1111) \doteq -2526.5$. The exact value is $|\mathbf{A}| = -2526.504$.

When the Gaussian elimination process has been completed with a useful pivotal strategy we can summarize it in an algorithmic form which can be easily implemented on a computer. The decomposition, the forward substitution and the backward substitution, are presented separately as self-contained processes. The values of the consecutive schemes are stored in a fixed array. This is possible, because the value of $a_{ij}^{(k-1)}$ is no longer needed as soon as either $l_{ij}$ or $a_{ij}^{(k)}$ is computed. Thus, the values of $l_{ij}$ replace those of $a_{ij}$, and the coefficients of the final equations are left unchanged in their places in the same way as in the previous example. At the end of the decomposition the meaning of the elements is therefore $a_{ij} = l_{ij}$ for $i > j$ and $a_{ij} = r_{ij}$ for $i \leqslant j$. The information about row exchanges is contained in the vector $\mathbf{p} = (p_1, p_2, \ldots, p_n)^{\mathrm{T}}$. The $k$th component is equal to the index of that row which has been exchanged with the $k$th row before the $k$th elimination step. No row exchange took place if $p_k = k$. In all the following algorithmic descriptions the assignment statements are understood dynamically, and empty loop statements are skipped.

The process of decomposition with relative maximal column pivoting strategy and the simultaneous computation of the determinant can be summarized as follows.

$$
\begin{aligned}
&\det = 1 \\
&\text{for } k = 1, 2, \ldots, n-1: \\
&\quad \max = 0; p_k = 0 \\
&\quad \text{for } i = k, k+1, \ldots, n: \\
&\quad\quad s = 0 \\
&\quad\quad \text{for } j = k, k+1, \ldots, n: \\
&\quad\quad\quad s = s + |a_{ij}| \\
&\quad\quad q = |a_{ik}|/s \\
&\quad\quad \text{if } q > \max: \\
&\quad\quad\quad \max = q; p_k = i \\
&\quad \text{if } \max = 0: \text{STOP} \\
&\quad \text{if } p_k \neq k: \\
&\quad\quad \det = -\det \\
&\quad\quad \text{for } j = 1, 2, \ldots, n: \\
&\quad\quad\quad h = a_{kj}; a_{kj} = a_{p_k, j}; a_{p_k, j} = h \\
&\quad \det = \det \times a_{kk} \\
&\quad \text{for } i = k+1, k+2, \ldots, n: \\
&\quad\quad a_{ik} = a_{ik}/a_{kk} \\
&\quad\quad \text{for } j = k+1, k+2, \ldots, n: \\
&\quad\quad\quad a_{ij} = a_{ij} - a_{ik} \times a_{kj} \\
&\det = \det \times a_{nn}
\end{aligned}
\qquad (1.43)
$$

Before the forward substitution, (1.26), we have to perform the required exchanges of the components of the constant vector **b**.

$$
\begin{aligned}
&\text{for } k = 1, 2, \ldots, n-1: \\
&\quad \text{if } p_k \neq k: \\
&\quad\quad h = b_k; b_k = b_{p_k}; b_{p_k} = h \\
&\text{for } i = 1, 2, \ldots, n: \\
&\quad c_i = b_i \\
&\quad \text{for } j = 1, 2, \ldots, i-1: \\
&\quad\quad c_i = c_i - a_{ij} \times c_j
\end{aligned}
\qquad (1.44)
$$

Finally, the backward substitution, (1.17), is described by

$$
\begin{aligned}
&\text{for } i = n, n-1, \ldots, 1: \\
&\quad s = c_i \\
&\quad \text{for } k = i+1, i+2, \ldots, n: \\
&\quad\quad s = s + a_{ik} \times x_k \\
&\quad x_i = -s/a_{ii}
\end{aligned}
\qquad (1.45)
$$

The algorithm (1.44) of the forward substitution allows the identification of the auxiliary vector **c** with **b**, since $b_i$ is no longer needed after $c_i$ has been computed. This is appropriate, because the given vector **b** is changed by the possible permutations anyway. A similar fact holds for the backward substitution (1.45), where the solution vector **x** can be identified with **c**, and hence with **b**! If we make both identifications, the given vector **b** is replaced by the desired solution vector **x** in the course of the algorithm.

### 1.1.3 Supplements

Certain applications require the simultaneous or consecutive solution of several systems of equations with the same matrix **A** but different constant vectors **b**. The three distinct steps (1.31) of the Gaussian algorithm are quite suitable for this. In fact, the decomposition **PA = LR** need only be performed once, because then all of the required values, together with the information about the row exchanges, are known for the forward and backward substitution. These processes can be applied to the different constant vectors.

If $m$ systems of equations with the constant vectors $\mathbf{b}_1, \mathbf{b}_2, \ldots, \mathbf{b}_m$ are to be solved simultaneously, we suitably combine them in the matrix

$$\mathbf{B} = (\mathbf{b}_1, \mathbf{b}_2, \ldots, \mathbf{b}_m) \in \mathbf{R}^{n \times m} \qquad (1.46)$$

Then we look for a matrix $\mathbf{X} \in \mathbf{R}^{n \times m}$ which solves the matrix equation

$$\mathbf{AX} + \mathbf{B} = \mathbf{0}. \qquad (1.47)$$

## 1.1 Gaussian Algorithm

The columns of $\mathbf{X}$ are the solution vectors $\mathbf{x}_\mu$ to the corresponding constant vectors $\mathbf{b}_\mu$. Many computer programs exist for the solution of (1.47). The amount of computational effort required to solve (1.47) is, if (1.32) and (1.35) are used,

$$Z = \tfrac{1}{3}(n^3 - n) + mn^2. \tag{1.48}$$

A special application of the technique mentioned is the *inversion* of a nonsingular matrix $\mathbf{A}$. The inverse of $\mathbf{X}$ is $\mathbf{A}^{-1}$ and satisfies the matrix equation

$$\mathbf{AX} = \mathbf{I} \quad \text{or} \quad \mathbf{AX} - \mathbf{I} = \mathbf{0}, \tag{1.49}$$

where $\mathbf{I}$ denotes the *identity matrix*. The computation of $\mathbf{A}^{-1}$ is thus reduced to the simultaneous solution of $n$ systems of equations with the same matrix $\mathbf{A}$. The total number of essential operations is given by

$$Z_{\text{Inv}} = \tfrac{4}{3}n^3 - \tfrac{1}{3}n. \tag{1.50}$$

In this operation count it has not been taken into account that even after row permutations of $\mathbf{I}$, there are still zeros above the ones. As a consequence, the process of forward substitution can be started in principle with the first element being one, thus reducing the number of operations. This possibility is hardly ever used in existing programs.

The computation of $\mathbf{A}^{-1}$ in the manner described requires storage space for the matrix $\mathbf{A}$ and additionally for $-\mathbf{I}$, where the inverse is generated successively. Therefore the storage requirement amounts to $2n^2$ places. In Section 1.4 we shall present a method which allows the calculation of the inverse within the array for $\mathbf{A}$.

The situation mentioned above is also met in connection with the *iterative improvement* of a computed solution. When we solve $\mathbf{Ax} + \mathbf{b} = \mathbf{0}$ numerically with the Gaussian algorithm, we only get an approximation $\tilde{\mathbf{x}}$ instead of the exact solution vector $\mathbf{x}$, because of the inevitable rounding errors. Substitution of $\tilde{\mathbf{x}}$ into the given equations yields a *residual vector* $\mathbf{r}$ instead of the zero vector

$$\mathbf{A}\tilde{\mathbf{x}} + \mathbf{b} = \mathbf{r}. \tag{1.51}$$

We now try to find the exact solution $\mathbf{x}$ by means of the computed approximation $\tilde{\mathbf{x}}$ using a correction approach

$$\mathbf{x} = \tilde{\mathbf{x}} + \mathbf{z}. \tag{1.52}$$

We determine the correction vector $\mathbf{z}$ such that the equations hold

$$\mathbf{Ax} + \mathbf{b} = \mathbf{A}(\tilde{\mathbf{x}} + \mathbf{z}) + \mathbf{b} = \mathbf{A}\tilde{\mathbf{x}} + \mathbf{Az} + \mathbf{b} = \mathbf{0}. \tag{1.53}$$

If we take equation (1.51) into account, it follows from (1.53) that the correction vector $\mathbf{z}$ has to satisfy the system of equations

$$\mathbf{Az} + \mathbf{r} = \mathbf{0} \tag{1.54}$$

with the same matrix $\mathbf{A}$, but with the new constant vector $\mathbf{r}$. Hence, we get the desired correction **z** by the process of the forward and backward substitution applied to the residual vector **r**.

*Example* 1.7 We illustrate the iterative improvement of an approximate solution with the following system of equations with four unknowns. The example is also chosen to motivate the subsequent investigations.

$$0.294\,12x_1 + 0.411\,76x_2 + 0.529\,41x_3 + 0.588\,24x_4 - 0.176\,42 = 0$$
$$0.428\,57x_1 + 0.571\,43x_2 + 0.714\,29x_3 + 0.642\,86x_4 - 0.214\,31 = 0$$
$$0.368\,42x_1 + 0.526\,32x_2 + 0.421\,05x_3 + 0.368\,42x_4 - 0.157\,92 = 0$$
$$0.384\,62x_1 + 0.538\,46x_2 + 0.461\,54x_3 + 0.384\,62x_4 - 0.153\,80 = 0$$

In the first step, only the triangular decomposition of the matrix **A** is performed by applying the relative maximal column pivoting strategy and calculating with five significant digits. The sums $s_i$ of the moduli of the matrix elements and the quotients $q_i$ are given to the right of the schemes.

| | $x_1$ | $x_2$ | $x_3$ | $x_4$ | $s_i$ | $q_i$ |
|---|---|---|---|---|---|---|
| | 0.294 12 | 0.411 76 | 0.529 41 | 0.588 24 | 1.8235 | 0.161 29 |
| | 0.428 57 | 0.571 43 | 0.714 29 | 0.642 86 | 2.3572 | 0.181 81 |
| | <u>0.368 42</u> | 0.526 32 | 0.421 05 | 0.368 42 | 1.6842 | 0.218 75 |
| | 0.384 62 | 0.538 46 | 0.461 54 | 0.384 62 | 1.7692 | 0.217 40 |

| | $x_1$ | $x_2$ | $x_3$ | $x_4$ | $s_i$ | $q_i$ |
|---|---|---|---|---|---|---|
| | 0.368 42 | 0.526 32 | 0.421 05 | 0.368 42 | — | — |
| | 1.1633 | −0.040 840 | 0.224 48 | 0.214 28 | 0.479 60 | 0.085 154 |
| | 0.798 33 | −0.008 4200 | 0.193 27 | 0.294 12 | 0.495 81 | 0.016 982 |
| | 1.0440 | <u>−0.011 020</u> | 0.021 960 | −0.000 01 | 0.032 980 | 0.334 14 |

| | $x_1$ | $x_2$ | $x_3$ | $x_4$ | $s_i$ | $q_i$ |
|---|---|---|---|---|---|---|
| | 0.368 42 | 0.526 32 | 0.421 05 | 0.368 42 | — | — |
| | 1.0440 | −0.011 020 | 0.021 960 | −0.000 01 | — | — |
| | 0.798 33 | 0.764 07 | 0.176 49 | 0.294 13 | 0.470 62 | 0.375 02 |
| | 1.1633 | 3.7060 | <u>0.143 10</u> | 0.214 32 | 0.357 42 | 0.400 37 |

## 1.1 Gaussian Algorithm

|  $x_1$  |  $x_2$  |  $x_3$  |  $x_4$  |
|---------|---------|---------|---------|
| 0.368 42 | 0.526 32 | 0.421 05 | 0.368 42 |
| 1.0440  | −0.011 020 | 0.021 960 | −0.000 01 |
| 1.1633  | 3.7060  | 0.143 10 | 0.214 32 |
| 0.798 33 | 0.764 07 | 1.2333  | 0.029 810 |

(1.55)

According to the three row exchanges the approximate value of the determinant is $|\mathbf{A}| \doteq (-1)^3 \times 0.368\,42 \times (-0.011\,020) \times 0.143\,10 \times 0.029\,810 \doteq 1.7319 \times 10^{-5}$. The three row exchanges have also to be performed for the constant vector **b**. Hence, the vector $\mathbf{Pb} = (-0.157\,92, -0.153\,80, -0.214\,31, -0.176\,42)^T$ is used for the forward substitution (1.44), yielding the vector $\mathbf{c} \doteq (-0.157\,92, 0.011\,070, -0.071\,625, 0.029\,527)^T$. Back substitution gives the approximate solution $\tilde{\mathbf{x}} \doteq (-7.9333, 4.9593, 1.9841, -0.990\,51)^T$. Substitution of $\tilde{\mathbf{x}}$ into the given equation yields the residual vector $\tilde{\mathbf{r}} \doteq (2, 3, -3, 7)^T \times 10^{-5}$, if we compute to five significant places. However, a computation with ten significant digits gives the residual vector, rounded to five decimal places

$$\mathbf{r} \doteq (2.3951, 7.1948, -4.5999, 5.0390)^T \times 10^{-5}.$$

The numerical results are quite different, revealing that almost all of the components of $\tilde{\mathbf{r}}$ are wrong, even in the first decimal place. Thus $\tilde{\mathbf{r}}$ is not an appropriate start to an iterative improvement. The residual vector $\mathbf{r} = \mathbf{A}\tilde{\mathbf{x}} + \mathbf{b}$ must always be computed with higher accuracy to make an iterative improvement meaningful (Wilkinson, 1969). Forward substitution for the permuted vector $\mathbf{Pr} = (-4.5999, 5.0390, 7.1948, 2.3951)^T \times 10^{-5}$ yields

$$\mathbf{c}_r \doteq (-4.5999, 9.8413, -23.926, 28.056)^T \times 10^{-5},$$

and by back substitution we get the correction vector

$$\mathbf{z} \doteq (-0.066\,142, 0.040\,360, 0.015\,768, -0.009\,4116)^T.$$

Since **z** is affected by errors analogously to $\tilde{\mathbf{x}}$, and we find that $\tilde{\mathbf{x}} + \mathbf{z} = \tilde{\tilde{\mathbf{x}}}$ is only another approximate solution, which will be a better approximation of the solution **x**, if certain conditions are satisfied. This is indeed true in our example, because

$$\tilde{\tilde{\mathbf{x}}} \doteq (-7.9994, 4.9997, 1.9999, -0.999\,92)^T$$

is a better approximation of $\mathbf{x} = (-8, 5, 2, -1)^T$. A second iterative improvement with the new residual vector $\mathbf{r} \doteq (4.7062, 6.5713, 5.0525, 5.3850)^T \times 10^{-5}$ yields the desired solution up to working accuracy.

We notice that the above example also reveals that small moduli of the components of the residual vector do not necessarily imply that $\tilde{\mathbf{x}}$ is an accurate approximation of **x**. Moreover, it may happen that residual vectors of smaller absolute components result in much larger corrections, as can be seen in the first step of the iterative improvement.

## 1.2 Accuracy, Error Estimates

We want to investigate the accuracy of a computed approximate solution $\tilde{\mathbf{x}}$ of the system $\mathbf{Ax} + \mathbf{b} = \mathbf{0}$ and in particular to look for the reasons which are responsible for the size of the differences. To derive the statements about the error $\mathbf{x} - \tilde{\mathbf{x}}$ we need a measure for the size of a vector and also an analogous measure for the magnitude of a matrix.

### 1.2.1 Norms

We only consider the important case of real vectors $\mathbf{x} \in \mathbf{R}^n$ and real matrices $\mathbf{A} \in \mathbf{R}^{n \times n}$.

*Definition 1.2  A vector norm $\|\mathbf{x}\|$ of a vector $\mathbf{x} \in \mathbf{R}^n$ is a real function of its components with the following three properties*

(a) $\|\mathbf{x}\| \geq 0$ *for all* $\mathbf{x}$, *and* $\|\mathbf{x}\| = 0$ *if and only if* $\mathbf{x} = \mathbf{0}$; (1.56)

(b) $\|c\mathbf{x}\| = |c| \cdot \|\mathbf{x}\|$ *for all* $c \in \mathbf{R}$ *and all* $\mathbf{x}$; (1.57)

(c) $\|\mathbf{x} + \mathbf{y}\| \leq \|\mathbf{x}\| + \|\mathbf{y}\|$ *for all* $\mathbf{x}, \mathbf{y}$ *(triangle inequality)* (1.58)

Examples of vector norms are

$$\|\mathbf{x}\|_\infty := \max_k |x_k| \quad \text{(maximum norm)} \tag{1.59}$$

$$\|\mathbf{x}\|_2 := \left( \sum_{k=1}^n x_k^2 \right)^{1/2} \quad \text{(Euclidean norm)} \tag{1.60}$$

$$\|\mathbf{x}\|_1 := \sum_{k=1}^n |x_k| \quad (L_1 - \text{norm}) \tag{1.61}$$

It is easy to see that the properties of a vector norm hold. The three vector norms are equivalent in the sense that they satisfy the almost obvious inequalities for all vectors $\mathbf{x} \in \mathbf{R}^n$:

$$\frac{1}{\sqrt{n}} \|\mathbf{x}\|_2 \leq \|\mathbf{x}\|_\infty \leq \|\mathbf{x}\|_2 \leq \sqrt{n} \|\mathbf{x}\|_\infty$$

$$\frac{1}{n} \|\mathbf{x}\|_1 \leq \|\mathbf{x}\|_\infty \leq \|\mathbf{x}\|_1 \leq n \|\mathbf{x}\|_\infty$$

$$\frac{1}{\sqrt{n}} \|\mathbf{x}\|_1 \leq \|\mathbf{x}\|_2 \leq \|\mathbf{x}\|_1 \leq \sqrt{n} \|\mathbf{x}\|_2$$

## 1.2 Accuracy, Error Estimates

**Definition 1.3** *A matrix norm* $\|\mathbf{A}\|$ *of a matrix* $\mathbf{A} \in \mathbf{R}^{n \times n}$ *is a real mapping of its elements with the four properties:*

(a) $\|\mathbf{A}\| \geq 0$ *for all* $\mathbf{A}$, *and* $\|\mathbf{A}\| = 0$, *if and only if* $\mathbf{A} = \mathbf{0}$; (1.62)

(b) $\|c\mathbf{A}\| = |c| \cdot \|\mathbf{A}\|$ *for all* $c \in \mathbf{R}$ *and all* $\mathbf{A}$; (1.63)

(c) $\|\mathbf{A} + \mathbf{B}\| \leq \|\mathbf{A}\| + \|\mathbf{B}\|$ *for all* $\mathbf{A}, \mathbf{B}$ (*triangle inequality*); (1.64)

(d) $\|\mathbf{A} \cdot \mathbf{B}\| \leq \|\mathbf{A}\| \cdot \|\mathbf{B}\|$. (1.65)

The property (1.65) restricts the matrix norms to the important class of *consistent norms*. Examples of matrix norms are

$$\|\mathbf{A}\|_T := n \cdot \max_{i,k} |a_{ik}| \quad \text{(total norm)} \tag{1.66}$$

$$\|\mathbf{A}\|_R := \max_i \sum_{k=1}^n |a_{ik}| \quad \text{(row sum norm)} \tag{1.67}$$

$$\|\mathbf{A}\|_C := \max_k \sum_{i=1}^n |a_{ik}| \quad \text{(column sum norm)} \tag{1.68}$$

$$\|\mathbf{A}\|_F := \left( \sum_{i,k=1}^n a_{ik}^2 \right)^{1/2} \quad \text{(Frobenius norm)} \tag{1.69}$$

The above matrix norms obviously satisfy the first three properties (1.62), (1.63) and (1.64). We will only prove the fourth property (1.65) for the case of the total norm. The verification of (1.65) for the other matrix norms proceeds analogously

$$\|\mathbf{A} \cdot \mathbf{B}\|_T = n \cdot \max_{i,k} \left| \sum_{j=1}^n a_{ij} b_{jk} \right| \leq n \cdot \max_{i,k} \sum_{j=1}^n |a_{ij}| \cdot |b_{jk}|$$

$$\leq n \cdot \max_{i,k} \sum_{j=1}^n \left\{ \max_{l,m} |a_{lm}| \right\} \cdot \left\{ \max_{r,s} |b_{rs}| \right\}$$

$$= n^2 \cdot \left\{ \max_{l,m} |a_{lm}| \right\} \cdot \left\{ \max_{r,s} |b_{rs}| \right\} = \|\mathbf{A}\|_T \cdot \|\mathbf{B}\|_T.$$

The above four matrix norms are also equivalent. For instance the following inequalities hold for all matrices $\mathbf{A} \in \mathbf{R}^{n \times n}$.

$$\frac{1}{n} \|\mathbf{A}\|_T \leq \|\mathbf{A}\|_{R,C} \leq \|\mathbf{A}\|_T \leq n \|\mathbf{A}\|_{R,C}$$

$$\frac{1}{n} \|\mathbf{A}\|_T \leq \|\mathbf{A}\|_F \leq \|\mathbf{A}\|_T \leq n \|\mathbf{A}\|_F.$$

In subsequent considerations matrices and vectors occur simultaneously, and therefore the used matrix norms and vector norms have to be related in such a way that we can suitably operate with them.

*Definition 1.4* A matrix norm $\|\mathbf{A}\|$ is called compatible with the vector norm $\|\mathbf{x}\|$ if

$$\|\mathbf{A}\mathbf{x}\| \leqslant \|\mathbf{A}\|\|\mathbf{x}\| \quad \text{for all } \mathbf{x} \in \mathbf{R}^n \text{ and all } \mathbf{A} \in \mathbf{R}^{n \times n}. \tag{1.70}$$

Combinations of compatible norms are, for example

$$\|\mathbf{A}\|_T \text{ or } \|\mathbf{A}\|_R \text{ are compatible with } \|\mathbf{x}\|_\infty; \tag{1.71}$$

$$\|\mathbf{A}\|_T \text{ or } \|\mathbf{A}\|_C \text{ are compatible with } \|\mathbf{x}\|_1; \tag{1.72}$$

$$\|\mathbf{A}\|_T \text{ or } \|\mathbf{A}\|_F \text{ are compatible with } \|\mathbf{x}\|_2. \tag{1.73}$$

The compatibility of pairs of norms is shown for two cases. Since

$$\|\mathbf{A}\mathbf{x}\|_\infty = \max_i \left( \left| \sum_{k=1}^n a_{ik} x_k \right| \right) \leqslant \max_i \left( \sum_{k=1}^n |a_{ik}| \cdot |x_k| \right)$$

$$\leqslant \max_i \left( \sum_{k=1}^n \left[ \max_{r,s} |a_{rs}| \right] \cdot \left[ \max_l |x_l| \right] \right) = \|\mathbf{A}\|_T \cdot \|\mathbf{x}\|_\infty$$

the total norm is compatible with the maximum norm. Likewise, the Frobenius norm is compatible with the Euclidean norm. By applying the Cauchy–Schwarz inequality we have

$$\|\mathbf{A}\mathbf{x}\|_2 = \left[ \sum_{i=1}^n \left( \sum_{k=1}^n a_{ik} x_k \right)^2 \right]^{1/2} \leqslant \left\{ \sum_{i=1}^n \left[ \left( \sum_{k=1}^n a_{ik}^2 \right) \left( \sum_{k=1}^n x_k^2 \right) \right] \right\}^{1/2}$$

$$= \left( \sum_{i=1}^n \sum_{k=1}^n a_{ik}^2 \right)^{1/2} \left( \sum_{k=1}^n x_k^2 \right)^{1/2} = \|\mathbf{A}\|_F \cdot \|\mathbf{x}\|_2.$$

For arbitrary compatible norms the right-hand side of the inequality (1.70) is in general larger than the left-hand side for all vectors $\mathbf{x} \neq \mathbf{0}$. Therefore for a given vector norm we define an appropriate matrix norm such that (1.70) holds as an equality for at least one nonzero vector.

*Definition 1.5* The value defined for a given vector norm by

$$\|\mathbf{A}\| := \max_{\mathbf{x} \neq \mathbf{0}} \frac{\|\mathbf{A}\mathbf{x}\|}{\|\mathbf{x}\|} = \max_{\|\mathbf{x}\|=1} \|\mathbf{A}\mathbf{x}\| \tag{1.74}$$

*is called the subordinate or natural matrix norm.*

*Theorem 1.6* The value defined by (1.74) is a matrix norm. It is compatible with the given vector norm and is the smallest matrix norm that is compatible with $\|\mathbf{x}\|$.

*Proof* We verify the properties of a matrix norm.

(a) For any $\mathbf{x} \neq \mathbf{0}$ we have $\|\mathbf{A}\mathbf{x}\| \geqslant 0$ for all $\mathbf{A} \in \mathbf{R}^{n \times n}$ and $\|\mathbf{x}\| > 0$. Hence it is true that $\max_{\mathbf{x} \neq \mathbf{0}} \|\mathbf{A}\mathbf{x}\|/\|\mathbf{x}\| \geqslant 0$. Moreover we have to show that from $\|\mathbf{A}\| = 0$ it follows that $\mathbf{A} = \mathbf{0}$. Let us assume the contrary, that is $\mathbf{A} \neq \mathbf{0}$, then there exists at

## 1.2 Accuracy, Error Estimates

least one element $a_{pq} \neq 0$. We now choose $\mathbf{x} = \mathbf{e}_q$, where $\mathbf{e}_q$ is the $q$th unit vector, for which we have $\mathbf{A}\mathbf{e}_q \neq \mathbf{0}$. For this vector the inequality $\|\mathbf{A}\mathbf{e}_q\|/\|\mathbf{e}_q\| > 0$ holds. However the maximum of the quotient (1.74) must be even larger, which is the desired contradiction.

(b) The second property of a vector norm guarantees

$$\|c\mathbf{A}\| := \max_{\|\mathbf{x}\|=1} \|c\mathbf{A}\mathbf{x}\| = \max_{\|\mathbf{x}\|=1} \{|c| \cdot \|\mathbf{A}\mathbf{x}\|\} = |c| \cdot \|\mathbf{A}\|.$$

(c) By using the triangle inequality of vector norms it follows that

$$\|\mathbf{A}+\mathbf{B}\| := \max_{\|\mathbf{x}\|=1} \|(\mathbf{A}+\mathbf{B})\mathbf{x}\| \leq \max_{\|\mathbf{x}\|=1} \{\|\mathbf{A}\mathbf{x}\| + \|\mathbf{B}\mathbf{x}\|\}$$

$$\leq \max_{\|\mathbf{x}\|=1} \|\mathbf{A}\mathbf{x}\| + \max_{\|\mathbf{x}\|=1} \|\mathbf{B}\mathbf{x}\| = \|\mathbf{A}\| + \|\mathbf{B}\|.$$

(d) To show the consistency of the norm we assume $\mathbf{A} \neq \mathbf{0}$. Otherwise, the inequality (1.65) obviously holds. Then we have

$$\|\mathbf{A}\cdot\mathbf{B}\| := \max_{\mathbf{x}\neq \mathbf{0}} \frac{\|\mathbf{A}\mathbf{B}\mathbf{x}\|}{\|\mathbf{x}\|} = \max_{\substack{\mathbf{x}\neq \mathbf{0} \\ \mathbf{B}\mathbf{x}\neq \mathbf{0}}} \frac{\|\mathbf{A}(\mathbf{B}\mathbf{x})\| \|\mathbf{B}\mathbf{x}\|}{\|\mathbf{B}\mathbf{x}\| \|\mathbf{x}\|}$$

$$\leq \max_{\mathbf{B}\mathbf{x}\neq \mathbf{0}} \frac{\|\mathbf{A}(\mathbf{B}\mathbf{x})\|}{\|\mathbf{B}\mathbf{x}\|} \cdot \max_{\mathbf{x}\neq \mathbf{0}} \frac{\|\mathbf{B}\mathbf{x}\|}{\|\mathbf{x}\|}$$

$$\leq \max_{\mathbf{y}\neq \mathbf{0}} \frac{\|\mathbf{A}\mathbf{y}\|}{\|\mathbf{y}\|} \cdot \max_{\mathbf{x}\neq \mathbf{0}} \frac{\|\mathbf{B}\mathbf{x}\|}{\|\mathbf{x}\|} = \|\mathbf{A}\| \cdot \|\mathbf{B}\|.$$

The compatibility of the defined matrix norm with the given vector norm is an immediate consequence of definition (1.74). The last statement of the theorem follows from the fact that there exists a vector $\mathbf{x} \neq \mathbf{0}$ such that $\|\mathbf{A}\mathbf{x}\| = \|\mathbf{A}\| \cdot \|\mathbf{x}\|$ holds.

According to Definition 1.5 the matrix norm $\|\mathbf{A}\|_\infty$ is given by

$$\|\mathbf{A}\|_\infty := \max_{\|\mathbf{x}\|_\infty=1} \|\mathbf{A}\mathbf{x}\|_\infty = \max_{\|\mathbf{x}\|_\infty=1} \left(\max_i \left|\sum_{k=1}^n a_{ik}x_k\right|\right)$$

$$= \max_i \left(\max_{\|\mathbf{x}\|_\infty=1} \left|\sum_{k=1}^n a_{ik}x_k\right|\right) = \max_i \sum_{k=1}^n |a_{ik}| = \|\mathbf{A}\|_R.$$

The modulus of the sum is largest for fixed index $i$, if $x_k = \text{sign}(a_{ik})$. According to Theorem 1.6 the row sum norm is the smallest matrix norm that is compatible with the maximum norm.

To derive the natural matrix norm $\|\mathbf{A}\|_2$ corresponding to the Euclidean vector norm $\|\mathbf{x}\|_2$, some fundamental knowledge of linear algebra is required

$$\|\mathbf{A}\|_2 := \max_{\|\mathbf{x}\|_2=1} \|\mathbf{A}\mathbf{x}\|_2 = \max_{\|\mathbf{x}\|_2=1} [(\mathbf{A}\mathbf{x})^T(\mathbf{A}\mathbf{x})]^{1/2} = \max_{\|\mathbf{x}\|_2=1} (\mathbf{x}^T\mathbf{A}^T\mathbf{A}\mathbf{x})^{1/2}$$

The matrix $A^TA$ in the last expression is obviously symmetric and positive semidefinite, since the corresponding quadratic form satisfies $Q(x) := x^T(A^TA)x \geq 0$ for all $x \neq 0$. As a consequence, the eigenvalues $\mu_i$ of $A^TA$ are real and nonnegative, and the $n$ eigenvectors $x_1, x_2, \ldots, x_n$ form an orthonormal base in $R^n$

$$A^TAx_i = \mu_i x_i \quad \mu_i \in R \quad \mu_i \geq 0 \quad x_i^T x_j = \delta_{ij} \tag{1.75}$$

There exists a unique representation of an arbitrary vector $x \in R^n$ as a linear combination of the eigenvectors $x_i$

$$x = \sum_{i=1}^{n} c_i x_i \tag{1.76}$$

and so we obtain, with respect to (1.75)

$$x^T A^T A x = \left(\sum_{i=1}^{n} c_i x_i\right)^T A^T A \left(\sum_{j=1}^{n} c_j x_j\right) = \left(\sum_{i=1}^{n} c_i x_i\right)^T \left(\sum_{j=1}^{n} c_j \mu_j x_j\right) = \sum_{i=1}^{n} c_i^2 \mu_i.$$

Let us number the eigenvalues $\mu_i$ such that $\mu_1 \geq \mu_2 \geq \cdots \geq \mu_n \geq 0$. As a consequence of $\|x\|_2 = 1$ we have $\sum_{i=1}^{n} c_i^2 = 1$, and hence the natural matrix norm is given by

$$\|A\|_2 = \max_{\|x\|_2 = 1} \left(\sum_{i=1}^{n} c_i^2 \mu_i\right)^{1/2} \leq \max_{\|x\|_2 = 1} \left(\mu_1 \sum_{i=1}^{n} c_i^2\right)^{1/2} = \sqrt{\mu_1}.$$

The maximum value $\sqrt{\mu_1}$ is taken for $x = x_1$ with $c_1 = 1, c_2 = \cdots = c_n = 0$. The result is therefore

$$\|A\|_2 := \max_{\|x\|_2 = 1} \|Ax\|_2 = \sqrt{\mu_1}, \tag{1.77}$$

where $\mu_1$ denotes the largest eigenvalue of $A^TA$. The matrix norm $\|A\|_2$, subordinate to the Euclidean vector norm is usually called the *spectral norm*. As stated by Theorem 1.6 it is the smallest matrix norm that is compatible with the Euclidean vector norm.

The name of the norm $\|A\|_2$ becomes clear in the special case of a symmetric matrix $A$. Let $\lambda_1, \lambda_2, \ldots, \lambda_n$ be the real eigenvalues of $A$. Then the eigenvalues $\mu_i$ of the matrix $A^TA = AA = A^2$ are known to be $\mu_i = \lambda_i^2 \geq 0$, and it follows from (1.77)

$$\|A\|_2 = |\lambda_1| \quad |\lambda_1| = \max_i |\lambda_i|. \tag{1.78}$$

Thus, the spectral norm of a symmetric matrix $A$ is given by its eigenvalue $\lambda_1$ of the largest modulus, i.e. by the spectral radius of $A$.

To prepare the subsequent application we determine the spectral norm of the inverse $A^{-1}$ of a nonsingular matrix $A$. According to (1.77) the norm is defined by $\|A^{-1}\|_2 = \sqrt{\psi_1}$, where $\psi_1$ is equal to the largest eigenvalue of $A^{-1^T}A^{-1} = (AA^T)^{-1}$. The eigenvalues of the inverse matrix $C^{-1}$ are the reciprocals of the

## 1.2 Accuracy, Error Estimates

eigenvalues of $\mathbf{C}$. Consequently, $\psi_1$ equals the reciprocal value of the smallest, positive eigenvalue of the positive definite matrix $\mathbf{AA}^T$. The last matrix is *similar* to $\mathbf{A}^T\mathbf{A}$, because $\mathbf{A}^{-1}(\mathbf{AA}^T)\mathbf{A} = \mathbf{A}^T\mathbf{A}$ holds, so that $\mathbf{AA}^T$ and $\mathbf{A}^T\mathbf{A}$ have the same eigenvalues. Therefore we conclude from this that

$$\|\mathbf{A}^{-1}\|_2 = 1/\sqrt{\mu_n}, \tag{1.79}$$

where $\mu_n$ denotes the smallest eigenvalue of the positive definite matrix $\mathbf{A}^T\mathbf{A}$. Moreover, for a symmetric, nonsingular matrix we have

$$\|\mathbf{A}^{-1}\|_2 = 1/|\lambda_n| \quad \mathbf{A}^T = \mathbf{A} \quad |\lambda_n| = \min_i |\lambda_i|. \tag{1.80}$$

### 1.2.2 Error estimates, condition

We now investigate two questions which concern the error of a computed approximation $\tilde{\mathbf{x}}$ of the solution $\mathbf{x}$ of the system $\mathbf{Ax} + \mathbf{b} = \mathbf{0}$. To begin with we consider what conclusions can be drawn from the size of the residual vector $\mathbf{r} = \mathbf{A}\tilde{\mathbf{x}} + \mathbf{b}$ with respect to the error $\mathbf{z} := \mathbf{x} - \tilde{\mathbf{x}}$. To do this, let $\|\mathbf{A}\|$ be an arbitrary matrix norm and $\|\mathbf{x}\|$ a compatible vector norm. From (1.54) the error vector $\mathbf{z}$ satisfies the system of linear equations $\mathbf{Az} + \mathbf{r} = \mathbf{0}$, and so we get from the relationships

$$\|\mathbf{b}\| = \|-\mathbf{Ax}\| \leqslant \|\mathbf{A}\|\|\mathbf{x}\| \quad \|\mathbf{z}\| = \|-\mathbf{A}^{-1}\mathbf{r}\| \leqslant \|\mathbf{A}^{-1}\|\|\mathbf{r}\| \tag{1.81}$$

the estimate of the relative error

$$\frac{\|\mathbf{z}\|}{\|\mathbf{x}\|} = \frac{\|\mathbf{x} - \tilde{\mathbf{x}}\|}{\|\mathbf{x}\|} \leqslant \|\mathbf{A}\|\|\mathbf{A}^{-1}\|\frac{\|\mathbf{r}\|}{\|\mathbf{b}\|} =: \kappa(\mathbf{A})\frac{\|\mathbf{r}\|}{\|\mathbf{b}\|}. \tag{1.82}$$

*Definition* 1.6 *The value* $\kappa(\mathbf{A}) := \|\mathbf{A}\| \cdot \|\mathbf{A}^{-1}\|$ *is called the condition number of the matrix* $\mathbf{A}$ *corresponding to the underlying matrix norm.*

The condition number $\kappa(\mathbf{A})$ is greater than or equal to one since

$$1 \leqslant \|\mathbf{I}\| = \|\mathbf{AA}^{-1}\| \leqslant \|\mathbf{A}\|\|\mathbf{A}^{-1}\| = \kappa(\mathbf{A}).$$

The estimate (1.82) has the practical meaning, that besides a small residual vector $\mathbf{r}$, related to the size of the constant vector $\mathbf{b}$, the condition number essentially determines the possible relative error of the approximation $\tilde{\mathbf{x}}$. We can only deduce a small relative error from a relative small residual vector if the condition number is small!

*Example* 1.8 We consider the system of linear equations of Example 1.7 and apply the error estimate (1.82). For simplicity we use the maximum norm $\|\mathbf{x}\|_\infty$ together with its subordinate norm $\|\mathbf{A}\|_\infty = \|\mathbf{A}\|_R$. To determine the condition number we need the inverse $\mathbf{A}^{-1}$

$$\mathbf{A}^{-1} \doteq \begin{pmatrix} 168.40 & -235.80 & -771.75 & 875.82 \\ -101.04 & 138.68 & 470.63 & -528.07 \\ -50.588 & 69.434 & 188.13 & -218.89 \\ 33.752 & -41.659 & -112.88 & 128.73 \end{pmatrix}$$

Hence we have $\|\mathbf{A}\|_\infty = 2.3572$, $\|\mathbf{A}^{-1}\|_\infty \doteq 2051.77$ and $\kappa_\infty(\mathbf{A}) \doteq 4836.4$. With the values of $\|\mathbf{x}\|_\infty = 8$, $\|\mathbf{r}\|_\infty = 7.1948 \times 10^{-5}$ and $\|\mathbf{b}\|_\infty = 0.21431$ the estimate (1.82) gives the bound of the absolute error $\|\mathbf{x} - \tilde{\mathbf{x}}\|_\infty \leqslant 1.624$. In fact, the error $\|\mathbf{x} - \tilde{\mathbf{x}}\|_\infty = 0.0667$ is much smaller.

As a consequence of the accuracy of the computer even the coefficients $a_{ik}$ and $b_i$ of the system of equations usually do not have an exact representation. If they are the result of a computation, they have already been affected by rounding errors. This is the motivation for studying the influence of errors in the given data on the solution $\mathbf{x}$, i.e. for investigating the *sensitivity* of the solution $\mathbf{x}$ to perturbations of the coefficients. Our next question is therefore: How large can the change $\Delta \mathbf{x}$ of the solution $\mathbf{x}$ to $\mathbf{Ax} + \mathbf{b} = \mathbf{0}$ be, if the matrix $\mathbf{A}$ and the constant vector $\mathbf{b}$ are perturbed by $\Delta \mathbf{A}$ and $\Delta \mathbf{b}$, respectively. We assume that $\Delta \mathbf{A}$ and $\Delta \mathbf{b}$ denote small perturbations and that the matrix $\mathbf{A} + \Delta \mathbf{A}$ is still nonsingular. Let $\mathbf{x} + \Delta \mathbf{x}$ be the solution of

$$(\mathbf{A} + \Delta \mathbf{A})(\mathbf{x} + \Delta \mathbf{x}) + (\mathbf{b} + \Delta \mathbf{b}) = \mathbf{0} \tag{1.83}$$

From this we obtain by rearranging terms

$$\mathbf{Ax} + \Delta \mathbf{A} \mathbf{x} + (\mathbf{A} + \Delta \mathbf{A})\Delta \mathbf{x} + \mathbf{b} + \Delta \mathbf{b} = \mathbf{0}$$
$$\Delta \mathbf{x} = -(\mathbf{A} + \Delta \mathbf{A})^{-1}(\Delta \mathbf{A} \mathbf{x} + \Delta \mathbf{b}) = -(\mathbf{I} + \mathbf{A}^{-1}\Delta \mathbf{A})^{-1}\mathbf{A}^{-1}(\Delta \mathbf{A} \mathbf{x} + \Delta \mathbf{b}).$$

For compatible norms the following estimate follows

$$\|\Delta \mathbf{x}\| \leqslant \|(\mathbf{I} + \mathbf{A}^{-1}\Delta \mathbf{A})^{-1}\| \, \|\mathbf{A}^{-1}\| (\|\Delta \mathbf{A}\| \, \|\mathbf{x}\| + \|\Delta \mathbf{b}\|)$$

and hence

$$\frac{\|\Delta \mathbf{x}\|}{\|\mathbf{x}\|} \leqslant \|(\mathbf{I} + \mathbf{A}^{-1}\Delta \mathbf{A})^{-1}\| \, \|\mathbf{A}^{-1}\| \left( \|\Delta \mathbf{A}\| + \frac{\|\Delta \mathbf{b}\|}{\|\mathbf{x}\|} \right). \tag{1.84}$$

In order to get an upper bound for the first norm on the right hand side we need the following lemma.

**Lemma 1.7** *If for an arbitrary matrix norm $\|\mathbf{B}\| < 1$, then $(\mathbf{I} + \mathbf{B})^{-1}$ exists. For a natural matrix norm (1.74) the following inequality holds*

$$\|(\mathbf{I} + \mathbf{B})^{-1}\| \leqslant \frac{1}{1 - \|\mathbf{B}\|}. \tag{1.85}$$

## 1.2 Accuracy, Error Estimates

*Proof* Let $\mathbf{x} \neq \mathbf{0}$ be an arbitrary vector. It follows from the triangle inequality for vector norms and from the assumption that

$$\|(\mathbf{I} + \mathbf{B})\mathbf{x}\| = \|\mathbf{x} + \mathbf{B}\mathbf{x}\| \geq \|\mathbf{x}\| - \|\mathbf{B}\mathbf{x}\| \geq \|\mathbf{x}\| - \|\mathbf{B}\|\|\mathbf{x}\|$$
$$= (1 - \|\mathbf{B}\|)\|\mathbf{x}\| > 0.$$

The homogeneous system of equations $(\mathbf{I} + \mathbf{B})\mathbf{x} = \mathbf{0}$ only has the trivial solution $\mathbf{x} = \mathbf{0}$, and the matrix $\mathbf{I} + \mathbf{B}$ is nonsingular.

Furthermore, since $\|\mathbf{I}\| = 1$ for any natural matrix norm we have

$$1 = \|\mathbf{I}\| = \|(\mathbf{I} + \mathbf{B})(\mathbf{I} + \mathbf{B})^{-1}\| = \|(\mathbf{I} + \mathbf{B})^{-1} + \mathbf{B}(\mathbf{I} + \mathbf{B})^{-1}\|$$
$$\geq \|(\mathbf{I} + \mathbf{B})^{-1}\| - \|\mathbf{B}(\mathbf{I} + \mathbf{B})^{-1}\|$$
$$\geq \|(\mathbf{I} + \mathbf{B})^{-1}\| - \|\mathbf{B}\|\|(\mathbf{I} + \mathbf{B})^{-1}\| = \|(\mathbf{I} + \mathbf{B})^{-1}\|(1 - \|\mathbf{B}\|).$$

From this the inequality (1.85) follows.

Now we set $\mathbf{B} = \mathbf{A}^{-1}\Delta\mathbf{A}$, assume that $\|\mathbf{A}^{-1}\Delta\mathbf{A}\| < 1$ and restrict the matrix norm to a subordinate norm, so that we can apply (1.85). From (1.84) due to (1.81) and the additional assumption $\|\mathbf{A}^{-1}\|\|\Delta\mathbf{A}\| < 1$ we obtain

$$\frac{\|\Delta\mathbf{x}\|}{\|\mathbf{x}\|} \leq \frac{\|\mathbf{A}\|\|\mathbf{A}^{-1}\|}{1 - \|\mathbf{A}^{-1}\Delta\mathbf{A}\|}\left(\frac{\|\Delta\mathbf{A}\|}{\|\mathbf{A}\|} + \frac{\|\Delta\mathbf{b}\|}{\|\mathbf{A}\|\|\mathbf{x}\|}\right)$$

$$\leq \frac{\|\mathbf{A}\|\|\mathbf{A}^{-1}\|}{1 - \|\mathbf{A}^{-1}\|\|\Delta\mathbf{A}\|}\left(\frac{\|\Delta\mathbf{A}\|}{\|\mathbf{A}\|} + \frac{\|\Delta\mathbf{b}\|}{\|\mathbf{b}\|}\right).$$

If we write $\|\mathbf{A}^{-1}\|\|\Delta\mathbf{A}\| = \kappa(\mathbf{A})\|\Delta\mathbf{A}\|/\|\mathbf{A}\| < 1$, the final result is

$$\boxed{\frac{\|\Delta\mathbf{x}\|}{\|\mathbf{x}\|} \leq \frac{\kappa(\mathbf{A})}{1 - \kappa(\mathbf{A})\dfrac{\|\Delta\mathbf{A}\|}{\|\mathbf{A}\|}}\left\{\frac{\|\Delta\mathbf{A}\|}{\|\mathbf{A}\|} + \frac{\|\Delta\mathbf{b}\|}{\|\mathbf{b}\|}\right\}} \qquad (1.86)$$

The error estimate (1.86) has only been proved here for a pair of norms, where the matrix norm is subordinate to the vector norm. However, the statement is also true for any arbitrary pair of compatible norms, as can be shown by a different reasoning (Bunse and Bunse–Gerstner 1984). The condition number $\kappa(\mathbf{A})$ of the coefficient matrix $\mathbf{A}$ is the critical quantity, which is responsible for the sensitivity of the solution $\mathbf{x}$ to variations $\Delta\mathbf{A}$ and $\Delta\mathbf{b}$. We want to explain the practical meaning and the numerical consequences of this error estimate. A floating point computation with $d$ significant digits may cause relative errors in the norms of the given data up to the size

$$\|\Delta\mathbf{A}\|/\|\mathbf{A}\| \approx 5 \times 10^{-d} \quad \|\Delta\mathbf{b}\|/\|\mathbf{b}\| \approx 5 \times 10^{-d}$$

On the assumption of a condition number $\kappa(\mathbf{A}) \approx 10^{\alpha}$ with $5 \times 10^{\alpha-d} \ll 1$, from

(1.86) we obtain the qualitative estimate

$$\|\Delta x\|/\|x\| \leqslant 10^{\alpha-d+1}.$$

This estimate of the sensitivity means that the size of $\|\Delta x\|$ can be as large as a unit in the $(d - \alpha - 1)$th decimal place of $\|x\|$, and we arrive at the following *rule of thumb*.

If a system of linear equations $Ax + b = 0$ is solved by a floating point computation with d decimal digits, <u>and</u> if the condition number $\kappa(A) \approx 10^\alpha$, then the computed solution $\tilde{x}$ is correct only up to $d - \alpha - 1$ decimal places, with respect to the component that is largest in absolute value. This is due to the inevitable input errors in the given data.

Although this rule is often too pessimistic, we should be aware of the following essential fact. Because the estimate (1.86) concerns the norms, the change $\Delta x$ can affect all components of $x$. If the values of the unknowns are very different in size, the smallest in absolute value may contain much larger relative errors, which can be so large that not even the sign of the computed unknown is correct.

*Example* 1.9 Let us consider a system of linear equations $Ax + b = 0$ with two unknowns and

$$A = \begin{pmatrix} 0.99 & 0.98 \\ 0.98 & 0.97 \end{pmatrix} \quad b = \begin{pmatrix} -1.97 \\ -1.95 \end{pmatrix} \quad x = \begin{pmatrix} 1 \\ 1 \end{pmatrix}.$$

The condition number of the symmetric matrix $A$ corresponding to the spectral norm is given by the quotient of the absolutely largest and smallest eigenvalues of $A$, i.e. $\kappa(A) = |\lambda_1|/|\lambda_2| \doteq 1.96005/0.000051019 \doteq 38418 \doteq 3.8 \times 10^4$. If the given values of the coefficients are assumed to be affected by rounding errors due to an accuracy of five significant digits, we expect to have, following the rule of thumb with $d = 5$ and $\alpha = 4$, no correct decimal digit. This is indeed true as can be seen from the perturbed system $(A + \Delta A)(x + \Delta x) + (b + \Delta b) = 0$ with

$$A + \Delta A = \begin{pmatrix} 0.990005 & 0.979996 \\ 0.979996 & 0.970004 \end{pmatrix} \quad b + \Delta b = \begin{pmatrix} -1.969967 \\ -1.950035 \end{pmatrix}$$

which has the solution

$$x + \Delta x \doteq \begin{pmatrix} 1.8072 \\ 0.18452 \end{pmatrix} \quad \text{and} \quad \Delta x \doteq \begin{pmatrix} 0.8072 \\ -0.81548 \end{pmatrix}.$$

The estimate (1.86) is realistic, for this constructed example. We get with $\|\Delta A\|_2 \doteq 8.531 \times 10^{-6}$, $\|A\|_2 \doteq 1.960$, $\|\Delta b\|_2 \doteq 4.810 \times 10^{-5}$, $\|b\|_2 \doteq 2.772$

$$\frac{\|\Delta x\|_2}{\|x\|_2} \leqslant \frac{3.842 \times 10^4}{1 - 0.1672}(4.353 \times 10^{-6} + 1.735 \times 10^{-5}) = 1.001,$$

whereas in reality we have $\|\Delta x\|_2/\|x\|_2 \doteq 0.8114$.

## 1.3 Systems with Special Properties

The estimate (1.86) for the relative error has another application, in the case where the given data is correct. The computed coefficients of the first reduced system can be interpreted as the exact values of a perturbed initial system. Thus the resulting triangular decomposition can be thought of being the exact decomposition of a perturbed initial matrix, so that $P(A + \Delta A) = \tilde{L}\tilde{R}$ holds, where $\tilde{L}$ and $\tilde{R}$ denote the triangular matrices obtained by means of a floating point computation. An analysis of the rounding errors yields estimates of the moduli of the elements of $\Delta A$. The idea of such a *backward error analysis* can be extended to the processes of the forward and backward substitution and leads to estimates for $\Delta b$. However, the theoretical results are in general too pessimistic and do not properly reflect practical experience (Stoer, 1983; Stummel and Hainer, 1982; Wilkinson, 1965; 1969). In general the actual changes $\Delta A$ and $\Delta b$ of a backward error computation are comparable in size with the inevitable input errors for small systems and for large systems they increase only by factors of small powers of ten. The estimate (1.86) is often too pessimistic due to its generality, and thus the above rule of thumb remains true, at least as a guiding principle, in connection with the backward error analysis.

On the basis of the preceding arguments we can add a heuristic statement concerning the iterative improvement of a solution. In order that each iteration step improves the approximate solution, it is necessary that at least one digit of the correction vector is correct, and this must hold in relation to the modulus of the largest component. The necessary condition for a floating point computation with $d$ digits is thus $\alpha < d - 1$. On this condition, each step of the iterative improvement will yield a further $(d - \alpha - 1)$ correct digits, and the sequence of approximate solutions converges. The values computed in Example 1.7 illustrate this fact nicely. Although the condition number corresponding to the spectral norm is $\kappa(A) \doteq 2.82 \times 10^3$, two digits of $\tilde{x}$ are correct, and each step of the iterative improvement yields an additional two correct digits.

If a program has to provide the user with a precise error bound of a computed approximate solution $\tilde{x}$, or if it has to be able to decide whether an iterative improvement is necessary or at all meaningful, the condition number $\kappa(A)$ must be known. For this purpose we need either to know the inverse $A^{-1}$ or the largest and the smallest eigenvalue of $A^T A$. In order to avoid these processes, which require too much computational effort compared with the solution of the system of equations, a method has been developed for the computation of a reasonable estimate of $\kappa(A)$ that requires only a small amount of work (Cline *et al.*, 1979; Forsythe and Moler, 1967).

### 1.3 Systems with Special Properties

Many applications require the solution of systems of linear equations with special properties or structures which, if taken into account, can reduce the amount of work and the storage requirements. We now consider some important special cases of systems of equations which will occur in the following chapters. We develop the appropriate algorithms and discuss the suitable computer implementations.

### 1.3.1 Symmetric, positive definite systems

The matrix $\mathbf{A}$ in $\mathbf{Ax} + \mathbf{b} = \mathbf{0}$ is often not only symmetric but also positive definite.

**Definition 1.7** *A symmetric matrix* $\mathbf{A} \in \mathbf{R}^{n \times n}$ *is called positive definite if the corresponding quadratic form is positive definite, i.e. if*

$$Q(\mathbf{x}) := \mathbf{x}^T \mathbf{A} \mathbf{x} = \sum_{i=1}^{n} \sum_{k=1}^{n} a_{ik} x_i x_k \begin{cases} \geq 0 & \text{for all } \mathbf{x} \in \mathbf{R}^n \\ = 0 & \text{only for } \mathbf{x} = \mathbf{0}. \end{cases} \tag{1.87}$$

**Theorem 1.8** *If* $\mathbf{A} \in \mathbf{R}^{n \times n}$ *is a symmetric, positive definite matrix* $\mathbf{A} \in \mathbf{R}^{n \times n}$ *then the elements satisfy the following conditions:*

(a) $a_{ii} > 0$ for $i = 1, 2, \ldots, n;$ \qquad (1.88)

(b) $a_{ik}^2 < a_{ii} a_{kk}$ for $i \neq k$; $i, k = 1, 2, \ldots, n;$ \qquad (1.89)

(c) there exists a $k$ such that $\max_{i,j} |a_{ij}| = a_{kk}.$ \qquad (1.90)

*Proof* We prove the first two properties by making special choices of $\mathbf{x} \neq \mathbf{0}$. In definition (1.87) with $\mathbf{x} = \mathbf{e}_i$, (1.88) follows immediately since $Q(\mathbf{x}) = a_{ii} > 0$, where $\mathbf{e}_i$ is the $i$th unit vector. If we choose $\mathbf{x} = \xi \mathbf{e}_i + \mathbf{e}_k$, where $\xi \in \mathbf{R}$, $i \neq k$, the quadratic form $Q(\mathbf{x})$ reduces to $a_{ii}\xi^2 + 2a_{ik}\xi + a_{kk} > 0$ for all $\xi \in \mathbf{R}$. The quadratic equation $a_{ii}\xi^2 + 2a_{ik}\xi + a_{kk} = 0$ has no real solution in $\xi$. Therefore its discriminant is $4a_{ik}^2 - 4a_{ii}a_{kk} < 0$, from which (1.89) follows. Finally, the assumption that the matrix element of largest absolute value is not a diagonal element contradicts (1.89).

A necessary and sufficient condition for the positive definiteness of a symmetric matrix can be derived by reducing a quadratic form to a sum of squares. We may assume $a_{11} > 0$, because otherwise $\mathbf{A}$ is not positive definite from Theorem 1.8. Hence we can complete to a full square all terms, in (1.87), which contain $x_1$

$$Q(\mathbf{x}) = a_{11} x_1^2 + 2 \sum_{i=2}^{n} a_{i1} x_1 x_i + \sum_{i=2}^{n} \sum_{k=2}^{n} a_{ik} x_i x_k$$

$$= \left( \sqrt{a_{11}} x_1 + \sum_{i=2}^{n} \frac{a_{i1}}{\sqrt{a_{11}}} x_i \right)^2 + \sum_{i=2}^{n} \sum_{k=2}^{n} \left( a_{ik} - \frac{a_{i1} a_{1k}}{a_{11}} \right) x_i x_k$$

$$= \left( \sum_{i=1}^{n} l_{i1} x_i \right)^2 + \sum_{i=2}^{n} \sum_{k=2}^{n} a_{ik}^{(1)} x_i x_k = \left( \sum_{i=1}^{n} l_{i1} x_i \right)^2 + Q^{(1)}(\mathbf{x}^{(1)}). \tag{1.91}$$

## 1.3 Systems with Special Properties

Here we have introduced the quantities

$$l_{11} = \sqrt{a_{11}} \quad l_{i1} = \frac{a_{i1}}{\sqrt{a_{11}}} = \frac{a_{i1}}{l_{11}} \quad (i = 2, 3, \ldots, n) \tag{1.92}$$

$$a_{ik}^{(1)} = a_{ik} - \frac{a_{i1}a_{1k}}{a_{11}} = a_{ik} - l_{i1}l_{k1} \quad (i, k = 2, 3, \ldots, n). \tag{1.93}$$

$Q^{(1)}(\mathbf{x}^{(1)})$ denotes the quadratic form of the $(n-1)$ variables $x_2, x_3, \ldots, x_n$ with the coefficients $a_{ik}^{(1)}$ (1.93), which are defined by the same formula that holds for the first elimination step of the Gaussian algorithm with the pivotal element $a_{11}$. It corresponds to the matrix of the reduced system of equations.

*Theorem 1.9* *The symmetric matrix* $\mathbf{A} = (a_{ik})$ *with* $a_{11} > 0$ *is positive definite if and only if the reduced matrix* $\mathbf{A}^{(1)} = (a_{ik}^{(1)}) \in \mathbf{R}^{(n-1) \times (n-1)}$, *whose elements* $a_{ik}^{(1)}$ *are given by* (1.93), *is positive definite.*

*Proof* (a) Necessity: Let $\mathbf{A}$ be positive definite. For each vector $\mathbf{x}^{(1)} := (x_2, x_3, \ldots, x_n)^T \neq \mathbf{0}$ we can determine the value $x_1$ such that $\sum_{i=1}^{n} l_{i1} x_i = 0$, since $a_{11} > 0$, and hence $l_{11} > 0$. From (1.91) we get for the corresponding vector $\mathbf{x} = (x_1, x_2, \ldots, x_n)^T \neq \mathbf{0}$ the relation $0 < Q^{(1)}(\mathbf{x}^{(1)})$, and as a consequence the matrix $\mathbf{A}^{(1)}$ is necessarily positive definite.
(b) Sufficiency: Let $\mathbf{A}^{(1)}$ be positive definite. Hence, for all $\mathbf{x} \neq \mathbf{0}$ we have according to (1.91) $Q(\mathbf{x}) \geq 0$. The equality $Q(\mathbf{x}) = 0$ can hold only if both terms of the sum vanish simultaneously. From $Q^{(1)}(\mathbf{x}^{(1)}) = 0$ it follows that $x_2 = x_3 = \cdots x_3 = 0$, and the first term vanishes only for $x_1 = 0$ since $l_{11} \neq 0$. Thus the matrix $\mathbf{A}$ is necessarily positive definite.

Theorem 1.9 has some immediate consequences which are significant for the solution of symmetric and positive definite systems of equations.

*Theorem 1.10* *A symmetric matrix* $\mathbf{A} = (a_{ik}) \in \mathbf{R}^{n \times n}$ *is positive definite if and only if Gaussian elimination with diagonal strategy is feasible with $n$ positive pivotal elements.*

*Proof* If $\mathbf{A}$ is positive definite, there necessarily exists a first pivotal element $a_{11} > 0$ in the diagonal. According to Theorem 1.9 the reduced matrix $\mathbf{A}^{(1)}$ is again positive definite, and hence $a_{22}^{(1)} > 0$ is the second admissible pivotal element. This is true for all subsequent, reduced matrices $\mathbf{A}^{(k)}, (k = 1, 2, \ldots, n-1)$, and in particular $a_{nn}^{(n-1)}$ is the last positive pivotal element.
Conversely, if $a_{11} > 0, a_{22}^{(1)} > 0, \ldots, a_{nn}^{(n-1)} > 0$ hold, the matrix $\mathbf{A}^{(n-1)} = (a_{nn}^{(n-1)})$ is positive definite, and therefore the matrices $\mathbf{A}^{(n-2)}, \mathbf{A}^{(n-3)}, \ldots, \mathbf{A}^{(1)}$, $\mathbf{A}$ must be positive definite by applying theorem 1.9 successively.

Theorem 1.10 ensures that the triangular decomposition is feasible for all symmetric and positive definite matrices **A** by using the Gaussian algorithm without row exchanges. Since the matrices of the reduced systems of equations are all symmetric, due to (1.93), this means that the amount of work for the decomposition is approximately halved.

*Theorem 1.11* *A symmetric matrix* $\mathbf{A} = (a_{ik}) \in \mathbf{R}^{n \times n}$ *is positive definite if and only if the reduction of the quadratic form* $Q(\mathbf{x})$ *to a sum of n squares*

$$Q(\mathbf{x}) = \sum_{i=1}^{n} \sum_{k=1}^{n} a_{ik} x_i x_k = \sum_{k=1}^{n} \left( \sum_{i=k}^{n} l_{ik} x_i \right)^2 \tag{1.94}$$

*is completely feasible in the field of real numbers.*

*Proof* The statement becomes obvious on the basis of theorem 1.10 if we define the following quantities which arise in the general $k$th reduction step:

$$l_{kk} = \sqrt{a_{kk}^{(k-1)}} \quad l_{ik} = \frac{a_{ik}^{(k-1)}}{l_{kk}} \quad (i = k+1, k+2, \ldots, n) \tag{1.95}$$

$$a_{ij}^{(k)} = a_{ij}^{(k-1)} - l_{ik} l_{jk} \quad (i, j = k+1, k+2, \ldots, n) \tag{1.96}$$

From Theorem 1.10 $a_{kk}^{(k-1)}$ in (1.95) are positive if and only if **A** is positive definite.

Now we define the left triangular matrix **L** with the elements $l_{ik}$, which were introduced in (1.92) and (1.95) for $i \geq k$

$$\mathbf{L} = \begin{pmatrix} l_{11} & 0 & 0 & \cdots & 0 \\ l_{21} & l_{22} & 0 & \cdots & 0 \\ l_{31} & l_{32} & l_{33} & \cdots & 0 \\ \vdots & \vdots & \vdots & & \vdots \\ l_{n1} & l_{n2} & l_{n3} & \cdots & l_{nn} \end{pmatrix}. \tag{1.97}$$

*Theorem 1.12* *The reduction of a positive definite quadratic form to a sum of squares,* (1.94), *achieves the decomposition of the corresponding matrix* **A** *into the product*

$$\mathbf{A} = \mathbf{L}\mathbf{L}^\mathrm{T}. \tag{1.98}$$

*Proof* Using (1.94), the quadratic form $Q(\mathbf{x})$ is represented in two different ways, which read with the left triangular matrix **L** (1.97) as follows

$$Q(\mathbf{x}) = \mathbf{x}^\mathrm{T} \mathbf{A} \mathbf{x} = (\mathbf{L}^\mathrm{T} \mathbf{x})^\mathrm{T} (\mathbf{L}^\mathrm{T} \mathbf{x}) = \mathbf{x}^\mathrm{T} \mathbf{L} \mathbf{L}^\mathrm{T} \mathbf{x}. \tag{1.99}$$

Equation (1.98) now follows from the uniqueness of the representation.

## 1.3 Systems with Special Properties

(1.98) is called the *Cholesky decomposition* of the symmetric, positive definite matrix **A**, proposed by the geodetic surveyor Cholesky (Benoit, 1924).

By means of the Cholesky decomposition, (1.98), systems of linear equations with symmetric and positive definite coefficient matrices can be solved as follows. We substitute $\mathbf{A} = \mathbf{LL}^T$ into $\mathbf{Ax} + \mathbf{b} = \mathbf{0}$ and get

$$\mathbf{LL}^T\mathbf{x} + \mathbf{b} = \mathbf{0} \quad \text{or equivalently} \quad \mathbf{L}(\mathbf{L}^T\mathbf{x}) + \mathbf{b} = \mathbf{0}. \tag{1.100}$$

We introduce the auxiliary vector $\mathbf{c} = -\mathbf{L}^T\mathbf{x}$, and thus the solution of $\mathbf{Ax} + \mathbf{b} = \mathbf{0}$ can be performed by *Cholesky's method* in three steps.

$$
\begin{array}{ll}
1.\ \mathbf{A} = \mathbf{LL}^T & \text{(Cholesky's decomposition)} \\
2.\ \mathbf{Lc} - \mathbf{b} = \mathbf{0} & \text{(Forward substitution} \to \mathbf{c}) \\
3.\ \mathbf{L}^T\mathbf{x} + \mathbf{c} = \mathbf{0} & \text{(Backward substitution} \to \mathbf{x})
\end{array}
\tag{1.101}
$$

It is clear from the Gaussian algorithm that solving a system of linear equations is a rational process. The Cholesky decomposition has the advantage of completely preserving the symmetry, but also requiring $n$ square roots.

We observe that according to (1.96) only the matrix elements $a_{ij}^{(k)}$ in and below the diagonal have to be computed. Therefore the $k$th reduction step requires a square root, $(n-k)$ divisions and $(1 + 2 + \cdots + (n-k)) = \frac{1}{2}(n-k+1)(n-k)$ multiplications. The complete decomposition requires a total of

$$Z_{\mathbf{LL}^T} = [(n-1) + (n-2) + \cdots + 1] + \tfrac{1}{2}[n(n-1) + (n-1)(n-2) + \cdots + 2\cdot 1]$$
$$= \tfrac{1}{2}n(n-1) + \tfrac{1}{2}[\tfrac{1}{6}n(n-1)(2n-1) + \tfrac{1}{2}n(n-1)] = \tfrac{1}{6}(n^3 + 3n^2 - 4n)$$

essential operations, if we ignore the $n$ square roots. The processes of forward and backward substitution require the same amount of work, because the diagonal elements $l_{ii}$ are in general not equal to 1 that is

$$Z_F = Z_B = \tfrac{1}{2}(n^2 + n)$$

multiplicative operations. Thus the amount of computational work required to solve a system of $n$ linear equations by applying Cholesky's method is

$$\boxed{Z_{\text{Cholesky}} = \tfrac{1}{6}n^3 + \tfrac{3}{2}n^2 + \tfrac{1}{3}n} \tag{1.102}$$

essential operations. For larger values of $n$ the number of operations is approximately halved with comparison to (1.36).

The algorithmic summary of the three solution steps (1.101) is as follows, on the assumption that only the elements $a_{ik}$ of **A** in and below the diagonal are given.

> for $k = 1, 2, \ldots, n$:
>   if $a_{kk} \leqslant 0$: STOP
>   $l_{kk} = \sqrt{a_{kk}}$
>   for $i = k + 1, k + 2, \ldots, n$:
>     $l_{ik} = a_{ik}/l_{kk}$
>     for $j = k + 1, k + 2, \ldots, i$:
>       $a_{ij} = a_{ij} - l_{ik} \times l_{jk}$

(1.103)

> for $i = 1, 2, \ldots, n$:
>   $s = b_i$
>   for $j = 1, 2, \ldots, i - 1$:
>     $s = s - l_{ij} \times c_j$
>   $c_i = s/l_{ii}$

(1.104)

> for $i = n, n - 1, \ldots, 1$:
>   $s = c_i$
>   for $k = i + 1, i + 2, \ldots, n$:
>     $s = s + l_{ki} \times x_k$
>   $x_i = -s/l_{ii}$

(1.105)

The matrix elements $a_{ik}$ are changed during the process (1.103). However, we observe that the value of $a_{ik}$ is needed for the last time when $l_{ik}$ is computed. Hence, the matrix **L** can be generated at the place of **A**. To achieve this in (1.103), we just identify the variable $l$ with $a$. Similarly, we may identify the vector **b** with **c** in (1.104), and in (1.105) the solution vector **x** is identifiable with **c**, so that the vector **b** is replaced by the solution **x**.

In order to take full advantage of the fact that we only work with the lower parts of the matrices **A** and **L**, when applying Cholesky's method the relevant matrix elements are stored row by row in a one-dimensional array (see Figure 1.1). The matrix element $a_{ik}$ is stored in the $r$th place of this array, where $r$ is given by $r = \frac{1}{2}i(i-1) + k$. The storage requirement is now only $S = \frac{1}{2}n(n+1)$, i.e. about half of that required for a usual arrangement of the matrix elements.

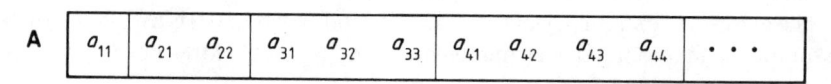

**A** | $a_{11}$ | $a_{21}$ | $a_{22}$ | $a_{31}$ | $a_{32}$ | $a_{33}$ | $a_{41}$ | $a_{42}$ | $a_{43}$ | $a_{44}$ | $\cdots$

Figure 1.1 Storage of the lower half of a symmetric, positive definite matrix

## 1.3 Systems with Special Properties

*Example* 1.10 Cholesky's method applied to

$$A = \begin{pmatrix} 5 & 7 & 3 \\ 7 & 11 & 2 \\ 3 & 2 & 6 \end{pmatrix} \quad b = \begin{pmatrix} 0 \\ 0 \\ -1 \end{pmatrix}$$

yields the following two reduced matrices, to five significant places

$$A^{(1)} = \begin{pmatrix} 1.2000 & -2.1999 \\ -2.1999 & 4.2001 \end{pmatrix} \quad A^{(2)} = (0.16680)$$

The left triangular matrix $L$, the vector $c$ of the forward substitution and the approximate solution $\tilde{x}$ are given by

$$L = \begin{pmatrix} 2.2361 & 0 & 0 \\ 3.1305 & 1.0954 & 0 \\ 1.3416 & -2.0083 & 0.40841 \end{pmatrix}$$

$$c = \begin{pmatrix} 0 \\ 0 \\ -2.4485 \end{pmatrix} \quad \tilde{x} = \begin{pmatrix} -18.984 \\ 10.991 \\ 5.9952 \end{pmatrix}.$$

Substitution of $\tilde{x}$ into the given equations results in the residual vector $r = (2.6, 3.4, 1.2)^T \times 10^{-3}$. The condition number corresponding to the spectral norm is $\kappa(A) \doteq 1.50 \times 10^3$. A step of the iterative improvement gives the correction $z \doteq (-15.99, 8.99, 4.80)^T \times 10^{-3}$, and thus the improved approximate solution coincides with the exact solution $x = (-19, 11, 6)^T$, within working accuracy. In this example the approximate solution $\tilde{x}$ is much more accurate than we would expect following our rule of thumb.

### 1.3.2 Banded systems

We call $A$ a *band matrix* if all the nonzero elements $a_{ik}$ are located in the diagonal and some adjacent off-diagonals. The symmetric and positive definite band matrices arise in several important applications.

*Definition* 1.8 *The bandwidth* $m$ *of a symmetric matrix* $A \in \mathbb{R}^{n \times n}$ *is defined to be the smallest integer* $m < n$ *such that*

$$a_{ik} = 0 \quad \text{for all } i \text{ and } k \text{ with } |i - k| > m. \tag{1.106}$$

The bandwidth $m$ indicates the number of off-diagonals on either side of the main diagonal which contain all the nonzero elements.

*Theorem* 1.13 *The left triangular matrix* $L$ *of the Cholesky decomposition* $A =$

$LL^T$ (1.98) *of a symmetric and positive definite band matrix* $A$ *which has a bandwidth* $m$ *has the same band structure, so that we have*

$$l_{ik} = 0 \text{ for all } i \text{ and } k \text{ with } i - k > m. \tag{1.107}$$

*Proof* It is sufficient to show that the first step of the reduction defined by (1.92) and (1.93) only generates nonzero elements in the first column of $L$ within the first $m$ off-diagonals, and that the reduced matrix $A^{(1)} = (a_{ik}^{(1)})$ is again a band matrix with the same bandwidth $m$. From (1.92) the first statement is obvious, since $l_{i1} = 0$ holds for all $i$ with $i - 1 > m$, because $a_{i1} = 0$. To show the second statement we only need to consider elements $a_{ik}^{(1)}$ below the diagonal because of symmetry. For an arbitrary index pair $(i, k)$, with $i \geqslant k \geqslant 2$ and $i - k > m$, we have $a_{ik} = 0$ from the assumption and, moreover, $l_{i1} = 0$ due to the first step and $i - 1 > i - k > m$. From (1.93) it follows that $a_{ik}^{(1)} = 0$ for all $i$, $k \geqslant 2$ with $|i - k| > m$.

According to Theorem 1.13 the Cholesky decomposition of a symmetric, positive definite band matrix $A$ can be completely carried out within the diagonal and the $m$ lower off-diagonals. Hence, the matrix $L$ can be generated exactly at the place of the essential part of the given matrix $A$. Moreover, it is clear that each step of the reduction only changes those elements of the band being within a triangular array that comprises, at most, the following $m$ rows. To make the situation clear in Figure 1.2 we illustrate the general $k$th step and the third to last step of the reduction in the case of a band matrix with $m = 4$.

The amount of work for a general reduction step consists of a square root for

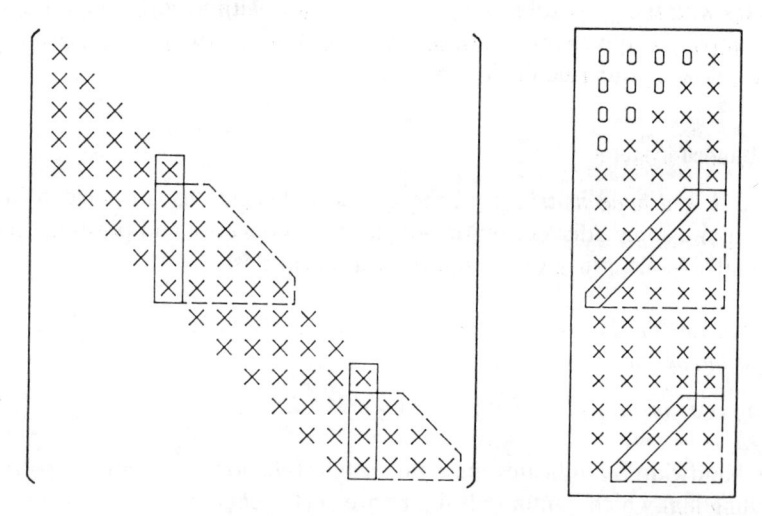

Figure 1.2 Reduction and storage of a symmetric, positive definite band matrix

## 1.3 Systems with Special Properties

$l_{kk}$, $m$ divisions for the values $l_{ik}$ and $\frac{1}{2}m(m+1)$ multiplications for the actual reduction of elements. The total amount of work for a Cholesky decomposition of a band matrix of order $n$ and of bandwidth $m$ is given by $n$ square roots and less than $\frac{1}{2}nm(m+3)$ essential operations. The amount of work required is proportional to the first power of the order $n$ and the square of the bandwidth $m$. The processes of the forward and backward substitution can both be done with less than $n(m+1)$ operations. As a consequence we obtain the estimate for the total amount of work to solve a system of $n$ linear equations with a symmetric and positive definite band matrix of bandwidth $m$

$$Z^{(\text{Band})}_{\text{Cholesky}} \leqslant \tfrac{1}{2}nm(m+3) + 2n(m+1) \tag{1.108}$$

In order to take full advantage of the bandstructure we store the lower part of a band matrix within a rectangular array of $n$ rows and $(m+1)$ columns, as shown in Figure 1.2. The off-diagonals of **A** are arranged columnwise in such a way that the $i$th row of **A** appears as the $i$th row of the array. The diagonal elements can be retrieved from the $(m+1)$th column, and the general element $a_{ik}$ of **A** with $\max(i-m, 1) \leqslant k \leqslant i$ can be found in the $(k-i+m+1)$th column. It is appropriate to set the undefined elements of the left upper triangle of the array equal to zero.

The algorithmic formulation of the Cholesky decomposition $\mathbf{A} = \mathbf{L}\mathbf{L}^{\mathrm{T}}$ of a band matrix **A** of order $n$ and bandwidth $m$, using the storage arrangement of Figure 1.2, is

$$
\begin{aligned}
&\text{for } k = 1, 2, \ldots, n: \\
&\quad \text{if } a_{k,m+1} \leqslant 0: \text{STOP} \\
&\quad l_{k,m+1} = \sqrt{a_{k,m+1}} \\
&\quad p = \min(k+m, n) \\
&\quad \text{for } i = k+1, k+2, \ldots, p: \\
&\qquad l_{i,k-i+m+1} = a_{i,k-i+m+1}/l_{k,m+1} \\
&\qquad \text{for } j = k+1, k+2, \ldots, i: \\
&\qquad\quad a_{i,j-i+m+1} = a_{i,j-i+m+1} - l_{i,k-i+m+1} \times l_{j,k-j+m+1}
\end{aligned}
$$

### 1.3.3 Tridiagonal systems

Systems of equations with a *tridiagonal* matrix **A** occur in several applications. Such systems can be treated in a simple manner since the general $i$th equation only contains three unknowns $x_{i-1}$, $x_i$ and $x_{i+1}$. In order to take into account this special structure of the matrix and to simplify the implementation on a computer, we start with a simple example of a system of equations for $n = 5$.

$$
\begin{array}{cccccc}
x_1 & x_2 & x_3 & x_4 & x_5 & 1
\end{array}
$$

$$
\left.\begin{array}{|ccccc|c|}
\hline
a_1 & b_1 & & & & d_1 \\
c_1 & a_2 & b_2 & & & d_2 \\
 & c_2 & a_3 & b_3 & & d_3 \\
 & & c_3 & a_4 & b_4 & d_4 \\
 & & & c_4 & a_5 & d_5 \\
\hline
\end{array}\right. \tag{1.109}
$$

To begin with, we assume that the Gaussian algorithm is feasible using diagonal strategy, i.e. without row exchanges, because **A** is known to be diagonally dominant or symmetric and positive definite, for instance. Then the triangular decomposition **A** = **LR** exists, and it is easy to see that **L** is a *bidiagonal* left triangular matrix and **R** is a *bidiagonal* right triangular matrix. Therefore we can directly use the following decomposition

$$
\begin{pmatrix}
a_1 & b_1 & & & \\
c_1 & a_2 & b_2 & & \\
 & c_2 & a_3 & b_3 & \\
 & & c_3 & a_4 & b_4 \\
 & & & c_4 & a_5
\end{pmatrix}
$$

$$
= \begin{pmatrix}
1 & & & & \\
l_1 & 1 & & & \\
 & l_2 & 1 & & \\
 & & l_3 & 1 & \\
 & & & l_4 & 1
\end{pmatrix} \cdot \begin{pmatrix}
m_1 & r_1 & & & \\
 & m_2 & r_2 & & \\
 & & m_3 & r_3 & \\
 & & & m_4 & r_4 \\
 & & & & m_5
\end{pmatrix}, \tag{1.110}
$$

and determine the unknown quantities $l_i$, $m_i$, $r_i$ by a simple comparison of corresponding matrix elements. We obtain the following equations

$$
\begin{aligned}
& & a_1 &= m_1 & b_1 &= r_1 \\
c_1 &= l_1 m_1 & a_2 &= l_1 r_1 + m_2 & b_2 &= r_2 \\
c_2 &= l_2 m_2 & a_3 &= l_2 r_2 + m_3 & b_3 &= r_3 \\
c_3 &= l_3 m_3 & a_4 &= l_3 r_3 + m_4 & b_4 &= r_4 \\
c_4 &= l_4 m_4 & a_5 &= l_4 r_4 + m_5 & &
\end{aligned} \tag{1.111}
$$

From (1.111) the unknowns are determined in the order $m_1$; $r_1$, $l_1$, $m_2$; $r_2$, $l_2$, $m_3$;...;$r_4$, $l_4$, $m_5$. Since $r_i = b_i$ for all $i$, we can summarize the algorithm for decomposing the tridiagonal matrix **A** (1.109) for general $n$:

## 1.3 Systems with Special Properties

$$\boxed{\begin{array}{l} m_1 = a_1 \\ \text{for } i = 1, 2, \ldots, n-1: \\ \quad l_i = c_i/m_i \\ \quad m_{i+1} = a_{i+1} - l_i \times b_i \end{array}} \qquad (1.112)$$

The forward substitution $\mathbf{Ly} - \mathbf{d} = \mathbf{0}$ can be summed up by the simple formulation

$$\boxed{\begin{array}{l} y_1 = d_1 \\ \text{for } i = 2, 3, \ldots, n: \\ \quad y_i = d_i - l_{i-1} \times y_{i-1} \end{array}} \qquad (1.113)$$

The back substitution $\mathbf{Rx} + \mathbf{y} = \mathbf{0}$ is formulated as

$$\boxed{\begin{array}{l} x_n = -y_n/m_n \\ \text{for } i = n-1, n-2, \ldots, 1: \\ \quad x_i = -(y_i + b_i \times x_{i+1})/m_i \end{array}} \qquad (1.114)$$

A computer program for solving a tridiagonal system of equations using the Gaussian algorithm with diagonal strategy essentially consists of three simple loop statements. The number of essential operations for the three steps together is

$$\boxed{Z_{\text{Gauss}}^{(\text{trid})} = 2(n-1) + (n-1) + 1 + 2(n-1) = 5n - 4} \qquad (1.115)$$

The amount of work is only proportional to the number of unknowns. So we can solve even large tridiagonal systems with a relatively small amount of effort.

The same holds true if we have to apply row exchanges in the course of the Gaussian algorithm. We explain the principle for the system (1.109) by applying the relative maximal column pivoting strategy. For the first elimination step the two elements $a_1$ and $c_1$ are the only pivot candidates. We define the two auxiliary quantities

$$\alpha := |a_1| + |b_1| \quad \beta := |c_1| + |a_2| + |b_2|. \qquad (1.116)$$

If $|a_1|/\alpha \geqslant |c_1|/\beta$, then $a_1$ is the pivotal element, otherwise a row exchange is needed. In this case, a nonzero element is produced at the place (1, 3) which is outside the original band. In order to have a uniform description of the elimination step we define the following quantities.

$$\begin{array}{ll} \text{If } a_1 \text{ is the pivot} & \begin{cases} r_1 := a_1 & s_1 := b_1 & t_1 := 0 & f_1 := d_1 \\ u := c_1 & v := a_2 & w := b_2 & z := d_2. \end{cases} \\ \text{If } c_1 \text{ is the pivot} & \begin{cases} r_1 := c_1 & s_1 := a_2 & t_1 := b_2 & f_1 := d_2 \\ u := a_1 & v := b_1 & w := 0 & z := d_1. \end{cases} \end{array} \qquad (1.117)$$

With these variables (1.109) becomes

$$
\begin{array}{ccccc|c}
x_1 & x_2 & x_3 & x_4 & x_5 & 1 \\
\hline
r_1 & s_1 & t_1 & & & f_1 \\
u & v & w & & & z \\
 & c_2 & a_3 & b_3 & & d_3 \\
 & & c_3 & a_4 & b_4 & d_4 \\
 & & & c_4 & a_5 & d_5 \\
\hline
\end{array}
\tag{1.118}
$$

The first elimination step now yields

$$
\begin{array}{c|cccc|c}
x_1 & x_2 & x_3 & x_4 & x_5 & 1 \\
\hline
r_1 & s_1 & t_1 & & & f_1 \\
\hline
l_1 & a_2' & b_2' & & & d_2' \\
 & c_2 & a_3 & b_3 & & d_3 \\
 & & c_3 & a_4 & b_4 & d_4 \\
 & & & c_4 & a_5 & d_5 \\
\hline
\end{array}
\tag{1.119}
$$

with the quantities

$$l_1 := u/r_1 \quad a_2' := v - l_1 s_1 \quad b_2' := w - l_1 t_1 \quad d_2' := z - l_1 f_1. \tag{1.120}$$

The same situation presents itself for the reduced system, because its matrix is again tridiagonal. So we can apply the same reasoning to the following steps. In order that the formulae (1.116) and (1.117) also hold for the last elimination step, we must define $b_n = 0$. A consequent continuation of the elimination ends with

$$
\begin{array}{ccccc|c}
x_1 & x_2 & x_3 & x_4 & x_5 & 1 \\
\hline
r_1 & s_1 & t_1 & & & f_1 \\
l_1 & r_2 & s_2 & t_2 & & f_2 \\
 & l_2 & r_3 & s_3 & t_3 & f_3 \\
 & & l_3 & r_4 & s_4 & f_4 \\
 & & & l_4 & r_5 & f_5 \\
\hline
\end{array}
\tag{1.121}
$$

The Gaussian algorithm, together with the relative maximal column pivoting strategy, for a tridiagonal system (1.109) can be formulated using the formulae (1.116), (1.117) and (1.120), whereby the forward substitution is included.

## 1.4 Exchange Step and Inversion of Matrices

> for $i = 1, 2, \ldots, n-1$:
> $\alpha = |a_i| + |b_i|; \beta = |c_i| + |a_{i+1}| + |b_{i+1}|$
> if $|a_i|/\alpha \geqslant |c_i|/\beta$
> then $r_i = a_i; s_i = b_i; t_i = 0; f_i = d_i;$
> $u = c_i; v = a_{i+1}; w = b_{i+1}; z = d_{i+1}$
> else $r_i = c_i; s_i = a_{i+1}; t_i = b_{i+1}; f_i = d_{i+1};$
> $u = a_i; v = b_i; w = 0; z = d_i$
> $l_i = u/r_i; a_{i+1} = v - l_i \times s_i$
> $b_{i+1} = w - l_i \times t_i; d_{i+1} = z - l_i \times f_i$
> $r_n = a_n; f_n = d_n$

(1.122)

The notation of the schemes (1.118) and (1.121) is used in the algorithmic formulation (1.122). We observe that the coefficients $a_i$, $b_i$, $c_{i-1}$, $d_i$ of the given equations are changed. Therefore we can identify the following variables in (1.122): $r_i = a_i, s_i = b_i, l_i = c_i, f_i = d_i$. With this system some storage space can be saved, and the formulation (1.122) can be simplified.

The unknowns $x_i$ are computed by back substitution described by

> $x_n = -f_n/r_n$
> $x_{n-1} = -(f_{n-1} + s_{n-1} \times x_n)/r_{n-1}$
> for $i = n-2, n-3, \ldots, 1$:
> $x_i = -(f_i + s_i \times x_{i+1} + t_i \times x_{i+2})/r_i$

(1.123)

The total number of multiplicative operations required to solve a general tridiagonal system of equations in $n$ unknowns is given by, if the two divisions for choosing the pivotal element are included,

$$Z_{\text{Gauss}}^{(\text{trid. gen})} = 5(n-1) + (n-1) + 3(n-1) = 9(n-1).$$

In comparison to (1.115) the amount of computational effort required is approximately doubled if pivoting is necessary. The left triangular matrix **L** of the decomposition $\mathbf{PA} = \mathbf{LR}$ is still bidiagonal, but **R** is a right triangular matrix in which two of the upper off-diagonals contain elements that are in general, nonzero.

### 1.4 Exchange Step and Inversion of Matrices

In the following we consider a fundamental operation of linear algebra. It will be the basis of an algorithm which allows the direct and practical inversion of nonsingular matrices. Moreover, the operation is the key for an algorithm to be developed in the next chapter.

### 1.4.1 Linear functions, exchange

Consider $m$ linear functions $y_i$, in $n$ variables $x_k$

$$y_i = \sum_{k=1}^{n} a_{ik} x_k + b_i \quad (i = 1, 2, \ldots, m), \tag{1.124}$$

which we represent again schematically for the special case $m = 3$ and $n = 4$.

$$
\begin{array}{c|cccc|c}
 & x_1 & x_2 & x_3 & x_4 & 1 \\
\hline
y_1 = & a_{11} & a_{12} & a_{13} & a_{14} & b_1 \\
y_2 = & a_{21} & a_{22} & a_{23} & a_{24} & b_2 \\
y_3 = & a_{31} & a_{32} & a_{33} & a_{34} & b_3
\end{array}
\tag{1.125}
$$

We call $x_1, x_2, \ldots, x_n$ the *independent variables* and $y_1, y_2, \ldots, y_m$ the *dependent variables*.

We now consider the $p$th linear function of (1.124). On the assumption that $a_{pq} \neq 0$ we can solve it for the variable $x_q$ and substitute the resulting expression into all the other linear functions

$$x_q = \frac{1}{a_{pq}} y_p - \sum_{\substack{k=1 \\ k \neq q}}^{n} \frac{a_{pk}}{a_{pq}} x_k - \frac{b_p}{a_{pq}} \tag{1.126}$$

$$y_i = \frac{a_{iq}}{a_{pq}} y_p + \sum_{\substack{k=1 \\ k \neq q}}^{n} \left( a_{ik} - \frac{a_{iq} a_{pk}}{a_{pq}} \right) x_k + \left( b_i - \frac{a_{iq} b_p}{a_{pq}} \right) \quad (i \neq p). \tag{1.127}$$

The equations (1.126) and (1.127) constitute $m$ new linear functions, where now $y_p$ plays the role of an independent and $x_q$ the role of a dependent variable. Since the variables $x_q$ and $y_p$ have been exchanged, in a sense, we call the algebraic operation an *exchange step*. Analogously to the preceding scheme, (1.125), we combine the new linear functions (1.126) and (1.127) in a new scheme. If we choose $p = 2$, $q = 3$ the scheme (1.125) results in

$$
\begin{array}{c|cccc|c}
 & x_1 & x_2 & y_2 & x_4 & 1 \\
\hline
y_1 = & a'_{11} & a'_{12} & a'_{13} & a'_{14} & b'_1 \\
x_3 = & a'_{21} & a'_{22} & a'_{23} & a'_{24} & b'_2 \\
y_3 = & a'_{31} & a'_{32} & a'_{33} & a'_{34} & b'_3
\end{array}
\tag{1.128}
$$

## 1.4 Exchange Step and Inversion of Matrices

The elements of the new scheme are defined by the formulae

$$a'_{pq} = \frac{1}{a_{pq}}$$

$$a'_{pk} = -\frac{a_{pk}}{a_{pq}} \quad (k \neq q) \quad b'_p = -\frac{b_p}{a_{pq}}$$

$$a'_{iq} = \frac{a_{iq}}{a_{pq}} \quad (i \neq p) \tag{1.129}$$

$$a'_{ik} = a_{ik} - \frac{a_{iq} a_{pk}}{a_{pq}} = a_{ik} + a_{iq} a'_{pk} \quad (i \neq p,\ k \neq q)$$

$$b'_i = b_i - \frac{a_{iq} b_p}{a_{pq}} = b_i + a_{iq} b'_p \quad (i \neq p)$$

The element $a_{pq}$ is called the *pivotal element*, it is in the intersection of the $y$ row and the $x$ column of the two variables which are to be exchanged. The corresponding $y$ row is called *pivotal row*, and the $x$ column is called *pivotal column*. We summarize the formulae (1.129) of an exchange step.

### Rule
1. *The pivotal element is replaced by its reciprocal value.*
2. *The other elements of the pivotal row have to be divided by the pivotal element followed by a change of sign.*
3. *The other elements of the pivotal column have to be divided by the pivotal element.*
4. *An element in the remaining part of the scheme is transformed by adding the product of the element that is in the same row in the pivotal column and the new element that is found in the same column in the pivotal row.*

*Example* 1.11 We perform an exchange step for the following linear functions with the pivot $a_{32}$

|       | $x_1$ | $x_2$ | $x_3$ | 1   |
|-------|-------|-------|-------|-----|
| $y_1 =$ | 3    | 7     | 4     | −13 |
| $y_2 =$ | −5   | 4     | 5     | 2   |
| $y_3 =$ | −1   | 2     | 3     | −6  |
|       | 0.5  |       | −1.5  | 3   |

→

|       | $x_1$ | $y_3$ | $x_3$ | 1   |
|-------|-------|-------|-------|-----|
| $y_1 =$ | 6.5  | 3.5   | −6.5  | 8   |
| $y_2 =$ | −3   | 2     | −1    | 14  |
| $x_2 =$ | 0.5  | 0.5   | −1.5  | 3   |

The elements of the pivotal row and column, as well as the variables to be exchanged, have been underlined to simplify the application of the above rule. For the same reason the new pivotal row, with the exception of the element of the pivotal column, has been written below the given scheme, forming the so-called *cellar row*.

### 1.4.2 Matrix inversion

The inversion of a nonsingular matrix $\mathbf{A} \in \mathbf{R}^{n \times n}$ is equivalent to the problem of solving $n$ *linear forms* in $n$ variables

$$y_i = \sum_{k=1}^{n} a_{ik} x_k \quad (i = 1, 2, \ldots, n) \quad \text{or} \quad \mathbf{y} = \mathbf{A}\mathbf{x} \tag{1.130}$$

for the independent variables $x_k$, so that we have

$$x_i = \sum_{k=1}^{n} \alpha_{ik} y_k \quad (i = 1, 2, \ldots, n) \quad \text{or} \quad \mathbf{x} = \mathbf{A}^{-1}\mathbf{y}. \tag{1.131}$$

We have at our disposal a means to solve this problem by applying an appropriate sequence of exchange steps, each of which exchanges just one independent $x$ variable for a dependent $y$ variable.

*Example* 1.12 We want to invert the matrix

$$\mathbf{A} = \begin{pmatrix} -3 & 5 & -4 \\ 2 & -6 & 12 \\ 1 & -2 & 2 \end{pmatrix}.$$

Three consecutive exchange steps will give us the desired result. The choice of the pivotal element, of the remaining $y$ rows and $x$ columns, becomes more restrictive during the progressive computation. The pivot is even uniquely determined for the last step.

|       | $x_1$ | $x_2$ | $x_3$ |
|-------|-------|-------|-------|
| $y_1 =$ | $\underline{-3}$ | 5 | $-4$ |
| $y_2 =$ | $\underline{2}$ | $-6$ | 12 |
| $y_3 =$ | $\underline{1}$ | $\underline{-2}$ | 2 |
|       | 2 | $-2$ |  |

|       | $y_3$ | $x_2$ | $x_3$ |
|-------|-------|-------|-------|
| $y_1 =$ | $-3$ | $\underline{-1}$ | 2 |
| $y_2 =$ | 2 | $\underline{-2}$ | 8 |
| $x_1 =$ | 1 | $\underline{2}$ | $-2$ |
|       | $-3$ |  | 2 |

|       | $y_3$ | $y_1$ | $x_3$ |
|-------|-------|-------|-------|
| $x_2 =$ | $-3$ | $-1$ | 2 |
| $y_2 =$ | 8 | 2 | $\underline{4}$ |
| $x_1 =$ | $-5$ | $-2$ | $\underline{2}$ |
|       | $-2$ | $-0.5$ |  |

|       | $y_3$ | $y_1$ | $y_2$ |
|-------|-------|-------|-------|
| $x_2 =$ | $-7$ | $-2$ | 0.5 |
| $x_3 =$ | $-2$ | $-0.5$ | 0.25 |
| $x_1 =$ | $-9$ | $-3$ | 0.5 |

## 1.4 Exchange Step and Inversion of Matrices

We obtain the inverse matrix $\mathbf{A}^{-1}$ from the last scheme, after rearranging its rows and columns

$$\mathbf{A}^{-1} = \begin{pmatrix} -3 & 0.5 & -9 \\ -2 & 0.5 & -7 \\ -0.5 & 0.25 & -2 \end{pmatrix}.$$

The described procedure is always feasible for nonsingular matrices due to the following theorem.

*Theorem 1.14* *If $\mathbf{A} \in \mathbf{R}^{n \times n}$ is nonsingular, there exists a pivotal element in the kth column for the kth step of the process of inversion.*

*Proof* The operations of the exchange steps can be related to those of the Gaussian alogrithm, so that we can apply those results. The statement is clear for the first exchange step, for otherwise the given matrix $\mathbf{A}$ would contain a first column with vanishing elements and would be singular, contrary to our assumption. An appropriate row permutation can produce a nonzero element at the place $(1, 1)$. An exchange step is now feasible with the pivotal element $a_{11}$. The elements which do not lie in the pivotal row and column are transformed according to (1.129) exactly by the same formulae holding for the matrix elements of the reduced equations (1.6). Hence, from Theorem 1.2, there exists a nonzero pivotal element among the last $(n-1)$ elements of the second column that can be used for the second exchange step. After a possible row exchange and an executed second exchange step the argumentation repeats itself in a completely analogous manner.

On the basis of the relationship between the exchange method for inverting a matrix and the Gaussian algorithm we can apply the pivotal strategies developed there in precisely the same manner. The determination of the pivot is simplified in a program if the necessary row exchanges are carried out explicitly. At the end of the process the $x$ variables appear in their correct order on the left-hand side of the scheme in contrast to the $y$ variables. The process yields the inverse $\mathbf{B}$ of the matrix $\mathbf{PA}$. Let us denote by $\mathbf{P}_k$ the permutation matrix describing the row exchange before the kth step of the inversion process. Then we have

$$\mathbf{B} = (\mathbf{P}_{n-1} \mathbf{P}_{n-2} \cdots \mathbf{P}_2 \mathbf{P}_1 \mathbf{A})^{-1} = \mathbf{A}^{-1} \mathbf{P}_1^{-1} \mathbf{P}_2^{-1} \cdots \mathbf{P}_{n-2}^{-1} \mathbf{P}_{n-1}^{-1},$$
$$\mathbf{A}^{-1} = \mathbf{B} \mathbf{P}_{n-1} \mathbf{P}_{n-2} \cdots \mathbf{P}_2 \mathbf{P}_1.$$

Hence we have to apply the row exchanges in the *opposite order* to the columns of $\mathbf{B}$ to produce $\mathbf{A}^{-1}$. The information on performed row exchanges can be generated in the components of a vector $\mathbf{p} \in \mathbf{R}^n$ as has been done in the formulation of the Gaussian algorithm. Moreover, we want to carry out the exchange steps using the storage space of $\mathbf{A}$. Therefore we have to change the order of the operations of the above rule in such a way that the original values, that are still needed, are available. With these explanations the detailed formulation of the process for inverting a nonsingular matrix $\mathbf{A} \in \mathbf{R}^{n \times n}$ is given in (1.132).

$$
\begin{aligned}
&\text{for } k = 1, 2, \ldots, n: \\
&\quad \max = 0;\ p_k = 0 \\
&\quad \text{for } i = k, k+1, \ldots, n: \\
&\quad\quad s = 0 \\
&\quad\quad \text{for } j = k, k+1, \ldots, n: \\
&\quad\quad\quad s = s + |a_{ij}| \\
&\quad\quad q = |a_{ik}|/s \\
&\quad\quad \text{if } q > \max: \\
&\quad\quad\quad \max = q;\ p_k = i \\
&\quad \text{if } \max = 0: \text{STOP} \\
&\quad \text{if } p_k \neq k: \\
&\quad\quad \text{for } j = 1, 2, \ldots, n: \\
&\quad\quad\quad h = a_{kj};\ a_{kj} = a_{p_k, j};\ a_{p_k, j} = h \\
&\quad \text{pivot} = a_{kk} \\
&\quad \text{for } j = 1, 2, \ldots, n: \\
&\quad\quad \text{if } j \neq k: \\
&\quad\quad\quad a_{kj} = -a_{kj}/\text{pivot} \\
&\quad\quad\quad \text{for } i = 1, 2, \ldots, n: \\
&\quad\quad\quad\quad \text{if } i \neq k:\ a_{ij} = a_{ij} + a_{ik} \times a_{kj} \\
&\quad \text{for } i = 1, 2, \ldots, n: \\
&\quad\quad a_{ik} = a_{ik}/\text{pivot} \\
&\quad a_{kk} = 1/\text{pivot} \\
&\text{for } k = n-1, n-2, \ldots, 1: \\
&\quad \text{if } p_k \neq k: \\
&\quad\quad \text{for } i = 1, 2, \ldots, n: \\
&\quad\quad\quad h = a_{ik};\ a_{ik} = a_{i, p_k};\ a_{i, p_k} = h
\end{aligned}
\qquad (1.132)
$$

A single exchange step for $n$ linear forms in $n$ variables requires $n^2$ essential operations, therefore the total number of such operations required to invert a nonsingular matrix $\mathbf{A} \in \mathbf{R}^{n \times n}$ is

$$Z_{\text{Inv}}^{(\text{AT})} = n^3. \qquad (1.133)$$

## 1.5 Exercises

**1.1.** Solve the system of equations by applying the Gaussian algorithm

$$
\begin{aligned}
2x_1 - 4x_2 + 6x_3 - 2x_4 - 3 &= 0 \\
3x_1 - 6x_2 + 10x_3 - 4x_4 + 2 &= 0
\end{aligned}
$$

## 1.5 Exercises

$$x_1 + 3x_2 + 13x_3 - 6x_4 - 3 = 0$$
$$5x_2 + 11x_3 - 6x_4 + 5 = 0.$$

What is the LR decomposition of the matrix?

**1.2.** Solve the system of linear equations

$$6.22x_1 + 1.42x_2 - 1.72x_3 + 1.91x_4 - 7.53 = 0$$
$$1.44x_1 + 5.33x_2 + 1.11x_3 - 1.82x_4 - 6.06 = 0$$
$$1.59x_1 - 1.23x_2 - 5.24x_3 - 1.42x_4 + 8.05 = 0$$
$$1.75x_1 - 1.69x_2 + 1.57x_3 + 6.55x_4 - 8.10 = 0$$

exploiting the diagonal dominance by means of the Gaussian algorithm and to five significant digits.

**1.3.** Compute the value of the determinant

$$\begin{vmatrix} 0.596 & 0.497 & 0.263 \\ 4.07 & 3.21 & 1.39 \\ 0.297 & 0.402 & 0.516 \end{vmatrix}$$

(a) Use the definition of a determinant, i.e. the rule of Sarrus, and compute first with full accuracy, and then with only three significant digits. In the second case the numerical result depends on the order of the operations. In order to explain the differences compare all intermediate results.

(b) Apply also the Gaussian algorithm with an accuracy of three significant digits using diagonal strategy and relative maximal column pivoting strategy.

**1.4.** Solve the system of linear equations

$$10x_1 + 14x_2 + 11x_3 - 1 = 0$$
$$13x_1 - 66x_2 + 14x_3 - 1 = 0$$
$$11x_1 - 13x_2 + 12x_3 - 1 = 0$$

by means of the Gaussian algorithm with relative maximal column pivoting strategy to five significant digits. What is the permutation matrix $\mathbf{P}$ and the matrices $\mathbf{L}$ and $\mathbf{R}$ of the decomposition? Compute the residuals with higher accuracy and perform one step of the iterative improvement. How large are the condition numbers of the matrix $\mathbf{A}$ corresponding to the total norm, the row sum norm and the Frobenius norm?

**1.5.** Verify the consistency $\|\mathbf{AB}\|_F \leq \|\mathbf{A}\|_F \|\mathbf{B}\|_F$ of the Frobenius norm.

**1.6.** Show that the Frobenius norm and the total norm are compatible with the Euclidean vector norm.

**1.7.** What is the matrix norm subordinate to the vector norm (1.61)?

**1.8.** Show that the condition numbers satisfy the following relations
(a) $\kappa(\mathbf{AB}) \leq \kappa(\mathbf{A}) \cdot \kappa(\mathbf{B})$ for all matrix norms;
(b) $\kappa(c\mathbf{A}) = \kappa(\mathbf{A})$ for all $c \in \mathbf{R}$;
(c) $\kappa_2(\mathbf{U}) = 1$ if $\mathbf{U}$ is an orthogonal matrix;
(d) $\kappa_2(\mathbf{A}) \leq \kappa_F(\mathbf{A}) \leq \kappa_T(\mathbf{A}) \leq n^2 \kappa_\infty(\mathbf{A})$;
(e) $\kappa_2(\mathbf{UA}) = \kappa_2(\mathbf{A})$ if $\mathbf{U}$ is an orthogonal matrix.

**1.9.** The system of linear equations

$$5x_1 + 7x_2 + 6x_3 + 5x_4 - 12 = 0$$
$$7x_1 + 10x_2 + 8x_3 + 7x_4 - 19 = 0$$
$$6x_1 + 8x_2 + 10x_3 + 9x_4 - 17 = 0$$
$$5x_1 + 7x_2 + 9x_3 + 10x_4 - 25 = 0$$

has a symmetric and positive definite matrix **A**. Solve it by means of the Gaussian algorithm (diagonal strategy and maximal column pivoting strategy) and by Cholesky's method to five significant digits. Then apply the iterative improvement method to the solution, compute the condition number of **A** in order to qualitatively explain the observation concerning the numerical results.

**1.10.** Decide whether the two symmetric matrices are positive definite

$$\mathbf{A} = \begin{pmatrix} 2 & -1 & 0 & -2 \\ -1 & 3 & -2 & 4 \\ 0 & -2 & 4 & -3 \\ -2 & 4 & -3 & 5 \end{pmatrix} \quad \mathbf{B} = \begin{pmatrix} 4 & -2 & 4 & -6 \\ -2 & 2 & -2 & 5 \\ 4 & -2 & 13 & -18 \\ -6 & 5 & -18 & 33 \end{pmatrix}.$$

**1.11.** Show that the Hilbert matrix

$$\mathbf{H} := \begin{pmatrix} 1 & \frac{1}{2} & \frac{1}{3} & \frac{1}{4} & \cdots \\ \frac{1}{2} & \frac{1}{3} & \frac{1}{4} & \frac{1}{5} & \cdots \\ \frac{1}{3} & \frac{1}{4} & \frac{1}{5} & \frac{1}{6} & \cdots \\ \frac{1}{4} & \frac{1}{5} & \frac{1}{6} & \frac{1}{7} & \cdots \\ \vdots & \vdots & \vdots & \vdots & \end{pmatrix} \in \mathbf{R}^{n \times n} \quad h_{ik} = \frac{1}{i+k-1}$$

of arbitrary order $n$ is positive definite. Then compute the inverses for $n = 3, 4, 5, \ldots, 12$, which are known to have integer valued elements, by using the Cholesky's method and the exchange method. Determine the growth of the condition number $\kappa(\mathbf{H})$ for increasing $n$ from your numerical results for $\mathbf{H}^{-1}$. Compute the condition number $\kappa_2(\mathbf{H})$ with the aid of methods from Chapter 6.

**1.12.** Solve the non-symmetric and tridiagonal system

$$-0.24x_1 + 1.76x_2 \qquad\qquad\qquad\qquad -1.28 = 0$$
$$-1.05x_1 + 1.26x_2 - 0.69x_3 \qquad\qquad\qquad -0.48 = 0$$
$$1.12x_2 - 2.12x_3 + 0.76x_4 \qquad\qquad -1.16 = 0$$
$$1.34x_3 + 0.36x_4 - 0.30x_5 \qquad +0.46 = 0$$
$$1.29x_4 + 1.05x_5 + 0.66x_6 + 0.66 = 0$$
$$0.96x_5 + 2.04x_6 + 0.57 = 0$$

by applying the relative maximal column pivoting strategy and to five significant digits. What are the matrices **L** and **R** of the row permuted matrix **A** of the system?

# 2

# Linear Programming

A branch of linear algebra is concerned with finding a solution of a class of extremal problems by finding the extremum value of a linear function of several variables, which have to satisfy some linear inequalities. To optimize processes in a certain sense, is a typical mathematical problem of operations research. Generally it is required to get the best from a conflict situation, where restrictive conditions have to be taken into account. In the following we confine ourselves to the fundamental treatment of simple problems of linear programming and refer you to Blum and Oettli (1975), Collatz and Wetterling (1971), Dantzig (1966), Glashoff and Gustafson (1978), Kall (1976) and Künzi et al. (1967) for more extensive treatment.

### 2.1 Introductory Examples, Graphical Solution

Two simple examples are shown to illustrate the problem of linear programming, and their graphical solution is used to motivate the computational procedure which is subsequently developed.

*Example* 2.1 We consider the problem that production might pose the owner of a small shoe factory. He plans to produce a ladies' and a men's model. His staff, consisting of 40 employees, and 10 machines should be used in an optimal way so that the profit of the production is maximized. The production of the shoes is subject to certain quite simple conditions concerning the available number of working hours of the employees and of the machine hours per month, as well as the available quantity of leather. The assumptions are given in Table 2.1.

The choice of the unknowns is obvious, namely

$x_1$ = number of ladies' shoes produced,
$x_2$ = number of men's shoes produced.

Table 2.1 The problem of production

|  | Ladies' shoe | Men's shoe | Available resources |
|---|---|---|---|
| Time of production (h) | 20 | 10 | 8000 |
| Machine handling (h) | 4 | 5 | 2000 |
| Leather required (dm$^2$) | 6 | 15 | 4500 |
| Net profit [£] | 16 | 32 | — |

The problem of optimization is mathematically formulated as follows:

$$\begin{aligned} 20x_1 + 10x_2 &\leqslant 8000 \\ 4x_1 + 5x_2 &\leqslant 2000 \\ 6x_1 + 15x_2 &\leqslant 4500 \\ x_1 &\geqslant 0 \\ x_2 &\geqslant 0 \\ 16x_1 + 32x_2 &= \text{Max!} \end{aligned} \quad (2.1)$$

The first three inequalities of (2.1) take into account the availability conditions, and the other two inequalities express the fact that the number of shoes produced cannot be negative. The last postulate of (2.1) corresponds to maximizing the profit. The system (2.1) represents a typical problem of optimization, in which a linear *objective function* has to be maximized under the condition that the unknowns have to satisfy a set of linear inequalities. The problem (2.1) is called a *linear program*.

The systematical treatment of a linear program requires a uniform formulation of the constraints, as linear functions, that must be nonnegative. For this reason we introduce the dependent variables $y_1, y_2, y_3$ for the first three linear functions. Moreover, we denote the objective function by $z$, and from (2.1) obtain the equivalent linear program

$$\begin{aligned} y_1 &= -20x_1 - 10x_2 + 8000 \geqslant 0 \\ y_2 &= -4x_1 - 5x_2 + 2000 \geqslant 0 \\ y_3 &= -6x_1 - 15x_2 + 4500 \geqslant 0 \\ x_1 &\geqslant 0 \\ x_2 &\geqslant 0 \\ z &= 16x_1 + 32x_2 = \text{Max!} \end{aligned} \quad (2.2)$$

The constraints of (2.2) now require that not only the independent variables $x_i$ but also the dependent variables $y_k$ are only allowed to take nonnegative values. All these conditions have to be fulfilled while maximizing the objective function.

## 2.1 Introductory Examples, Graphical Solution

As the linear program (2.2) has only two unknowns, it can be solved graphically in the $(x_1, x_2)$ plane. The set of points $P(x_1, x_2)$ with the coordinates $x_1, x_2$ satisfying the linear inequality

$$y = ax_1 + bx_2 + c \geq 0 \tag{2.3}$$

consists of a half plane of the $(x_1, x_2)$ plane, including its boundary. The boundary of the half plane is given by the equation of the straight line $ax_1 + bx_2 + c = 0$. Having determined the boundary, we find the corresponding half plane of a given inequality by substituting the coordinates $(x_1, x_2)$ of a point not lying on the boundary, e.g. the origin, and testing whether the inequality (2.3) is satisfied. The five inequalities of (2.2) define five half planes. Their boundaries are given in Figure 2.1. A point P, with coordinates $(x_1, x_2)$ is called *feasible* for the linear program (2.2), if all inequalities are satisfied. Hence, a feasible point must belong to the intersection of the five half planes, i.e. the *feasible domain* which is shaded in Figure 2.1. In the case of two unknowns the feasible domain is a *convex polygon* since it is the intersection of convex half planes.

The objective function is linear in the two unknowns $x_1$ and $x_2$, and its level lines, $z = $ constant, are given by a set of parallel straight lines. The level line $z = 3200$ is drawn in Figure 2.1, and the arrow indicates the direction in which the value of $z$ increases. The objective function is maximized within all feasible points, the point D, which is at greatest distance from the drawn level line in the direction of the arrow being the solution. The *solution point* of the linear program (2.2) is a *corner* of the feasible domain. If $x_1 = 250$ (ladies' shoes) and $x_2 = 200$ (men's shoes), the obtained profit $z_{max} = 10\,400$ is maximum.

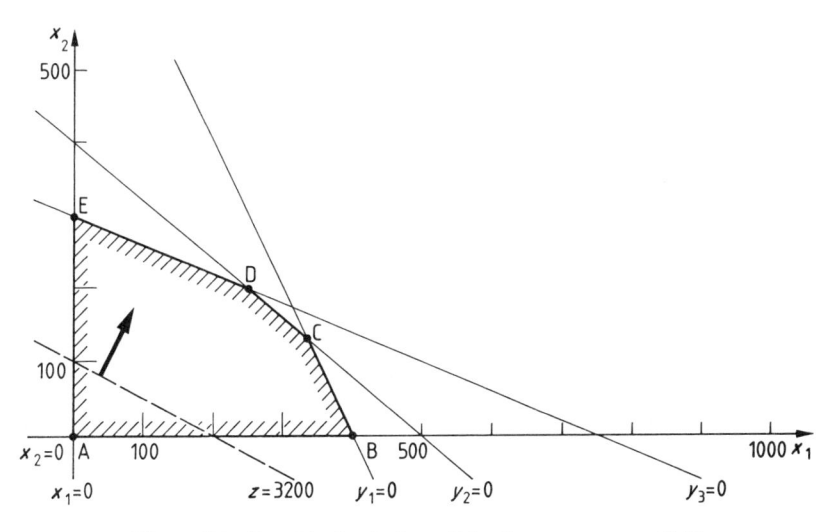

Figure 2.1 Graphical solution of the linear program (2.2)

The graphical solution in Figure 2.1 gives the additional information that the solution point D with $y_2 = 0$ and $y_3 = 0$, lies on the boundary of the two half planes defined by $y_2 \geqslant 0$ and $y_3 \geqslant 0$, respectively, whereas $y_1 > 0$. This means that for our problem the available time for machine handling and quantity of leather are completely consumed, whereas the existing capacity of working hours of the employees is not consumed, because $y_1 = 1000$. The first two mentioned boundary conditions are called the *essential restrictions*, since they determine the best possible production.

*Example 2.2* We consider a typical problem of a railway enterprise. In two shunting yards, A and B, there are 18 and 12 empty luggage vans, respectively. In three stations R, S and T the enterprise needs 11, 10 and 9 luggage vans, respectively. The distances in kilometres between the shunting yards and the stations are given in Table 2.2. The luggage vans should be directed in such a way that the total distance is minimized.

In view of the mathematical treatment of the problem, we notice that the total number of required vans equals that of the available ones. Therefore two unknowns suffice to describe the transport problem. From several possibilities we choose the unknowns as follows:

$x_1$ = number of vans from A to R
$x_2$ = number of vans from A to S.

The number of vans to be directed along the other sections can be expressed by these unknowns, and the number of idle kilometres is given by

$$z = 5x_1 + 4x_2 + 9(18 - x_1 - x_2) + 7(11 - x_1) + 8(10 - x_2)$$
$$+ 10(x_1 + x_2 - 9)$$
$$= -x_1 - 3x_2 + 229.$$

As the number of luggage vans, which go along the six sections from the shunting yards to the stations, cannot be negative, we obtain the following linear program

$$\begin{array}{ll}
A \to T: & y_1 = -x_1 - x_2 + 18 \geqslant 0 \\
B \to R: & y_2 = -x_1 \phantom{- x_2} + 11 \geqslant 0 \\
B \to S: & y_3 = \phantom{-x_1} - x_2 + 10 \geqslant 0 \\
B \to T: & y_4 = x_1 + x_2 - 9 \geqslant 0 \\
A \to R: & x_1 \phantom{+ x_2 - 99} \geqslant 0 \\
A \to S: & \phantom{x_1 +} x_2 \phantom{- 99} \geqslant 0 \\
& z = -x_1 - 3x_2 + 229 = \text{Min}!
\end{array} \quad (2.4)$$

## 2.1 Introductory Examples, Graphical Solution

Table 2.2 Distance table

|   | R | S | T  |
|---|---|---|----|
| A | 5 | 4 | 9  |
| B | 7 | 8 | 10 |

In the linear program (2.4) we have already introduced the dependent variables $y_1$ to $y_4$. The graphical solution of (2.4) is shown in Figure 2.2. The six inequalities define six half planes, whose intersection is the shaded *feasible domain*. Moreover, the level line of the objective function for $z = 226$ is drawn, together with an arrow which shows in the direction of decreasing $z$. The *solution point* of the linear program (2.4) is the *corner* L of the feasible domain, since it is at the greatest distance from the drawn level line in the direction of the arrow. Hence the enterprise has to direct $x_1 = 8$ vans from A to R, $x_2 = 10$ vans from A to S, $y_1 = 0$ vans from A to T, $y_2 = 3$ vans from B to R, $y_3 = 0$ vans from B to S and $y_4 = 9$ vans from B to T yielding a minimum number of idle kilometres of $z_{\min} = 191$ km.

The two examples have in common that the unique solution point is a *corner* of the feasible domain. This is usually true for general linear programs of two unknowns. This fact can also be seen to be true in the case of three unknowns by an intuitive argument because the feasible domain is a convex polyhedron of three-dimensional space, whose boundary surfaces are given by those planes that are defined by the vanishing value of either an independent variable, $x_i$, or a dependent variable, $y_k$. The level surfaces $z = $ constant are now parallel planes, and the solution point will usually be a corner of the polyhedron.

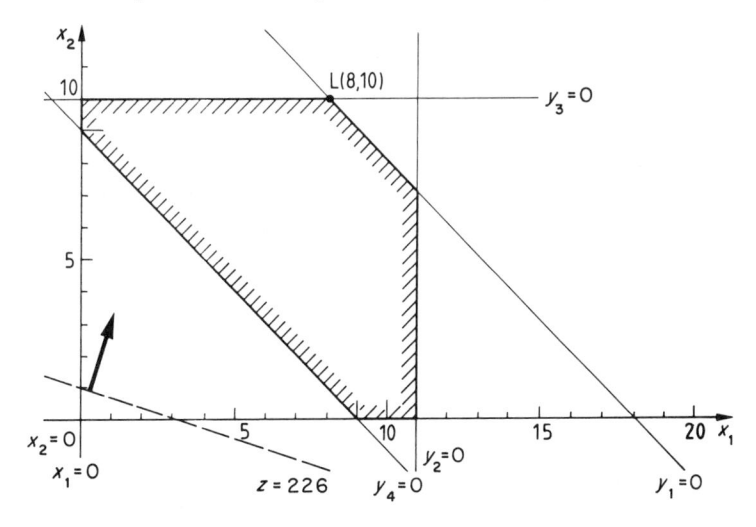

Figure 2.2 Graphical solution of the transport problem

If the level lines $z = $ constant are parallel to one of the boundary straight lines, it is obvious that a whole *edge* of the polygon may become the solution set of a linear program, because the objective function takes on the same extremal value for all points of the edge. The solution is not unique in this case. However, the two endpoints of the edge, being corners of the polygon, are two special solution points of the linear program, because all the other points of the solution set are a *convex linear combination* of the two end points. Analogously the solution set of a linear program in three unknowns may consist either of an edge or, more generally, of the boundary and the interior of a convex polygon of the surface of the polyhedron. The last case occurs if the level planes $z = $ constant are parallel to a boundary surface of the polyhedron. In any case the corners of this polygon are special solution points of the linear program, from which every other point of the solution set is obtained by a convex linear combination.

The geometrical consideration shows that the solution point, or special solution points, of a linear program must be among the corners of the feasible domain. The algorithm subsequently developed will have to determine a sequence of corners of the feasible domain in such a systematic way, that it ends up with the solution corner. This procedure requires, of course, the knowledge of a starting corner with which the algorithm can be initialized. At this point the two examples differ. In Example 2.1 the origin with $x_1 = 0$ and $x_2 = 0$ represents a feasible corner which can be used as starting corner for the process. However, in Example 2.2 the origin does not belong to the feasible domain, and no obvious starting corner is available. This situation will require special treatment.

## 2.2 The Simplex Algorithm

To motivate the procedure we look at Example 2.1 and Figure 2.1. Each corner of the polygon is characterized by the fact that two of the involved variables vanish, whereas all the other variables take on a nonnegative value. According to our goal of producing a sequence of corners, an obvious method consists of proceeding from a known corner to an adjacent corner, along an edge of the polygon. To obtain some insight we consider the linear program (2.2) and the representative transition from the known starting corner A to the neighbouring corner B along the $x_1$ axis (see Figure 2.1). The situation in the two corners is described as follows

$$\begin{array}{llllll} \text{corner A:} & \underline{x_1 = 0,} & x_2 = 0, & \underline{y_1 > 0,} \ y_2 > 0, \ y_3 > 0; & z = 0 \\ \text{corner B:} & x_1 > 0, & x_2 = 0, & \underline{y_1 = 0,} \ y_2 > 0, \ y_3 > 0; & z = 6400. \end{array} \quad (2.5)$$

The variable $x_2$ has kept its value zero along the edge of the polygon, $x_1$ has increased from zero to a positive value, whereas $y_1$ has decreased from a positive value to zero. All the other variables involved are still positive. We see that the

## 2.2 The Simplex Algorithm

variables $x_1$ and $y_1$ have interchanged their roles. This statement leads us to the obvious conclusion that the transition from one corner to an adjacent one can be done by an appropriate exchange step applied to the given linear functions of the linear program. In order to approach the solution point systematically, we require that the value of the objective function increases in each step, thus avoiding the procedure that will produce the same corner in a later step.

We want to derive the rules for the proper choice of the pivotal element, of an exchange step, for the linear program

$$y_i = \sum_{k=1}^{n} a_{ik} x_k + c_i \geq 0 \quad (i = 1, 2, \ldots, m)$$
$$x_k \geq 0 \quad (k = 1, 2, \ldots, n) \tag{2.6}$$
$$z = \sum_{k=1}^{n} b_k x_k + d = \text{Max!}$$

The $n$ unknowns $x_1, x_2, \ldots, x_n$ have to be determined in such a way that they satisfy the $n$ linear inequalities, are nonnegative and maximize the objective function $z$.

The origin $x_1 = x_2 = \cdots = x_n = 0$ is a feasible corner of (2.6) if and only if

$$c_i \geq 0 \quad (i = 1, 2, \ldots, m) \tag{2.7}$$

We assume (2.7) in the following, so that a starting corner is known, however assumption (2.7) will be dropped later on in Section 2.4.

We write down the $m$ linear inequalities and the objective function of (2.6) in a scheme that indicates the most important rows and columns

|         | $x_1$    | $\cdots$ | $x_q$    | $\cdots$ | $x_n$    | 1        |
|---------|----------|----------|----------|----------|----------|----------|
| $y_1 =$ | $a_{11}$ | $\cdots$ | $a_{1q}$ | $\cdots$ | $a_{1n}$ | $c_1$    |
| $\vdots$| $\vdots$ |          | $\vdots$ |          | $\vdots$ | $\vdots$ |
| $y_i =$ | $a_{i1}$ | $\cdots$ | $a_{iq}$ | $\cdots$ | $a_{in}$ | $c_i$    |
| $\vdots$| $\vdots$ |          | $\vdots$ |          | $\vdots$ | $\vdots$ |
| $y_p =$ | $a_{p1}$ | $\cdots$ | $a_{pq}$ | $\cdots$ | $a_{pn}$ | $c_p$    |
| $\vdots$| $\vdots$ |          | $\vdots$ |          | $\vdots$ | $\vdots$ |
| $y_m =$ | $a_{m1}$ | $\cdots$ | $a_{mq}$ | $\cdots$ | $a_{mn}$ | $c_m$    |
| $z =$   | $b_1$    | $\cdots$ | $b_q$    | $\cdots$ | $b_n$    | $d$      |

(2.8)

From assumption (2.7) the independent variables $x_1, x_2, \ldots, x_n$ which appear above the scheme (2.8) characterize a feasible corner if they are set equal to zero. They are called *basis variables* of (2.8). The dependent variables $y_1, y_2, \ldots, y_m$ to the left of the scheme, which take on nonnegative values for the feasible corner, are called *nonbasis variables*. Finally the quantity $d$ in the bottom right-hand corner of the scheme (2.8) represents the value of the objective function $z$ for the feasible corner of the linear program.

After these preliminary considerations we apply an exchange step to (2.8) with the pivotal element $a_{pq}$ following the rules of Section 1.4, in order to exchange the variables $x_q$ and $y_p$. We look for the conditions to be satisfied by the pivotal element, so that the resulting scheme

$$
\begin{array}{c|ccccc|c}
 & x_1 & \cdots & y_p & \cdots & x_n & 1 \\
\hline
y_1 = & a'_{11} & \cdots & a'_{1q} & \cdots & a'_{1n} & c'_1 \\
\vdots & \vdots & & \vdots & & \vdots & \vdots \\
y_i = & a'_{i1} & \cdots & a'_{iq} & \cdots & a'_{in} & c'_i \\
\vdots & \vdots & & \vdots & & \vdots & \vdots \\
x_q = & a'_{p1} & \cdots & a'_{pq} & \cdots & a'_{pn} & c'_p \\
\vdots & \vdots & & \vdots & & \vdots & \vdots \\
y_m = & a'_{m1} & \cdots & a'_{mq} & \cdots & a'_{mn} & c'_m \\
z = & b'_1 & \cdots & b'_q & \cdots & b'_n & d' \\
\end{array}
\qquad (2.9)
$$

corresponds to a feasible corner if we set the basis variables $x_1, \ldots, y_p, \ldots, x_n$ of (2.9) equal to zero. This is true if and only if the *nonbasis variables* $y_1, \ldots, x_q, \ldots, y_m$ on the left-hand side of (2.9) are nonnegative, that is if the conditions

$$c'_i \geq 0 \quad (i = 1, 2, \ldots, m) \qquad (2.10)$$

hold. Moreover, the value of the objective function should increase leading to the weakened requirement

$$d' \geq d. \qquad (2.11)$$

On the basis of the rules (1.129) of an exchange step, we have the following selected equations

$$c'_p = -\frac{c_p}{a_{pq}} \qquad (2.12)$$

$$c'_i = c_i - \frac{a_{iq} c_p}{a_{pq}} \quad (i = 1, 2, \ldots, m;\, i \neq p) \qquad (2.13)$$

$$d' = d - \frac{b_q c_p}{a_{pq}} \qquad (2.14)$$

## 2.2 The Simplex Algorithm

Due to the stronger conditions $c_i > 0$, $i = 1, 2, \ldots, m$, and the requirement (2.10) the condition for the pivotal element follows from (2.12)

$$\boxed{a_{pq} < 0.} \tag{2.15}$$

If $c_p = 0$, then the pivot $a_{pq}$ has to be negative.

The condition (2.10) combines, according to (2.13), the elements of the pivotal column with those of the last column of the scheme (2.8). We have for the new element $c'_i$ ($i \neq p$), following (2.7) and (2.15),

$$c'_i = c_i - \frac{a_{iq}c_p}{a_{pq}} \geq c_i \geq 0 \quad \text{if } a_{iq} \geq 0,$$

and we see that the condition (2.10) is satisfied for those $c'_i$ having nonnegative elements $a_{iq}$. Hence, we need to consider (2.13) combined with (2.10) only for the negative elements $a_{iq}$, among which the pivot will be found. We divide those inequalities, to be satisfied by $c'_i$, by the positive value $(-a_{iq})$ to obtain the conditions

$$-\frac{c_i}{a_{iq}} + \frac{c_p}{a_{pq}} \geq 0 \quad \text{for all } i \neq p \text{ with } a_{iq} < 0,$$

or

$$\boxed{\frac{c_p}{a_{pq}} \geq \frac{c_i}{a_{iq}} \quad \text{for all } i \neq p \text{ with } a_{iq} < 0.} \tag{2.16}$$

From (2.7) and (2.15) the pivotal row must have, according to (2.16), the largest of the negative quotients $Q_i := c_i/a_{iq}$ which are computed with the negative elements of the pivotal column.

Finally, from (2.11) and (2.14) due to $c_p \geq 0$ and $a_{pq} < 0$, we obtain the condition

$$\boxed{b_q \geq 0} \tag{2.17}$$

which has to be satisfied by the pivotal column.

We summarize the conditions (2.15), (2.16) and (2.17) as two rules for the choice of the pivotal element.

**Rule 1** *The pivotal column must be determined so that its element $b_q$ in the row of the objective function is positive or possibly zero.*

**Rule 2** *In the chosen pivotal column the pivot $a_{pq}$ must be negative. The pivotal row* is then defined by the largest, i.e. in absolute value the smallest, quotient $Q_i := c_i/a_{iq}$, which are computed with the negative elements $a_{iq}$ of the pivotal column.

The two rules are the basis of the *simplex algorithm*. It allows the systematic computation of corners of the feasible convex domain in $\mathbf{R}^n$, which is called a simplex. The value of the objective function increases in each step.

Rule 1 does not fix the pivotal column uniquely, if there exist several positive elements in the last row. For a computer program we need additional rules. For instance, we can simply choose the first column with a positive element $b_q$ as the pivotal column, or the largest of the positive elements of the last row determines the pivotal column.

If the pivotal column is chosen with a vanishing $b_q$, the objective function will not increase, due to (2.14). This exceptional situation will be treated in Section 2.3.

The pivotal row is not uniquely determined by Rule 2, whenever several quotients $Q_i$ yield the same greatest value. Let $p$ and $j$ be two different indices such that

$$\max_{i, a_{iq} < 0} \frac{c_i}{a_{iq}} = \frac{c_p}{a_{pq}} = \frac{c_j}{a_{jq}} \quad p \neq j. \tag{2.18}$$

For an exchange step with the pivot $a_{pq}$ it follows from (2.13) that we have $c'_j = 0$. If we set all basis variables equal to zero, at least one nonbasis variable will also vanish. This situation is called *degeneration*, and will be considered in the following section in more detail.

As soon as all elements $b_j^*$ of the last row are negative or zero, no pivotal column can be found according to Rule 1, which would guarantee an increase of the objective function. The simplex algorithm breaks down. The objective function has the representation

$$z = b_1^* \xi_1 + b_2^* \xi_2 + \cdots + b_n^* \xi_n + d^* = \text{Max}!, \tag{2.19}$$

where $\xi_i$ denote $n$ basis variables among the $(n + m)$ variables $x_k$ and $y_i$ of the given linear program (2.6) and where $b_j^* \leq 0$ for $j = 1, 2, \ldots, n$. From the conditions $\xi_i \geq 0$, the maximal value of $z$ is equal to $d^*$ which is obtained if all basis variables are set equal to zero. This characterizes a corner of the feasible simplex in which the objective function is maximized. Hence, a solution point of (2.6) is found, for which the values of the nonbasis variables are given by the corresponding $c_j^* \geq 0$ of the last column. The unknowns $x_k$ are indeed a solution of the linear program (2.6), because of the *reversibility* of the exchange steps. This means that linear functions represent just the same relationships between the involved variables after a sequence of one or several exchange steps.

If in (2.19) $b_j^* < 0$ for all $j$, the objective function $z$ attains its maximal value if and only if all basis variables are $\xi_1 = \xi_2 = \cdots = \xi_n = 0$. Their values, and therefore also those of the nonbasis variables, are uniquely determined in this

## 2.2 The Simplex Algorithm

case. On the basis of the reversibility of the exchange steps the solution of (2.6) is unique.

*Example 2.3* The linear program (2.2) of example 2.1 satisfies the assumption (2.7). We solve it by means of the simplex algorithm and subsequently give the geometrical interpretation of the computational steps. The *simplex scheme* for (2.2) is

|       | $x_1$ | $x_2$ | 1    | $Q_i =$ |
|-------|-------|-------|------|---------|
| $y_1 =$ | $\underline{-20}$ | $-10$ | 8000 | $-400$ |
| $y_2 =$ | $-4$  | $-5$  | 2000 | $-500$ |
| $y_3 =$ | $-6$  | $-15$ | 4500 | $-750$ |
| $z =$ | $\underline{16}$ | 32 | 0 | |
|       | | $-\dfrac{1}{2}$ | 400 | |

(2.20)

Setting the basis variables $x_1 = x_2 = 0$ the scheme (2.20) corresponds to the origin A of Figure 2.1.

We now choose the first column as the pivotal column for several reasons. To the right of the scheme (2.20) the quotients $Q_i$ are given, from which Rule 2 fixes the pivotal row. Following the rules of Section 1.4 we use the bottom row to obtain

|       | $y_1$ | $x_2$ | 1 | $Q_i =$ |
|-------|-------|-------|-----|---------|
| $x_1 =$ | $-\dfrac{1}{20}$ | $-\dfrac{1}{2}$ | 400 | $-800$ |
| $y_2 =$ | $\dfrac{1}{5}$ | $\underline{-3}$ | 400 | $-133\dfrac{1}{3}$ |
| $y_3 =$ | $\dfrac{3}{10}$ | $-12$ | 2100 | $-175$ |
| $z =$ | $-\dfrac{4}{5}$ | $\underline{24}$ | 6400 | |
|       | $\dfrac{1}{15}$ | | $\dfrac{400}{3}$ | |

(2.21)

The scheme (2.21) with the basis variables $y_1$ and $x_2$ can be identified with the corner point B. The exchange step performed corresponds to the transition from the starting corner A to the neighbouring corner B along the edge $x_2 = 0$ of the polygon. The value of the objective function has increased from zero to 6400. The nonbasis variables are strictly positive.

Rule 1 of the simplex algorithm (2.21), uniquely determines the second column as the pivotal column. According to Rule 2 the second row becomes the pivotal row. The result of the exchange step is

|  | $y_1$ | $y_2$ | 1 | $Q_i =$ |
|---|---|---|---|---|
| $x_1 =$ | $-\dfrac{1}{12}$ | $\dfrac{1}{6}$ | $\dfrac{1000}{3}$ | $-4000$ |
| $x_2 =$ | $\dfrac{1}{15}$ | $-\dfrac{1}{3}$ | $\dfrac{400}{3}$ | — |
| $y_3 =$ | $-\dfrac{1}{2}$ | $4$ | $500$ | $-1000$ |
| $z =$ | $\dfrac{4}{5}$ | $-8$ | $9600$ | |
|  | $8$ | $1000$ | | |

(2.22)

The two basis variables $y_1$ and $y_2$ of (2.22) correspond to corner C of Figure 2.1, where the objective function takes on the value 9600. Although the two unknowns $x_1$ and $x_2$ have been exchanged and become nonbasis variables in (2.22), the solution has not been found yet, because there exists a positive element in the last row of (2.22), and hence the simplex algorithm has to be continued with the first column as the pivotal column. Among the elements $a_{i1}$ only the two negative ones are candidates for pivots, so that only the corresponding $Q_i$ are computed and given. The third exchange step yields

|  | $y_3$ | $y_2$ | 1 |
|---|---|---|---|
| $x_1 =$ | $\dfrac{1}{6}$ | $-\dfrac{1}{2}$ | $250$ |
| $x_2 =$ | $-\dfrac{2}{15}$ | $\dfrac{1}{5}$ | $200$ |
| $y_1 =$ | $-2$ | $8$ | $1000$ |
| $z =$ | $-\dfrac{8}{5}$ | $-\dfrac{8}{5}$ | $10400$ |

(2.23)

## 2.2 The Simplex Algorithm

With the scheme (2.23), corresponding to the solution point D, the simplex algorithm breaks off. If we set the basis variables $y_3 = y_2 = 0$ from (2.23) we read the required solution $x_1 = 250$, $x_2 = 200$, the value of the nonbasis variable $y_1 = 1000$ and the maximal value of the objective function $z_{max} = 10400$.

The number of exchange steps of the simplex algorithm is not known *a priori*, because the number of corner points from the starting corner to the solution corner does not only depend on the given linear program but also on the choice of the pivotal columns. However, the simplex algorithm produces the solution in a finite number of steps on certain restrictive assumptions.

**Theorem 2.1** *A solution of the linear program (2.6) is determined in a finite number of simplex steps, if*
   (a) *the elements $c_i$ of the last column in all simplex schemes are strictly positive, i.e. there is no degeneration;*
   (b) *it is possible to choose a pivot according to Rules 1 and 2 with a strictly positive $b_q$ in the last row.*

*Proof* With $n$ unknowns $x_k$ and $m$ dependent variables $y_i$ of the linear program (2.6) in $\mathbf{R}^n$ there exist $\binom{n+m}{n}$ points which are defined by the vanishing of $n$ of the $(n + m)$ variables involved. Among these not necessarily different points (degeneration!) several points, in general, will not belong to the feasible domain of (2.6). It follows that the number of corners of the feasible domain is at most $\binom{n+m}{n}$ and hence finite.

The assumptions (a) and (b) guarantee that the value of $d$ strictly increases for each exchange step due to (2.14). The sequence of values of the objective function therefore increases strictly monotonically. Since each simplex scheme can be identified with a corner of the feasible domain, by setting the $n$ basis variables equal to zero, we conclude that each possible corner can occur at most once in the course of the simplex algorithm complying with the given assumptions. Therefore after a finite number of steps one must encounter that for the elements of the last row $b_k^* \leq 0$, $k = 1, 2, \ldots, n$. The simplex algorithm stops, and the corresponding corner is a solution point with a maximal value of the objective function because of the previous considerations.

An immediate consequence of Theorem 2.1 and of the representation, (2.19), of the objective function is the following theorem.

**Theorem 2.2** *A linear program (2.6) has a unique solution if the hypothesis (a) and (b) of Theorem 2.1 hold, and the additional condition $b_k^* < 0$, $k = 1, 2, \ldots, n$, for the last simplex scheme is satisfied.*

The simplex algorithm is now summarized on the assumptions of Theorem 2.1. For an appropriate implementation on a computer, the elements of the simplex schemes are combined in the matrix $\mathbf{A} = (a_{ik})$ with $(m+1)$ rows and $(n+1)$ columns, such that we have $b_k = a_{m+1,k}$, $(k = 1, 2, \ldots, n)$, $c_i = a_{i,n+1}$, $(i = 1, 2, \ldots, m)$ and $d = a_{m+1,n+1}$. In order to be able to assign the corresponding values to the proper unknowns $x_k$ and to the dependent variables $y_i$ after the end of the simplex algorithm, the information on the performed exchange steps is needed. This can be done, for example, by means of two auxiliary vectors of $n$ and $m$ components, respectively. They contain the information on the basis and

$$
\begin{aligned}
&\text{for } k = 1, 2, \ldots, n: ba_k = k \\
&\text{for } i = 1, 2, \ldots, m: nb_i = -i \\
\text{PIV:}\quad &q = 0;\ \max = 0 \\
&\text{for } k = 1, 2, \ldots, n: \\
&\quad \text{if } a_{m+1,k} > \max: \\
&\quad\quad q = k;\ \max = a_{m+1,k} \\
&\text{if } q = 0:\ \text{go to SOL} \\
&p = 0;\ \max = -10^{50} \\
&\text{for } i = 1, 2, \ldots, m: \\
&\quad \text{if } a_{iq} < 0: \\
&\quad\quad \text{quot} = a_{i,n+1}/a_{iq} \\
&\quad\quad \text{if quot} > \max: \\
&\quad\quad\quad p = i;\ \max = \text{quot} \\
&\text{if } p = 0:\ \text{STOP} \\
\text{ES:}\quad &h = nb_p;\ nb_p = ba_q;\ ba_q = h;\ \text{pivot} = a_{pq} \\
&\text{for } k = 1, 2, \ldots, n+1: \\
&\quad \text{if } k \neq q: \\
&\quad\quad a_{pk} = -a_{pk}/\text{pivot} \\
&\quad\quad \text{for } i = 1, 2, \ldots, m+1: \\
&\quad\quad\quad \text{if } i \neq p:\ a_{ik} = a_{ik} + a_{iq} \times a_{pk} \\
&\text{for } i = 1, 2, \ldots, m+1: \\
&\quad a_{iq} = a_{iq}/\text{pivot} \\
&a_{pq} = 1/\text{pivot} \\
&\text{go to PIV} \\
\text{SOL:}\quad &\text{for } k = 1, 2, \ldots, n: \\
&\quad j = |ba_k| \\
&\quad \text{if } ba_k > 0 \text{ then } x_j = 0,\ \text{else } y_j = 0 \\
&\text{for } i = 1, 2, \ldots, m: \\
&\quad j = |nb_i| \\
&\quad \text{if } nb_i > 0 \text{ then } x_j = a_{i,n+1},\ \text{else } y_j = a_{i,n+1} \\
&z = a_{m+1,m+1}
\end{aligned}
$$

(2.24)

nonbasis variables of the present scheme. In the following program the two vectors are denoted by **ba** and **nb**, the unknown $x_k$ is identified by the index $k$ and the variable $y_i$ by the negative index $-i$. The corresponding components have to be exchanged for each simplex step. The pivotal column is defined by the largest, positive element $b_k$. If Rule 2 cannot determine a pivotal element, the hypothesis (b) of Theorem 2.1 is not fulfilled. This indicates an exceptional case, and the process ends with STOP.

## 2.3 Supplements to the Simplex Algorithm

While deriving the rules of the simplex algorithm, possible exceptional cases have been pointed out which will now be analysed in more detail. We consider three independent situations whose combinations are possible.

### 2.3.1 Degeneracy

In the course of the simplex algorithm as soon as one encounters the situation that the maximum of the quotients, considered in Rule 2 for the determination of the pivotal row, is attained by more than one quotient $Q_i$, (2.18), an immediate consequence is that after the exchange step at least one coefficient $c_j$ vanishes. In addition to the $n$ basis variables at least one nonbasis variable is equal to zero. If we set one of the $(n+m)$ involved variables equal to zero this corresponds geometrically to a hyperplane in the $\mathbf{R}^n$. It represents the boundary of a half space, and therefore this situation means that the corner of the feasible simplex is the intersection of more than $n$ hyperplanes. Therefore we call it a *degenerate corner* and speak of *degeneracy* of the simplex scheme.

In the case of two unknowns a degenerate corner of the convex polygon in $\mathbf{R}^2$ can only result from more than two boundary lines passing through the corner. However, because only two of them are boundary pieces of the feasible domain, the inequalities corresponding to the other straight lines are *redundant*, because they will be satisfied as soon as the first two are.

A degenerate corner of the feasible polyhedron in $\mathbf{R}^3$ can be generated in two ways which are combinable. Firstly, the analogous situation to $\mathbf{R}^2$ may arise, that typically three boundary planes define the corner and that additional, but redundant planes, pass through the corner. The second situation, that the degenerate corner is the intersection of at least four boundary planes, is more frequent. This occurs for instance in the peak of a pyramid, where there are more than three edges of the feasible polyhedron emerging from such a degenerate corner.

An immediate consequence of the presence of a degenerate corner for the simplex algorithm will be that one of the rows, that contains a vanishing value $c_j$ in the last column of the simplex scheme, may become a pivotal row in a subsequent step. If this happens, we shall have $d' = d$ after such an exchange step from $c_p = 0$ and (2.14). The value of the objective function does not increase. Furthermore, the exchanged basis and nonbasis variables are still zero after this step, and the other variables are unchanged because $c'_i = c_i$ holds for all $i \neq p$, due to (2.13). The resulting simplex scheme must therefore be identified with the same degenerate corner. Hence, we have performed a *stationary step* with which no new corner has been produced.

If the degeneration of the simplex scheme is higher, such that several nonbasis variables vanish, the simplex algorithm may perform an indefinite sequence of stationary steps, called a *cycle*. To avoid such cycles, special techniques have been developed which guarantee an advance by considering appropriate perturbations of the linear program (Kall, 1976; Künzi et al., 1967).

*Example* 2.4 We want to illustrate the degeneracy with the linear program (2.25) of three unknowns and give the geometrical interpretation of the steps in Figure 2.3. The feasible domain of the linear program consists of a cuboid with a superimposed pyramid.

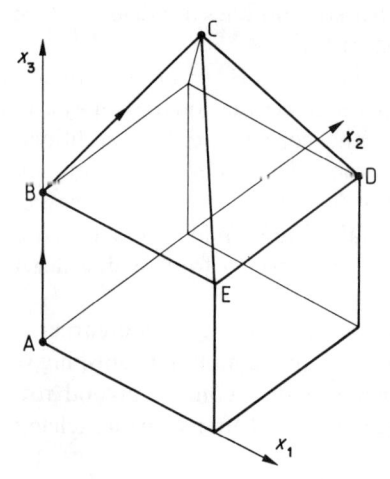

$$\begin{aligned}
y_1 &= -x_1 &&&& + 2 \geq 0 \\
y_2 &= && -x_2 && + 2 \geq 0 \\
y_3 &= -x_1 && && -x_3 + 3 \geq 0 \\
y_4 &= && -x_2 & -x_3 + 3 \geq 0 \\
y_5 &= x_1 && && -x_3 + 1 \geq 0 \\
y_6 &= && x_2 & -x_3 + 1 \geq 0 \\
&& x_1 \geq 0 \quad x_2 \geq 0 \quad x_3 \geq 0 \\
z &= x_1 + 2x_2 + 3x_3 = \text{Max!}
\end{aligned}$$
(2.25)

Figure 2.3 Solution steps of the linear program (2.25)

## 2.3 Supplements to the Simplex Algorithm

The pivotal column will be determined by the greatest element $b_q > 0$. In the case of equal maximal quotients $Q_i$ the row with the smaller index is taken as pivotal row. This strategy is applied in the program (2.24). The first simplex step leads to the degenerate simplex scheme, (2.26), corresponding to the degenerate corner B in Figure 2.3.

|        | $x_1$ | $x_2$ | $x_3$ | 1 |        | $x_1$ | $x_2$ | $y_5$ | 1 |
|--------|-------|-------|-------|---|--------|-------|-------|-------|---|
| $y_1=$ | $-1$  | 0     | 0     | 2 | $y_1=$ | $\underline{-1}$ | 0 | 0 | 2 |
| $y_2=$ | 0     | $-1$  | 0     | 2 | $y_2=$ | 0     | $-1$  | 0     | 2 |
| $y_3=$ | $-1$  | 0     | $\underline{-1}$ | 3 | $y_3=$ | $-2$ | 0 | 1 | 2 |
| $y_4=$ | 0     | $-1$  | $\underline{-1}$ | 3 | $y_4=$ | $-1$ | $-1$ | 1 | 2 |
| $\underline{y_5=}$ | 1 | 0 | $\underline{-1}$ | 1 | $x_3=$ | 1 | 0 | $-1$ | 1 |
| $y_6=$ | 0     | 1     | $\underline{-1}$ | 1 | $y_6=$ | $\underline{-1}$ | 1 | 1 | 0 |
| $z=$   | 1     | 2     | $\underline{3}$ | 0 | $z=$   | 4 | 2 | $-3$ | 3 |
|        | 1     | 0     | 1     |   |        | 1 | 1 | 0 |   |

(2.26)

For the second step the first column is chosen as the pivotal column. This choice leads to a stationary step because $c_6 = 0$. In the case of a different choice, namely the second column, a nonstationary step with a pivot in the second row would immediately lead to the (degenerate) corner C. However, with the chosen pivot, the resulting simplex scheme corresponds again to the corner B. For the third exchange step, the degeneracy is no longer a problem, since now a nonstationary simplex step is possible corresponding to a transition to the degenerate corner C.

|        | $y_6$ | $x_2$ | $y_5$ | 1 |        | $y_6$ | $y_3$ | $y_5$ | 1 |
|--------|-------|-------|-------|---|--------|-------|-------|-------|---|
| $y_1=$ | 1     | $\underline{-1}$ | $-1$ | 2 | $y_1=$ | 0 | 0.5 | $-0.5$ | 1 |
| $y_2=$ | 0     | $\underline{-1}$ | 0    | 2 | $y_2=$ | $-1$ | 0.5 | 0.5 | 1 |
| $\underline{y_3=}$ | 2 | $\underline{-2}$ | $-1$ | 2 | $x_2=$ | 1 | $-0.5$ | $-0.5$ | 1 |
| $y_4=$ | 1     | $\underline{-2}$ | 0    | 2 | $y_4=$ | $\underline{-1}$ | $\underline{1}$ | $\underline{1}$ | 0 |
| $x_3=$ | $-1$  | 1     | 0     | 1 | $x_3=$ | 0 | $-0.5$ | $-0.5$ | 2 |
| $x_1=$ | $-1$  | $\underline{1}$ | 1 | 0 | $x_1=$ | 0 | $-0.5$ | $-0.5$ | 2 |
| $z=$   | $-4$  | $\underline{6}$ | 1 | 3 | $z=$   | $\underline{2}$ | $-3$ | $-2$ | 9 |
|        | 1     | $-0.5$ | 1 |   |        | 1 | 1 | 0 |   |

(2.27)

The rules of the simplex algorithm require another exchange step, with a pivot in the first column which is again stationary, so that the resulting scheme corresponds again to the corner C.

|         | $y_4$ | $y_3$ | $y_5$ | 1 |
|---------|------|------|------|---|
| $y_1 =$ | 0    | 0.5  | $-0.5$ | 1 |
| $y_2 =$ | 1    | $-0.5$ | $-0.5$ | 1 |
| $x_2 =$ | $-1$ | 0.5  | 0.5  | 1 |
| $y_6 =$ | $-1$ | 1    | 1    | 0 |
| $x_3 =$ | 0    | $-0.5$ | $-0.5$ | 2 |
| $x_1 =$ | 0    | $-0.5$ | 0.5  | 1 |
| $z =$   | $-2$ | $-1$ | 0    | 9 |
|         | 0    | 1    | 2    |   |

(2.28)

Here the simplex algorithm ends because $b_k \leq 0$, $(k = 1, 2, 3)$. The scheme is degenerate. By setting the basis variables equal to zero we get the unknowns to be $x_1 = 1, x_2 = 1, x_3 = 2$ with a maximal value of the objective function $z_{max} = 9$. The dependent variables are $y_1 = 1$, $y_2 = 1$, $y_3 = y_4 = y_5 = y_6 = 0$.

### 2.3.2 Nonunique solution

As soon as the elements of the last row of a simplex scheme satisfy $b_k \leq 0$, $k = 1, 2, \ldots, n$, the value of the objective function can no longer be increased. A solution of the linear program is obtained by setting the current basis variables equal to zero. If some of the $b_k$ vanish, the solution is, in general, not unique. Let $b_q$ be equal to zero, and let the simplex scheme be nondegenerate with $c_i > 0$, $i = 1, 2, \ldots, m$. Moreover, we assume that Rule 2, of the simplex algorithm, provides a pivotal element in the $q$th column, so that an exchange step can be performed. On these assumptions the exchange of a basis variable with a nonbasis variable is equivalent to a transition from one corner to a different one, whereby the value of the objective function is not changed. Obviously we have got another solution of the linear program with the same maximal value of the objective function. With these two different solutions each convex linear combination is also a solution of the linear program. Geometrically this means that the straight line joining the two corners of the simplex belongs to the set of solution points of the linear program.

If several $b_k = 0$, there are correspondingly many exchange steps possible, which produce as many solutions that are not necessarily pairwise different. The dimension of the convex set of solutions is at most equal to the number of

## 2.3 Supplements to the Simplex Algorithm

vanishing $b_k$. However, it is a difficult combinatorial problem to find the number of all possible optimal corners of the feasible domain. In the case of three unknowns, the set of optimal solution points may consist of a convex polygon with several solution corners, if the level planes, $z = $ constant, are parallel to a boundary plane of the polyhedron.

**Example 2.5** In the simplex scheme (2.28), the element $b_3 = 0$, whereas the other $b_k$ are strictly negative. We choose the third column to be pivotal. Rule 2 determines the first row as the pivotal row, and the exchange step yields the following, again degenerate, simplex scheme

|       | $y_4$ | $y_3$ | $y_1$ | 1 |
|-------|------|------|------|---|
| $y_5 =$ | 0  | 1  | $-2$ | 2 |
| $y_2 =$ | 1  | $-1$ | 1  | 0 |
| $x_2 =$ | $-1$ | 1  | $-1$ | 2 |
| $y_6 =$ | $-1$ | 2  | $-2$ | 2 |
| $x_3 =$ | 0  | $-1$ | 1  | 1 |
| $x_1 =$ | 0  | 0  | $-1$ | 2 |
| $z =$  | $-2$ | $-1$ | 0  | 9 |

(2.29)

From this we derive the different solution $x_1^* = 2$, $x_2^* = 2$, $x_3^* = 1$, $z_{\max}^* = 9$, $y_1^* = y_2^* = y_3^* = y_4^* = 0$, $y_5^* = 2$, $y_6^* = 2$. It corresponds to the corner D of the polyhedron of Figure 2.3. In addition to the two solution corners $C(x_1, x_2, x_3)$ and $D(x_1^*, x_2^*, x_3^*)$, all points $P(\tilde{x}_1, \tilde{x}_2, \tilde{x}_3)$ on the joining edge are solution points, that is

$$\tilde{x}_1 = \lambda x_1 + (1 - \lambda) x_1^* \quad \tilde{x}_2 = \lambda x_2 + (1 - \lambda) x_2^* \quad \tilde{x}_3 = \lambda x_3 + (1 - \lambda) x_3^* \quad 0 \leq \lambda \leq 1.$$

For all these points P, we have $\tilde{z} = z_{\max} = 9$. The solution edge is parallel to the level planes $z = x_1 + 2x_2 + 3x_3 = $ constant.

### 2.3.3 Unbounded objective function

In the course of the simplex algorithm it may happen that after choosing the pivotal column with $b_q > 0$ that all the elements $a_{iq}$ are nonnegative for $i = 1, 2, \ldots, m$. Therefore there does not exist a pivotal element according to Rule 2 satisfying the condition (2.15). We want to understand the consequences of this situation for the given linear program. To do this we consider the simplex scheme (2.30), with the basis variables $\xi_1, \xi_2, \ldots, \xi_n$ and the nonbasis variables $\eta_1, \eta_2, \ldots, \eta_m$ and $c_i \geq 0$ for $i = 1, 2, \ldots, m$.

|        | $\xi_1$   | $\cdots$ | $\xi_q$   | $\cdots$ | $\xi_n$   | 1     |
|--------|-----------|----------|-----------|----------|-----------|-------|
| $\eta_1 =$ | $a_{11}$ | $\cdots$ | $a_{1q}$ | $\cdots$ | $a_{1n}$ | $c_1$ |
| $\vdots$ | $\vdots$ | | $\vdots$ | | $\vdots$ | $\vdots$ |
| $\eta_i =$ | $a_{i1}$ | $\cdots$ | $a_{iq}$ | $\cdots$ | $a_{in}$ | $c_i$ |
| $\vdots$ | $\vdots$ | | $\vdots$ | | $\vdots$ | $\vdots$ |
| $\eta_m =$ | $a_{m1}$ | $\cdots$ | $a_{mq}$ | $\cdots$ | $a_{mn}$ | $c_m$ |
| $z =$ | $b_1$ | $\cdots$ | $b_q$ | $\cdots$ | $b_n$ | $d$ |

(2.30)

The general $i$th inequality with $i = 1, 2, \ldots, m$

$$\eta_i = a_{i1}\xi_1 + \cdots + a_{iq}\xi_q + \cdots + a_{in}\xi_n + c_i \geqslant 0$$

is satisfied for $\xi_1 = \cdots = \xi_{q-1} = \xi_{q+1} = \cdots = \xi_n = 0$ and an arbitrary large, positive $\xi_q$. As $b_q > 0$ has been assumed, the objective function to be maximized can take on arbitrarily large values. The given linear inequalities do not bound the value of the objective function, and as a consequence the linear program does not have a finite solution. Whenever this situation occurs, in practice it is most likely that the problem has been formulated incorrectly.

## 2.4 General Linear Programs

In this section we mainly present methods for the treatment of linear programs for which the origin is not a feasible point. We start with a special case that is fundamental to the development of a method and a later application.

### 2.4.1 Treatment of free variables

In certain problems of linear programming only some of the unknowns have to be nonnegative. The other unknowns are called *free variables*. Since these free variables have to be regarded as foreign bodies of a linear program, it is quite natural to eliminate them.

To derive the rules for the elimination of a free variable, as usual we assume $c_i \geqslant 0$, $i = 1, 2, \ldots, m$. This assumption will be satisfied in the following applications. Let $x_q$ be a free variable to be eliminated from the simplex scheme (2.8). After the exchange step has been performed, the resulting row corresponding to the exchanged variable $x_q$ will usually be left out of the new scheme (2.9). However, if the value of the free variable is sought, the corresponding row is carried over but will no longer be examined in connection with the following considerations. For the remaining rows, we maintain the condition (2.10) $c_i' \geqslant 0$. Moreover, the value $d$ should increase in the weak sense of (2.11), so that $d' \geqslant d$.

The pivotal column has already been fixed by the free variable $x_q$ to be

## 2.4 General Linear Programs

eliminated. The decision of in which row the pivot $a_{pq}$ has to be taken, so that the conditions

$$c'_i = c_i - \frac{a_{iq}c_p}{a_{pq}} \geq 0 \quad (i \neq p) \tag{2.31}$$

$$d' = d - \frac{b_q c_p}{a_{pq}} \geq d \tag{2.32}$$

are satisfied, requires a distinction concerning the value $b_q$.

**Case 1** Let $b_q > 0$. Since $c_p \geq 0$ is assumed, the pivotal element $a_{pq}$ must be negative in the case $c_p > 0$, in order to satisfy the condition (2.32). Also the same condition should hold in the exceptional case $c_p = 0$. This is the same situation we encountered while deriving the rules of the simplex algorithm. The condition (2.31) is satisfied if we have for the quotient

$$Q_p = \frac{c_p}{a_{pq}} = \max_{a_{iq} < 0} \frac{c_i}{a_{iq}}.$$

**Case 2** Let $b_q < 0$. If $c_p > 0$, then the condition (2.32) requires that the pivot $a_{pq}$ is positive and this same condition also holds if $c_p = 0$. The condition (2.31) therefore is always satisfied for $a_{iq} \leq 0$. It is critical for $a_{iq} > 0$. We divide the inequality (2.31) by $a_{iq}$ and obtain

$$\frac{c_i}{a_{iq}} \geq \frac{c_p}{a_{pq}} \quad \text{for all } i \neq p \quad \text{with } a_{iq} > 0. \tag{2.33}$$

From (2.33) the pivotal row is now defined by the smallest of the nonnegative quotients $Q_i$.

**Case 3** If $b_q = 0$ the sign of the pivot $a_{pq}$ is not fixed by (2.32). This case can be regarded as a limiting case of either of the two previous cases. In the following we shall treat it as Case 2.

We summarize these conditions with the following rule.

**Rule for the elimination of a free variable.** *If the element $b_q$ of the pivotal column, which is fixed by the free variable $x_q$ to be eliminated, is positive (negative or zero), we have to compute the quotients $Q_i = c_i/a_{iq}$ with the negative (positive) elements $a_{iq}$. The pivotal row is determined by the smallest in absolute value of these quotients.*

### 2.4.2 Coordinate shift method

We consider the situation where the origin $x_1 = x_2 = \cdots = x_n = 0$ is not a feasible point of the linear program (2.6), because the assumption, (2.7), $c_i \geq 0$,

($i = 1, 2, \ldots, m$) is not satisfied. As a consequence, the simplex algorithm of Section 2.2 cannot be applied immediately, since no starting corner is available.

For many practical applications it is easy to find $n$ values $\bar{x}_1, \bar{x}_2, \ldots, \bar{x}_n$ so that the linear constraints

$$y_i = \sum_{k=1}^{n} a_{ik}\bar{x}_k + c_i \geq 0 \quad (i = 1, 2, \ldots, m)$$

$$\bar{x}_k \geq 0 \quad (k = 1, 2, \ldots, n) \tag{2.34}$$

hold. We call such an $n$-tuple a *feasible solution* of the linear program (2.6). With these values we use the substitutions

$$x_k = \bar{x}_k + \xi_k \quad (k = 1, 2, \ldots, n). \tag{2.35}$$

Geometrically this is equivalent to shifting the origin of the coordinate system of $\mathbf{R}^n$ into the feasible point $\bar{P}(\bar{x}_1, \bar{x}_2, \ldots, \bar{x}_n)$. From the given linear program (2.6), after substitution of (2.35) we obtain

$$y_i = \sum_{k=1}^{n} a_{ik}\xi_k + \bar{c}_i \geq 0 \quad (i = 1, 2, \ldots, m)$$

$$x_k = \xi_k + \bar{x}_k \geq 0 \quad (k = 1, 2, \ldots, n) \tag{2.36}$$

$$z = \sum_{k=1}^{n} b_k \xi_k + \bar{d} = \text{Max!}$$

From (2.34), the constants of the linear program (2.36) satisfy

$$\bar{c}_i = \sum_{k=1}^{n} a_{ik}\bar{x}_k + c_i \geq 0 \quad (i = 1, 2, \ldots, m),$$

$$\bar{x}_k \geq 0 \quad (k = 1, 2, \ldots, n)$$

so that the condition (2.7) holds. All the unknowns $\xi_k$ are free variables which have to be successively eliminated by applying the rule of Section 2.4.1. The linear program (2.36) contains, in comparison with (2.6), $n$ additional linear inequalities, so that the corresponding simplex scheme, including the objective function, will have $(m + n + 1)$ rows. As the values of the free variables $\xi_k$ are of no interest in this context, the corresponding pivotal row is dropped after each elimination step. After the complete elimination of the free variables the simplex scheme will comprise only $(m + 1)$ rows. The original unknowns $x_k$, and the dependent variables $y_i$ are arranged at the upper and left boundary of the scheme. Due to the fact, that the condition (2.31) holds for the remaining coefficients $c_i$ for each elimination step, the resulting simplex scheme corresponds to a feasible corner of

## 2.4 General Linear Programs

the simplex. Hence, the usual simplex algorithm can now be applied to find the required solution.

*Example 2.6* We consider the linear program (2.4) of the transport problem, for which the origin is not feasible. A possible, feasible solution is $\bar{x}_1 = 6$, $\bar{x}_2 = 4$. As all inequalities are strictly satisfied by this pair of values, the chosen feasible point $\bar{P}(6, 4)$ lies in the interior of the feasible domain. The substitutions $x_1 = 6 + \xi_1$, $x_2 = 4 + \xi_2$ leads to the equivalent linear program in which we now maximize the negative objective function.

$$
\begin{aligned}
y_1 &= -\xi_1 - \xi_2 + 8 \geqslant 0 \\
y_2 &= -\xi_1 \phantom{-\xi_2} + 5 \geqslant 0 \\
y_3 &= \phantom{-\xi_1} -\xi_2 + 6 \geqslant 0 \\
y_4 &= \xi_1 + \xi_2 + 1 \geqslant 0 \\
x_1 &= \xi_1 \phantom{+\xi_2} + 6 \geqslant 0 \\
x_2 &= \phantom{\xi_1 +} \xi_2 + 4 \geqslant 0 \\
z^* &= \xi_1 + 3\xi_2 - 211 = \text{Max!}
\end{aligned}
$$

|        | $\xi_1$ | $\xi_2$ | 1    |
|--------|---------|---------|------|
| $y_1 =$ | $-1$    | $-1$    | 8    |
| $y_2 =$ | $-1$    | 0       | 5    |
| $y_3 =$ | 0       | $-1$    | 6    |
| $y_4 =$ | 1       | 1       | 1    |
| $x_1 =$ | 1       | 0       | 6    |
| $x_2 =$ | 0       | 1       | 4    |
| $z^* =$ | 1       | 3       | $-211$ |
|        | 0       | 5       |      |

(2.37)

The successive elimination of the free variables $\xi_1$ and $\xi_2$, by applying the above rule, yields the following schemes where the pivotal row has been systematically dropped.

|        | $y_2$ | $\xi_2$ | 1    |
|--------|-------|---------|------|
| $y_1 =$ | 1     | $-1$    | 3    |
| $y_3 =$ | 0     | $-1$    | 6    |
| $y_4 =$ | $-1$  | 1       | 6    |
| $x_1 =$ | $-1$  | 0       | 11   |
| $x_2 =$ | 0     | 1       | 4    |
| $z^* =$ | $-1$  | 3       | $-206$ |
|        | 1     | 3       |      |

|        | $y_2$ | $y_1$ | 1    |
|--------|-------|-------|------|
| $y_3 =$ | $-1$  | 1     | 3    |
| $y_4 =$ | 0     | $-1$  | 9    |
| $x_1 =$ | $-1$  | 0     | 11   |
| $x_2 =$ | 1     | $-1$  | 7    |
| $z^* =$ | 2     | $-3$  | $-197$ |
|        | 1     | 3     |      |

(2.38)

With the second scheme (2.38) the goal has been reached. By setting the two basis variables $y_1$ and $y_2$ equal to zero a feasible corner is found, and the usual simplex algorithm can begin. Another exchange step that exchanges the variable $y_2$ yields the final scheme (2.39). From this we find the optimal solution $x_1 = 8$, $x_2 = 10$, and, moreover, $y_1 = 0$, $y_2 = 3$, $y_3 = 0$, $y_4 = 9$ with $z^*_{max} = -191$ or $z_{min} = -z^*_{max} = 191$.

|       | $y_3$ | $y_1$ | 1    |
|-------|-------|-------|------|
| $y_2 =$ | $-1$  | 1     | 3    |
| $y_4 =$ | 0     | $-1$  | 9    |
| $x_1 =$ | 1     | $-1$  | 8    |
| $x_2 =$ | $-1$  | 0     | 10   |
| $z^* =$ | $-2$  | $-1$  | $-191$ |

(2.39)

In Figure 2.4 the solution steps of the coordinate shift method are interpreted in such a way that the independent variables of each scheme are set equal to zero. Thus the starting scheme, (2.37), corresponds to the origin A of the $(\xi_1, \xi_2)$ coordinate system. The elimination of the first free variable $\xi_1$ leads to the boundary point B on the edge with $y_2 = 0$, and the second elimination step produces the feasible corner C with coordinates $x_1 = 11$, $x_2 = 7$. The simplex algorithm generates the solution corner with another step.

The coordinate shift method is a very simple concept, but it has certain disadvantages. First of all, a feasible solution must be known. The method is not systematic enough, because it requires some previous knowledge. Secondly, the given linear program is generally enlarged by $n$ linear inequalities, so that the first simplex scheme contains $(m + n + 1)$ rows, which are subsequently reduced to the original $(n + 1)$ rows by the elimination process. This additional requirement of storage space is not satisfactory. Therefore several procedures exist in order to

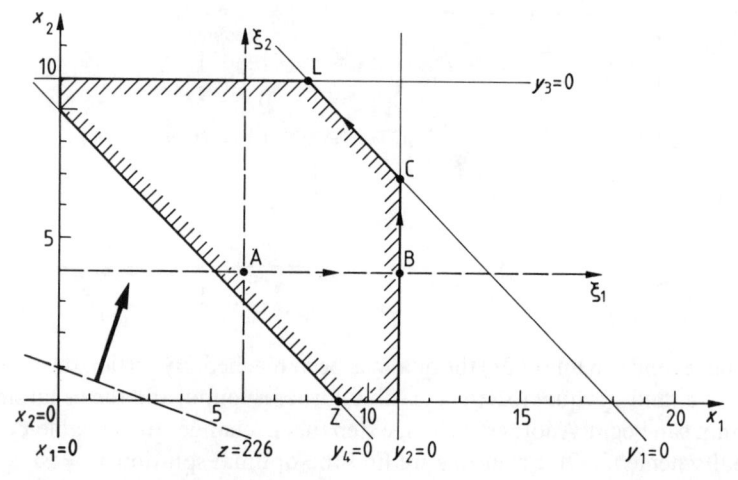

Figure 2.4 Graphical interpretation of the solution steps of the coordinate shift method

## 2.4 General Linear Programs

determine a feasible solution in a systematic way with a minimum amount of additional effort and storage.

### 2.4.3 The two-phase method

In order to determine a feasible solution of a linear program, (2.6), not satisfying the assumption (2.7), an auxiliary linear program is considered and solved in a first step. To do this we extend the given linear program (2.6) by a variable $x_0$ and consider the problem

$$y_i = x_0 + \sum_{k=1}^{n} a_{ik}x_k + c_i \geq 0, \quad (i = 1, 2, \ldots, m)$$
$$x_k \geq 0, \quad (k = 1, 2, \ldots, n) \qquad (2.40)$$
$$h = x_0 \qquad = \text{Min!}$$

We can immediately find a feasible solution of (2.40) just by choosing $x_0$ sufficiently large, namely

$$\bar{x}_0 = -\min_i(c_i) \quad \bar{x}_1 = \bar{x}_2 = \cdots = \bar{x}_n = 0. \qquad (2.41)$$

This feasible solution can be used as the new origin of a coordinate system in $\mathbf{R}^{n+1}$. Because $\bar{x}_k = 0$ holds for $k = 1, 2, \ldots, n$, the additional variable $x_0$ is therefore the only free variable.

If $x_0^*, x_1^*, x_2^*, \ldots, x_n^*$ is an optimal solution of (2.40), with $h_{\min} = x_0^* \leq 0$, then the inequalities

$$\sum_{k=1}^{n} a_{ik}x_k^* + c_i \geq 0 \quad (i = 1, 2, \ldots, m)$$
$$x_k^* \geq 0 \quad (k = 1, 2, \ldots, n) \qquad (2.42)$$

are obviously satisfied. Hence $x_1^*, x_2^*, \ldots, x_n^*$ is a feasible solution of the given linear program (2.6).

If, however, $h_{\min} = x_0^* > 0$ is the minimal value of the auxiliary objective function, then the feasible domain of (2.6) is empty. We can see this by an indirect argument. If we assume $\tilde{x}_1, \tilde{x}_2, \ldots, \tilde{x}_n$ to be a feasible solution of the linear program (2.6), it follows that $\tilde{x}_0 = 0, \tilde{x}_1, \tilde{x}_2, \ldots, \tilde{x}_n$ is a feasible solution of (2.40) with an even smaller value of the auxiliary objective function. This contradicts $h_{\min} > 0$.

The treatment of a linear program (2.6) with a nonfeasible origin will consist of two different solution steps. We start by solving the auxiliary linear program (2.40) to get a feasible solution of (2.6), and then on its basis compute the desired solution. Therefore we call this procedure a *two-phase method*. As soon as we have found a feasible solution in the first step we could, in principle, apply the coordinate shift method to determine the desired optimal solution. However, we shall proceed in a different way by taking into account that the auxiliary linear program is closely related to the given linear program. As soon as a feasible solution has been computed it will be much more advantageous to modify the resulting simplex scheme of the first phase so that it can directly be used as the starting scheme for the second phase. We shall now describe the steps in more detail.

We extend the given linear program, (2.6), by the variable $x_0$, for which we immediately make the substitution

$$x_0 = \bar{x}_0 + \xi_0 \quad \bar{x}_0 = -\min_i(c_i), \tag{2.43}$$

such that the new coefficients

$$\bar{c}_i = c_i + \bar{x}_0 \geq 0 \quad (i = 1, 2, \ldots, m) \tag{2.44}$$

satisfy the condition of Section 2.4.1 for the elimination of the free variable $\xi_0$ in the linear program (2.45)

$$\begin{aligned} h^* &= -\xi_0 & -\bar{x}_0 = \text{Max!} \\ y_i &= \xi_0 + \sum_{k=1}^n a_{ik}x_k + \bar{c}_i \geq 0 \quad (i = 1, 2, \ldots, m) \\ & x_k \geq 0 \quad (k = 1, 2, \ldots, n) \\ \left( z \right. &= \left. \sum_{k=1}^n b_k x_k + d = \text{Max!} \right) \end{aligned} \tag{2.45}$$

In view of the algorithmic implementation, the auxiliary objective function $h^* = -h$ to be maximized has been put at the top of (2.45), and the actual objective function $z$ is added in parentheses at the bottom because it will simply be carried along with the first phase. To solve the linear program (2.45) we use a simplex scheme in which the objective function $h^*$ appears in the uppermost (zeroth) row and the free variable $\xi_0$ in the leftmost (zeroth) column. Thereby the simplex scheme of the given linear program is extended by just one row and one column.

We start with the elimination of the free variable $\xi_0$. Although its value is of no immediate interest, we do not drop the resulting pivotal row, because it will be used later on. Next the simplex algorithm is applied with respect to the auxiliary objective function. It should be obvious that neither the row of the exchanged free variable, $\xi_0$, nor the last row of the objective function $z$ is allowed to be the pivotal row.

The first solution step can be stopped as soon as the auxiliary objective function, $h^*$, takes on a nonnegative value. Indeed, if we set the current $(n + 1)$ basis variables equal to zero we obtain a feasible solution of (2.6) according to (2.42). However this feasible solution point, of the auxiliary program (2.45), is a corner of the corresponding simplex in $\mathbf{R}^{n+1}$.

We now want to find a corner of the given linear program (2.6) in order to start

## 2.4 General Linear Programs

the second solution phase. For this the number of unknowns must be reduced by one so that we return to the appropriate $n$-dimensional space of (2.6). However this is a subspace of $\mathbf{R}^{n+1}$ of the auxiliary program defined by $x_0 = 0$, i.e. by $\xi_0 = -\bar{x}_0$. This reduction can be achieved by a suitable exchange step as can be seen from the geometrical interpretation of the simplex steps.

During the simplex step, in which the objective function $h^*$ changes its sign, or becomes zero, the hyperplane $x_0 = 0$ is crossed or just reached on an edge of the simplex in $\mathbf{R}^{n+1}$. This edge is characterized by setting all those basis variables equal to zero which were basis variables before and after this exchange step. The penetrating point of the edge, with the hyperplane $x_0 = 0$ in $\mathbf{R}^{n+1}$, is a corner of the feasible domain of the given linear program and is, therefore, the desired starting corner for the usual simplex algorithm of the second phase. The corresponding simplex scheme can be derived by an exchange step which exchanges the last variable entering the basis against the free variable $\xi_0$. Afterwards $\xi_0$ is replaced by the $-\bar{x}_0$ thus changing the constants of the linear inequalities, and finally the column of the variable $\xi_0$ is dropped from the scheme. We summarize the results below.

*Stopping criterion for the auxiliary program* The simplex algorithm, for the auxiliary linear program (2.45), has to be stopped as soon as the value of the objective function $h^*$, i.e. the element in the right-hand upper corner, is no longer negative.

*Reduction to the given linear program* (1) With the pivotal column last used and the row of the free variable $\xi_0$ and exchange step is performed. (2) In the resulting scheme we add the $(-\bar{x}_0)$ multiple of the obtained pivotal column to the last column. (3) Afterwards the row of the auxiliary objective function as well as the column of the variable $\xi_0$ are dropped.

The exchange step of the reduction is always possible with a nonzero pivot, as can be seen by a geometrical argument. The step is performed despite the resulting signs of the $c_i$ values of the last column. These coefficients must become nonnegative in the second step because the simplex scheme necessarily corresponds to a feasible corner point.

The two-phase method for the solution of a linear program (2.6) not satisfying (2.7) consists of the following steps

> 1. Extension of the simplex scheme to the auxiliary program (2.45). Set $\bar{x}_0 = -\min(c_i)$; $\bar{c}_i = c_i + \bar{x}_0$; extending the original scheme.
> 2. Elimination of the free variable $\xi_0$.
> 3. Simplex algorithm for the auxiliary objective function $h^*$. Stop as soon as $h^* \geq 0$.
> 4. Reduction of the simplex scheme.
> 5. Simplex algorithm for the objective function $z$.

(2.46)

The implementation of the two-phase method on a computer requires a two-dimensional array of $(m+2)$ rows and $(n+2)$ columns for the first phase. The storage requirement is only increased by $(m+n+3)$ places, in comparison with that of the given linear program (2.6).

*Example 2.7* The linear program (2.4) of the transport problem becomes, with $x_0 = 9 + \xi_0$, the extended form

$$
\begin{aligned}
h^* = -\xi_0 \quad &- 9 = \text{Max!} \\
y_1 = \xi_0 - x_1 - x_2 &+ 27 \geq 0 \\
y_2 = \xi_0 - x_1 \quad &+ 20 \geq 0 \\
y_3 = \xi_0 \quad - x_2 &+ 19 \geq 0 \\
y_4 = \xi_0 + x_1 + x_2 \quad &\geq 0 \\
x_1 \quad &\geq 0 \\
x_2 \quad &\geq 0 \\
(z^* = x_1 + 3x_2 &- 229 = \text{Max!})
\end{aligned}
\tag{2.47}
$$

Starting with the corresponding simplex scheme the elimination of the free variable $\xi_0$ yields

|       | $\xi_0$ | $x_1$ | $x_2$ | 1    |       | $y_4$ | $x_1$ | $x_2$ | 1    |
|-------|---------|-------|-------|------|-------|-------|-------|-------|------|
| $h^*=$ | $\underline{-1}$ | 0 | 0 | 9 | $h^*=$ | $-1$ | $\underline{1}$ | 1 | $-9$ |
| $y_1=$ | $\underline{1}$ | $-1$ | $-1$ | 27 | $y_1=$ | 1 | $\underline{-2}$ | $-2$ | 27 |
| $y_2=$ | $\underline{1}$ | $-1$ | 0 | 20 | $y_2=$ | 1 | $\underline{-2}$ | $-1$ | 20 |
| $y_3=$ | $\underline{1}$ | 0 | $-1$ | 19 | $y_3=$ | 1 | $\underline{-1}$ | $-2$ | 19 |
| $y_4=$ | $\underline{1}$ | 1 | 1 | 0 | $\xi_0=$ | 1 | $\underline{-1}$ | $-1$ | 0 |
| $z^*=$ | $\underline{0}$ | 1 | 3 | $-229$ | $z^*=$ | 0 | $\underline{1}$ | 3 | $-229$ |
|       | $-1$ | $-1$ | 0 |    |       | 0.5 | $-0.5$ | 10 |    |

$$\tag{2.48}$$

The scheme on the right of (2.48) is the starting scheme of the simplex algorithm with respect to the auxiliary objective function $h^*$ in the first row. The pivotal column is, as usual, fixed by Rule 1 and the pivotal element by Rule 2. The corresponding exchange step has already produced a positive value for $h^*$, and the first phase ends with

## 2.5 Discrete Chebyshev Approximation

|        | $y_4$ | $y_2$ | $x_2$ | 1    |
|--------|-------|-------|-------|------|
| $h^* =$ | −0.5  | −0.5  | 0.5   | 1    |
| $y_1 =$ | 0     | 1     | −1    | 7    |
| $x_1 =$ | 0.5   | −0.5  | −0.5  | 10   |
| $y_3 =$ | 0.5   | 0.5   | −1.5  | 9    |
| $\xi_0 =$ | 0.5 | 0.5   | −0.5  | −10  |
| $z^* =$ | 0.5   | −0.5  | 2.5   | −219 |
|        | −1    |       | 1     | 20   |

|        | $y_4$ | $\xi_0$ | $x_2$ | 1    |
|--------|-------|---------|-------|------|
| $y_1 =$ | −1   | 2       | 0     | 27   |
| $x_1 =$ | 1    | −1      | −1    | 0    |
| $y_3 =$ | 0    | 1       | −1    | 19   |
| $y_2 =$ | −1   | 2       | 1     | 20   |
| $z^* =$ | 1    | −1      | 2     | −229 |

(2.49)

Following the rule of reduction of the linear program the exchange step that interchanges $y_2$ and $\xi_0$ is added. The superfluous row of the auxiliary objective function has already been dropped. With $\xi_0 = -\bar{x}_0 = -9$ we obtain the simplex scheme (2.50), which corresponds to the corner Z of the feasible domain (see Figure 2.4).

|        | $y_4$ | $x_2$ | 1    |
|--------|-------|-------|------|
| $y_1 =$ | −1   | 0     | 9    |
| $x_1 =$ | 1    | −1    | 9    |
| $y_3 =$ | 0    | −1    | 10   |
| $y_2 =$ | −1   | 1     | 2    |
| $z^* =$ | 1    | 2     | −220 |

|        | $y_3$ | $y_1$ | 1    |
|--------|-------|-------|------|
| $x_1 =$ | 1    | −1    | 8    |
| $x_2 =$ | −1   | 0     | 10   |
| $y_4 =$ | 0    | −1    | 9    |
| $y_2 =$ | −1   | 1     | 3    |
| $z^* =$ | −2   | −1    | −191 |

(2.50)

After three steps of the usual simplex algorithm we finally obtain the simplex scheme on the right-hand side of (2.50). From this we derive the solution $x_1 = 8$, $x_2 = 10$, $y_1 = 0$, $y_2 = 3$, $y_3 = 0$, $y_4 = 9$ and $z^*_{max} = -191$.

## 2.5 Discrete Chebyshev Approximation

An application of linear programming is approximating a continuous function $y = f(x)$ on an interval by a polynomial $P_n(x)$ of degree $n$ such that the maximum absolute value of the differences $P_n(x_k) - f(x_k)$, at $N$ discrete abscissae $x_k$, ($k =$

$1, 2, \ldots, N$) is a minimum. Such a problem arises, for instance, if a function $f(x)$ that is defined by a complicated expression, is approximated by a polynomial, that it is easy to evaluate. As the maximum absolute value of the errors has to be minimized, we also speak of a problem of *Chebyshev approximation*.

Given the abscissae, $x_k$, and corresponding function values, $y_k = f(x_k)$, ($k = 1, 2, \ldots, N$) the polynomial

$$P_n(x) = a_0 + a_1 x + a_2 x^2 + \cdots + a_n x^n \quad n < N - 1 \tag{2.51}$$

of degree $n$ is sought so that we have

$$\max_k |P_n(x_k) - y_k| = \text{Min!} \tag{2.52}$$

To solve this problem we introduce

$$P_n(x_k) - y_k = r_k \quad (k = 1, 2, \ldots, N) \tag{2.53}$$

the *residuals* $r_k$ of the abscissae $x_k$, so that (2.52) becomes

$$\max_k |r_k| = \text{Min!} \tag{2.54}$$

Among the residuals $r_k$ there exists at least one with the largest absolute value. Let us denote it by $H = \max_k |r_k| > 0$. Hence we have the inequalities

$$|P_n(x_k) - y_k| \leq H \quad (k = 1, 2, \ldots, N). \tag{2.55}$$

The requirement (2.52) is equivalent to minimizing the value $H$ as an upper bound of the absolute values of the residuals. Now we replace $P_n(x_k)$ in (2.55) by the expression (2.51) and subsequently divide the inequalities (2.55) by the positive quantity $H$ to obtain the new inequalities

$$\left| \sum_{j=0}^{n} \left( \frac{a_j}{H} \right) x_k^j - \left( \frac{1}{H} \right) y_k \right| \leq 1 \quad (k = 1, 2, \ldots, N). \tag{2.56}$$

With the unknowns

$$\xi_1 = \frac{a_0}{H} \quad \xi_2 = \frac{a_1}{H} \quad \xi_3 = \frac{a_2}{H}, \ldots \quad \xi_{n+1} = \frac{a_n}{H} \quad \xi_{n+2} = \frac{1}{H} \tag{2.57}$$

the conditions (2.56) read as follows

$$|\xi_1 + x_k \xi_2 + x_k^2 \xi_3 + \cdots + x_k^n \xi_{n+1} - y_k \xi_{n+2}| \leq 1 \quad (k = 1, 2, \ldots, N). \tag{2.58}$$

Each inequality of (2.58), for the absolute value of an expression, is equivalent to two inequalities. Moreover, the variable $\xi_{n+2}$ equals the reciprocal value of the nonnegative quantity $H$ which has to be minimized. Thus for the unknowns $\xi_1$,

## 2.5 Discrete Chebyshev Approximation

$\xi_2, \ldots, \xi_{n+2}$ we obtain the following linear program

$$\xi_1 + x_k\xi_2 + x_k^2\xi_3 + \cdots + x_k^n\xi_{n+1} - y_k\xi_{n+2} \leqslant 1$$
$$\xi_1 + x_k\xi_2 + x_k^2\xi_3 + \cdots + x_k^n\xi_{n+1} - y_k\xi_{n+2} \geqslant -1 \quad (k = 1, 2, \ldots, N)$$
$$\xi_{n+2} \geqslant 0$$
$$\xi_{n+2} = \text{Max!}$$

Therefore, the normal form is

$$\boxed{\begin{aligned}
\eta_k &= -\xi_1 - x_k\xi_2 - x_k^2\xi_3 - \cdots - x_k^n\xi_{n+1} + y_k\xi_{n+2} + 1 \geqslant 0 \\
\eta'_k &= \xi_1 + x_k\xi_2 + x_k^2\xi_3 + \cdots + x_k^n\xi_{n+1} - y_k\xi_{n+2} + 1 \geqslant 0
\end{aligned} \quad (k = 1, 2, \ldots, N)}$$
$$\xi_{n+2} \geqslant 0$$
$$\zeta = \xi_{n+2} = \text{Max!}$$

(2.59)

The linear program (2.59), for the $(n+2)$ unknowns $\xi_1, \xi_2, \ldots, \xi_{n+2}$, contains $2N$ linear inequalities satisfying the obvious relations $\eta_k + \eta'_k = 2$, $k = 1, 2, \ldots, N$. As only $\xi_{n+2}$ has to fulfil a sign condition, $\xi_1, \xi_2, \ldots, \xi_{n+1}$ are free variables. The constants of the linear constraints (2.59) are positive. The hypotheses of Section 2.4.1 for the elimination of the $(n+1)$ free variables are satisfied. Due to the very special objective function we shall have the case $b_q = 0$. The values of the free variables $\xi_j$ are needed to compute the desired coefficients $a_0, a_1, \ldots, a_n$ of the polynomial $P_n(x)$ by means of (2.57). Therefore, the rows corresponding to the eliminated free variables must be carried over in the schemes, but they must be excluded from being a pivotal row during the elimination steps and during the simplex algorithm.

The approximation problem considered in (2.51) and (2.52) can obviously be generalized by replacing the polynomial $P_n(x)$ by a function $F_n(x)$ with the representation

$$F_n(x) = a_0\psi_0(x) + a_1\psi_1(x) + \cdots + a_n\psi_n(x) \tag{2.60}$$

where the $(n+1)$ functions $\psi_0(x), \psi_1(x), \ldots, \psi_n(x)$ are supposed to be linearly independent. In the above procedure the powers $x^j$ are simply replaced by the functions $\psi_j(x)$.

*Example 2.8* We consider only a simple problem so that we can give the complete solution. We look for a linear function $y = a_0 + a_1 x$ such that the maximum of the absolute values of the residuals is a minimum from four given points. The abscissae $x_k$ and the ordinates $y_k$ of the points are given in (2.61),

| $k$   | 1 | 2 | 3    | 4 |
|-------|---|---|------|---|
| $x_k$ | 1 | 2 | 3    | 4 |
| $y_k$ | 1 | 3 | 3.25 | 4 |

(2.61)

The corresponding simplex scheme for the unknowns $\xi_1, \xi_2$ and $\xi_3$ with the eight inequalities (2.59) is

|          | $\underline{\xi_1}$ | $\xi_2$ | $\xi_3$ | 1 |
|----------|---------|---------|---------|---|
| $\eta_1 =$ | $-1$ | $-1$ | $1$ | $1$ |
| $\eta_2 =$ | $-1$ | $-2$ | $3$ | $1$ |
| $\eta_3 =$ | $-1$ | $-3$ | $3.25$ | $1$ |
| $\eta_4 =$ | $-1$ | $-4$ | $4$ | $1$ |
| $\eta'_1 =$ | $\underline{1}$ | $1$ | $-1$ | $1$ |
| $\eta'_2 =$ | $1$ | $2$ | $-3$ | $1$ |
| $\eta'_3 =$ | $1$ | $3$ | $-3.25$ | $1$ |
| $\eta'_4 =$ | $1$ | $4$ | $-4$ | $1$ |
| $\zeta =$ | $\underline{0}$ | $0$ | $1$ | $0$ |
|          | $-1$ | $1$ | $-1$ |   |

The elimination of the two free variables happens in the two following schemes.

|          | $\eta'_1$ | $\underline{\xi_2}$ | $\xi_3$ | 1 |
|----------|---------|---------|---------|---|
| $\eta_1 =$ | $-1$ | $0$ | $0$ | $2$ |
| $\eta_2 =$ | $-1$ | $\underline{-1}$ | $2$ | $2$ |
| $\eta_3 =$ | $-1$ | $-2$ | $2.25$ | $2$ |
| $\eta_4 =$ | $-1$ | $-3$ | $3$ | $2$ |
| $\xi_1 =$ | $1$ | $-1$ | $1$ | $-1$ |
| $\eta'_2 =$ | $1$ | $1$ | $-2$ | $0$ |
| $\eta'_3 =$ | $1$ | $\underline{2}$ | $-2.25$ | $0$ |
| $\eta'_4 =$ | $1$ | $3$ | $-3$ | $0$ |
| $\zeta =$ | $0$ | $\underline{0}$ | $1$ | $0$ |
|          | $-1$ | $2$ | $0$ |   |

|          | $\eta'_1$ | $\eta'_2$ | $\xi_3$ | 1 |
|----------|---------|---------|---------|---|
| $\eta_1 =$ | $-1$ | $0$ | $0$ | $2$ |
| $\eta_2 =$ | $0$ | $-1$ | $0$ | $2$ |
| $\eta_3 =$ | $1$ | $-2$ | $\underline{-1.75}$ | $2$ |
| $\eta_4 =$ | $2$ | $-3$ | $\underline{-3}$ | $2$ |
| $\xi_1 =$ | $2$ | $-1$ | $\underline{-1}$ | $-1$ |
| $\xi_2 =$ | $-1$ | $1$ | $2$ | $0$ |
| $\eta'_3 =$ | $-1$ | $2$ | $1.75$ | $0$ |
| $\eta'_4 =$ | $-2$ | $3$ | $3$ | $0$ |
| $\zeta =$ | $0$ | $0$ | $\underline{1}$ | $0$ |
|          | $2/3$ | $-1$ |  | $2/3$ |

Both elimination steps do not change the value of the objective function. They

## 2.5 Discrete Chebyshev Approximation

lead to a degenerate simplex scheme to which the simplex algorithm is applied. The degeneracy causes no problem in this simple example because the corresponding rows cannot become pivotal rows. Two simplex steps give the solution scheme (2.62)

|            | $\eta_1'$ | $\eta_2'$ | $\eta_4$ | 1     |
|------------|-----------|-----------|----------|-------|
| $\eta_1 =$ | $-1$      | 0         | 0        | 2     |
| $\eta_2 =$ | 0         | $-1$      | 0        | 2     |
| $\eta_3 =$ | $-1/6$    | $-1/4$    | $7/12$   | $5/6$ |
| $\xi_3 =$  | $2/3$     | $-1$      | $-1/3$   | $2/3$ |
| $\xi_1 =$  | $4/3$     | 0         | $1/3$    | $-5/3$|
| $\xi_2 =$  | $1/3$     | $-1$      | $-2/3$   | $4/3$ |
| $\eta_3' =$| $1/6$     | $1/4$     | $-7/12$  | $7/6$ |
| $\eta_4' =$| 0         | 0         | $-1$     | 2     |
| $\zeta =$  | $2/3$     | $-1$      | $-1/3$   | $2/3$ |
|            | 0         | 0         |          | 2     |

|            | $\eta_1$  | $\eta_2'$ | $\eta_4$ | 1     |
|------------|-----------|-----------|----------|-------|
| $\eta_1' =$| $-1$      | 0         | 0        | 2     |
| $\eta_2 =$ | 0         | $-1$      | 0        | 2     |
| $\eta_3 =$ | $1/6$     | $-1/4$    | $7/12$   | $1/2$ |
| $\xi_3 =$  | $-2/3$    | $-1$      | $-1/3$   | 2     |
| $\xi_1 =$  | $-4/3$    | 0         | $1/3$    | 1     |
| $\xi_2 =$  | $-1/3$    | $-1$      | $-2/3$   | 2     |
| $\eta_3' =$| $-1/6$    | $1/4$     | $-7/12$  | $3/2$ |
| $\eta_4' =$| 0         | 0         | $-1$     | 2     |
| $\zeta =$  | $-2/3$    | $-1$      | $-1/3$   | 2     |

(2.62)

From (2.62) we obtain the interesting values $\zeta_{\max} = 1/H = 2$, $\xi_1 = 1$, $\xi_2 = 2$ from which we obtain the coefficients $a_0 = \xi_1 H = 0.5$ and $a_1 = \xi_2 H = 1$ and $H = \max|r_k| = 0.5$. In Figure 2.5 the straight line of best approximation, in the Chebyshev sense, is drawn together with the residuals $r_k$. It is typical of the solution line that there exist three residuals of the same magnitude $|r_1| = |r_2| = |r_4| = 0.5$, whereas $|r_3| < 0.5$. Moreover, the residuals of largest absolute value equioscillate with increasing abscissae. We say they form an *alternante* (Rutishauser, 1976; Werner and Schaback, 1972).

*Example* 2.9 In the Chebyshev sense, we want to approximate the function $f(x) = \cos(\frac{1}{2}\pi x)$ on the interval $[0, 1]$ by a polynomial $P_4(x)$ of degree four with respect to the 11 equidistant abscissae $x_k = (k-1)/10$, $(k = 1, 2, \ldots, 11)$. The corresponding linear program with 22 linear constraints for the six unknowns $\xi_1$, $\xi_2, \ldots, \xi_6$ has been set up and solved with the aid of a computer program. After the elimination of the five free variables, 11 simplex steps were needed. The resulting polynomial is

$$P_4(x) = 0.999\,896\,0847 + 0.004\,960\,1071x - 1.271\,044\,5207x^2$$
$$+ 0.093\,307\,1199x^3 + 0.172\,985\,1242x^4. \qquad (2.63)$$

The maximum absolute value of the residuals at the discrete abscissae, $x_k$, is

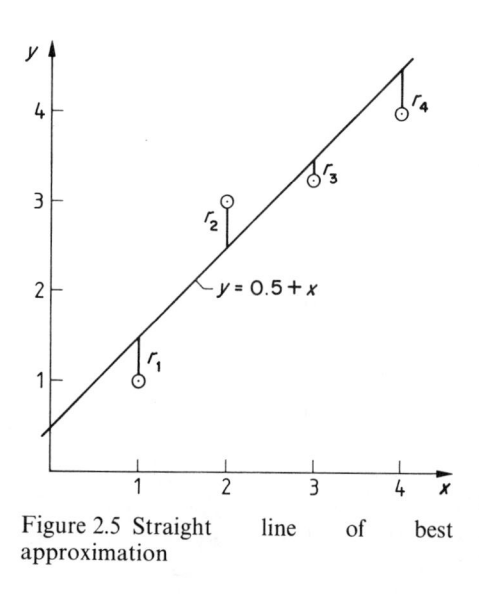

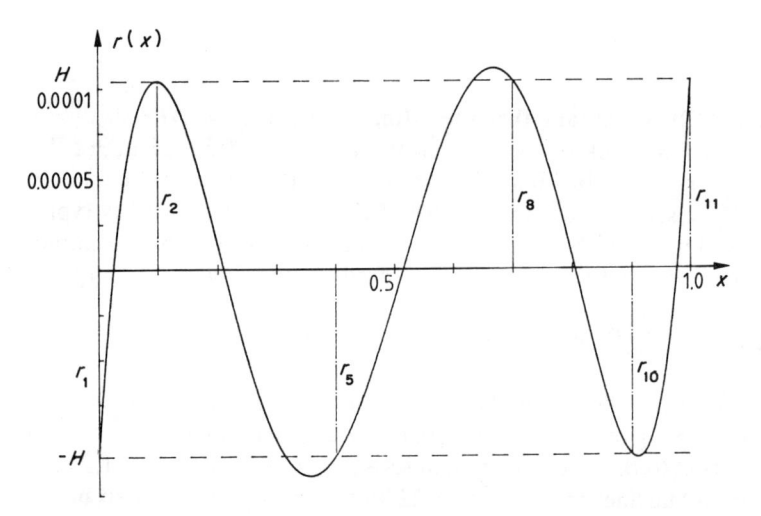

Figure 2.5 Straight line of best approximation

Figure 2.6 Error function of the Chebyshev approximation

$H = \max |r_k| = 0.000\,1039$ and is taken on at $x_1, x_2, x_5, x_8, x_{10}, x_{11}$ with alternating sign. Figure 2.6 shows the error function $r(x) := P_4(x) - \cos(\tfrac{1}{2}\pi x)$. Its maximum absolute value is only slightly larger than $H$. Hence we can say that $P_4(x)$, (2.63), approximates the given function within an absolute accuracy of about four decimal places.

## 2.6 Exercises

**2.1.** A factory produces the products X and Y. The production of a unit requires a certain number of working days, machine hours and materials, $A$ and $B$, for which all have limited capacities available each month. According to an analysis of the market it is decided to produce at most 30 units more of product Y than X. On the following assumptions what is the production with a maximum net profit.

|  | Product X | Product Y | Available resources |
|---|---|---|---|
| Working days | 1 | 5 | 300 |
| Machine hours | 2 | 2 | 200 |
| Material $A(m^2)$ | 2 | 1 | 170 |
| Material $B(l)$ | 5 | 2 | 420 |
| Net profit | 20 | 30 |  |

Solve the problem graphically and by means of the simplex algorithm, give the interpretation of the steps and discuss the result.

**2.2.** Solve the linear program

$$\begin{array}{ll} x_1 \leq 2 & -x_1 + 2x_2 + 2x_3 \leq 6 \\ x_2 \leq 2 & 2x_1 - x_2 + 2x_3 \leq 6 \\ x_3 \leq 2 & x_1 \geq 0, x_2 \geq 0, x_3 \geq 0 \\ x_1 + x_2 + x_3 \leq 5 & z = x_1 + x_2 + x_3 = \text{Max!} \end{array}$$

by means of the simplex algorithm and determine the complete set of solution corners. With these, represent the nonunique solution of the linear program. Geometrically construct the three-dimensional feasible domain, cube with cut corners, as an aid and interpret the simplex steps.

**2.3.** Which exceptional cases are encountered when the linear program

$$\begin{array}{ll} x_1 - 5x_2 \leq 5 & -2x_1 + x_2 \leq 4 \\ 2x_1 - 5x_2 \leq 10 & x_1 \geq 0, x_2 \geq 0 \\ x_1 - x_2 \leq 8 & z = x_1 - x_2 + 4 = \text{Max!} \end{array}$$

is solved by the simplex algorithm? For an interpretation of the situations and the steps a figure is helpful.

**2.4.** Solve the following linear program that does not satisfy (2.7)

$$x_1 + x_2 \leq 18 \qquad -x_1 + 2x_2 \geq 2$$
$$x_1 \leq 10 \qquad x_1 \geq 0, x_2 \geq 0$$
$$x_2 \leq 11 \qquad z = x_1 + x_2 + 5 = \text{Max}!$$
$$x_1 + x_2 \geq 9$$

by means of a coordinate shift, with the feasible solution $\bar{x}_1 = 5, \bar{x}_2 = 6$ and by means of the two-phase method. In both cases give a graphical interpretation of the steps.

**2.5.** In three tram depots A, B and C are 18, 12 and 8 trams ready. At the places U, V and W 13, 15 and 10 extra trams are required. According to the distance table

|   | U | V | W |
|---|---|---|---|
| A | 5 | 8 | 11 |
| B | 8 | 12 | 14 |
| C | 6 | 10 | 13 |

all trams are to be directed in such a way that the total number of idle kilometres is minimal. What is the optimal solution?

*Hint*: Note that the number of available trams is equal to the number of required ones. Therefore four unknowns are needed for the mathematical formulation of the problem. It is suggested to choose the unknowns $x_1 = $ number of trams from A to U, $x_2 = $ number of trams from A to V, $x_3 = $ number of trams from B to U, $x_4 = $ number of trams from C to V. Solve the resulting linear program, with the origin nonfeasible, by means of a coordinate shift and the two-phase method. In the first case a coordinate shift with respect to two variables is sufficient.

**2.6.** Approximate the function $y = \sin(x)$ on the interval $[0, \pi/4]$ by a polynomial $P(x) = a_1 x + a_3 x^3$ that is odd like $\sin(x)$. Determine the coefficients $a_1$ and $a_3$ by a discrete Chebyshev approximation for the abscissae $x_k = k\pi/24$, $(k = 1, 2, \ldots, 6)$. What is the linear program and the approximating polynomial? Also draw the graph of the error function $r(x) := P(x) - \sin(x)$ on the interval $[0, \pi/4]$ and the residuals at the discrete values $x_k$.

**2.7.** Determine the approximating polynomials $P_2(x) = a_0 + a_1 x + a_2 x^2$ and $P_4(x) = a_0 + a_1 x + a_2 x^2 + a_3 x^3 + a_4 x^4$ of degree two and four, respectively, in the sense of a discrete Chebyshev approximation, for the function $y = e^x$ on the interval $[0, 1]$ using $N = 6$ and $N = 11$ equidistant abscissae $x_k$. How large are the corresponding maxima $H$ of the absolute values of the residuals? Set up and solve the corresponding linear programs by means of a computer program and represent the error functions graphically.

# 3

# Interpolation

We consider the problem of approximating a real-valued function $f(x)$ on the assumption that the function values are only known at discrete points. We may then approximately evaluate between the discrete points. Interpolation polynomials are the most simple and common means of approximation. Although the interpolation polynomial is uniquely determined by the given interpolation points there exist different representations with specific advantages, depending on the actual application. We shall see that interpolation by polynomials is not always adequate. Therefore we shall also consider the interpolation problem for rational functions. Finally we shall present a useful procedure for determining a smooth, i.e. twice continuously differentiable, piecewise polynomial interpolating function that has many practical applications.

## 3.1 Existence and Uniqueness of Polynomial Interpolation

Let $(x_i, y_i)$ be $(n+1)$ support points, with discrete, pairwise different *support abscissae* $x_i$, $(i = 0, 1, \ldots, n)$ and arbitrary support ordinates $y_i$, $(i = 0, 1, \ldots, n)$. We seek a polynomial of degree $n$

$$P_n(x) = a_0 + a_1 x + \cdots + a_n x^n, \tag{3.1}$$

satisfying the interpolation conditions

$$P_n(x_i) = y_i \quad (i = 0, 1, \ldots, n). \tag{3.2}$$

The following theorem on existence and uniqueness serves as the basis for all subsequent considerations about polynomial interpolation.

**Theorem 3.1** *If $(x_i, y_i)$ are $(n+1)$ arbitrary points, $(i = 0, 1, \ldots, n)$, with pairwise different abscissae, $x_i \neq x_j$ for all $i \neq j$, then there exists a unique interpolation polynomial $P_n(x)$ satisfying (3.2), whose degree is at most n.*

*Proof* (a) We show the existence of the interpolation polynomial by construction. Consider the special polynomials

$$L_i(x) := \prod_{\substack{j=0 \\ j \neq i}}^{n} \frac{(x-x_j)}{(x_i-x_j)} = \frac{(x-x_0)\cdots(x-x_{i-1})(x-x_{i+1})\cdots(x-x_n)}{(x_i-x_0)\cdots(x_i-x_{i-1})(x_i-x_{i+1})\cdots(x_i-x_n)}$$

$$(i = 0, 1, \ldots, n). \qquad (3.3)$$

These *Lagrange-polynomials* $L_i(x)$ are exactly of degree $n$ and possess the obvious property

$$L_i(x_k) = \delta_{ik} = \begin{cases} 1, & \text{if } i = k \\ 0, & i \neq k \end{cases}. \qquad (3.4)$$

Then the polynomial

$$P_n(x) := \sum_{i=0}^{n} y_i L_i(x) \qquad (3.5)$$

has the required interpolation properties. From (3.4) we have

$$P_n(x_k) = \sum_{i=0}^{n} y_i L_i(x_k) = \sum_{i=0}^{n} y_i \delta_{ik} = y_k \quad \text{for } k = 0, 1, \ldots, n.$$

The degree of $P_n(x)$, (3.5), is less than or equal to $n$ because it is a linear combination of polynomials of degree $n$.

(b) We now show the uniqueness of the interpolation polynomial. Let $P_n(x)$ and $Q_n(x)$ be two polynomials of degree at most $n$ satisfying the interpolation conditions

$$P_n(x_k) = Q_n(x_k) = y_k \quad (k = 0, 1, \ldots, n). \qquad (3.6)$$

(3.6) implies that $D(x) := P_n(x) - Q_n(x)$ is a polynomial of degree at most $n$ with the $(n+1)$ pairwise different roots $x_0, x_1, \ldots, x_n$. According to the fundamental theorem of algebra we must have $D(x) \equiv 0$ or $P_n(x) = Q_n(x)$.

## 3.2 Lagrange Interpolation

### 3.2.1 Technique of computation

The representation (3.5) of the interpolation polynomial $P_n(x)$ is called the *Langrange interpolation formula*. It solves the interpolation problem by an explicit expression. However, it is not appropriate for practical applications, for example the computation of an interpolating value, because the evaluation of the expression (3.5) would take too long. Therefore after substituting (3.3) into (3.5) we arrange it in a different way, on the assumption that $x \neq x_i$ for $i = 0, 1, \ldots, n$.

$$P_n(x) = \sum_{i=0}^{n} y_i \prod_{\substack{j=0 \\ j \neq i}}^{n} \frac{x-x_j}{x_i-x_j} = \sum_{i=0}^{n} y_i \frac{1}{x-x_i} \left( \prod_{\substack{j=0 \\ j \neq i}}^{n} \frac{1}{x_i-x_j} \right) \prod_{k=0}^{n} (x-x_k). \qquad (3.7)$$

## 3.2 Lagrange Interpolation

At first we define the *support coefficients*

$$\lambda_i := \prod_{\substack{j=0 \\ j \neq i}}^{n} \frac{1}{x_i - x_j} = 1 \Big/ \prod_{\substack{j=0 \\ j \neq i}}^{n} (x_i - x_j) \quad (i = 0, 1, \ldots, n) \tag{3.8}$$

which only depend on the given support abscissae $x_k$. Furthermore we introduce the auxiliary quantities

$$\mu_i := \frac{\lambda_i}{x - x_i} \quad (i = 0, 1, \ldots, n), \tag{3.9}$$

which depend on the value of $x$. With these definitions, (3.7) takes on the simpler form

$$P_n(x) = \left( \sum_{i=0}^{n} \mu_i y_i \right) \prod_{k=0}^{n} (x - x_k) \tag{3.10}$$

in which we want to eliminate the product factor. The representation (3.10) is valid for arbitrary $y_i$ values and hence for the special values $y_i = 1, (i = 0, 1, \ldots, n)$. For these function values $P_n(x) = 1$ is obviously the unique solution of this interpolation problem due to Theorem 3.1. Hence, (3.10) implies the identity

$$1 = \left( \sum_{i=0}^{n} \mu_i \right) \prod_{k=0}^{n} (x - x_k)$$

for all $x$, and hence

$$\prod_{k=0}^{n} (x - x_k) = \frac{1}{\sum_{i=0}^{n} \mu_i}. \tag{3.11}$$

From (3.10) and (3.11) we deduce the final representation

$$P_n(x) = \frac{\sum_{i=0}^{n} \mu_i y_i}{\sum_{i=0}^{n} \mu_i} \tag{3.12}$$

We call (3.12) the *barycentric formula* of the Lagrange interpolation for the evaluation of $P_n(x)$ for the value $x$, because it is formed as a weighted mean of the function values $y_i$ with the weights $\mu_i$. However, the signs of the $\mu_i$ are different for $n \geqslant 2$.

The formulae (3.8), (3.9) and (3.12) define the computational technique of the Lagrange interpolation. First, the $(n+1)$ support coefficients $\lambda_i$ are computed from the given support abscissae $x_k$. Then for each new value $x$ the weights $\mu_i$ can be computed and at the same time the two sums of (3.12) can be formed. At first sight the computation of the support coefficients $\lambda_i$ requires $n(n+1)$ essential operations. However, this amount can be halved by a simple observation (Werner, 1984). Let us assume that the $(n+1)$ support coefficients $\lambda_i^{(n)}$ for the support abscissae $x_0, x_1, \ldots, x_n$ are known. If we now add a $(n+2)$th abscissa $x_{n+1}$, for the corresponding coefficients $\lambda_i^{(n+1)}$ and the known values $\lambda_i^{(n)}$ we obtain the relationships

$$\lambda_i^{(n+1)} = \lambda_i^{(n)}/(x_i - x_{n+1}) \quad (i = 0, 1, \ldots, n), \tag{3.13}$$

which allow us to obtain the first $(n+1)$ support coefficients $\lambda_i^{(n+1)}$ from the $\lambda_i^{(n)}$ by one division each. The remaining coefficient is determined by means of the following theorem.

*Theorem 3.2* The $(n+1)$ support coefficients $\lambda_i^{(n)}$ for $(n+1)$ pairwise different support abscissae $x_0, x_1, \ldots, x_n$ satisfy the equation

$$\sum_{i=0}^{n} \lambda_i^{(n)} = 0 \quad (n \geqslant 1). \tag{3.14}$$

*Proof* The interpolation polynomial $P_n(x)$ has the following representation from (3.8)

$$P_n(x) = \sum_{i=0}^{n} y_i \lambda_i^{(n)} \prod_{\substack{j=0 \\ j \neq i}}^{n} (x - x_j). \tag{3.15}$$

From (3.15) the coefficient $a_n$ of $x^n$ is given by

$$a_n = \sum_{i=0}^{n} y_i \lambda_i^{(n)}. \tag{3.16}$$

For the special choice $y_i = 1$ for all $i = 0, 1, \ldots, n$ we again have $P_n(x) = 1$ and hence $a_n = 0$ for all $n \geqslant 1$. From this (3.14) follows.

The recursive computation of the support coefficients $\lambda_i^{(k)}$ for increasing $k$ requires starting values for a certain index $k$. For $k = 1$ and the abscissae $x_0$ and $x_1$ we have $\lambda_0^{(1)} = 1/(x_0 - x_1)$, $\lambda_1^{(1)} = 1/(x_1 - x_0) = -\lambda_0^{(1)}$. Equation (3.13) shows that the starting value is $\lambda_0^{(0)} = 1$. Therefore we can summarize the computation

## 3.2 Lagrange Interpolation

of the coefficients $\lambda_i^{(n)}$ for given support abscissae $x_0, x_1, \ldots, x_n$ as follows.

$$
\begin{aligned}
&\text{Start:} \quad \lambda_0^{(0)} = 1 \\
&\text{Recursion: for } k = 1, 2, \ldots, n: \\
&\qquad \text{for } i = 0, 1, \ldots, k-1: \\
&\qquad\qquad \lambda_i^{(k)} = \lambda_i^{(k-1)}/(x_i - x_k) \\
&\qquad \lambda_k^{(k)} = -\sum_{i=0}^{k-1} \lambda_i^{(k)}
\end{aligned}
\qquad (3.17)
$$

According to (3.17) the amount of computational effort required is only

$$Z_{\text{Supp Coeff}} = 1 + 2 + \cdots + n = \tfrac{1}{2}n(n+1) \qquad (3.18)$$

divisions. We note that the support coefficients $\lambda_i^{(k)}$ can be built up on a computer in a vector of $(n+1)$ components. Moreover, the described procedure allows us to increase the number of support abscissae successively in a simple way, which is useful for certain practical applications.

*Example* 3.1 Let the support abscissae be $x_0 = 0$, $x_1 = 1.5$, $x_2 = 2.5$, $x_3 = 4.5$ and the support ordinates be $y_0 = 1$, $y_1 = 2$, $y_2 = 2$, $y_3 = 1$. We seek the interpolating value for $x = 2.1$. The recursive computation of the support coefficients is given in the following scheme to five significant digits.

| k \ i | 0 | 1 | 2 | 3 |
|---|---|---|---|---|
| 0 | 1.0000 | | | |
| 1 | −0.66667 | 0.66667 | | |
| 2 | 0.26667 | −0.66667 | 0.40000 | |
| 3 | −0.059260 | 0.22222 | −0.20000 | 0.037040 |

From the $\lambda_i^{(3)}$ we obtain $\mu_0 \doteq -0.028\,219$, $\mu_1 \doteq 0.370\,37$, $\mu_2 \doteq 0.500\,00$, $\mu_3 \doteq -0.015\,433$ and hence $P_3(2.1) \doteq 1.6971/0.826\,72 \doteq 2.0528$.

In the case of *equidistant support abscissae* the support coefficients are defined by a simple expression. We arrange the abscissae in increasing order with step $h$

$$x_0, x_1 = x_0 + h, \ldots, x_j = x_0 + jh, \ldots, x_n = x_0 + nh.$$

For the $i$th coefficient from (3.8) we obtain

$$\lambda_i = \frac{1}{(x_i - x_0)\cdots(x_i - x_{i-1})(x_i - x_{i+1})\cdots(x_i - x_n)}$$

$$= \frac{1}{(-1)^{n-i}h^n[i(i-1)\cdots 1][1\cdot 2\cdots(n-i)]} = \frac{(-1)^{n-i}}{h^n n!}\binom{n}{i}. \tag{3.19}$$

As a common factor of the weights $\mu_i$ obviously cancels in the formula (3.12), instead of (3.19), we may use the equivalent support coefficients defined by the alternating *binomial coefficients*:

$$\lambda_i^* = (-1)^i \binom{n}{i} \quad (i = 0, 1, \ldots, n). \tag{3.20}$$

For a given $n$ the $\lambda_i^*$ are generated by means of the recursion

$$\lambda_0^* = 1 \quad \lambda_i^* = -\lambda_{i-1}^* \cdot \frac{n-i+1}{i} \quad (i = 1, 2, \ldots, n)$$

with only $2n$ essential operations.

*Example* 3.2 From a table of $\sin(x)$ we take the function values at equidistant support abscissae rounded to five decimal places.

| $x_i =$ | 20° | 30° | 40° | 50° |
|---|---|---|---|---|
| $y_i =$ | 0.342 02 | 0.500 00 | 0.642 79 | 0.766 04 |

We seek the interpolated function value for $x = 36°$ by means of cubic interpolation. The modified support coefficients are $\lambda_0^* = 1$, $\lambda_1^* = -3$, $\lambda_2^* = 3$, $\lambda_3^* = -1$, and so we obtain the weights $\mu_0^* = 0.0625$, $\mu_1^* = -0.5$, $\mu_2^* = -0.75$, $\mu_3^* \doteq 0.071\,429$ and the interpolated function value $P_3(36°) \doteq -0.655\,99/(-1.1161) \doteq 0.587\,75$. As we used five-digit arithmetic, the interpolated value is affected by rounding errors and differs by four units in the last decimal place from the exact value $\sin(36°) \doteq 0.587\,79$.

### 3.2.2 Applications

Interpolation polynomials, for tabulated functions, serve as a basis for the approximate computation of derivatives of functions. Formulae derived in this way for the *numerical differentiation* can be applied for the approximate computation of derivatives of analytically defined functions. The representation of the interpolation polynomial by the Lagrange formula, (3.5) or (3.15), is well

## 3.2 Lagrange Interpolation

suited for deriving the rules of differentiation. If we differentiate (3.15) $n$ times, then the expression

$$\frac{d^n P_n(x)}{dx^n} = \sum_{i=0}^{n} y_i \lambda_i n! \approx f^{(n)}(x) \tag{3.21}$$

is a possible approximation of the $n$th derivative $f^{(n)}(x)$.

**Theorem 3.3** *Let $f(x)$ be at least $n$ times continuously differentiable on the interval $[a, b]$, where $a = \min_i(x_i)$, $b = \max_i(x_i)$, then there exists a $\xi \in (a, b)$ such that*

$$f^{(n)}(\xi) = \sum_{i=0}^{n} y_i \lambda_i n!. \tag{3.22}$$

*Proof* We consider the function $g(x) := f(x) - P_n(x)$, where $P_n(x)$ is the interpolation polynomial for the support abscissae $x_i$, $(i = 0, 1, \ldots, n)$ with the function values $y_i = f(x_i)$. Therefore $g(x)$ has at least the $(n+1)$ zeros $x_0, x_1, \ldots, x_n$ from the interpolation property. We repeatedly apply Rolle's theorem to $g(x)$ and deduce the existence of a point $\xi$ in the interior of the smallest interval containing all support abscissae such that $g^{(n)}(\xi) = f^{(n)}(\xi) - P_n^{(n)}(\xi) = 0$ holds. From (3.21) we derive the assertion (3.22). Figure 3.1 illustrates the situation for $n = 3$.

The *rule of numerical differentiation*, (3.21), for approximating the $n$th derivative of a given, tabular, function is usually only applied to equidistant points $x_i = x_0 + ih$, $(i = 0, 1, \ldots, n)$. According to (3.19) the rule (3.21) becomes

$$f^{(n)}(x) \approx \frac{1}{h^n} \left[ (-1)^n y_0 + (-1)^{n-1} \binom{n}{1} y_1 \right.$$
$$\left. + (-1)^{n-2} \binom{n}{2} y_2 + \cdots - \binom{n}{n-1} y_{n-1} + y_n \right]. \tag{3.23}$$

Thus the approximation of the $n$th derivative $f^{(n)}(x)$ is computed as a linear combination of the function values $y_i$ multiplied by the binomial coefficients of

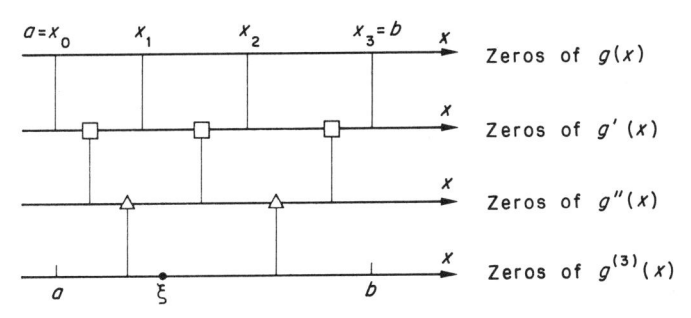

Figure 3.1 Existence of zeros according to Rolle's theorem

alternating signs, where the last one is positive, and divided by $h^n$. We call the expression (3.23) the *n*th *difference quotient* of the $(n+1)$ functions values $y_i$. For $n = 1, 2$ and 3 the corresponding rules of numerical differentiation are

$$f'(x) \approx \frac{y_1 - y_0}{h} \qquad \text{1. difference quotient}$$

$$f''(x) \approx \frac{y_2 - 2y_1 + y_0}{h^2} \qquad \text{2. difference quotient} \qquad (3.24)$$

$$f^{(3)}(x) \approx \frac{y_3 - 3y_2 + 3y_1 - y_0}{h^3} \qquad \text{3. difference quotient}$$

The point $\xi$ for which the *n*th difference quotient is equal to the *n*th derivative of a function $f(x)$, that is at least $n$ times continuously differentiable, is usually near to the middle point $x_M = \frac{1}{2}(x_0 + x_n)$. In the case of purely tabulated functions a difference quotient (3.24) will usually best approximate the corresponding derivative in the middle of the interval.

The *p*th derivative of a function may also be approximated by using an interpolation polynomial $P_n(x)$ of degree $n$ strictly greater than $p$. Because then the *p*th derivative $P_n^{(p)}(x)$ is no longer constant, we have to specify the value $x$. We illustrate the procedure for the first derivate $f'(x)$ on the basis of a quadratic interpolation polynomial. From (3.15) and (3.19) we have

$$P_2(x) = \frac{1}{2h^2}[y_0(x - x_1)(x - x_2)$$

$$- 2y_1(x - x_0)(x - x_2) + y_2(x - x_0)(x - x_1)] \qquad (3.25)$$

$$P'_2(x) = \frac{1}{2h^2}[y_0(2x - x_1 - x_2) - 2y_1(2x - x_0 - x_2) + y_2(2x - x_0 - x_1)] \qquad (3.26)$$

For the first derivative at the points $x_0$ and $x_1$ we obtain from (3.26)

$$f'(x_0) \approx \frac{1}{2h}(-3y_0 + 4y_1 - y_2) \qquad (3.27)$$

$$f'(x_1) \approx \frac{1}{2h}(-y_0 + y_2) \qquad (3.28)$$

## 3.2 Lagrange Interpolation

The expression (3.28) is called *central difference quotient*. It represents the slope of the line passing through the two points $(x_0, y_0)$ and $(x_2, y_2)$. It is known that this slope is equal to the slope of the tangent of the interpolating parabola at the middle support point $(x_1, y_1)$. The central difference quotient (3.28) is in fact a better approximation of the first derivative than the first difference quotient (3.24).

In a completely analogous way the first and second derivatives may be approximated at certain points by means of cubic interpolation polynomials. The analogous calculation yields the following selected formulae of numerical differentiation.

$$f'(x_0) \approx \frac{1}{6h}(-11y_0 + 18y_1 - 9y_2 + 2y_3)$$

$$f'(x_1) \approx \frac{1}{6h}(-2y_0 - 3y_1 + 6y_2 - y_3)$$

$$f'(x_M) \approx \frac{1}{24h}(y_0 - 27y_1 + 27y_2 - y_3) \qquad (3.29)$$

$$f''(x_M) \approx \frac{1}{2h^2}(y_0 - y_1 - y_2 + y_3)$$

$$x_M = \tfrac{1}{2}(x_0 + x_3)$$

*Example 3.3* To exhibit a difficulty of the numerical differentiation, we consider the problem of computing the second derivative of the function $f(x) = \sinh(x)$ at $x = 0.6$ with the second difference quotient (3.24). We shall do this for a sequence of step lengths $h$ tending to zero. We use function values rounded to nine significant figures, whereas the second difference quotient is evaluated in floating point arithmetic with ten digits. The following table contains the different values of $h$, $x_0$, $x_2$, $y_0$, $y_2$ and the resulting approximation of $f''(0.6) = \sinh(0.6) = 0.636\,653\,582$.

| $h$ | $x_0$ | $x_2$ | $y_0$ | $y_2$ | $f''(x_1) \approx$ |
|---|---|---|---|---|---|
| 0.1 | 0.50 | 0.7 | 0.521 095 305 | 0.758 583 702 | 0.637 184 300 |
| 0.01 | 0.59 | 0.61 | 0.624 830 565 | 0.648 540 265 | 0.636 660 000 |
| 0.001 | 0.599 | 0.601 | 0.635 468 435 | 0.637 839 366 | 0.637 000 000 |
| 0.000 1 | 0.599 9 | 0.600 1 | 0.636 535 039 | 0.636 772 132 | 0.700 000 000 |
| 0.000 01 | 0.599 99 | 0.600 01 | 0.636 641 728 | 0.636 665 437 | 10.000 000 00 |

With decreasing step size $h$ the cancellation of digits becomes more and more catastrophic so that a wrong approximate value results for the smallest $h = 0.00001$.

Numerical differentiation is in general a dangerous process that, due to the cancellation for small step sizes $h$, produces results with increasing relative errors. The limiting process as $h \to 0$ is not feasible for numerical reasons unless an arbitrarily high precision is used.

However, if we require as accurate values as possible for the derivatives, computed in arithmetic of a given precision, the method of *extrapolation* is useful. To explain the principle we need an analysis of the error. For this we must assume that the function $f(x)$, whose $p$th derivative is to be computed by numerical differentiation, is arbitrarily many times continuously differentiable on the closed interval considered, and that it admits convergent Taylor series expansions.

Let us start with the first difference quotient (3.24) where we substitute the Taylor series for $y_1 = f(x_1) = f(x_0 + h)$

$$y_1 = f(x_0) + hf'(x_0) + \frac{h^2}{2!}f''(x_0) + \frac{h^3}{3!}f^{(3)}(x_0) + \frac{h^4}{4!}f^{(4)}(x_0) + \cdots$$

and obtain with $y_0 = f(x_0)$

$$\frac{y_1 - y_0}{h} = f'(x_0) + \frac{h}{2!}f^{(2)}(x_0) + \frac{h^2}{3!}f^{(3)}(x_0) + \frac{h^3}{4!}f^{(4)}(x_0) + \cdots. \quad (3.30)$$

The difference quotient approximates $f'(x_0)$ with an error that can be expressed by a power series in $h$.

For the central difference quotient (3.28) with the Taylor series we obtain

$$y_2 = f(x_2) = f(x_1 + h) = f(x_1) + hf'(x_1)$$
$$+ \frac{h^2}{2!}f''(x_1) + \frac{h^3}{3!}f^{(3)}(x_1) + \frac{h^4}{4!}f^{(4)}(x_1) + + \cdots$$

$$y_0 = f(x_0) = f(x_1 - h) = f(x_1) - hf'(x_1)$$
$$+ \frac{h^2}{2!}f''(x_1) - \frac{h^3}{3!}f^{(3)}(x_1) + \frac{h^4}{4!}f^{(4)}(x_1) - + \cdots$$

$$\frac{1}{2h}(y_2 - y_0) = f'(x_1) + \frac{h^2}{3!}f^{(3)}(x_1)$$
$$+ \frac{h^4}{5!}f^{(5)}(x_1) + \frac{h^6}{7!}f^{(7)}(x_1) + \cdots \quad (3.31)$$

Hence, the central difference quotient, (3.28), approximates the first derivative with an error having a series expansion that has only *even powers* of $h$. The error of (3.31) is of *second order*, whereas the error of the first difference quotient (3.30) is

## 3.2 Lagrange Interpolation

of *first order*. Thus the approximation error of the central difference quotient is smaller, at least for sufficiently small $h$.

Likewise for the second difference quotient we obtain the result

$$\frac{y_2 - 2y_1 + y_0}{h^2} = f''(x_1) + \frac{2h^2}{4!} f^{(4)}(x_1)$$

$$+ \frac{2h^4}{6!} f^{(6)}(x_1) + \frac{2h^6}{8!} f^{(8)}(x_1) + \cdots. \tag{3.32}$$

The three results (3.30), (3.31) and (3.32) all have in common that a *computable quantity* $B(t)$ depending on a parameter $t$ approximates a *desired value* $A$ with an error, that has a series expansion, so that we have with constant coefficients $a_1, a_2, \ldots$.

$$B(t) = A + a_1 t + a_2 t^2 + a_3 t^3 + \cdots + a_n t^n + \cdots. \tag{3.33}$$

For numerical or computational reasons it is often impossible to determine the quantity $B(t)$ for a small parameter $t$ for that $B(t)$ is an acceptable approximation of $A$. The *principle of extrapolation* consists of computing the values $B(t_k)$ for some parameters $t_0 > t_1 > t_2 > \cdots > t_n > 0$ and then of evaluating the corresponding interpolation polynomials $P_k(t)$ for the new value $t = 0$ outside the interpolation interval. With increasing $k$ the values $P_k(0)$ represent better approximations of the desired value $B(0) = A$. The extrapolation is stopped as soon as two successive extrapolated values $P_k(0)$ coincide within a prescribed accuracy. The extrapolation process may be performed by means of the Lagrange interpolation, as in (3.12). As the new value is $t = 0$, the weights (3.9) are simply given by $\mu_i = \lambda_i/t_i$, where we omit the common factor $(-1)$. We summarize the process of extrapolation complementing (3.17) as follows.

$$
\begin{array}{l}
\text{Start: Input of } t_0, y_0 := B(t_0) \\
\quad \lambda_0 = 1 \\
\text{Extrapolation: for } k = 1, 2, \ldots, n: \\
\quad \text{Input of } t_k, y_k := B(t_k) \\
\quad s = 0 \\
\quad \text{for } i = 0, 1, \ldots, k-1: \\
\quad\quad \lambda_i := \lambda_i/(t_i - t_k) \\
\quad\quad s := s + \lambda_i \\
\quad \lambda_k = -s \\
\quad s = 0;\ z = 0 \\
\quad \text{for } i = 0, 1, \ldots, k: \\
\quad\quad \mu = \lambda_i/t_i \\
\quad\quad s := s + \mu;\ z := z + \mu \times y_i \\
\quad P_k(0) = z/s
\end{array} \tag{3.34}
$$

*Example* 3.4 We try to compute the second derivative of the function $f(x) = \sinh(x)$ at $x = 0.6$ as accurately as possible by means of the extrapolation method. From (3.32) our parameter is $t = h^2$. The computable values $B(t_k) := (y_2 - 2y_1 + y_0)/t_k$ must not have too large a relative error. Therefore the step size used cannot be too small. The essential numerical values are given in the following table. The given function values $y_i = \sinh(x_i)$ are rounded to nine significant digits, but otherwise a precision of ten significant digits is used.

| $k$ | $h_k$ | $t_k = h_k^2$ | $B(t_k)$ | $P_k(0)$ |
|---|---|---|---|---|
| 0 | 0.30 | 0.09   | 0.641 442 8889 |              |
| 1 | 0.20 | 0.04   | 0.638 778 6000 | 0.636 647 17 |
| 2 | 0.15 | 0.0225 | 0.637 848 2222 | 0.636 653 64 |
| 3 | 0.10 | 0.01   | 0.637 184 3000 | 0.636 653 53 |

The extrapolated values $P_k(0)$ are indeed much better approximations than the starting values $B(t_k)$. The last value $P_3(0)$ has already been affected by the error in $B(t_3)$.

### 3.3 Error Estimates

We raise the question of how accurately does an interpolation polynomial approximate a given function. This question of the magnitude of the interpolation error can of course only be answered if the function $f(x)$ satisfies the hypothesis of being continuously differentiable for a sufficient number of times. We will specialize the general error estimate for some cases and will discuss the consequences for the approximation of functions.

**Theorem 3.4** *Let $f(x)$ be a real-valued, $(n+1)$ times continuously differentiable function on the bounded interval $[a,b]$. Let $P_n(x)$ be the interpolation polynomial of the $(n+1)$ pairwise different support abscissae $x_0, x_1, \ldots, x_n$, with $\min_i(x_i) = a$, $\max_i(x_i) = b$, and the functions values $y_i = f(x_i)$. Then for all $\bar{x} \in [a,b]$ we have*

$$f(\bar{x}) - P_n(\bar{x}) = \frac{f^{(n+1)}(\xi)}{(n+1)!} \prod_{i=0}^{n} (\bar{x} - x_i), \tag{3.35}$$

*where $\xi \in (a,b)$ is a point depending on $\bar{x}$.*

*Proof* If $\bar{x}$ coincides with any one of the support abscissae $x_i$, the assertion (3.35) is obvious from the interpolation property. Therefore we may assume $\bar{x} \neq x_i$ for all $i = 0, 1, \ldots, n$. For the fixed abscissa $\bar{x} \in [a, b]$ we consider the function

$$F(x) := f(x) - P_n(x) - g(\bar{x}) \prod_{i=0}^{n} (x - x_i) \tag{3.36}$$

## 3.3 Error Estimates

where the constant $g(\bar{x})$ is chosen in such a way that $F(\bar{x}) = 0$, that is

$$g(\bar{x}) = \frac{f(\bar{x}) - P_n(\bar{x})}{\prod_{i=0}^{n}(\bar{x} - x_i)}. \tag{3.37}$$

In the interval $[a, b]$ the function $F(x)$ has, at least, $(n + 2)$ pairwise different zeros $x_0, x_1, \ldots, x_n, \bar{x}$. From the hypothesis for $f(x)$, it follows that $F(x)$ is a real, $(n + 1)$ times continuously differentiable function on the interval $[a, b]$. We apply Rolle's theorem to $F(x)$ repeatedly to deduce the existence of a point $\xi \in (a, b)$ such that $F^{(n+1)}(\xi) = 0$. Figure 3.2 shows the situation for $n = 3$.

If we take into account that $P_n(x)$ is a polynomial of at most degree $n$,

$$F^{(n+1)}(x) = f^{(n+1)}(x) - g(\bar{x})(n + 1)!$$

and since $F^{(n+1)}(\xi) = 0$ we obtain the equation

$$g(\bar{x}) = f^{(n+1)}(\xi)/(n + 1)!.$$

Together with (3.37) the assertion of the theorem follows.

From Theorem 3.4 we can derive an upper bound for the interpolation error whenever the absolute value of the $(n + 1)$th derivative of the function $f(x)$ can easily be estimated for all $x$ in the interval of interpolation. From (3.35) we obviously have

$$|f(x) - P_n(x)| \leqslant \frac{\max_{\xi \in [a,b]} |f^{(n+1)}(\xi)|}{(n+1)!} \left| \prod_{i=0}^{n}(x - x_i) \right|. \tag{3.38}$$

For *equidistant support abscissae*, and for small $n$, it is easy to derive explicit estimates of the interpolation error, because the maximum of the last factor of (3.38) is computable. For the following we define the bounds

$$\max_{\xi \in [a,b]} |f^{(m)}(\xi)| \leqslant M_m \quad (m = 2, 3, 4, \ldots). \tag{3.39}$$

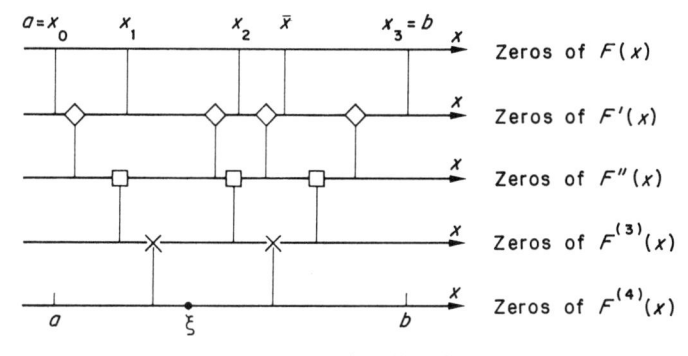

Figure 3.2 Application of Rolle's theorem to $F(x)$

For *linear interpolation* ($n = 1$) the estimate (3.38) is

$$|f(x) - P_1(x)| \leq \frac{M_2}{2}|(x - x_0)(x - x_1)| \quad x_1 = x_0 + h.$$

The quadratic function $(x - x_0)(x - x_1)$ attains its largest absolute value on $[x_0, x_1]$ at the midpoint $x_M = \frac{1}{2}(x_0 + x_1) = x_0 + \frac{1}{2}h$ for symmetry reasons. The maximum value is $\frac{1}{4}h^2$. Hence we have

$$|f(x) - P_1(x)| \leq \tfrac{1}{8}h^2 M_2 \quad x \in [x_0, x_1] \quad x_1 = x_0 + h \tag{3.40}$$

In the case of *quadratic interpolation* ($n = 2$) we have to determine the maximum in absolute value of $(x - x_0)(x - x_1)(x - x_2)$ on $[x_0, x_2]$. Without loss of generality, we choose the equidistant support abscissae to be $x_0 = -h$, $x_1 = 0$, $x_2 = h$, because the maximum is not affected by a shift. The auxiliary function $g(x) := x(x^2 - h^2)$ has its extrema at the points $z_{1,2} = \pm h/\sqrt{3}$ so that $\max_{[-h,h]} |g(x)| = (2\sqrt{3}/9)h^3$. Thus we obtain the error estimate

$$|f(x) - P_2(x)| \leq \frac{\sqrt{3}}{27} M_3 h^3 \doteq 0.0624 M_3 h^3 \quad x \in [x_0, x_2] \tag{3.41}$$

For *cubic interpolation* ($n = 3$) we have to distinguish between the interval in which the interpolating value $x$ lies in order to get the best possible estimates. With the choice $x_0 = -\frac{3}{2}h$, $x_1 = -\frac{1}{2}h$, $x_2 = \frac{1}{2}h$, $x_3 = \frac{3}{2}h$ the auxiliary function is $g(x) = (x^2 - \frac{1}{4}h^2)(x^2 - \frac{9}{4}h^2)$. The maximum in absolute value of $g(x)$ on the middle interval is attained for $x = 0$, so that the estimate holds

$$|f(x) - P_3(x)| \leq \tfrac{3}{128} M_4 h^4 \doteq 0.0234 M_4 h^4 \quad x \in [x_1, x_2] \tag{3.42}$$

For the two other intervals we get by means of the extremal points of $g(x)$ about twice as large an estimate of the interpolation error

$$|f(x) - P_3(x)| \leq \tfrac{1}{24} M_4 h^4 \doteq 0.0417 M_4 h^4 \quad x \in [x_0, x_1] \cup [x_2, x_3] \tag{3.43}$$

*Example* 3.5 How large is the interpolation error if we interpolate the sine function cubically at given equidistant support abscissae with step size $h = 10° = \pi/18$? Since $f^{(4)}(x) = \sin(x)$ we have $M_4 = \max |f^{(4)}(x)| = 1$. If we use the

## 3.3 Error Estimates

interpolation formula for the middle interval, from (3.42) we find

$$|\sin(x) - P_3(x)| \leq \tfrac{3}{128} \cdot \left(\frac{\pi}{18}\right)^4 \doteq 2.2 \times 10^{-5}.$$

The interpolation error is thus at most about two units in the fifth decimal place. In Example 3.2 the difference was larger due to rounding errors. A calculation with higher precision confirms the error bound with $P_3(36°) \doteq 0.5877733$ as compared with $\sin(36°) \doteq 0.5877853$.

The distribution of the support abscissae $x_0, x_1, \ldots, x_n$ in the interval of interpolation has a decisive influence upon the quality of approximation, because the error function is primarily defined by the term $\varphi(x) := \prod_{i=0}^{n}(x - x_i)$. In the case of equidistant support abscissae $x_i = x_0 + ih$ and large values $n$, the function $\varphi(x)$ oscillates a lot towards the ends of the interval $[x_0, x_n]$, also the absolute maximum is much larger there than the values of $|\varphi(x)|$ on the intervals in the middle. This typical situation is shown in Figure 3.3 for $n = 9$. Although the point $\xi$ is dependent on the new value $x$, the interpolation error behaves at least qualitatively in the same way. An improvement of this situation can be achieved by a different choice of the support abscissae so that they are denser towards the ends of the interval of interpolation. For instance, we can take the extremal points of the $n$th Chebyshev polynomial $T_n(x)$ transformed to the interval $[a, b]$

$$x_k^* = \frac{a+b}{2} + \frac{b-a}{2} \cos\left(\frac{n-k}{n}\pi\right) \quad (k = 0, 1, \ldots, n). \tag{3.44}$$

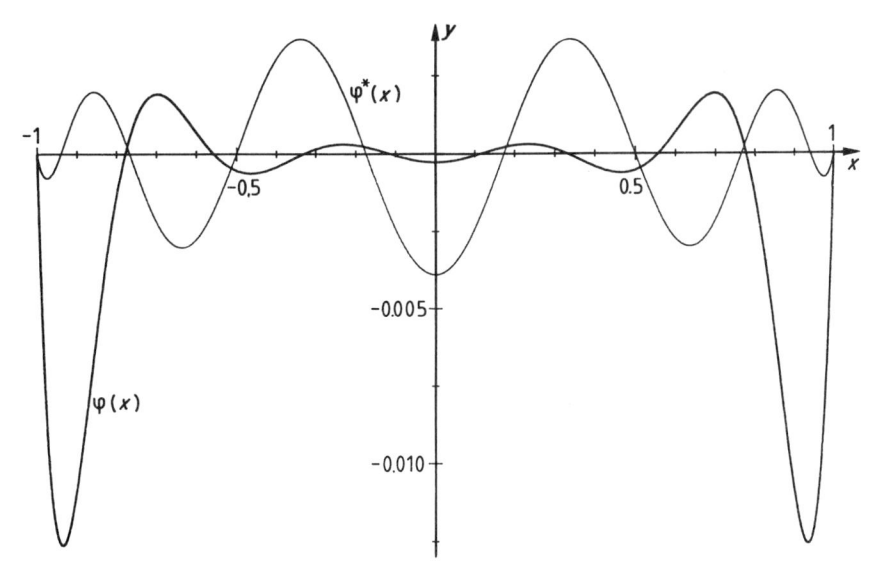

Figure 3.3 Graph of $\varphi(x)$ for equidistant and non-equidistant support abscissae

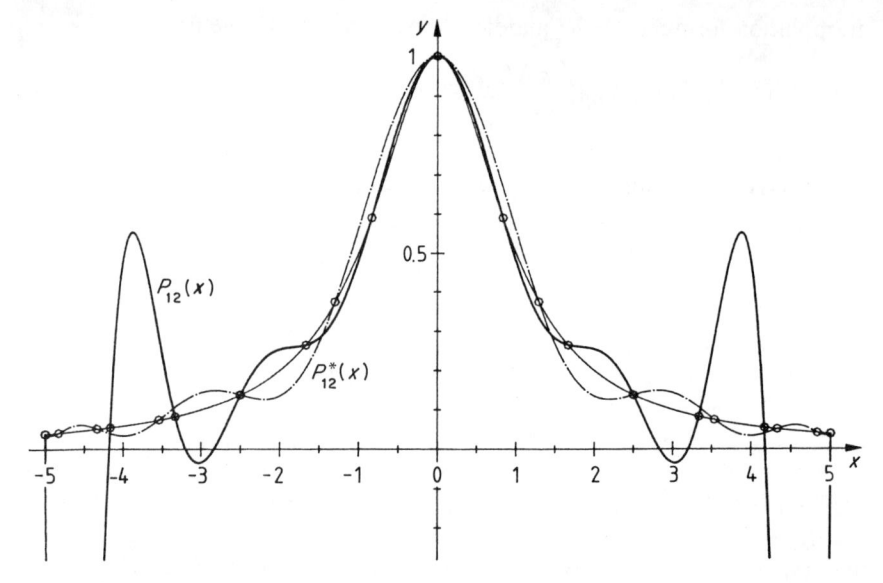

Figure 3.4 Interpolation in Runge's example

The function $\varphi^*(x) := \prod_{k=0}^{n}(x - x_k^*)$ is less variable (see Figure 3.3), and moreover we have

$$\max_{x \in [a,b]} |\varphi^*(x)| \leqslant \max_{x \in [a,b]} |\varphi(x)|.$$

*Example* 3.6 A classic example to illustrate the facts mentioned above was given by Runge (Runge, 1901; Epperson, 1987). For the function $f(x) = 1/(1 + x^2)$ defined on the interval $[-5, 5]$ we consider the sequence of interpolating polynomials $P_n(x)$ for the equidistant support abscissae $x_k = -5 + (10/n)k$. For $|x| > 3.63$ the sequence $P_n(x)$ is not pointwise convergent. However, the interpolation polynomials $P_n^*(x)$, corresponding to the nonequidistant support abscissae $x_k^*$ (3.44), converge towards the function $f(x)$ on the interval $[-5, 5]$ for increasing $n$. The comparison is shown in Figure 3.4 for the case $n = 12$.

## 3.4 Newton Interpolation

In addition to the representations (3.1) and (3.5) of the interpolation polynomial $P_n(x)$ *Newton's interpolation formula* is introduced so that the addition of a further support point is simplified. For the $(n + 1)$ pairwise different support abscissae $x_0, x_1, \ldots, x_n$ we use the following representation

$$P_n(x) = c_0 + c_1(x - x_0) + c_2(x - x_0)(x - x_1)$$
$$+ c_3(x - x_0)(x - x_1)(x - x_2) + \cdots + c_n(x - x_0)(x - x_1)\cdots(x - x_{n-1}).$$

(3.45)

## 3.4 Newton Interpolation

The unknown coefficients $c_0, c_1, \ldots, c_n$ can be successively derived from the set of interpolation conditions

$$\begin{aligned}
P_n(x_0) &= c_0 &&= y_0 \\
P_n(x_1) &= c_0 + c_1(x_1 - x_0) &&= y_1 \\
P_n(x_2) &= c_0 + c_1(x_2 - x_0) + c_2(x_2 - x_0)(x_2 - x_1) &&= y_2
\end{aligned} \quad (3.46)$$

etc.

because the system of linear equations (3.46) is of left triangular structure. However, we want to derive the general formula of computation and therefore we need some auxiliary considerations.

The coefficient $c_n$ in (3.45) is equal to the coefficient of $x^n$ of the interpolation polynomial, and hence it is uniquely defined by the $(n+1)$ pairs $(x_i, y_i)$, $(i = 0, 1, \ldots, n)$, for instance by formula (3.16). Similarly this holds for all other coefficients $c_k$, because $c_k$ is the coefficient of the highest power $x^k$ of the interpolation polynomial $P_k(x)$ corresponding to the $(k+1)$ points $(x_i, y_i)$, $i = 0, 1, \ldots, k$.

**Definition 3.1** Let

$$c_k := f[x_0, x_1, \ldots, x_k] \quad (k = 0, 1, \ldots, n), \quad (3.47)$$

be the values $c_k$ of (3.45) that are uniquely defined by the $(k+1)$ points $(x_i, y_i)$, $i = 0, 1, \ldots, k$. The notation expresses the dependence on the corresponding support abscissae.

In the following we shall consider interpolation polynomials for given subsets of the $(n+1)$ support abscissae. We denote $P^*_{i_0 i_1 \cdots i_k}(x)$ by the interpolation polynomial of the $(k+1)$ support abscissae $x_{i_0}, x_{i_1}, \ldots, x_{i_k}$, where $\{i_0, i_1, \ldots, i_k\}$ are $(k+1)$ pairwise different integers from the set $\{0, 1, \ldots, n\}$, so that we have

$$P^*_{i_0 i_1 \cdots i_k}(x_{i_j}) = y_{i_j} \quad (j = 0, 1, \ldots, k). \quad (3.48)$$

With this new notation we have

$$P^*_k(x) \equiv y_k \quad (k = 0, 1, \ldots, n). \quad (3.49)$$

**Theorem 3.5** For $1 \leq k \leq n$ we have the recursion formula

$$P^*_{i_0 i_1 \cdots i_k}(x) = \frac{(x - x_{i_0}) P^*_{i_1 i_2 \cdots i_k}(x) - (x - x_{i_k}) P^*_{i_0 i_1 \cdots i_{k-1}}(x)}{(x_{i_k} - x_{i_0})}. \quad (3.50)$$

*Proof* The interpolation polynomials $P^*_{i_1 i_2 \cdots i_k}(x)$ and $P^*_{i_0 i_1 \cdots i_{k-1}}(x)$ have degrees not exceeding $(k-1)$. So we have to show that the right-hand side of (3.50) is in fact the interpolation polynomial for the points $x_{i_0}, x_{i_1}, \ldots, x_{i_k}$. From the interpolation property we obviously have

$$P^*_{i_0 i_1 \cdots i_k}(x_{i_0}) = y_{i_0} \quad \text{and} \quad P^*_{i_0 i_1 \cdots i_k}(x_{i_k}) = y_{i_k}$$

and furthermore for $j = 1, 2, \ldots, k - 1$

$$P^*_{i_0 i_1 \cdots i_k}(x_{i_j}) = \frac{(x_{i_j} - x_{i_0})y_{i_j} - (x_{i_j} - x_{i_k})y_{i_j}}{(x_{i_k} - x_{i_0})} = y_{i_j}.$$

Theorem 3.1 now guarantees that the right-hand side of (3.50) is the uniquely determined interpolation polynomial $P^*_{i_0 i_1 \cdots i_k}(x)$.

As an immediate consequence of Theorem 3.5 we get the recursion formula for the quantities (3.47) by comparing the coefficients of $x^k$

$$f[x_{i_0}, x_{i_1}, \ldots, x_{i_k}] = \frac{f[x_{i_1}, x_{i_2}, \ldots, x_{i_k}] - f[x_{i_0}, x_{i_1}, \ldots, x_{i_{k-1}}]}{x_{i_k} - x_{i_0}}. \quad (3.51)$$

Thus we have found the desired recursion formula for the coefficients $c_k$. With regard to (3.51) $f[x_{i_0}, x_{i_1}, \ldots, x_{i_k}]$ is called the *kth divided difference* corresponding to the support abscissae $x_{i_0}, x_{i_1}, \ldots, x_{i_k}$. The application of the recursion (3.51) is conveniently implemented by using the *divided difference scheme*. The initial values are $f[x_i] = y_i$, $i = 0, 1, \ldots, n$.

$$
\begin{array}{c|ccccc}
x_0 & f[x_0] & & & & \\
    & & f[x_0, x_1] & & & \\
x_1 & f[x_1] & & f[x_0, x_1, x_2] & & \\
    & & f[x_1, x_2] & & f[x_0, x_1, x_2, x_3] & \\
x_2 & f[x_2] & & f[x_1, x_2, x_3] & & f[x_0, x_1, x_2, x_3, x_4] \\
    & & f[x_2, x_3] & & f[x_1, x_2, x_3, x_4] & \\
x_3 & f[x_3] & & f[x_2, x_3, x_4] & & \\
    & & f[x_3, x_4] & & & \\
x_4 & f[x_4] & & & & \\
\end{array}
\quad (3.52)
$$

The scheme is usually computed columnwise for consecutive sequences of support abscissae, for instance using the formulae

$$f[x_0, x_1] = \frac{f[x_1] - f[x_0]}{x_1 - x_0}, \quad f[x_2, x_3] = \frac{f[x_3] - f[x_2]}{x_3 - x_2},$$

$$f[x_2, x_3, x_4] = \frac{f[x_3, x_4] - f[x_2, x_3]}{x_4 - x_2}, \ldots \quad (3.53)$$

According to (3.47) the desired coefficients $c_k$ of Newton's interpolation formula (3.45) are found in the uppermost diagonal of the divided difference scheme (3.52). In connection with interpolation we only need the values of the uppermost diagonal. They can be computed by successive evaluation of the columns of (3.52)

## 3.4 Newton Interpolation

and then storing them as components $c_0, c_1, \ldots, c_n$ of a vector $\mathbf{c} \in \mathbf{R}^{n+1}$. In order not to lose a value that is still required it is necessary to compute the columns from below. At the end of the following program the vector $\mathbf{c}$ contains the desired coefficients of (3.45)

$$
\begin{array}{l}
\text{for } i = 0, 1, \ldots, n: \\
\quad c_i = y_i \\
\text{for } k = 1, 2, \ldots, n: \\
\quad \text{for } i = n, n-1, \ldots, k: \\
\quad\quad c_i := (c_i - c_{i-1})/(x_i - x_{i-k})
\end{array}
\tag{3.54}
$$

The amount of computational effort required to compute the coefficients $c_k$ by means of the divided difference scheme consists of

$$Z_{\text{div. diff.}} = n + (n-1) + \cdots + 2 + 1 = \tfrac{1}{2}n(n+1) \tag{3.55}$$

divisions. It is equal to that needed for the calculation of the support coefficients of Lagrange interpolation (3.18).

The evaluation of an interpolating value for the new value $x$, from (3.45), is performed most efficiently by means of a Horner-like scheme, that is shown for the case $n = 4$ as follows

$$\begin{aligned}
P_4(x) &= c_0 + c_1(x - x_0) + c_2(x - x_0)(x - x_1) + c_3(x - x_0)(x - x_1)(x - x_2) \\
&\quad + c_4(x - x_0)(x - x_1)(x - x_2)(x - x_3) \\
&= c_0 + (x - x_0)(c_1 + (x - x_1)\{c_2 + (x - x_2)[c_3 + (x - x_3)c_4]\}).
\end{aligned}$$

The interpolation polynomial $P_n(x)$, (3.45), can thus be evaluated at a new point $x$ by the following algorithm which starts with the innermost brackets

$$
\begin{array}{l}
p = c_n \\
\text{for } k = n-1, n-2, \ldots, 0: \\
\quad p = c_k + (x - x_k) \times p
\end{array}
\tag{3.56}
$$

For each new value $x$, $n$ multiplications are required, that is less than half the number of operations that were required for the Lagrange interpolation. If the same interpolation polynomial $P_n(x)$ has to be evaluated at many points, Newton's interpolation formula is most suitable.

*Example 3.7* For the four support abscissae $x_0 = 0$, $x_1 = 1.5$, $x_2 = 2.5$, $x_3 = 4.5$ with the corresponding function values $y_0 = 1$, $y_1 = 2$, $y_2 = 2$, $y_3 = 1$ the divided difference scheme is used to five significant digits

| $x_0 = 0$   | 1 |          |          |          |
|-------------|---|----------|----------|----------|
|             |   | 0.666 67 |          |          |
| $x_1 = 1.5$ | 2 |          | −0.266 67|          |
|             |   | 0        |          | 0.022 22 |
| $x_2 = 2.5$ | 2 |          | −0.166 67|          |
|             |   | −0.500 00|          |          |
| $x_3 = 4.5$ | 1 |          |          |          |

Newton's interpolation polynomial is

$$P_3(x) = 1 + x\{0.666\,67 + (x - 1.5)[-0.266\,67 + (x - 2.5)0.022\,222]\}.$$

Its value at the new argument $x = 2.1$ is $P_3(2.1) \doteq 2.0528$.

For equidistant support abscissae $x_i = x_0 + ih$, $i = 0, 1, \ldots, n$. Newton's interpolation formula (3.45) becomes an easily remembered representation. The $k$th divided difference of the scheme (3.52) simplifies because the denominators are constant for each column. Setting $f[x_i] = y_i$ we have

$$f[x_i, x_{i+1}] = \frac{y_{i+1} - y_i}{h} =: \frac{1}{h}\Delta^1_{i+1/2}, \qquad \text{first differences}$$

$$f[x_i, x_{i+1}, x_{i+2}] = \frac{1}{2h^2}(\Delta^1_{i+3/2} - \Delta^1_{i+1/2}) =: \frac{1}{2h^2}\Delta^2_{i+1}, \quad \text{second differences}$$

$$\vdots$$

$$f[x_i, x_{i+1}, \ldots, x_{i+k}] =: \frac{1}{k!h^k}\Delta^k_{i+k/2}, \qquad k\text{th differences}$$

Accordingly, we can compute the coefficients $c_k$, of (3.45), by means of the difference scheme

| $x_0$ | $y_0$ |               |               |               |               |        |
|-------|-------|---------------|---------------|---------------|---------------|--------|
|       |       | $\Delta^1_{0.5}$ |               |               |               |        |
| $x_1$ | $y_1$ |               | $\Delta^2_{1.0}$ |               |               |        |
|       |       | $\Delta^1_{1.5}$ |               | $\Delta^3_{1.5}$ |               |        |
| $x_2$ | $y_2$ |               | $\Delta^2_{2.0}$ |               | $\Delta^4_{2.0}$ | (3.57) |
|       |       | $\Delta^1_{2.5}$ |               | $\Delta^3_{2.5}$ |               |        |
| $x_3$ | $y_3$ |               | $\Delta^2_{3.0}$ |               |               |        |
|       |       | $\Delta^1_{3.5}$ |               |               |               |        |
| $x_4$ | $y_4$ |               |               |               |               |        |

## 3.4 Newton Interpolation

The interpolation formula (3.45) now becomes

$$P_n(x) = y_0 + \Delta_{0.5}^1 \frac{x - x_0}{h} + \Delta_{1.0}^2 \frac{(x - x_0)(x - x_1)}{2h^2}$$

$$+ \Delta_{1.5}^3 \frac{(x - x_0)(x - x_1)(x - x_2)}{3!h^3} + \cdots \qquad (3.58)$$

$$+ \Delta_{n/2}^n \frac{(x - x_0)(x - x_1)\cdots(x - x_{n-1})}{n!h^n}.$$

If we define the *relative distance* between $x$ and $x_0$ by

$$t := \frac{x - x_0}{h} \qquad (3.59)$$

with

$$\frac{(x - x_0)(x - x_1)\cdots(x - x_{i-1})}{i!h^i} = \frac{t(t-1)\cdots(t-i+1)}{1 \cdot 2 \cdots i} = \binom{t}{i}$$

we obtain the *Newton–Gregory interpolation* formula

$$P_n(x) = y_0 + \binom{t}{1}\Delta_{0.5}^1 + \binom{t}{2}\Delta_{1.0}^2 + \cdots + \binom{t}{n}\Delta_{n/2}^n; \quad t = \frac{x - x_0}{h} \qquad (3.60)$$

The required values of (3.60) are in the uppermost diagonal of the difference scheme (3.57). Taking account of the recursion formula for the binomial coefficients, we can evaluate $P_n(x)$ by a Horner-like algorithm if we define $c_0 = y_0$, $c_i = \Delta_{i/2}^i$, $i = 1, 2, \ldots, n$.

$$\boxed{\begin{array}{l} t = (x - x_0)/h; \; p = c_n \\ \text{for } i = n-1, n-2, \ldots, 0: \\ \quad p = p \times (t - i)/(i + 1) + c_i \end{array}} \qquad (3.61)$$

The number of essential operations of (3.61) is $2n + 1$, which is about double that of (3.56).

**Example 3.8** With the values of Example 3.2, from a table of $\sin(x)$, the difference scheme is

| | | | | |
|---|---|---|---|---|
| 20° | 0.342 02 | | | |
| | | 0.157 98 | | |
| 30° | 0.500 00 | | −0.015 19 | |
| | | 0.142 79 | | −0.004 35 |
| 40° | 0.642 79 | | −0.019 54 | |
| | | 0.123 25 | | |
| 50° | 0.766 04 | | | |

The interpolated value for $x = 36°$ with the relative difference $t = 1.6$ is computed using (3.61) to be $P_3(36°) \approx 0.58778$ to five significant figures.

Newton's interpolation formula has been derived on the assumption that the support abscissae $x_0, x_1, \ldots, x_n$ are pairwise different. It can be extended to the case of partially coinciding support abscissae. For this purpose we have to define those divided differences with vanishing denominators by means of limiting processes. For this reason we assume that to be interpolated $f(x)$ can be continuously differentiated a sufficient number of times. To begin with we consider the case where two adjacent points $x_k$ and $x_{k+1} = x_k + h$ coincide with $h \to 0$, and define the corresponding first divided difference by

$$f[x_k, x_k] := \lim_{h \to 0} \frac{f[x_k + h] - f[x_k]}{x_k + h - x_k} = f'(x_k).$$

To treat the case of three coinciding abscissae we start with three different and equidistant abscissae $x_k$, $x_{k+1} = x_k + h$, $x_{k+2} = x_k + 2h$ and perform the limiting process as $h \to 0$. Thus we define the corresponding second divided difference by the limit

$$f[x_k, x_k, x_k] := \lim_{h \to 0} \frac{f[x_{k+1}, x_{k+2}] - f[x_k, x_{k+1}]}{x_{k+2} - x_k}$$

$$= \lim_{h \to 0} \frac{f[x_{k+2}] - 2f[x_{k+1}] + f[x_k]}{2h^2} = \tfrac{1}{2} f''(x_k).$$

It can be shown that the $m$th divided difference for $(m + 1)$ coinciding support abscissae is given by

$$f[x_k, x_k, \ldots, x_k] := \frac{1}{m!} f^{(m)}(x_k) \quad (m = 1, 2, \ldots). \tag{3.62}$$

To see this we have to take account of the rule of numerical differentiation (3.23), the representation of the $m$th divided difference for equidistant points and of Theorem 3.3 when $h \to 0$.

As a consequence of (3.62) in the scheme of divided differences (3.52) we have to replace all $m$th divided differences with $(m + 1)$ coinciding support abscissae by the $m$th derivative of the function divided by $m!$. The remaining entries of the scheme are computed according to the usual rules (3.51). It is certainly appropriate to arrange the coinciding support abscissae in groups while forming the scheme of divided differences. On this assumption it is guaranteed that all denominators of (3.51) are nonzero when the missing entries of the scheme are computed.

In the following we consider two special situations, which can be generalized in an obvious way to show the most essential points. In the first special case of six

## 3.4 Newton Interpolation

support abscissae we assume that we have two triples of coinciding abscissae. The divided difference scheme is

$$
\begin{array}{l|lllllll}
x_0 & f[x_0] = c_0 \\
 & & f'(x_0) = c_1 \\
x_0 & f[x_0] & & \tfrac{1}{2}f''(x_0) = c_2 \\
 & & f'(x_0) & & f[x_0,x_0,x_0,x_1] = c_3 \\
x_0 & f[x_0] & & f[x_0,x_0,x_1] & & f[x_0,x_0,x_0,x_1,x_1] = c_4 \\
 & & f[x_0,x_1] & & f[x_0,x_0,x_1,x_1] & & f[x_0,x_0,x_0,x_1,x_1,x_1] = c_5 \\
x_1 & f[x_1] & & f[x_0,x_1,x_1] & & f[x_0,x_0,x_1,x_1,x_1] \\
 & & f'(x_1) & & f[x_0,x_1,x_1,x_1] \\
x_1 & f[x_1] & & \tfrac{1}{2}f''(x_1) \\
 & & f'(x_1) \\
x_1 & f[x_1]
\end{array}
$$

Newton's interpolation polynomial (3.45) is given by

$$P_5(x) = c_0 + c_1(x - x_0) + c_2(x - x_0)^2 + c_3(x - x_0)^3 + c_4(x - x_0)^3(x - x_1)$$
$$+ c_5(x - x_0)^3(x - x_1)^2. \quad (3.63)$$

It has the following interpolation properties

$$\begin{array}{lll} P_5(x_0) = f(x_0) & P_5'(x_0) = f'(x_0) & P_5''(x_0) = f''(x_0) \\ P_5(x_1) = f(x_1) & P_5'(x_1) = f'(x_1) & P_5''(x_1) = f''(x_1). \end{array} \quad (3.64)$$

The first three equations are obvious from the special values of $c_0$, $c_1$, $c_2$. The remaining properties (3.64) can be verified by an elementary calculation. Hence the polynomial $P_5(x)$, (3.63), not only interpolates the function values at the triple of coinciding points $x_0$ and $x_1$ but also the first and second derivative. We speak of *Hermite interpolation* in this case.

**Example 3.9** Hermite interpolation of degree five for the sine function at the support abscissae $x_0 = \pi/6 \doteq 0.523\,598\,78$ and $x_1 = \pi/3 \doteq 1.047\,1976$ is used to determine the interpolating value at the point $x = 2\pi/9 \doteq 0.698\,131\,70$. A calculation to eight significant figures yields the divided difference scheme

|       |             |            |             |              |             |             |
|-------|-------------|------------|-------------|--------------|-------------|-------------|
| $x_0$ | 0.500 000 00 |            |             |              |             |             |
|       |             | 0.866 025 40 |           |              |             |             |
| $x_0$ | 0.500 000 00 |            | −0.250 000 00 |            |             |             |
|       |             | 0.866 025 40 |           | −0.131 562 94 |           |             |
| $x_0$ | 0.500 000 00 |            | −0.318 886 20 |            | 0.027 727 450 |           |
|       |             | 0.699 056 96 |           | −0.117 044 88 |           | 0.005 858 5674 |
| $x_1$ | 0.866 025 40 |            | −0.380 170 76 |            | 0.030 794 989 |           |
|       |             | 0.500 000 00 |           | −0.100 920 66 |           |             |
| $x_1$ | 0.866 025 40 |            | −0.433 012 70 |            |             |             |
|       |             | 0.500 000 00 |           |              |             |             |
| $x_1$ | 0.866 025 40 |            |             |              |             |             |

The Hermite interpolation polynomial (3.63) is

$$P_5(x) = 0.5 + 0.86602540(x - x_0) - 0.25(x - x_0)^2$$
$$- 0.13156294(x - x_0)^3 + 0.027727450(x - x_0)^3(x - x_1)$$
$$+ 0.0058585674(x - x_0)^3(x - x_1)^2$$

and the interpolating value $P_5(0.69813170) \doteq 0.64278739$ represents a very accurate approximation of $\sin(2\pi/9) \doteq 0.64278761$.

Another important situation is that of two coinciding support abscissae. The corresponding Hermite interpolation polynomial is now defined by the function values and the first derivatives at the support abscissae. This interpolation problem is encountered with the numerical integration of ordinary differential equations, if interpolated values are needed in the case of a change of the step size (see Section 9.2). In the case of three pairwise coinciding support abscissae $x_0, x_1, x_2$ the following scheme of divided differences has to be formed.

| | | | | | |
|---|---|---|---|---|---|
| $x_0$ | $f[x_0] = c_0$ | | | | |
| | $f'(x_0) = c_1$ | | | | |
| $x_0$ | $f[x_0]$ | $f[x_0, x_0, x_1] = c_2$ | | | |
| | $f[x_0, x_1]$ | | $f[x_0, x_0, x_1, x_1] = c_3$ | | |
| $x_1$ | $f[x_1]$ | $f[x_0, x_1, x_1]$ | | $f[x_0, x_0, x_1, x_1, x_2] = c_4$ | |
| | $f'(x_1)$ | | $f[x_0, x_1, x_1, x_2]$ | | $f[x_0, x_0, x_1, x_1, x_2, x_2] = c_5$ |
| $x_1$ | $f[x_1]$ | $f[x_1, x_1, x_2]$ | | $f[x_0, x_1, x_1, x_2, x_2]$ | |
| | $f[x_1, x_2]$ | | $f[x_1, x_1, x_2, x_2]$ | | |
| $x_2$ | $f[x_2]$ | $f[x_1, x_2, x_2]$ | | | |
| | $f'(x_2)$ | | | | |
| $x_2$ | $f[x_2]$ | | | | |

The Hermite interpolation polynomial of degree five has the following representation, if appropriate terms are combined.

$$P_5(x) = [c_0 + c_1(x - x_0)] + [c_2 + c_3(x - x_1)](x - x_0)^2$$
$$+ [c_4 + c_5(x - x_2)](x - x_0)^2(x - x_1)^2. \quad (3.65)$$

It allows a simple evaluation and has the following properties

$$P_5(x_i) = f(x_i) \quad P'_5(x_i) = f'(x_i) \quad (i = 0, 1, 2). \quad (3.66)$$

## 3.5 Aitken–Neville Interpolation

The Lagrange and Newton interpolations are aimed at the explicit representation of the interpolation polynomial in order to compute interpolating values, for instance. There exist certain applications where exactly one interpolation value is sought for given data points, and support abscissae often have to be added according to a prescribed rule. In these cases it is more suitable to solve this problem by a direct approach avoiding using the explicit representation of the polynomial.

## 3.5 Aitken–Neville Interpolation

### 3.5.1 Aitken's and Neville's algorithms

The value of the interpolation polynomial $P_n(x)$ for a prescribed point $x$ is computed by the Aitken–Neville procedure recursively on the basis of Theorem 3.5. The recursion formula (3.50) is now used in such a way that the numerical values of certain interpolation polynomials are systematically computed so that finally $P_n(x) = P^*_{012\ldots n}(x)$ results. Aitken proposed applying (3.50) in such a way that the values of the following arrangement are computed columnwise

$$
\begin{array}{c|llllll}
x_0 & y_0 = P^*_0 \\
x_1 & y_1 = P^*_1 & P^*_{01} \\
x_2 & y_2 = P^*_2 & P^*_{02} & P^*_{012} \\
x_3 & y_3 = P^*_3 & P^*_{03} & P^*_{013} & P^*_{0123} \\
x_4 & y_4 = P^*_4 & P^*_{04} & P^*_{014} & P^*_{0124} & P^*_{01234} = P_4(x)
\end{array}
\qquad (3.67)
$$

The evaluation is done following the formulae

$$P^*_{0i} = \frac{(x - x_0)P^*_i - (x - x_i)P^*_0}{x_i - x_0} \quad (i = 1, 2, 3, 4) \tag{3.68}$$

$$P^*_{01i} = \frac{(x - x_1)P^*_{0i} - (x - x_i)P^*_{01}}{x_i - x_1} \quad (i = 2, 3, 4) \tag{3.69}$$

$$P^*_{012i} = \frac{(x - x_2)P^*_{01i} - (x - x_i)P^*_{012}}{x_i - x_2} \quad (i = 3, 4) \tag{3.70}$$

To understand the last two formulae we note that the interpolation polynomial $P^*_{i_0 i_1 \cdots i_k}(x)$ and hence also its value for $x$ does not depend on the order of the support abscissae. Hence we have $P^*_{01i} = P^*_{10i}$ in (3.69), $P^*_{012i} = P^*_{201i}$ and $P^*_{012} = P^*_{201}$ in (3.70).

The computational scheme (3.67) proposed by Aitken is rarely used in practice, and it has been replaced by the widely used algorithm of Neville. Here, the recursion (3.50) is consequently applied to consecutively indexed support abscissae. This is similar to forming the divided difference scheme (3.52), so that the following scheme is produced columnwise

$$
\begin{array}{c|llllll}
x_0 & y_0 = P^*_0 \\
x_1 & y_1 = P^*_1 & P^*_{01} \\
x_2 & y_2 = P^*_2 & P^*_{12} & P^*_{012} \\
x_3 & y_3 = P^*_3 & P^*_{23} & P^*_{123} & P^*_{0123} \\
x_4 & y_4 = P^*_4 & P^*_{34} & P^*_{234} & P^*_{1234} & P^*_{01234} = P_4(x)
\end{array}
\qquad (3.71)
$$

Any value appearing in (3.71) arises from the value to the left and the one that is above that, according to the formulae

$$P_{23}^* = \frac{(x-x_2)P_3^* - (x-x_3)P_2^*}{x_3 - x_2} = P_3^* + \frac{x-x_3}{x_3-x_2}(P_3^* - P_2^*) \qquad (3.72)$$

$$P_{123}^* = \frac{(x-x_1)P_{23}^* - (x-x_3)P_{12}^*}{x_3 - x_1} = P_{23}^* + \frac{x-x_3}{x_3-x_1}(P_{23}^* - P_{12}^*) \qquad (3.73)$$

$$P_{1234}^* = \frac{(x-x_1)P_{234}^* - (x-x_4)P_{123}^*}{x_4 - x_1} = P_{234}^* + \frac{x-x_4}{x_4-x_1}(P_{234}^* - P_{123}^*) \qquad (3.74)$$

The recursion formula (3.50) has already been arranged in the given examples (3.72), (3.73) and (3.74) in a suitable form that allows a more efficient evaluation requiring only two essential operations. As a consequence we can say, that any value of the Neville scheme, (3.71), results from the value to the left of it by adding a multiple of the difference between the last mentioned value and that above it. The multiplier depends on the new value $x$, and the difference of the indices of the points used in the denominator increases from column to column by one.

*Example* 3.10 For the four abscissae $x_0 = 0$, $x_1 = 1.4$, $x_2 = 2.6$, $x_3 = 3.9$ and the corresponding function values $y_0 = 0.4$, $y_1 = 1.5$, $y_2 = 1.8$, $y_3 = 2.6$ the Neville scheme for the new value $x = 2.0$ is, to five significant figures

| 0   | 0.4 |        |        |        |
|-----|-----|--------|--------|--------|
| 1.4 | 1.5 | 1.9714 |        |        |
| 2.6 | 1.8 | 1.6500 | 1.7242 |        |
| 3.9 | 2.6 | 1.4308 | 1.5974 | 1.6592 |

All the numbers appearing in the Neville scheme are significant as being the values of certain interpolation polynomials, and thus have a concrete meaning which is shown in Figure 3.5 with the data used.

For an implementation of the Neville algorithm on a computer we only need a vector $\mathbf{p}$ of $(n+1)$ components that successively holds the values of the columns. $P_{i-k,i-k+1,\ldots,i}^*$, $i = k, k+1, \ldots, n$, of the $k$th column of the Neville scheme (3.71) is the component $p_i$. On this assumption the columns of (3.71) must be computed from below in order to keep the values still required available. The *Neville algorithm* is thus

$$\begin{array}{l} \text{for } i = 0, 1, \ldots, n: \\ \quad p_i = y_i \\ \text{for } k = 1, 2, \ldots, n: \\ \quad \text{for } i = n, n-1, \ldots, k: \\ \quad\quad p_i = p_i + (x - x_i) \times (p_i - p_{i-1})/(x_i - x_{i-k}) \end{array} \qquad (3.75)$$

## 3.5 Aitken–Neville Interpolation

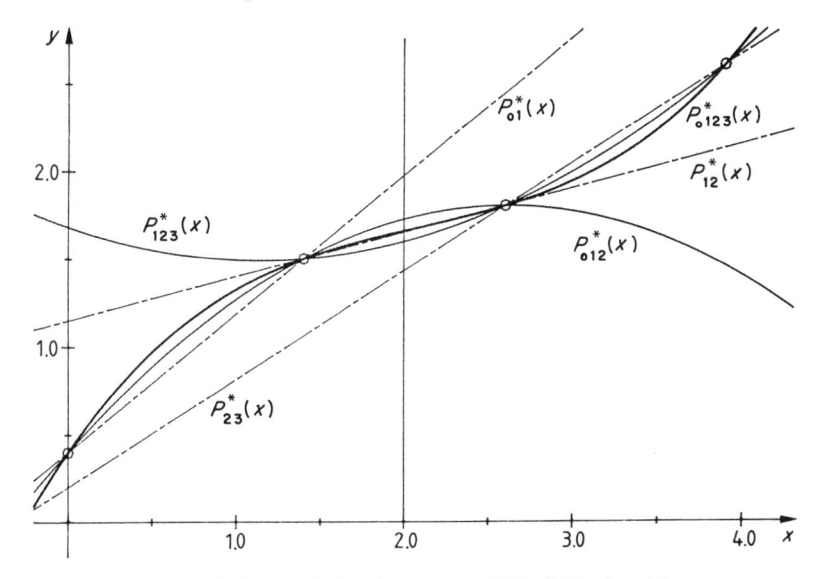

Figure 3.5 Interpolation by means of Neville's algorithm

After completion of the program (3.75), $p_n$ is the desired value $P_n(x)$ for the prescribed value $x$. The total amount of work required to evaluate an interpolating value for $(n+1)$ given data points consists of

$$Z_{\text{Neville}} = 2[n + (n-1) + \cdots 2 + 1] = n(n+1) \tag{3.76}$$

essential operations. It should be noted that the Neville algorithm (3.75) is extremely simple and compact. Therefore it is very popular and we accept the moderately increased effort with respect to Newton's interpolation.

### 3.5.2 Extrapolation and the Romberg scheme

The most important application of the Neville algorithm is extrapolation. The problem and its fundamental solution have been given in Section 3.2.2. As a single extrapolated value has to be computed, the Neville algorithm is certainly adequate. Again we denote the parameter by $t$ and by $B(t)$ the computable quantity. Our algorithm (3.75) of the Neville scheme is simplified in the case of extrapolation for $t = 0$, and we have

$$p_i^{(k)} = p_i^{(k-1)} - \frac{t_i}{t_i - t_{i-k}} (p_i^{(k-1)} - p_{i-1}^{(k-1)}). \tag{3.77}$$

In the formula (3.77) we have used $p_i^{(k)} := P^*_{i-k, i-k+1, \ldots, i}$. The support abscissae form a monotonically decreasing sequence of positive parameters $t_0 > t_1 > t_2 > \cdots > t_n$. Instead of (3.77) we can write the following equivalent recursion formulae

$$p_i^{(k)} = p_i^{(k-1)} + \frac{t_i}{t_{i-k} - t_i}(p_i^{(k-1)} - p_{i-1}^{(k-1)})$$
$$= p_i^{(k-1)} + \frac{1}{(t_{i-k}/t_i) - 1}(p_i^{(k-1)} - p_{i-1}^{(k-1)}) \qquad \begin{array}{l}(i = k, k+1, \ldots, n; \\ k = 1, 2, \ldots, n).\end{array} \qquad (3.78)$$

The first form of the recursion (3.78) is suitable, if the $t_i$ form an irregular sequence, whereas the second form is useful in connection with special parameter sequences.

If the number of parameter values $t_i$ is prescribed, the Neville algorithm for the extrapolation for $t = 0$ is analogous to the formulation (3.75).

$$\boxed{\begin{array}{l} \text{for } i = 0, 1, \ldots, n: \\ \quad \text{Input of } t_i, B(t_i); p_i = B(t_i) \\ \text{for } k = 1, 2, \ldots, n: \\ \quad \text{for } i = n, n-1, \ldots, k: \\ \quad\quad p_i = p_i + t_i \times (p_i - p_{i-1})/(t_{i-k} - t_i) \end{array}} \qquad (3.79)$$

After completion of the algorithm (3.79) $p_k$ corresponds to the value in the uppermost diagonal of the Neville scheme in the $k$th column, and hence represents the extrapolated value according to the first $(k+1)$ quantitites $B(t_0), \ldots, B(t_k)$. These are precisely the same values that are produced by the extrapolation using the Lagrange process (3.34).

Usually the process of extrapolation should be continued only as long as an extrapolated value agrees with the previous one within a given tolerance. For this purpose the Neville scheme must be built up *row-wise*. If we want to work again with a single vector containing the values of the $i$th row, some care is needed so that numerical values that are still required are not destroyed too early. Two auxiliary variables $h$ and $d$ serve this purpose in the following algorithmic description where $p_k = P^*_{i-k, i-k+1, \ldots, i}$ now denotes the $k$th value of the $i$th row.

$$\boxed{\begin{array}{l} \text{for } i = 0, 1, \ldots, n: \\ \quad \text{Input of } t_i, B(t_i); h = B(t_i) \\ \text{for } k = 1, 2, \ldots, i: \\ \quad d = h - p_{k-1}; p_{k-1} = h \\ \quad h = p_{k-1} + t_i \times d/(t_{i-k} - t_i) \\ \quad p_i = h \end{array}} \qquad (3.80)$$

**Example 3.11** The approximate computation of $\pi$ by means of the circumference of regular polygons with $n$ corners which are inscribed in a circle of unit diameter

## 3.5 Aitken–Neville Interpolation

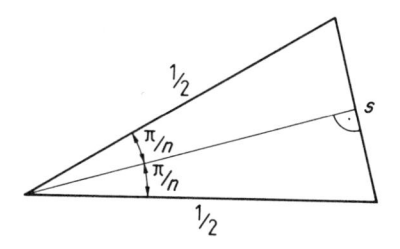

Figure 3.6 Partial triangle of a regular polygon with $n$ corners

is a typical application of extrapolation with a surprising result. First we must analyze the error between the computable circumferences and $\pi$ as a function of the regular polygon with $n$ corners before we can apply the method of extrapolation. The length $s$ of a side of the polygon is $s = \sin(\pi/n)$, see Figure 3.6, and the circumference is therefore $U_n = n \sin(\pi/n)$. From the Taylor series expansion of $\sin(x)$ we have

$$U_n = n \left[ \left(\frac{\pi}{n}\right) - \frac{1}{3!}\left(\frac{\pi}{n}\right)^3 + \frac{1}{5!}\left(\frac{\pi}{n}\right)^5 - \frac{1}{7!}\left(\frac{\pi}{n}\right)^7 + - \cdots \right]$$

$$= \pi - \frac{\pi^3}{3!}\left(\frac{1}{n}\right)^2 + \frac{\pi^5}{5!}\left(\frac{1}{n}\right)^4 - \frac{\pi^7}{7!}\left(\frac{1}{n}\right)^6 + - \cdots. \quad (3.81)$$

If we set $t = (1/n)^2$, $U_n$ represents the computable quantity $B(t)$ and $\pi$ is the desired quantity $A$. Therefore it should be possible to obtain accurate approximations of $\pi$ from some simple circumferences $U_n$. However, we should not use any trigonometric functions since they implicitly use the number $\pi$. The following circumferences are elementary: $U_2 = 2$, $U_3 = 3\sqrt{3}/2 = \sqrt{6.75}$, $U_4 = 2\sqrt{2} = \sqrt{8}$, $U_6 = 3$, $U_8 = 4\sqrt{2-\sqrt{2}} = \sqrt{32 - \sqrt{512}}$. With $t_0 = 1/4$, $t_1 = 1/9$, $t_2 = 1/16$, $t_3 = 1/36$, $t_4 = 1/64$ Neville's extrapolation process yields, to precision of 10 significant figures,

| | | | | | |
|---|---|---|---|---|---|
| 1/4 | 2.000 000 000 | | | | |
| 1/9 | 2.598 076 211 | 3.076 537 180 | | | |
| 1/16 | 2.828 427 125 | 3.124 592 586 | 3.140 611 055 | | |
| 1/36 | 3.000 000 000 | 3.137 258 300 | 3.141 480 205 | 3.141 588 849 | |
| 1/64 | 3.061 467 459 | 3.140 497 049 | 3.141 576 632 | 3.141 592 411 | 3.141 592 648 |

The result of the extrapolation, applied to the five quite rough approximations of the starting column, is amazing, because it only differs by 6 units in the last decimal place with the rounded value of $\pi \doteq 3.141\,592\,654$.

The series expansions (3.31), (3.32) and (3.81) of the error contain only *even* powers of the step size $h$, or of $(1/n)$. This situation occurs quite frequently, and then the step sizes $h_i$ will often form a *geometric sequence* with the quotient $q = \frac{1}{2}$ for practical reasons. Hence, the parameter values $t_i = h_i^2$ form a geometric sequence with the quotient $q = 1/4$, and we have $t_i = t_0/4^i$, $i = 0, 1, 2, \ldots$. As $t_{i-k}/t_i = 4^k$, the recursion formula (3.78) simplifies to

$$p_i^{(k)} = p_i^{(k-1)} + \frac{1}{4^k - 1}(p_i^{(k-1)} - p_{i-1}^{(k-1)}). \tag{3.82}$$

We note that the factor $1/(4^k - 1)$ is constant for the $k$th column. This factor decreases rapidly with increasing $k$, so that a small correction of the difference of the values $p_i^{(k-1)}$ and $p_{i-1}^{(k-1)}$ is added to $p_i^{(k-1)}$. Thus we see that the *Romberg scheme* is a special case of the Neville scheme, which was introduced originally in connection with the approximation of integrals (Bauer *et al.*, 1963; Bulirsch, 1964; Romberg, 1955), see also Section 8.1.4.

*Example* 3.12 We seek an accurate approximation of the first derivative of the function $f(x) = \sinh(x)$ for $x = 0.6$ by means of the central difference quotient, (3.28), in combination with extrapolation. We evaluate the central difference quotient for the step sizes $h_0 = 0.40$, $h_1 = 0.20$, $h_2 = 0.10$, $h_3 = 0.05$, so that the Romberg scheme can be applied. A calculation to ten significant figures yields

| $h_i$ | $B(h_i)$ | $p_i^{(1)}$ | $p_i^{(2)}$ | $p_i^{(3)}$ |
|---|---|---|---|---|
| 0.40 | 1.217 331 490 | | | |
| 0.20 | 1.193 384 140 | 1.185 401 690 | | |
| 0.10 | 1.187 441 980 | 1.185 461 260 | 1.185 465 231 | |
| 0.05 | 1.185 959 220 | 1.185 464 967 | 1.185 465 214 | 1.185 465 214 |

The last two numbers in the last row coincide indicating convergence of the extrapolated values. The result shows an error of four units in the last decimal place of $\sinh'(0.6) = \cosh(0.6) \doteq 1.185\,465\,218$. This error is due to the inaccuracies of the starting column whose values are only correct to the first nine digits.

### 3.5.3 Inverse interpolation

The problem of finding the value $x$ of a function $f(x)$ defined by $(n + 1)$ points $(x_i, y_i)$, $(i = 0, 1, \ldots, n)$ such that for a given value $y$ we have $f(x) = y$ can be solved by *inverse interpolation*. We simply interpolate the inverse function $x = f^{-1}(y)$ by essentially interchanging the roles of $x$ and $y$. However, the procedure is only meaningful on the assumption that the function $f(x)$ is monotonic in the interval of interpolation. Then the Neville algorithm is suitable. In contrast to the usual polynomial interpolation we denote the interpolation polynomial of the inverse

## 3.5 Aitken–Neville Interpolation

problem by $x = Q_n(y)$. Furthermore we define

$$q_i^{(k)} := Q_{i-k, i-k+1, \ldots, i}^* \quad (i = k, k+1, \ldots, n; k = 0, 1, \ldots, n). \tag{3.83}$$

With this notation, the Neville algorithm with the starting values

$$y_i = f(x_i) \quad q_i^{(0)} = Q_i^* = x_i \quad (i = 0, 1, \ldots, n) \tag{3.84}$$

becomes

$$q_i^{(k)} = q_i^{(k-1)} + \frac{y - y_i}{y_i - y_{i-k}} (q_i^{(k-1)} - q_{i-1}^{(k-1)}) \quad \begin{matrix} (i = k, k+1, \ldots, n; \\ k = 1, 2, \ldots, n) \end{matrix} \tag{3.85}$$

*Example* 3.13 The function $f(x) = \sinh(x)$ is tabulated with a step size $h = 0.1$. We seek $x$ such that $\sinh(x) = 2$. It is known that $1.4 < x < 1.5$. We apply cubic inverse interpolation with the support abscissae $x_0 = 1.3$, $x_1 = 1.4$, $x_2 = 1.5$, $x_3 = 1.6$. Using a precision of seven significant digits, the Neville scheme for the new value $y = 2$ is as follows

| $y_i = \sinh(x_i)$ | $x_i = q_i^{(0)}$ | $q_i^{(1)}$ | $q_i^{(2)}$ | $q_i^{(3)}$ |
|---|---|---|---|---|
| 1.698 382 | 1.3 | | | |
| 1.904 302 | 1.4 | 1.446 473 | | |
| 2.129 279 | 1.5 | 1.442 537 | 1.443 718 | |
| 2.375 568 | 1.6 | 1.447 509 | 1.443 547 | 1.443 642 ≈ $x$ |

Compared with the exact value $x = \operatorname{arcsinh}(2) \doteq 1.443\,635$ the cubic inverse interpolation yields a very good approximation. Once again, it should be pointed out that all numbers appearing in the Neville scheme have the meaning of interpolating values corresponding to appropriate support abscissae. They all represent approximations of the desired value.

Inverse interpolation has a special application to solving nonlinear equations $f(x) = 0$ in one unknown. We assume that we know an interval $I = [x_0, x_1]$ containing a zero of the continuous function $f(x)$, that is $f(x_0) \cdot f(x_1) < 0$, and that $f(x)$ is monotonic on $I$. The idea is now to determine a point $x_2$ by inverse interpolation such that $y = 0$ and then to determine the corresponding function value $y_2 = f(x_2)$. If $y_2 \neq 0$ we compute in a second step a point $x_3$, by inverse interpolation, from the three data pairs $(x_0, y_0), (x_1, y_1), (x_2, y_2)$ with $y = 0$ and the corresponding $y_3 = f(x_3)$. If $y_3 \neq 0$ inverse cubic interpolation yields $x_4$, and we determine $y_4 = f(x_4)$. Although the degree of the polynomials of the inverse interpolation could be increased without limit, it will be bounded. If, at most, cubic inverse interpolation is used, then the last four data points are involved for the subsequent interpolation step. For practical reasons, the Neville scheme is built up rowwise and continued downwards. The procedure is schematically shown in (3.86), and the steps are marked by arrows

$$\text{Start} \begin{cases} y_0 = f(x_0) & x_0 = q_0^{(0)} \\ y_1 = f(x_1) & x_1 = q_1^{(0)} \xrightarrow{1.} q_1^{(1)} =: x_2 \end{cases}$$

$$y_2 = f(x_2) \xleftarrow{1.} x_2 = q_2^{(0)} \xrightarrow{2.} q_2^{(1)} \xrightarrow{2.} q_2^{(2)} =: x_3 \qquad (3.86)$$

$$y_3 = f(x_3) \xleftarrow{2.} x_3 = q_3^{(0)} \xrightarrow{3.} q_3^{(1)} \xrightarrow{3.} q_3^{(2)} \xrightarrow{3.} q_3^{(3)} =: x_4$$

$$y_4 = f(x_4) \xleftarrow{3.} x_4 = q_4^{(0)} \xrightarrow{4.} q_4^{(1)} \xrightarrow{4.} q_4^{(2)} \xrightarrow{4.} q_4^{(3)} =: x_5$$

$$y_5 = f(x_5) \xleftarrow{4.} x_5 = q_5^{(0)} \cdots$$

*Example* 3.14 We determine the zero of $f(x) = \sinh(x) - 2$ which lies in the interval $[1.3, 1.6]$. If we use cubic inverse interpolation and a precision of ten digits is used we obtain the following scheme

| $y_i = f(x_i)$ | $x_i = q_i^{(0)}$ | $q_i^{(1)}$ | $q_i^{(2)}$ | $q_i^{(3)}$ |
|---|---|---|---|---|
| −0.301 617 562 | 1.300 000 000 | | | |
| 0.375 567 953 | 1.600 000 000 | 1.433 619 616 | | |
| −0.022 296 198 | 1.433 619 616 | 1.442 943 527 | 1.443 687 787 | |
| 0.000 116 976 | 1.443 687 787 | 1.443 635 240 | 1.433 635 456 | 1.443 635 476 |
| 0.000 000 002 | 1.443 635 476 | 1.443 635 475 | 1.443 635 475 | 1.443 635 475 |
| 0.000 000 000 | 1.443 635 475 = x | | | |

## 3.6 Rational Interpolation

### 3.6.1 Formulation of the problem, difficulties

We denote by *rational interpolation* the problem of constructing a rational function

$$R(x) = \frac{p_0 + p_1 x + \cdots + p_\zeta x^\zeta}{q_0 + q_1 x + \cdots + q_\nu x^\nu} = \frac{P_\zeta(x)}{Q_\nu(x)} \qquad (3.87)$$

with prescribed degrees $\zeta$ and $\nu$ of the polynomials of the numerator and denominator, so that $R(x)$ satisfies for $(n+1)$ pairwise different support abscissae $x_0, x_1, \ldots, x_n$ and given function values $y_0, y_1, \ldots, y_n$ the following conditions

$$R(x_i) = y_i \quad (i = 0, 1, \ldots, n). \qquad (3.88)$$

As we can multiply numerator and denominator of (3.87) by an arbitrary nonzero number a coefficient can be normalized. Hence the rational function $R(x)$ has $\zeta + \nu + 1$ free coefficients. The number of interpolation conditions (3.88)

## 3.6 Rational Interpolation

should coincide with the number of unknown coefficients, so we must have

$$\zeta + \nu = n. \tag{3.89}$$

The degrees of the numerator and denominator polynomial need not be exact because it may happen that the coefficients $p_\zeta$ and/or $p_\nu$ of the highest powers vanish.

In contrast to the polynomial interpolation it cannot be shown that a rational function $R(x)$ exists that solves the interpolation problem for arbitrarily given data. The conditions (3.88) are equivalent to a system of homogeneous equations

$$p_0 + p_1 x_i + \cdots + p_\zeta x_i^\zeta - y_i(q_0 + q_1 x_i + \cdots + q_\nu x_i^\nu) = 0 \quad (i = 0, 1, \ldots, n) \tag{3.90}$$

with $(n+1)$ equations for the $\zeta + \nu + 2 = n + 2$ unknowns $p_0, p_1, \ldots, p_\zeta, q_0, q_1, \ldots, q_\nu$. This system always has a nontrivial solution, but the corresponding rational function $R(x)$ does not necessarily satisfy the interpolation conditions. It is possible that the denominator polynomial has a zero for one of the given support abscissae $x_i$. As a consequence of (3.90) the numerator polynomial must also have the same zero, so both polynomials have a common linear factor $(x - x_i)$ which cancels. The resulting rational function $\tilde{R}(x)$ will in general no longer satisfy the interpolation condition (3.88) for $x_i$. As $\tilde{R}(x_i) \neq y_i$ we speak of an *inaccessible point*.

**Example 3.15** We seek $R(x)$, (3.87), for the abscissae $x_0 = -1$, $x_1 = 1$, $x_2 = 2$ and the ordinate values $y_0 = 2$, $y_1 = 3$, $y_2 = 3$ and prescribe $\zeta = \nu = 1$. For

$$R(x) = \frac{p_0 + p_1 x}{q_0 + q_1 x}$$

the homogeneous linear system (3.90) is

$$p_0 - p_1 - 2q_0 + 2q_1 = 0$$
$$p_0 + p_1 - 3q_0 - 3q_1 = 0$$
$$p_0 + 2p_1 - 3q_0 - 6q_1 = 0.$$

It has the rank $r = 3$ as can be verified by the Gaussian algorithm. The only nontrivial solution, determined up to a common factor, is $p_0 = 3$, $p_1 = 3$, $q_0 = 1$, $q_1 = 1$. Thus we have

$$R(x) = \frac{3 + 3x}{1 + x} = \frac{3(1 + x)}{(1 + x)} = 3 = \tilde{R}(x).$$

The denominator polynomial of $R(x)$ has the zero $x_0 = -1$, and in this situation $\tilde{R}(x_0) = 3 \neq 2 = y_0$. The interpolation problem has no solution for the prescribed degrees $\zeta$ and $\nu$.

For the same interpolation data, but with $\zeta = 0$ and $\nu = 2$ the rational interpolation problem with

$$R(x) = \frac{p_0}{q_0 + q_1 x + q_2 x^2}$$

and the homogeneous system of equations

$$\begin{aligned} p_0 - 2q_0 + 2q_1 - 2q_2 &= 0 \\ p_0 - 3q_0 - 3q_1 - 3q_2 &= 0 \\ p_0 - 3q_0 - 6q_1 - 12q_2 &= 0 \end{aligned}$$

of rank $r = 3$ has the proper solution

$$R(x) = \frac{36}{14 - 3x + x^2}.$$

In spite of the mentioned difficulties, rational interpolation is more adequate and yields better results, compared with polynomial interpolation, in all cases where the function to be interpolated has a pole or its graph has an inclined, or even a horizontal asymptote.

### 3.6.2 Special interpolation problem, Thiele's continued fraction

Below we only consider a special rational interpolation problem for which the degrees of the numerator and the denominator polynomial are either equal or differ by one. We assume the existence of the interpolating rational function $R(x)$.

More precisely we look for the rational function, (3.87), $R(x) = P_\zeta(x)/Q_\nu(x)$ for $(n + 1)$ pairwise different support abscissae $x_0, x_1, \ldots, x_n$ and function values $y_0 = f(x_0)$, $y_1 = f(x_1), \ldots, y_n = f(x_n)$ satisfying

(a) $R(x_i) = y_i$, $(i = 0, 1, \ldots, n)$; (3.91)

(b) $\left.\begin{aligned} \zeta &= \nu = \tfrac{1}{2}n, \quad \text{if } n \text{ is even,} \\ \zeta &= \tfrac{1}{2}(n+1), \quad \nu = \tfrac{1}{2}(n-1), \quad \text{if } n \text{ is odd.} \end{aligned}\right\}$ (3.92)

To solve this special problem, we need the *inverse divided differences* that play a similar role as the divided differences did in Newton's interpolation.

**Definition 3.2** *For the $(n + 1)$ pairwise different support abscissae $x_0, x_1, \ldots, x_n$ and the corresponding function values $y_0 = f(x_0)$, $y_1 = f(x_1), \ldots, y_n = f(x_n)$ we define*
(a) *the first inverse divided differences*

$$\varphi_1(x_i, x_0) := \frac{x_i - x_0}{f(x_i) - f(x_0)} \quad (i = 1, 2, \ldots, n); \tag{3.93}$$

(b) *the second inverse divided differences*

$$\varphi_2(x_i, x_1, x_0) := \frac{x_i - x_1}{\varphi_1(x_i, x_0) - \varphi_1(x_1, x_0)} \quad (i = 2, 3, \ldots, n); \tag{3.94}$$

## 3.6 Rational Interpolation

(c) the kth inverse divided differences

$$\varphi_k(x_i, x_{k-1}, \ldots, x_0) := \frac{x_i - x_{k-1}}{\varphi_{k-1}(x_i, x_{k-2}, \ldots, x_0) - \varphi_{k-1}(x_{k-1}, x_{k-2}, \ldots, x_0)},$$

$$(i = k, k+1, \ldots, n; k = 2, 3, \ldots, n). \quad (3.95)$$

In contrast to the divided differences (3.47), and (3.51) the order of the support abscissae is essential in the definition of the inverse divided differences. It should be noted that the subtrahends of both the numerator and the denominator are constant for the $k$th inverse divided difference. The corresponding values are arranged in the *inverse divided differences scheme*

| | | | | | |
|---|---|---|---|---|---|
| $x_0$ | $f(x_0)$ | | | | |
| $x_1$ | $f(x_1)$ | $\varphi_1(x_1, x_0)$ | | | |
| $x_2$ | $f(x_2)$ | $\varphi_1(x_2, x_0)$ | $\varphi_2(x_2, x_1, x_0)$ | | |
| $x_3$ | $f(x_3)$ | $\varphi_1(x_3, x_0)$ | $\varphi_2(x_3, x_1, x_0)$ | $\varphi_3(x_3, x_2, x_1, x_0)$ | |
| $x_4$ | $f(x_4)$ | $\varphi_1(x_4, x_0)$ | $\varphi_2(x_4, x_1, x_0)$ | $\varphi_3(x_4, x_2, x_1, x_0)$ | $\varphi_4(x_4, x_3, x_2, x_1, x_0)$ |

$$(3.96)$$

The third inverse divided difference, doubly underlined, results from the four underlined quantities. It may happen that a denominator is equal to zero when the numerical entries of the scheme (3.96) are computed. This obviously occurs when the two $k$th inverse divided differences $\varphi_k(x_k, x_{k-1}, \ldots, x_0)$ and $\varphi_k(x_i, x_{k-1}, \ldots, x_0)$ are equal. Moreover, by means of the solution to be developed it can be shown that the $m$th inverse divided differences are constant for a certain $m$ if the function values $f(x_i)$ are produced by a rational function $f(x)$. In this case the $(m+1)$th column of (3.96) necessarily contains 'infinite' numbers. A computer program should take this possible situation into account.

**Example 3.16** For some support abscissae and function values of the rational function $f(x) = (x^2 + 3x + 5)/(x - 2)$ we obtain the following inverse divided differences scheme, to six significant figures

| | | | | | | |
|---|---|---|---|---|---|---|
| 0 | $-2.50$ | | | | | |
| 1 | $-9.00$ | $-0.153\,846$ | | | | |
| 3 | 23.00 | $0.117\,647$ | $7.366\,67$ | | | |
| 4 | 16.50 | $0.210\,526$ | $8.233\,34$ | $1.153\,84$ | | |
| 6 | 14.75 | $0.347\,826$ | $9.966\,67$ | $1.153\,85$ | $2 \times 10^5 \approx \infty$ | |
| 7 | 15.00 | $0.400\,000$ | $10.833\,3$ | $1.153\,86$ | $1.5 \times 10^5 \approx \infty$ | |

Now we replace the support abscissa $x_i$ by the variable $x$ in the defining equations of the inverse divided differences and then solve (3.93), (3.94) and so on successively for $f(x)$, $\varphi_1(x, x_0)$ and so on

$$f(x) = f(x_0) + \frac{x - x_0}{\varphi_1(x, x_0)}$$

$$\varphi_1(x, x_0) = \varphi_1(x_1, x_0) + \frac{x - x_1}{\varphi_2(x, x_1, x_0)}$$

$$\varphi_2(x, x_1, x_0) = \varphi_2(x_2, x_1, x_0) + \frac{x - x_2}{\varphi_3(x, x_2, x_1, x_0)}$$

$$\varphi_3(x, x_2, x_1, x_0) = \varphi_3(x_3, x_2, x_1, x_0) + \frac{x - x_3}{\varphi_4(x, x_3, x_2, x_1, x_0)}$$
$$\vdots$$

We successively substitute this sequence of identities for $x$ to obtain the following, new identity which can be generalized in an obvious way

$$f(x) = f(x_0) + \cfrac{x - x_0}{\varphi_1(x_1, x_0) + \cfrac{x - x_1}{\varphi_2(x_2, x_1, x_0) + \cfrac{x - x_2}{\varphi_3(x_3, x_2, x_1, x_0) + \cfrac{x - x_3}{\varphi_4(x, x_3, x_2, x_1, x_0)}}}} \tag{3.97}$$

The known inverse divided differences of the uppermost diagonal of the scheme (3.96) appear in this *continued fraction*. Now we imagine that the last term $\varphi_4(x, x_3, x_2, x_1, x_0)$, in (3.97), is also substituted, and then we drop the quotient $(x - x_4)/\varphi_5(x, x_4, \ldots, x_0)$. In this way we have got the finite *Thiele's continued fraction* (for $n = 4$)

$$R(x) := f(x_0) + \cfrac{x - x_0}{\varphi_1(x_1, x_0) + \cfrac{x - x_1}{\varphi_2(x_2, x_1, x_0) + \cfrac{x - x_2}{\varphi_3(x_3, x_2, x_1, x_0) + \cfrac{x - x_3}{\varphi_4(x_4, x_3, x_2, x_1, x_0)}}}}$$

(3.98)

**Theorem 3.6** *Thiele's continued fraction $R(x)$, (3.98), solves the special rational interpolation problem (3.87), (3.91), (3.92).*

## 3.6 Rational Interpolation

*Proof* The interpolation property $R(x_i) = f(x_i)$, $i = 0, 1, \ldots, n$, of the continued fraction (3.98) is obvious if we generalize the identity (3.97) for arbitrary $n$. It remains to verify the property (3.92) concerning the degrees of the polynomials of the numerator and the denominator, if the continued fraction (3.98) is formally transformed into the representation (3.87). To simplify the notation we introduce the coefficients

$$c_0 := f(x_0) \quad c_1 := \varphi_1(x_1, x_0), \ldots, c_k := \varphi_k(x_k, x_{k-1}, \ldots, x_0). \tag{3.99}$$

We show property (3.92) by induction on $m$.

*Induction basis:* For $m = 0$ we have $R_0(x) = c_0 = P_0(x)/Q_0(x)$ with $\zeta = \nu = 0$, $P_0(x) = c_0$, $Q_0(x) = 1$ so that (3.92) holds. For $m = 1$ we have

$$R_1(x) = c_0 + \frac{x - x_0}{c_1} = \frac{x + (c_0 c_1 - x_0)}{c_1} = \frac{P_1(x)}{Q_0(x)}$$

a rational function with $\zeta = 1$, $\nu = 0$.

*Induction hypothesis:* For $k = 0, 1, \ldots, m$ we assume that (3.92) holds for

$$R_k(x) = c_0 + \cfrac{x - x_0}{c_1 + \cfrac{x - x_1}{c_2 + \cfrac{\ddots}{+ \cfrac{x - x_{k-2}}{c_{k-1} + \cfrac{x - x_{k-1}}{c_k}}}}} = \frac{P_\zeta(x)}{Q_\nu(x)}, \quad \zeta + \nu = k. \tag{3.100}$$

*Induction step:* We show that the property (3.92) is also valid for $R(x)$ for $m + 1$. In the continued fraction

$$R_{m+1}(x) = c_0 + \cfrac{x - x_0}{c_1 + \cfrac{x - x_1}{c_2 + \cfrac{\ddots}{+ \cfrac{x - x_{m-1}}{c_m + \cfrac{x - x_m}{c_{m+1}}}}}}$$

the denominator of $(x - x_0)$ is itself a continued fraction of the same structure as $R_m(x)$ and is therefore, due to the induction hypothesis, a rational function satisfying the property (3.92). Hence we have

$$R_{m+1}(x) = c_0 + \frac{x - x_0}{\frac{P_\zeta(x)}{Q_\nu(x)}} = \frac{c_0 P_\zeta(x) + (x - x_0)Q_\nu(x)}{P_\zeta(x)} \qquad (3.101)$$

where $\zeta = \nu = \tfrac{1}{2}m$ if $m$ is even and $\zeta = \tfrac{1}{2}(m+1)$, $\nu = \tfrac{1}{2}(m-1)$ if $m$ is odd. Now if $(m+1)$ is odd, the degree of the numerator polynomial of (3.101) is one greater than the degree of the denominator polynomial. Otherwise the degrees are equal.

Thiele's continued fraction (3.98) represents the explicit solution of the special rational interpolation problem. Next we consider the problem of evaluating the continued fraction for a new given value $x$. To do this we want to compute the partial continued fractions $R_0(x), R_1(x), R_2(x), \ldots$ successively so that it will be easy to extend the continued fraction in the case of additional support abscissae. To derive the recursion formula we start with the first three simple cases and introduce the *values* of the corresponding numerators and denominators. We use the coefficients $c_k$ (3.99) for brevity

$$k = 0: \quad R_0(x) = c_0 =: \frac{p_0}{q_0} \qquad p_0 := c_0 \quad q_0 := 1 \qquad (3.102)$$

$$k = 1 \quad R_1(x) = c_0 + \frac{x - x_0}{c_1} = \frac{c_0 c_1 + (x - x_0)}{c_1} =: \frac{p_1}{q_1}$$

$$p_1 := c_0 c_1 + (x - x_0) \quad q_1 := c_1 \qquad (3.103)$$

$$k = 2 \quad R_2(x) = c_0 + \frac{x - x_0}{c_1 + \frac{x - x_1}{c_2}}$$

$$= \frac{c_2[c_0 c_1 + (x - x_0)] + (x - x_1)c_0}{c_2 \cdot [c_1] + (x - x_1) \cdot 1} =: \frac{p_2}{q_2}$$

$$p_2 := c_2 p_1 + (x - x_1)p_0 \quad q_2 = c_2 q_1 + (x - x_1)q_0 \qquad (3.104)$$

We have used the values $p_1, q_1, p_0, q_0$ to define the quantities $p_2$ and $q_2$ in (3.104). We now sketch the idea of the general induction step by means of the next step where we compute $R_3(x)$. In order to get the value of

$$R_3(x) = c_0 + \frac{x - x_0}{c_1 + \frac{x - x_1}{c_2 + \frac{x - x_2}{c_3}}}$$

## 3.6 Rational Interpolation

In $R_2(x)$ we replace the coefficient $c_2$ by $c_2 + (x - x_2)/c_3$, substituting this expression into (3.104) and multiplying by $c_3$ we obtain

$$p_3 := \{c_2 c_3 + (x - x_2)\} p_1 + c_3(x - x_1) p_0$$
$$= c_3[c_2 p_1 + (x - x_1) p_0] + (x - x_2) p_1$$
$$q_3 := \{c_2 c_3 + (x - x_2)\} q_1 + c_3(x - x_1) q_0$$
$$= c_3[c_2 q_1 + (x - x_1) q_0] + (x - x_2) q_1$$

We recognize the contents of the square brackets to be $p_2$ and $q_2$, respectively, and so we can conclude that the general recursion formulae are

$$\left. \begin{array}{l} p_k = c_k p_{k-1} + (x - x_{k-1}) p_{k-2} \\ q_k = c_k q_{k-1} + (x - x_{k-1}) q_{k-2} \end{array} \right\} \quad (k = 2, 3, \ldots, n), \tag{3.105}$$

with the initial values given by (3.102) and (3.103).

The procedure of the rational interpolation by means of Thiele's continued fraction consists of two parts. The inverse divided difference scheme, (3.96), is computed from which only the entries $c_k$ of the uppermost diagonal are required for the subsequent evaluation of the continued fraction. The *rowwise* computation of the scheme (3.96) is described by the following algorithm

$$\boxed{\begin{array}{l} \text{for } i = 0, 1, \ldots, n: \\ \quad \text{Input of } x_i, f(x_i); h = f(x_i) \\ \quad \text{for } k = 1, 2, \ldots, i: \\ \quad\quad h = (x_i - x_{k-1})/(h - c_{k-1}) \\ \quad c_i = h \end{array}} \tag{3.106}$$

The evaluation of the partial continued fractions in ascending order is performed with the recursion formulae (3.105) terminating with the desired value. It has the following algorithmic description:

$$\boxed{\begin{array}{ll} \text{Start:} & p_0 := c_0, q_0 := 1, \qquad\qquad\qquad R_0(x) = p_0/q_0 \\ & p_1 := c_0 \times c_1 + (x - x_0), q_1 := c_1, R_1(x) = p_1/q_1 \\ \text{Recursion:} & \text{for } k = 2, 3, \ldots, n: \\ & \quad p_k = c_k \times p_{k-1} + (x - x_{k-1}) \times p_{k-2} \\ & \quad q_k = c_k \times q_{k-1} + (x - x_{k-1}) \times q_{k-2} \\ & \quad R_k(x) = p_k/q_k \end{array}} \tag{3.107}$$

The two algorithms (3.106) and (3.107) can, of course, be appropriately combined in order to apply them to the method of *extrapolation*. The sequence of partial continued fractions $R_k(x)$ gives insight into the convergence which can be used to define a stopping criterion.

*Example* 3.17 Thiele's continued fraction belonging to the inverse divided difference scheme is

| $x_i$ | $f_i$ | $\varphi_1$ | $\varphi_2$ | $\varphi_3$ |
|---|---|---|---|---|
| 1 | 1 | | | |
| 2 | $-1$ | $-\dfrac{1}{2}$ | | |
| 3 | 2 | 2 | $\dfrac{2}{5}$ | |
| 4 | $-2$ | $-1$ | $-4$ | $\dfrac{5}{22}$ |

$$R(x) = 1 + \cfrac{x-1}{-\dfrac{1}{2} + \cfrac{x-2}{\dfrac{2}{5} + \cfrac{x-3}{-\dfrac{5}{22}}}}$$

$$= \frac{-11x^2 + 53x - 56}{8x - 22} = \frac{P_2(x)}{Q_1(x)}.$$

For the new value $x = 2.4$ we obtain the values

| $k$ | $p_k$ | $q_k$ | $R_k(x)$ |
|---|---|---|---|
| 0 | 1 | 1 | 1 |
| 1 | 0.9 | $-0.5$ | $-1.8$ |
| 2 | 0.76 | 0.2 | 3.8 |
| 3 | $-0.712\,727$ | $0.254\,545$ | $-2.8 = R(2.4)$ |

*Example* 3.18 We want to show the superiority of rational interpolation in comparison with polynomial interpolation in the presence of a pole by means of the function $f(x) = 0.1/(x\sqrt{5x+1})$. We assume that we know the function values for $x_0 = 0.01$, $x_1 = 0.02$, $x_2 = 0.03$, $x_3 = 0.04$, $x_4 = 0.05$. We seek the interpolated value for $x = 0.024$. The inverse divided difference scheme and the recursive computation of the partial continued fractions, according to (3.107), is given in the following two tables of ten digits

| $x_i$ | $f_i$ | $\varphi_1$ | $\varphi_2$ | $\varphi_3$ | $\varphi_4$ |
|---|---|---|---|---|---|
| 0.01 | $9.759\,000\,729 = c_0$ | | | | |
| 0.02 | $4.767\,312\,946$ | $-0.002\,003\,330\,423 = c_1$ | | | |
| 0.03 | $3.108\,349\,361$ | $-0.003\,007\,224\,239$ | $-9.961\,212\,870 = c_2$ | | |
| 0.04 | $2.282\,177\,323$ | $-0.004\,012\,399\,166$ | $-9.954\,860\,962$ | $1.574\,330\,107 = c_3$ | |
| 0.05 | $1.788\,854\,382$ | $-0.005\,018\,728\,422$ | $-9.948\,935\,434$ | $1.629\,004\,623$ | $0.182\,900\,5674 = c_4$ |

| $k$ | $p_k$ | $q_k$ | $R_k(0.024)$ |
|---|---|---|---|
| 0 | $9.759\,000\,729$ | $1.000\,000\,000$ | $9.759\,000\,729$ |
| 1 | $-0.005\,550\,503\,060$ | $-0.002\,003\,330\,423$ | $2.770\,637\,832$ |
| 2 | $0.094\,325\,745\,44$ | $0.023\,955\,600\,79$ | $3.937\,523\,683$ |
| 3 | $0.148\,533\,163\,9$ | $0.037\,726\,043\,54$ | $3.937\,151\,897$ |
| 4 | $0.025\,657\,588\,03$ | $0.006\,516\,825\,156$ | $3.937\,130\,031$ |

## 3.7 Spline Interpolation

In fact Thiele's continued fraction provides an excellent approximation of $f(0.024) \doteq 3.937\,129\,927$, whereas the interpolation polynomial of degree four yields the poor value of $P_4(0.024) \doteq 3.8886$.

*Example 3.19* Once more we approximately compute $\pi$ by means of the circumferences of regular polygons of $n$ corners inscribed in a circle of unit diameter (see Example 3.11). The extrapolation by rational functions may often provide a better extrapolated value. A calculation to ten significant figures yields the inverse divided difference scheme from which the partial continued fractions result.

| $h_i$ | $f_i$ | $\varphi_1$ | $\varphi_2$ | $\varphi_3$ | $\varphi_4$ |
|---|---|---|---|---|---|
| 1/4  | $2.000\,000\,000 = c_0$ | | | | |
| 1/9  | 2.598 076 211 | $-0.232\,226\,0714 = c_1$ | | | |
| 1/16 | 2.828 427 125 | $-0.226\,332\,5214$ | $-8.248\,188\,460 = c_2$ | | |
| 1/36 | 3.000 000 000 | $-0.222\,222\,2222$ | $-8.330\,126\,900$ | $0.423\,759\,8644 = c_3$ | |
| 1/64 | 3.061 467 459 | $-0.220\,802\,8122$ | $-8.358\,920\,115$ | $0.423\,320\,6846$ | $27.671\,531\,75 = c_4$ |

| $k$ | $p_k$ | $q_k$ | $R_k$ |
|---|---|---|---|
| 0 | 2.000 000 000 | 1.000 000 000 | 2.000 000 000 |
| 1 | $-0.714\,452\,1428$ | $-0.232\,226\,0714$ | 3.076 537 180 |
| 2 | 5.670 713 697 | 1.804 333 291 | 3.142 830 499 |
| 3 | 2.447 674 126 | 0.779 118 1602 | 3.141 595 526 |
| 4 | 67.573 372 47 | 21.509 272 54 | 3.141 592 648 |

The extrapolated approximation for $\pi$ is by chance the same as we got for polynomial extrapolation in Example 3.11.

### 3.7 Spline Interpolation

We consider the problem of approximating a function $f(x)$, which is defined only by its function values at given support abscissae, by an interpolation function that is at least continuously differentiable. If the number of support abscissae is large, it is often meaningless to take the corresponding interpolation polynomial as the solution, especially if the support abscissae are equidistant. Therefore interpolation polynomials of low degree are usually used which are suitably associated with the intervals containing the new value. The interpolation polynomials defined in this way will, in general, be different in adjacent intervals, so that the resulting interpolating function composed piecewise of polynomials is continuous at all interior support abscissae but not continuously differentiable. For certain applications, however, only a smooth, that is continuously differentiable, interpolating function is adequate. The smooth interpolation problem was probably considered for the first time by Rutishauser (1960). We start by characterizing the desired smooth interpolating function by certain properties which are based on the model of thin splines and thus the computation of the socalled spline function is developed. For extensive treatments of the theory and of applications see Ahlberg et al. (1967), Böhmer (1974), Boor (1978) and Späth (1973).

### 3.7.1 Characterization of the spline function

Let $x_0 < x_1 < \cdots < x_n$ be $(n+1)$ pairwise different support abscissae, and corresponding function values $y_0, y_1, \ldots, y_n$ be given. We seek an interpolating function $s(x)$ that is at least continuously differentiable. To do this, we need additional assumptions. For this purpose we think of the mechanical model of a thin, homogeneous lath or spline passing through the pivoted points $(x_i, y_i)$ no exterior forces acting on it. The bending curve of the lath should be the solution $s(x)$ of the interpolation problem.

Following physical extremal principles, the lath takes on a form that minimizes its deformation energy. For a thin, homogeneous lath, on certain simplifying assumptions and disregarding physical and geometrical constants, the energy is given by the integral expression

$$E = \frac{1}{2} \int_{x_0}^{x_n} s''(x)^2 \, dx.$$

The desired interpolating spline function is now defined by the solution of the following *variational problem*.

(a) the function $s(x)$ satisfies the interpolation conditions

$$s(x_i) = y_i \quad (i = 0, 1, \ldots, n). \tag{3.108}$$

(b) The function $s(x)$ must be at least continuously differentiable on $[x_0, x_n]$.

(c) On each subinterval $[x_{i-1}, x_i]$ $s(x)$ must be four times continuously differentiable.

(d) $s(x)$ minimizes the integral

$$J = \frac{1}{2} \int_{x_0}^{x_n} s''(x)^2 \, dx. \tag{3.109}$$

Classical calculus of variations (Clegg, 1970; Funk, 1970) gives necessary conditions that the function $s(x)$ must satisfy in solving the variational problem. The first variation of the functional (3.109) must vanish for all admissible variations of the solution function. Thus we must have

$$\delta J = \int_{x_0}^{x_n} s''(x) \delta s''(x) \, dx = 0. \tag{3.110}$$

## 3.7 Spline Interpolation

The support abscissae $x_i$ will certainly play a special role when we have to take into account the conditions (a) and (b). Therefore we write the integral of the first variation as a sum of integrals over the $n$ subintervals and then apply partial integration. Since the operations of variation and differentiation commute, from (3.110) after two integrations by parts we obtain

$$\delta J = \sum_{i=1}^{n} \int_{x_{i-1}}^{x_i} s''(x)\delta s''(x)\,dx$$
$$= \sum_{i=1}^{n} \left( s''(x)\delta s'(x)\Big|_{x_{i-1}}^{x_i} - s^{(3)}(x)\delta s(x)\Big|_{x_{i-1}}^{x_i} \right. \tag{3.111}$$
$$\left. + \int_{x_{i-1}}^{x_i} s^{(4)}(x)\delta s(x)\,dx \right) = 0.$$

As the condition (a), the first variation $\delta s(x)$ has to be zero at all support abscissae $x_i$. As a consequence all terms $\delta s(x_i)$ of the second sum vanish. Moreover, all admissible variations $\delta s(x)$ are assumed to be continuously differentiable a sufficient number of times. The continuity condition (b) for the first derivative of $s(x)$ at the interior support abscissae implies the continuity of $\delta s'(x)$ at $x_i$, $i = 1, 2, \ldots, n-1$. Due to $\delta s'(x_i - 0) = \delta s'(x_i + 0) = \delta s'(x_i)$ from (3.111) we obtain

$$\delta J = s''(x_n - 0)\delta s'(x_n) - s''(x_0 + 0)\delta s'(x_0)$$
$$- \sum_{i=1}^{n-1} [s''(x_i + 0) - s''(x_i - 0)]\delta s'(x_i)$$
$$+ \sum_{i=1}^{n} \int_{x_{i-1}}^{x_i} s^{(4)}(x)\delta s(x)\,dx = 0. \tag{3.112}$$

The first variation must vanish for arbitrary admissible variations $\delta s(x)$. If we apply the well known technique of appropriate restrictions of the admissible variations, we obtain, in addition to the properties (a) and (b), the following necessary conditions for $s(x)$:

$$s^{(4)}(x) = 0 \qquad \text{for all } x \neq x_0, x_1, \ldots, x_n \tag{3.113}$$
$$s''(x_i + 0) = s''(x_i - 0) \qquad \text{for } i = 1, 2, \ldots, n-1 \tag{3.114}$$
$$s''(x_0) = 0 \quad \text{and} \quad s''(x_n) = 0. \tag{3.115}$$

The desired spline function $s(x)$ is characterized by the following properties:

(1) Because of (3.113) $s(x)$ is a cubic polynomial in each subinterval $(x_i, x_{i+1})$. The interpolating spline function is piecewise composed of polynomials of degree three.

(2) Not only the first, but also the second derivative of $s(x)$ is continuous at the interior support abscissae because of (3.114).

(3) The second derivative is zero at the end-points $x_0$ and $x_n$.

These three properties uniquely define the interpolating spline function $s(x)$, which will be seen from the constructive algorithm below. The third property $s''(x_0) = s''(x_n) = 0$ is a so-called natural condition, which is a consequence of the variational problem. Therefore the resulting function $s(x)$ is called a *natural cubic spline function*.

The cubic spline interpolation has several generalizations and extensions of a purely mathematical nature for which a physical interpretation is missing. For instance, the order of the derivative of the integrand of the variational integral can obviously be increased. Then the conditions (b) and (c) have to be adapted. So we can define a spline function for any integer $p \geqslant 2$ as solution of the following variational problem.

(a) The function $s(x)$ satisfies the interpolation conditions (3.108).

(b) The function $s(x)$ is at least $(p-1)$ times continuously differentiable on $[x_0, x_n]$.

(c) $s(x)$ is $2p$ times continuously differentiable on each subinterval $[x_{i-1}, x_i]$.

(d) $s(x)$ minimizes the integral

$$J = \frac{1}{2} \int_{x_0}^{x_n} s^{(p)}(x)^2 \, dx.$$

The variational problem uniquely characterizes the interpolating spline functions. It is a polynomial of degree $(2p-1)$ on each interval with continuous derivatives up to order $(2p-2)$ at all interior support abscissae, and it has vanishing derivatives $s^{(p)}(x), s^{(p+1)}(x), \ldots, s^{(2p-2)}(x)$ at the two end points as natural conditions. In the case $p = 3$ the additional necessary conditions for the *natural quintic spline function* are derived from the variational formulation

$$s^{(6)}(x) = 0 \qquad \text{for all } x \neq x_0, x_1, \ldots, x_n,$$

$$\left.\begin{array}{l} s^{(3)}(x_i + 0) = s^{(3)}(x_i - 0) \\ s^{(4)}(x_i + 0) = s^{(4)}(x_i - 0) \end{array}\right\} \quad \text{for } i = 1, 2, \ldots, n-1.$$

$$s^{(3)}(x_0) = s^{(4)}(x_0) = s^{(3)}(x_n) = s^{(4)}(x_n) = 0.$$

In comparison with the cubic spline functions, the quintic spline functions possess higher differentiability properties. However, in general the spline functions of higher degree oscillate more than the cubic ones (Sauer and Szabo, 1968, Späth, 1973). Therefore they are less suitable for interpolation purposes.

Finally we can completely abandon the variational background and use, instead of polynomials, other functions that are more appropriate for the special problem in view by directly requiring analogous conditions of continuity for the derivatives (Späth, 1973).

### 3.7.2 Computation of the cubic spline function

The properties (1), (2) and (3) define the cubic interpolating function $s(x)$, and the necessary steps for its construction are almost obvious: for the subinterval

## 3.7 Spline Interpolation

$[x_i, x_{i+1}]$ of length

$$h_i = x_{i+1} - x_i \tag{3.116}$$

we choose the following cubic polynomial of a suitable form

$$s_i(x) = a_i(x - x_i)^3 + b_i(x - x_i)^2 + c_i(x - x_i) + d_i. \tag{3.117}$$

Its value and the first two derivatives at the end points of the interval are given by

$$s_i(x_i) \quad = \quad d_i = y_i \tag{3.118}$$

$$s_i(x_{i+1}) = a_i h_i^3 + b_i h_i^2 + c_i h_i + d_i = y_{i+1} \tag{3.119}$$

$$s_i'(x_i) \quad = \quad c_i \tag{3.120}$$

$$s_i'(x_{i+1}) = 3a_i h_i^2 + 2b_i h_i + c_i \tag{3.121}$$

$$s_i''(x_i) \quad = \quad 2b_i \quad\quad = y_i'' \tag{3.122}$$

$$s_i''(x_{i+1}) = 6a_i h_i + 2b_i \quad\quad = y_{i+1}'' \tag{3.123}$$

In order to satisfy the interpolation and continuity conditions of the first and second derivatives at the support abscissae, it is most convenient to express the coefficients $a_i$, $b_i$, $c_i$ and $d_i$ by means of the given function values $y_i$ and $y_{i+1}$ and the unknown second derivatives $y_i''$ and $y_{i+1}''$ at both ends of the interval $[x_i, x_{i+1}]$. From (3.118), (3.119), (3.122) and (3.123) we obtain

$$\boxed{\begin{aligned} a_i &= \frac{1}{6h_i}(y_{i+1}'' - y_i'') \\ b_i &= \tfrac{1}{2} y_i'' \\ c_i &= \frac{1}{h_i}(y_{i+1} - y_i) - \tfrac{1}{6} h_i (y_{i+1}'' + 2 y_i'') \\ d_i &= y_i \end{aligned}} \tag{3.124}$$

In addition to the given function values $y_k$, as soon as the second derivatives $y_k''$ are also known at all support abscissae, the cubic polynomials $s_i(x)$ are uniquely determined and can be evaluated on each subinterval. We observe that these values that completely describe the spline function, guarantee the interpolation property, together with the continuity of the second derivative of the function at the interior support abscissae. Hence, it remains to satisfy the continuity of the first derivative at the interior support abscissae. After substitution of the expressions (3.124) for $a_i, b_i, c_i$ into (3.121) we get the first derivative at the end of the interval

$$s_i'(x_{i+1}) = \frac{1}{h_i}(y_{i+1} - y_i) + \tfrac{1}{6} h_i (2 y_{i+1}'' + y_i'')$$

130  *Interpolation*

and by decreasing the index $i$ by one

$$s'_{i-1}(x_i) = \frac{1}{h_{i-1}}(y_i - y_{i-1}) + \tfrac{1}{6}h_{i-1}(2y''_i + y''_{i-1}). \tag{3.125}$$

The condition $s'_{i-1}(x_i) = s'_i(x_i)$ for an interior support abscissa $x_i$ leads, from (3.125), (3.120) and (3.124), to the equation

$$\frac{1}{h_{i-1}}(y_i - y_{i-1}) + \tfrac{1}{6}h_{i-1}(2y''_i + y''_{i-1}) = \frac{1}{h_i}(y_{i+1} - y_i) - \tfrac{1}{6}h_i(y''_{i+1} + 2y''_i)$$

which, after multiplication by 6, becomes

$$h_{i-1}y''_{i-1} + 2(h_{i-1} + h_i)y''_i + h_i y''_{i+1}$$
$$- \frac{6}{h_i}(y_{i+1} - y_i) + \frac{6}{h_{i-1}}(y_i - y_{i-1}) = 0. \tag{3.126}$$

This condition must hold for all interior support abscissae $x_i$, $(i = 1, 2, \ldots, n-1)$, and from this we get for the $(n-1)$ unknowns $y''_1, y''_2, \ldots, y''_{n-1}$, of a natural cubic spline function, a system of $(n-1)$ linear equations. For $n = 5$ the system is as follows

| $y''_1$ | $y''_2$ | $y''_3$ | $y''_4$ | | 1 |
|---|---|---|---|---|---|
| $2(h_0 + h_1)$ | $h_1$ | | | | $\dfrac{6}{h_0}(y_1 - y_0) - \dfrac{6}{h_1}(y_2 - y_1) + h_0 y''_0$ |
| $h_1$ | $2(h_1 + h_2)$ | $h_2$ | | | $\dfrac{6}{h_1}(y_2 - y_1) - \dfrac{6}{h_2}(y_3 - y_2)$ |
| | $h_2$ | $2(h_2 + h_3)$ | $h_3$ | | $\dfrac{6}{h_2}(y_3 - y_2) - \dfrac{6}{h_3}(y_4 - y_3)$ |
| | | $h_3$ | $2(h_3 + h_4)$ | | $\dfrac{6}{h_3}(y_4 - y_3) - \dfrac{6}{h_4}(y_5 - y_4) + h_4 y''_5$ |

$$\tag{3.127}$$

For purposes that will become clear later, we have combined the quantities $y''_0$ and $y''_5$ with the constants of the first and the last equation of the system (3.127). For a natural spline they are equal to zero, of course.

The system of equations (3.127) has a symmetric, *tridiagonal* matrix which is clearly *diagonally dominant*. From Theorem 1.5 the system can be solved by means of the Gaussian algorithm using the *diagonal strategy*, that is the algorithm of Section 1.3.3 is applicable. Moreover, the procedure can be simplified

## 3.7 Spline Interpolation

somewhat by using the symmetry. We deduce the essential fact, that the system of $(n-1)$ equations (3.127) for the $(n-1)$ unknown second derivatives $y_1'', y_2'', \ldots, y_{n-1}''$ has a unique solution, so that the existence and uniqueness of the cubic spline function $s(x)$ follows.

In the case of equidistant support abscissae $x_i = x_0 + ih$, we can divide each equation (3.126) by the constant length $h$ of the intervals. The system (3.127) is simplified to

| $y_1''$ | $y_2''$ | $y_3''$ | $y_4''$ | 1 |
|---|---|---|---|---|
| 4 | 1 | | | $-\frac{6}{h^2}(y_2 - 2y_1 + y_0) + y_0''$ |
| 1 | 4 | 1 | | $-\frac{6}{h^2}(y_3 - 2y_2 + y_1)$ |
| | 1 | 4 | 1 | $-\frac{6}{h^2}(y_4 - 2y_3 + y_2)$ |
| | | 1 | 4 | $-\frac{6}{h^2}(y_5 - 2y_4 + y_3) + y_5''$ |

(3.128)

The matrix of the system (3.128) has quite a special structure. The constants of the equations are essentially the second difference quotients, (3.24), of the given function values.

The procedure for determining the cubic spline function $s(x)$ is now obvious. From the given data points $(x_i, y_i)$, $i = 0, 1, \ldots, n$, the interval lengths $h_i$ are determined and the system of equations (3.127) is built up. After its solution by means of the algorithm (1.112), (1.113) and (1.114) for the unknowns $y_k$, the coefficients $a_i$, $b_i$, $c_i$ and $d_i$ of the cubic polynomials $s_i(x)$ of the subintervals $[x_i, x_{i+1}]$ are computed. Now we can evaluate the spline function $s(x)$ for any $x \in [x_0, x_n]$. The amount of computational effort required is only proportional to $n$, if $(n + 1)$ data points are given, because this is true for each of the steps mentioned. Even for larger values of $n$, the required effort is reasonably small.

Finally the *condition number* of the symmetric, tridiagonal and positive definite matrices is not large, so that the numerical solution of the systems is good, and no iterative improvement is necessary. In order to estimate the condition number corresponding to the spectral norm of the matrix **A** of the system (3.127), we need an upper bound for the largest eigenvalue $\lambda_{max}$ and a lower bound for the smallest eigenvalue $\lambda_{min}$ of **A**. *Gershgorin's theorem* (Maess, 1985; Schmeisser and Schirmeier, 1976; Stummel and Hainer, 1982; Wilkinson, 1965; Young and

Gregory, 1973) provides these bounds, and we have

$$\kappa(\mathbf{A}) = \frac{\lambda_{\max}}{\lambda_{\min}} \leqslant \frac{\max_i \{2h_0 + 3h_1, 3(h_i + h_{i+1}), 3h_{n-2} + 2h_{n-1}\}}{\min_i \{2h_0 + h_1, h_i + h_{i+1}, h_{n-2} + 2h_{n-1}\}}. \tag{3.129}$$

The estimate (3.129) of the condition number is essentially determined by the ratio between the largest and smallest interval length. If for example $\max(h_i):\min(h_i) = 10:1$, from (3.129) we have $\kappa(\mathbf{A}) \leqslant 30$. In case of equidistant support abscissae we get the bound $\kappa(\mathbf{A}) \leqslant 3$. It should be noted that the given upper bound of the condition number does not depend on the number of points.

*Example* 3.20 The table (3.130) contains the given data for 15 equidistant support abscissae $x_k$, together with the corresponding function values $y_k$. The matrix of the system is given by (3.128). Therefore in (3.130) only the constant terms of the 13 linear equations are given. The last column contains the resulting values of the second derivatives complemented by the vanishing values of the natural cubic spline function $s(x)$. It is shown in Figure 3.7, and its graph is quite satisfactory. For reasons of comparison the graph of the interpolation polynomial $P_{14}(x)$ of degree 14 is drawn as a dashed line, within the limits of the diagram. Once again this shows the extreme oscillatory, and hence, unsatisfactory behaviour of the interpolation polynomial.

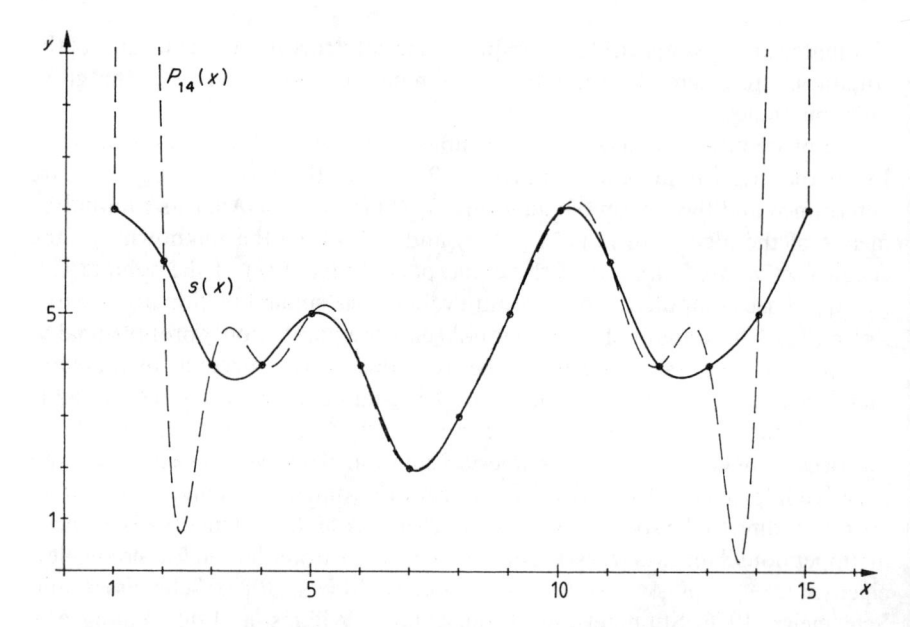

Figure 3.7 Natural cubic spline interpolation

## 3.7 Spline Interpolation

| $k$ | $x_k$ | $y_k$ | Constants | $y_k''$ | |
|---|---|---|---|---|---|
| 0  | 1  | 7 | —    | 0         | |
| 1  | 2  | 6 | 6    | −2.306 437 | |
| 2  | 3  | 4 | −12  | 3.225 748 | |
| 3  | 4  | 4 | −6   | 1.403 445 | |
| 4  | 5  | 5 | 12   | −2.839 529 | |
| 5  | 6  | 4 | 6    | −2.045 331 | |
| 6  | 7  | 2 | −18  | 5.020 853 | |
| 7  | 8  | 3 | −6   | −0.038 080 | (3.130) |
| 8  | 9  | 5 | 0    | 1.131 468 | |
| 9  | 10 | 7 | 18   | −4.487 791 | |
| 10 | 11 | 6 | 6    | −1.180 305 | |
| 11 | 12 | 4 | −12  | 3.209 010 | |
| 12 | 13 | 4 | −6   | 0.344 264 | |
| 13 | 14 | 5 | −6   | 1.413 934 | |
| 14 | 15 | 7 | —    | 0         | |

### 3.7.3 General cubic spline functions

The natural boundary conditions $s''(x_0) = s''(x_n) = 0$ are not always adequate. Indeed this is true if the function to be approximated has a large curvature at the ends of the interval. Thus the two natural boundary conditions are replaced by two other conditions, so that the spline function is still uniquely determined.

The most obvious modification would be to prescribe appropriate values of the second derivatives $y_0''$ and $y_n''$ at the support abscissae $x_0$ and $x_n$. The behaviour of the resulting *general cubic spline function* can in fact be influenced favourably, but the influence of these values extends only over a few of the adjacent intervals. Prescribed values for $y_0''$ and $y_n''$ do not have a decisive influence on the computational procedure, because only the constant terms of the first and the last equations of the system (3.127), or (3.128), are changed. The systems of linear equations for the remaining unknown second derivatives still have a unique solution, so that the general cubic spline function is unique, too.

It is often difficult to provide appropriate values $y_0''$ and $y_n''$ because they do not have a direct intuitive meaning. One attempts to overcome this difficulty by requiring that the second derivative of the end point is equal to a given positive multiple of the second derivative of the adjacent interior support abscissa (Späth, 1973). Hence we require $y_0'' = \alpha y_1''$, $y_n'' = \beta y_{n-1}''$, which changes only the diagonal elements of the first and last equation of the systems (3.127), or (3.128), so that the diagonal dominance is maintained. With the values $\alpha = \beta = 1$ or $\alpha = \beta = 0.5$ the resulting spline functions are usually quite satisfactory.

A different approach to the construction of a general cubic spline function is based on the condition that the polynomials of the two adjacent subintervals at the boundary of $[x_0, x_n]$ should coincide (Boor, 1978). This can simply be satisfied by the requirement that in addition the third derivatives of the two adjoining polynomials are equal at the support abscissae $x_1$ and $x_{n-1}$, respectively. The two conditions $s_0'''(x_1) = s_1'''(x_1)$ and $s_{n-2}'''(x_{n-1}) = s_{n-1}'''(x_{n-1})$ lead, from (3.117) and (3.124), to the two linear equations

$$h_1 y_0'' - (h_0 + h_1) y_1'' + h_0 y_2'' = 0$$
$$h_{n-1} y_{n-2}'' - (h_{n-2} + h_{n-1}) y_{n-1}'' + h_{n-2} y_n'' = 0. \tag{3.131}$$

As the values $y_0''$ and $y_n''$ are to be considered as unknown, the $(n-1)$ equations (3.126) have to be extended by the two equations (3.131) to yield a system of $(n+1)$ linear equations for the unknowns $y_0'', y_1'', \ldots, y_{n-1}'', y_n''$. Unfortunately, the system is no longer tridiagonal because the first and last equation are no longer symmetric or diagonally dominant. However, it can be solved by the Gaussian algorithm using the diagonal strategy, whereby the first and last equations require special treatment in order to take advantage of the special structure of the matrix and the reduced system.

In certain applications it is adequate to prescribe the values of the first derivatives at the end points. In this case we have to consider $y_0''$ and $y_n''$ again as unknowns. The new conditions $s_0'(x_0) = y_0'$ and $s_{n-1}'(x_n) = y_n'$, on the basis of (3.120), (3.121) and (3.124), after multiplication by $-6$ and $6$ respectively, yield the two equations

$$2h_0 y_0'' + h_0 y_1'' - \frac{6}{h_0}(y_1 - y_0) + 6 y_0' = 0 \tag{3.132}$$

$$h_{n-1} y_{n-1}'' + 2h_{n-1} y_n'' + \frac{6}{h_{n-1}}(y_n - y_{n-1}) - 6 y_n' = 0. \tag{3.133}$$

If we add (3.132) as the first and (3.133) as the last equation to the system (3.127), we get a system of $(n+1)$ equations for the unknowns $y_0'', y_1'', \ldots, y_n''$, whose matrix is tridiagonal, symmetric and diagonally dominant and hence positive definite. For $n = 4$ equidistant support abscissae and after division of all equations by the length $h$ of the interval, it is given by

| $y_0''$ | $y_1''$ | $y_2''$ | $y_3''$ | $y_4''$ | | 1 |
|---|---|---|---|---|---|---|
| 2 | 1 | | | | | $-\frac{6}{h^2}(y_1 - y_0) + \frac{6}{h} y_0'$ |
| 1 | 4 | 1 | | | | $-\frac{6}{h^2}(y_2 - 2y_1 + y_0)$ |

## 3.7 Spline Interpolation

$$\begin{vmatrix} 1 & 4 & 1 & & \\ & 1 & 4 & 1 & \\ & & 1 & 2 & \end{vmatrix} \quad \begin{matrix} -\frac{6}{h^2}(y_3 - 2y_2 + y_1) \\ -\frac{6}{h^2}(y_4 - 2y_3 + y_2) \\ \frac{6}{h^2}(y_4 - y_3) - \frac{6}{h}y'_4 \end{matrix} \tag{3.134}$$

*Example* 3.21 A hysteresis curve is defined by measured values at nonequidistant support abscissae according to (3.135). We want to represent it by a cubic spline function. We know that for physical reasons the first derivative at the origin is $y'_0 = 0.00125664$. At the last point we fix the value $y'_8 = 0.0001$.

| k | $x_k$ | $y_k$ | $y''_k$ |
|---|-------|-------|---------|
| 0 | 0     | 0     | 0.022 181 |
| 1 | 8.2   | 0.5   | −0.000 665 |
| 2 | 14.7  | 1.0   | −0.010 253 |
| 3 | 17.0  | 1.1   | −0.006 909 |
| 4 | 21.1  | 1.2   | −0.000 613 |
| 5 | 35.0  | 1.4   | −0.000 691 |
| 6 | 54.1  | 1.5   | −0.000 040 |
| 7 | 104   | 1.6   | −0.000 014 |
| 8 | 357   | 1.7   | 0.000 004 |

(3.135)

According to (3.127), (3.132) and (3.133) the system of linear equations for the nine unknowns $y''_0, y''_1, \ldots, y''_8$ is

| $y''_0$ | $y''_1$ | $y''_2$ | $y''_3$ | $y''_4$ | $y''_5$ | $y''_6$ | $y''_7$ | $y''_8$ | 1 |
|---|---|---|---|---|---|---|---|---|---|
| 16.4 | 8.2 |  |  |  |  |  |  |  | −0.358 314 |
| 8.2 | 29.4 | 6.5 |  |  |  |  |  |  | −0.095 685 |
|  | 6.5 | 17.6 | 2.3 |  |  |  |  |  | 0.200 668 |
|  |  | 2.3 | 12.8 | 4.1 |  |  |  |  | 0.114 530 |
|  |  |  | 4.1 | 36.0 | 13.9 |  |  |  | 0.060 010 |
|  |  |  |  | 13.9 | 66.0 | 19.1 |  |  | 0.054 917 |
|  |  |  |  |  | 19.1 | 138 | 49.9 |  | 0.019 390 |
|  |  |  |  |  |  | 49.9 | 605.8 | 253 | 0.009 653 |
|  |  |  |  |  |  |  | 253 | 506 | 0.001 772 |

The resulting unknowns are given in (3.135). The spline function has a turning point near $x_2 = 8.2$, a property that is even desired.

As the interval lengths $h_i$ are quite different in size, we determine the condition number of the matrix by using Gershgorin's theorem which yields the bounds $\lambda_{\min} \geqslant 6.4$ and $\lambda_{\max} \leqslant 908.7$, so we have $\kappa \leqslant 142$. In fact we have $\lambda_{\max} \doteq 816$, $\lambda_{\min} \doteq 10.1$ and therefore $\kappa \doteq 81$. The condition number is small, and the numerical solution of the system presents no problems.

### 3.7.4 Periodic cubic spline interpolation

If a periodic function which is at least twice continuously differentiable is to be represented on its interval of periodicity by an interpolating *periodic spline function*, the natural conditions (3.115) must be replaced by conditions that take into account the periodicity. The support abscissae are arranged in increasing order $x_0 < x_1 < \cdots < x_n$ and $x_n = x_0 + T$, where $T$ denotes the period. On these assumptions we have $y_0 = y_n$, $y'_0 = y'_n$ and $y''_0 = y''_n$. Our unknowns are now the second derivatives $y''_i$, $(i = 0, 1, \ldots, n-1)$, and we have to satisfy the condition of continuity of the first derivative at the $n$ support abscissae $x_0, x_1, \ldots, x_{n-1}$. The conditions (3.126) give a system of $n$ linear equations for the unknowns $y''_0$, $y''_1, \ldots, y''_{n-1}$. When set up, we must take into account the relationships $h_{-1} = x_0 - x_{-1} = x_n - x_{n-1} = h_{n-1}, y''_{-1} = y''_{n-1}, y_{-1} = y_{n-1}$, due to the periodicity. For $n = 5$ we have the system

| $y''_0$ | $y''_1$ | $y''_2$ | $y''_3$ | $y''_4$ | 1 |
|---|---|---|---|---|---|
| $2(h_4 + h_0)$ | $h_0$ | | | $h_4$ | $-\frac{6}{h_0}(y_1 - y_0) + \frac{6}{h_4}(y_0 - y_4)$ |
| $h_0$ | $2(h_0 + h_1)$ | $h_1$ | | | $-\frac{6}{h_1}(y_2 - y_1) + \frac{6}{h_0}(y_1 - y_0)$ |
| | $h_1$ | $2(h_1 + h_2)$ | $h_2$ | | $-\frac{6}{h_2}(y_3 - y_2) + \frac{6}{h_1}(y_2 - y_1)$ |
| | | $h_2$ | $2(h_2 + h_3)$ | $h_3$ | $-\frac{6}{h_3}(y_4 - y_3) + \frac{6}{h_2}(y_3 - y_2)$ |
| $h_4$ | | | $h_3$ | $2(h_3 + h_4)$ | $-\frac{6}{h_4}(y_5 - y_4) + \frac{6}{h_3}(y_4 - y_3)$ |

(3.136)

The matrix of the system (3.136) is *symmetric*, but no longer tridiagonal. It is still *diagonally dominant*, so that the system can be solved by the Gaussian algorithm with the diagonal strategy due to Theorem 1.5. As the matrix has positive diagonal elements, it is positive definite (Schwarz, 1972), and so the Cholesky's method is better. The two nonzero elements in the corners of the matrix require further consideration. It is obvious that the first reduction step generates a reduced matrix with the same structure. This is true for all subsequent steps, and

## 3.7 Spline Interpolation

we deduce that the left triangular matrix **L** of the Cholesky decomposition $\mathbf{A} = \mathbf{L}\mathbf{L}^T$ has nonzero elements in its last row in addition to the subdiagonal. In order to take the quite special structure of the two matrices **A** and **L** into account we write them as follows

$$\begin{pmatrix} a_1 & b_1 & & & c \\ b_1 & a_2 & b_2 & & \\ & b_2 & a_3 & b_3 & \\ & & b_3 & a_4 & b_4 \\ c & & & b_4 & a_5 \end{pmatrix} = \begin{pmatrix} l_1 & & & & \\ m_1 & l_2 & & & \\ & m_2 & l_3 & & \\ & & m_3 & l_4 & \\ e_1 & e_2 & e_3 & m_4 & l_5 \end{pmatrix} \begin{pmatrix} l_1 & m_1 & & & e_1 \\ & l_2 & m_2 & & e_2 \\ & & l_3 & m_3 & e_3 \\ & & & l_4 & m_4 \\ & & & & l_5 \end{pmatrix}$$

(3.137)

The unknown quantities $l_i$, $m_i$ and $e_i$ are determined by comparing corresponding matrix elements. We get the following equations

$$\begin{aligned} a_1 &= l_1^2 & b_1 &= m_1 l_1 & c &= e_1 l_1 \\ a_2 &= m_1^2 + l_2^2 & b_2 &= m_2 l_2 & 0 &= e_1 m_1 + e_2 l_2 \\ a_3 &= m_2^2 + l_3^2 & b_3 &= m_3 l_3 & 0 &= e_2 m_2 + e_3 l_3 \\ a_4 &= m_3^2 + l_4^2 & b_4 &= e_3 m_3 + m_4 l_4 & & \\ a_5 &= e_1^2 + e_2^2 + e_3^2 + m_4^2 + l_5^2 & & & & \end{aligned}$$

From this set of equations, the unknown elements of **L** are computed in the order $l_1; m_1, e_1, l_2; m_2, e_2, l_3; \ldots, m_4, l_5$. Therefore, the corresponding algorithm for the decomposition (3.137) for a general $n$ is

$$\boxed{\begin{aligned} & l_1 = \sqrt{a_1}\,; e_1 = c/l_1\,; s = 0 \\ & \text{for } i = 1, 2, \ldots, n-2: \\ & \quad m_i = b_i/l_i \\ & \quad \text{if } i \neq 1: e_i = -e_{i-1} \times m_{i-1}/l_i \\ & \quad l_{i+1} = \sqrt{a_{i+1} - m_i^2} \\ & \quad s = s + e_i^2 \\ & m_{n-1} = (b_{n-1} - e_{n-2} \times m_{n-2})/l_{n-1} \\ & l_n = \sqrt{a_n - m_{n-1}^2 - s} \end{aligned}}$$

(3.138)

The forward substitution $\mathbf{L}\mathbf{y} - \mathbf{d} = \mathbf{0}$ is described by

$$\boxed{\begin{aligned} & y_1 = d_1/l_1\,; s = 0 \\ & \text{for } i = 2, 3, \ldots, n-1: \\ & \quad y_i = (d_i - m_{i-1} \times y_{i-1})/l_i \\ & \quad s = s + e_{i-1} \times y_{i-1} \\ & y_n = (d_n - m_{n-1} \times y_{n-1} - s)/l_n \end{aligned}}$$

(3.139)

The backward substitution $\mathbf{L}^T\mathbf{x} + \mathbf{y} = \mathbf{0}$ is formulated as

$$\begin{aligned} x_n &= -y_n/l_n \\ x_{n-1} &= -(y_{n-1} + m_{n-1} \times x_n)/l_{n-1} \\ \text{for } i &= n-2, n-3, \ldots, 1: \\ x_i &= -(y_i + m_i \times x_{i+1} + e_i \times x_n)/l_i \end{aligned} \tag{3.140}$$

*Example* 3.22 A periodic function of period $T = 16$ is known by values at nonequidistant support abscissae according to (3.141). The coefficients $a_k$ and $b_k$ as well as the constants $d_k$ of the system of linear equations $\mathbf{A}\mathbf{y}'' + \mathbf{d} = \mathbf{0}$ (3.136) for the unknowns $y_0'', y_1'', \ldots, y_5''$ are given in (3.141), together with the resulting values of $y_k''$. The corresponding periodic spline function $s(x)$ and the function $f(x) = 2.5[\cos(2\pi x/16) - \sin(4\pi x/16)] + 5$, that has been used to define the function values $y_k$, are shown in Figure 3.8. Even with these few support points the two curves agree quite well.

| $k$ | $x_k$ | $y_k$ | $a_k$ | $b_k$ | $d_k$ | $y_k''$ |
|---|---|---|---|---|---|---|
| 0 | 1.0   | 5.541 932 | —    | —    | —          | 0.819 490  |
| 1 | 2.5   | 4.079 227 | 8.0  | 1.50 | −3.380 638 | 1.555 705  |
| 2 | 5.25  | 5.900 182 | 8.5  | 2.75 | −9.823 814 | −1.683 242 |
| 3 | 9.5   | 0.611 627 | 14.0 | 4.25 | 11.439 189 | 1.846 591  |
| 4 | 12.0  | 5.000 000 | 13.5 | 2.50 | −17.998 290| 0.089 237  |
| 5 | 14.5  | 9.388 373 | 10.0 | 2.50 | 0          | −2.203 537 |
| 6 | 17.0  | 5.541 932 | 10.0 | —    | 19.763 553 | 0.819 490  |

(3.141)

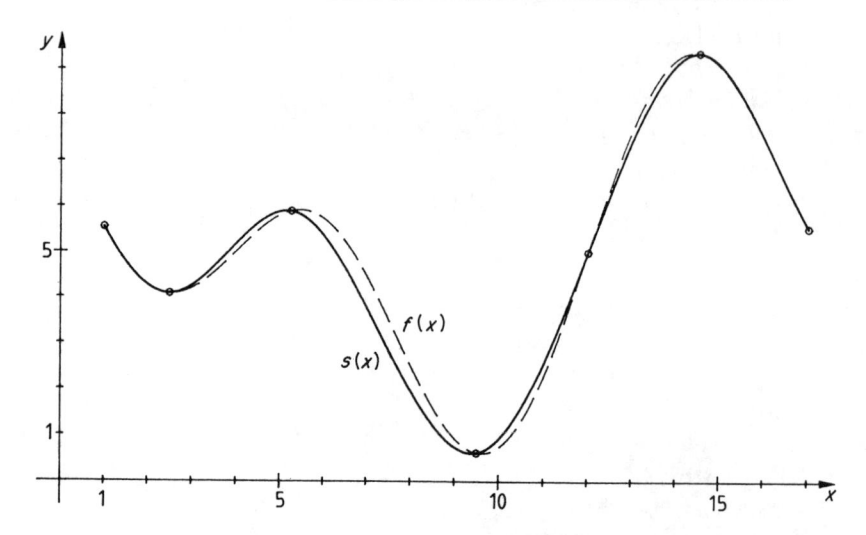

Figure 3.8 Periodic cubic spline interpolation

## 3.7 Spline Interpolation

### 3.7.5 Smooth two-dimensional curves

An application of cubic spline interpolation is the construction of smooth curves in the plane passing through $(n + 1)$ given points with coordinates $(x_k, y_k)$, where a general representation of the form $y = f(x)$ is not possible. Therefore we have to use the parametric representation

$$x = x(t) \quad y = y(t), \tag{3.142}$$

where $t$ denotes the parameter. The parameter values $t_0, t_1, \ldots, t_n$ corresponding to the $(n + 1)$ given points can be assumed to be ordered by magnitude. Hence we determine the two spline functions that interpolate the tabulated functions $(t_k, x_k)$ and $(t_k, y_k)$, $(k = 0, 1, \ldots, n)$, which, by means of the parametric representation (3.142), will describe the desired curve.

The arc length of the curve would be the most appropriate parameter $t$. However, it is not known *a priori*, and therefore the parameter values $t_k$ are usually chosen to be the known distances of consecutive points so that we have

$$t_0 = 0 \quad t_k = t_{k-1} + \sqrt{(x_k - x_{k-1})^2 + (y_k - y_{k-1})^2} \quad (k = 1, 2, \ldots, n). \tag{3.143}$$

*Example* 3.23 The profile of a body in connection with stream problems is fixed by nine points with coordinates given in (3.144). The parameter values $t_k$, according to (3.143), are rounded to three decimal places, and the resulting second derivatives $x_k''$ and $y_k''$ of the two cubic spline functions are given in (3.144).

The result of this smooth, parametric representation of the profile is shown in Figure 3.9.

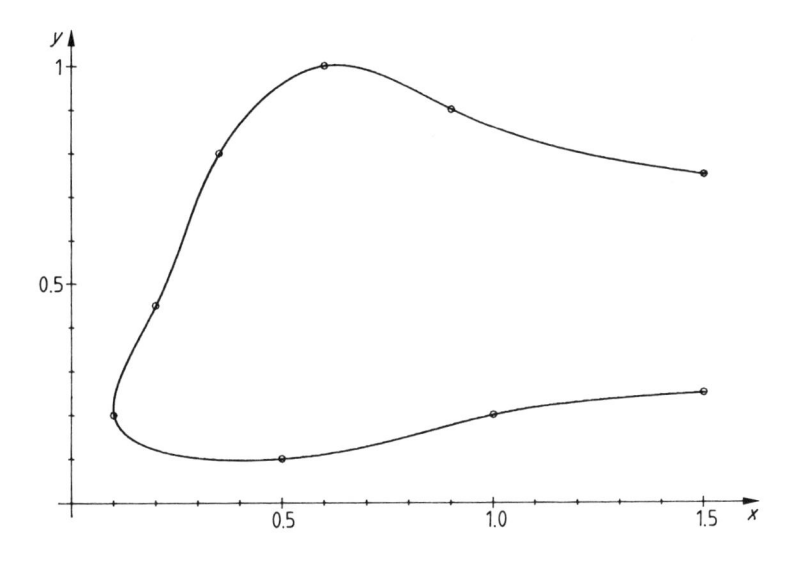

Figure 3.9 Two-dimensional representation of a curve

| k | $x_k$ | $y_k$ | $t_k$ | $x_k''$ | $y_k''$ |
|---|---|---|---|---|---|
| 0 | 1.50 | 0.75 | 0     | 0         | 0         |
| 1 | 0.90 | 0.90 | 0.618 | 0.026 366 | 1.029 327 |
| 2 | 0.60 | 1.00 | 0.935 | 0.251 299 | −4.686 680 |
| 3 | 0.35 | 0.80 | 1.255 | 2.119 916 | −0.018 687 |
| 4 | 0.20 | 0.45 | 1.636 | −1.918 770 | −0.630 358 |
| 5 | 0.10 | 0.20 | 1.905 | 6.768 642 | 2.862 123 |
| 6 | 0.50 | 0.10 | 2.317 | −1.602 547 | 0.931 113 |
| 7 | 1.00 | 0.20 | 2.827 | 0.446 407 | −0.520 770 |
| 8 | 1.50 | 0.25 | 3.330 | 0         | 0         |

(3.144)

*Example* 3.24 The problem of constructing and drawing a smooth curve in the plane defined by a sequence of given points may arise in connection with the numerical integration of a system of two first-order ordinary differential equations (see Chapter 9). The computed discrete solution points in the phase plane, arising after a certain number of integration steps, are to be interpolated by a smooth phase curve in order to get a graphical representation on the screen. Figure 3.10 shows the solution points $P_0, P_1, \ldots, P_{30}$ and the approximation of the desired phase curve defined by the two natural cubic spline functions.

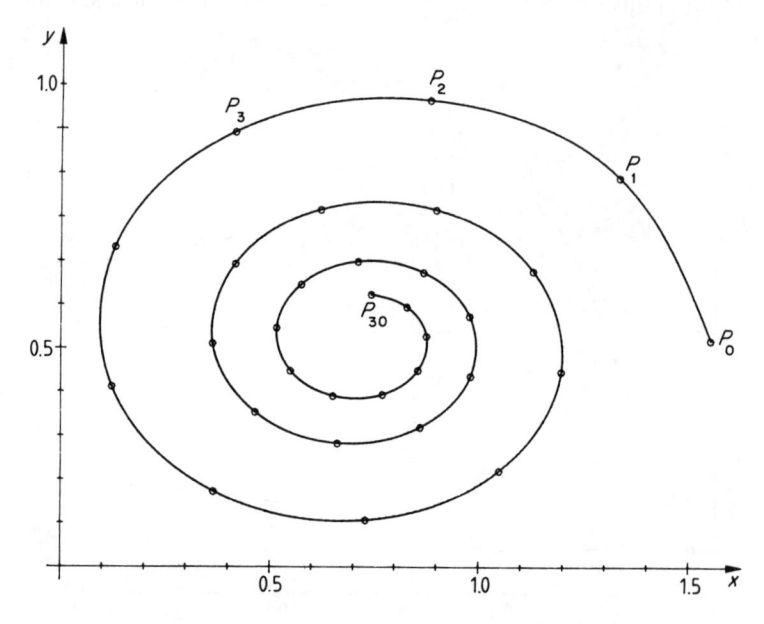

Figure 3.10 Construction of a phase curve

## 3.8 Exercises

**3.1.** Derive the formulae (3.29) for the approximate computation of the first and second derivatives at the points $x_0$, $x_1$ and $x_M = (x_0 + x_3)/2$ by means of the cubic interpolation polynomial $P_3(x)$ corresponding to the equidistant support abscissae $x_0$, $x_1 = x_0 + h$, $x_2 = x_0 + 2h$, $x_3 = x_0 + 3h$. With these formulae compute the approximations of the first and second derivative of the function $f(x) = e^{-x}$ at the point $x_M = 0.4$ for the sequence of step sizes $h = 0.1, 0.01, 0.001, \ldots, 10^{-8}$. Compare the values with the exact derivatives.

**3.2.** Consider the function $f(x) = \log(x) - 2(x-1)/x$ and the nonequidistant support abscissae $x_0 = 1$, $x_1 = 2$, $x_2 = 4$, $x_3 = 8$, $x_4 = 10$. Compute the interpolated values by means of the interpolation polynomial $P_4(x)$ at the points $x = 2.9$ and $x = 5.25$. Apply the Lagrange and the Newton interpolation formula. How large are the interpolation errors and the error bounds?

**3.3.** A function $f(x)$ that is six times continuously differentiable is approximated by an interpolation polynomial $P_5(x)$ corresponding to equidistant support abscissae $x_0$, $x_1 = x_0 + h, \ldots, x_5 = x_0 + 5h$. What bounds for the maximum interpolation error can be given for each of the intervals? Choose the support abscissae suitably to simplify the derivation of the bounds. What are the error bounds for

(a) $f(x) = \sin(x)$ $\quad x \in [0, \pi/2]$ $\quad h = \pi/10$;
(b) $g(x) = e^{-3x}$ $\quad x \in [0, 5]$ $\quad h = 1$?

**3.4.** Determine the error function $r(x) = f(x) - P_{10}(x)$ for $f(x) = e^{-3x}$ in the interval $[0, 5]$ corresponding to the 11 equidistant support abscissae $x_k = 0.5k$, $(k = 0, 1, \ldots, 10)$. Use a computer program and apply Lagrange's and Newton's formula to evaluate $P_{10}(x)$. Draw the error function graphically.

**3.5.** Apply the method of extrapolation to determine the first derivative of the function $f(x) = e^{-x}$ as accurately as possible at the point $x = 0.4$. Use the first difference quotient and the central difference quotient and the step sizes $h_0 = 0.2$, $h_1 = 0.15$, $h_2 = 0.10$, $h_3 = 0.05$, $h_4 = 0.02$. What are the extrapolated values for linear, quadratic, ... interpolation? Apply the algorithms of the Lagrange, Newton, and Neville interpolation.

**3.6.** The number $\pi$ can be approximately computed by means of the circumferences $V_n$ of regular polygons with $n$ corners that are circumscribed to a circle of unit diameter. Show that an analogous representation of the error holds for $V_n$. Then compute by extrapolation (Neville algorithm and rational interpolation) a good approximation of $\pi$ from the circumferences $V_3$, $V_4$, $V_6$, $V_8$, and $V_{12}$ for their calculation no trigonometric functions should be used.

**3.7.** Use Hermite interpolation of degree three and five to compute the interpolated value at the point $x = 0.34$ on the assumption that the function values, and the derivatives, of $f(x) = e^{-x}$ at the support abscissae $x_0 = 0.3$ and $x_1 = 0.4$ are known. How large is the actual error of interpolation. Derive error bounds for the Hermite interpolation in these two cases.

**3.8.** Apply the inverse interpolation method to determine the smallest positive root of the equation $f(x) = \cos(x)\cosh(x) + 1 = 0$ that lies in the interval $[1.8, 1.9]$.

**3.9.** Take the function values of $f(x) = \cot(x)$ for the support abscissae $x_0 = 1°$, $x_1 = 2°$,

$x_2 = 3°$, $x_3 = 4°$, $x_4 = 5°$. Compute the interpolated value at the point $x = 2.5°$ by the Neville algorithm and by rational interpolation. Explain the differences between the interpolation errors.

**3.10.** Determine the natural cubic spline function $s(x)$ for the function $f(x) = 1/(1 + x^2)$ in the interval $[-5, 5]$

(a) for the six equidistant support abscissae $x_k = -5 + 2k$, $(k = 0, 1, \ldots, 5)$;
(b) for the 11 equidistant support abscissae $x_k = -5 + k$, $(k = 0, 1, \ldots, 10)$.

Represent both cubic spline functions graphically and give the maximum errors.

**3.11.** Given the function $f(x) = 4 + 3\cos(x) - 2\sin(2x) + \sin(5x)$ of period $2\pi$. Determine the periodic spline function $s(x)$ and represent it together with $f(x)$ graphically. Use the support abscissae (a) $x_k = 2\pi k/5$, $(k = 0, 1, \ldots, 5)$; (b) $x_k = \pi k/5$, $(k = 0, 1, \ldots, 10)$; (c) $x_k = \pi k/6$, $(k = 0, 1, \ldots, 12)$.

**3.12.** The complete profile of a body is described by the 12 points $P_0, P_1, \ldots, P_{10}, P_{11} = P_0$ in the plane. The profile has a peak at the point $P_0$. The coordinates of the points $P_k(x_k, y_k)$ are given as follows

| k | 0 | 1 | 2 | 3 | 4 | 5 | 6 | 7 | 8 | 9 | 10 | 11 |
|---|---|---|---|---|---|---|---|---|---|---|----|----|
| $x_k$ | 25 | 19 | 13 | 9 | 5 | 2.2 | 1 | 3 | 8 | 13 | 18 | 25 |
| $y_k$ | 5 | 7.5 | 9.1 | 9.4 | 9 | 7.5 | 5 | 2.1 | 2 | 3.5 | 4.5 | 5 |

Find the parametric representation of the curve on the basis of two spline functions, and draw the resulting profile.

**3.13.** Give an approximation of an ellipse with the half axis $a = 3$ and $b = 1$ by a smooth two-dimensional curve. Choose 4, 8 and 12 points on the ellipse and determine the parametric representations by periodic cubic spline functions.

# 4

# Approximation of Functions

In the following we consider the problem of approximating a function $f(x)$ defined explicitly on a given interval by another function $g(x)$ in the quadratic mean. The approximating function $g(x)$ should be an element of a linear function space that is chosen appropriately according to the properties of the given function $f(x)$. Orthogonal basis functions are the most suitable. We start by treating the approximation of periodic functions by finite Fourier series, and then consider two systems of orthogonal polynomials, namely the Chebyshev and the Legendre polynomials. The first mentioned polynomials are important in connection with the approximation of functions on computers.

## 4.1 Fourier Series

Let $f: \mathbf{R} \to \mathbf{R}$ be a piecewise continuous function of period $2\pi$, that is

$$f(x + 2\pi) = f(x) \quad \text{for all } x \in \mathbf{R}. \tag{4.1}$$

The function $f(x)$ is allowed to have discontinuities of the form, that for a point $x_0$ the limits $y_0^-$ and $y_0^+$

$$\lim_{h \to +0} f(x_0 - h) = y_0^- \quad \lim_{h \to +0} f(x_0 + h) = y_0^+ \tag{4.2}$$

exist and are finite. We aim to approximate the function $f(x)$ by a linear combination of the $(2\pi)$ periodic trigonometric functions

$$1, \cos(x), \sin(x), \cos(2x), \sin(2x), \ldots, \cos(nx), \sin(nx) \tag{4.3}$$

in the form

such that in the *quadratic mean* we have

$$\|g_n(x) - f(x)\|_2 := \left( \int_{-\pi}^{\pi} [g_n(x) - f(x)]^2 \, dx \right)^{1/2} = \text{Min!} \quad (4.5)$$

$$g_n(x) = \tfrac{1}{2}a_0 + \sum_{k=1}^{n} [a_k \cos(kx) + b_k \sin(kx)] \quad (4.4)$$

The difference between the approximating function $g_n(x)$ and the given function $f(x)$ has to be minimal with respect to the $L_2$ norm (4.5). The coefficients $a_k$ and $b_k$ of (4.4) can be determined from the condition (4.5) by means of the following theorem.

**Theorem 4.1** *The trigonometric functions* (4.3) *form a system of pairwise orthogonal functions with respect to the interval* $[-\pi, \pi]$. *The following relationships hold*

$$\int_{-\pi}^{\pi} \cos(jx)\cos(kx)\,dx = \begin{cases} 0 & \text{for all } j \neq k \\ 2\pi & \text{for } j = k = 0 \\ \pi & \text{for } j = k > 0 \end{cases} \quad (4.6)$$

$$\int_{-\pi}^{\pi} \sin(jx)\sin(kx)\,dx = \begin{cases} 0 & \text{for all } j \neq k,\ j, k > 0 \\ \pi & \text{for } j = k > 0 \end{cases} \quad (4.7)$$

$$\int_{-\pi}^{\pi} \cos(jx)\sin(kx)\,dx = 0 \quad \text{for all } j \geq 0, k > 0 \quad (4.8)$$

*Proof* Well known trigonometric identities give

$$\int_{-\pi}^{\pi} \cos(jx)\cos(kx)\,dx = \frac{1}{2} \int_{-\pi}^{\pi} \{\cos[(j+k)x] + \cos[(j-k)x]\}\,dx. \quad (4.9)$$

For $j \neq k$ from (4.9) we get

$$\frac{1}{2} \left( \frac{1}{j+k} \sin[(j+k)x] + \frac{1}{j-k} \sin[(j-k)x] \right) \Big|_{-\pi}^{\pi} = 0$$

and hence the first relationship of (4.6) is verified. For $j = k > 0$ from (4.9) we obtain

$$\frac{1}{2} \left( \frac{1}{j+k} \sin[(j+k)x] + x \right) \Big|_{-\pi}^{\pi} = \pi$$

## 4.1 Fourier Series

the third case of (4.6). The second relationship of (4.6) is obvious. The statement (4.7) is proved analogously because of the identity

$$\int_{-\pi}^{\pi} \sin(jx)\sin(kx)\,dx = -\frac{1}{2}\int_{-\pi}^{\pi}\{\cos[(j+k)x] - \cos[(j-k)x]\}\,dx$$

whereas (4.8) follows from the fact that the integrand is odd, as a product of an even and an odd function, so that the integral is zero.

It is obvious that the relationships of orthogonality, (4.6) to (4.8), hold for an arbitrary interval of length $2\pi$ because of the periodicity of the trigonometric functions.

From the orthogonality relationships (4.6) to (4.8) we obtain the following representation of the square of the $L_2$-norm of the difference $g_n(x) - f(x)$

$$\|g_n(x) - f(x)\|_2^2 = \int_{-\pi}^{\pi}\left(\frac{1}{2}a_0 + \sum_{k=1}^{n}[a_k\cos(kx) + b_k\sin(kx)] - f(x)\right)^2 dx$$

$$= \int_{-\pi}^{\pi}\left(\frac{1}{2}a_0 - f(x)\right)^2 dx + 2\sum_{k=1}^{n}\int_{-\pi}^{\pi}\left(\frac{1}{2}a_0 - f(x)\right)$$
$$\cdot [a_k\cos(kx) + b_k\sin(kx)]\,dx$$

$$+ \sum_{k=1}^{n}\sum_{j=1}^{n}\int_{-\pi}^{\pi}[a_k\cos(kx) + b_k\sin(kx)]$$
$$\cdot [a_j\cos(jx) + b_j\sin(jx)]\,dx \qquad (4.10)$$

$$= \frac{\pi}{2}a_0^2 - a_0\int_{-\pi}^{\pi}f(x)\,dx + \int_{-\pi}^{\pi}f(x)^2\,dx$$

$$- 2\sum_{k=1}^{n}a_k\int_{-\pi}^{\pi}f(x)\cos(kx)\,dx$$

$$- 2\sum_{k=1}^{n}b_k\int_{-\pi}^{\pi}f(x)\sin(kx)\,dx + \pi\sum_{k=1}^{n}(a_k^2 + b_k^2) =: F.$$

The necessary condition for a minimum of the quadratic function $F$ of the $(2n+1)$ variables $a_0, a_1, \ldots, a_n, b_1, \ldots, b_n$ is that its first partial derivatives with respect to the variables are equal to zero.

$$\frac{\partial F}{\partial a_0} = \pi a_0 - \int_{-\pi}^{\pi} f(x)\,dx = 0$$

$$\frac{\partial F}{\partial a_k} = -2\int_{-\pi}^{\pi} f(x)\cos(kx)\,dx + 2\pi a_k = 0 \quad (k=1,2,\ldots,n)$$

$$\frac{\partial F}{\partial b_k} = -2\int_{-\pi}^{\pi} f(x)\sin(kx)\,dx + 2\pi b_k = 0 \quad (k=1,2,\ldots,n)$$

Hence the desired coefficients $a_k$ and $b_k$ are given by

$$a_k = \frac{1}{\pi} \int_{-\pi}^{\pi} f(x) \cos(kx)\,dx \quad (k = 0, 1, \ldots, n)$$

$$b_k = \frac{1}{\pi} \int_{-\pi}^{\pi} f(x) \sin(kx)\,dx \quad (k = 1, 2, \ldots, n) \tag{4.11}$$

The values $a_k$ and $b_k$ defined by (4.11) are called the *Fourier coefficients* of the $(2\pi)$ periodic function $f(x)$, and the function $g_n(x)$, (4.4), is called the *Fourier polynomial*. The Fourier coefficients and the corresponding Fourier polynomials are defined for arbitrary large $n$, so that formally we can define the infinite *Fourier series*

$$g(x) := \tfrac{1}{2}a_0 + \sum_{k=1}^{\infty} [a_k \cos(kx) + b_k \sin(kx)] \tag{4.12}$$

for an arbitrary, $(2\pi)$ periodic and piecewise continuous function $f(x)$. Without proof we cite (Courant and Hilbert, 1968; Smirnow, 1972) the following theorem.

*Theorem 4.2 If $f(x)$ is a $(2\pi)$ periodic, piecewise continuous function whose first derivative is piecewise continuous, then the corresponding Fourier series $g(x)$, (4.12), converges to*

*(a) the value $f(x_0)$, if $f(x)$ is continuous at the point $x_0$,*

*(b) $\tfrac{1}{2}\{y_0^- + y_0^+\}$, where $y_0^-$ and $y_0^+$ are the limits (4.2), if $f(x)$ has a discontinuity at the point $x_0$.*

The hypotheses of Theorem 4.2 exclude piecewise continuous functions whose graph have a vertical tangent as is the case for example with $f(x) = \sqrt{|x|}$ at $x = 0$.

*Example* 4.1 The $(2\pi)$ periodic function $f(x)$ is defined on its basic interval $[-\pi, \pi]$ by

$$f(x) = |x| \quad -\pi \leqslant x \leqslant \pi. \tag{4.13}$$

Due to its graph, see Figure 4.1, it is called a roof function. It is a continuous and even function, and hence all coefficients $b_k = 0$. Moreover, the evaluation of the $a_k$ can be simplified.

$$a_0 = \frac{1}{\pi} \int_{-\pi}^{\pi} |x|\,dx = \frac{2}{\pi} \int_0^{\pi} x\,dx = \pi$$

$$a_k = \frac{2}{\pi} \int_0^{\pi} x \cos(kx)\,dx = \frac{2}{\pi}\left(\frac{1}{k} x \sin(kx)\Big|_0^{\pi} - \frac{1}{k}\int_0^{\pi} \sin(kx)\,dx\right)$$

$$= \frac{2}{\pi k^2} \cos(kx)\Big|_0^{\pi} = \frac{2}{\pi k^2}[(-1)^k - 1] \quad k > 0$$

## 4.1 Fourier Series

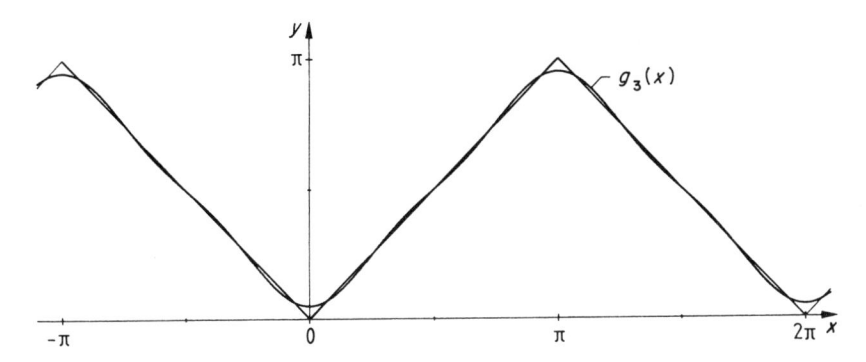

Figure 4.1 Roof function and approximating Fourier polynomial

Therefore the corresponding Fourier series is

$$g(x) = \frac{1}{2}\pi - \frac{4}{\pi}\left(\frac{\cos(x)}{1^2} + \frac{\cos(3x)}{3^2} + \frac{\cos(5x)}{5^2} + \cdots\right). \tag{4.14}$$

In Figure 4.1 the Fourier polynomial $g_3(x)$ is shown. It already approximates the function $f(x)$ quite well. The function $f(x)$ satisfies the assumptions of Theorem 4.2, and hence we get from (4.14) for $x = 0$ the value of the series

$$\frac{1}{1^2} + \frac{1}{3^2} + \frac{1}{5^2} + \frac{1}{7^2} + \cdots = \frac{\pi^2}{8}.$$

*Example 4.2* The $(2\pi)$ periodic function $f(x)$ is defined on the interval $(0, 2\pi)$ by

$$f(x) = x^2 \qquad 0 < x < 2\pi. \tag{4.15}$$

It is discontinuous at all points $x_k = 2\pi k$, $(k \in \mathbf{Z})$, but it satisfies the hypotheses of Theorem 4.2. The corresponding Fourier coefficients are calculated by means of integration by parts.

$$a_0 = \frac{1}{\pi}\int_0^{2\pi} x^2 \, dx = \frac{8\pi^2}{3}$$

$$a_k = \frac{1}{\pi}\int_0^{2\pi} x^2 \cos(kx) \, dx = \frac{4}{k^2} \quad (k = 1, 2, \ldots)$$

$$b_k = \frac{1}{\pi}\int_0^{2\pi} x^2 \sin(kx) \, dx = -\frac{4\pi}{k} \quad (k = 1, 2, \ldots)$$

The Fourier series is therefore

$$g(x) = \frac{4\pi^2}{3} + \sum_{k=1}^{\infty}\left(\frac{4}{k^2}\cos(kx) - \frac{4\pi}{k}\sin(kx)\right). \tag{4.16}$$

In Figure 4.2 the Fourier polynomial $g_4(x)$ illustrates the approximation of $f(x)$. The discontinuities cause a worsening of local convergence.

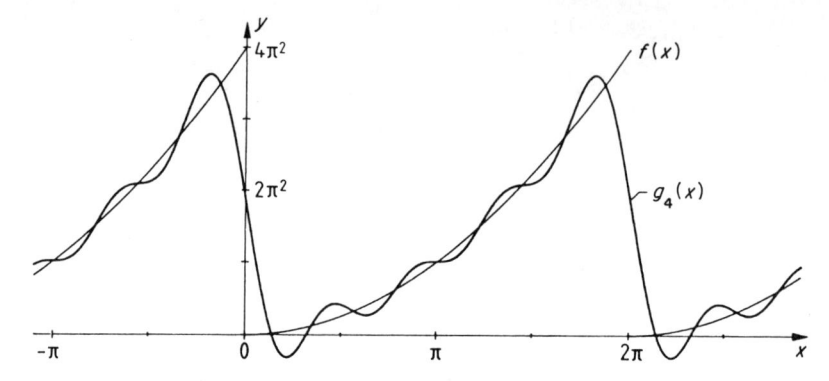

Figure 4.2 Periodic function with discontinuities

To compute the Fourier coefficients numerically and approximately we choose the *trapezoidal rule*, because it yields an approximation of the integral that is shift invariant for equidistant integration nodes of a given step length. Moreover, it has quite special properties, if a periodic function is approximately integrated over an interval of periodicity (see Section 8.2.1). If the interval $[0, 2\pi]$ is divided into $N$ subintervals of equal length $h$, we get the integration nodes $x_j$

$$h = \frac{2\pi}{N} \quad x_j = hj = \frac{2\pi}{N} j \quad (j = 0, 1, 2, \ldots, N) \tag{4.17}$$

and from the trapezoidal rule (8.4)

$$a_k = \frac{1}{\pi} \int_0^{2\pi} f(x) \cos(kx)\, dx$$

$$\approx \frac{1}{\pi} \frac{2\pi}{2N} \left( f(x_0) \cos(kx_0) + 2 \sum_{j=1}^{N-1} f(x_j) \cos(kx_j) + f(x_N) \cos(kx_N) \right).$$

As $f(x)\cos(kx)$ is $2\pi$ periodic we obtain for the $a_k$, and similarly for the $b_k$, the approximating values

$$\begin{aligned} a_k^* &:= \frac{2}{N} \sum_{j=1}^{N} f(x_j) \cos(kx_j) \quad (k = 0, 1, 2, \ldots) \\ b_k^* &:= \frac{2}{N} \sum_{j=1}^{N} f(x_j) \sin(kx_j) \quad (k = 1, 2, 3, \ldots) \end{aligned} \tag{4.18}$$

Fourier polynomials, which are defined by the coefficients $a_k^*$ and $b_k^*$ in place of $a_k$ and $b_k$, have specific properties to be derived next. As a preliminary step we need the discrete orthogonality relationships of the trigonometric functions.

## 4.1 Fourier Series

**Theorem 4.3** *For the discrete support abscissae $x_j$, (4.17), we have*

$$\sum_{j=1}^{N} \cos(kx_j) = \begin{cases} 0 & \text{if } k/N \notin \mathbf{Z} \\ N & \text{if } k/N \in \mathbf{Z} \end{cases} \tag{4.19}$$

$$\sum_{j=1}^{N} \sin(kx_j) = 0 \quad \text{for all } k \in \mathbf{Z}. \tag{4.20}$$

*Proof* We consider the complex sum

$$S := \sum_{j=1}^{N} [\cos(kx_j) + i\sin(kx_j)] = \sum_{j=1}^{N} e^{ikx_j} = \sum_{j=1}^{N} e^{ijkh} \tag{4.21}$$

which is nothing but a finite geometric series with the (complex) quotient $q := e^{ikh} = e^{2\pi i k/N}$. For its evaluation we have to distinguish two cases.

If $k/n \notin \mathbf{Z}$, then $q \neq 1$ and we have

$$S = e^{ikh} \frac{e^{ikhN} - 1}{e^{ikh} - 1} = e^{ikh} \frac{e^{2\pi k i} - 1}{e^{ikh} - 1} = 0 \quad k/N \notin \mathbf{Z}.$$

If, however, $k/N \in \mathbf{Z}$, then $q = 1$ and $S = N$. From these two results the assertions (4.19) and (4.20) follow by taking the real and imaginary parts of the sum $S$ (4.21).

**Theorem 4.4** *The trigonometric functions (4.3) satisfy the following discrete orthogonality relationships with respect to the equidistant support abscissae $x_j$ given in (4.17):*

$$\sum_{j=1}^{N} \cos(kx_j)\cos(lx_j) = \begin{cases} 0, & \text{if } \frac{k+l}{N} \notin \mathbf{Z} \text{ and } \frac{k-l}{N} \notin \mathbf{Z} \\ \frac{N}{2}, & \text{if either } \frac{k+l}{N} \in \mathbf{Z} \text{ or } \frac{k-l}{N} \in \mathbf{Z} \\ N, & \text{if } \frac{k+l}{N} \in \mathbf{Z} \text{ and } \frac{k-l}{N} \in \mathbf{Z} \end{cases} \tag{4.22}$$

$$\sum_{j=1}^{N} \sin(kx_j)\sin(lx_j) = \begin{cases} 0, & \text{if } \frac{k+l}{N} \notin \mathbf{Z} \text{ and } \frac{k-l}{N} \notin \mathbf{Z} \\ & \text{or } \frac{k+l}{N} \in \mathbf{Z} \text{ and } \frac{k-l}{N} \in \mathbf{Z} \\ -\frac{N}{2}, & \text{if } \frac{k+l}{N} \in \mathbf{Z} \text{ and } \frac{k-l}{N} \notin \mathbf{Z} \\ \frac{N}{2}, & \text{if } \frac{k+l}{N} \notin \mathbf{Z} \text{ and } \frac{k-l}{N} \in \mathbf{Z} \end{cases} \tag{4.23}$$

$$\sum_{j=1}^{N} \cos(kx_j)\sin(lx_j) = 0 \quad \text{for all } k,l \in \mathbf{N}_0 \tag{4.24}$$

*Proof* To verify the relationships (4.22) to (4.24) we first use the identities for trigonometric functions

$$\cos(kx_j)\cos(lx_j) = \tfrac{1}{2}\{\cos[(k+l)x_j] + \cos[(k-l)x_j]\}$$
$$\sin(kx_j)\sin(lx_j) = \tfrac{1}{2}\{\cos[(k-l)x_j] - \cos[(k+l)x_j]\}$$
$$\cos(kx_j)\sin(lx_j) = \tfrac{1}{2}\{\sin[(k+l)x_j] - \sin[(k-l)x_j]\}$$

and then apply the statements of Theorem 4.3.

**Theorem 4.5** Let $N = 2n$, $n \in \mathbf{N}$. The special Fourier polynomial

$$g_n^*(x) := \tfrac{1}{2}a_0^* + \sum_{k=1}^{n-1} [a_k^* \cos(kx) + b_k^* \sin(kx)] + \tfrac{1}{2}a_n^* \cos(nx) \qquad (4.25)$$

with the coefficients (4.18) is the unique, interpolating Fourier polynomial corresponding to the support abscissae $x_j$, (4.17), with the function values $f(x_j)$, $j = 1, 2, \ldots, N$.

*Proof* Because of the periodicity of $f(x)$ we have $f(x_0) = f(x_N)$. Therefore $N$ interpolation conditions have to be satisfied. A Fourier polynomial with as many coefficients is

$$g_n(x) = \tfrac{1}{2}\alpha_0 + \sum_{k=1}^{n-1} [\alpha_k \cos(kx) + \beta_k \sin(kx)] + \tfrac{1}{2}\alpha_n \cos(nx). \qquad (4.26)$$

We first show that the system of $N$ linear equations for the $N$ unknown coefficients $\alpha_k$ and $\beta_k$

$$\tfrac{1}{2}\alpha_0 + \sum_{k=1}^{n-1} [\alpha_k \cos(kx_j) + \beta_k \sin(kx_j)] + \tfrac{1}{2}\alpha_n \cos(nx_j) = f(x_j)$$
$$(j = 1, 2, \ldots, N) \quad (4.27)$$

has a unique solution. The columns of the coefficient matrix of the system of equations (4.27) are nonzero and pairwise orthogonal, due to the discrete relationships (4.22) to (4.24). Hence the matrix is nonsingular, and the system of equations, (4.27), has a unique solution.

The unknowns $\alpha_k$ and $\beta_k$ can be determined from (4.27) by means of the discrete othogonality relationships as follows. We multiply the $j$th equation by $\cos(lx_j)$, $0 \leq l \leq n$, where $l$ is fixed, and then sum all of the equations to obtain

$$\tfrac{1}{2}\alpha_0 \sum_{j=1}^{N} \cos(lx_j) + \sum_{k=1}^{n-1} \left( \alpha_k \sum_{j=1}^{N} \cos(kx_j)\cos(lx_j) + \beta_k \sum_{j=1}^{N} \sin(kx_j)\cos(lx_j) \right)$$
$$+ \tfrac{1}{2}\alpha_n \sum_{j=1}^{N} \cos(nx_j)\cos(lx_j) = \sum_{j=1}^{N} f(x_j)\cos(lx_j).$$

From (4.22) and (4.24) the left-hand side reduces to the term $\tfrac{1}{2}N\alpha_l$, so that $\alpha_l = a_l^*$ holds for all $l = 0, 1, \ldots, n$. If we multiply the $j$th equation of (4.27) by $\sin(lx_j)$,

## 4.1 Fourier Series

$1 \leq l \leq n-1$, where $l$ is fixed, and sum all of the equations again, we obtain by a similar argument $\beta_l = b_l^*$, $l = 1, 2, \ldots, n-1$.

In addition to Theorem 4.5 it should be pointed out, that the function value $f(x_j)$ in the case of a discontinuity at $x_j$, must be defined by the arithmetic mean of the two limits $y_j^-$ and $y_j^+$ (4.2) to be in agreement with Theorem 4.2. If $N$ is odd, it is also possible to define an interpolating Fourier polynomial with the coefficients $a_k^*$ and $b_k^*$ (Young and Gregory, 1973).

*Example* 4.3 For the roof function $f(x)$, (4.13), for $N = 8$, $h = \pi/4$, $x_j = \pi j/4$, ($j = 1, 2, \ldots, 8$) we get the values

$$a_0^* = \pi \quad a_1^* \doteq -1.34076 \quad a_2^* = 0$$
$$a_3^* \doteq -0.230038 \quad a_4^* = 0 \quad b_1^* = b_2^* = b_3^* = 0.$$

The interpolating Fourier polynomial $g_4^*(x)$ is

$$g_4^*(x) \doteq 1.57080 - 1.34076 \cos(x) - 0.230038 \cos(3x)$$

and is shown in Figure 4.3. In comparison with Figure 4.1 the approximation of the roof function is different, of course, because of the interpolation property.

**Theorem 4.6** *Let $N = 2n$, $n \in \mathbf{N}$. The Fourier polynomial*

$$g_m^*(x) := \tfrac{1}{2}a_0^* + \sum_{k=1}^{m} [a_k^* \cos(kx) + b_k^* \sin(kx)] \tag{4.28}$$

*of degree $m < n$ with the coefficients (4.18) approximates the function $f(x)$ in such a way that the sum of squares of the differences*

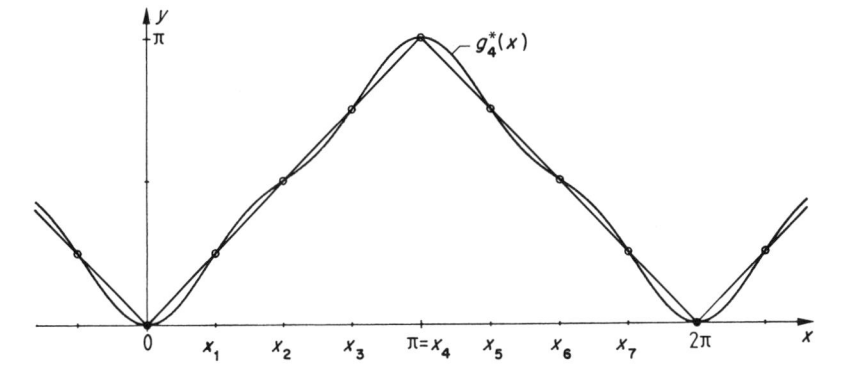

$$F := \sum_{j=1}^{N} [g_m^*(x_j) - f(x_j)]^2 \tag{4.29}$$

Figure 4.3 Roof function and the interpolating Fourier polynomial

is a minimum. This is called approximation in the sense of the discrete quadratic mean with respect to the $N$ support abscissae $x_j$ (4.17).

*Proof* We consider the Fourier polynomial

$$g_m(x) := \tfrac{1}{2}\alpha_0 + \sum_{k=1}^{m} [\alpha_k \cos(kx) + \beta_k \sin(kx)] \quad m < n = \frac{N}{2}, \quad (4.30)$$

whose coefficients $\alpha_0, \alpha_1, \ldots, \alpha_m, \beta_1, \ldots, \beta_m$ are to be determined such that the sum of the squares of the errors at the $N$ support abscissae $x_j$ attains a minimum.

$$F := \sum_{j=1}^{N} \left( \tfrac{1}{2}\alpha_0 + \sum_{k=1}^{m} [\alpha_k \cos(kx_j) + \beta_k \sin(kx_j)] - f(x_j) \right)^2 = \text{Min!} \quad (4.31)$$

A necessary condition for a minimum of $F$ is that the partial derivatives of $F$, with respect to the variables $\alpha_k$ and $\beta_k$, are equal to zero. The following conditions result if we take into account the orthogonality relationships (4.22) to (4.24).

$$\frac{\partial F}{\partial \alpha_0} = \sum_{j=1}^{N} \left( \tfrac{1}{2}\alpha_0 + \sum_{k=1}^{m} [\alpha_k \cos(kx_j) + \beta_k \sin(kx_j)] - f(x_j) \right)$$

$$= \frac{N}{2}\alpha_0 - \sum_{j=1}^{N} f(x_j) = 0 \Rightarrow \alpha_0 = \frac{2}{N} \sum_{j=1}^{N} f(x_j).$$

For $l = 1, 2, \ldots, m$ they are

$$\frac{\partial F}{\partial \alpha_l} = 2 \sum_{j=1}^{N} \left( \tfrac{1}{2}\alpha_0 + \sum_{k=1}^{m} [\alpha_k \cos(kx_j) + \beta_k \sin(kx_j)] - f(x_j) \right) \cos(lx_j)$$

$$= 2\left( \alpha_l \frac{N}{2} - \sum_{j=1}^{N} f(x_j) \cos(lx_j) \right) = 0 \Rightarrow \alpha_l = \frac{2}{N} \sum_{j=1}^{N} f(x_j) \cos(lx_j)$$

$$\frac{\partial F}{\partial \beta_l} = 2\left( \beta_l \frac{N}{2} - \sum_{j=1}^{N} f(x_j) \sin(lx_j) \right) = 0 \Rightarrow \beta_l = \frac{2}{N} \sum_{j=1}^{N} f(x_j) \sin(lx_j).$$

Consequently, we have in fact $\alpha_k = a_k^*$, $(k = 0, 1, \ldots, m)$ and $\beta_k = b_k^*$, $(k = 1, 2, \ldots, m)$ for the desired Fourier polynomial (4.30).

The statement of Theorem 4.6 means that the Fourier polynomial $g_m^*(x)$ (4.28) is the solution of an approximation problem of the least squares method (see Chapter 7) as it approximates the given periodic function $f(x)$ in such a way that the sum of the squares of the residuals $r_j := g_m^*(x_j) - f(x_j), j = 1, 2, \ldots, N$, attains a minimum. The solution of this problem can be given explicitly because of the discrete orthogonality properties.

*Example 4.4* For the $(2\pi)$ periodic continued function $f(x) = x^2, 0 < x < 2\pi$, for which we define the value $f(0) = f(2\pi) = 2\pi^2$ at the points of discontinuity, we

## 4.1 Fourier Series

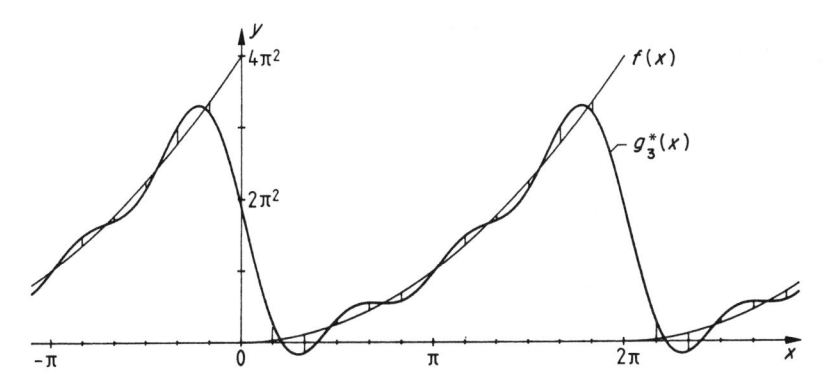

Figure 4.4 Approximation in the sense of the discrete quadratic mean

obtain for $N = 12$ the coefficients

$$a_0^* \doteq 26.410\,330 \quad a_1^* \doteq 4.092\,652 \quad a_2^* \doteq 1.096\,623 \quad a_3^* \doteq 0.548\,311,$$
$$b_1^* \doteq -12.277\,955 \quad b_2^* \doteq -5.698\,219 \quad b_3^* \doteq -3.289\,868.$$

The corresponding Fourier polynomial $g_3^*(x)$ is represented, together with $f(x)$, in Figure 4.4. Moreover, the residuals $r_j$ are shown.

The Fourier coefficients $a_k$ and $b_k$, (4.11), of a function $f(x)$ are approximated by the values $a_k^*$ and $b_k^*$, (4.18). The following theorem provides an error bound.

**Theorem 4.7** *Let $f(x)$ be a $(2\pi)$ periodic function satisfying the hypotheses of Theorem 4.2, and let $N = 2n$, $n \in \mathbf{N}$. If $f(x_j)$ is defined by the arithmetic mean of the two limits, (4.2), at a point of discontinuity, then the coefficients $a_k^*$ and $b_k^*$ obey the following relationships*

$$\boxed{\begin{aligned} a_k^* &= a_k + a_{N-k} + a_{N+k} + a_{2N-k} + a_{2N+k} + \cdots, \quad (k = 0, 1, \ldots, n) \\ b_k^* &= b_k - b_{N-k} + b_{N+k} - b_{2N-k} + b_{2N+k} + \cdots, \quad (k = 1, 2, \ldots, n-1) \end{aligned}} \quad \begin{aligned} (4.32) \\ (4.33) \end{aligned}$$

*Proof* From the assumptions on $f(x)$, the infinite Fourier series (4.12) converges to $f(x_j)$ at the support abscissae $x_j$. After substitution into (4.18) we obtain

$$a_k^* = \frac{2}{N} \sum_{j=1}^{N} \left( \tfrac{1}{2} a_0 + \sum_{l=1}^{\infty} [a_l \cos(lx_j) + b_l \sin(lx_j)] \right) \cos(kx_j)$$

$$= \frac{2}{N} \left[ \tfrac{1}{2} a_0 \sum_{j=1}^{N} \cos(kx_j) + \sum_{l=1}^{\infty} \left( a_l \sum_{j=1}^{N} \cos(kx_j) \cos(lx_j) \right.\right.$$

$$\left.\left. + b_l \sum_{j=1}^{N} \cos(kx_j) \sin(lx_j) \right) \right].$$

The discrete orthogonality relationships (4.22) and (4.24) give us three cases which have to be treated separately

$$k = 0 \quad a_0^* = \frac{2}{N}\left(\frac{N}{2}a_0 + N(a_N + a_{2N} + a_{3N} + \cdots)\right)$$

$$= a_0 + 2(a_N + a_{2N} + \cdots)$$

$$1 \leqslant k < n \quad a_k^* = \frac{2}{N}\left(\frac{N}{2}a_k + \frac{N}{2}(a_{N-k} + a_{N+k} + a_{2N-k} + a_{2N+k} + \cdots)\right)$$

$$= a_k + a_{N-k} + a_{N+k} + a_{2N-k} + a_{2N+k} + \cdots$$

$$k = n \quad a_n^* = \frac{2}{N}[N(a_n + a_{3n} + a_{5n} + \cdots)] = 2(a_n + a_{3n} + a_{5n} + \cdots)$$

The three results can be combined in the uniform relation (4.32). For $b_k^*$ we get

$$b_k^* = \frac{2}{N}\left[\tfrac{1}{2}a_0 \sum_{j=1}^{N} \sin(kx_j)\right.$$
$$\left. + \sum_{l=1}^{\infty}\left(a_l \sum_{j=1}^{N} \cos(lx_j)\sin(kx_j) + b_l \sum_{j=1}^{N} \sin(lx_j)\sin(kx_j)\right)\right]$$

$$= \frac{2}{N}\left(\frac{N}{2}(b_k - b_{N-k} + b_{N+k} - b_{2N-k} + b_{2N+k} - \cdots)\right)$$

which is the relationship (4.33).

By means of the relationships (4.32) and (4.33) for a given $N$ we can estimate the errors

$$|a_k^* - a_k| \leqslant \sum_{\mu=1}^{\infty}(|a_{\mu N-k}| + |a_{\mu N+k}|) \quad (0 \leqslant k \leqslant n) \tag{4.34}$$

$$|b_k^* - b_k| \leqslant \sum_{\mu=1}^{\infty}(|b_{\mu N-k}| + |b_{\mu N+k}|) \quad (1 \leqslant k \leqslant n-1) \tag{4.35}$$

if it is known how the Fourier coefficients $a_k$ and $b_k$ tend to zero for a given function $f(x)$. Conversely, on this assumption the value of $N$ can be estimated such that for a prescribed $\varepsilon > 0$ we have $|a_k^* - a_k| \leqslant \varepsilon$ and $|b_k^* - b_k| \leqslant \varepsilon$ for $k = 0, 1, \ldots, m < n$.

**Example 4.5** The Fourier coefficients of the roof function $f(x)$ (4.13) were found in Example 4.1 to be

$$a_0 = \pi \quad a_k = \begin{cases} -\dfrac{4}{\pi k^2} & \text{if } k \text{ odd} \\ 0 & \text{if } k > 0 \text{ even} \end{cases} \tag{4.36}$$

## 4.2 Evaluation of Fourier Coefficients

For $N = 8$ we have

$$a_0^* = a_0 + 2(a_8 + a_{16} + a_{24} + a_{32} + \cdots) = a_0,$$

$$a_1^* = a_1 + a_7 + a_9 + a_{15} + a_{17} + \cdots = a_1 - \frac{4}{\pi}\left(\frac{1}{49} + \frac{1}{81} + \frac{1}{225} + \frac{1}{289} + \cdots\right).$$

We want to estimate the value of $N$ such that the first few nonzero coefficients $a_k$ are approximated by the corresponding $a_k^*$ with a maximum absolute error of $\varepsilon = 10^{-6}$. For $k$ odd with $k \ll N$ we have

$$a_k^* - a_k = -\frac{4}{\pi}\left(\frac{1}{(N-k)^2} + \frac{1}{(N+k)^2} + \frac{1}{(2N-k)^2} + \frac{1}{(2N+k)^2} + \cdots\right)$$

$$= -\frac{4}{\pi}\left(\frac{2N^2 + 2k^2}{(N^2 - k^2)^2} + \frac{8N^2 + 2k^2}{(4N^2 - k^2)^2} + \cdots\right)$$

$$\approx -\frac{8}{\pi}\left(\frac{1}{N^2} + \frac{1}{(2N)^2} + \frac{1}{(3N)^2} + \cdots\right)$$

$$= -\frac{8}{\pi N^2}\left(\frac{1}{1^2} + \frac{1}{2^2} + \frac{1}{3^2} + \cdots\right) = -\frac{8}{\pi N^2} \cdot \frac{\pi^2}{6} = -\frac{4\pi}{3N^2}.$$

The sum of the reciprocals of the squares of integers is found from the Fourier series (4.16) if we set $x = 0$. From the condition $|a_k^* - a_k| \leq \varepsilon = 10^{-6}$ it follows that $N > 2046$.

### 4.2 Efficient Evaluation of Fourier Coefficients

We used the formulae (4.18) for the approximate calculation of the Fourier coefficients $a_k$ and $b_k$ of a $(2\pi)$ periodic function $f(x)$. The sums (4.18) also occur in connection with the *discrete Fourier transformation* which is important in physics, electrical engineering, statistics and image processing. In these applications $N$ is usually quite large, and it is very important that the sums are evaluated with the least amount of computational effort possible. The direct evaluation of the sums requires for the calculation of the $N = 2n$ coefficients, $a_k^*$, $k = 0, 1, \ldots, n$, and $b_k^*$, $k = 1, 2, \ldots, n-1$, about $N^2$ multiplications and $N^2$ evaluations of trigonometric functions. For large $N$, say $N \geq 1000$, the computational effort becomes prohibitively large. In the following we shall present two algorithms for an efficient performance of the discrete Fourier transformation. We also refer to the *Reinsch algorithm* (Sauer and Szabo, 1968; Stoer, 1983) which yields the numerically stable Fourier coefficients, somewhat less efficiently, but with smaller storage requirements.

#### 4.2.1 Runge's algorithm

In 1903 Runge (Runge, 1903, 1905; Runge and König, 1924) proposed a scheme for computing the sums

$$a'_k := \sum_{j=1}^{N} f(x_j) \cos(kx_j) \quad k = 0, 1, 2, \ldots, \tfrac{1}{2}N \tag{4.37}$$
$$b'_k := \sum_{j=1}^{N} f(x_j) \sin(kx_j) \quad k = 1, 2, \ldots, \tfrac{1}{2}N - 1$$

with $x_j = (2\pi/N)j$, on the assumption

$$N = 4m \quad m \in \mathbf{N} \tag{4.38}$$

that considerably reduces the number of operations by exploiting certain properties of the trigonometric functions involved. The idea is to consider two terms of the sums with the index values $j$ and $N - j$. The corresponding cos and sin values fulfil the following relations because of periodicity and symmetry:

$$\cos(kx_{N-j}) = \cos\left(k(N-j)\frac{2\pi}{N}\right) = \cos(2\pi k - kx_j) = \cos(kx_j) \tag{4.39}$$

$$\sin(kx_{N-j}) = \sin\left(k(N-j)\frac{2\pi}{N}\right) = \sin(2\pi k - kx_j) = -\sin(kx_j) \tag{4.40}$$

In the sums (4.37) certain pairs of terms can be combined, whereby we have to take into account that for $j = N/2 = 2m$ and $j = N = 4m$ the terms have no counterparts.

On the other hand we have $\sin(kx_{2m}) = \sin[k(2\pi/4m)2m] = \sin(k\pi) = 0$ and $\sin(kx_N) = \sin[k(2\pi/N)N] = \sin(2k\pi) = 0$. Hence, because of (4.39) and (4.40), from (4.37) we obtain

$$a'_k = \sum_{j=1}^{2m-1} [f(x_j) + f(x_{N-j})] \cos(kx_j)$$
$$+ f(x_{2m}) \cos(k\pi) + f(x_{4m}) \cos(2k\pi) \tag{4.41}$$
$$b'_k = \sum_{j=1}^{2m-1} [f(x_j) - f(x_{N-j})] \sin(kx_j).$$

We see that the number of operations can be halved by introducing sums and differences of the given function values $f(x_j)$. Therefore in the first step we define the following quantities, taking the periodicity of $f(x)$ into account

$$\begin{array}{l} s_0 = f(x_0) \quad s_{2m} = f(x_{2m}) \\ s_j = f(x_j) + f(x_{N-j}) \quad d_j = f(x_j) - f(x_{N-j}) \quad (j = 1, 2, \ldots, 2m-1) \end{array} \tag{4.42}$$

## 4.2 Evaluation of Fourier Coefficients

The process (4.42) is called a *first convolution* of the given data $f(x_j)$. The sums (4.37) have the new forms

$$a'_k = \sum_{j=0}^{2m} s_j \cos(kx_j) \quad b'_k = \sum_{j=1}^{2m-1} d_j \sin(kx_j). \tag{4.43}$$

Again we can collect pairs of terms in these sums with an odd number of terms, because the following relationships now hold

$$\cos(kx_{2m-j}) = \cos\left(k(2m-j)\frac{2\pi}{4m}\right) = \cos(k\pi - kx_j) = (-1)^k \cos(kx_j) \tag{4.44}$$

$$\sin(kx_{2m-j}) = \sin\left(k(2m-j)\frac{2\pi}{4m}\right) = \sin(k\pi - kx_j) = (-1)^{k+1} \sin(kx_j). \tag{4.45}$$

Due to (4.44) and (4.45) we have to distinguish between two cases for a further treatment of (4.43). For $k$ even, with $0 < k < 2m$, from (4.43) we get

$$\left. \begin{array}{l} a'_k = \sum_{j=0}^{m-1} (s_j + s_{2m-j}) \cos(kx_j) + s_m \cos(kx_m) \\ b'_k = \sum_{j=1}^{m-1} (d_j - d_{2m-j}) \sin(kx_j) + d_m \sin(kx_m) \end{array} \right\} \quad (k = 2, 4, \ldots, 2m-2) \tag{4.46}$$

For later use we note that $\sin(kx_m) = \sin[k(2\pi/4m)m] = \sin[k(\pi/2)] = 0$ for $k$ even. For odd $k$ with $1 \leq k \leq 2m-1$ we have

$$\left. \begin{array}{l} a'_k = \sum_{j=0}^{m-1} (s_j - s_{2m-j}) \cos(kx_j) + s_m \cos(kx_m) \\ b'_k = \sum_{j=1}^{m-1} (d_j + d_{2m-j}) \sin(kx_j) + d_m \sin(kx_m) \end{array} \right\} \quad (k = 1, 3, \ldots, 2m-1) \tag{4.47}$$

Here $\cos(kx_m) = 0$. Hence we form other sums and differences with the quantities $s_j$ and $d_j$, (4.42), by a *second convolution* according to

$$\boxed{\begin{array}{ll} ss_j = s_j + s_{2m-j} & ds_j = s_j - s_{2m-j} \quad (j = 0, 1, \ldots, m-1) \\ sd_j = d_j + d_{2m-j} & dd_j = d_j - d_{2m-j} \quad (j = 1, 2, \ldots, m-1) \\ ss_m = s_m & sd_m = d_m \end{array}} \tag{4.48}$$

The sums $a'_0$ and $a'_{2m}$ are special cases of (4.46), because $\cos(0) = 1$ and $\cos(2mx_j) = \cos(j\pi) = (-1)^j$, so that we can summarize the computation of the coefficients $a^*_k$ and $b^*_k$ as follows

$$a_0^* = \frac{2}{N} \sum_{j=0}^{m} ss_j \quad a_{2m}^* = \frac{2}{N} \sum_{j=0}^{m} (-1)^j ss_j$$

$$\left.\begin{aligned} a_k^* &= \frac{2}{N} \sum_{j=0}^{m} ss_j \cos(kx_j) \\ b_k^* &= \frac{2}{N} \sum_{j=1}^{m-1} dd_j \sin(kx_j) \end{aligned}\right\} \quad (k = 2, 4, \ldots, 2m-2)$$

$$\left.\begin{aligned} a_k^* &= \frac{2}{N} \sum_{j=0}^{m-1} ds_j \cos(kx_j) \\ b_k^* &= \frac{2}{N} \sum_{j=1}^{m} sd_j \sin(kx_j) \end{aligned}\right\} \quad (k = 1, 3, \ldots, 2m-1)$$

(4.49)

The *Runge algorithm* requires the two preliminary convolutions, (4.42) and (4.48), which are completed by about $N$ additions and subtractions each. In order to determine the amount of work needed to perform (4.49) we notice that because of the periodicity of $\cos(x)$ and $\sin(x)$ only the function values

$$\gamma_j := \cos(x_j) \quad \sigma_j := \sin(x_j) \quad (j = 0, 1, \ldots, N-1) \tag{4.50}$$

at the support abscissae $x_j$ are required, since all values $kx_j$ can be reduced to these. Therefore we assume that these $2N$ function values are available in tabulated form. Taking hypothesis (4.38) into account, the computation of this table requires only $m$ function values of one of the two trigonometric functions. The remaining entries of the table are obtained in a simple way. Finally, the number of multiplications to compute the $a_k^*$ and $b_k^*$ by means of the Runge scheme, (4.49), is

$$Z_{\text{Runge}} \simeq \tfrac{1}{4} N^2. \tag{4.51}$$

This effort corresponds to the performance of (4.49) on a computer. It can be reduced further for special values of $N$, if we take into account that only a few different cos and sin function values are involved, with some even equal to one or zero. Special schemes for computation by hand were developed for $N = 12, 24, 36$ and 72 which allowed a performance of the discrete Fourier transformation with an absolute minimum number of multiplications (Jordan-Engeln, 1982; Runge and König, 1924; Terebesi, 1930; Willers, 1957).

*Example* 4.6 For the $(2\pi)$ periodic continued function $f(x) = x^2, 0 < x < 2\pi$, with the points of discontinuity $x = 2\pi k$, $(k \in \mathbf{Z})$, the numerical values of Runge's algorithm are given in (4.52) for $N = 12$ to six significant figures.

4.2 Evaluation of Fourier Coefficients

| $f(x_j)=f_j$ | 1st convolution | 2nd convolution | $a_k^*, b_k^*$ |
|---|---|---|---|
| $f_0\ =19.7392$ | $s_0=\ \ \ \ 19.7392$ | $ss_0=\ \ \ \ 29.6088$ | $a_0^*=\ \ \ \ 26.4103$ |
| $f_1\ =\ \ 0.274156$ | $s_1=\ \ \ \ 33.4470$ | $ss_1=\ \ \ \ 53.7345$ | $a_2^*=\ \ \ \ \ \ 1.09662$ |
| $f_2\ =\ \ 1.09662$ | $s_2=\ \ \ \ 28.5122$ | $ss_2=\ \ \ \ 50.4447$ | $a_4^*=\ \ \ \ \ \ 0.365550$ |
| $f_3\ =\ \ 2.46740$ | $s_3=\ \ \ \ 24.6740$ | $ss_3=\ \ \ \ 24.6740$ | $a_6^*=\ \ \ \ \ \ 0.274167$ |
| $f_4\ =\ \ 4.38649$ | $s_4=\ \ \ \ 21.9325$ | $ds_0=\ \ \ \ \ \ 9.86960$ | $a_1^*=\ \ \ \ \ \ 4.09267$ |
| $f_5\ =\ \ 6.85389$ | $s_5=\ \ \ \ 20.2875$ | $ds_1=\ \ \ \ 13.1595$ | $a_3^*=\ \ \ \ \ \ 0.548317$ |
| $f_6\ =\ \ 9.86960$ | $s_6=\ \ \ \ \ \ 9.86960$ | $ds_2=\ \ \ \ \ \ 6.57970$ | $a_5^*=\ \ \ \ \ \ 0.293825$ |
| $f_7\ =13.4336$ | $d_1=-32.8986$ | $sd_1=-39.4783$ | $b_1^*=-12.2780$ |
| $f_8\ =17.5460$ | $d_2=-26.3190$ | $sd_2=-39.4785$ | $b_3^*=\ \ -3.28985$ |
| $f_9\ =22.2066$ | $d_3=-19.7392$ | $sd_3=-19.7392$ | $b_5^*=\ \ -0.881505$ |
| $f_{10}=27.4156$ | $d_4=-13.1595$ | $dd_1=-26.3189$ | $b_2^*=\ \ -5.69822$ |
| $f_{11}=33.1728$ | $d_5=\ \ -6.57971$ | $dd_2=-13.1595$ | $b_4^*=\ \ -1.89940$ |

(4.52)

The obvious question arises of whether it is possible to reduce the computational effort once more by an analogous continuation of the convolution process on the assumption $N=8m$ or even more generally $N=2^\gamma$. Runge noticed that trigonometric relationships like (4.44) and (4.45) do not admit further simplifications of the sums (4.49). However, the number of multiplications (4.51) can be halved again if convolutions with the Fourier coefficients are used (Zurmühl, 1965), but the resulting algorithm is not systematic enough.

### 4.2.2 The fast Fourier transform

An extremely efficient algorithm exists for the special case where $N$ is a power of 2, if we consider a complex Fourier transformation. In the following we study the evaluation of the trigonometric sums

$$a_k' := \sum_{j=0}^{N-1} f(x_j)\cos(kx_j) \quad \left(k=0,1,2,\ldots,\frac{N}{2}\right)$$

$$b_k' := \sum_{j=0}^{N-1} f(x_j)\sin(kx_j) \quad \left(k=1,2,\ldots,\frac{N}{2}-1\right)$$

(4.53)

where $x_j = 2\pi j/n$. The periodicity now causes the summation index to run from 0 to $N-1$ in (4.53). Two subsequent function values are taken together to define the following $n = N/2$ complex numbers

$$y_j := f(x_{2j}) + if(x_{2j+1}) \quad (j = 0, 1, \ldots, n-1) \quad n = \frac{N}{2} \tag{4.54}$$

For these complex values $y_j$ we define the *discrete, complex Fourier transformation* of order $n$

$$c_k := \sum_{j=0}^{n-1} y_j \exp(-ijk(2\pi/n)) = \sum_{j=0}^{n-1} y_j w_n^{jk} \quad (k = 0, 1, \ldots, n-1)$$

with $\tag{4.55}$

$$w_n := \exp(-i(2\pi/n)) = \cos\left(\frac{2\pi}{n}\right) - i \sin\left(\frac{2\pi}{n}\right).$$

The complex quantity $w_n$ is equal to an $n$th root of unity. The connection between the complex Fourier transforms $c_k$ and our desired real quantities $a'_k$ and $b'_k$ is given by the following theorem.

**Theorem 4.8** *The real valued trigonometric sums $a'_k$ and $b'_k$ are related to the complex Fourier transforms $c_k$ by the relationships*

$$a'_k - ib'_k = \frac{1}{2}(c_k + \bar{c}_{n-k}) + \frac{1}{2i}(c_k - \bar{c}_{n-k})\exp(-ik\pi/n) \tag{4.56}$$

$$a'_{n-k} - ib'_{n-k} = \frac{1}{2}(\bar{c}_k + c_{n-k}) + \frac{1}{2i}(\bar{c}_k - c_{n-k})\exp(ik\pi/n) \tag{4.57}$$

*for $k = 0, 1, \ldots, n$, defining $b'_0 = b'_n = 0$ and $c_n = c_0$.*

**Proof** The first term of the sum (4.56) is equal to

$$\tfrac{1}{2}(c_k + \bar{c}_{n-k}) = \tfrac{1}{2}\sum_{j=0}^{n-1}(y_j w_n^{jk} + \overline{\bar{y}_j w_n^{j(n-k)}}) = \tfrac{1}{2}\sum_{j=0}^{n-1}(y_j + \bar{y}_j)w_n^{jk}$$

and the expression in the parentheses of the second term is

$$\frac{1}{2i}(c_k - \bar{c}_{n-k}) = \frac{1}{2i}\sum_{j=0}^{n-1}(y_j w_n^{jk} - \overline{\bar{y}_j w_n^{j(n-k)}}) = \frac{1}{2i}\sum_{j=0}^{n-1}(y_j - \bar{y}_j)w_n^{jk}.$$

If we use definition (4.54) we obtain

$$\frac{1}{2}(c_k + \bar{c}_{n-k}) + \frac{1}{2i}(c_k - \bar{c}_{n-k})\exp(-ik\pi/n)$$

## 4.2 Evaluation of Fourier Coefficients

$$= \sum_{j=0}^{n-1} \{f(x_{2j})\exp(-ijk2\pi/n) + f(x_{2j+1})\exp[-ik(2j+1)\pi/n]$$

$$= \sum_{j=0}^{n-1} \{f(x_{2j})[\cos(kx_{2j}) - i\sin(kx_{2j})]$$

$$+ f(x_{2j+1})[\cos(kx_{2j+1}) - i\sin(kx_{2j+1})]\} = a'_k - ib'_k.$$

This verifies (4.56). Equation (4.57) can be shown by substituting $n-k$ for $k$.

For an efficient computation of the values $a'_k$ and $b'_k$ for fixed index $k$, simple relationships for the sums and differences of the pairs $a'_k, a'_{n-k}, b'_k$ and $b'_{n-k}$ respectively, can be derived by adding and subtracting equations (4.56) and (4.57). Only eight real essential operations are required to compute the four Fourier coefficients $a^*_k, a^*_{n-k}, b^*_k, b^*_{n-k}$, (4.18), from two complex Fourier transforms $c_k$ and $c_{n-k}$, (Sauer and Szabo, 1968; Stoer, 1983).

Next we wish to present the fundamental idea of an efficient algorithm for performing the discrete Fourier transformation (4.55). As $w_n$ is an $n$th root of unity, the powers $w_n^{jk}$ lie on the unit circle of the complex plane and form the $n$ corners of a regular polygon. The exponents can be reduced to mod $n$. After these preliminaries we consider the case $n = 4$. The transformation (4.55) is a linear mapping with a matrix $\mathbf{W}_4 \in \mathbf{C}^{4\times 4}$.

$$\begin{pmatrix} c_0 \\ c_1 \\ c_2 \\ c_3 \end{pmatrix} = \begin{pmatrix} 1 & 1 & 1 & 1 \\ 1 & w^1 & w^2 & w^3 \\ 1 & w^2 & 1 & w^2 \\ 1 & w^3 & w^2 & w^1 \end{pmatrix} \begin{pmatrix} y_0 \\ y_1 \\ y_2 \\ y_3 \end{pmatrix} \quad w = w_4 \quad \mathbf{c} = \mathbf{W}_4 \mathbf{y} \tag{4.58}$$

In (4.58) we permute the second and third components of the vector $\mathbf{c}$ and the corresponding rows in $\mathbf{W}_4$. The resulting matrix can clearly be factorized into the product of two special matrices, and from (4.58) we find

$$\begin{pmatrix} \tilde{c}_0 \\ \tilde{c}_1 \\ \tilde{c}_2 \\ \tilde{c}_3 \end{pmatrix} = \begin{pmatrix} c_0 \\ c_2 \\ c_1 \\ c_3 \end{pmatrix} = \left(\begin{array}{cc|cc} 1 & 1 & 1 & 1 \\ 1 & w^2 & 1 & w^2 \\ \hline 1 & w & w^2 & w^3 \\ 1 & w^3 & w^2 & w^1 \end{array}\right) \begin{pmatrix} y_0 \\ y_1 \\ y_2 \\ y_3 \end{pmatrix}$$

$$= \left(\begin{array}{cc|cc} 1 & 1 & 0 & 0 \\ 1 & w^2 & 0 & 0 \\ \hline 0 & 0 & 1 & 1 \\ 0 & 0 & 1 & w^2 \end{array}\right) \left(\begin{array}{cc|cc} 1 & 0 & 1 & 0 \\ 0 & 1 & 0 & 1 \\ \hline 1 & 0 & w^2 & 0 \\ 0 & w^1 & 0 & w^3 \end{array}\right) \begin{pmatrix} y_0 \\ y_1 \\ y_2 \\ y_3 \end{pmatrix}. \tag{4.59}$$

The first matrix of the factorization, subdivided into four submatrices of order two, is block diagonal, the two matrices on the diagonal being identical. The second matrix consists of four diagonal matrices of order two, among which the unit matrix appears twice. According to (4.59) we execute the linear transformation in two steps. First we multiply the vector y by the second matrix of the factorization in order to get the vector z whose components are given, if we take into account $w^2 = -1$, $w^3 = -w^1$, by

$$z_0 = y_0 + y_2 \qquad z_1 = y_1 + y_3 \tag{4.60}$$

$$z_2 = (y_0 - y_2)w^0 \qquad z_3 = (y_1 - y_3)w^1. \tag{4.61}$$

In view of the subsequent generalization, we write in (4.61) the trivial multiplication by $w^0 = 1$. In the second step we multiply the auxiliary vector z by the first matrix of the factorization to get

$$\tilde{c}_0 = c_0 = z_0 + z_1 \qquad \tilde{c}_1 = c_2 = z_0 + w^2 z_1 \tag{4.62}$$

$$\tilde{c}_2 = c_1 = z_2 + z_3 \qquad \tilde{c}_3 = c_3 = z_2 + w^2 z_3. \tag{4.63}$$

The sets of formulae (4.62) and (4.63) are identical. In view of the subsequent generalization, the trivial multiplication by $w^2 = -1$ is explicitly written in these equations. If we take into account $w_4^2 = w_2^1$ we notice that (4.62) and (4.63) are each complex Fourier transformations of order two. Thus the given Fourier transformation (4.58) of order four has been reduced by means of (4.60) and (4.61) to two Fourier transformations of order two.

The reduction of a complex Fourier transformation of an even order to two Fourier transformations each of half order is always feasible. This can be shown by analogous factorization of the row permuted matrix. In its place, we use an algebraic approach. Let $n = 2m$, $m \in \mathbf{N}$. Then for the complex Fourier transforms $c_k$ (4.55) with even indices $k = 2l$, $(l = 0, 1, \ldots, m-1)$ we have

$$c_{2l} = \sum_{j=0}^{2m-1} y_j w_n^{2lj} = \sum_{j=0}^{m-1} (y_j + y_{m+j}) w_n^{2lj} = \sum_{j=0}^{m-1} (y_j + y_{m+j})(w_n^2)^{lj}.$$

We have used the identity $w_n^{2l(m+j)} = w_n^{2lj} w_n^{2lm} = w_n^{2lj}$. With the $m$ auxiliary values

$$\boxed{z_j := y_j + y_{m+j} \quad (j = 0, 1, \ldots, m-1)} \tag{4.64}$$

and because $w_n^2 = w_m$ the coefficients

$$\boxed{c_{2l} = \sum_{j=0}^{m-1} z_j w_m^{jl} \quad (l = 0, 1, \ldots, m-1)} \tag{4.65}$$

## 4.2 Evaluation of Fourier Coefficients

are the Fourier transforms of order $m$ of the auxiliary values $z_j$ (4.64). For $c_k$ with odd indices $k = 2l+1$, $(l = 0, 1, \ldots, m-1)$ we have

$$c_{2l+1} = \sum_{j=0}^{2m-1} y_j w_n^{(2l+1)j} = \sum_{j=0}^{m-1} (y_j w_n^{(2l+1)j} + y_{m+j} w_n^{(2l+1)(m+j)})$$

$$= \sum_{j=0}^{m-1} (y_j - y_{m+j}) w_n^{(2l+1)j} = \sum_{j=0}^{m-1} [(y_j - y_{m+j}) w_n^j] w_n^{2lj}.$$

With the additional $m$ auxiliary values

$$z_{m+j} := (y_j - y_{m+j}) w_n^j \quad (j = 0, 1, \ldots, m-1) \tag{4.66}$$

the $m$ coefficients

$$c_{2l+1} = \sum_{j=0}^{m-1} z_{m+j} w_m^{jl} \quad (l = 0, 1, \ldots, m-1) \tag{4.67}$$

are the Fourier transforms of order $m$ of the auxiliary values $z_{m+j}$ (4.66).

The reduction of a complex Fourier transformation of the order $n = 2m$ to two complex Fourier transformations of the order $m$ each requires, because of (4.66), $m$ complex multiplications as the amount of essential effort. If the order $n$ is a power of two, $n = 2^\gamma$, $\gamma \in \mathbf{N}$, the two Fourier transformations of order $m$ can be reduced again to two Fourier transformations each of half order, so that a systematical reduction is possible. In the case $n = 32 = 2^5$ this process can be formally described as follows

$$\text{FT}_{32} \xrightarrow{16} 2(\text{FT}_{16}) \xrightarrow{2 \times 8} 4(\text{FT}_8) \xrightarrow{4 \times 4} 8(\text{FT}_4) \xrightarrow{8 \times 2} 16(\text{FT}_2) \xrightarrow{16 \times 1} 32(\text{FT}_1) \tag{4.68}$$

where $\text{FT}_k$ denotes a Fourier transformation of order $k$, and where the number of complex multiplications of the corresponding reduction steps are indicated. As the Fourier transforms of order one are identical to the quantities to be transformed, the values that are obtained after the last reduction step are the desired Fourier transforms $c_k$.

Hence a complex Fourier transformation (4.55) of order $n = 2^\gamma$, $\gamma \in \mathbf{N}$ can be reduced to $n$ Fourier transformations of order one by $\gamma$ reduction steps. Each reduction step obviously requires $\frac{1}{2}n$ complex multiplications, and therefore the total number of complex multiplications is given by

$$Z_{\text{FT}_n} = \tfrac{1}{2} n \gamma = \tfrac{1}{2} n \log_2 n \tag{4.69}$$

The amount of computational effort required increases only about linearly with the order $n$, and therefore the sketched method is called the *fast Fourier transformation* (FFT). It is attributed to Cooley and Tukey (1965), but had been previously proposed by Good (1960). The high efficiency of the fast Fourier transformation is illustrated in (4.70), where for several orders $n$ the number $n^2$ of complex multiplications required by the direct evaluation of the sums (4.55) is compared with the effort $Z_{FT_n}$ (4.69).

| $\gamma$ | 5 | 6 | 8 | 9 | 10 | 11 | 12 |
|---|---|---|---|---|---|---|---|
| $n$ | 32 | 64 | 256 | 512 | 1024 | 2048 | 4096 |
| $n^2$ | 1024 | 4096 | 65536 | $2.62 \times 10^5$ | $1.05 \times 10^6$ | $4.19 \times 10^6$ | $1.68 \times 10^7$ |
| $Z_{FT_n}$ | 80 | 192 | 1024 | 2304 | 5120 | 11264 | 24576 |
| factor | 12.8 | 21.3 | 64 | 114 | 205 | 372 | 683 |

(4.70)

The amount of computational effort required by the FFT is reduced by the factor in the last row in (4.70). The FFT is also much better than the Runge's algorithm for evaluating the real trigonometric sums (4.53). As $N = 2n$, the number of *real* multiplications, due to (4.51), is about $\frac{1}{4}N^2 = n^2$, so that the factor of reduction for the FFT is still quite essential.

A possible algorithmic implementation of the fast Fourier transformation on a computer, following the previous considerations, transforms the given $y$ values successively into the desired $c$ values. The corresponding steps are outlined in Table 4.1 for the case $n = 16 = 2^4$, where $w := \exp(-i2\pi/16) = \exp(-i\pi/8) = \cos(\pi/8) - i\sin(\pi/8)$. The auxiliary values $z_j$, (4.64), and $z_{m+j}$, (4.66), are again denoted by $y_i$ and $y_{m+j}$. The assignment statements are to be understood in the sense that the quantities to the right always mean the values *before* the corresponding step.

In the first step sums and differences must be simultaneously computed according to (4.64) and (4.66), whereby the differences have to be multiplied by powers of $w$. The first group of eight $y$ values will produce the Fourier transforms $c_k$ with even indices, and the second group the other $c_k$. The corresponding $c$ values are given to clarify the process. In the second step for each of the two groups we correspondingly have to compute sums and differences where, due to $w_8 = w_{16}^2 = w^2$, only even powers of $w$ are the multipliers. Moreover, the Fourier transforms $c_k$ have to be permuted accordingly within each group. In the third step the four groups are treated analogously with the root of unity $w_4 = w_{16}^4 = w^4$ and the Fourier transforms $c_k$ are permuted according to (4.59). The fourth and final step executes the final reduction with $w_2 = w_{16}^8 = w^8$.

The resulting $y$ values are essentially the desired $c$ values. However, we have to solve the problem of the correct assignment. The key to the problem is found by

## 4.2 Evaluation of Fourier Coefficients

**Table 4.1** An algorithmic implementation of the FFT

| | ① →2FT$_8$ | | ② →4FT$_4$ | | ③ →8FT$_2$ | | ④ →16FT$_1$ |
|---|---|---|---|---|---|---|---|
| $y_0$ | $y_0 := y_0 + y_8$ | $c_0$ | $y_0 := y_0 + y_4$ | $c_0$ | $y_0 := y_0 + y_2$ | $c_0$ | $y_0 := y_0 + y_1 = c_0$ |
| $y_1$ | $y_1 := y_1 + y_9$ | $c_2$ | $y_1 := y_1 + y_5$ | $c_4$ | $y_1 := y_1 + y_3$ | $c_8$ | $y_1 := (y_0 - y_1)w^0 = c_8$ |
| $y_2$ | $y_2 := y_2 + y_{10}$ | $c_4$ | $y_2 := y_2 + y_6$ | $c_8$ | $y_2 := (y_0 - y_2)w^0$ | $c_4$ | $y_2 := y_2 + y_3 = c_4$ |
| $y_3$ | $y_3 := y_3 + y_{11}$ | $c_6$ | $y_3 := y_3 + y_7$ | $c_{12}$ | $y_3 := (y_1 - y_3)w^4$ | $c_{12}$ | $y_3 := (y_2 - y_3)w^0 = c_{12}$ |
| $y_4$ | $y_4 := y_4 + y_{12}$ | $c_8$ | $y_4 := (y_0 - y_4)w^0$ | $c_2$ | $y_4 := y_4 + y_6$ | $c_2$ | $y_4 := y_4 + y_5 = c_2$ |
| $y_5$ | $y_5 := y_5 + y_{13}$ | $c_{10}$ | $y_5 := (y_1 - y_5)w^2$ | $c_6$ | $y_5 := y_5 + y_7$ | $c_{10}$ | $y_5 := (y_4 - y_5)w^0 = c_{10}$ |
| $y_6$ | $y_6 := y_6 + y_{14}$ | $c_{12}$ | $y_6 := (y_2 - y_6)w^4$ | $c_{10}$ | $y_6 := (y_4 - y_6)w^0$ | $c_6$ | $y_6 := y_6 + y_7 = c_6$ |
| $y_7$ | $y_7 := y_7 + y_{15}$ | $c_{14}$ | $y_7 := (y_3 - y_7)w^6$ | $c_{14}$ | $y_7 := (y_5 - y_7)w^4$ | $c_{14}$ | $y_7 := (y_6 - y_7)w^0 = c_{14}$ |
| $y_8$ | $y_8 := (y_0 - y_8)w^0$ | $c_1$ | $y_8 := y_8 + y_{12}$ | $c_1$ | $y_8 := y_8 + y_{10}$ | $c_1$ | $y_8 := y_8 + y_9 = c_1$ |
| $y_9$ | $y_9 := (y_1 - y_9)w^1$ | $c_3$ | $y_9 := y_9 + y_{13}$ | $c_5$ | $y_9 := y_9 + y_{11}$ | $c_9$ | $y_9 := (y_8 - y_9)w^0 = c_9$ |
| $y_{10}$ | $y_{10} := (y_2 - y_{10})w^2$ | $c_5$ | $y_{10} := y_{10} + y_{14}$ | $c_9$ | $y_{10} := (y_8 - y_{10})w^0$ | $c_5$ | $y_{10} := y_{10} + y_{11} = c_5$ |
| $y_{11}$ | $y_{11} := (y_3 - y_{11})w^3$ | $c_7$ | $y_{11} := y_{11} + y_{15}$ | $c_{13}$ | $y_{11} := (y_9 - y_{11})w^4$ | $c_{13}$ | $y_{11} := (y_{10} - y_{11})w^0 = c_{13}$ |
| $y_{12}$ | $y_{12} := (y_4 - y_{12})w^4$ | $c_9$ | $y_{12} := (y_8 - y_{12})w^0$ | $c_3$ | $y_{12} := y_{12} + y_{14}$ | $c_3$ | $y_{12} := y_{12} + y_{13} = c_3$ |
| $y_{13}$ | $y_{13} := (y_5 - y_{13})w^5$ | $c_{11}$ | $y_{13} := (y_9 - y_{13})w^2$ | $c_7$ | $y_{13} := y_{13} + y_{15}$ | $c_{11}$ | $y_{13} := (y_{12} - y_{13})w^0 = c_{11}$ |
| $y_{14}$ | $y_{14} := (y_6 - y_{14})w^6$ | $c_{13}$ | $y_{14} := (y_{10} - y_{14})w^4$ | $c_{11}$ | $y_{14} := (y_{12} - y_{14})w^0$ | $c_7$ | $y_{14} := y_{14} + y_{15} = c_7$ |
| $y_{15}$ | $y_{15} := (y_7 - y_{15})w^7$ | $c_{15}$ | $y_{15} := (y_{11} - y_{15})w^6$ | $c_{15}$ | $y_{15} := (y_{13} - y_{15})w^4$ | $c_{15}$ | $y_{15} := (y_{14} - y_{15})w^0 = c_{15}$ |

Table 4.2 Binary representations of the indices

| $j$ | $y_i$ | $c_k^{(1)}$ | $c_k^{(2)}$ | $c_k^{(3)}$ | $c_k^{(4)}$ | $k$ |
|---|---|---|---|---|---|---|
| 0  | 0000 | 0000 | 0000 | 0000 | 0000 | 0  |
| 1  | 000L | 00L0 | 0L00 | L000 | L000 | 8  |
| 2  | 00L0 | 0L00 | L000 | 0L00 | 0L00 | 4  |
| 3  | 00LL | 0LL0 | LL00 | LL00 | LL00 | 12 |
| 4  | 0L00 | L000 | 00L0 | 00L0 | 00L0 | 2  |
| 5  | 0L0L | L0L0 | 0LL0 | L0L0 | L0L0 | 10 |
| 6  | 0LL0 | LL00 | L0L0 | 0LL0 | 0LL0 | 6  |
| 7  | 0LLL | LLL0 | LLL0 | LLL0 | LLL0 | 14 |
| 8  | L000 | 000L | 000L | 000L | 000L | 1  |
| 9  | L00L | 00LL | 0L0L | L00L | L00L | 9  |
| 10 | L0L0 | 0L0L | L00L | 0L0L | 0L0L | 5  |
| 11 | L0LL | 0LLL | LL0L | LL0L | LL0L | 13 |
| 12 | LL00 | L00L | 00LL | 00LL | 00LL | 3  |
| 13 | LL0L | L0LL | 0LLL | L0LL | L0LL | 11 |
| 14 | LLL0 | LL0L | L0LL | 0LLL | 0LLL | 7  |
| 15 | LLLL | LLLL | LLLL | LLLL | LLLL | 15 |

studying the binary representation of the indices. In Table 4.2 the binary representation of the indices of the $c$ values, after the reduction steps, are given.

From Table 4.2 we gather that the binary representations in the last column are the reverse of those in the first column. A thorough proof of this observation is based on the fact that the permutation of the indices of the first step corresponds to a cyclic permutation of the $\gamma$ binary digits. In the second step the permutation of the indices within each group is equivalent to a cyclic permutation of the first $(\gamma - 1)$ binary digits, because now the last binary digit is fixed. The argument continues in a similar manner. Therefore the unique assignment of the resulting $y_j$ to the desired $c_k$ is achieved by means of a *bit reversal* of the binary representation of $j$. It can easily be seen that it is sufficient to exchange the values $y_j$ and $y_k$ only if $k > j$ in order to obtain the correctly ordered $c_k$.

A computer program of the above algorithm of the FFT is given by Schwarz (1977). Different implementations can be found, for example, in Brigham (1974), Cooley and Tukey (1965), Sauer and Szabo (1968) and Singleton (1968). The process of bit reversal can be eliminated by a different organization of the steps, but the corresponding implementation requires additional storage space for $n$ complex values (G-AE, 1967).

*Example* 4.7 The fast Fourier transformation is applied to compute the approximate Fourier coefficients $a_k^*$ and $b_k^*$ of the $(2\pi)$ periodic function $f(x) = x^2$, $0 < x < 2\pi$ of Example 4.2. For $N = 16$ support abscissae a FFT of order $n = 8$ has to be performed. We denote by $y_j^{(0)}$ the complex values that are built from the given

## 4.2 Evaluation of Fourier Coefficients

Table 4.3 Fast Fourier transformation

| $j$ | $y_j^{(0)}$ | $y_j^{(1)}$ | $y_j^{(2)}$ |
|---|---|---|---|
| 0 | 19.739 21 + 0.154 21i | 29.608 81 + 12.645 43i | 54.282 82 + 42.562 67i |
| 1 | 0.616 85 + 1.387 91i | 16.038 11 + 20.047 63i | 51.815 42 + 62.301 88i |
| 2 | 2.467 40 + 3.855 31i | 24.674 01 + 29.917 24i | 4.934 80 − 17.271 81i |
| 3 | 5.551 65 + 7.556 42i | 35.777 32 + 42.254 24i | −22.206 61 + 19.739 21i |
| 4 | 9.869 60 + 12.491 22i | 9.869 60 − 12.337 01i | −12.337 01 + 7.402 20i |
| 5 | 15.421 26 + 18.659 72i | −22.681 31 − 1.744 72i | −24.426 02 + 34.894 32i |
| 6 | 22.206 61 + 26.061 92i | −22.206 61 + 19.739 21i | 32.076 21 − 32.076 21i |
| 7 | 30.225 66 + 34.697 83i | −1.744 72 + 36.639 04i | −38.383 75 + 20.936 59i |

| $j, k$ | $y_j^{(3)}$ | $c_k$ | $a_k^*$ | $b_k^*$ |
|---|---|---|---|---|
| 0 | 106.098 25 + 104.864 55i | 106.098 25 + 104.864 55i | 26.370 349 | — |
| 1 | 2.467 40 − 19.739 21i | −36.763 03 + 42.296 52i | 4.051 803 | −12.404 463 |
| 2 | −17.271 81 + 2.467 40i | −17.271 81 + 2.467 40i | 1.053 029 | −5.956 833 |
| 3 | 27.141 41 − 37.011 02i | −6.307 54 − 11.139 62i | 0.499 622 | −3.692 727 |
| 4 | −36.763 03 + 42.296 52i | 2.467 40 − 19.739 21i | 0.308 425 | −2.467 401 |
| 5 | 12.089 02 − 27.492 12i | 12.089 02 − 27.492 12i | 0.223 063 | −1.648 665 |
| 6 | −6.307 54 − 11.139 62i | 27.141 41 − 37.011 02i | 0.180 671 | −1.022 031 |
| 7 | 70.459 97 − 53.012 81i | 70.459 97 − 53.012 81i | 0.160 314 | −0.490 797 |
| 8 | — | — | 0.154 213 | — |

real function values $f(x_l)$. Table 4.3 gives the rounded values of the steps of the FFT and the Fourier coefficients $a_k^*$ and $b_k^*$ resulting from the Fourier transforms $c_k$.

An efficient performance of the complex Fourier transformation (4.55) is also possible for a more general order $n$ (Boor, 1980; Brigham, 1974; Winograd, 1978). Let $n = p \cdot m$ be the product of a prime number $p$ and an integer $m$. In this case we collect the Fourier transforms $c_k$ whose indices are $k \equiv \mu \pmod{p}$ for a fixed $\mu = 0, 1, \ldots, p − 1$. With $k = lp + \mu$, $(l = 0, 1, \ldots, m − 1)$ we can write

$$c_{lp+\mu} = \sum_{j=0}^{pm-1} y_j w_n^{j(lp+\mu)} = \sum_{j=0}^{m-1} \left( \sum_{\sigma=0}^{p-1} y_{j+\sigma m} w_n^{(j+\sigma m)(lp+\mu)} \right). \quad (4.71)$$

For the powers of $w_n$ in (4.71) we have

$$w_n^{(j+\sigma m)(lp+\mu)} = w_n^{jlp} w_n^{\sigma lmp} w_n^{(j+\sigma m)\mu} = w_n^{(j+\sigma m)\mu} (w_n^p)^{jl}$$

and hence from (4.71) it follows that

$$c_{lp+\mu} = \sum_{j=0}^{m-1} \left( \sum_{\sigma=0}^{p-1} y_{j+\sigma m} w_n^{(j+\sigma m)\mu} \right) w_m^{jl} \quad (l = 0, 1, \ldots, m − 1). \quad (4.72)$$

Now for each $\mu \in \{0, 1, \ldots, p-1\}$ we introduce the $m$ auxiliary values

$$z_{j+m\mu} := \sum_{\sigma=0}^{p-1} y_{j+\sigma m} w_n^{(j+\sigma m)\mu} \quad (j = 0, 1, \ldots, m-1), \tag{4.73}$$

so that (4.72) implies

$$c_{lp+\mu} = \sum_{j=0}^{m-1} z_{j+m\mu} w_m^{jl} \quad (l = 0, 1, \ldots, m-1). \tag{4.74}$$

From (4.74) the $m$ values $c_k$ with the indices $k = lp + \mu$ are the Fourier transforms of order $m$ of the auxiliary values $z_{j+m\mu}$, for each fixed $\mu \in \{0, 1, \ldots, p-1\}$. Conversely to (4.73) we now consider the auxiliary values $z_{j+m\mu}$ for fixed $j$ and varying $\mu = 0, 1, \ldots, p-1$. Then (4.73) has the new representation

$$z_{j+m\mu} = \left( \sum_{\sigma=0}^{p-1} y_{j+\sigma m} w_n^{\sigma m \mu} \right) w_n^{j\mu} = \left( \sum_{\sigma=0}^{p-1} y_{j+\sigma m} w_p^{\sigma \mu} \right) w_n^{j\mu} \quad (\mu = 0, 1, \ldots, p-1). \tag{4.75}$$

In the parentheses of (4.75) we recognize the formulae of a Fourier transformation of order $p$ for all the $p$ tuples of values $y_j, y_{j+m}, \ldots, y_{j+(p-1)m}$. Hence the $p$ values $z_j$, $z_{j+m}, \ldots, z_{j+(p-1)m}$ are essentially, disregarding the multiplier $w_n^{j\mu}$, defined by a Fourier transformation of order $p$.

Consequently, a complex Fourier transformation of order $n = p \cdot m$ can be reduced to $p$ Fourier transformations of order $m$, (4.74), which have to be applied to the $m$-tuples of auxiliary values $z_{j+m\mu}$ (4.73). These values in their turn are obtained according to (4.75) as the result of $m$ Fourier transformations of order $p$. If we take into account that in (4.75) for $\mu = 0$ we only have to compute a sum for $z_j$, and that for $\sigma = 0$ the multiplier of the first term is equal to one, we see that such a reduction step requires $p(p-1)m$ complex multiplications. Therefore the amount of computational effort $Z_n$ of a Fourier transformation of the order $n = p \cdot m$ is given by

$$Z_n = Z_{p \cdot m} = p \cdot Z_m + p(p-1)m \quad p > 2. \tag{4.76}$$

If $m$ is not prime, the Fourier transformations of order $m$ can again be reduced to one of smaller order. If the factorization into primes is

$$n = p_1^{q_1} \cdot p_2^{q_2} \cdots p_r^{q_r} \quad p_i \text{ prime} \quad q_i \in \mathbf{N} \tag{4.77}$$

## 4.3 Orthogonal Polynomials

Table 4.4 Multiplications of a FFT of general order

| Step | $p$ | Complex multiplications | Number of FT | Order of the FT |
|---|---|---|---|---|
| 1 | 2 | $1 \times 1 \times 180 = 180$ | 2 | 180 |
| 2 | 2 | $2 \times 1 \times 90 = 180$ | 4 | 90 |
| 3 | 2 | $4 \times 1 \times 45 = 180$ | 8 | 45 |
| 4 | 3 | $8 \times 6 \times 15 = 720$ | 24 | 15 |
| 5 | 3 | $24 \times 6 \times 5 = 720$ | 72 | 5 |
| 6 | 5 | $72 \times 20 \times 1 = 1440$ | 360 | 1 |
|   |   | Total: 3420 |   |   |

then the systematic reduction to Fourier transformations of order one is achieved after $q_1 + q_2 + \cdots + q_r$ steps.

*Example* 4.8 We consider a complex Fourier transformation of order $n = 360 = 2^3 \times 3^2 \times 5$. In Table 4.4 the number of the complex multiplications of the reduction steps are collected together. In comparison with the direct evaluation of the sums requiring $n^2 = 360^2 = 129\,600$ multiplications, the fast Fourier transformation is about 28 times more efficient.

It is not necessary to factorize the given order $n$ completely into its primes. The reduction of the order can be done using an arbitrary divisor of $n$, thus many variants arise (Brigham, 1974). For example, the amount of work is further reduced if in place of $p = 2$ the reduction step can be done with 4, 8 or 16 and the very special values of the powers of $w$ involved are taken into account. For computer programs of the general fast Fourier transformation, see Boor (1980) and Singleton (1968).

### 4.3 Orthogonal Polynomials

In the following we consider two systems of orthogonal polynomials which are of importance for certain applications. We build up the fundamental properties, that will be needed later. The special properties of the *Chebyshev polynomials* make them suitable for the approximation of functions, whereby a relationship to a Fourier series will be recognized. The *Legendre polynomials* will mainly be used to derive a certain class of quadrature rules in Chapter 8.

#### 4.3.1 The Chebyshev polynomials

To define the Chebyshev polynomials we start with the trigonometric identity

$$\cos[(n+1)\varphi] + \cos[(n-1)\varphi] = 2\cos(\varphi)\cos(n\varphi), \quad n \in \mathbf{N}. \tag{4.78}$$

We verify the relationship (4.78) by means of $\cos(\varphi) = \frac{1}{2}(e^{i\varphi} - e^{-i\varphi})$:

$$\cos[(n+1)\varphi] + \cos[(n-1)\varphi] = \tfrac{1}{2}\{\exp[i(n+1)\varphi]$$
$$+ \exp[-i(n+1)\varphi] + \exp[i(n-1)\varphi]$$
$$+ \exp[-i(n-1)\varphi]\}$$
$$= \tfrac{1}{2}[\exp(in\varphi) + \exp(-in\varphi)][\exp(i\varphi)$$
$$+ \exp(-i\varphi)] = 2\cos(\varphi)\cos(n\varphi).$$

*Theorem 4.9* For $n \in \mathbf{N}_0$, $\cos(n\varphi)$ can be represented as a polynomial of degree n in $\cos(\varphi)$.

*Proof* The statement is obvious for $n = 0$ and $n = 1$. From (4.78) we successively have

$$\cos(2\varphi) = 2\cos^2(\varphi) - 1$$
$$\cos(3\varphi) = 2\cos(\varphi)\cos(2\varphi) - \cos(\varphi) = 4\cos^3(\varphi) - 3\cos(\varphi) \qquad (4.79)$$
$$\cos(4\varphi) = 2\cos(\varphi)\cos(3\varphi) - \cos(2\varphi) = 8\cos^4(\varphi) - 8\cos^2(\varphi) + 1$$

An obvious proof by induction, with respect to n, shows the assertion.

The nth *Chebyshev polynomial* $T_n(x)$ is now defined on the basis of Theorem 4.9 by

$$\boxed{\cos(n\varphi) =: T_n(\cos(\varphi)) = T_n(x) \quad x = \cos(\varphi) \quad x \in [-1, 1] \quad n \in \mathbf{N}_0.} \qquad (4.80)$$

At first sight the range of definition is restricted to the interval $[-1, 1]$, but it can easily be extended to $\mathbf{R}$. In the following we shall consider the polynomials $T_n(x)$ only on the interval $[-1, 1]$ which is of most interest for our purposes. From (4.79) the first few polynomials are given by

$$T_0(x) = 1 \quad T_1(x) = x \quad T_2(x) = 2x^2 - 1$$
$$T_3(x) = 4x^3 - 3x \quad T_4(x) = 8x^4 - 8x^2 + 1.$$

The definition (4.80) implies the obvious property

$$\boxed{|T_n(x)| \leq 1 \quad \text{for} \quad x \in [-1, 1] \quad n \in \mathbf{N}_0.} \qquad (4.81)$$

The polynomial $T_n(x)$ takes on the extreme values $\pm 1$ if $n\varphi = k\pi$, $k = 0, 1, \ldots, n$. Hence the $(n+1)$ *extreme points* of $T_n(x)$ are

$$\boxed{x_k^{(e)} = \cos\left(\frac{k\pi}{n}\right) \quad (k = 0, 1, 2, \ldots, n) \quad n \geq 1.} \qquad (4.82)$$

## 4.3 Orthogonal Polynomials

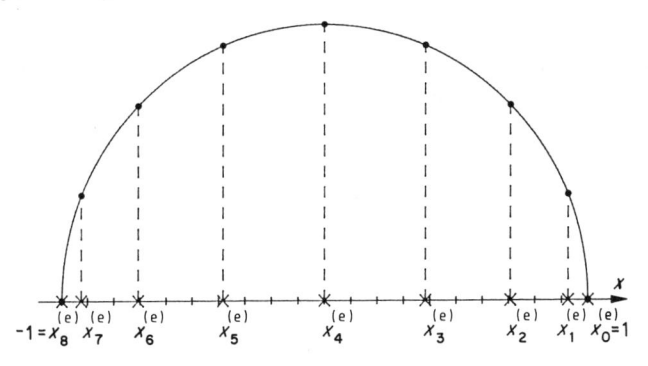

Figure 4.5 Extremal points of $T_8(x)$

They are not equidistant in the interval $[-1, 1]$, but are more densely distributed towards the ends of the interval. They can be constructed geometrically as projections of regularly distributed points on the half circle of unit radius. The construction is shown in Figure 4.5 for $n = 8$. The extreme points $x_k^{(e)}$ are symmetric with respect to the origin. Due to the oscillating behaviour of $\cos(n\varphi)$ the sign at the extreme function values alternates if $x$, and consequently $\varphi$, runs through the interval monotonically.

Between two consecutive extreme points of $T_n(x)$ there necessarily exists a zero. From the condition $\cos(n\varphi) = 0$ it follows that $n\varphi = (2k-1)\pi/2$, $k = 1, 2, \ldots, n$. Consequently the $n$ roots of $T_n(x)$ are

$$x_k = \cos\left(\frac{2k-1}{n}\frac{\pi}{2}\right) \quad (k = 1, 2, \ldots, n) \quad n \geq 1. \tag{4.83}$$

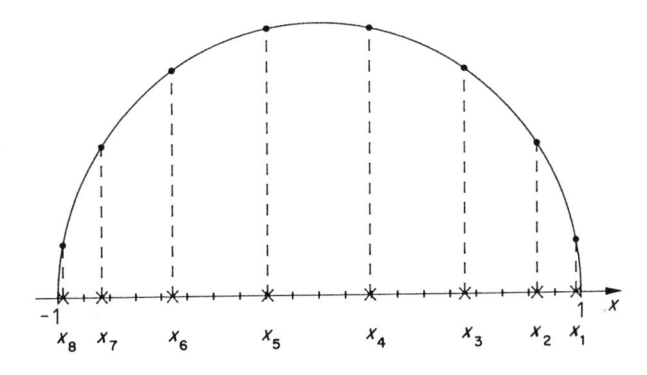

Figure 4.6 Zeros of $T_8(x)$

The zeros of $T_n(x)$ are real, simple and are located in the interior of the interval $[-1, 1]$ symmetrically about the origin. They are also denser towards the ends of the interval. Their geometrical construction is shown in Figure 4.6 for $n = 8$. From their importance in connection with the approximation of functions, they are called the *Chebyshev abscissae* of the $n$th Chebyshev polynomial.

The Chebyshev polynomials satisfy the recursion formula, due to (4.78) and (4.80),

$$T_{n+1}(x) = 2xT_n(x) - T_{n-1}(x) \quad n \geq 1 \quad T_0(x) = 1 \quad T_1(x) = x. \quad (4.84)$$

By means of (4.84) it is easy to see that the coefficient of $x^n$ of the polynomial $T_n(x)$ is equal to $2^{n-1}$ for all $n \geq 1$. Moreover, from (4.84) it can be proved that

$$T_n(-x) = (-1)^n T_n(x) \quad n \geq 0. \quad (4.85)$$

Hence according to the parity of $n$ $T_n(x)$ is an even or an odd polynomial. For this reason the Chebyshev polynomials $T_2(x)$ to $T_{10}(x)$ are shown in Figure 4.7 only on the interval $[0, 1]$.

**Theorem 4.10** *The polynomials $T_n(x)$, $(n = 0, 1, 2, ...)$ form a system of orthogonal polynomials on the interval $[-1, 1]$ with respect to the weight function $w(x) = 1/\sqrt{1-x^2}$. The following relationships hold*

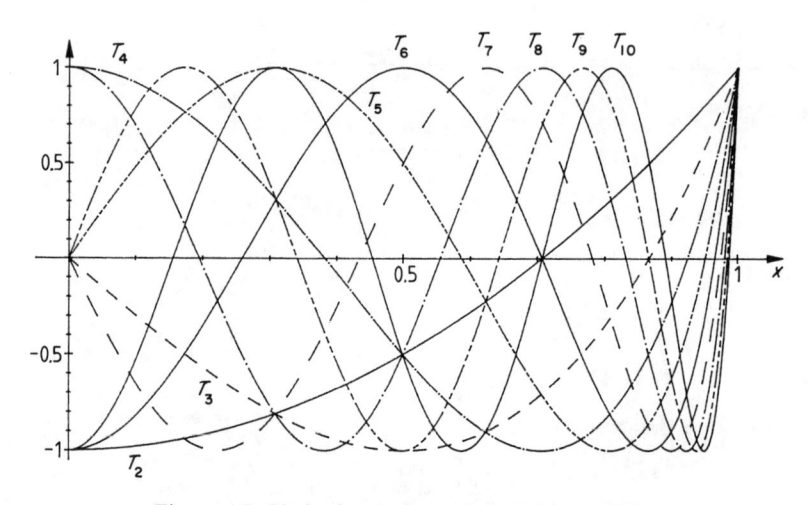

Figure 4.7 Chebyshev polynomials $T_2(x)$ to $T_{10}(x)$

## 4.3 Orthogonal Polynomials

$$\int_{-1}^{1} T_k(x)T_j(x)\frac{dx}{\sqrt{1-x^2}} = \begin{cases} 0 & \text{if } k \neq j \\ \frac{1}{2}\pi & \text{if } k = j > 0 \\ \pi & \text{if } k = j = 0 \end{cases} \quad k, j \in \mathbf{N}_0 \qquad (4.86)$$

*Proof* We substitute $x = \cos(\varphi)$, $T_k(x) = \cos(k\varphi)$, $T_j(x) = \cos(j\varphi)$ and $dx = -\sin(\varphi)d\varphi$ to obtain

$$\int_{-1}^{1} T_k(x)T_j(x)\frac{dx}{\sqrt{1-x^2}} = -\int_{\pi}^{0} \cos(k\varphi)\cos(j\varphi)\frac{\sin(\varphi)d\varphi}{\sin(\varphi)}$$

$$= \int_{0}^{\pi} \cos(k\varphi)\cos(j\varphi)d\varphi = \frac{1}{2}\int_{-\pi}^{\pi} \cos(k\varphi)\cos(j\varphi)d\varphi.$$

If we apply the relationships (4.6) to the last integral we obtain (4.86).

Now we consider the problem of approximating a continuous function $f(x)$ on the interval $[-1, 1]$ in the mean square by a polynomial of degree $n$, $g_n(x)$, having the following expansion

$$g_n(x) = \tfrac{1}{2}c_0 T_0(x) + \sum_{k=1}^{n} c_k T_k(x) \qquad (4.87)$$

The coefficients $c_0, c_1, \ldots, c_n$ are to be determined in such a way that the difference between the approximating function $g_n(x)$ and the given function $f(x)$ is a minimum with respect to the weighted $L_2$ norm, with the weight function $w(x)$. From (4.86) this leads to the requirement

$$\Phi := \int_{-1}^{1} [g_n(x) - f(x)]^2 \frac{dx}{\sqrt{1-x^2}}$$

$$= \int_{-1}^{1} \left(\tfrac{1}{2}c_0 T_0(x) + \sum_{k=1}^{n} c_k T_k(x) - f(x)\right)^2 \frac{dx}{\sqrt{1-x^2}}$$

$$= \frac{\pi}{4}c_0^2 + \sum_{k=1}^{n} \frac{\pi}{2}c_k^2 - c_0 \int_{-1}^{1} \frac{f(x)dx}{\sqrt{1-x^2}}$$

$$- 2\sum_{k=1}^{n} c_k \int_{-1}^{1} f(x)T_k(x)\frac{dx}{\sqrt{1-x^2}} + \int_{-1}^{1} \frac{f(x)^2 dx}{\sqrt{1-x^2}} = \text{Min!}$$

As quadratic function of the $(n+1)$ variables $c_0, c_1, \ldots, c_n$, the necessary conditions for a minimum of $\Phi$ are

$$\frac{\partial \Phi}{\partial c_0} = \frac{\pi}{2} c_0 - \int_{-1}^{1} \frac{f(x) \, dx}{\sqrt{1-x^2}} = 0$$

$$\frac{\partial \Phi}{\partial c_k} = \pi c_k - 2 \int_{-1}^{1} f(x) T_k(x) \frac{dx}{\sqrt{1-x^2}} = 0 \quad (k=1,2,\ldots,n).$$

Hence the desired coefficients $c_k$ are given by

$$c_k = \frac{2}{\pi} \int_{-1}^{1} f(x) T_k(x) \frac{dx}{\sqrt{1-x^2}} \quad (k=0,1,2,\ldots,n). \tag{4.88}$$

The integral (4.88) is much simpler if the substitution $x = \cos \varphi$ is made and (4.80) is taken into account. After some elementary adaptations from (4.88) we obtain

$$c_k = \frac{2}{\pi} \int_0^{\pi} f(\cos \varphi) \cos(k\varphi) \, d\varphi \quad (k=0,1,\ldots,n). \tag{4.89}$$

However, $F(\varphi) := f(\cos \varphi)$ is obviously an even, $(2\pi)$ periodic function of $\varphi$, and therefore we finally get

$$c_k = \frac{1}{\pi} \int_{-\pi}^{\pi} f(\cos \varphi) \cos(k\varphi) \, d\varphi \quad (k=0,1,2,\ldots,n). \tag{4.90}$$

*Conclusion* The coefficients $c_k$ of the expansion (4.87) of the approximating function $g_n(x)$ are equal to the Fourier coefficients $a_k$, (4.11), of the even, $(2\pi)$ periodic function $F(\varphi) := f(\cos \varphi)$.

The approximation problem is thus reduced to the computation of Fourier coefficients, and the methods of Section 4.1 can be applied. The coefficients $c_k$ and the corresponding polynomial $g_n(x)$ are defined for arbitrarily large $n$. So for every continuous function on $[-1, 1]$ we can form the infinite Chebyshev expansion

$$g(x) = \tfrac{1}{2} c_0 T_0(x) + \sum_{k=1}^{\infty} c_k T_k(x). \tag{4.91}$$

The question of convergence of (4.91) can be answered by means of Theorem 4.2. We observe that the $(2\pi)$ periodic function $F(\varphi)$ is continuous as the composition of two continuous functions. Moreover, if the given function $f(x)$ has a piecewise

## 4.3 Orthogonal Polynomials

continuous first derivative on $[-1, 1]$, then so has $F(\varphi)$. On this assumption the expansion (4.91) converges to $f(x)$ for all $x \in [-1, 1]$.

The approximation error $f(x) - g_n(x)$ can be estimated on an additional hypothesis.

*Theorem 4.11* If the series $\sum_{k=1}^{\infty} |c_k|$ is convergent, where $c_k$ are the expansion coefficients of (4.91) corresponding to a continuous function $f(x)$ with a piecewise continuous first derivative, then the Chebyshev expansion converges uniformly to $f(x)$ for all $x \in [-1, 1]$, and we have

$$|f(x) - g_n(x)| \leq \sum_{k=n+1}^{\infty} |c_k| \quad \text{for } x \in [-1, 1]. \tag{4.92}$$

*Proof* According to the hypothesis for $f(x)$, the infinite expansion (4.91) converges pointwise to $f(x)$ for $|x| \leq 1$. To show the uniform convergence we use the fact that for each $\varepsilon > 0$ there exists a $N$ such that

$$\sum_{k=n}^{\infty} |c_k| < \varepsilon \quad \text{for all } n > N. \tag{4.93}$$

So for every $n > N$, $|x| \leq 1$ and because $|T_k(x)| \leq 1$

$$|f(x) - g_n(x)| = \left| \sum_{k=n+1}^{\infty} c_k T_k(x) \right| \leq \sum_{k=n+1}^{\infty} |c_k| < \varepsilon \tag{4.94}$$

This proves the error estimate (4.92).

*Example 4.9* We seek an approximating polynomial $g_n(x)$ for the function $f(x) = e^x$ on $[-1, 1]$. Following (4.89) the coefficients $c_k$ of the Chebyshev expansion are given by

$$c_k = \frac{2}{\pi} \int_0^\pi e^{\cos \varphi} \cos(k\varphi) \, d\varphi = 2I_k(1) \quad (k = 0, 1, 2, \ldots), \tag{4.95}$$

where $I_k(\bar{x})$ denotes the modified $k$th Bessel function (Abramowitz and Stegun, 1971). The appropriate series expansion of $I_k(x)$ yields

$c_0 \doteq 2.532\,131\,7555 \quad c_1 \doteq 1.130\,318\,2080 \quad c_2 \doteq 0.271\,495\,3395$
$c_3 \doteq 0.044\,336\,8498 \quad c_4 \doteq 0.005\,474\,2404 \quad c_5 \doteq 0.000\,542\,9263$
$c_6 \doteq 0.000\,044\,9773 \quad c_7 \doteq 0.000\,003\,1984 \quad c_8 \doteq 0.000\,000\,1992$
$c_9 \doteq 0.000\,000\,0110 \quad c_{10} \doteq 0.000\,000\,0006$

(4.96)

The coefficients $c_k$ converge rapidly towards zero, and the corresponding series is uniformly convergent. The function $f(x)$ is arbitrarily many times continuously differentiable, so that the assumptions of Theorem 4.11 are satisfied. For $g_6(x)$, from (4.92), we get the error estimate

$$|f(x) - g_6(x)| \leq 3.409 \times 10^{-6} \quad \text{for all } |x| \leq 1.$$

The coefficients $c_k$ of the expansion (4.87) rarely exist in an explicit formula, and therefore they must be computed approximately. As the $c_k$ are identical to the Fourier coefficients $a_k$ of the $(2\pi)$ periodic function $F(\varphi) = f(\cos \varphi)$, the formulae (4.18) are at hand. There we replace $x$ by $\varphi$ and obtain

$$c_k^* = \frac{2}{N} \sum_{j=1}^{N} F(\varphi_j) \cos(k\varphi_j) = \frac{2}{N} \sum_{j=0}^{N-1} f(\cos \varphi_j) \cos(k\varphi_j)$$

$$\varphi_j = \frac{2\pi}{N} j \quad N \in \mathbf{N} \quad \left( k = 0, 1, 2, \ldots, \left[\frac{N}{2}\right] \right). \tag{4.97}$$

The approximations $c_k^*$ can be efficiently computed for small $N$ by Runge's algorithm and for large $N$ by the fast Fourier transformation. In Runge's algorithm we need not compute the differences $d_j$, (4.42), and the quantities derived from them, because $F(\varphi)$ is an even function and hence all coefficients $b_k^*$ are known to be zero. Some more involved measures are required to take the same fact into account for the FFT.

For a direct computation of some of the coefficients $c_k^*$ for $N = 2m$, $m \in \mathbf{N}$, not too large the summation (4.97) can be simplified. As $F(\varphi)$ is an even function the relationship $F(\varphi_j) = F(\varphi_{N-j})$ holds. So essentially we can apply the first convolution, (4.41), of Runge's method to obtain from (4.97)

$$c_k^* = \frac{2}{N} \left( f(\cos \varphi_0) \cos(k\varphi_0) + 2 \sum_{j=1}^{m-1} f(\cos \varphi_j) \cos(k\varphi_j) \right.$$

$$\left. + f(\cos \varphi_m) \cos(k\varphi_m) \right)$$

$$= \frac{2}{2m} \left( f(1) + 2 \sum_{j=1}^{m-1} f(\cos \varphi_j) \cos(k\varphi_j) + f(-1) \cos(k\pi) \right)$$

and hence the representation

$$\boxed{c_k^* = \frac{2}{m} \left\{ \frac{1}{2}(f(1) + f(-1) \cos(k\pi)) + \sum_{j=1}^{m-1} f\left[\cos\left(\frac{j\pi}{m}\right)\right] \cos\left(\frac{kj\pi}{m}\right) \right\}}$$

$$(k = 0, 1, 2, \ldots, n), n \leq m.$$

(4.98)

The formula (4.98) can be regarded as the summed trapezoidal rule, (8.4), for the approximate evaluation of the integral (4.89) if the interval $[0, \pi]$ is divided into $m$ subintervals.

## 4.3 Orthogonal Polynomials

On the assumption that the continuous function $f(x)$ also possesses a piecewise continuous first derivative on the interval $[-1, 1]$, the approximations $c_k^*$ and the exact coefficients $c_k$ satisfy, due to (4.32), the relation

$$c_k^* = c_k + c_{N-k} + c_{N+k} + c_{2N-k} + c_{2N+k} + \cdots,$$
$$(k = 0, 1, \ldots, n \leqslant m) \quad N = 2m. \quad (4.99)$$

If the asymptotic behaviour of the coefficients $c_k$ for a given function $f(x)$ is known, then the error $|c_k^* - c_k|$ can be estimated for a prescribed $N$, or conversely, the required value of $N$ can be determined so that the error is sufficiently small.

*Example* 4.10 The coefficients $c_k$ of the Chebyshev expansion of the function $f(x) = e^x$ rapidly converge to zero. As $c_{11} \doteq 2.50 \times 10^{-11}$, $c_{12} \doteq 1.04 \times 10^{-12}$ and $|c_k| < 10^{-13}$ for all $k \geqslant 13$, the relationship (4.99) implies that (4.98) yields the first eleven coefficients to ten significant places even for $m = 12$, that is $N = 24$. More precisely we have the error estimate

$$|c_k^* - c_k| \leqslant |c_{24-k}| + |c_{24+k}| + \cdots < 10^{-13} \quad \text{for } k = 0, 1, \ldots, 10.$$

Finally we consider the problem of evaluating a polynomial $g_n(x)$ in the representation (4.87), in an efficient and numerically stable way. The obvious method for determining $g_n(x)$ for a given $x$ consists of successively evaluating the Chebyshev polynomials by means of the recursion formula (4.84) and building up the partial sums. However, this procedure is unstable. It is better to reduce the finite sum (4.87) backwards by successively eliminating the last term using the recursion (4.84). We develop the algorithm in the case $n = 5$

$$g_5(x) = \tfrac{1}{2}c_0 T_0 + c_1 T_1 + c_2 T_2 + c_3 T_3 + c_4 T_4 + c_5 T_5$$
$$= \tfrac{1}{2}c_0 T_0 + c_1 T_1 + c_2 T_2 + (c_3 - c_5) T_3 + (c_4 + 2xc_5) T_4$$

With the substitution $d_4 := c_4 + 2xc_5$ the subsequent step yields

$$g_5(x) = \tfrac{1}{2}c_0 T_0 + c_1 T_1 + (c_2 - d_4) T_2 + (c_3 + 2xd_4 - c_5) T_3.$$

Now we define $d_3 := c_3 + 2xd_4 - c_5$ and by an analogous continuation with the correspondingly defined quantities obtain

$$g_5(x) = \tfrac{1}{2}c_0 T_0 + (c_1 - d_3) T_1 + (c_2 + 2xd_3 - d_4) T_2$$
$$= (\tfrac{1}{2}c_0 - d_2) T_0 + (c_1 + 2xd_2 - d_3) T_1 = (\tfrac{1}{2}c_0 - d_2) T_0 + d_1 T_1.$$

As $T_0(x) = 1$ and $T_1(x) = x$, finally

$$g_5(x) = \tfrac{1}{2}c_0 + xd_1 - d_2 = \tfrac{1}{2}[(c_0 + 2xd_1 - d_2) - d_2] = \tfrac{1}{2}(d_0 - d_2).$$

Hence we have to compute the values $d_{n-1}, d_{n-2}, \ldots, d_0$ recursively from the given coefficients $c_n, c_{n-1}, \ldots, c_0$ of the Chebyshev expansion. The value of the polynomial $g_n(x)$ is then given by half of the difference of $d_0$ and $d_2$. Clenshaw's algorithm (Clenshaw, 1955) for the evaluation of $g_n(x)$ for a prescribed $x$ can be summarized as follows.

$$\boxed{\begin{aligned} &d_n = c_n \quad y = 2 \times x \quad d_{n-1} = c_{n-1} + y \times c_n \\ &\text{for } k = n-2, n-3, \ldots, 0: \\ &\quad d_k = c_k + y \times d_{k+1} - d_{k+2} \\ &g_n(x) = (d_0 - d_2)/2 \end{aligned}} \quad (4.100)$$

The amount of computational effort required to evaluate $g_n(x)$ consists of only $(n+2)$ multiplications. The algorithm (4.100) is stable, as it can be shown that the total error of $g_n(x)$ is at most equal to the sum of absolute values of the rounding errors of the computed values $d_k$ (Fox and Parker, 1968).

### 4.3.2 Chebyshev interpolation

To approximate a given function with the aid of Chebyshev polynomials, an interpolation polynomial is often as suitable as the truncated Chebyshev series expansion. As motivation we need the following theorem.

*Theorem 4.12* *Among all polynomials $P_n(x)$ of degree $n \geq 1$, whose coefficient of $x^n$ is equal to one, the polynomial $T_n(x)/2^{n-1}$ has the smallest maximum norm on the interval $[-1, 1]$, that is we have*

$$\min_{P_n(x)} \left( \max_{x \in [-1,1]} |P_n(x)| \right) = \max_{x \in [-1,1]} \left| \left( \frac{1}{2^{n-1}} T_n(x) \right) \right| = \frac{1}{2^{n-1}}. \quad (4.101)$$

*Proof* We show the minimax property (4.101) of the Chebyshev polynomial $T_n(x)$ by an indirect argument. We assume that there is a polynomial $P_n(x)$ whose coefficient of the highest power $x^n$ is equal to one such that $|P_n(x)| < 1/2^{n-1}$ holds for all $x \in [-1, 1]$. This assumption implies that the following inequalities hold at the $(n+1)$ extremum points $x_k^{(e)}$ (4.82) of $T_n(x)$:

$$P_n(x_0^{(e)}) < T_n(x_0^{(e)})/2^{n-1} = 1/2^{n-1},$$
$$P_n(x_1^{(e)}) > T_n(x_1^{(e)})/2^{n-1} = -1/2^{n-1},$$
$$P_n(x_2^{(e)}) < T_n(x_2^{(e)})/2^{n-1} = 1/2^{n-1}, \text{ etc.}$$

Consequently the difference polynomial

$$Q(x) := P_n(x) - T_n(x)/2^{n-1}$$

## 4.3 Orthogonal Polynomials

takes on values of alternating signs at the $(n+1)$ extremal points $x_0^{(e)} > x_1^{(e)} > \cdots > x_n^{(e)}$. For reasons of continuity $Q(x)$ must have at least $n$ different zeros. However, the polynomials $P_n(x)$ and $T_n(x)/2^{n-1}$ have the same coefficient of $x^n$, and hence the degree of $Q(x)$ is at most equal to $(n-1)$. This contradicts the fundamental theorem of algebra.

The minimax property of the Chebyshev polynomials has an important application in connection with polynomial interpolation. If $f(x)$ can be continuously differentiated $(n+1)$ times and is approximated on the interval $[-1, 1]$ by an interpolation polynomial $P_n(x)$, then because of (3.38) and with $M_{n+1} := \max_{-1 \leq \xi \leq 1} |f^{(n+1)}(\xi)|$ we have the error estimate

$$|f(x) - P_n(x)| \leq \frac{M_{n+1}}{(n+1)!} |(x-x_0)(x-x_1)\cdots(x-x_n)| \quad x \in [-1, 1].$$

The $(n+1)$ support abscissae $x_0, x_1, \ldots, x_n$ should be chosen such that

$$\max_{-1 \leq x \leq 1} |(x-x_0)(x-x_1)\cdots(x-x_n)| = \text{Min!}$$

holds. The function $\varphi(x) := (x-x_0)(x-x_1)\cdots(x-x_n)$ is a polynomial of degree $(n+1)$ whose coefficient of the highest power is one. Due to Theorem 4.12 its maximum norm on $[-1, 1]$ is a minimum if and only if the $(n+1)$ support abscissae coincide with the $(n+1)$ zeros of $T_{n+1}(x)$. In this case we have $\max |\varphi(x)| = 2^{-n}$. The interpolation polynomial $P_n^*(x)$ whose support abscissae $x_k$ are equal to the *Chebyshev abscissae* of the $(n+1)$th Chebyshev polynomial provides the smallest possible bound for the interpolation error

$$\boxed{|f(x) - P_n^*(x)| \leq \frac{M_{n+1}}{2^n \cdot (n+1)!} \quad x \in [-1, 1].} \tag{4.102}$$

The result (4.102) does not mean, however, that the interpolation polynomial $P_n^*(x)$ is the polynomial of the best approximation with respect to the maximum norm, as the interpolation error is given by

$$f(x) - P_n^*(x) = \frac{f^{(n+1)}(\xi)}{2^n \cdot (n+1)!} T_{n+1}(x) \quad x \in [-1, 1]$$

where $\xi$ depends on $x$.

In Sections 3.2 and 3.4 several representations of the interpolation polynomial have been extensively treated. However, now it is more suitable to represent the desired polynomial, corresponding to the Chebyshev abscissae, as the following linear combination of Chebyshev polynomials

$$P_n^*(x) = \tfrac{1}{2}\gamma_0 T_0(x) + \sum_{k=1}^{n} \gamma_k T_k(x). \tag{4.103}$$

The unknown coefficients $\gamma_0, \gamma_1, \ldots, \gamma_n$ are determined from the $(n+1)$ interpolation conditions

$$\tfrac{1}{2}\gamma_0 T_0(x_l) + \sum_{k=1}^{n} \gamma_k T_k(x_l) = f(x_l) \quad (l = 1, 2, \ldots, n+1) \tag{4.104}$$

for the Chebyshev abscissae $x_l = \cos\left(\dfrac{2l-1}{n+1} \cdot \dfrac{\pi}{2}\right)$ of $T_{n+1}(x)$. To do this we need a discrete orthogonality property of the Chebyshev polynomials.

**Theorem 4.13** Let $x_l$, $1 \leq l \leq n+1$ be $(n+1)$ zeros of $T_{n+1}(x)$. Then we have

$$\sum_{l=1}^{n+1} T_k(x_l) T_j(x_l) = \begin{cases} 0 & \text{if } k \neq j \\ \tfrac{1}{2}(n+1) & \text{if } k = j > 0 \\ n+1 & \text{if } k = j = 0 \end{cases} \quad 0 \leq k, \, j \leq n \tag{4.105}$$

*Proof* From (4.80) and (4.83) the function values of the Chebyshev polynomials at the Chebyshev abscissae are

$$T_k(x_l) = \cos(k \arccos(x_l)) = \cos\left(k \frac{(2l-1)}{(n+1)} \frac{\pi}{2}\right)$$

$$= \cos[k(l - \tfrac{1}{2})h] \quad h := \frac{\pi}{n+1}.$$

We again use trigonometric identities to obtain

$$\sum_{l=1}^{n+1} T_k(x_l) T_j(x_l) = \sum_{l=1}^{n+1} \cos[kh(l - \tfrac{1}{2})] \cos[jh(l - \tfrac{1}{2})]$$

$$= \tfrac{1}{2} \sum_{l=1}^{n+1} \{\cos[(k-j)h(l - \tfrac{1}{2})] + \cos[(k+j)h(l - \tfrac{1}{2})]\}$$

$$= \tfrac{1}{2} \operatorname{Re} \left\{ \sum_{l=1}^{n+1} \exp[i(k-j)h(l - \tfrac{1}{2})] + \sum_{l=1}^{n+1} \exp[i(k+j)h(l - \tfrac{1}{2})] \right\}. \tag{4.106}$$

The two sums are finite geometric series with the corresponding quotients $q_1 = \exp[i(k-j)h]$ and $q_2 = \exp[i(k+j)h]$. We first treat the case $k \neq j$. The inequalities $0 < |k-j| \leq n$ and $0 < k+j < 2n$ hold as $0 \leq k \leq n$ and $0 \leq j \leq n$. Therefore $q_1 \neq 1$ and $q_2 \neq 1$ since

$$\frac{\pi}{n+1} \leq |(k-j)h| \leq \frac{n\pi}{n+1}, \quad \frac{\pi}{n+1} \leq (k+j)h < \frac{2\pi n}{n+1}.$$

## 4.3 Orthogonal Polynomials

Hence the first sum is

$$\sum_{l=1}^{n+1} \exp[i(k-j)h(l-\tfrac{1}{2})] = \exp(\tfrac{1}{2}i(k-j)h)\frac{\exp[i(k-j)h(n+1)]-1}{\exp[i(k-j)h]-1}$$

$$= \frac{(-1)^{k-j}-1}{2i\sin\left(\tfrac{1}{2}(k-j)\dfrac{\pi}{n+1}\right)}$$

purely imaginary or vanishes. The same holds for the second sum, so that the real part of (4.106) is equal to zero in any case, and the first row of (4.105) is shown.

For $k = j > 0$ the first sum of (4.106) is equal to $(n+1)$, whereas the second is zero. For $k = j = 0$ all terms of the two sums are equal to one. From this the other two assertions of (4.105) follow.

The relationships (4.105) mean that the columns of the matrix of the system of linear equations, (4.104), are pairwise orthogonal. For this reason the unknowns $\gamma_0, \gamma_1, \ldots, \gamma_n$ follow explicitly from (4.104). We multiply the $l$th equation of (4.104) by $T_j(x_l)$, where $j$ is a fixed index $0 \leq j \leq n$. Then we add all $(n+1)$ equations and from (4.105), after a change of indices for $\gamma_k$, we get the representation

$$\boxed{\begin{aligned}\gamma_k &= \frac{2}{n+1}\sum_{l=1}^{n+1} f(x_l)T_k(x_l) \\ &= \frac{2}{n+1}\sum_{l=1}^{n+1} f\left[\cos\left(\frac{(2l-1)}{(n+1)}\frac{\pi}{2}\right)\right]\cos\left(k\frac{(2l-1)}{(n+1)}\frac{\pi}{2}\right) \quad (k=0,1,\ldots,n).\end{aligned}}$$

(4.107)

The coefficients $\gamma_k$ of the expansion of the interpolation polynomial $P_n^*(x)$ corresponding to the Chebyshev abscissae of $T_{n+1}(x)$ differ from the coefficients $c_k^*$ (4.98). With these another polynomial of degree $n$

$$g_n^*(x) := \tfrac{1}{2}c_0^* T_0(x) + \sum_{k=1}^{n-1} c_k^* T_k(x) + \tfrac{1}{2}c_n^* T_n(x) \tag{4.108}$$

can be formed. In the special case $n = m = N/2$ it interpolates the corresponding function values at the $(n+1)$ extremal points $x_j^{(e)} = \cos(j\pi/n)$ of $T_n(x)$. In this case the endpoints $\pm 1$ of the interval are support abscissae of the interpolation polynomial $g_n^*(x)$. The error estimate (4.102) is not valid for $g_n^*(x)$.

*Example 4.11* For the function $f(x) = e^x$ the coefficients $\gamma_k$ of the interpolation

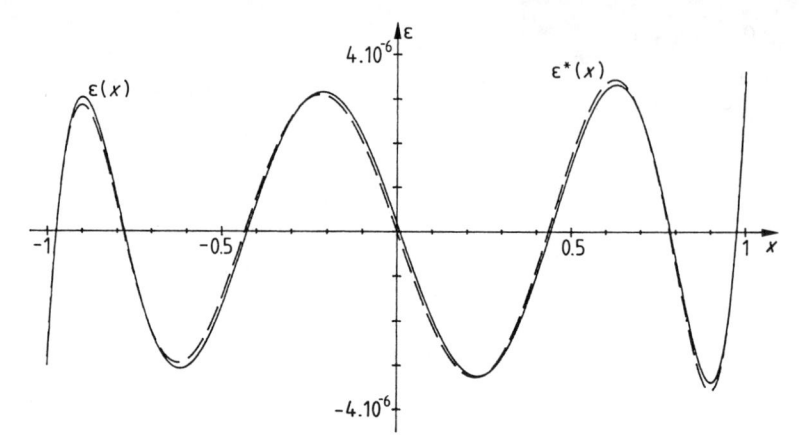

Figure 4.8 Error of approximation for $e^x$, Chebyshev polynomials

polynomial $P_6^*(x)$ corresponding to the Chebyshev abscissae are given by

$\gamma_0 \doteq 2.532\,131\,7555 \quad \gamma_1 \doteq 1.130\,318\,2080 \quad \gamma_2 \doteq 0.271\,495\,3395$
$\gamma_3 \doteq 0.044\,336\,8498 \quad \gamma_4 \doteq 0.005\,474\,2399 \quad \gamma_5 \doteq 0.000\,542\,9153$
$\gamma_6 \doteq 0.000\,044\,7781.$

In comparison with the coefficients $c_k$ (4.96) only the last three coefficients $\gamma_4, \gamma_5, \gamma_6$ differ from $c_4, c_5, c_6$ in the ten decimal places. The bound of the maximal interpolating error of $P_6^*(x)$, because $M_7 = \max_{-1 \leq x \leq 1} |f^{(7)}(x)| = \max_{-1 \leq x \leq 1} |e^x| \doteq 2.7183$ according to (4.102) is

$|e^x - P_6^*(x)| \leq 8.43 \times 10^{-6} \quad \text{for all } |x| \leq 1.$

The maximum of the interpolation error is of course overestimated. The maximal absolute error is actually $3.620 \times 10^{-6}$. The interpolation polynomial $P_6^*(x)$ yields as good an approximation as $g_6(x)$. In Figure 4.8 the error functions $\varepsilon^*(x) := e^x - P_6^*(x)$ and $\varepsilon(x) := e^x - g_6(x)$ are shown.

### 4.3.3 Legendre polynomials

In this section $P_n(x)$ denotes the $n$th *Legendre polynomial* defined by

$$P_n(x) := \frac{1}{2^n \cdot n!} \frac{d^n}{dx^n}[(x^2-1)^n] \quad n \in \mathbf{N}_0. \tag{4.109}$$

As the expression in the brackets is a polynomial of exactly degree $2n$, the $n$th derivative is a polynomial of exactly degree $n$.

## 4.3 Orthogonal Polynomials

**Theorem 4.14** *The Legendre polynomials $P_n(x)$, $n = 0, 1, 2, \ldots$ form an orthogonal system with respect to the interval $[-1, 1]$. The following relationships hold*

$$\int_{-1}^{1} P_m(x) P_n(x) \, dx = \begin{cases} 0 & \text{if } m \neq n \\ \dfrac{2}{2n+1} & \text{if } m = n \end{cases} \quad m, n \in \mathbf{N}_0. \tag{4.110}$$

*Proof* We start by showing the property of orthogonality of the Legendre polynomials. Without any restriction we can assume that $m < n$. Integration by parts gives

$$\begin{aligned}
I_{m,n} &:= 2^m \cdot m! \, 2^n \cdot n! \int_{-1}^{1} P_m(x) P_n(x) \, dx \\
&= \int_{-1}^{1} \frac{d^m}{dx^m} [(x^2 - 1)^m] \cdot \frac{d^n}{dx^n} [(x^2 - 1)^n] \, dx \\
&= \frac{d^m}{dx^m} [(x^2 - 1)^m] \cdot \frac{d^{n-1}}{dx^{n-1}} [(x^2 - 1)^n] \Big|_{-1}^{1} \\
&\quad - \int_{-1}^{1} \frac{d^{m+1}}{dx^{m+1}} [(x^2 - 1)^m] \cdot \frac{d^{n-1}}{dx^{n-1}} [(x^2 - 1)^n] \, dx.
\end{aligned} \tag{4.111}$$

Now we observe that the polynomial $(x^2 - 1)^n$ has zeros of order $n$ for $x = \pm 1$. Hence we have

$$\frac{d^{n-k}}{dx^{n-k}} [(x^2 - 1)^n] = 0 \quad \text{for } x = \pm 1 \text{ and for } k = 1, 2, \ldots, n. \tag{4.112}$$

Thus from (4.111) after $(n - 1)$ more integrations by parts we get

$$I_{m,n} = (-1)^n \int_{-1}^{1} \frac{d^{m+n}}{dx^{m+n}} [(x^2 - 1)^m] \cdot (x^2 - 1)^n \, dx. \tag{4.113}$$

On our assumption $m + n > 2m$, and hence the first factor of the integrand vanishes, and we have $I_{m,n} = 0$.

The second part of the assertion (4.110) follows from (4.113) which also holds for $m = n$. As

$$\frac{d^{2n}}{dx^{2n}} [(x^2 - 1)^n] = (2n)!$$

from (4.113) after $n$ integrations by parts we obtain

$$I_{n,n} = (-1)^n (2n)! \int_{-1}^{1} (x - 1)^n (x + 1)^n \, dx$$

$$= (-1)^n (2n)! \left( (x-1)^n \frac{1}{n+1} (x+1)^{n+1} \Big|_{-1}^{1} \right.$$
$$\left. - \frac{n}{n+1} \int_{-1}^{1} (x-1)^{n-1}(x+1)^{n+1} \, dx \right)$$
$$\vdots$$
$$= (-1)^{2n}(2n!) \frac{n(n-1)(n-2)\cdots 1}{(n+1)(n+2)(n+3)\cdots(2n)}$$
$$\cdot \int_{-1}^{1} (x+1)^{2n} \, dx = (n!)^2 \cdot \frac{2^{2n+1}}{2n+1}.$$

Taking (4.111) into account, the second statement of (4.110) follows.

The definition (4.109) of the Legendre polynomials obviously implies that $P_n(x)$ is an even or odd function of $x$ corresponding to the parity of $n$. This follows from the fact that $(x^2 - 1)^n$ is even and that the derivative of an even function is odd and vice versa. Hence we have

$$\boxed{P_n(-x) = (-1)^n P_n(x) \quad n \in \mathbf{N}_0.} \tag{4.114}$$

**Theorem 4.15** *In the open interval* $(-1, 1)$, *the Legendre polynomial* $P_n(x)$, $n \geq 1$, *has n simple zeros.*

*Proof* We start from the fact that $(x^2 - 1)^n$ has zeros of order $n$ for $x = \pm 1$, then apply Rolle's theorem $n$ times. With each step the multiplicity of the zeros at $x = \pm 1$ decreases by one, and an additional zero appears in the interior of the interval. Hence there exist at least $n$ pairwise different zeros within $(-1, 1)$. However a polynomial of degree $n$ has exactly $n$ zeros, and the assertion follows.

The zeros of $P_n(x)$ cannot be defined for general $n$ by an explicit formula as is the case for those of the Chebyshev polynomials. The zeros are tabulated in Abramowitz and Stegun (1971) and Schmeisser and Schirmeier (1976).

**Theorem 4.16** *Three consecutive Legendre polynomials satisfy the recursion formula*

$$\boxed{\begin{aligned} P_{n+1}(x) &= \frac{2n+1}{n+1} x P_n(x) - \frac{n}{n+1} P_{n-1}(x) \quad (n = 1, 2, \ldots) \\ P_0(x) &= 1 \quad P_1(x) = x. \end{aligned}} \tag{4.115}$$

## 4.3 Orthogonal Polynomials

*Proof* The first two polynomials $P_0(x) = 1$ and $P_1(x) = x$ follow immediately from the definition (4.109). First we show the existence of a recursion formula of the form (4.115) between three consecutive Legendre polynomials and in a second step verify the specific coefficients. Let $a_n$ denote the coefficient of the highest power of $P_n(x) = a_n x^n + \cdots$. Then a relationship

$$P_{n+1}(x) - \frac{a_{n+1}}{a_n} x P_n(x) = \sum_{i=0}^{n-1} c_i P_i(x) \quad n \geq 1 \tag{4.116}$$

must hold, because the left-hand side of (4.116) is a polynomial of degree at most $(n-1)$ as it is the difference of either two even or odd polynomials with the same coefficients of the highest power. Hence there must exist a linear combination of the Legendre polynomials $P_0(x), P_1(x), \ldots, P_{n-1}(x)$ that represents this difference. From (4.116) it follows that for each fixed index $j = 0, 1, \ldots, n-2$

$$\int_{-1}^{1} P_j(x) P_{n+1}(x) \, dx - \frac{a_{n+1}}{a_n} \int_{-1}^{1} P_j(x) x P_n(x) \, dx = \sum_{i=0}^{n-1} c_i \int_{-1}^{1} P_j(x) P_i(x) \, dx. \tag{4.117}$$

Both integrals on the left-hand side of the equation (4.117) are zero from (4.110). To see this we take into account that $xP_j(x)$ is a polynomial of degree less than $n$ and can be represented by a linear combination of Legendre polynomials of degree less than $n$. On the right-hand side the only nonvanishing term is $2c_j/(2j+1)$ due to (4.110). Consequently the relationship (4.116) holds for $c_0 = c_1 = \cdots = c_{n-2} = 0$ and $c_{n-1} \neq 0$.

In order to determine the coefficients of the recursion we need the coefficients of $x^n$ and $x^{n-2}$ of $P_n(x)$. According to (4.109) they are

$$P_n(x) = \frac{1}{2^n \cdot n!} \left\{ \frac{d^n}{dx^n} \left[ x^{2n} - \binom{n}{1} x^{2n-2} + \cdots \right] \right\}$$

$$= \frac{(2n)!}{2^n \cdot (n!)^2} x^n - \frac{n \cdot (2n-2)!}{2^n \cdot n! \cdot (n-2)!} x^{n-2} + \cdots. \tag{4.118}$$

From (4.118) we obtain

$$\frac{a_{n+1}}{a_n} = \frac{[2(n+1)]! \cdot 2^n \cdot (n!)^2}{2^{n+1} \cdot [(n+1)!]^2 \cdot (2n)!} = \frac{(2n+2)(2n+1)}{2(n+1)(n+1)} = \frac{2n+1}{n+1}.$$

If we compare the coefficients of $x^{n-1}$ in (4.116) we get

$$-\frac{(n+1) \cdot (2n)!}{2^{n+1} \cdot (n+1)! \cdot (n-1)!} + \frac{2n+1}{n+1} \frac{n \cdot (2n-2)!}{2^n \cdot n! \cdot (n-2)!}$$

$$= c_{n-1} \cdot \frac{(2n-2)!}{2^{n-1} \cdot [(n-1)!]^2},$$

and by a simple calculation

$$c_{n-1} = -\frac{n}{n+1}.$$

By means of the recursion formula (4.115) some further Legendre polynomials are

$$P_2(x) = \tfrac{1}{2}(3x^2 - 1) \quad P_3(x) = \tfrac{1}{2}(5x^3 - 3x)$$
$$P_4(x) = \tfrac{1}{8}(35x^4 - 30x^2 + 3) \quad P_5(x) = \tfrac{1}{8}(63x^5 - 70x^3 + 15x)$$
$$P_6(x) = \tfrac{1}{16}(231x^6 - 315x^4 + 105x^2 - 5).$$

By an induction proof with respect to $n$ it can be shown that the Legendre polynomials have the property

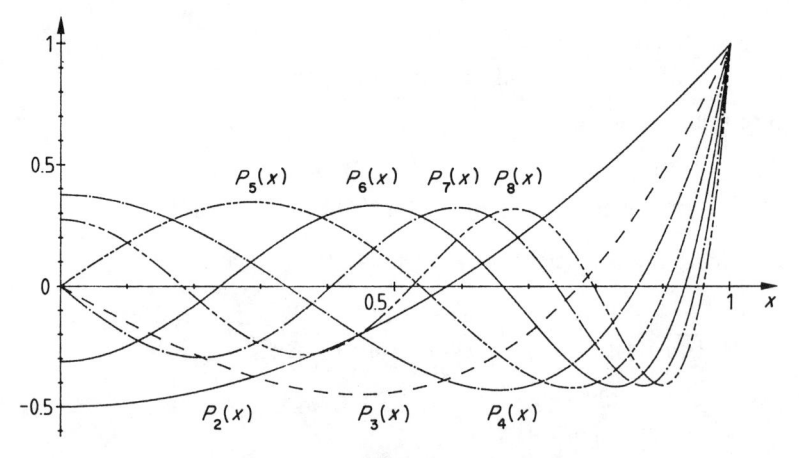

$$P_n(1) = 1 \quad P_n(-1) = (-1)^n \quad (n = 0, 1, 2, \ldots). \tag{4.119}$$

In figure 4.9 the Legendre polynomials $P_2(x)$ to $P_8(x)$ are shown only in the interval $[0, 1]$, due to (4.114). Without proof we mention the fact that the Legendre polynomials are bounded in absolute value by one for $x \in [-1, 1]$.

The Legendre polynomials are suitable for approximating a continuous function $f(x)$ in $[-1, 1]$ by a polynomial of degree $n$

$$g_n(x) = \sum_{k=0}^{n} c_k P_k(x) \tag{4.120}$$

in the mean square so that we have

$$\int_{-1}^{1} [g_n(x) - f(x)]^2 \, dx = \text{Min!} \tag{4.121}$$

Figure 4.9 Legendre polynomials $P_2(x)$ to $P_8(x)$

## 4.3 Orthogonal Polynomials

Analogous to the corresponding problem with Chebyshev polynomials, because of (4.110), the coefficients $c_k$ of the expansion (4.120) are given by

$$c_k = \frac{2k+1}{2} \int_{-1}^{1} f(x) P_k(x) \, dx \quad (k = 0, 1, \ldots, n). \tag{4.122}$$

The analytic evaluation of the integrals (4.122) is in general quite laborious, and the resulting formulae lead to severe cancellation in the course of their numerical evaluation. For an approximate computation of the integrals the *Gaussian quadrature rules* are most adequate, which in turn are based on the Legendre polynomials (see Section 8.4).

*Example* 4.12 The function $f(x) = e^x$ is to be approximated by a polynomial $g_6(x)$ (4.120) in the mean square on the interval $[-1, 1]$. Its coefficients $c_k$ (4.122) are given by

$$c_k = \frac{2k+1}{2} \int_{-1}^{1} e^x P_k(x) \, dx \quad (k = 0, 1, 2, \ldots, 6). \tag{4.123}$$

In order to compute these integrals we determine the following auxiliary integrals

$$I_n := \int_{-1}^{1} x^n e^x \, dx \quad (n = 0, 1, 2, \ldots, 6), \tag{4.124}$$

from which $c_k$ will follow by simple linear combinations. Integration by parts yields the recursion formula

$$I_n = x^n e^x \Big|_{-1}^{1} - n \int_{-1}^{1} x^{n-1} e^x \, dx = \left( e - (-1)^n \frac{1}{e} \right) - n I_{n-1} \quad n \geq 1,$$

hence

$$I_n = \begin{cases} \left( e - \dfrac{1}{e} \right) - n I_{n-1} & \text{if } n \text{ is even} \\ \left( e + \dfrac{1}{e} \right) - n I_{n-1} & \text{if } n \text{ is odd} \end{cases} \tag{4.125}$$

For numerical purposes the recursion (4.125) is highly unstable for large values of $n$, because an initial error of $I_0$ will be enlarged by a factor $n!$ to that of $I_n$. However, the recursion formula (4.125) is numerically stable if it is applied in the opposite order

$$I_{n-1} = \begin{cases} \left[ \left( e - \dfrac{1}{e} \right) - I_n \right] \Big/ n & \text{if } n \text{ is even} \\ \left[ \left( e + \dfrac{1}{e} \right) - I_n \right] \Big/ n & \text{if } n \text{ is odd} \end{cases} \tag{4.126}$$

Table 4.6 Integrals and coefficients of the expansion

| $k =$ | $I_k$ | $c_k$ |
|---|---|---|
| 0 | 2.350 402 387 29 | 1.175 201 1936 |
| 1 | 0.735 758 882 34 | 1.103 638 3235 |
| 2 | 0.878 884 622 60 | 0.357 814 3506 |
| 3 | 0.449 507 401 82 | 0.070 455 6337 |
| 4 | 0.552 372 779 99 | 0.009 965 1281 |
| 5 | 0.324 297 369 69 | 0.001 099 5861 |
| 6 | 0.404 618 169 13 | 0.000 099 4543 |

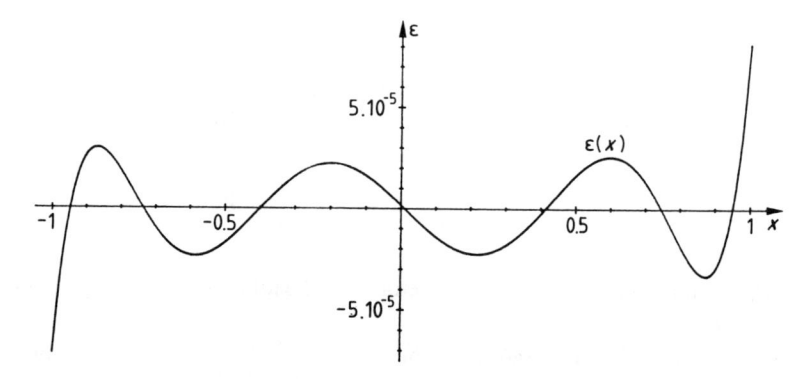

Figure 4.10 Error of approximation for $e^x$, approximation in the quadratic mean

started for a sufficiently large $N$ with $I_N = 0$. In order to compute the desired integrals with an accuracy of 14 decimal places $N = 24$ is sufficient. Table 4.6 contains the values of the integrals $I_0$ to $I_6$ and the resulting coefficients $c_0$ to $c_6$. The error function $\varepsilon(x) := e^x - g_6(x)$ is shown in Figure 4.10. In comparison with the corresponding approximation by Chebyshev polynomials the absolute value of the error at the ends of the interval is more than twice as large, whereas it is of comparable size in the interior of the interval. The different behaviour of the error function is due to the weight function.

## 4.4 Exercises

**4.1.** What are the Fourier series of the following $(2\pi)$ periodic functions that are of some importance in electrical engineering?

(a) $f_1(x) = |\sin(x)|$ (commuted sine function);

(b) $f_2(x) = \begin{cases} \sin(x) & 0 \leqslant x \leqslant \pi \\ 0 & \pi \leqslant x \leqslant 2\pi \end{cases}$ (rectified sine function);

## 4.4 Exercises

(c) $f_3(x) = \begin{cases} 1 & 0 < x < \pi \\ -1 & \pi < x < 2\pi \end{cases}$ (rectangular function);

(d) $f_4(x) = x - \pi$, $0 < x < 2\pi$ (sawtooth function).

From the resulting Fourier series derive for appropriately chosen $x$, the sums of special infinite series by applying Theorem 4.2. To study the convergence of the Fourier series draw the graphs of the Fourier polynomials $g_n(x)$ for $3 \leqslant n \leqslant 8$.

**4.2.** Compute the approximate Fourier coefficients (4.18) of the functions $f_i(x)$, ($i = 1, 2, 3, 4$) of Exercise 4.1 for $N = 8, 12, 16, 24, 32, 64$. Apply Runge's algorithm and the fast Fourier transformation whenever this is possible. Verify the relationships (4.32) and (4.33) with the resulting coefficients. How large must $N$ be chosen such that the approximations $a_k^*$ and $b_k^*$ differ from the Fourier coefficients $a_k$ and $b_k$ by less than $10^{-6}$?

**4.3.** With the approximate Fourier coefficients $a_k^*$ and $b_k^*$ of the functions $f_i(x)$ of Exercise 4.1, for $N = 16$, represent the corresponding Fourier polynomials $g_4^*(x)$ and $g_8^*(x)$ together with the functions $f_i(x)$ graphically. What can be said about the approximations?

**4.4.** Let $N = 2n + 1$, $n \in \mathbb{N}$. Show that

$$g_n^*(x) := \tfrac{1}{2}a_0^* + \sum_{k=1}^{n} [a_k^* \cos(kx) + b_k^* \sin(kx)]$$

with the coefficients (4.18) is the unique interpolating Fourier polynomial corresponding to the support abscissae $x_j$ (4.17) and the functions values $f(x_j)$, $j = 0, 1, \ldots, N$.

**4.5.** Approximate the function $f(x) = \sin(\tfrac{1}{2}\pi x)$ on the interval $[-1, 1]$ by an expansion in terms of Chebyshev polynomials $g_n(x)$, (4.87), for $n = 3, 5, 7, 9, 11$. Compute the coefficients $c_k$ approximately as Fourier coefficients with $N = 32$ and estimate the error of the coefficients $c_k^*$. What are the error bounds $|g_n(x) - f(x)|$? What can be said about the qualitative behaviour of the error functions $\varepsilon_n(x) := g_n(x) - f(x)$?

**4.6.** Determine the interpolating polynomials $P_n^*(x)$ for $n = 3, 5, 7, 9, 11$ for the function $f(x) = \sin(\tfrac{1}{2}\pi x)$ on the interval $[-1, 1]$ corresponding to the Chebyshev abscissae. What are the bounds for the maximal interpolation errors? Compare them with the error bounds of Exercise 4.5 and draw the graphs of the error functions $\varepsilon_n^*(x) := P_n^*(x) - f(x)$.

**4.7.** Determine the interpolating polynomials $P_n^*(x)$ for $n = 6, 8, 10, 12, 14, 16$ for the function $f(x) = 1/(1 + 25x^2)$ on the interval $[-1, 1]$ (Runge's example) corresponding to the Chebyshev abscissae and draw the graphs of the approximations. Observe the convergence behaviour of the polynomials $P_n^*(x)$ with increasing $n$.

**4.8.** The function $f(x) = \sin(\tfrac{1}{2}\pi x)$ is to be approximated by polynomials of degree $n$, ($n = 3, 5, 7, 9$) in the mean square on the interval $[-1, 1]$ as an expansion in terms of Legendre polynomials. The coefficients $c_k$ of the expansion can be computed by means of a recursion formula. Are the resulting formulae suitable for the numerical evaluation? Compare the coefficients of the approximating polynomials $g_n(x)$ with those of the truncated Taylor series $t_n(x)$ of $f(x)$. Finally draw the graphs of the error functions $\varepsilon_n(x) := g_n(x) - f(x)$ and $\delta_n(x) := t_n(x) - f(x)$ and determine the ratios of the maximal error sizes.

**4.9.** The Laguerre polynomials $L_n(x)$ can be defined by

$$L_n(x) := e^x \frac{d^n}{dx^n}(x^n e^{-x}) \quad (n = 0, 1, 2, \ldots).$$

(a) Give the Laguerre polynomials $L_0(x), L_1(x), \ldots, L_6(x)$ explicitly. Show that $L_n(x)$ is a polynomial of exact degree $n$ and that for general $n$ we have the representation

$$L_n(x) = n! \sum_{j=0}^{n} \binom{n}{j} \frac{(-x)^j}{j!} \quad (n = 0, 1, 2, \ldots).$$

(b) Verify the orthogonality of $L_n(x)$:

$$\int_0^\infty e^{-x} L_i(x) L_j(x) \, dx = 0 \quad \text{for all } i \neq j, \; i, j \in \mathbf{N}_0.$$

What is the constant of normalization?

(c) Show that $L_n(x)$ has $n$ simple zeros in the interval $(0, \infty)$.

(d) Show that the following recursion holds

$$L_{n+1}(x) = (2n + 1 - x) L_n(x) - n^2 L_{n-1}(x) \quad (n = 1, 2, 3, \ldots).$$

# 5

# Nonlinear Equations

A frequent problem in applied mathematics is the approximate solution of nonlinear equations. For this purpose we use iterative methods that produce the desired solution as the limit of a sequence of approximations. We shall formulate the Banach fixed point theorem which is the theoretical basis for the study of the convergence properties. The convergence behaviour of the sequence of approximate values towards the solution is of paramount importance for the usefulness and the efficiency of a method. With this point of view, some methods for determining a solution of a nonlinear equation in one unknown are developed and analyzed. The considerations will be generalized to systems. The determination of the zeros of polynomials will be a special application of the methods.

## 5.1 The Banach Fixed Point Theorem

A large class of iterative methods have the form

$$x^{(k+1)} = F(x^{(k)}) \quad (k=0,1,2,\ldots), \tag{5.1}$$

where $x^{(k)}$ can be a real number, a vector or a function with certain properties and $F(x)$ is a mapping of the corresponding set into itself. Given the initial value $x^{(0)}$, equation (5.1) defines a sequence of iterates, $x^{(k)}$, with the aim to solve the equation

$$x = F(x). \tag{5.2}$$

The desired solution $x$ is called the *fixed point* of the mapping $F(x)$, and therefore we call (5.2) a fixed point equation. The iteration scheme (5.1) is called a *fixed point iteration* or a *method of successive approximation*. Because the scheme (5.1) only requires the element $x^{(k)}$ to define the subsequent element $x^{(k+1)}$, and because it does not depend on $k$, the procedure is often called a one-stage, stationary iterative method.

For a uniform theoretical analysis of iterative schemes, (5.1), we introduce the concept of a Banach space.

**Definition 5.1** *A Banach space $B$ is a vector space over the field $\mathbf{K}(=\mathbf{R}, \mathbf{C})$, equipped with a norm, with the properties*

$\|x\| \geqslant 0$ \qquad for all $x \in B$,

$\|x\| = 0 \Leftrightarrow x = 0$,

$\|\gamma x\| = |\gamma| \cdot \|x\|$ \qquad for all $\gamma \in \mathbf{K}, x \in B$,

$\|x + y\| \leqslant \|x\| + \|y\|$ \qquad for all $x, y \in B$,

*and each Cauchy sequence $x^{(k)}$ of $B$ converges in $B$.*

The field of the real numbers $\mathbf{R}$ with the norm defined by the absolute value is a Banach space. More generally, the $n$-dimensional real vector space $\mathbf{R}^n$ with an arbitrary vector norm $\|\mathbf{x}\|$ is a Banach space. Finally, the set of continuous functions over a bounded interval $I = [a, b]$, that is $C(I)$, with the norm

$$\|f\| := \max_{x \in I} |f(x)| \quad f(x) \in C(I) \tag{5.3}$$

is a Banach space.

**Definition 5.2** *We consider a closed subset $A \subset B$ of a Banach space $B$ and a mapping $F: A \to A$ of $A$ onto itself. The mapping $F$ is called Lipschitz continuous on $A$, with the Lipschitz constant $0 < L < \infty$, if*

$$\|F(x) - F(y)\| \leqslant L\|x - y\| \quad \text{for all } x, y \in A \tag{5.4}$$

*The mapping is a contraction on $A$ if the Lipschitz constant of (5.4) satisfies $L < 1$.*

**Theorem 5.1** (Banach fixed point theorem). *Let $A$ be a closed subset of a Banach space $B$ and $F: A \to A$ a contraction mapping ($L < 1$). Then we have*

   (a) *The mapping $F$ has a unique fixed point $s \in A$.*

   (b) *For arbitrary initial values $x^{(0)} \in A$ the sequence $x^{(k)}$ defined by (5.1) converges to the fixed point $s$.*

   (c) *The error estimate holds*

$$\|s - x^{(k)}\| \leqslant \frac{L^{k-1}}{1 - L} \|x^{(l+1)} - x^{(l)}\| \quad \text{for } 0 \leqslant l < k. \tag{5.5}$$

*Proof* We first prepare an estimate of the norm of the difference of consecutive

## 5.1 The Banach Fixed Point Theorem

elements of the sequence $x^{(k)}$

$$\|x^{(k+1)} - x^{(k)}\| = \|F(x^{(k)}) - F(x^{(k-1)})\| \leq L\|x^{(k)} - x^{(k-1)}\|$$
$$= L\|F(x^{(k-1)}) - F(x^{(k-2)})\| \leq L^2 \|x^{(k-1)} - x^{(k-2)}\| = \cdots$$

From this we deduce

$$\|x^{(k+1)} - x^{(k)}\| \leq L^{k-l} \|x^{(l+1)} - x^{(l)}\| \quad \text{for } 0 \leq l \leq k. \tag{5.6}$$

In order to show the existence of a fixed point we first show that the sequence of iterates $x^{(k)}$, (5.1), is convergent for an arbitrary $x^{(0)} \in A$. The $x^{(k)}$ form a Cauchy sequence, because for every $m \geq 1$ and $k \geq 1$ and using (5.6) with $l = 0$, we have

$$\|x^{(k+m)} - x^{(k)}\| = \|x^{(k+m)} - x^{(k+m-1)} + x^{(k+m-1)} - + \cdots - x^{(k)}\|$$
$$\leq \sum_{\mu=k}^{k+m-1} \|x^{(\mu+1)} - x^{(\mu)}\|$$
$$\leq L^k(L^{m-1} + L^{m-2} + \cdots + L + 1)\|x^{(1)} - x^{(0)}\|$$
$$= L^k \frac{1-L^m}{1-L} \|x^{(1)} - x^{(0)}\|. \tag{5.7}$$

From the hypothesis $L < 1$, given any $\varepsilon > 0$ we can find an $N \in \mathbb{N}$ such that for $k \geq N$ and $m \geq 1$ the relation $\|x^{(k+m)} - x^{(k)}\| < \varepsilon$ holds. Hence the sequence $x^{(k)}$ has a limit value in the closed subset $A$

$$s := \lim_{k \to \infty} x^{(k)} \quad s \in A.$$

However, the mapping $F$ is Lipschitz continuous, hence we have

$$F(s) = F\left(\lim_{k \to \infty} x^{(k)}\right) = \lim_{k \to \infty} F(x^{(k)}) = \lim_{k \to \infty} x^{(k+1)} = s.$$

Hence, we have not only proved the existence of a fixed point $s \in A$, but also showed the convergence of the sequence $x^{(k)}$ to a fixed point.

The uniqueness of the fixed point is proved by contradiction. We assume that $s_1 \in A$ and $s_2 \in A$ are two fixed points of the mapping $F$ with $\|s_1 - s_2\| > 0$. Since $s_1 = F(s_1)$ and $s_2 = F(s_2)$ we have $\|s_1 - s_2\| = \|F(s_1) - F(s_2)\| \leq L\|s_1 - s_2\|$. This leads to the contradiction $L \geq 1$.

Now we apply the estimate (5.6) in (5.7) for a fixed integer $l < k$ and obtain

$$\|x^{(k+m)} - x^{(k)}\| \leq L^{k-l}\left(\frac{1-L^m}{1-L}\right) \|x^{(l+1)} - x^{(l)}\| \quad m \geq 1. \tag{5.8}$$

We keep $k$ and $l$ fixed in (5.8) and let $m$ tend to infinity. Because $\lim_{m \to \infty} x^{(k+m)} = s$ and $L < 1$ the error estimate (5.5) follows from (5.8).

Two useful estimates can be derived from (5.5) which one can apply if the

Lipschitz constant $L$ is approximately known. For $l = 0$ we get the *a priori error estimate*

$$\|s - x^{(k)}\| \leqslant \frac{L^k}{1-L} \|x^{(1)} - x^{(0)}\| \quad (k = 1, 2, \ldots). \tag{5.9}$$

This allows us to predict an absolute bound for the error $x^{(k)} - s$ as soon as the first iterate $x^{(1)}$ has been computed from the initial value $x^{(0)}$. At the same time from (5.9) we learn that the norm of the error $x^{(k)} - s$ decreases at least like a geometric sequence with the quotient $L$. The smaller the Lipschitz constant $L$ the better the convergence of the sequence $x^{(k)}$ to $s$.

For $l = k - 1$ from (5.5) we get the *a posteriori error estimate*

$$\|s - x^{(k)}\| \leqslant \frac{L}{1-L} \|x^{(k)} - x^{(k-1)}\| \quad (k = 1, 2, \ldots). \tag{5.10}$$

After the $k$th iteration step has been completed, the error $x^{(k)} - s$ of the latest computed approximation can be estimated.

The statement of Theorem 5.1 has only a local character in many practical situations, because the closed subset $A$ may be quite small. Moreover, it is usually difficult to describe the set $A$ quantitatively for a given mapping, and we must often be content to know that $A$ merely exists.

*Example* 5.1 We seek the solution of the nonlinear equation

$$x = e^{-x} =: F(x) \quad x \in \mathbf{R}, \tag{5.11}$$

that is the fixed point of the mapping $F: \mathbf{R} \to \mathbf{R}_+$. In order to be able to apply the fixed point theorem we explicitly need to know that there exists a closed interval $A$ of the Banach space $B = \mathbf{R}$ for which the hypotheses are satisfied. For example $A := [0.5, 0.69]$ is mapped onto itself by the mapping $F$ (5.11). The mean value theorem implies that the Lipschitz constant $L$ of the continuously differentiable function $F$ is given by

$$L = \max_{x \in A} |F'(x)| = \max_{x \in A} |-e^{-x}| = e^{-0.5} \doteq 0.606\,531 < 1.$$

Hence, $F$ is a contraction mapping on $A$, and there exists a unique fixed point $s$ in $A$. The result of the fixed point iteration is summarized partly in Table 5.1 for the initial value $x^{(0)} = 0.55 \in A$ to eight significant figures.

## 5.2 Behaviour and Order of Convergence

Table 5.1 Fixed point iteration

| k | $x^{(k)}$ | k | $x^{(k)}$ | k | $x^{(k)}$ |
|---|---|---|---|---|---|
| 0 | 0.550 000 00 | 10 | 0.567 083 94 | 20 | 0.567 143 09 |
| 1 | 0.576 949 81 | 11 | 0.567 176 95 | 21 | 0.567 143 40 |
| 2 | 0.561 608 77 | 12 | 0.567 124 20 | 22 | 0.567 143 23 |
| 3 | 0.570 290 86 | 13 | 0.567 154 12 | 23 | 0.567 143 32 |
| 4 | 0.565 360 97 | 14 | 0.567 137 15 | 24 | 0.567 143 27 |

From the first two values $x^{(0)}$ and $x^{(1)}$ we can estimate the number, $k$, of iteration steps that are required so that the error satisfies $|s - x^{(k)}| \leqslant \varepsilon = 10^{-6}$. The *a priori* error estimate yields

$$k \geqslant \log\left(\frac{\varepsilon(1-L)}{|x^{(1)} - x^{(0)}|}\right) \Big/ \log(L) \doteq 22.3 \tag{5.12}$$

a slight overestimate as can be seen from Table 5.1.

For the iterate $x^{(12)}$ the *a priori* estimate (5.9) yields $|s - x^{(12)}| \leqslant 1.70 \times 10^{-4}$, whereas (5.10) provides the better *a posteriori* error bound $|s - x^{(12)}| \leqslant 8.13 \times 10^{-5}$. The fixed point is $s \doteq 0.567 143 29$ and hence the error is in fact $|s - x^{(12)}| \doteq 1.91 \times 10^{-5}$. For $x^{(22)}$ from (5.10) we obtain the realistic error bound $|s - x^{(22)}| \leqslant 2.6 \times 10^{-7}$, which is about four times too large. This shows that fewer than 22 iterations are required to attain the desired accuracy.

### 5.2 Behaviour and Order of Convergence

We want to analyse in more detail the convergence behaviour of the sequence $x^{(k)}$, of (5.1), to $s$. On the assumptions of the Banach fixed point theorem the relationship

$$\|s - x^{(k+1)}\| = \|F(s) - F(x^{(k)})\| \leqslant L\|s - x^{(k)}\| \quad (k = 0, 1, 2, \ldots). \tag{5.13}$$

holds, so that the norm of the error

$$\varepsilon^{(k)} := x^{(k)} - s \quad (k = 0, 1, 2, \ldots) \tag{5.14}$$

decreases at least by the factor $L < 1$ in each iteration step. In the following we mainly consider the case $B = \mathbf{R}$ and a mapping $F: \mathbf{R} \to \mathbf{R}$.

**Theorem 5.2** *Let $I = [a, b] \subset \mathbf{R}$ be a closed interval, and let $F: I \to I$ be a Lipschitz continuous, contraction mapping on $I$, that has a continuous first derivative. Moreover, we assume $F'(x) \neq 0$ for all $x \in I$. Then we have for the error $\varepsilon^{(k)}$*

$$\lim_{k \to \infty} \frac{\varepsilon^{(k+1)}}{\varepsilon^{(k)}} = F'(s). \tag{5.15}$$

*Proof* We first show that $x^{(k)} \neq s$ for all $k > 0$ if $s \neq x^{(0)} \in I$. This means that the fixed point iteration cannot produce the fixed point $s$ in a finite number of steps. We verify this statement by an indirect argument. Assume the existence of an index $k > 0$ for which $x^{(k-1)} \neq s$ and $x^{(k)} = s$ hold. Because $F(x^{(k)}) = x^{(k)}$ from the mean value theorem we obtain

$$0 = F(x^{(k-1)}) - F(x^{(k)}) = (x^{(k-1)} - x^{(k)})F'(x^*)$$

since $(x^{(k-1)} - x^{(k)}) \neq 0$, this implies that $F'(x^*) = 0$ for a $x^* \in I$ contradicting the hypothesis.

Now we apply the mean value theorem again and get

$$\varepsilon^{(k+1)} = x^{(k+1)} - s = F(x^{(k)}) - F(s)$$
$$= F(s + \varepsilon^{(k)}) - F(s) = \varepsilon^{(k)} F'(s + \theta_k \varepsilon^{(k)})$$

where $0 < \theta_k < 1$. As $\varepsilon^{(k)} \neq 0$, this is equivalent to

$$\frac{\varepsilon^{(k+1)}}{\varepsilon^{(k)}} = F'(s + \theta_k \varepsilon^{(k)}) \quad (k = 0, 1, 2, \ldots).$$

The hypotheses on $F(x)$ imply $\lim_{k \to \infty} x^{(k)} = s$ and $\lim_{k \to \infty} \varepsilon^{(k)} = 0$. The assertion (5.15) then follows from the continuity of $F'(x)$ on $I$.

For sufficiently large $k$ the error $\varepsilon^{(k)}$ decreases as a geometric sequence with the quotient $q = F'(s)$, and asymptotically we have

$$\varepsilon^{(k+1)} \approx q \varepsilon^{(k)} \quad |q| < 1, k \gg 1. \tag{5.16}$$

As the error $\varepsilon^{(k+1)}$ is approximately proportional to $\varepsilon^{(k)}$, the sequence $x^{(k)}$ is said to be *linearly convergent*, and $|q|$ is called the *convergence quotient* or *convergence factor*. In the case of linear convergence the same number of iteration steps are asymptotically required to reduce the magnitude of $\varepsilon^{(k)}$ by one tenth. In other words this means that after $m$ further iterations, another decimal place of the approximate value is correct. From $\varepsilon^{(k+m)} \approx q^m \varepsilon^{(k)}$ the condition for $m$ follows

$$m \geq \frac{-1}{\log_{10}|q|} \quad q = F'(s). \tag{5.17}$$

The convergence factor $|q|$ is crucial for the convergence behaviour, and the following table provides some insight.

| $\|q\| =$ | 0.316 | 0.562 | 0.681 | 0.750 | 0.794 | 0.891 | 0.944 | 0.9716 |
|---|---|---|---|---|---|---|---|---|
| $m =$ | 2 | 4 | 6 | 8 | 10 | 20 | 40 | 80 |

(5.18)

For efficiency the convergence factor should be as small as possible so that, in the case of linear convergence, the sequence of $x^{(k)}$ converges as fast as possible.

## 5.2 Behaviour and Order of Convergence

To illustrate that the statement (5.17) is asymptotically valid we consider Example 5.1. There the convergence quotient is $q = F'(s) = -e^{-s} = -s \doteq -0.5671$. From (5.18) about four iterations are needed to gain an extra correct decimal place. This is equivalent to saying that twelve iterations produce about three extra correct decimal places. This qualitative statement is confirmed by the values $x^{(0)}$, $x^{(12)}$, $x^{(24)}$.

We shall become acquainted with fixed point iterations for which the hypotheses of Theorem 5.2 are not satisfied because $F'(s) = 0$. In this case the convergence is characterized by the following theorem.

**Theorem 5.3** *Let $F: I \to I$ be a Lipschitz continuous contraction mapping of the closed interval $I = [a, b]$ onto itself, where $F(x)$ is twice continuously differentiable. Moreover, let $F'(s) = 0$ and $F''(x) \neq 0$ for all $x \in I$. Then the error $\varepsilon^{(k)}$ satisfies*

$$\lim_{k \to \infty} \frac{\varepsilon^{(k+1)}}{\varepsilon^{(k)^2}} = \tfrac{1}{2} F''(s). \tag{5.19}$$

*Proof* We apply the Taylor series expansion with remainder term, and obtain

$$\varepsilon^{(k+1)} = x^{(k+1)} - s = F(x^{(k)}) - F(s) = F(s + \varepsilon^{(k)}) - F(s)$$
$$= F(s) + \varepsilon^{(k)} F'(s) + \tfrac{1}{2} \varepsilon^{(k)^2} F''(s + \theta_k \varepsilon^{(k)}) - F(s) = \tfrac{1}{2} \varepsilon^{(k)^2} F''(s + \theta_k \varepsilon^{(k)})$$

with $0 < \theta_k < 1$. The hypothesis $F''(x) \neq 0$ for all $x \in I$, as before, implies that for $s \neq x^{(0)} \in I$ the iterates $x^{(k)}$ for all $k > 0$ are not equal to $s$, so that $\varepsilon^{(k)} \neq 0$. Hence we obtain

$$\frac{\varepsilon^{(k+1)}}{\varepsilon^{(k)^2}} = \tfrac{1}{2} F''(s + \theta_k \varepsilon^{(k)}) \quad (k = 0, 1, 2, \ldots). \tag{5.20}$$

Because of the hypothesis on $F(x)$ it follows that $\lim_{k \to \infty} \varepsilon^{(k)} = 0$, and from the continuity of $F''(x)$ on $I$ the assertion (5.19) follows from (5.20).

From the assumptions of Theorem 5.3 and for sufficiently large $k$, the following relationship is approximately valid

$$\varepsilon^{(k+1)} \approx K \varepsilon^{(k)^2} \quad K = \tfrac{1}{2} F''(s) \quad k \gg 1. \tag{5.21}$$

Hence, $\varepsilon^{(k+1)}$ is now proportional to the square of $\varepsilon^{(k)}$ with a constant of proportionality $K$ that is independent of $k$. Therefore we call such a sequence, $x^{(k)}$, *quadratically convergent*. The quadratic convergence of a sequence $x^{(k)}$ to $s$ means that for $|K| \approx 1$ the number of digits coinciding with $s$ is doubled in each step. If the magnitude of $K$ is large compared to one, this qualitative statement is true in a weaker sense.

**Definition 5.3** *An iterative method is said to possess the order of convergence $p \geq 1$, if the generated sequence $x^{(k)}$ converges to the limit $s$ such that*

$$\limsup_{k\to\infty} \frac{\|x^{(k+1)} - s\|}{\|x^{(k)} - s\|^p} = K \quad \text{where } 0 < K < \infty \text{ and } K < 1 \text{ for } p = 1. \quad (5.22)$$

In addition to the order of convergence $p$, the *asymptotic error constant* $K$ defined by (5.22) characterizes the behaviour of the sequence $x^{(k)}$. A high order of convergence in connection with a small error constant means a rapid convergence, at least asymptotically. The special cases $p = 1$ (linear convergence) and $p = 2$ (quadratic convergence) have been analysed in Theorems 5.2 and 5.3, and the error constants have also been given there. There exist iterative methods whose order of convergence is not an integer.

If the sequence $x^{(k)}$ is known to be linearly convergent to $s$, it can be used to construct a sequence that converges more rapidly. We assume $k$ to be sufficiently large so that (5.16) holds in the sense of a good approximation for two consecutive index values

$$x^{(k+1)} - s \approx q(x^{(k)} - s)$$
$$x^{(k+2)} - s \approx q(x^{(k+1)} - s). \quad (5.23)$$

In (5.23) $s$ and $q$ are unknown quantities. The unknown quotient $q$ can easily be eliminated from these two approximately valued equations, and so we get

$$(x^{(k+1)} - s)^2 \approx (x^{(k)} - s)(x^{(k+2)} - s),$$

and solve for $s$

$$s \approx \frac{x^{(k+2)} x^{(k)} - x^{(k+1)^2}}{x^{(k+2)} - 2x^{(k+1)} + x^{(k)}}. \quad (5.24)$$

Three consecutive iterates $x^{(k)}$, $x^{(k+1)}$, $x^{(k+2)}$ allow us to compute the limit $s$ according to (5.24), at least approximately. We assume that the denominator of (5.24) is nonzero, and thus we understand the expression to the right to be the $k$th term of a new sequence. In the literature two different representations exist. We choose the form which will be the more adequate for the definition of a new method, derived from the fixed point iteration. In order to eliminate the problem of cancellation of digits in the numerator of (5.24), we transform the expression as follows

$$z^{(k)} := \frac{x^{(k)} x^{(k+2)} - x^{(k+1)^2}}{x^{(k+2)} - 2x^{(k+1)} + x^{(k)}}$$

$$= \frac{x^{(k)}(x^{(k+2)} - 2x^{(k+1)} + x^{(k)}) - x^{(k+1)^2} + 2x^{(k+1)} x^{(k)} - x^{(k)^2}}{x^{(k+2)} - 2x^{(k+1)} + x^{(k)}}$$

$$\boxed{z^{(k)} := x^{(k)} - \frac{(x^{(k+1)} - x^{(k)})^2}{x^{(k+2)} - 2x^{(k+1)} + x^{(k)}} \quad (k = 0, 1, 2, \ldots)} \quad (5.25)$$

## 5.2 Behaviour and Order of Convergence

The formula (5.25) for the calculation of the new terms $z^{(k)}$ can easily be remembered if we introduce the first and second differences

$$\Delta^1_{k+\frac{1}{2}} := x^{(k+1)} - x^{(k)} \tag{5.26}$$

$$\Delta^2_{k+1} := x^{(k+2)} - 2x^{(k+1)} + x^{(k)} = \Delta^1_{k+\frac{3}{2}} - \Delta^1_{k+\frac{1}{2}} \tag{5.27}$$

Thus from (5.25) we get the computational scheme

$$\boxed{z^{(k)} = x^{(k)} - \frac{\left(\Delta^1_{k+\frac{1}{2}}\right)^2}{\Delta^2_{k+1}} \quad (k = 0, 1, 2, \ldots),} \tag{5.28}$$

which is called *Aitken's $\Delta^2$ process*. The sequence $z^{(k)}$ converges to $s$ more rapidly than the original sequence.

**Theorem 5.4** *Let $x^{(k)}$ be a linearly convergent sequence with $\varepsilon^{(k+1)} \approx q\varepsilon^{(k)}$, $|q| < 1$. Then the sequence $z^{(k)}$ obtained by Aitken's $\Delta^2$ process (5.28) satisfies*

$$\lim_{k \to \infty} \frac{z^{(k)} - s}{x^{(k)} - s} = 0. \tag{5.29}$$

*Proof* The asymptotically valid relationship of the error $\varepsilon^{(k)}$ is expressed in the form

$$\varepsilon^{(k+1)} = (q + \delta_k)\varepsilon^{(k)} \quad \text{with } \lim_{k \to \infty} \delta_k = 0 \tag{5.30}$$

and hence

$$\varepsilon^{(k+2)} = (q + \delta_{k+1})\varepsilon^{(k+1)} = (q + \delta_{k+1})(q + \delta_k)\varepsilon^{(k)}.$$

For the first and second differences we obtain

$$\Delta^1_{k+\frac{1}{2}} = (x^{(k+1)} - s) - (x^{(k)} - s) = \varepsilon^{(k+1)} - \varepsilon^{(k)} = (q - 1 + \delta_k)\varepsilon^{(k)}$$

$$\Delta^2_{k+1} = \varepsilon^{(k+2)} - 2\varepsilon^{(k+1)} + \varepsilon^{(k)}$$

$$= [q^2 - 2q + 1 + (q\delta_k + q\delta_{k+1} + \delta_k\delta_{k+1} - 2\delta_k)]\varepsilon^{(k)}$$

$$=: [(q-1)^2 + \delta'_k]\varepsilon^{(k)} \quad \text{with } \lim_{k \to \infty} \delta'_k = 0. \tag{5.31}$$

As $|q| < 1$ and $\varepsilon^{(k)} \neq 0$ it follows that for sufficiently large $k$, at least theoretically, the denominator of (5.28) fulfils $\Delta^2_{k+1} \neq 0$. With these expressions the error $\eta^{(k)} := z^{(k)} - s$ is given by

$$\eta^{(k)} = \varepsilon^{(k)} - \frac{[(q-1) + \delta_k]^2}{[(q-1)^2 + \delta'_k]}\varepsilon^{(k)} = \frac{\delta'_k - 2(q-1)\delta_k - \delta_k^2}{(q-1)^2 + \delta'_k}\varepsilon^{(k)}. \tag{5.32}$$

If we take (5.30) and (5.31) into account, the assertion (5.29) is a consequence of (5.32).

Table 5.2 $\Delta^2$ process of Aitken, fixed point iteration

| $k$ | $x^{(k)}$ | $\Delta^1_{k+\frac{1}{2}}$ | $\Delta^2_{k+1}$ | $z^{(k)}$ | $\eta^{(k)}$ |
|---|---|---|---|---|---|
| 0 | 0.55000000 | 0.02694981 | −0.04229085 | 0.56717375 | 0.00003046 |
| 1 | 0.57694981 | −0.01534104 | 0.02402313 | 0.56715311 | 0.00000982 |
| 2 | 0.56160877 | 0.00868209 | −0.01361198 | 0.56714644 | 0.00000315 |
| 3 | 0.57029086 | −0.00492989 | 0.00772394 | 0.56714430 | 0.00000101 |
| 4 | 0.56536097 | 0.00279405 | −0.00437929 | 0.56714361 | 0.00000032 |
| 5 | 0.56815502 | −0.00158524 | 0.00248411 | 0.56714340 | 0.00000011 |
| 6 | 0.56656978 | 0.00089887 | −0.00140873 | 0.56714332 | 0.00000003 |
| 7 | 0.56746865 | −0.00050986 | 0.00079901 | 0.56714330 | 0.00000001 |
| 8 | 0.56695879 | 0.00028915 | −0.00045315 | 0.56714329 | |
| 9 | 0.56724794 | −0.00016400 | | | |
| 10 | 0.56708394 | | | | |

*Example* 5.2  We illustrate the effect of the $\Delta^2$ process of Aitken by the example of the fixed point iteration for solving $x = e^{-x}$. We use a precision of eight significant figures and give the values in Table 5.2. The faster convergence of the sequence $z^{(k)}$ is obvious. From the computed errors $\eta^{(k)}$ we can determine an approximation of the convergence factor of the sequence $z^{(k)}$

$$q_z = \lim_{k \to \infty} \frac{\eta^{(k+1)}}{\eta^{(k)}} \doteq 0.32 < |q_x| \doteq 0.567. \tag{5.33}$$

We see that already $z^{(0)}$ is a good approximation of $s$, and from (5.33) and (5.18) the element $z^{(8)}$ coincides with $s$ within the given precision.

The $\Delta^2$ process of Aitken is applicable to an arbitrary linearly convergent sequence to improve the convergence. For instance it can be useful to compute an approximation of the sum $s = \sum_{j=0}^{\infty} a_j$ of an infinite, convergent series. If the series satisfies the quotient criterion $\lim_{j \to \infty} a_{j+1}/a_j = q, |q| < 1$, so that the terms of the series decrease for sufficiently large index $j$ as a geometric sequence, then the partial sums $x^{(k)} := \sum_{j=0}^{k} a_j$ are a linearly convergent sequence. Hence the sequence $z^{(k)}$ converges more rapidly to $s$ according to Theorem 5.4. As the $z^{(k)}$ also usually form a linearly convergent sequence, it is possible to apply Aitken's $\Delta^2$ process again to obtain an even more rapidly convergent sequence. However, the procedure is limited for numerical reasons because the computation of the first and second differences together with the terms of the new sequence, suffers from rounding errors.

*Example* 5.3  The series

$$s = \sum_{j=0}^{\infty} \frac{1}{\cosh(j)} \quad \text{with} \lim_{j \to \infty} a_{j+1}/a_j = e^{-1}$$

satisfies the hypotheses for an application of Aitken's $\Delta^2$ process to the partial

## 5.2 Behaviour and Order of Convergence

Table 5.3 Summation of a series, $\Delta^2$ process

| k | $x^{(k)}$ | $\Delta^1_{k+\frac{1}{2}} = a_{k+1}$ | $z^{(k)}$ | $\zeta^{(k)}$ |
|---|---|---|---|---|
| 0 | 1.000 0000 | 0.648 054 27 | 2.098 6844 | 2.071 1234 |
| 1 | 1.648 0543 | 0.265 802 23 | 2.072 4491 | 2.071 1213 |
| 2 | 1.913 8565 | 0.099 327 927 | 2.071 1872 | 2.071 1213 |
| 3 | 2.013 1844 | 0.036 618 993 | 2.071 1246 | 2.071 1213 |
| 4 | 2.049 8034 | 0.013 475 282 | 2.071 1215 | |
| 5 | 2.063 2787 | 0.004 957 4739 | 2.071 1213 | |
| 6 | 2.068 2362 | 0.001 823 7624 | | |
| 7 | 2.070 0600 | | | |
| ⋮ | ⋮ | | | |
| 18 | 2.071 1213 | | | |

sums $x^{(k)}$. Moreover, we have $\Delta^1_{k+\frac{1}{2}} = a_{k+1}$. Table 5.3 contains the most significant numerical values where a precision of eight significant digits has been used. Aitken's $\Delta^2$ process is applied to the sequence of the $z^{(k)}$ again. Within the used arithmetic precision the value of $s$ is obtained with little effort.

Aitken's $\Delta^2$ process is also the basis for defining a given fixed point iteration scheme with linear convergence, another iterative method that has a higher order of convergence. The value $z^{(0)}$, which is obtained from three consecutive iterates $x^{(0)}$, $x^{(1)}$ and $x^{(2)}$ using formula (5.25), is usually a much better approximation of $s$ than $x^{(2)}$. Therefore it is quite natural to take $z^{(0)}$ as the initial value for two subsequent steps of the fixed point iteration and to apply Aitken's $\Delta^2$ process to this triplet of iterates. The combination of the given fixed point iteration $x^{(k+1)} = F(x^{(k)})$ with Aitken's $\Delta^2$ process yields the *Steffensen method*

$$x^{(k+1)} = x^{(k)} - \frac{[F(x^{(k)}) - x^{(k)}]^2}{F(F(x^{(k)})) - 2F(x^{(k)}) + x^{(k)}} =: \tilde{F}(x^{(k)}) \quad (k = 0, 1, 2, \ldots)$$

(5.34)

The computational scheme (5.34) is again a fixed point iteration but the mapping $\tilde{F}(x)$ is a composition of $F(x)$. The quotient in (5.34) takes for a fixed point $s$ the indefinite expression 0/0, but from the proof of Theorem 5.4 it is clear that if $x^{(k)}$ tends to $s$ then the numerator converges to zero faster than the denominator. This remark is essential for the numerical evaluation of (5.34). Steffensen's method is performed best according to the algorithmic description:

$$\begin{aligned} &y := F(x^{(k)}) \quad \Delta^1 := y - x^{(k)} \quad \text{Test: } \Delta^1 = 0 \\ &z := F(y) \quad \Delta^2 := (z - y) - \Delta^1 \\ &x^{(k+1)} = x^{(k)} - (\Delta^1)^2/\Delta^2 \; (= \tilde{F}(x^{(k)})) \end{aligned}$$

(5.35)

A single iteration step (5.35) requires two evaluations of the function $F(x)$. However, this additional effort pays off due to the following theorem.

**Theorem 5.5** *If the order of convergence of the fixed point iteration $x^{(k+1)} = F(x^{(k)})$ is one, then the order of convergence of Steffensen's method is at least two.*

**Proof** Let the function $F(x)$ be sufficiently many times continuously differentiable. From (5.34) the error $\varepsilon^{(k)} = x^{(k)} - s$ satisfies the relationship

$$\varepsilon^{(k+1)} = \varepsilon^{(k)} - \frac{[F(s+\varepsilon^{(k)}) - s - \varepsilon^{(k)}]^2}{F(F(s+\varepsilon^{(k)})) - 2F(s+\varepsilon^{(k)}) + s + \varepsilon^{(k)}} =: \varepsilon^{(k)} - \frac{Z}{N}. \quad (5.36)$$

We expand the numerator and denominator of (5.36) into Taylor series at the point $s$ and use $F(s) = s$. To simplify the notation we omit the index $k$. The Taylor series

$$F(s+\varepsilon) = s + \varepsilon F'(s) + \tfrac{1}{2}\varepsilon^2 F''(s) + O(\varepsilon^3)$$
$$F(F(s+\varepsilon)) = F(s + \varepsilon F'(s) + \tfrac{1}{2}\varepsilon^2 F''(s) + O(\varepsilon^3))$$
$$= s + \varepsilon F'(s)^2 + \tfrac{1}{2}\varepsilon^2 F'(s)F''(s)[1 + F'(s)] + O(\varepsilon^3)$$

yield for the numerator and denominator

$$Z = [(F'(s) - 1) + \tfrac{1}{2}\varepsilon F''(s)]^2 \varepsilon^2 + O(\varepsilon^4)$$
$$N = [(F'(s) - 1)^2 + \tfrac{1}{2}\varepsilon(F'(s)^2 + F'(s) - 2)F''(s)]\varepsilon + O(\varepsilon^3).$$

These expressions are substituted into (5.36), and after some further manipulation we obtain

$$\varepsilon^{(k+1)} = \frac{\tfrac{1}{2}F'(s)F''(s)[F'(s) - 1]\varepsilon^{(k)^2} + O(\varepsilon^{(k)^3})}{[F'(s) - 1]^2 + \tfrac{1}{2}F''(s)[F'(s)^2 + F'(s) - 2]\varepsilon^{(k)} + O(\varepsilon^{(k)^2})}$$

$$= \frac{1}{2}\frac{F'(s)F''(s) + O(\varepsilon^{(k)})}{[F'(s) - 1] + O(\varepsilon^{(k)})}(\varepsilon^{(k)})^2. \quad (5.37)$$

The fixed point iteration $x^{(k+1)} = F(x^{(k)})$ is assumed to be linearly convergent, $F'(s) \neq 1$ and $F'(s) \neq 0$. If $F''(s) \neq 0$ from (5.37) due to $\varepsilon^{(k)} \to 0$ we obtain

$$\lim_{k \to \infty} \frac{|\varepsilon^{(k+1)}|}{|\varepsilon^{(k)}|^2} = \frac{1}{2}\left|\frac{F'(s)F''(s)}{F'(s) - 1}\right| =: K \quad 0 < K < \infty, \quad (5.38)$$

so that the order of convergence is $p = 2$ in this case, otherwise it is higher.

**Example 5.4** Steffensen's method for solving $x = F(x) = e^{-x}$ works as follows if ten significant digit arithmetic is used (see Table 5.4).

The asymptotic error constant, (5.38), is $K \doteq 0.1026$. This means that for this example the number of correct digits is more than doubled in each iteration step.

## 5.3 Equations in One Unknown

Table 5.4 Steffensen's method

| $k$ | $x^{(k)}$ | $F(x^{(k)}) = y$ | $F(y) = z$ | $\varepsilon^{(k)} = x^{(k)} - s$ |
|---|---|---|---|---|
| 0 | 0.400 000 0000 | 0.670 320 0460 | 0.511 544 8337 | −0.167 143 2904 |
| 1 | 0.570 295 3502 | 0.565 358 4353 | 0.568 156 4629 | 0.003 152 0598 |
| 2 | 0.567 144 3082 | 0.567 142 7132 | 0.567 143 6178 | 0.000 001 0178 |
| 3 | 0.567 143 2904 $= s$ | | | |

Steffenson's method has the interesting property that it generates a quadratically convergent sequence $x^{(k)}$ even in those cases where the conditions of convergence are not satisfied for the given fixed point iteration. Also the result (5.38) remains valid if $|F'(s)| > 1$, so that the quadratic convergence is guaranteed asymptotically. In these cases, however, the initial value $x^{(0)}$ must be a sufficiently good approximation of $s$, because the interval for which $\tilde{F}(x)$ is a contraction is usually quite small.

### 5.3 Equations in One Unknown

We consider the problem of finding the solution of the equation

$$\boxed{f(x) = 0} \qquad (5.39)$$

where $f: \mathbf{R} \to \mathbf{R}$ is a continuous nonlinear function. We treat mainly the case of real solutions of (5.39), but will give indications of how complex solutions may be found.

#### 5.3.1 Bisection, regula falsi, secant method

The *bisection method* locates the real solution of (5.39) within a sequence of intervals of systematically smaller lengths. We assume an interval $I = [a, b]$ is known such that $f(a) \cdot f(b) < 0$ holds. Due to the continuity there is a solution $s$ in the interior of $I$. Next the function $f(\mu)$ is evaluated at the midpoint $\mu = (a + b)/2$. If $f(\mu) \neq 0$, then its sign decides in which of the two subintervals $[a, \mu]$ and $[\mu, b]$ the desired solution $s$ is located. The length of the new interval containing the solution $s$ has been halved, and the procedure is continued until $s$ is located in an interval of sufficiently small length. Let us denote by $L_0 := b - a$ the length of the initial interval, then the lengths $L_k := (b - a)/2^k$, $(k = 0, 1, 2, \ldots)$ tend to zero. The middle point $x^{(k)}$ of the interval after $k$ bisections is an approximation of $s$ with the a priori error estimate

$$|x^{(k)} - s| \leq \frac{b - a}{2^{k+1}} \quad (k = 0, 1, 2, \ldots). \qquad (5.40)$$

This error bound decreases as a geometric sequence with the quotient $q = 0.5$, and hence the order of convergence is $p = 1$.

The initial assumption of the *regula falsi method* (or *false position method*) is the same as before, namely there exists an interval $I = [x^{(0)}, x^{(1)}]$ with $f(x^{(0)}) \cdot f(x^{(1)}) < 0$. The two initial values $x^{(0)}$ and $x^{(1)}$ are supposed to be appropriate approximations of the desired solution $s$. Instead of using the middle point of the interval as the subsequent test point, $x^{(2)}$ is determined to be the zero of the linear interpolation function corresponding to the support abscissae $x^{(0)}$ and $x^{(1)}$ with the function values $y_0 := f(x^{(0)})$ and $y_1 := f(x^{(1)})$. The interpolation polynomial $P_1(x)$ is

$$P_1(x) = y_0 + (x - x^{(0)}) \frac{y_1 - y_0}{x^{(1)} - x^{(0)}},$$

from which we obtain

$$x^{(2)} = x^{(0)} - y_0 \frac{x^{(1)} - x^{(0)}}{y_1 - y_0} = \frac{x^{(0)} y_1 - x^{(1)} y_0}{y_1 - y_0}. \tag{5.41}$$

We then proceed with $x^{(2)}$ in a similar manner to the bisection method. The sign of $y_2 := f(x^{(2)})$ determines the interval in which $s$ is located. We summarize the computational scheme of the regula falsi method as follows

$$
\begin{aligned}
&\text{START:} \quad &&\text{Input of } x^{(0)} < x^{(1)} \\
& &&y_0 = f(x^{(0)}), \; y_1 = f(x^{(1)}) \text{ with } y_0 y_1 < 0 \\
&\text{ITERATION:} &&\text{for } k = 1, 2, \ldots: \\
& &&x^{(k+1)} = (x^{(k-1)} \times y_k - x^{(k)} \times y_{k-1})/(y_k - y_{k-1}) \\
& &&y_{k+1} = f(x^{(k+1)}) \\
& &&\text{if } y_{k+1} = 0 \text{ then } s = x^{(k+1)}; \text{ STOP} \\
& &&\text{if } y_{k+1} \times y_{k-1} < 0 \\
& &&\text{then } x^{(k)} := x^{(k-1)}; \; y_k := y_{k-1} \\
& &&\text{else } (x^{(k+1)}, y_{k+1}) \leftrightarrow (x^{(k)}, y_k)
\end{aligned}
$$

(5.42)

The algorithm (5.42) is designed so that $x^{(k-1)} < x^{(k)}$ holds for all $k > 1$, and thus the interval $I_k := [x^{(k-1)}, x^{(k)}]$ contains the desired solution $s$ in its interior. For numerical reasons, the test $y_{k+1} = 0$ will hardly ever be satisfied. It has to be replaced by a bound for the magnitude of $y_{k+1}$ that is appropriate for the given function $f(x)$. For a suitable implementation we only need six values $x^{(k-1)}$, $x^{(k)}$, $x^{(k+1)}$, $y_{k-1}$, $y_k$, $y_{k+1}$ for which no subscripted variables must be used.

The lengths of the intervals $I_k$ produced by the false position method need not tend to zero as can be seen from Figure 5.1 where the sequences $x^{(k)}$ are illustrated for the two cases of a convex and concave function $f(x)$.

## 5.3 Equations in One Unknown

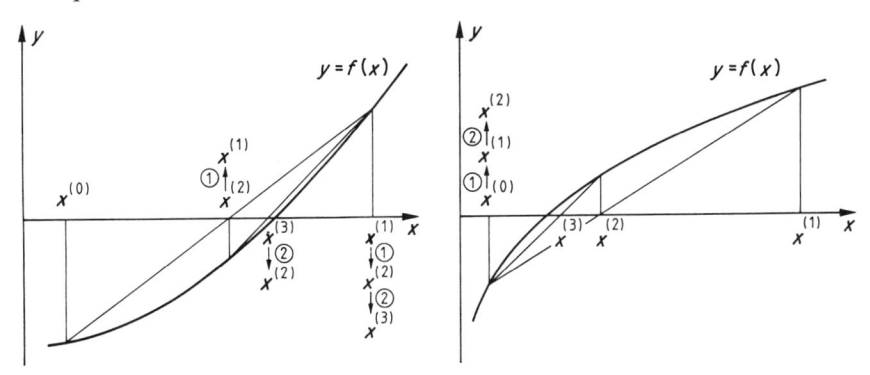

Figure 5.1 Regula falsi for convex and concave functions

**Theorem 5.6** *If $f'(s) \neq 0$ and $f''(s) \neq 0$, then the order of convergence of the regula falsi method is one.*

*Proof* Let the function $f(x)$ be at least three times continuously differentiable. Then according to (5.42), and because of $f(s) = 0$ we have

$$\varepsilon^{(k+1)} = x^{(k+1)} - s = \frac{(x^{(k-1)} - s)f(x^{(k)}) - (x^{(k)} - s)f(x^{(k-1)})}{f(x^{(k)}) - f(x^{(k-1)})}$$

$$= \frac{\varepsilon^{(k-1)}f(s + \varepsilon^{(k)}) - \varepsilon^{(k)}f(s + \varepsilon^{(k-1)})}{f(s + \varepsilon^{(k)}) - f(s + \varepsilon^{(k-1)})}$$

$$= \frac{\varepsilon^{(k-1)}(\varepsilon^{(k)}f'(s) + \tfrac{1}{2}\varepsilon^{(k)2}f''(s) + \cdots) - \varepsilon^{(k)}(\varepsilon^{(k-1)}f'(s) + \tfrac{1}{2}\varepsilon^{(k-1)2}f''(s) + \cdots)}{(\varepsilon^{(k)}f'(s) + \tfrac{1}{2}\varepsilon^{(k)2}f''(s) + \cdots) - (\varepsilon^{(k-1)}f'(s) + \tfrac{1}{2}\varepsilon^{(k-1)2}f''(s) + \cdots)}$$

$$= \frac{\tfrac{1}{2}\varepsilon^{(k-1)}\varepsilon^{(k)}(\varepsilon^{(k)} - \varepsilon^{(k-1)})f''(s) + \cdots}{(\varepsilon^{(k)} - \varepsilon^{(k-1)})(f'(s) + \tfrac{1}{2}(\varepsilon^{(k)} + \varepsilon^{(k-1)})f''(s) + \cdots)} = \frac{\tfrac{1}{2}\varepsilon^{(k-1)}\varepsilon^{(k)}f''(s) + \cdots}{f'(s) + \cdots}$$

As $x^{(k-1)} < s < x^{(k)}$ the errors $\varepsilon^{(k-1)}$ and $\varepsilon^{(k)}$ have opposite signs, and, moreover, we have $\varepsilon^{(k)} - \varepsilon^{(k-1)} > 0$. Therefore for sufficiently small $|\varepsilon^{(k-1)}|$ and $|\varepsilon^{(k)}|$ we have, in the sense of an asymptotically valid analysis of convergence

$$\varepsilon^{(k+1)} \approx \frac{f''(s)}{2f'(s)} \varepsilon^{(k)} \varepsilon^{(k-1)}. \tag{5.43}$$

The qualitative error relationship (5.43) does not imply quadratic convergence. Since the function $f(x)$ is either convex or concave in a sufficiently small neighbourhood of $s$, from the hypothesis $f''(s) \neq 0$, one of the end points of the interval is fixed and is only renamed in the course of the procedure. Therefore $\varepsilon^{(k+1)}$ is proportional to one of the two preceding errors $\varepsilon^{(k)}$ or $\varepsilon^{(k-1)}$. The error

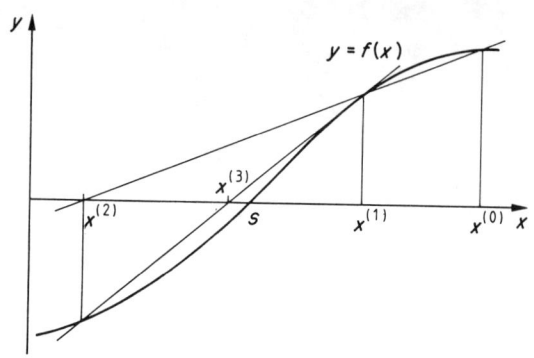

Figure 5.2 Method of the secant

constant $K$, in the case of a concave function, is given by

$$K = \left| \frac{f''(s)}{2f'(s)} \varepsilon^{(k-1)} \right| \quad \varepsilon^{(k-1)} = x^{(0)} - s,$$

and the sequence $x^{(k)}$ converges linearly.

The *secant method* is a modification of the regula falsi method where the condition that the desired solution, $s$, is located between two approximations is dropped. For two given approximate values $x^{(0)}$ and $x^{(1)}$, where $s$ does not necessarily lie between them, the value $x^{(2)}$ is determined analogously to the regula falsi by the intersection point of the secant with the $x$ axis. In spite of the signs of the function values $f(x^{(0)})$, $f(x^{(1)})$ and $f(x^{(2)})$ the subsequent approximation $x^{(3)}$ is determined by means of the linear interpolation polynomial corresponding to the support abscissae $x^{(1)}$ and $x^{(2)}$ (see Figure 5.2).

The iterate $x^{(k+1)}$ from $x^{(k)}$ and $x^{(k-1)}$ can, in principle, be computed by the formulae (5.41) or (5.42). It is now more appropriate to transform that expression into

$$x^{(k+1)} = x^{(k)} - y_k \frac{x^{(k)} - x^{(k-1)}}{y_k - y_{k-1}} \quad (k = 1, 2, \ldots) \tag{5.44}$$

The computational scheme (5.44) assumes that $y_k = f(x^{(k)}) \neq y_{k-1} = f(x^{(k-1)})$. The method does not belong to the class of fixed point iterations (5.1) because the information at the points $x^{(k)}$ and $x^{(k-1)}$ are required to compute $x^{(k+1)}$. The secant method is therefore called a *two-stage iterative method*.

**Theorem 5.7** *If $f'(s) \neq 0$ and $f''(s) \neq 0$, then the order of convergence of the secant method is $p = (1 + \sqrt{5})/2 \doteq 1.618$.*

## 5.3 Equations in One Unknown

*Proof* For sufficiently small errors $|\varepsilon^{(k-1)}|$ and $|\varepsilon^{(k)}|$ the relationship (5.43) remains valid for the secant method, but with the essential difference that both values $\varepsilon^{(k-1)}$ and $\varepsilon^{(k)}$ change accordingly with increasing $k$. If we introduce the constant $C := |f''(s)/(2f'(s))|$, then for sufficiently large $k$ we can write

$$|\varepsilon^{(k+1)}| \approx C|\varepsilon^{(k)}| \cdot |\varepsilon^{(k-1)}|. \tag{5.45}$$

According to Definition 5.3 of the order of convergence, we seek a solution of the approximately valid nonlinear difference equation (5.45) having the form

$$|\varepsilon^{(k)}| = K|\varepsilon^{(k-1)}|^p \quad K > 0, p \geq 1. \tag{5.46}$$

We consider the relationship (5.45) as an equation and substitute (5.46) into it to obtain

$$K|\varepsilon^{(k)}|^p = K \cdot K^p|\varepsilon^{(k-1)}|^{p^2} = C \cdot K|\varepsilon^{(k-1)}|^{p+1}. \tag{5.47}$$

The last equation of (5.47) can only be valid for all (sufficiently large) values of $k$ if

$$K^p = C \quad \text{and} \quad p^2 = p + 1. \tag{5.48}$$

The positive solution $p = (1 + \sqrt{5})/2 \doteq 1.618$ of the quadratic equation is thus equal to the order of convergence of the secant method. The error constant is $K = C^{1/p} \doteq C^{0.618}$.

The practical meaning of the *superlinear* convergence of the secant method is illustrated in Table 5.5. It contains for some constants $C$, and different initial values $|\varepsilon^{(0)}|$ and $|\varepsilon^{(1)}|$, the resulting sequence of terms $|\varepsilon^{(k)}|$ showing the fast convergence to zero.

Theorem 5.7 on the order of convergence of the secant method only states how the iterates $x^{(k)}$ tend to $s$ if convergence is assumed. The theorem does not give any indications concerning the choice of the initial values $x^{(0)}$ and $x^{(1)}$ so that a convergent sequence of iterates is produced. There exist some theoretical results that guarantee that iterates $x^{(k)}$ remain within a certain neighbourhood of $s$ if $x^{(0)}$ and $x^{(1)}$ are chosen in it (Törnig, 1979). However, $s$ is unknown, and such results are

Table 5.5 The meaning of the superlinear convergence

| $k$ | $C=1$ $|\varepsilon^{(k)}|$ | $C=1$ $|\varepsilon^{(k)}|$ | $C=5$ $|\varepsilon^{(k)}|$ | $C=5$ $|\varepsilon^{(k)}|$ | $C=5$ $|\varepsilon^{(k)}|$ |
|---|---|---|---|---|---|
| 0 | $1.00 \times 10^{-1}$ | $1.00 \times 10^{-1}$ | $1.00 \times 10^{-1}$ | $1.00 \times 10^{-1}$ | $5.00 \times 10^{-2}$ |
| 1 | $8.00 \times 10^{-2}$ | $5.00 \times 10^{-2}$ | $5.00 \times 10^{-2}$ | $2.00 \times 10^{-2}$ | $1.00 \times 10^{-1}$ |
| 2 | $8.00 \times 10^{-3}$ | $5.00 \times 10^{-3}$ | $2.50 \times 10^{-2}$ | $1.00 \times 10^{-2}$ | $2.50 \times 10^{-2}$ |
| 3 | $6.40 \times 10^{-4}$ | $2.50 \times 10^{-4}$ | $6.25 \times 10^{-3}$ | $1.00 \times 10^{-3}$ | $1.25 \times 10^{-2}$ |
| 4 | $5.12 \times 10^{-6}$ | $1.25 \times 10^{-6}$ | $7.81 \times 10^{-4}$ | $5.00 \times 10^{-5}$ | $1.56 \times 10^{-3}$ |
| 5 | $3.28 \times 10^{-9}$ | $3.13 \times 10^{-10}$ | $2.44 \times 10^{-5}$ | $2.50 \times 10^{-7}$ | $9.77 \times 10^{-5}$ |
| 6 | $1.68 \times 10^{-14}$ | $3.91 \times 10^{-16}$ | $9.54 \times 10^{-8}$ | $6.25 \times 10^{-11}$ | $7.63 \times 10^{-7}$ |

often not very helpful for practical applications. Therefore one usually proceeds as follows. An interval $I = [a,b]$ is determined by tabulating the function $f(x)$ such that $f(a) \cdot f(b) < 0$. Then the secant method is applied with $x^{(0)} = a$ and $x^{(1)} = b$ and the iterates $x^{(k)}$ are tested to see if $x^{(k)} \in I$ for $k \geqslant 2$ and whether $|f(x^{(k)})|$ is decreasing. If this is not true two different initial values are chosen.

*Example* 5.5 To determine the smallest positive solution of the transcendental equation

$$f(x) = \cos(x)\cosh(x) + 1 = 0 \tag{5.49}$$

we apply the three methods. The initial interval $I = [1.8, 1.9]$ is used in all three cases for which $f(1.8) \doteq 0.29398$ and $f(1.9) \doteq -0.10492$. The results of the bisection method, to 10 significant digits are partly summarized in Table 5.6. The iteration has been stopped as soon as $|f(\mu)| \leqslant 10^{-9}$. The length of the last interval is $L_{24} \doteq 6 \times 10^{-9}$ so that the error of the last middle point $\mu$ is bounded by $|\mu - s| \leqslant 3 \times 10^{-9}$.

In comparison with the bisection method the regula falsi method is much more efficient. The function $f(x)$ is convex on the interval considered thus the upper

Table 5.6 Bisection method

| $k$ | $a$ | $b$ | $\mu$ | $f(\mu)$ |
|---|---|---|---|---|
| 0 | 1.8 | 1.9 | 1.85 | $>0$ |
| 1 | 1.85 | 1.9 | 1.875 | $>0$ |
| 2 | 1.875 | 1.9 | 1.8875 | $<0$ |
| 3 | 1.875 | 1.8875 | 1.88125 | $<0$ |
| 4 | 1.875 | 1.88125 | 1.878125 | $<0$ |
| 5 | 1.875 | 1.878125 | 1.8765625 | $<0$ |
| ⋮ | ⋮ | ⋮ | ⋮ | ⋮ |
| 21 | 1.875 104 048 | 1.875 104 096 | 1.875 104 072 | $<0$ |
| 22 | 1.875 104 048 | 1.875 104 072 | 1.875 104 060 | $>0$ |
| 23 | 1.875 104 060 | 1.875 104 072 | 1.875 104 066 | $>0$ |
| 24 | 1.875 104 066 | 1.875 104 072 | $1.875\,104\,069 \doteq s$ | |

Table 5.7 Regula falsi method

| $k$ | $x^{(k-1)}$ | $x^{(k)}$ | $x^{(k+1)}$ | $f(x^{(k+1)})$ |
|---|---|---|---|---|
| 1 | 1.80 | 1.90 | 1.873 697 942 | $5.8127 \times 10^{-3}$ |
| 2 | 1.873 697 942 | 1.90 | 1.875 078 665 | $1.0512 \times 10^{-4}$ |
| 3 | 1.875 078 665 | 1.90 | 1.875 103 609 | $1.9021 \times 10^{-6}$ |
| 4 | 1.875 103 609 | 1.90 | 1.875 104 061 | $3.19 \times 10^{-8}$ |
| 5 | 1.875 104 061 | 1.90 | 1.875 104 068 | $2.9 \times 10^{-9}$ |

## 5.3 Equations in One Unknown

Table 5.8 Secant method

| k | $x^{(k)}$ | $f(x^{(k)})$ | $|\varepsilon^{(k)}|$ |
|---|-----------|--------------|------------------------|
| 0 | 1.80          | $2.939\,755\,852 \times 10^{-1}$  | $7.5104 \times 10^{-2}$ |
| 1 | 1.90          | $-1.049\,169\,460 \times 10^{-1}$ | $2.4896 \times 10^{-2}$ |
| 2 | 1.873 697 942 | $5.812\,736\,4 \times 10^{-3}$    | $1.4061 \times 10^{-3}$ |
| 3 | 1.875 078 664 | $1.051\,126 \times 10^{-4}$       | $2.5405 \times 10^{-5}$ |
| 4 | 1.875 104 095 | $-1.09 \times 10^{-7}$            | $2.6 \times 10^{-8}$    |
| 5 | 1.875 104 069 | $-1.0 \times 10^{-9}$             | $(4.8 \times 10^{-13})$ |

bound of the interval is not changed. The solution $s$ is found with only five iteration steps. The relevant numbers are given in Table 5.7.

The secant method yields the solution after four iteration steps according to Table 5.8.

Theoretically the computed values $x^{(3)}$ should be identical for the regula falsi and the secant method. However, the different formulae yield slightly different results due to rounding effects. The secant method requires the smallest number of function evaluations. The superlinear convergence has not fully appeared yet, due to the restricted precision of ten digits. The constant $C$ of (5.45) is $C = |f''(s)/(2f'(s))| \doteq 0.734\,10$, so that according to $|\varepsilon^{(3)}|$ and $|\varepsilon^{(4)}|$ the iterate $x^{(5)}$ would have coincided with the solution $s$ within 12 decimal places if a higher precision had been used.

Whether the regula falsi is preferred to the secant method, or vice versa, depends on the practical situation and on whether we want to include the desired solution between lower and upper bounds. The false position method has been generalized to the *Pegasus method*, which has an order of convergence $p \doteq 1.642$ (Dowell and Jarrat, 1972).

### 5.3.2 Newton's method

If the function $f(x)$ of the given equation $f(x) = 0$ is continuously differentiable and moreover, if the first derivative can be computed without too much effort, then *Newton's method* consists of linearizing the function $f(x)$ at the approximating point $x^{(k)}$ and defining the abscissa of the intersection point between the tangent and the $x$ axis to be the next iterate $x^{(k+1)}$ (see Figure 5.3).
From the equation of the tangent

$$y = f(x^{(k)}) + (x - x^{(k)})f'(x^{(k)})$$

we obtain the iteration scheme

$$x^{(k+1)} = x^{(k)} - \frac{f(x^{(k)})}{f'(x^{(k)})} \quad (k = 0, 1, 2, \ldots). \tag{5.50}$$

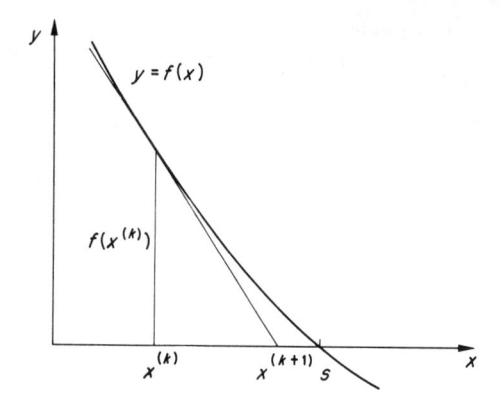

Figure 5.3 Newton's method

Newton's method belongs to the class of fixed point iterations with the function

$$F(x) := x - \frac{f(x)}{f'(x)} \quad \text{with } F(s) = s. \tag{5.51}$$

The condition and order of convergence of Newton's method are formulated in the following theorem.

*Theorem 5.8* If the function $f(x)$ is three times continuously differentiable on an interval $I_1 = [a, b]$, with $a < s < b$, and $f'(s) \neq 0$, that is s is a simple zero of $f(x)$, then there exists an interval $I = [s - \delta, s + \delta]$, $\delta > 0$, for which $F: I \to I$ is a contraction mapping, and for any initial value $x^{(0)} \in I$ the sequence $x^{(k)}$ produced by Newton's method, (5.50), converges at least quadratically to s.

*Proof* To prove the contraction property of the mapping $F$, (5.51), in a nonempty neighbourhood of s, we show that the absolute value of the first derivative of $F$ is smaller than one in a neighbourhood of s. The first derivative is

$$F'(x) = 1 - \frac{f'(x)^2 - f(x)f''(x)}{f'(x)^2} = \frac{f(x)f''(x)}{f'(x)^2}, \tag{5.52}$$

hence $F'(s) = 0$ because $f(s) = 0$ and the assumed properties of differentiability of $f(x)$. For reasons of continuity there exists $\delta > 0$ such that

$$-1 < F'(x) < 1 \quad \text{for all } x \in [s - \delta, s + \delta] =: I$$

The hypotheses of the Banach fixed point theorem are thus satisfied for the interval $I$, and hence the convergence of the sequence $x^{(k)}$ for any $x^{(0)} \in I$ is shown.

From Theorem 5.3 the order of convergence of Newton's method is $p = 2$ if

## 5.3 Equations in One Unknown

$F''(s) \neq 0$. From (5.52) we obtain

$$F''(x) = \frac{f'(x)^2 f''(x) + f(x)f'(x)f^{(3)}(x) - 2f(x)f''(x)^2}{f'(x)^3}, \qquad (5.53)$$

and hence $F''(s) = f''(s)/f'(s)$. Consequently $p = 2$ if and only if $f''(s) \neq 0$, otherwise the order of convergence of Newton's method is even higher.

The interval $I$ whose existence has been shown in Theorem 5.8 for which convergence is guaranteed, can be arbitrarily small. The practical determination of $I$ on the basis of (5.52) can, however, be quite tedious.

Newton's method appears favourable because of the quadratic convergence of the sequence $x^{(k)}$, (5.50). However, we should compare the amount of computational efforts required by Newton's method and the secant method. To do this we assume that the evaluation of the function value $f(x)$ requires a comparable amount of effort to the evaluation of the first derivative. An iteration step of Newton's method needs the simultaneous evaluation of $f(x^{(k)})$ and $f'(x^{(k)})$. Therefore we can perform two steps of the secant method with the same amount of computational effort. From (5.48) we then have the asymptotic relationship

$$|\varepsilon_S^{(k+2)}| \approx K_S |\varepsilon_S^{(k+1)}|^p \approx K_S^{p+1} |\varepsilon_S^{(k)}|^{p^2}$$

$$= \left|\frac{f''(s)}{2f'(s)}\right|^p \cdot |\varepsilon_S^{(k)}|^{p^2} =: K_{DS} |\varepsilon_S^{(k)}|^{p^2} \qquad (5.54)$$

Therefore the order of convergence of a *double step* of the secant method is $p_{DS} = p_S^2 = p_S + 1 \doteq 2.618$ and hence larger than the order of convergence of Newton's method. From this point of view the secant method is more efficient. The error constants $K_{DS}$ of a double step and $K_N$ of Newton's method are related by $K_{DS} \doteq (K_N)^{1.618}$.

*Example* 5.6 The $m$th root of a positive number $a$ is a solution of the equation

$$f(x) = x^m - a = 0 \quad a > 0. \qquad (5.55)$$

Newton's method yields, with $f'(x) = mx^{m-1}$, the iteration scheme

$$x^{(k+1)} = x^{(k)} - \frac{(x^{(k)})^m - a}{m(x^{(k)})^{m-1}} = \frac{a + (m-1)(x^{(k)})^m}{m(x^{(k)})^{m-1}}$$

or

$$\boxed{x^{(k+1)} = \frac{1}{m}\left(\frac{a}{(x^{(k)})^{m-1}} + (m-1)x^{(k)}\right) \quad (k = 0, 1, 2, \ldots).} \qquad (5.56)$$

From this general scheme we derive the following special cases

$$x^{(k+1)} = \frac{1}{2}\left(\frac{a}{x^{(k)}} + x^{(k)}\right) \qquad \text{square root} \qquad (5.57)$$

$$x^{(k+1)} = \frac{1}{3}\left(\frac{a}{(x^{(k)})^2} + 2x^{(k)}\right) \qquad \text{cubic root} \qquad (5.58)$$

$$x^{(k+1)} = (2 - ax^{(k)})x^{(k)} \qquad m = -1, \text{ reciprocal value} \qquad (5.59)$$

From $f''(s) = m(m-1)s^{m-2} \neq 0$ the order of convergence is $p = 2$. The formula (5.57) is usually used on computers to compute the square root. In order to bound the error constant $K := |\frac{1}{2}f''(s)/f'(s)| = 1/(2\sqrt{a})$ the given $a$ is multiplied by an even power of the basis of the number system of the computer to obtain an auxiliary value $a'$ within a prescribed interval. Then an initial value $x^{(0)}$ depending on $a'$ is chosen such that a certain number of iteration steps produces the square root to the full precision of the computer.

The iteration scheme (5.59) was the key for the first computers that made division possible by reducing this arithmetic operation to multiplications and subtractions. The division of two numbers was executed by means of the reciprocal value of the denominator.

*Example* 5.7 The computation of the smallest positive solution of the equation $f(x) = \cos(x)\cosh(x) + 1 = 0$ by means of Newton's method with $f'(x) = \cos(x)\sinh(x) - \sin(x)\cosh(x)$ proceeds, to ten significant figures, according to Table 5.9.

The iteration was stopped after four steps because the correction of $x^{(3)}$ had a sufficiently small absolute value. Four evaluations of both $f(x)$ and $f'(x)$ were required. Since the two evaluations require the same amount of work the secant method is more efficient for computing the solution $s$. The corrections obviously form a quadratically convergent sequence with limit zero.

The value of the first derivative does not change much in Newton's method during the last iteration steps. To take this observation into account and at the

Table 5.9 Newton's method

| $k$ | $x^{(k)}$ | $f(x^{(k)})$ | $f'(x^{(k)})$ | $f(x^{(k)})/f'(x^{(k)})$ |
|---|---|---|---|---|
| 0 | 1.800 000 000 | $2.939\,755\,852 \times 10^{-1}$ | $-3.694\,673\,552$ | $7.956\,740\,46 \times 10^{-2}$ |
| 1 | 1.879 567 405 | $-1.853\,046\,7 \times 10^{-2}$ | $-4.165\,295\,668$ | $-4.448\,775\,90 \times 10^{-3}$ |
| 2 | 1.875 118 629 | $-6.025\,3 \times 10^{-5}$ | $-4.138\,222\,447$ | $-1.456\,011\,6 \times 10^{-5}$ |
| 3 | 1.875 104 069 | $-1.0 \times 10^{-9}$ | $-4.138\,133\,987$ | $-2.416\,5 \times 10^{-10}$ |
| 4 | $1.875\,104\,069 = s$ | | | |

## 5.3 Equations in One Unknown

Table 5.10 Simplified Newton's Method

| k | $x^{(k)}$ | $f(x^{(k)})$ | $f'(x^{(0)})$ | $f(x^{(k)})/f'(x^{(0)})$ |
|---|---|---|---|---|
| 0 | 1.880 000 000 | $-2.033\,2923 \times 10^{-2}$ | $-4.167\,932\,940$ | $-4.878\,419 \times 10^{-3}$ |
| 1 | 1.875 121 581 | $-7.2469 \quad \times 10^{-5}$ | | $-1.738\,728 \times 10^{-5}$ |
| 2 | 1.875 104 194 | $-5.19 \quad\quad \times 10^{-7}$ | | $-1.245\,222 \times 10^{-7}$ |
| 3 | 1.875 104 069 | $-1.0 \quad\quad \times 10^{-9}$ | | $-2.399 \quad \times 10^{-10}$ |
| 4 | $1.875\,104\,069 = s$ | | | |

same time to reduce the number of evaluations involved in the first derivative, one often applies the *simplified Newton's method*

$$x^{(k+1)} = x^{(k)} - \frac{f(x^{(k)})}{f'(x^{(0)})} \quad (k = 0, 1, 2, \ldots). \tag{5.60}$$

The first derivative is computed just once for a (good!) initial value $x^{(0)}$ and kept fixed for the subsequent iterations. The order of convergence of the simplified scheme (5.60) is, however, reduced to $p = 1$, but the error constant $K := |F'(s)| = |1 - f'(s)/f'(x^{(0)})|$ is often quite small.

*Example* 5.8 For a good initial approximation the simplified Newton's method works satisfactorily for determining the smallest positive solution of $f(x) = \cos(x)\cosh(x) + 1 = 0$ (see Table 5.10). The error constant here is $K \doteq 0.007\,15$.

### 5.3.3 Interpolation methods

The regula falsi and the secant method are two representatives of schemes that determine $x^{(k+1)}$ by means of the linear interpolation polynomial to the two given values $x^{(k-1)}$ and $x^{(k)}$. In order to approximate the function $f(x)$ better an interpolation polynomial of higher degree is used, from whose zeros the subsequent iterate is determined. *Muller's method* (Muller, 1956) uses quadratic interpolation. For this reason three approximate values $x^{(k-2)}$, $x^{(k-1)}$, $x^{(k)}$ with the corresponding function values $y_i := f(x^{(i)})$, $(i = k-2, k-1, k)$ are required (see Figure 5.4). The interpolation polynomial $P_2(x)$ corresponding to the three support points is used in this context in the following appropriate form

$$P_2(x) = A(x - x^{(k)})^2 + B(x - x^{(k)}) + C \tag{5.61}$$

because we shall be interested in that zero of $P_2(x)$ being closest to $x^{(k)}$. The

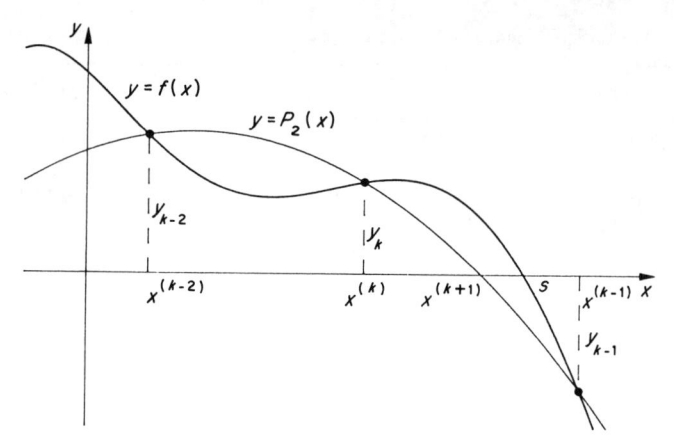

Figure 5.4 Muller's method

coefficients $A$, $B$ and $C$ are determined from the interpolating conditions

$$A(x^{(k-2)} - x^{(k)})^2 + B(x^{(k-2)} - x^{(k)}) + C = y_{k-2}$$
$$A(x^{(k-1)} - x^{(k)})^2 + B(x^{(k-1)} - x^{(k)}) + C = y_{k-1} \qquad (5.62)$$
$$C = y_k$$

We introduce the auxiliary quantities

$$h_2 := x^{(k-2)} - x^{(k)} \quad h_1 := x^{(k-1)} - x^{(k)}$$
$$d_2 := y_{k-2} - y_k \quad d_1 := y_{k-1} - y_k \qquad (5.63)$$

and get

$$A = \frac{h_1 d_2 - h_2 d_1}{h_2 h_1 (h_2 - h_1)} \quad B = \frac{h_2^2 d_1 - h_1^2 d_2}{h_2 h_1 (h_2 - h_1)} \quad C = y_k. \qquad (5.64)$$

From (5.61) we thus obtain in general a quadratic equation for the desired difference $h := x^{(k+1)} - x^{(k)}$

$$Ah^2 + Bh + C = 0 \quad \text{with } C \neq 0. \qquad (5.65)$$

From (5.63) and (5.64) the exceptional case $A = B = 0$ occurs if and only if $y_{k-2} = y_{k-1} = y_k$. We do not consider this case in the following. If $A = 0$ and $B \neq 0$, then the three points define a straight line, and hence there is a single solution of (5.65). If $A \neq 0$ and $B = 0$, (5.65) has two solutions of the same absolute value, and we can choose one of them. In summary

## 5.3 Equations in One Unknown

$$h = \begin{cases} -C/B, & \text{if } A = 0 \text{ and } B \neq 0 \\ \sqrt{-C/A}, & \text{if } A \neq 0 \text{ and } B = 0 \\ \dfrac{-2\,\text{sgn}(B)C}{|B| + \sqrt{B^2 - 4AC}} & \end{cases} \qquad (5.66)$$

The formulae (5.63), (5.64) and (5.66) define the essential elements of an iteration step of Muller's method. The iterate $x^{(k+1)}$ is computed with the information at the three points $x^{(k-2)}$, $x^{(k-1)}$ and $x^{(k)}$ and has thus the general form $x^{(k+1)} = F(x^{(k-2)}, x^{(k-1)}, x^{(k)})$, of a so-called *three-stage iteration method*. The derivation of the relationship that holds for four consecutive errors $\varepsilon^{(k)} = x^{(k)} - s$ is involved (Young and Gregory, 1973). It is found to be

$$\varepsilon^{(k+1)} \approx \frac{f^{(3)}(s)}{6f'(s)} \varepsilon^{(k)} \varepsilon^{(k-1)} \varepsilon^{(k-2)} \quad (k \gg 1). \qquad (5.67)$$

An analogous technique that was applied to determine the order of convergence of the secant method shows that Muller's method has *superlinear convergence* with $p \doteq 1.839$. The order of convergence of Muller's method is, as could have been expected, higher than that of the secant method.

Muller's method was originally proposed to compute all the zeros of a polynomial (Muller, 1956). It is applicable, in this context, without restrictions even if complex zeros occur. It has the advantage over Newton's method that it produces, even with real initial values, complex iterates and hence convergent sequences $x^{(k)}$. This is because the quadratic equation (5.65), for $h$, possesses complex solutions, as soon as for real $A$, $B$, $C$ the discriminant $B^2 - AC$ is negative. Thus this method automatically indicates if a given equation $f(x) = 0$ may have complex solutions.

Table 5.11 Muller's method

| $k$ | $x^{(k)}$ | $A$ | $B$ | $C = f(x^{(k)})$ | $h$ |
|---|---|---|---|---|---|
| 0 | 1.800 000 000 | | | $2.939\,76 \times 10^{-1}$ | |
| 1 | 1.900 000 000 | | | $-1.049\,17 \times 10^{-1}$ | |
| 2 | 1.850 000 000 | $-2.980\,82$ | $-3.988\,93$ | $1.019\,81 \times 10^{-1}$ | $2.509\,55 \times 10^{-2}$ |
| 3 | 1.875 095 505 | $-3.037\,52$ | $-4.138\,55$ | $3.543\,76 \times 10^{-5}$ | $8.562\,76 \times 10^{-6}$ |
| 4 | 1.875 104 068 | $-3.019\,39$ | $-4.138\,14$ | $2.9 \quad \times 10^{-9}$ | $7.008 \quad \times 10^{-10}$ |
| 5 | 1.875 104 069 $= s$ | | | | |

*Example* 5.9 Muller's method is quite efficient for computing the smallest positive solution of the equation $f(x) = \cos(x) \cdot \cosh(x) + 1 = 0$. Table 5.11 illustrates the procedure where a precision of ten significant digits is used. The coefficients $A$, $B$ and $C$ of the quadratic equation (5.65) are only given to six significant digits. The first two initial values $x^{(0)}$ and $x^{(1)}$ are the same as for the secant method, so that $f(x^{(0)})f(x^{(1)}) < 0$. $x^{(2)}$ is defined arbitrarily as the arithmetic mean. Five function evaluations are sufficient to produce the solution.

On the assumption that the function $f(x)$ is monotonic in a neighbourhood of the desired zero the *inverse interpolation method* can be successfully applied. The corresponding algorithm is outlined in Section 3.5.3.

## 5.4 Equations in Several Unknowns

We now seek the solutions of a given system of $n$ nonlinear equations in $n$ unknowns. The problem is formulated, in general, by means of a continuous, nonlinear function $\mathbf{f}: \mathbf{R}^n \to \mathbf{R}^n$ as follows. We want to find $\mathbf{x} \in \mathbf{R}^n$ such that

$$\boxed{\mathbf{f}(\mathbf{x}) = \mathbf{0}} \quad (5.68)$$

To simplify the notation we shall present the fundamental ideas and principles for a system of two nonlinear equations in two unknowns. The generalization will be obvious.

### 5.4.1 Fixed point iteration and convergence

We consider two nonlinear equations in two unknowns which are assumed to have *fixed point form*

$$x = F(x, y) \quad y = G(x, y) \quad (5.69)$$

This situation will serve as the basis for discussing the convergence behaviour of subsequently presented iteration methods. Each pair of values $(s, t)$ with the property

$$s = F(s, t) \quad t = G(s, t) \quad (5.70)$$

is called a *fixed point* of the mapping $\mathbf{F}: \mathbf{R}^2 \to \mathbf{R}^2$ defined by

$$\mathbf{F}(\mathbf{x}) := \begin{bmatrix} F(x, y) \\ G(x, y) \end{bmatrix} \quad \mathbf{x} := \begin{bmatrix} x \\ y \end{bmatrix} \in \mathbf{R}^2. \quad (5.71)$$

In order to apply the fixed point theorem of Banach a closed domain $A \subset \mathbf{R}^2$, for example a closed rectangle or a closed disc, must be determined such that the

## 5.4 Equations in Several Unknowns

mapping $F: \mathbf{R}^2 \to \mathbf{R}^2$ satisfies

$$F: A \to A \quad \text{with} \quad \|F(x) - F(x^*)\| \leq L \|x - x^*\| \quad L < 1 \quad x, x^* \in A \tag{5.72}$$

for an arbitrary vector norm. On the additional hypothesis that the functions $F(x, y)$ and $G(x, y)$ have continuous partial derivatives in $A$, sufficient criterions for the Lipschitz condition (5.72) can be given. The *Jacobian* of the mapping $F: \mathbf{R}^2 \to \mathbf{R}^2$ is given by

$$J(x, y) := \begin{bmatrix} \dfrac{\partial F}{\partial x} & \dfrac{\partial F}{\partial y} \\ \dfrac{\partial G}{\partial x} & \dfrac{\partial G}{\partial y} \end{bmatrix}_{(x,y)} \tag{5.73}$$

If for any matrix norm $\|J(x, y)\| \leq L < 1$ holds for all $x \in A$, then (5.72) is satisfied for a compatible vector norm. By means of sophisticated methods of linear algebra it can be shown that the following condition is necessary and sufficient. Let $\lambda_i(\mathbf{B})$ be the eigenvalues of a matrix $\mathbf{B} \in \mathbf{R}^{n \times n}$. Then we call

$$\rho(\mathbf{B}) := \max_i |\lambda_i(\mathbf{B})| \tag{5.74}$$

the *spectral radius* of the matrix $\mathbf{B}$. The mapping $F: A \to A$ is a contraction if and only if for all $x \in A$ the spectral radius of the Jacobian satisfies $\rho(\mathbf{J}) < 1$.

Usually it is difficult to determine the domain $A$ of a given mapping $F$ explicitly for which the hypotheses of Banach's fixed point theorem are satisfied. We now analyze the order of convergence of the iterates on the assumption that the above conditions hold. To reveal the asymptotically valid behaviour of the error vector

$$\varepsilon^{(k)} := \mathbf{x}^{(k)} - \mathbf{s} = \begin{bmatrix} x^{(k)} - s \\ y^{(k)} - t \end{bmatrix} = \begin{bmatrix} \varepsilon^{(k)} \\ \delta^{(k)} \end{bmatrix} \tag{5.75}$$

we assume that $\varepsilon^{(k)}$ and $\delta^{(k)}$ are small in absolute value and that the functions $F(x, y)$ and $G(x, y)$ are at least twice continuously differentiable. Then the following expansions hold

$$\begin{aligned}
\varepsilon^{(k+1)} &= x^{(k+1)} - s = F(x^{(k)}, y^{(k)}) - F(s, t) \\
&= F(s + \varepsilon^{(k)}, t + \delta^{(k)}) - F(s, t) \\
&= \varepsilon^{(k)} F_x(s, t) + \delta^{(k)} F_y(s, t) \\
&\quad + \tfrac{1}{2} \varepsilon^{(k)2} F_{xx} + \varepsilon^{(k)} \delta^{(k)} F_{xy} + \tfrac{1}{2} \delta^{(k)2} F_{yy} + \cdots \\
\delta^{(k+1)} &= \varepsilon^{(k)} G_x(s, t) + \delta^{(k)} G_y(s, t) \\
&\quad + \tfrac{1}{2} \varepsilon^{(k)2} G_{xx} + \varepsilon^{(k)} \delta^{(k)} G_{xy} + \tfrac{1}{2} \delta^{(k)2} G_{yy} + \cdots
\end{aligned} \tag{5.76}$$

From the representations (5.76) it follows that if $\mathbf{J}(s, t) \neq \mathbf{0}$ then the approximate relationship holds

$$\varepsilon^{(k+1)} \approx \mathbf{J}(s,t)\varepsilon^{(k)} \quad (k \gg 1). \tag{5.77}$$

Thus the sequence $\mathbf{x}^{(k)}$ converges *linearly* to the fixed point $\mathbf{s}$.

However, if the Jacobian $\mathbf{J}(s,t)$ is the zero matrix from (5.76) with the *Hessian matrices* of the functions $F(x,y)$ and $G(x,y)$

$$\mathbf{H}_F := \begin{bmatrix} F_{xx} & F_{xy} \\ F_{xy} & F_{yy} \end{bmatrix} \quad \mathbf{H}_G := \begin{bmatrix} G_{xx} & G_{xy} \\ G_{xy} & G_{yy} \end{bmatrix}, \tag{5.78}$$

both evaluated at the fixed point $\mathbf{s}$ we get the relationship

$$\varepsilon^{(k+1)} = \tfrac{1}{2}\varepsilon^{(k)^T}\mathbf{H}_F \varepsilon^{(k)} + \cdots \quad \delta^{(k+1)} = \tfrac{1}{2}\varepsilon^{(k)^T}\mathbf{H}_G \varepsilon^{(k)} + \cdots \tag{5.79}$$

If one of the Hessian matrices (5.78) evaluated at $\mathbf{s}$ is a nonzero matrix, then (5.79) means the iterates $\mathbf{x}^{(k)}$ *converge quadratically*.

### 5.4.2 Newton's method

To determine a solution $(s,t)$ of a system of two nonlinear equations

$$\boxed{f(x,y) = 0 \quad g(x,y) = 0} \tag{5.80}$$

the Newton's iteration or one of its modifications is well suited. In the following functions $f(x,y)$ and $g(x,y)$ are assumed to have at least continuous first partial derivatives. Let $\mathbf{x}^{(0)} = (x^{(0)}, y^{(0)})^T$ be an appropriate initial approximation of the desired solution $\mathbf{s} = (s,t)^T$ of (5.80). Then we use in the general $k$th step the substitution

$$s = x^{(k)} + \xi^{(k)} \quad t = y^{(k)} + \eta^{(k)} \tag{5.81}$$

where the (small) *correction terms* $\xi^{(k)}$ and $\eta^{(k)}$ serve to *linearize* the given system as follows

$$\begin{aligned}
f(s,t) &= f(x^{(k)} + \xi^{(k)}, y^{(k)} + \eta^{(k)}) \approx f(x^{(k)}, y^{(k)}) + \xi^{(k)} f_x(x^{(k)}, y^{(k)}) \\
&\quad + \eta^{(k)} f_y(x^{(k)}, y^{(k)}) = 0 \\
g(s,t) &= g(x^{(k)} + \xi^{(k)}, y^{(k)} + \eta^{(k)}) \approx g(x^{(k)}, y^{(k)}) + \xi^{(k)} g_x(x^{(k)}, y^{(k)}) \\
&\quad + \eta^{(k)} g_y(x^{(k)}, y^{(k)}) = 0
\end{aligned} \tag{5.82}$$

Thus we have obtained a system of linear equations for the corrections $\xi^{(k)}$ and $\eta^{(k)}$ that with the vectors reads

$$\mathbf{x}^{(k)} := \begin{bmatrix} x^{(k)} \\ y^{(k)} \end{bmatrix} \quad \xi^{(k)} := \begin{bmatrix} \xi^{(k)} \\ \eta^{(k)} \end{bmatrix} \quad \mathbf{f}^{(k)} := \begin{bmatrix} f(x^{(k)}, y^{(k)}) \\ g(x^{(k)}, y^{(k)}) \end{bmatrix}$$

$$\boxed{\boldsymbol{\Phi}(x^{(k)}, y^{(k)})\xi^{(k)} + \mathbf{f}^{(k)} = \mathbf{0},} \tag{5.83}$$

## 5.4 Equations in Several Unknowns

where we have introduced, in contrast to (5.73), the Jacobian

$$\Phi(x,y) := \begin{bmatrix} f_x & f_y \\ g_x & g_y \end{bmatrix}_{(x,y)} \tag{5.84}$$

corresponding to the functions $f(x,y)$ and $g(x,y)$ of the system (5.80). On the assumption that the Jacobian $\Phi$ (5.84) is nonsingular for the approximation $\mathbf{x}^{(k)}$, the system of linear equations (5.83) has a unique solution $\xi^{(k)}$. The resulting correction vector $\xi^{(k)}$ does not yield the solution $\mathbf{s}$, of course, but only a new approximation $\mathbf{x}^{(k+1)} = \mathbf{x}^{(k)} + \xi^{(k)}$, that has the *formal* representation

$$\mathbf{x}^{(k+1)} = \mathbf{x}^{(k)} - \Phi^{-1}(\mathbf{x}^{(k)})\mathbf{f}^{(k)} =: \mathbf{F}(\mathbf{x}^{(k)}) \quad (k = 0, 1, 2, \ldots). \tag{5.85}$$

The computational scheme (5.85) belongs to the class of fixed point iterations and (5.85) can be seen as a direct generalization of (5.50). The explicit formula should not mislead one into computing the inverse of $\Phi$, but the correction vector $\xi^{(k)}$ should be determined by solving (5.83) by means of the Gaussian algorithm (see Section 1.1).

In the case of a system of two nonlinear equations, (5.80), the computational scheme (5.85) is in its explicit form

$$\begin{aligned} x^{(k+1)} &= x^{(k)} + \zeta^{(k)} = x^{(k)} + \left.\frac{g \cdot f_y - f \cdot g_y}{f_x \cdot g_y - f_y \cdot g_x}\right|_{(x^{(k)}, y^{(k)})} =: F(x^{(k)}, y^{(k)}) \\ y^{(k+1)} &= y^{(k)} + \eta^{(k)} = y^{(k)} + \left.\frac{f \cdot g_x - g \cdot f_x}{f_x \cdot g_y - f_y \cdot g_x}\right|_{(x^{(k)}, y^{(k)})} =: G(x^{(k)}, y^{(k)}) \end{aligned} \tag{5.86}$$

**Theorem 5.9** *If the functions $f(x,y)$ and $g(x,y)$ of (5.80) are three times continuously differentiable within a domain A that contains the solution $\mathbf{s}$ in its interior and the Jacobian $\Phi(s,t)$, (5.84), is nonsingular, then Newton's method has an order of convergence of at least $p = 2$.*

*Proof* We show that the Jacobian $\mathbf{J}(s,t)$ of the fixed point iteration (5.85) is equal to the zero matrix. This fact can be easily verified in the case of two equations from the explicit representation (5.86) just by partial differentiation of $F(x,y)$ and $G(x,y)$ and substitution of the solution $(s,t)$. However, we present a proof that can be generalized to systems of $n$ nonlinear equations in an obvious manner. The Jacobian $\mathbf{J}(x,y)$ is

$$\mathbf{J}(x,y) = \begin{bmatrix} 1 + \xi_x & \xi_y \\ \eta_x & 1 + \eta_y \end{bmatrix}_{(x,y)}, \tag{5.87}$$

where $\xi(x,y)$ and $\eta(x,y)$ denote the corrections depending on $x$ and $y$. They are defined as the solution of the system of linear equations

$$f_x(x,y)\xi(x,y) + f_y(x,y)\eta(x,y) + f(x,y) = 0$$
$$g_x(x,y)\xi(x,y) + g_y(x,y)\eta(x,y) + g(x,y) = 0 \tag{5.88}$$

The partial derivatives of the corrections that are needed in (5.87) are obtained by an implicit differentiation of the equations (5.88) with respect to $x$ and $y$. Differentiation with respect to $x$ yields the following relationships in which the variables $x$ and $y$ are omitted

$$f_{xx}\xi + f_x\xi_x + f_{xy}\eta + f_y\eta_x + f_x = 0$$
$$g_{xx}\xi + g_x\xi_x + g_{xy}\eta + g_y\eta_x + g_x = 0 \tag{5.89}$$

Now we substitute $x = s$ and $y = t$ into (5.88). Since $f(s,t) = g(s,t) = 0$ and from the nonsingularity of the Jacobian $\Phi(s,t)$ we have $\xi(s,t) = \eta(s,t) = 0$. Hence from (5.89), with $x = s$ and $y = t$, we get the system of homogeneous equations

$$f_x \cdot (1 + \xi_x) + f_y \cdot \eta_x = 0$$
$$g_x \cdot (1 + \xi_x) + g_y \cdot \eta_x = 0 \tag{5.90}$$

for the unknowns $1 + \xi_x(s,t)$ and $\eta_x(s,t)$. It only has the trivial solution because $\Phi(s,t)$ is assumed to be nonsingular. This argument shows that the elements of the first column of $\mathbf{J}(s,t)$, (5.87), vanish. Differentiation of (5.88) with respect to $y$ yields in a completely analogous manner the same result for the second column of $\mathbf{J}(s,t)$. Consequently the order of convergence of Newton's method is greater than one. Whether its order of convergence is $p = 2$ or even higher depends on the Hessian matrices (5.78). Their elements can be determined by implicit differentiation of the equations (5.89) and their analogous ones.

*Example* 5.10 We seek the solution of the quadratic equations

$$f(x,y) = x^2 + y^2 + 0.6y - 0.16 = 0$$
$$g(x,y) = x^2 - y^2 + x - 1.6y - 0.14 = 0 \tag{5.91}$$

for which $s > 0$ and $t > 0$. The first partial derivatives are

$$f_x = 2x \quad f_y = 2y + 0.6 \quad g_x = 2x + 1 \quad g_y = -2y - 1.6.$$

We choose the initial guesses $x^{(0)} = 0.6$, $y^{(0)} = 0.25$ for which (5.83) becomes

$$1.2\xi^{(0)} + 1.1\eta^{(0)} + 0.4125 = 0$$
$$2.2\xi^{(0)} - 2.1\eta^{(0)} + 0.3575 = 0 \tag{5.92}$$

with the resulting corrections $\xi^{(0)} \doteq -0.254\,960$, $\eta^{(0)} \doteq -0.096\,862$ and the new approximations $x^{(1)} \doteq 0.345\,040$, $y^{(1)} \doteq 0.153\,138$. The corresponding system of equations is, with rounded values

$$0.690\,081\xi^{(1)} + 0.906\,275\eta^{(1)} + 0.074\,3867 = 0$$
$$1.690\,081\xi^{(1)} - 1.906\,275\eta^{(1)} + 0.055\,6220 = 0. \tag{5.93}$$

## 5.4 Equations in Several Unknowns

Table 5.12 Newton's method for a system of nonlinear equations

| k | $x^{(k)}$ | $y^{(k)}$ | $\xi^{(k)}$ | $\eta^{(k)}$ | $\|\varepsilon^{(k)}\|$ |
|---|---|---|---|---|---|
| 0 | 0.600 000 0000 | 0.250 000 0000 | $-2.549\,60 \times 10^{-1}$ | $-9.686\,23 \times 10^{-2}$ | $3.531 \times 10^{-1}$ |
| 1 | 0.345 040 4858 | 0.153 137 6518 | $-6.750\,94 \times 10^{-2}$ | $-3.067\,47 \times 10^{-2}$ | $8.050 \times 10^{-2}$ |
| 2 | 0.277 531 0555 | 0.122 462 9827 | $-5.645\,94 \times 10^{-3}$ | $-2.798\,60 \times 10^{-3}$ | $6.347 \times 10^{-3}$ |
| 3 | 0.271 885 1108 | 0.119 664 3843 | $-4.060\,23 \times 10^{-5}$ | $-2.100\,56 \times 10^{-5}$ | $4.572 \times 10^{-5}$ |
| 4 | 0.271 844 5085 | 0.119 643 3787 | $-2.1579 \times 10^{-9}$ | $-1.1043 \times 10^{-9}$ | $2.460 \times 10^{-9}$ |
| 5 | 0.271 884 5063 | 0.119 643 3776 | | | |

The continuation of Newton's method is summarized in Table 5.12 to ten significant figures.

The solution steps of Newton's method can be interpreted geometrically. The interpretation of the first two steps is shown in Figure 5.5. The two linear equations (5.82) are the equations of the traces $s_f$ and $s_g$ of the two tangent planes of the surfaces $z = f(x, y)$ and $z = g(x, y)$ at the point $P(x^{(k)}, y^{(k)})$, related to a local $(\xi^{(k)}, \eta^{(k)})$ coordinate system with origin $P(x^{(k)}, y^{(k)})$. The intersection of the two traces yields the iterated point $P(x^{(k+1)}, y^{(k+1)})$.

The evaluation of the Jacobian $\Phi(x^{(k)})$ may require a large amount of computation, because for a general system of $n$ nonlinear equations a total of $n^2$ partial derivatives have to be calculated. Therefore we give briefly simplifying modifications.

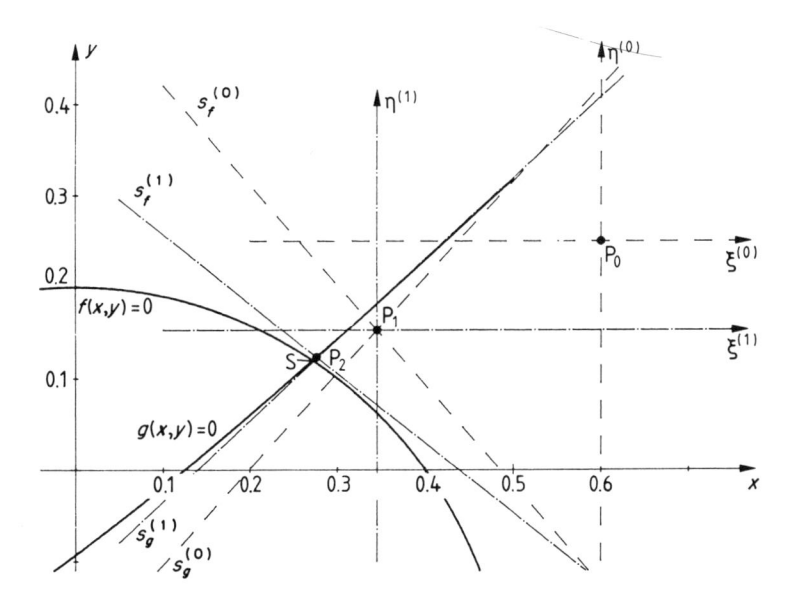

Figure 5.5 Newton's method for a system of equations

The *simplified Newton's method* uses a constant Jacobian $\Phi(\mathbf{x}^{(0)})$, which is evaluated for a good initial vector $\mathbf{x}^{(0)}$, and the corrections $\boldsymbol{\xi}^{(k)}$ are determined not from (5.83) but as solutions of the system of equations

$$\Phi(\mathbf{x}^{(0)})\boldsymbol{\xi}^{(k)} + \mathbf{f}^{(k)} = \mathbf{0} \quad (k = 0, 1, 2, \ldots). \tag{5.94}$$

The amount of computational effort required is considerably reduced because the LR-decomposition of the Jacobian must only be carried out once for the Gaussian algorithm, then the correction vector $\boldsymbol{\xi}^{(k)}$ is obtained from (5.94) by the processes of forward and backward substitution in each iteration step. However, it is obvious that the simplified method (5.94) only has the order of convergence $p = 1$ because the Jacobian $\mathbf{J}(s, t)$ (5.87) is no longer the zero matrix. However, a similar analysis to the proof of Theorem 5.9 shows that for a good initial approximation $\mathbf{x}^{(0)}$, the matrix elements of $\mathbf{J}(s, t)$ are small in absolute value, and hence the sequence $\mathbf{x}^{(k)}$ converges relatively rapidly due to (5.77).

*Example* 5.11 The system of equations (5.91) is now treated by means of the simplified Newton's method with the better initial approximations $x^{(0)} = 0.3$ and $y^{(0)} = 0.1$. With the Jacobian

$$\Phi(x^{(0)}, y^{(0)}) = \begin{pmatrix} 0.6 & 0.8 \\ 1.6 & -1.8 \end{pmatrix} \tag{5.95}$$

the sequence of iterates $\mathbf{x}^{(k)}$ converges linearly with a convergence factor $q \approx 0.06$, so that the Euclidian norm of the error $\boldsymbol{\varepsilon}^{(k)}$ decreases in each step by a factor of approximately 15. Therefore only seven iteration steps are required to obtain the solution to ten significant figures (see Table 5.13).

If the same initial values are used as in Example 5.10, the simplified Newton's method has a slow linear convergence with a convergence quotient $q \approx 0.45$. In this case 27 iteration steps are required to produce an approximation with the same accuracy.

Table 5.13 Simplified Newton's method for a system

| $k$ | $x^{(k)}$ | $y^{(k)}$ | $\xi^{(k)}$ | $\eta^{(k)}$ | $\|\boldsymbol{\varepsilon}^{(k)}\|$ |
|---|---|---|---|---|---|
| 0 | 0.300 000 0000 | 0.100 000 0000 | $-2.711\,86 \times 10^{-2}$ | $2.033\,90 \times 10^{-2}$ | $3.433 \times 10^{-2}$ |
| 1 | 0.272 881 3559 | 0.120 338 9831 | $-9.854\,95 \times 10^{-4}$ | $-6.972\,48 \times 10^{-4}$ | $1.249 \times 10^{-3}$ |
| 2 | 0.271 895 8608 | 0.119 641 7355 | $-4.814\,41 \times 10^{-5}$ | $2.926\,48 \times 10^{-6}$ | $5.138 \times 10^{-5}$ |
| 3 | 0.271 847 7167 | 0.119 644 6620 | $-3.032\,56 \times 10^{-6}$ | $-1.254\,83 \times 10^{-6}$ | $3.458 \times 10^{-6}$ |
| 4 | 0.271 844 6841 | 0.119 643 4072 | $-1.672\,46 \times 10^{-7}$ | $-2.644\,07 \times 10^{-8}$ | $1.802 \times 10^{-7}$ |
| 5 | 0.271 844 5169 | 0.119 643 3808 | $-9.932\,2 \times 10^{-9}$ | $-3.050\,8 \times 10^{-9}$ | $1.107 \times 10^{-8}$ |
| 6 | 0.271 844 5070 | 0.119 643 3777 | $-6.101\,7 \times 10^{-10}$ | $-4.237\,3 \times 10^{-11}$ | $7.07 \times 10^{-10}$ |
| 7 | 0.271 844 5064 | 0.119 643 3777 | | | |

## 5.4 Equations in Several Unknowns

A different approach avoiding the evaluation of the Jacobian $\Phi(\mathbf{x}^{(k)})$, at least partially, is based on the idea of reducing the problem of solving a system of $n$ nonlinear equations in $n$ unknowns into the successive solution of nonlinear equations in a single unknown. We consider the system of $n$ equations

$$f_i(x_1, x_2, \ldots, x_n) = 0 \quad (i = 1, 2, \ldots, n). \tag{5.96}$$

We assume that they are ordered such that

$$\frac{\partial f_i(x_1, x_2, \ldots, x_n)}{\partial x_i} \neq 0 \quad \text{for } i = 1, 2, \ldots, n \tag{5.97}$$

This only means that the $i$th equation depends on the $i$th unknown $x_i$. We denote by $\mathbf{x}^{(k)} = (x_1^{(k)}, x_2^{(k)}, \ldots, x_n^{(k)})^T$ the $k$th approximation vector. Analogous to the *method of successive relaxation* (or *Gauss–Seidel method*) for the iterative solution of systems of linear equations (Schwarz et al., 1972; Stummel and Hainer, 1982) the $i$th component of the $(k+1)$-th approximation vector is found to be the solution of the $i$th equation as follows

$$f_i(x_1^{(k+1)}, \ldots, x_{i-1}^{(k+1)}, x_i^{(k+1)}, x_{i+1}^{(k)}, \ldots, x_n^{(k)}) = 0 \quad (i = 1, 2, \ldots, n) \tag{5.98}$$

Note that in (5.98) for the first $(i-1)$ variables the known components of $\mathbf{x}^{(k+1)}$ are substituted, whereas for the last $(n-i)$ variables the corresponding components of $\mathbf{x}^{(k)}$ are used. For each index $i$, (5.98) represents an equation for the single unknown $x_i^{(k+1)}$ which is determined by means of Newton's method. To perform this so-called *inner iteration* we only need the partial derivatives $\partial f_i/\partial x_i$, that is the $n$ diagonal elements of the Jacobian $\Phi$, that are to be evaluated for changing values. It is quite natural to use $x_i^{(k)}$ as initial value for this iteration. Thus we can give the following algorithmic description of the *nonlinear successive relaxation method* for a system (5.96) using an initial vector $\mathbf{x}^{(0)}$.

for $k = 0, 1, 2, \ldots$:
$s = 0$
for $i = 1, 2, \ldots, n$:
$\xi_i = x_i^{(k)}$
INNIT: $\Delta\xi_i = \dfrac{f_i(x_1^{(k+1)}, \ldots, x_{i-1}^{(k+1)}, \xi_i, x_{i+1}^{(k)}, \ldots, x_n^{(k)})}{\dfrac{\partial f_i(x_1^{(k+1)}, \ldots, x_{i-1}^{(k+1)}, \xi_i, x_{i+1}^{(k)}, \ldots, x_n^{(k)})}{\partial x_i}}$
$\xi_i = \xi_i - \Delta\xi_i$
if $|\Delta\xi_i| > \text{tol}_1$ then go to INNIT
$s = s + |\xi_i - x_i^{(k)}|; \; x_i^{(k+1)} = \xi_i$
if $s < \text{tol}_2$ then STOP
(5.99)

The nonlinear successive relaxation method based on Newton's method for solving the $i$th equation, (5.98), admits several possible modifications. For instance, we do not need to solve the $i$th equation for $x_i^{(k+1)}$ exactly because this value is only a new approximation. Therefore the inner iteration is reduced to a single step of the Newton correction for computing $x_i^{(k+1)}$, and we obtain *Newton's successive relaxation method* with the computational scheme

$$x_i^{(k+1)} = x_i^{(k)} - \frac{f_i(x_1^{(k+1)},\ldots,x_{i-1}^{(k+1)},x_i^{(k)},x_{i+1}^{(k)},\ldots,x_n^{(k)})}{\dfrac{\partial f_i(x_1^{(k+1)},\ldots,x_{i-1}^{(k+1)},x_i^{(k)},x_{i+1}^{(k)},\ldots,x_n^{(k)})}{\partial x_i}} \quad (i=1,2,\ldots,n).$$

(5.100)

The linear convergence of Newton's successive relaxation method can usually be improved by multiplying the correction of the $i$th component $x_i^{(k)}$, given in (5.100), by an appropriately chosen constant called the *relaxation factor* $\omega \in (0,2)$. The motivation to do this is given by the *method of successive overrelaxation* (SOR) widely used for the iterative solution of systems of linear equations (Schwarz et al., 1972; Törnig, 1979). In a natural way this leads to the *SOR–Newton method*

$$x_i^{(k+1)} = x_i^{(k)} - \omega\frac{f_i(x_1^{(k+1)},\ldots,x_{i-1}^{(k+1)},x_i^{(k)},\ldots,x_n^{(k)})}{\dfrac{\partial f_i(x_1^{(k+1)},\ldots,x_{i-1}^{(k+1)},x_i^{(k)},\ldots,x_n^{(k)})}{\partial x_i}} \quad (i=1,2,\ldots,n).$$

(5.101)

The SOR–Newton scheme is successfully applied to solving large systems of nonlinear equations where the $i$th equation involves only a few of the unknowns. This situation occurs in connection with the numerical treatment of nonlinear boundary value problems by means of finite differences or finite elements. For some important special cases results exist on the proper choice of the relaxation factor $\omega$ to achieve optimal convergence (Ortega and Rheinboldt, 1970; Törnig, 1979).

### 5.5 Zeros of Polynomials

To compute the zeros of a polynomial of degree $n$

$$P_n(x) = a_0 x^n + a_1 x^{n-1} + a_2 x^{n-2} + \cdots + a_{n-1} x + a_n \quad a_0 \neq 0 \tag{5.102}$$

Newton's method is suitable, efficient and simple to implement. We mainly consider polynomials with real coefficients $a_j$. However, most of the considerations remain valid without restrictions for polynomials with complex coefficients.

## 5.5 Zeros of Polynomials

Newton's method requires the evaluation of the polynomial $P_n(x)$ and of its first derivative for a prescribed value $x$. Hence we first consider the problem of an efficient and simple computation of these two values. The algorithm to be developed is based on the division with remainder of a polynomial by a linear factor

$$P_n(x) = (x - p)P_{n-1}(x) + R \quad p \text{ given.} \tag{5.103}$$

Let the quotient polynomial $P_{n-1}(x)$ be

$$P_{n-1}(x) = b_0 x^{n-1} + b_1 x^{n-2} + b_2 x^{n-3} + \cdots + b_{n-2} x + b_{n-1} \tag{5.104}$$

and the remainder is defined to be $R = b_n$ which will be seen to be reasonable. From (5.103) we have

$$a_0 x^n + a_1 x^{n-1} + a_2 x^{n-2} + \cdots + a_{n-2} x^2 + a_{n-1} x + a_n$$
$$= (x - p)(b_0 x^{n-1} + b_1 x^{n-2} + b_2 x^{n-3} + \cdots + b_{n-2} x + b_{n-1}) + b_n$$

and by comparing coefficients

$$a_0 = b_0 \quad a_1 = b_1 - pb_0 \quad a_2 = b_2 - pb_1, \ldots, a_{n-1} = b_{n-1} - pb_{n-2}$$
$$a_n = b_n - pb_{n-1}.$$

Thus the coefficients $b_j$ can be computed recursively by the following algorithm

$$\boxed{b_0 = a_0 \quad b_j = a_j + pb_{j-1} \quad (j = 1, 2, \ldots, n) \quad R = b_n} \tag{5.105}$$

The practical meaning of the division algorithm (5.105) becomes clear from the following theorem.

**Theorem 5.10** *The value of the polynomial $P_n(x)$ for $x = p$ is equal to the remainder $R$ after division of $P_n(x)$ by $(x - p)$.*

*Proof* From (5.103) it follows that for $x = p$

$$\boxed{P_n(p) = R.} \tag{5.106}$$

The simple algorithm (5.105) yields the value of the polynomial $P_n(x)$ for a given value $x$ with the amount of computational effort required being $n$ multiplications and additions. In order to get a similar computational scheme for computing the first derivative we differentiate (5.103) with respect to $x$ and obtain the relationship

$$P'_n(x) = P_{n-1}(x) + (x - p)P'_{n-1}(x). \tag{5.107}$$

This implies for $x = p$

$$P'_n(p) = P_{n-1}(p). \tag{5.108}$$

The value of the first derivative of $P_n(x)$, for a prescribed $x = p$, is equal to the value of the quotient polynomial $P_{n-1}(x)$ for $x = p$. From Theorem 5.10 its value can be computed by means of the division algorithm. We set

$$P_{n-1}(x) = (x - p)P_{n-2}(x) + R_1$$

and

$$P_{n-2}(x) = c_0 x^{n-2} + c_1 x^{n-3} + \cdots + c_{n-3} x + c_{n-2}, \quad c_{n-1} = R_1$$

then the coefficients $c_j$ are given by the recursion

$$c_0 = b_0 \quad c_j = b_j + p c_{j-1} \quad (j = 1, 2, \ldots, n-1) \quad R_1 = c_{n-1}. \tag{5.109}$$

Thus the first derivative $P'_n(p)$ can be computed from (5.108) and (5.109) by another $(n-1)$ multiplications.

Although the higher derivatives are not needed in connection with Newton's method, it should be pointed out that they can be computed similarly by the division algorithm. Differentiation of (5.107) yields the identity

$$P''_n(x) = 2P'_{n-1}(x) + (x - p)P''_{n-1}(x) \tag{5.110}$$

and hence by applying (5.108) accordingly we get

$$P''_n(p) = 2P'_{n-1}(p) = 2P_{n-2}(p). \tag{5.111}$$

It can be shown by successive differentiation of (5.110) and by recursive use of previous results that the $m$th derivative can be computed by

$$P_n^{(m)}(p) = m! P_{n-m}(p) \quad (m = 1, 2, \ldots, n) \tag{5.112}$$

where $P_{n-m}(x)$ is the quotient polynomial after $m$ steps of the division algorithm. The result (5.112) allows us to evaluate all $n$ derivatives of a polynomial $P_n(x)$ for a prescribed $x = p$ in an efficient way. This procedure is applied, for instance, when we seek the expansion of a given polynomial $P_n(x)$ in powers of $(x - p)$.

The quantities that are generated by the division algorithm are arranged in the *Horner scheme* that is especially useful for calculations by hand and that also gives some insight into an appropriate implementation. The Horner scheme, with

## 5.5 Zeros of Polynomials

the intermediate quantities, is shown in (5.113) for the first two division steps for the case $n = 6$.

$$P_6(x): \quad \begin{array}{ccccccc} a_0 & a_1 & a_2 & a_3 & a_4 & a_5 & a_6 \\ & pb_0 & pb_1 & pb_2 & pb_3 & pb_4 & pb_5 \end{array}$$

$$P_5(x): \quad \begin{array}{cccccc|c} b_0 & b_1 & b_2 & b_3 & b_4 & b_5 & b_6 = P_6(p) \\ & pc_0 & pc_1 & pc_2 & pc_3 & pc_4 & \end{array} \quad (5.113)$$

$$P_4(x): \quad \begin{array}{ccccc|c} c_0 & c_1 & c_2 & c_3 & c_4 & c_5 = P_6'(p) \end{array}$$

Each row is shortened by one entry corresponding to the decreasing degree of the quotient polynomials.

If we are only interested in the value of a polynomial of degree $n > 2$ and of its first derivative for a given $p$, this can be done on a computer without indexed variables $b$ and $c$. We observe that in the Horner scheme the value of $b_{j-1}$ is no longer needed as soon as $c_{j-1}$ and then $b_j$ have been computed. Thus the corresponding algorithm reads as follows

$$\boxed{\begin{array}{l} b = a_0;\ c = b;\ b = a_1 + p \times b \\ \text{for } j = 2, 3, \ldots, n: \\ \quad c = b + p \times c;\ b = a_j + p \times b \\ P_n(p) = b;\ P_n'(p) = c \end{array}} \quad (5.114)$$

**Example 5.12** We want to compute a zero of the polynomial of degree five

$$P_5(x) = x^5 - 5x^3 + 4x + 1$$

for which the approximate value $x^{(0)} = 2$ is known. The Horner scheme is

$$\begin{array}{r|rrrrrr} & 1 & 0 & -5 & 0 & 4 & 1 \\ 2) & & 2 & 4 & -2 & -4 & 0 \\ \hline & 1 & 2 & -1 & -2 & 0 & 1 = P_5(2) \\ 2) & & 2 & 8 & 14 & 24 & \\ \hline & 1 & 4 & 7 & 12 & 24 = P_5'(2) \end{array}$$

With the resulting values, Newton's method yields the iterate $x^{(1)} = 2 - \frac{1}{24} \doteq 1.958\,33$. We continue the process using a precision of six significant figures. The next Horner scheme for $x^{(1)}$ reads

|  | 1 | 0 | −5 | 0 | 4 | 1 |
|---|---|---|---|---|---|---|
| 1.958 33) |  | 1.958 33 | 3.835 06 | −2.281 34 | −4.467 62 | −0.915 754 |
|  | 1 | 1.958 33 | −1.164 94 | −2.281 34 | −0.467 620 | 0.084 246 |
| 1.958 33) |  | 1.958 33 | 7.670 11 | 12.739 3 | 20.480 2 |  |
|  | 1 | 3.916 66 | 6.505 17 | 10.458 0 | 20.012 6 |  |

Hence the second iterate is $x^{(2)} \doteq 1.958\,33 - 0.004\,209\,65 \doteq 1.954\,12$. Another iteration step produces $x^{(3)} \doteq 1.954\,08$, which is the first zero rounded to six digits.

If $z_1$ is a zero of $P_n(x)$, then $(x - z_1)$ divides $P_n(x)$ so that a known zero can be removed. We simply divide $P_n(x)$ by the linear factor $(x - z_1)$ and thus get the quotient polynomial $P_{n-1}(x)$ whose zeros are the remaining ones of $P_n(x)$. The coefficients of $P_{n-1}(x)$ are computed by means of the division algorithm (5.105), and the resulting remainder $R$ may be used for checking purposes. We continue with $P_{n-1}(x)$, determine a zero $z_2$, divide it by the linear factor $(x - z_2)$ and thus successively reduce the degree of the polynomial for which a further zero may be computed.

The described successive *deflation* of zeros may have an unfavourable effect on the zeros of the polynomials of lower degree subsequently computed for two reasons. First of all, the computed zero is, in general, only an approximate value of the exact zero, and secondly the computed coefficients $b_j$ of $P_{n-1}(x)$ inevitably suffer from rounding errors. Therefore we determine a zero $z_2^*$ of a perturbed polynomial $P_{n-1}^*(x)$. If the sensitivity of $z_2$ to perturbations of the coefficients of the exact quotient polynomial $P_{n-1}(x)$ is high, then $z_2^*$ can be quite different from the exact zero $z_2$ of $P_n(x)$ (see Section 6.1 for more details of a sensitivity analysis). The effect outlined can be so disastrous on the last computed zeros that they cannot even be considered as approximations! When the deflation method is applied nevertheless, the zeros should be determined with increasing magnitude because the resulting errors of the polynomial coefficients are smaller (Wilkinson, 1969). Moreover, an iterative improvement of the approximations with respect to the original polynomial $P_n(x)$ is advisable.

The described difficulties can be avoided by a simple modification of Newton's method. The already known zeros of $P_n(x)$ are removed *implicitly* so that we can work with the original and unchanged coefficients of $P_n(x)$. We denote by $z_1, z_2, \ldots, z_n$ the zeros of $P_n(x)$. Thus we have

$$P_n(x) = \sum_{j=0}^{n} a_j x^{n-j} = a_0 \prod_{j=1}^{n} (x - z_j) \quad a_0 \neq 0, \tag{5.115}$$

$$P_n'(x) = a_0 \sum_{i=1}^{n} \prod_{\substack{j=1 \\ j \neq i}}^{n} (x - z_j) \quad \frac{P_n'(x)}{P_n(x)} = \sum_{i=1}^{n} \frac{1}{x - z_i}. \tag{5.116}$$

## 5.5 Zeros of Polynomials

The correction of an iterate $x^{(k)}$ defined by Newton's method is now recognized as the reciprocal of the formal expression (5.116) for the value $x = x^{(k)}$. Let us assume that the $m$ zeros $z_1, z_2, \ldots, z_m$, $(1 \leqslant m < n)$ have already been computed, so that after their deflation the polynomial

$$P_{n-m}(x) := P_n(x) \bigg/ \prod_{i=1}^{m}(x-z_i) = a_0 \prod_{j=m+1}^{n}(x-z_j) \tag{5.117}$$

is, in principle, used for the determination of the following zero. However, from (5.116) we have the expression,

$$\frac{P'_{n-m}(x)}{P_{n-m}(x)} = \sum_{j=m+1}^{n} \frac{1}{x-z_j} = \frac{P'_n(x)}{P_n(x)} - \sum_{i=1}^{m} \frac{1}{x-z_i} \tag{5.118}$$

and hence the modified computational scheme of Newton's method with *implicit deflation* of approximately known zeros $z_1, z_2, \ldots, z_m$ is

$$\boxed{x^{(k+1)} = x^{(k)} - \frac{1}{\dfrac{P'_n(x^{(k)})}{P_n(x^{(k)})} - \sum_{i=1}^{m} \dfrac{1}{(x^{(k)} - z_i)}} \quad (k=0,1,2,\ldots).} \tag{5.119}$$

The amount of computational effort required by an iteration step (5.119) for the $(m+1)$th zero consists of $Z_N = 2n + m + 1$ essential operations and increases with each computed zero.

*Example 5.13* We determine a second zero of the polynomial of Example 5.12 by using the technique of the implicit deflation of the approximately known zero $z_1 \doteq 1.95408$. Again we use a precision of six significant digits and get numbers that are given in Table 5.14. The same initial value $x^{(0)} = 2$ is chosen intentionally to illustrate that the sequence of iterates $x^{(k)}$ converges in fact to a different zero $z_2$.

Newton's method can also be directly applied to the situation where the

Table 5.14 Newton's method with implicit deflation

| $k$ | $x^{(k)}$ | $P_5(x^{(k)})$ | $P'_5(x^{(k)})$ | $(x^{(k)} - z_1)^{-1}$ | $\Delta x^{(k)}$ |
|---|---|---|---|---|---|
| 0 | 2.00000 | 1.00000 | 24.000 | 21.7770 | $-4.49843 \times 10^{-1}$ |
| 1 | 1.55016 | $-2.47327$ | $-3.17298$ | $-2.47574$ | $-2.66053 \times 10^{-1}$ |
| 2 | 1.28411 | $-0.959150$ | $-7.13907$ | $-1.49260$ | $-1.11910 \times 10^{-1}$ |
| 3 | 1.17220 | $-0.151390$ | $-7.17069$ | $-1.27897$ | $-2.05572 \times 10^{-2}$ |
| 4 | 1.15164 | $-0.00466$ | $-7.09910$ | $-1.24620$ | $-6.55884 \times 10^{-4}$ |
| 5 | $1.15098 \doteq z_2$ | | | | |

coefficients of the polynomial are complex or where the coefficients are real, but pairs of conjugate complex zeros exist. All steps must be carried out in complex arithmetic. The initial value must be complex in the second case because otherwise the sequence of iterates $x^{(k)}$ would be real.

Complex arithmetic can, however, be avoided completely in the case of polynomials with real coefficients. If $z_1 = u + iv, v \neq 0$, is a complex zero of $P_n(x)$, then it is known that the complex conjugate value $z_2 = u - iv$ is also a zero of $P_n(x)$. Hence the product of the two linear factors

$$(x - z_1)(x - z_2) = (x - u - iv)(x - u + iv) = x^2 - 2ux + (u^2 + v^2) \quad (5.120)$$

is a *quadratic divisor* of $P_n(x)$ whose coefficients are real due to (5.120). Due to this observation our new aim will be to find a real quadratic divisor of $P_n(x)$. If such a quadratic divisor has been found either a pair of conjugate complex zeros or a pair of real zeros of $P_n(x)$ has been determined. The remaining $(n-2)$ zeros are subsequently determined as zeros of the quotient polynomial $P_{n-2}(x)$.

We start with the division algorithm for a quadratic factor. Let the given polynomial be

$$P_n(x) = a_0 x^n + a_1 x^{n-1} + a_2 x^{n-2} + \cdots$$
$$+ a_{n-2} x^2 + a_{n-1} x + a_n \quad a_j \in \mathbf{R} \quad a_0 \neq 0.$$

The quadratic factor is defined by

$$x^2 - px - q \quad p, q \in \mathbf{R}. \quad (5.121)$$

We seek the coefficients of the quotient polynomial

$$P_{n-2}(x) = b_0 x^{n-2} + b_1 x^{n-3} + b_2 x^{n-4} + \cdots + b_{n-4} x^2 + b_{n-3} x + b_{n-2} \quad (5.122)$$

so that we have

$$P_n(x) = (x^2 - px - q) P_{n-2}(x) + b_{n-1}(x - p) + b_n. \quad (5.123)$$

The linear remainder polynomial $R_1(x) = b_{n-1}(x - p) + b_n$ in (5.123) is defined in this unconventional form in order to achieve a resulting computational scheme for the coefficients $b_j$ that will be systematically valid for all indices. Comparing coefficients leads to the following relationships:

$$
\begin{array}{ll}
x^n & a_0 = b_0 \\
x^{n-1} & a_1 = b_1 - p b_0 \\
x^{n-2} & a_2 = b_2 - p b_1 - q b_0 \\
\vdots & \\
x^{n-j} & a_j = b_j - p b_{j-1} - q b_{j-2} \quad (j = 2, 3, \ldots, n).
\end{array}
$$

The desired coefficients of $P_{n-2}(x)$ and of the remainder $R_1(x)$ are thus

## 5.5 Zeros of Polynomials

computable by the simple recursive algorithm

$$b_0 = a_0 \quad b_1 = a_1 + pb_0$$
$$b_j = a_j + pb_{j-1} + qb_{j-2} \quad (j = 2, 3, \ldots, n) \tag{5.124}$$

A factor $x^2 - px - q$ is a divisor of $P_n(x)$ if and only if the remainder polynomial $R_1(x)$ is zero, that is if the two coefficients of (5.123) $b_{n-1} = b_n = 0$. For given values of $p$ and $q$ the coefficients of $P_{n-2}(x)$, as well as those of $R_1(x)$, are uniquely determined. They can be considered as functions of the two variables $p$ and $q$. The problem formulated above of finding a quadratic divisor of $P_n(x)$ is thus equivalent to solving the system of nonlinear equations

$$b_{n-1}(p, q) = 0 \quad b_n(p, q) = 0 \tag{5.125}$$

for the unknowns $p$ and $q$. The functions $b_{n-1}(p, q)$ and $b_n(p, q)$ are defined by the algorithm (5.124). We are only interested in the real solutions of (5.125), and want to compute them by means of Newton's method. To do this we need the partial derivatives of the two functions $b_{n-1}(p, q)$ and $b_n(p, q)$ with respect to $p$ and $q$. They can be evaluated by means of a single recursion formula, analogous to (5.124). From (5.124) we find that

$$\frac{\partial b_j}{\partial p} = b_{j-1} + p \frac{\partial b_{j-1}}{\partial p} + q \frac{\partial b_{j-2}}{\partial p} \quad (j = 2, 3, \ldots, n) \tag{5.126}$$

and in addition

$$\frac{\partial b_0}{\partial p} = 0 \quad \frac{\partial b_1}{\partial p} = b_0. \tag{5.127}$$

The structure of (5.126) suggests the definition of the quantities

$$c_{j-1} := \frac{\partial b_j}{\partial p} \quad (j = 1, 2, \ldots, n). \tag{5.128}$$

Thus from (5.126), because of (5.127), we get the computational scheme

$$c_0 = b_0 \quad c_1 = b_1 + pc_0$$
$$c_j = b_j + pc_{j-1} + qc_{j-2} \quad (j = 2, 3, \ldots, n-1). \tag{5.129}$$

Partial differentiation of (5.124) with respect to $q$ yields

$$\frac{\partial b_j}{\partial q} = b_{j-2} + p \frac{\partial b_{j-1}}{\partial q} + q \frac{\partial b_{j-2}}{\partial q} \quad (j = 2, 3, \ldots, n), \tag{5.130}$$

where we have to take account of the additional relationships

$$\frac{\partial b_0}{\partial q} = 0 \quad \frac{\partial b_1}{\partial q} = 0 \quad \frac{\partial b_2}{\partial q} = b_0 \quad \frac{\partial b_3}{\partial q} = b_1 + p\frac{\partial b_2}{\partial q}.$$

Hence it is obvious that with the identification

$$\frac{\partial b_j}{\partial q} =: c_{j-2} \quad (j = 2, 3, \ldots, n) \tag{5.131}$$

these partial derivatives satisfy the recursion formula (5.129). In particular the derivatives that are required for Newton's method from (5.128) and (5.131), are found to be

$$\frac{\partial b_{n-1}}{\partial p} = c_{n-2} \quad \frac{\partial b_{n-1}}{\partial q} = c_{n-3} \quad \frac{\partial b_n}{\partial p} = c_{n-1} \quad \frac{\partial b_n}{\partial q} = c_{n-2}. \tag{5.132}$$

With these preliminaries all the elements are known that are required to perform a Newton's iteration step for solving the system (5.125). The relationships (5.132) define the matrix elements of the Jacobian $\Phi$, and from the general scheme (5.86), after corresponding substitutions we get

$$p^{(k+1)} = p^{(k)} + \frac{b_n c_{n-3} - b_{n-1} c_{n-2}}{c_{n-2}^2 - c_{n-1} c_{n-3}} \quad q^{(k+1)} = q^{(k)} + \frac{b_{n-1} c_{n-1} - b_n c_{n-2}}{c_{n-2}^2 - c_{n-1} c_{n-3}}$$

$$\tag{5.133}$$

The $b$ and $c$ values in (5.133) have to be computed for the iterates $p^{(k)}$ and $q^{(k)}$. The denominator $c_{n-2}^2 - c_{n-1} c_{n-3}$ is equal to the determinant of the Jacobian $\Phi$. If the determinant vanishes, the formula (5.133) cannot be used. In this case the iterates $p^{(k)}$ and $q^{(k)}$ are simply changed for example by adding random numbers. Moreover, it may happen that the determinant is not zero but small in absolute value. This may cause quite large iterates $p^{(k+1)}$ and $q^{(k+1)}$ that cannot be useful approximations of the coefficients of a quadratic divisor of the given polynomial. The absolute sizes of the zeros of a polynomial $P_n(x)$ (5.102) are bounded, for instance, by

$$|z_i| \leqslant R := \max\left(\left|\frac{a_n}{a_0}\right|, 1 + \left|\frac{a_1}{a_0}\right|, 1 + \left|\frac{a_2}{a_0}\right|, \ldots, 1 + \left|\frac{a_{n-1}}{a_0}\right|\right). \tag{5.134}$$

## 5.5 Zeros of Polynomials

The bound (5.134) can be derived from the so-called companion matrix (Wilkinson and Reinsch, 1971) whose characteristic polynomial is equal to $P_n(x)/a_0$. This matrix essentially contains the coefficients of $P_n(x)/a_0$ in the last column, elements equal to one in the subdiagonal and zeros otherwise. The row sum norm (1.67) is a bound of the absolute values of the eigenvalues and hence of the zeros of $P_n(x)$. According to (5.120) and (5.121) we have $q = -(u^2 + v^2) = -|z_1|^2$ in the case of a pair of complex conjugate zeros or $|q| = |z_1| \cdot |z_2|$ in the case of two real zeros. Therefore the test $|q^{(k+1)}| \leq R^2$ is reasonable. If it is not satisfied the computed iterates $p^{(k+1)}$ and $q^{(k+1)}$ are reduced accordingly.

The *Bairstow method* for determining a quadratic divisor $x^2 - px - q$ of a polynomial $P_n(x)$ with real coefficients and of degree $n > 2$ consists of the following steps of a single iteration: (1) for given approximations $p^{(k)}$ and $q^{(k)}$ we compute the values $b_j$ by the recursion (5.124) and the values $c_j$ by (5.129). (2) If the determinant of the Jacobian $\Phi$ is nonzero the iterates $p^{(k+1)}$ and $q^{(k+1)}$ are computed according to (5.133) and the test mentioned above is applied. (3) The iteration is stopped as soon as the values $b_{n-1}$ and $b_n$ are sufficiently small in absolute value with respect to $a_{n-1}$ and $a_n$, from which they are computed according to (5.124). We propose the condition

$$|b_{n-1}| + |b_n| \leq \varepsilon(|a_{n-1}| + |a_n|). \tag{5.135}$$

Here $\varepsilon > 0$ is a tolerance that must be larger than the smallest positive number $\delta$ for which $1 + \delta \neq 1$ holds on the computer. (4) The two zeros of the quadratic divisor are determined. This is the first time when a complex number arises in the case of a pair of complex conjugate zeros. (5) The deflation of the two zeros is performed by division of $P_n(x)$ by the quadratic divisor. The coefficients of the quotient polynomial are given by (5.124). An implicit deflation seems to be impossible.

Bairstows method has the order of convergence $p = 2$, because the system of nonlinear equations (5.125) is treated by Newton's method. The quantities produced by the division algorithm are arranged in a *double rowed Horner scheme* shown in (5.136) for the case $n = 6$

$$\begin{array}{c|ccccccc}
 & a_0 & a_1 & a_2 & a_3 & a_4 & a_5 & a_6 \\
q) & & & qb_0 & qb_1 & qb_2 & qb_3 & qb_4 \\
p) & & pb_0 & pb_1 & pb_2 & pb_3 & pb_4 & pb_5 \\
\hline
 & b_0 & b_1 & b_2 & b_3 & b_4 & b_5 & b_6 \\
q) & & & qc_0 & qc_1 & qc_2 & qc_3 & \\
p) & & pc_0 & pc_1 & pc_2 & pc_3 & pc_4 & \\
\hline
 & c_0 & c_1 & c_2 & \boxed{c_3} & c_4 & c_5 & \\
\end{array} \tag{5.136}$$

Table 5.15 Bairstow's method

| $k =$ | 0 | 1 | 2 | 3 | 4 |
|---|---|---|---|---|---|
| $p^{(k)} =$<br>$q^{(k)} =$ | $-2$<br>$-5$ | $-1.7681159$<br>$-4.6923077$ | $-1.7307716$<br>$-4.7281528$ | $-1.7320524$<br>$-4.7320527$ | $-1.7320508$<br>$-4.7320508$ |
| $b_4 =$<br>$b_5 =$ | $-4$<br>$16$ | $0.1714180\,0$<br>$2.274299\,0$ | $0.0457260$<br>$-0.0455520$ | $-0.000029$<br>$0.000077$ | $-1 \times 10^{-6}$<br>$0$ |
| $c_2 =$<br>$c_3 =$<br>$c_4 =$ | $13$<br>$0$<br>$-69$ | $10.066549$<br>$5.072222\,0$<br>$-56.032203$ | $9.4534922$<br>$6.9160820$<br>$-56.621988$ | $9.4641203$<br>$6.9281770$<br>$-56.784711$ | $9.4641012$<br>$6.9282040$<br>$-56.784610$ |
| $\Delta p^{(k)} =$<br>$\Delta q^{(k)} =$ | $0.23188406$<br>$0.30769231$ | $0.037344320$<br>$-0.035845122$ | $-0.0012808444$<br>$-0.0038998894$ | $1.5880 \times 10^{-6}$<br>$1.9017 \times 10^{-6}$ | $1.1835 \times 10^{-8}$<br>$9.6999 \times 10^{-8}$ |

**Example 5.14** For the polynomial of degree five
$$P_5(x) = x^5 - 2x^4 + 3x^3 - 12x^2 + 18x - 12$$
the double rowed Horner scheme (5.136) for the initial values $p^{(0)} = -2$ and $q^{(0)} = -5$ is

|     |     | 1  | −2 | 3   | −12 | 18  | −12 |
|-----|-----|----|----|-----|-----|-----|-----|
| −5) |     |    |    | −5  | 20  | −30 | 20  |
| −2) |     |    | −2 | 8   | −12 | 8   | 8   |
|     |     | 1  | −4 | 6   | −4  | −4  | 16  |
| −5) |     |    |    | −5  | 30  | −65 |     |
| −2) |     |    | −2 | 12  | −26 | 0   |     |
|     |     | 1  | −6 | 13  | 0   | −69 |     |

The essential numbers of Bairstow's method are summarized in Table 5.15 where a precision of eight significant digits is used.

The resulting quadratic divisor $x^2 + 1.7320508x + 4.7320507$ has a pair of complex conjugate zeros $z_{1,2} \doteq -0.86602540 \pm 1.9955076i$. With the quotient polynomial $P_3(x) = x^3 - 3.7320508x^2 + 4.7320509x - 2.5358990$ we can find another quadratic divisor $x^2 - 1.7320506x + 1.2679494$ by means of Bairstow's method which has a second pair of complex conjugate zeros $z_{3,4} = 0.86602530 \pm 0.71968714i$. The last zero is $z_5 = 2$.

## 5.6 Exercises

**5.1.** A fixed point iteration $x^{(k+1)} = F(x^{(k)})$ is defined by $F(x) = 1 + 1/x + (1/x)^2$.
  (a) Verify that $F(x)$ satisfies the hypotheses of the Banach fixed point theorem for the interval $[1.75, 2.0]$. How large is the Lipschitz constant $L$?
  (b) For the initial value $x^{(0)} = 1.825$ compute the iterates up to $x^{(20)}$. What *a priori* and *a posteriori* error bounds can be given? How large is the convergence quotient $|q|$ and the number of iterations to gain another correct digit?
  (c) Apply Aitken's $\Delta^2$-process to the sequence $x^{(k)}$ and approximately determine the convergence factor of the Aitken sequence $z^{(k)}$.
  (d) Use Steffensen's method to compute the fixed point $s$.

**5.2.** The nonlinear equation $2x - \sin(x) = 0.5$ can be transformed into the fixed point form $x = 0.5\sin(x) + 0.25 =: F(x)$. Find an interval for which $F(x)$ satisfies the hypotheses of the Banach fixed point theorem. Then proceed as in Exercise 5.1 to find the fixed point $s$.

**5.3.** Compute the positive zero of $F(x) = e^{2x} - \sin(x) - 2$
  (a) by means of the regula falsi method, $x^{(0)} = 0$, $x^{(1)} = 1$;
  (b) by means of the secant method, $x^{(0)} = 0.25$, $x^{(1)} = 0.35$;

(c) by means of Newton's method, $x^{(0)} = 0.25$;
  (d) by means of the simplified Newton's method, $x^{(0)} = 0.25$;
  (e) by means of Muller's method, $x^{(0)} = 0$, $x^{(1)} = 1$.

**5.4.** Use the same methods as in Exercise 5.3 to determine the smallest positive solutions of the following equations:
  (a) $\tan(x) - \tanh(x) = 0$;
  (b) $\tan(x) - x - 4x^3 = 0$;
  (c) $(1 + x^2)\tan(x) - x(1 - x^2) = 0$.

**5.5.** To solve an equation $f(x) = 0$ the iteration scheme

$$x^{(k+1)} = x^{(k)} - \frac{f(x^{(k)})f'(x^{(k)})}{f'(x^{(k)})^2 - 0.5 f(x^{(k)}) f''(x^{(k)})} \quad (k = 0, 1, 2, \ldots)$$

can be applied on the assumption that $f(x)$ is at least twice continuously differentiable. Show that the order of convergence is, at least, $p = 3$. What are the resulting iteration schemes for computing $\sqrt{a}$ from the equation $f(x) = x^2 - a = 0$ and of $\sqrt[3]{a}$ from $f(x) = x^3 - a = 0$? With these methods compute $\sqrt{2}$ and $\sqrt[3]{2}$ with the initial value $x^{(0)} = 1$.

**5.6.** The system of linear equations

$$\begin{aligned} 8x_1 + x_2 - 2x_3 - 8 &= 0 \\ x_1 + 18x_2 - 6x_3 + 10 &= 0 \\ 2x_1 + x_2 + 16x_3 + 2 &= 0 \end{aligned}$$

with diagonally dominant matrix can be transformed into the following fixed point form:

$$\begin{aligned} x_1 &= \phantom{-}0.2x_1 - 0.1x_2 + 0.2x_3 + 0.8 \\ x_2 &= -0.05x_1 + 0.1x_2 + 0.3x_3 - 0.5 \\ x_3 &= -0.1x_1 - 0.05x_2 + 0.2x_3 - 0.1 \end{aligned}$$

Verify the equivalence of the two systems. Then show that the linear mapping is a contraction for the whole of $\mathbf{R}^3$ and determine the Lipschitz constant $L$ and the spectral radius of the Jacobian. With the initial vector $\mathbf{x}^{(0)} = (1, -0.5, -0.3)^T$ perform, at least, ten iteration steps and find the *a priori* and the *a posteriori* error bounds. Verify the convergence factor of the computed iterated vectors.

**5.7.** Solve the system of nonlinear equations

$$\begin{aligned} e^{xy} + x^2 + y - 1.2 &= 0 \\ x^2 + y^2 + x - 0.55 &= 0 \end{aligned}$$

For the initial values $x^{(0)} = 0.6$, $y^{(0)} = 0.5$ apply Newton's method and give the geometrical interpretation of the first two steps. Then solve the system by means of the simplified Newton's method but with the better initial values $x^{(0)} = 0.4$, $y^{(0)} = 0.25$. What is the convergence quotient of the linearly convergent sequence derived from the spectral radius of the corresponding Jacobian matrix?

**5.8.** To solve the system of nonlinear equations

$$\begin{aligned} 4.72 \sin(2x) - 3.14 e^y - 0.495 &= 0 \\ 3.61 \cos(3x) + \sin(y) - 0.402 &= 0 \end{aligned}$$

## 5.6 Exercises

use the initial values $x^{(0)} = 1.5$ and $y^{(0)} = -4.7$. Apply Newton's method, the simplified Newton's method, Newton's successive relaxation method and the SOR–Newton method for different values of $\omega$.

**5.9.** Determine the zeros of the polynomials

$$P_5(x) = x^5 - 2x^4 - 13x^3 + 14x^2 + 24x - 1$$
$$P_8(x) = x^8 - x^7 - 50x^6 + 62x^5 + 650x^4 - 528x^3 - 2760x^2 + 470x + 2185$$

by means of Newton's method in combination with the implicit deflation of computed zeros.

**5.10.** Use Bairstow's method and Muller's method to compute the (complex) zeros of the polynomials

$$P_6(x) = 3x^6 + 9x^5 + 9x^4 + 5x^3 + 3x^2 + 8x + 5$$
$$P_8(x) = 3x^8 + 2x^7 + 6x^6 + 4x^5 + 7x^4 + 3x^3 + 5x^2 + x + 8.$$

# 6

# Eigenvalue Problems

We consider the problem of computing the eigenvalues and eigenvectors of a real, square matrix **A**. This problem particularly occurs in physics and engineering sciences in the treatment of vibration problems. An eigenvalue problem must also be solved in statistics to analyze variances. We have already encountered the eigenvalue problem in Chapter 1, where the determination of the condition number of a matrix, corresponding to the spectral norm, requires the computation of the largest and the smallest eigenvalue of a certain matrix. In a similar way the eigenvalues of matrices are crucial for the stability of methods for solving systems of ordinary differential equations or of partial differential equations (see Chapters 9 and 10).

We shall restrict ourselves to the treatment of a few methods for determining eigenvalues that are usually available as library routines on computers. More extensive details of eigenvalue methods can be found, for instance, in Bathe and Wilson (1976), Bunse and Bunse–Gerstner (1984), Golub and Van Loan (1983), Jennings (1977), Parlett (1980), Stewart (1973), Törnig (1979), Wilkinson (1965) and Zurmühl and Falk (1984; 1986).

## 6.1 The Characteristic Polynomial, Difficulties

Let **A** be a real, square matrix of order $n$. We seek its eigenvalues $\lambda_k$ and the corresponding eigenvectors $\mathbf{x}_k$ such that

$$\mathbf{A}\mathbf{x}_k = \lambda_k \mathbf{x}_k \quad \text{or} \quad (\mathbf{A} - \lambda_k \mathbf{I})\mathbf{x}_k = \mathbf{0} \tag{6.1}$$

As the eigenvector $\mathbf{x}_k$ must be nonzero, the determinant of the matrix $\mathbf{A} - \lambda_k \mathbf{I}$ must necessarily vanish. Hence the eigenvalue $\lambda_k$ must be a zero of the *characteristic polynomial*

$$P(\lambda) := (-1)^n |\mathbf{A} - \lambda \mathbf{I}| = \lambda^n + p_{n-1}\lambda^{n-1} + p_{n-2}\lambda^{n-2} + \cdots + p_1 \lambda + p_0. \tag{6.2}$$

## 6.1 The Characteristic Polynomial, Difficulties

The characteristic polynomial is of exact degree $n$ and thus has $n$ zeros, counting multiplicities. Hence the problem of determining the eigenvalues seems to have been already solved because we only need to compute the coefficients of the characteristic polynomial and then to determine the zeros by means of a known method of Section 5.5. The eigenvectors $\mathbf{x}_k$ corresponding to the computed eigenvalues $\lambda_k$ are obtained as the nontrivial solutions of the systems of homogeneous equations (6.1).

However, this theoretical procedure for solving the problem is inadequate for numerical reasons as soon as the order $n$ of the matrix is large. Once again we have to take into account the finite precision of every computation. Each method suffers from rounding errors for computing the coefficients $p_j$ of the characteristic polynomial $P(\lambda)$ from the given numerical elements of the matrix $\mathbf{A}$. The resulting values $p_j^*$ have an error. In the most favourable case, the $p_j^*$ will be equal to computer numbers closest to the exact values $p_j$. Consequently, it is realistic to assume that all computed coefficients $p_j^*$ possess a relative error $\varepsilon$, that is at most equal to $5 \times 10^{-d}$ if a precision of $d$ digits is used. The exact characteristic polynomial $P(\lambda)$ is thus replaced by

$$P^*(\lambda) = \lambda^n + p_{n-1}^* \lambda^{n-1} + p_{n-2}^* \lambda^{n-2} + \cdots + p_1^* \lambda + p_0^* \tag{6.3}$$

whose zeros $\lambda_k^*$ serve as approximations of the eigenvalues of the given matrix.

We now want to investigate the sensitivity of the zeros $\lambda_k$ of the polynomial $P(\lambda)$ to slight changes of the coefficients. For this purpose we introduce the quantities $q_j$

$$p_j^* = p_j + \varepsilon q_j \quad (j = 0, 1, \ldots, n-1) \tag{6.4}$$

and define the corresponding polynomial

$$Q(\lambda) := q_{n-1} \lambda^{n-1} + q_{n-2} \lambda^{n-2} + \cdots + q_1 \lambda + q_0 \tag{6.5}$$

whose degree is at most equal to $(n-1)$. Thus we can write the polynomial $P^*(\lambda)$ of (6.3) as

$$P^*(\lambda) = P(\lambda) + \varepsilon Q(\lambda). \tag{6.6}$$

In (6.6) $\varepsilon$ is considered to be a small parameter, and $Q(\lambda)$ represents the polynomial of perturbation. The zeros $\lambda_k^*$ of $P^*(\lambda)$ depend continuously on $\varepsilon$, but not in a continuously differentiable way. In order to simplify the sensitivity analysis we only consider the case of a *simple zero* $\lambda_k$ of $P(\lambda)$, so that we have

$$P(\lambda_k) = 0 \quad \text{and} \quad P'(\lambda_k) \neq 0. \tag{6.7}$$

For sufficiently small $\varepsilon$, the zero $\lambda_k^*$ can be well approximated by using Newton's correction, (5.53), for the initial guess $\lambda_k$. The correction term is given by

$$\Delta \lambda_k = -\frac{P^*(\lambda_k)}{P^{*'}(\lambda_k)} = -\frac{P(\lambda_k) + \varepsilon Q(\lambda_k)}{P'(\lambda_k) + \varepsilon Q'(\lambda_k)} = \frac{-\varepsilon Q(\lambda_k)}{P'(\lambda_k) + \varepsilon Q'(\lambda_k)}. \tag{6.8}$$

In addition we assume that $\varepsilon$ is small enough so that

$$|\varepsilon Q'(\lambda_k)| \ll |P'(\lambda_k)|$$

holds. Due to (6.7) this condition can be satisfied. Thus we obtain the desired qualitative result

$$\boxed{\lambda_k^* \approx \lambda_k - \varepsilon \frac{Q(\lambda_k)}{P'(\lambda_k)}} \tag{6.9}$$

The change of the zero $\lambda_k$ depends on the value of the perturbation polynomial $Q(\lambda_k)$ and on the value of the derivative $P'(\lambda_k)$ of the given polynomial $P(\lambda)$. The quotient $Q(\lambda_k)/P'(\lambda_k)$ in (6.9) represents the amplifying factor of the inevitable relative error $\varepsilon$. These quotients that depend on $\lambda_k$ can be called the *condition numbers* of the zeros $\lambda_k$. The condition number of a certain zero $\lambda_k$ is, in general, large if the derivative $P'(\lambda_k)$ is small in absolute value. This occurs in the case of closely neighbouring zeros, and so (6.9) explains the obvious fact that close zeros are very sensitive to small changes of the coefficients.

*Example* 6.1 We consider the polynomial of degree $n = 12$ with the zeros, $\lambda_1 = 1$, $\lambda_2 = 2, \ldots, \lambda_{12} = 12$

$$P(\lambda) = \prod_{k=1}^{12} (\lambda - k) = \lambda^{12} - 78\lambda^{11} + 2717\lambda^{10} - + \cdots - 6\,926\,634\lambda^7$$

$$+ \cdots + 479\,001\,600.$$

For reasons of clarity and simplicity we analyze the sensitivity of the zeros $\lambda_k$ if a single coefficient $p_j$ is changed by the value $\varepsilon p_j$. The perturbation polynomial (6.5) is given by

$$Q(\lambda) = p_j \lambda^j \quad (0 \leqslant j \leqslant 11).$$

The first derivative of the polynomial $P(\lambda)$ at the zero $\lambda_k = k$ is equal to

$$P'(\lambda_k) = P'(k) = (-1)^{12-k}(k-1)!(12-k)!. \tag{6.10}$$

Following (6.9) the condition numbers, depending on $k$ and $j$, are given by

$$C_{k,j} := \left| \frac{Q(\lambda_k)}{P'(\lambda_k)} \right| = \frac{|p_j| k^j}{(k-1)!(12-k)!} \quad \begin{array}{l} (k = 1, 2, \ldots, 12; \\ j = 0, 1, \ldots, 11). \end{array} \tag{6.11}$$

To illustrate the quite different, partly large sensitivities of the zeros $\lambda_k$ the condition numbers $C_{k,j}$ are summarized in Table 6.1 for some index combinations $k$ and $j$.

In this example the smallest zeros $\lambda_1$ to $\lambda_3$ are the least sensitive ones, whereas $\lambda_6$ to $\lambda_{10}$ have much larger condition numbers. The largest condition number is

## 6.2 Jacobi Methods

Table 6.1 Condition numbers $C_{k,j}$

| j \ k | 1 | 3 | 6 | 8 | 9 | 10 | 12 |
|---|---|---|---|---|---|---|---|
| 0 | $1.20 \times 10$ | $6.60 \times 10^2$ | $5.54 \times 10^3$ | $3.96 \times 10^3$ | $1.98 \times 10^3$ | $6.60 \times 10^2$ | $1.20 \times 10^1$ |
| 3 | $3.54 \times 10^1$ | $5.26 \times 10^4$ | $3.54 \times 10^6$ | $5.99 \times 10^6$ | $4.26 \times 10^6$ | $1.95 \times 10^6$ | $6.12 \times 10^4$ |
| 6 | $1.13 \times 10^0$ | $4.52 \times 10^4$ | $2.43 \times 10^7$ | $9.75 \times 10^7$ | $9.88 \times 10^7$ | $6.20 \times 10^7$ | $3.37 \times 10^6$ |
| 7 | $1.74 \times 10^{-1}$ | $2.09 \times 10^4$ | $2.24 \times 10^7$ | $1.20 \times 10^8$ | $1.37 \times 10^8$ | $9.54 \times 10^7$ | $6.22 \times 10^6$ |
| 8 | $1.88 \times 10^{-2}$ | $6.78 \times 10^3$ | $1.46 \times 10^7$ | $1.04 \times 10^8$ | $1.33 \times 10^8$ | $1.03 \times 10^8$ | $8.07 \times 10^6$ |
| 11 | $1.95 \times 10^{-6}$ | $1.90 \times 10^1$ | $3.28 \times 10^5$ | $5.54 \times 10^6$ | $1.01 \times 10^7$ | $1.07 \times 10^7$ | $1.45 \times 10^6$ |

$C_{9,7} \doteq 1.37 \times 10^8$. If the coefficient $p_7 = -6\,926\,634$ is changed by a unit in the tenth decimal place to $p_7^* = -6\,926\,634.001$ due to rounding errors the relative error is $\varepsilon \doteq 1.444 \times 10^{-10}$. If the signs of $Q(\lambda_9) = p_7 \lambda_9^7$ and of $P'(\lambda_9)$ according to (6.10) are taken into account, our qualitative result (6.9) yields

$$\lambda_9^* \approx \lambda_9 - 1.444 \times 10^{-10} \times 1.37 \times 10^8 \doteq 8.980.$$

The large change of the zero $\lambda_9$ that is caused by the small perturbation of the coefficient $p_7$ can be verified by a computation using high precision.

The high sensitivity of zeros that is illustrated in Example 6.1 is, in general, even greater for increasing degree of the polynomial. It may happen that pairs of real zeros become complex conjugate pairs of zeros (Wilkinson, 1969). Therefore the characteristic polynomial has to be definitely avoided as a means of computing eigenvalues. That is why in the following we will only be considering methods that do not require the explicit computation of the coefficients of the characteristic polynomial.

### 6.2 Jacobi Methods

In this section we consider the problem of computing all the eigenvalues and corresponding eigenvectors of a real *symmetric matrix* **A**. The numerical treatment of this problem is simplified because the eigenvalues, $\lambda_k$, of a symmetric matrix of order $n$ are real and the eigenvectors, $\mathbf{x}_k$, form a complete system of $n$ orthonormal vectors, even in the case of multiple eigenvalues.

The motivation of the following procedures is given by the *theorem of principal axes* stating that each symmetric matrix **A** can be transformed by means of an *orthogonal* matrix **U** into a similar diagonal matrix **D**, that is

$$\mathbf{U}^{-1}\mathbf{A}\mathbf{U} = \mathbf{D}. \tag{6.12}$$

The columns of **U** are equal to the eigenvectors $\mathbf{x}_k$, and the diagonal elements of **D** are the eigenvalues $\lambda_k$ of **A**. This can be seen from (6.12) which is equivalent to $\mathbf{AU} = \mathbf{UD}$. This relation summarizes the $n$ equations (6.1) in matrix form.

Two matrices **A** and **B** are called *similar* if there exists a nonsingular matrix **C**

such that
$$\mathbf{C}^{-1}\mathbf{A}\mathbf{C} = \mathbf{B} \tag{6.13}$$

The transition from the matrix $\mathbf{A}$ to the matrix $\mathbf{B}$ according to (6.13) is called a *similarity transformation*. We note the essential fact that similar matrices have the same eigenvalues. In fact, their characteristic polynomials are identical, because we have

$$P_B(\lambda) = |\lambda \mathbf{I} - \mathbf{B}| = |\lambda \mathbf{I} - \mathbf{C}^{-1}\mathbf{A}\mathbf{C}| = |\mathbf{C}^{-1}(\lambda \mathbf{I} - \mathbf{A})\mathbf{C}|$$
$$= |\mathbf{C}^{-1}| \cdot |\lambda \mathbf{I} - \mathbf{A}| \cdot |\mathbf{C}| = |\lambda \mathbf{I} - \mathbf{A}| = P_A(\lambda).$$

We have used the relationships for determinants $|\mathbf{XY}| = |\mathbf{X}| \cdot |\mathbf{Y}|$ and $|\mathbf{C}^{-1}| = |\mathbf{C}|^{-1}$.

### 6.2.1 Elementary rotations

In order to transform a given matrix into a specified form we often use transformation matrices that are as simple as possible so that the effect of a multiplication is obvious. The following orthogonal matrices belong to this class of matrices

$$\mathbf{U}(p,q;\varphi) := \begin{pmatrix} 1 & & & & & & & & \\ & \ddots & & & & & & & \\ & & 1 & & & & & & \\ & & & \cos\varphi & & \sin\varphi & & & \\ & & & & 1 & & & & \\ & & & & & \ddots & & & \\ & & & & & & 1 & & \\ & & & -\sin\varphi & & \cos\varphi & & & \\ & & & & & & & 1 & \\ & & & & & & & & \ddots \\ & & & & & & & & & 1 \end{pmatrix} \begin{matrix} \\ \\ \\ \leftarrow p \\ \\ \\ \\ \leftarrow q \\ \\ \\ \end{matrix} \quad \begin{matrix} u_{ii} = 1, i \neq p,q \\ u_{pp} = u_{qq} = \cos\varphi \\ u_{pq} = \sin\varphi \\ u_{qp} = -\sin\varphi \\ u_{ij} = 0 \text{ otherwise} \end{matrix}$$

$$(6.14)$$

The orthogonality of the matrix $\mathbf{U}(p,q;\varphi)$ is obvious so that from $\mathbf{U}^T\mathbf{U} = \mathbf{I}$ we have $\mathbf{U}^{-1} = \mathbf{U}^T$. If we interpret $\mathbf{U}(p,q;\varphi)$ as a representation matrix of a linear transformation in $\mathbf{R}^n$ it corresponds to a rotation by the angle $-\varphi$ in the two-dimensional plane defined by the $p$th and $q$th coordinate directions. The pair $(p,q)$ with $1 \leq p < q \leq n$ is called the *rotation index pair* and $\mathbf{U}(p,q;\varphi)$ is called a $(p,q)$ *rotation matrix*.

The effect of a similarity transformation of a matrix $\mathbf{A}$ by a $(p,q)$ rotation matrix (6.14) into $\mathbf{A}'' = \mathbf{U}^{-1}\mathbf{A}\mathbf{U} = \mathbf{U}^T\mathbf{A}\mathbf{U}$ is analyzed in two steps. The multiplication of

## 6.2 Jacobi Methods

the matrix $\mathbf{A}$ from the left by $\mathbf{U}^T$ producing $\mathbf{A}' = \mathbf{U}^T\mathbf{A}$ causes only a linear combination of the $p$th and $q$th rows of $\mathbf{A}$, whereas all the other matrix elements remain unchanged. The elements of $\mathbf{A}' = \mathbf{U}^T\mathbf{A}$ are given by the formulae

$$\left.\begin{aligned} a'_{pj} &= a_{pj}\cos\varphi - a_{qj}\sin\varphi \\ a'_{qj} &= a_{pj}\sin\varphi + a_{qj}\cos\varphi \\ a'_{ij} &= a_{ij} \quad \text{for } i \neq p,q \end{aligned}\right\} \quad (j = 1,2,\ldots,n) \tag{6.15}$$

The subsequent multiplication of the matrix $\mathbf{A}'$ from the right by $\mathbf{U}$ resulting in $\mathbf{A}'' = \mathbf{A}'\mathbf{U}$ causes only a linear combination of the $p$th and $q$th columns of $\mathbf{A}'$, whereas all the other columns are unchanged. The elements of $\mathbf{A}'' = \mathbf{A}'\mathbf{U} = \mathbf{U}^T\mathbf{A}\mathbf{U}$ are given by

$$\left.\begin{aligned} a''_{ip} &= a'_{ip}\cos\varphi - a'_{iq}\sin\varphi \\ a''_{iq} &= a'_{ip}\sin\varphi + a'_{iq}\cos\varphi \\ a''_{ij} &= a'_{ij} \quad \text{for } j \neq p,q \end{aligned}\right\} \quad (i = 1,2,\ldots,n) \tag{6.16}$$

To summarize we can state that a similarity transformation of the matrix $\mathbf{A}$ by a $(p,q)$ rotation matrix $\mathbf{U}$ only changes the elements of the $p$th and $q$th rows and columns as is shown in Figure 6.1.

The matrix elements of the four cross points are transformed according to (6.15) and (6.16). Before deriving the corresponding formulae we state the fact that the symmetry of a matrix is maintained if an orthogonal similarity transformation is applied. Since $\mathbf{A}^T = \mathbf{A}$ and $\mathbf{U}^{-1} = \mathbf{U}^T$ we have

$$\mathbf{A}''^T = (\mathbf{U}^{-1}\mathbf{A}\mathbf{U})^T = (\mathbf{U}^T\mathbf{A}\mathbf{U})^T = \mathbf{U}^T\mathbf{A}^T\mathbf{U} = \mathbf{U}^T\mathbf{A}\mathbf{U} = \mathbf{A}''.$$

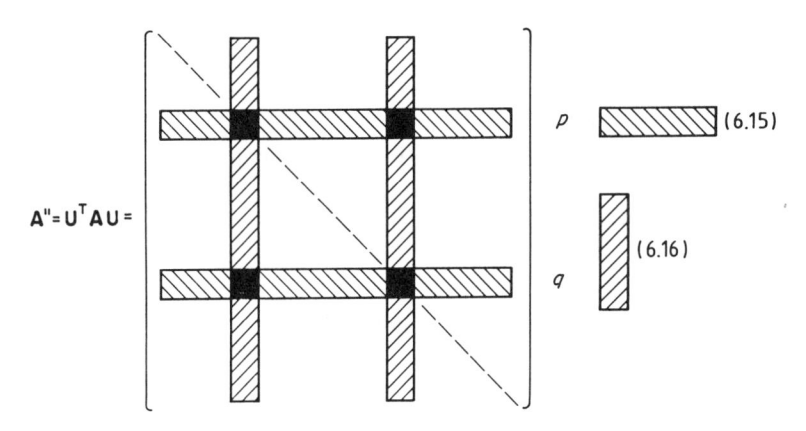

Figure 6.1 The effect of a similarity transformation

After substitution of (6.15) into (6.16) the elements of the cross points are defined by

$$a''_{pp} = a_{pp}\cos^2\varphi - 2a_{pq}\cos\varphi\sin\varphi + a_{qq}\sin^2\varphi \tag{6.17}$$
$$a''_{qq} = a_{pp}\sin^2\varphi + 2a_{pq}\cos\varphi\sin\varphi + a_{qq}\cos^2\varphi \tag{6.18}$$
$$a''_{pq} = a''_{qp} = (a_{pp} - a_{qq})\cos\varphi\sin\varphi + a_{pq}(\cos^2\varphi - \sin^2\varphi). \tag{6.19}$$

Such elementary orthogonal similarity transformations were used by Jacobi in 1846 (Jacobi, 1846) to successively diagonalize a symmetric matrix. Therefore they are called *Jacobi rotations* or equivalently $(p, q)$ rotations to emphasize the rotation index pair.

If we take into account the symmetry we can perform a Jacobi rotation with the matrix elements in and below the diagonal. Hence, according to Figure 6.1 we have to compute a total of $2(n-2)$ elements following (6.15) and (6.16) together with the three elements of the cross points. As $a''_{pp} + a''_{qq} = a_{pp} + a_{qq}$ such a step requires slightly more than $4n$ multiplications.

### 6.2.2 The classical Jacobi method

From the theorem of principal axes each symmetric matrix **A** can be diagonalized by means of an orthogonal similarity transformation. The idea of Jacobi consists of performing this transformation by an appropriate sequence of rotations.

A natural strategy is to annihilate the pair of off-diagonal elements $a_{pq} = a_{qp}$ of largest absolute value by a single transformation step. This can be achieved by a $(p, q)$ rotation, which means that $a_{pq}$ and $a_{qp}$ are located in the two off-diagonal cross points, see Figure 6.1. To avoid unnecessary indices we consider a typical step of the sequence of transformations. From (6.19) we obtain the condition for the determination of the angle $\varphi$

$$(a_{pp} - a_{qq})\cos\varphi\sin\varphi + a_{qp}(\cos^2\varphi - \sin^2\varphi) = 0. \tag{6.20}$$

Here we use $a_{qp}$ taking into account that we can work with the lower half of the matrix **A**. Using the trigonometric identities $\sin(2\varphi) = 2\cos(\varphi)\sin(\varphi)$ and $\cos(2\varphi) = \cos^2(\varphi) - \sin^2(\varphi)$ the condition (6.20) becomes

$$\cot(2\varphi) = \frac{\cos^2\varphi - \sin^2\varphi}{2\cos\varphi\sin\varphi} = \frac{a_{qq} - a_{pp}}{2a_{qp}} =: \Theta. \tag{6.21}$$

Since $a_{qp} \neq 0$, the quotient $\Theta$ has a well defined finite value. However, the Jacobi rotation does not require the angle $\varphi$ but the values of $\cos\varphi$ and $\sin\varphi$. They can be computed in a numerically safe way as follows. With $t := \tan\varphi$ from (6.21) we get

$$\frac{1-t^2}{2t} = \Theta \quad \text{or} \quad t^2 + 2\Theta t - 1 = 0 \tag{6.22}$$

## 6.2 Jacobi Methods

a quadratic equation in $t$ with the two solutions

$$t_{1,2} = -\Theta \pm \sqrt{\Theta^2 + 1} = \frac{1}{\Theta \pm \sqrt{\Theta^2 + 1}}. \tag{6.23}$$

We choose the solution of smaller absolute value and define

$$t = \tan \varphi = \begin{cases} \dfrac{1}{\Theta + \text{sgn}(\Theta)\sqrt{\Theta^2 + 1}} & \text{for } \Theta \neq 0 \\ 1 & \text{for } \Theta = 0 \end{cases} \tag{6.24}$$

This choice is made in order to avoid cancellation of digits when the denominator of (6.24) is computed. Moreover, we have $-1 < \tan \varphi \leqslant 1$ and hence $-\pi/4 < \varphi \leqslant \pi/4$. From the known value of $t = \tan \varphi$ we get the desired quantities

$$\cos \varphi = \frac{1}{\sqrt{1 + t^2}} \quad \sin \varphi = t \cdot \cos \varphi \tag{6.25}$$

Now all elements of the Jacobi rotation are fixed, we can perform the similarity transformation $\mathbf{A}'' = \mathbf{U}^T \mathbf{A} \mathbf{U}$ according to formulae (6.15) and (6.16). The formulae of the diagonal elements of the transformed matrix $\mathbf{A}''$ can be simplified substantially if the condition (6.20) for determining the angle $\varphi$ is taken into account. From (6.17) because of (6.21) we get

$$a''_{pp} = a_{pp} - 2a_{qp} \cos \varphi \sin \varphi + (a_{qq} - a_{pp}) \sin^2 \varphi$$

$$= a_{pp} - a_{qp} \left( 2\cos \varphi \sin \varphi - \frac{\cos^2 \varphi - \sin^2 \varphi}{\cos \varphi \sin \varphi} \sin^2 \varphi \right) = a_{pp} - a_{qp} \tan \varphi.$$

Hence we get

$$a''_{pp} = a_{pp} - a_{qp} \tan \varphi \quad a''_{qq} = a_{qq} + a_{qp} \tan \varphi \tag{6.26}$$

The formulae (6.26) do not only reduce the number of multiplications, but also the rounding errors of the diagonal elements. Similarly formulae (6.15) and (6.16) can be changed into a form that is less affected by rounding errors (Rutishauser, 1966). This is especially true for rotation angles $\varphi$ that are small in absolute value. In order to derive the modified formulae of the transformed matrix elements we use the identity

$$\cos \varphi = \frac{\cos \varphi + \cos^2 \varphi}{1 + \cos \varphi} = \frac{1 + \cos \varphi - \sin^2 \varphi}{1 + \cos \varphi} = 1 - \frac{\sin^2 \varphi}{1 + \cos \varphi}.$$

We define the quantity

$$r := \frac{\sin\varphi}{1+\cos\varphi}\left(=\tan\left(\frac{\varphi}{2}\right)\right) \qquad (6.27)$$

and from (6.15) and (6.16) obtain the representations

$$\left.\begin{array}{l} a'_{pj} = a_{pj} - \sin\varphi(a_{qj} + ra_{pj}) \\ a'_{qj} = a_{qj} + \sin\varphi(a_{pj} - ra_{qj}) \end{array}\right\} \quad (j=1,2,\ldots,n) \qquad (6.28)$$

$$\left.\begin{array}{l} a''_{ip} = a'_{ip} - \sin\varphi(a'_{iq} + ra'_{ip}) \\ a''_{iq} = a'_{iq} + \sin\varphi(a'_{ip} - ra'_{iq}) \end{array}\right\} \quad (i=1,2,\ldots,n) \qquad (6.29)$$

We observe that the number of multiplications is unchanged.

The *classical Jacobi method* generates, with $\mathbf{A}^{(0)} = \mathbf{A}$, a sequence of orthogonally similar matrices

$$\mathbf{A}^{(k)} = \mathbf{U}_k^T \mathbf{A}^{(k-1)} \mathbf{U}_k \quad (k=1,2,\ldots) \qquad (6.30)$$

such that in the $k$th step the off-diagonal element $a_{qp}^{(k-1)}$ of largest absolute value of the matrix $\mathbf{A}^{(k-1)}$

$$|a_{qp}^{(k-1)}| = \max_{i>j} |a_{ij}^{(k-1)}| \qquad (6.31)$$

is annihilated by a Jacobi rotation $\mathbf{U}_k = \mathbf{U}(p,q;\varphi)$. Although the off-diagonal elements are eliminated by a certain Jacobi rotation, in general through one of the subsequent transformation steps they become nonzero, but we have the following theorem.

*Theorem 6.1* *The sequence of similar matrices $\mathbf{A}^{(k)}$ (6.30) of the classical Jacobi method converges to a diagonal matrix $\mathbf{D}$.*

*Proof* We consider the sum of squares of the off-diagonal elements of the matrix $\mathbf{A}^{(k)}$

$$S(\mathbf{A}^{(k)}) = \sum_{i=1}^{n} \sum_{\substack{j=1 \\ j\neq i}}^{n} (a_{ij}^{(k)})^2 \quad (k=1,2,\ldots) \qquad (6.32)$$

We will show that $S(\mathbf{A}^{(k)})$ is a monotonically decreasing sequence converging to

## 6.2 Jacobi Methods

zero. To show this, we start by analyzing the change of $S(\mathbf{A})$ if a general Jacobi rotation, with the rotation index pair $(p, q)$, is applied to $\mathbf{A}$. We split up the sum $S(\mathbf{A}'') = S(\mathbf{U}^T\mathbf{A}\mathbf{U})$ into partial sums as follows

$$S(\mathbf{A}'') = \sum_{\substack{i=1 \\ j \neq i}}^{n} \sum_{j=1}^{n} a_{ij}''^2 = \sum_{\substack{i=1 \\ i \neq p,q}}^{n} \sum_{\substack{j=1 \\ j \neq i,p,q}}^{n} a_{ij}''^2 + \sum_{\substack{i=1 \\ i \neq p,q}}^{n} (a_{ip}''^2 + a_{iq}''^2)$$

$$+ \sum_{\substack{j=1 \\ j \neq p,q}}^{n} (a_{pj}''^2 + a_{qj}''^2) + 2a_{qp}''^2 \tag{6.33}$$

Only the elements of the $p$th and $q$th rows and columns are changed, so that $a_{ij}'' = a_{ij}$ holds for all $i \neq p, q$ and all $j \neq i, p, q$. From the orthogonality of the transformation the following relationships hold

$$a_{ip}''^2 + a_{iq}''^2 = a_{ip}^2 + a_{iq}^2 \quad (i \neq p, q),$$
$$a_{pj}''^2 + a_{qj}''^2 = a_{pj}^2 + a_{qj}^2 \quad (j \neq p, q).$$

Therefore from (6.33) for a general Jacobi rotation we get

$$S(\mathbf{A}'') = S(\mathbf{U}^T\mathbf{A}\mathbf{U}) = (S(\mathbf{A}) - 2a_{qp}^2) + 2a_{qp}''^2. \tag{6.34}$$

In the $k$th step of the classical Jacobi method, $a_{qp}^{(k)} = 0$ holds, and thus (6.34) reduces to

$$S(\mathbf{A}^{(k)}) = S(\mathbf{A}^{(k-1)}) - 2a_{qp}^{(k-1)2} \quad (k = 1, 2, \ldots) \tag{6.35}$$

The values $S(\mathbf{A}^{(k)})$ form, with increasing $k$, a strictly monotonically decreasing sequence as long as $\max_{i \neq j}|a_{ij}^{(k-1)}| \neq 0$. The decrease of the value $S(\mathbf{A}^{(k-1)})$ obtained by the $k$th Jacobi rotation is even maximal. It remains to show that $S(\mathbf{A}^{(k)})$ converges to zero. From (6.31) we have

$$S(\mathbf{A}^{(k-1)}) \leqslant (n^2 - n)a_{qp}^{(k-1)2},$$

and hence from (6.35) we get the estimate for $S(\mathbf{A}^{(k)})$

$$S(\mathbf{A}^{(k)}) = S(\mathbf{A}^{(k-1)}) - 2a_{qp}^{(k-1)2} \leqslant \left(1 - \frac{2}{n^2 - n}\right)S(\mathbf{A}^{(k-1)}). \tag{6.36}$$

The inequality (6.36) does not depend on $a_{qp}^{(k-1)}$ thus its recursive application yields

$$S(\mathbf{A}^{(k)}) \leqslant \left(1 - \frac{2}{n^2 - n}\right)^k S(\mathbf{A}^{(0)}). \tag{6.37}$$

For $n = 2$ we have $1 - 2/(n^2 - n) = 0$ in accordance with the fact that a single Jacobi rotation is sufficient, in this case, to diagonalize a matrix of order two so that $S(\mathbf{A}^{(1)}) = 0$. For $n > 2$ we have $0 < 1 - 2/(n^2 - n) < 1$ and thus

$$\lim_{k \to \infty} S(\mathbf{A}^{(k)}) = 0.$$

As far as the convergence of the diagonal elements is concerned we have

$$a_{ii}^{(k)} = \begin{cases} a_{ii}^{(k-1)} & \text{for } i \neq p \text{ and } i \neq q \\ a_{pp}^{(k-1)} - a_{qp}^{(k-1)} \tan\varphi & \text{for } i = p \\ a_{qq}^{(k-1)} + a_{qp}^{(k-1)} \tan\varphi & \text{for } i = q \end{cases}$$

where $\varphi$ denotes the rotation angle in the $k$th step. As $|\tan\varphi| \leq 1$, we have in each case

$$|a_{ii}^{(k)} - a_{ii}^{(k-1)}| \leq |a_{qp}^{(k-1)}| \leq \sqrt{S(\mathbf{A}^{(k-1)})} \leq \sqrt{\left(1 - \frac{2}{n^2 - n}\right)^{k-1}} \cdot \sqrt{S(\mathbf{A}^{(0)})}.$$

Therefore the series $a_{ii}^{(0)} + \sum_{k=1}^{\infty}(a_{ii}^{(k)} - a_{ii}^{(k-1)})$ is even absolutely convergent according to the comparison test.

The product of the rotation matrices

$$\mathbf{V}_k := \mathbf{U}_1 \mathbf{U}_2 \cdots \mathbf{U}_k \quad (k = 1, 2, \ldots) \tag{6.38}$$

is an orthogonal matrix for which

$$\mathbf{A}^{(k)} = \mathbf{U}_k^T \mathbf{U}_{k-1}^T \cdots \mathbf{U}_2^T \mathbf{U}_1^T \mathbf{A}^{(0)} \mathbf{U}_1 \mathbf{U}_2 \cdots \mathbf{U}_{k-1} \mathbf{U}_k = \mathbf{V}_k^T \mathbf{A} \mathbf{V}_k \tag{6.39}$$

holds. Following Theorem 6.1 the matrix $\mathbf{A}^{(k)}$ represents, for sufficiently large $k$, a diagonal matrix $\mathbf{D}$ to any given accuracy, and hence we can write

$$\mathbf{A}^{(k)} = \mathbf{V}_k^T \mathbf{A} \mathbf{V}_k \approx \mathbf{D}.$$

The diagonal elements of $\mathbf{A}^{(k)}$ are approximations of the eigenvalues $\lambda_j$ of $\mathbf{A}$. The columns of $\mathbf{V}_k$ are approximations of the corresponding orthonormal eigenvectors $\mathbf{x}_j$. In particular we get a complete system of orthonormal approximations of the eigenvectors, even in the case of multiple eigenvalues.

Due to the estimate (6.37) the sequence $S(\mathbf{A}^{(k)})$ decreases, at least, like a geometric sequence with the quotient $q = 1 - 2/(n^2 - n)$, that is the convergence is at least linear. Hence it is possible to estimate the number of Jacobi rotations required to fulfil the condition

$$S(\mathbf{A}^{(k)})/S(\mathbf{A}^{(0)}) \leq \varepsilon^2. \tag{6.40}$$

Let us denote by $N = \frac{1}{2}(n^2 - n)$ the number of off-diagonal elements of the lower part of the matrix $\mathbf{A}$. Then, due to (6.37), (6.40) is certainly satisfied if we have

$$\left(1 - \frac{1}{N}\right)^k \leq \varepsilon^2.$$

From this we can estimate $k$ if we use the natural logarithm and assume $N$ to be large

$$k \geq \frac{2\log(\varepsilon)}{\log\left(1 - \frac{1}{N}\right)} \approx 2N \log\left(\frac{1}{\varepsilon}\right) = (n^2 - n)\log\left(\frac{1}{\varepsilon}\right). \tag{6.41}$$

## 6.2 Jacobi Methods

As a Jacobi rotation requires about $4n$ multiplications from (6.41), with $\varepsilon = 10^{-\alpha}$, we obtain the following estimate of the number of multiplications required to diagonalize a symmetric matrix $\mathbf{A}$ of order $n$ by means of the classical Jacobi method

$$Z_{\text{Jacobi}} \approx 9.21\alpha(n^3 - n^2). \tag{6.42}$$

The amount of computational work required increases with $n$ cubed and the constant $9.21\alpha$ depends on the prescribed accuracy. However, the estimate (6.42) is too pessimistic because the sequence of the values $S(\mathbf{A}^{(k)})$ converges *quadratically* to zero as soon as the off-diagonal elements are sufficiently small in absolute value, relative to the minimal difference of two eigenvalues (Henrici, 1958; Schönhage, 1961; 1964). The result of Schönhage (1961, 1964) states that if the eigenvalues satisfy the relationship

$$\min_{i \neq j} |\lambda_i - \lambda_j| = 2\delta > 0 \tag{6.43}$$

and if in the course of the classical Jacobi method

$$S(\mathbf{A}^{(k)}) < \tfrac{1}{4}\delta^2 \tag{6.44}$$

holds, then after $N = \tfrac{1}{2}(n^2 - n)$ subsequent Jacobi rotations we have

$$S(\mathbf{A}^{(k+N)}) \leqslant (\tfrac{1}{2}n - 1)S(\mathbf{A}^{(k)})^2/\delta^2. \tag{6.45}$$

It can be shown that a statement on the quadratic convergence similar to that of (6.45) also holds in the case of double eigenvalues (Schönhage, 1961; Schröder, 1964). The asymptotic quadratic convergence reduces the total amount of computational work considerably. It is still proportional to $n^3$, but with a much smaller factor of proportionality than that of (6.42).

The difference between the diagonal elements $a_{ii}^{(k)}$ of the matrix $\mathbf{A}^{(k)}$ and the eigenvalues $\lambda_j$ of $\mathbf{A}$ can be estimated. A proof of the following theorem is given by Henrici (1958).

*Theorem 6.2* Let the eigenvalues $\lambda_j$ of the symmetric matrix $\mathbf{A} = \mathbf{A}^{(0)}$ be ordered by magnitude so that $\lambda_1 \leqslant \lambda_2 \leqslant \cdots \leqslant \lambda_n$ holds. Likewise we denote by $d_j^{(k)}$ the diagonal elements $a_{ii}^{(k)}$ ordered such that $d_1^{(k)} \leqslant d_2^{(k)} \leqslant \cdots \leqslant d_n^{(k)}$ holds. Then the eigenvalues $\lambda_j$ satisfy the error estimates

$$|d_j^{(k)} - \lambda_j| \leqslant \sqrt{S(\mathbf{A}^{(k)})} \quad (j = 1, 2, \ldots, n;\ k = 0, 1, 2, \ldots). \tag{6.46}$$

In order to avoid the computation of $S(\mathbf{A}^{(k)})$ we estimate the sum by means of the off-diagonal element $a_{qp}^{(k)}$ of largest absolute value and so get $S(\mathbf{A}^{(k)}) \leqslant (n^2 - n)(a_{qp}^{(k)})^2 < n^2(a_{qp}^{(k)})^2$. This yields, instead of (6.46), the following error estimate that is in general much less rigorous

$$|d_j^{(k)} - \lambda_j| < n|a_{qp}^{(k)}| \quad (j = 1, 2, \ldots, n;\ k = 0, 1, 2, \ldots). \tag{6.47}$$

It can be used as a simple stopping criterion of the method.

The simultaneous computation of the eigenvectors of the given matrix $\mathbf{A}$ has already been given by (6.38) and (6.39). The $j$th column of $\mathbf{V}_k$ is an approximation of the normed eigenvector corresponding to the eigenvalue approximation $a_{jj}^{(k)}$. The resulting approximations of the eigenvectors always form a system of pairwise orthogonal and normed vectors, at least within the numerical accuracy. This is also true in the case of multiple eigenvalues. The convergence of the matrix sequence $\mathbf{V}_k$ obeys more complicated rules than the convergence of the diagonal elements $a_{jj}^{(k)}$ towards the eigenvalues. Therefore the eigenvectors are usually approximated less accurately (Wilkinson, 1965).

The sequence of matrices $\mathbf{V}_k$ is computed recursively by successive multiplication by the rotation matrices according to

$$\mathbf{V}_0 = \mathbf{I} \quad \mathbf{V}_k = \mathbf{V}_{k-1}\mathbf{U}_k \quad (k = 1, 2, \ldots). \tag{6.48}$$

The storage of the matrices $\mathbf{V}_k$ requires an array of a full $(n \times n)$ matrix because they are not symmetric. If $\mathbf{U}_k$ is a $(p, q)$ rotation matrix, the multiplication $\mathbf{V}_{k-1}\mathbf{U}_k$ is equivalent to a linear combination of the $p$th and $q$th columns of $\mathbf{V}_{k-1}$ according to formulae (6.16) or (6.29). The amount of computational effort required for this operation consists of $4n$ essential operations. The total amount of work is doubled if the eigenvectors are simultaneously computed.

### 6.2.3 Cyclic Jacobi method

The search process of the classical Jacobi method for determining the off-diagonal element of largest absolute value requires $N = \frac{1}{2}(n^2 - n)$ comparison operations. A Jacobi rotation requires only $4n$ multiplications thus the amount of work required for the search process compares unfavourably with that of the proper transformation, at least for large order $n$. Therefore it is more appropriate to annihilate the $N$ off-diagonal elements below the diagonal in a systematic and fixed order in a cycle of $N$ rotations. For the *special cyclic Jacobi method* the sequence of the rotation index pairs to be fixed is

$$(1, 2), (1, 3), \ldots, (1, n), (2, 3), (2, 4), \ldots, (2, n), (3, 4), \ldots, (n-1, n) \tag{6.49}$$

so that the off-diagonal elements of the lower part are annihilated columnwise once per cycle. A $(p, q)$ rotation is not performed if the current element $a_{qp}$ is zero. The values of $\cos \varphi$ and $\sin \varphi$ are determined in the same way as in the classical Jacobi method so that the angle $\varphi$ satisfies $-\pi/4 < \varphi \leqslant \pi/4$.

*Theorem 6.3 The matrix sequence $\mathbf{A}^{(k)}$, produced by the special cyclic Jacobi method, (6.49), with the rotation angles $\varphi_k$, defined by (6.24), converges to a diagonal matrix $\mathbf{D}$.*

The proof of convergence is far less elementary in comparison with the classical Jacobi method. Although the sequence of values $S(\mathbf{A}^{(k)})$ is only nonincreasing

## 6.2 Jacobi Methods

now, it is shown by Forsythe and Henrici (1960) and Henrici (1958) using a sophisticated argument taking into account the previous history of the off-diagonal elements that $S(\mathbf{A}^{(k)})$ tends to zero. Moreover, it can be shown on the assumptions (6.43) and (6.44) that the matrix sequence $\mathbf{A}^{(k)}$ converges quadratically to $\mathbf{D}$, in the sense that after a full cycle of $N$ rotations the following estimate holds (Henrici, 1958; Schönhage, 1961; Schröder, 1964; Wilkinson, 1962)

$$S(\mathbf{A}^{(k+N)}) \leqslant S(\mathbf{A}^{(k)})^2/(2\delta^2). \tag{6.50}$$

The fundamental idea of the proof of the asymptotically quadratic convergence is simple and will be outlined here. We assume the eigenvalues to be simple, so that (6.43) holds. From (6.44) and (6.46) the diagonal elements of $\mathbf{A}^{(k)}$ satisfy

$$|a_{ii}^{(k)} - a_{jj}^{(k)}| > \delta \quad \text{for all } i \neq j. \tag{6.51}$$

In addition we assume that the off-diagonal elements fulfil the condition

$$\max_{i>j} |a_{ij}^{(k)}| = \varepsilon \ll \delta. \tag{6.52}$$

Then the rotation angle $\varphi$ of a $(p, q)$ rotation satisfies, because of (6.51) and (6.21), the condition $|\varphi| < \pi/4$ and the inequality

$$|\sin \varphi| \leqslant |\varphi| \leqslant |\tan \varphi| \leqslant \tfrac{1}{2}|\tan(2\varphi)| = \frac{|a_{qp}|}{|a_{qq} - a_{pp}|} < \frac{\varepsilon}{\delta}.$$

An arbitrary element of the lower part, say $a_{pj}$, that is not situated in the cross point of the rotation, is changed according to (6.28) in absolute value at most by

$$|a'_{pj} - a_{pj}| = |\sin \varphi||a_{qj} + ra_{pj}| < \frac{\varepsilon}{\delta}\left(\varepsilon + \frac{\varepsilon}{\delta}\varepsilon\right) = \frac{\varepsilon^2}{\delta} + O(\varepsilon^3) \tag{6.53}$$

that is by an amount that is proportional to $\varepsilon^2$. Now we study the changes of a certain matrix element of the lower half during a cycle (6.49) on the assumption that all of the cited conditions are fulfilled at the beginning of the cycle. We consider a typical element $a_{52}$. It is first changed by the rotations with the index pairs $(1, 2), (1, 5), (2, 3)$ and $(2, 4)$ according to (6.53) at most by values of the size $\varepsilon^2$, so that its value is still $a_{52} = O(\varepsilon)$. Then the $(2, 5)$ rotation changes its value to zero. Subsequently its value will be changed by the rotations with the index pairs $(2, 6), (2, 7), \ldots, (2, n), (3, 5), (4, 5), (5, 6), \ldots, (5, n)$. This is a total of $(n - 5) + 2 + (n - 5) = 2n - 8$ rotations. Each of them can increase $a_{52}$ in absolute value at most by $\varepsilon^2/\delta$. In general an arbitrary matrix element satisfies after a full cycle

$$|a_{ij}^{(k+N)}| < (2n - 4)\frac{\varepsilon^2}{\delta} \quad (i > j) \tag{6.54}$$

and this shows the quadratic convergence of the off-diagonal elements, and hence of $S(\mathbf{A}^{(k)})$ to zero.

We now give an algorithmic description of the special cyclic Jacobi method. It is assumed that the matrix elements of **A** are stored in and below the diagonal. For reasons of clarity the usual indexing of the matrix elements is used. The quantity $\varepsilon$ represents the absolute tolerance of the desired accuracy of the eigenvalues. The iteration is stopped as soon as $S(\mathbf{A}^{(k)}) < \varepsilon^2$. Furthermore $\delta$ denotes the smallest positive number of the computer such that $1 + \delta \neq 1$ holds. After termination of the program the diagonal elements $a_{ii}$ represent the approximate eigenvalues, and the matrix **V** contains, columnwise, the corresponding orthonormal approximations of the eigenvectors. A Jacobi rotation is skipped whenever $|a_{qp}| < \varepsilon^2$.

```
START:   for i = 1, 2, ..., n:
             for j = 1, 2, ..., n:
                 v_ij = 0
             v_ii = 1
CYCLE:   sum = 0
         for i = 2, 3, ..., n:
             for j = 1, 2, ..., i − 1:
                 sum = sum + a_ij²
TEST:    if 2 × sum < ε²: STOP
         for p = 1, 2, ..., n − 1:
             for q = p + 1, p + 2, ..., n:
                 if |a_qp| ≥ ε² then
                     Θ = (a_qq − a_pp)/(2 × a_qp); t = 1
                     if |Θ| > δ then t = 1/(Θ + sgn(Θ)√(Θ² + 1))
                     c = 1/√(1 + t²); s = c × t; r = s/(1 + c)
                     a_pp = a_pp − t × a_qp; a_qq = a_qq + t × a_qp; a_qp = 0
                     for j = 1, 2, ..., p − 1:
                         g = a_qj + r × a_pj; h = a_pj − r × a_qj
                         a_pj = a_pj − s × g; a_qj = a_qj + s × h
                     for i = p + 1, p + 2, ..., q − 1:
                         g = a_qi + r × a_ip; h = a_ip − r × a_qi
                         a_ip = a_ip − s × g; a_qi = a_qi + s × h
                     for i = q + 1, q + 2, ..., n:
                         g = a_iq + r × a_ip; h = a_ip − r × a_iq
                         a_ip = a_ip − s × g; a_iq = a_iq + s × h
                     for i = 1, 2, ..., n:
                         g = v_iq + r × v_ip; h = v_ip − r × v_iq
                         v_ip = v_ip − s × g; v_iq = v_iq + s × h
         go to CYCLE
```

It is observed that after a few cycles, the cyclic Jacobi method becomes quadratically convergent so that usually about six to eight cycles are needed to

## 6.3 Transformation Methods

diagonalize a given matrix **A**, up to the desired accuracy. As a cycle consists of $N = \frac{1}{2}(n^2 - n)$ rotations and each rotation requires about $8n$ multiplications, including the computation of the eigenvectors, the number of essential operations can be estimated by

$$Z_{\text{cycl Jac}} \approx 32n^3. \tag{6.55}$$

The computation of all the eigenvalues and eigenvectors, by means of the cyclic Jacobi method, requires a much larger amount of work than the solution of a system of linear equations. Although we shall describe in the following more efficient methods, the Jacobi method is very popular. This can be explained by its simplicity, clarity, numerical stability and easy implementation.

*Example 6.2* We compute the eigenvalues and eigenvectors of the matrix

$$\mathbf{A}^{(0)} = \begin{pmatrix} 20 & -7 & 3 & -2 \\ -7 & 5 & 1 & 4 \\ 3 & 1 & 3 & 1 \\ -2 & 4 & 1 & 2 \end{pmatrix} \quad \text{with } S(\mathbf{A}^{(0)}) = 160$$

by means of the cyclic Jacobi method. After a complete cycle of six rotations, the lower part of the resulting matrix $\mathbf{A}^{(6)}$ is

$$\mathbf{A}^{(6)} \doteq \begin{pmatrix} 23.523\,089 & & & \\ -0.009\,053 & -0.437\,554 & & \\ -0.238\,471 & -1.397\,689 & 6.174\,371 & \\ 0.151\,640 & 0.931\,475 & 0 & 0.740\,095 \end{pmatrix}$$

The sum of squares of the off-diagonal elements is $S(\mathbf{A}^{(6)}) \doteq 5.802\,252$. The three following cycles yield $S(\mathbf{A}^{(12)}) \doteq 1.387\,334 \times 10^{-2}$, $S(\mathbf{A}^{(18)}) \doteq 1.094\,265 \times 10^{-9}$ and $S(\mathbf{A}^{(24)}) \doteq 3.7645 \times 10^{-31}$. The fast, quadratic convergence is obvious in this example and becomes effective quite early because the eigenvalues $\lambda_1 \doteq 23.527\,386$, $\lambda_2 \doteq -1.160\,950$, $\lambda_3 \doteq 6.460\,515$, $\lambda_4 \doteq 1.173\,049$ taken from the diagonal of $\mathbf{A}^{(24)}$ are well separated with $\min_{i \neq j} |\lambda_i - \lambda_j| \doteq 2.334$. The matrix $\mathbf{V}_{24}$ with the approximate eigenvectors is

$$\mathbf{V}_{24} \doteq \begin{pmatrix} 0.910\,633 & 0.172\,942 & 0.260\,705 & 0.269\,948 \\ -0.370\,273 & 0.674\,951 & 0.587\,564 & 0.249\,212 \\ 0.107\,818 & -0.116\,811 & 0.549\,910 & -0.819\,957 \\ -0.148\,394 & -0.707\,733 & 0.533\,292 & 0.438\,967 \end{pmatrix}.$$

## 6.3 Transformation Methods

The numerical treatment of the eigenvalue problem is simplified if the given matrix **A** is first reduced to an appropriate form by means of an orthogonal similarity transformation. We shall treat this transformation in the sense of a preparatory step and mainly consider the case of a nonsymmetric matrix **A**. If the same transformation is applied to a symmetric matrix a simpler form results. In both cases the resulting matrices will be the starting point for certain procedures.

### 6.3.1 Transformation into Hessenberg form

We intend to reduce any given matrix **A** of order $n$ to an upper *Hessenberg matrix*

$$\mathbf{H} = \begin{pmatrix} h_{11} & h_{12} & h_{13} & h_{14} & \cdots & h_{1,n-1} & h_{1n} \\ h_{21} & h_{22} & h_{23} & h_{24} & \cdots & h_{2,n-1} & h_{2n} \\ 0 & h_{32} & h_{33} & h_{34} & \cdots & h_{3,n-1} & h_{3n} \\ 0 & 0 & h_{43} & h_{44} & \cdots & h_{4,n-1} & h_{4n} \\ \vdots & \vdots & \vdots & \vdots & & \vdots & \vdots \\ 0 & 0 & 0 & 0 & \cdots & h_{n,n-1} & h_{nn} \end{pmatrix} \tag{6.56}$$

by means of a similarity transformation so that the elements satisfy

$$h_{ij} = 0 \quad \text{for all } i > j + 1. \tag{6.57}$$

The desired transformation is achieved by means of an appropriate sequence of Jacobi rotations that annihilate a matrix element in each step. In this process of elimination we pay attention to the fact that a matrix element that has been annihilated will no longer be changed by subsequent rotations. In contrast to the Jacobi method, the rotation index pairs are chosen so that the element to be eliminated is not situated in the off-diagonal cross point of the changed rows and columns. To distinguish them, they are called *Givens rotations* (Parlett, 1980). There exist several possibilities to perform the desired transformation. It seems quite natural to eliminate the matrix elements below the subdiagonal columnwise in the order

$$a_{31}, a_{41}, \ldots, a_{n1}, a_{42}, a_{52}, \ldots, a_{n2}, a_{53}, \ldots, a_{n,n-2} \tag{6.58}$$

If we choose the corresponding rotation index pairs

$$(2,3), (2,4), \ldots, (2,n), (3,4), (3,5), \ldots, (3,n), (4,5), \ldots, (n-1,n) \tag{6.59}$$

and determine the rotation angles $\varphi$ suitably, the desired result can be achieved. To eliminate the current element $a_{ij} \neq 0$ with $i \geq j + 2$ we apply, according to (6.58) and (6.59), a $(j+1, i)$ rotation. The element $a_{ij}$ is changed only by the row operation of this rotation. The condition $a'_{ij} = 0$ leads us, because of (6.15) and $p = j + 1 < i = q$, to the equation

$$a'_{ij} = a_{j+1,j} \sin \varphi + a_{ij} \cos \varphi = 0. \tag{6.60}$$

Together with the identity $\cos^2 \varphi + \sin^2 \varphi = 1$ the values of $\cos \varphi$ and $\sin \varphi$ are

## 6.3 Transformation Methods

determined apart from a common sign. We restrict the rotation angle $\varphi$ to the interval $[-\pi/2, \pi/2]$ for a reason that will become clear. Therefore the desired values are defined as follows:

$$\text{if } a_{j+1,j} \neq 0 \quad \cos\varphi = \frac{|a_{j+1,j}|}{\sqrt{a_{j+1,j}^2 + a_{ij}^2}} \quad \sin\varphi = \frac{-\operatorname{sgn}(a_{j+1,j})a_{ij}}{\sqrt{a_{j+1,j}^2 + a_{ij}^2}}$$
$$\text{if } a_{j+1,j} = 0 \quad \cos\varphi = 0 \quad \sin\varphi = 1 \qquad (6.61)$$

Now we have to verify that the sequence of rotations, defined by (6.59), transforms a matrix $\mathbf{A}$ into the Hessenberg form (6.56). We have to show that the generated vanishing matrix elements remain zero when subsequent rotations are applied. This is obvious for the first column because each of the $(2, i)$ rotations only changes the second and $i$th rows and columns, thus eliminating the desired element $a_{i1}$. We show by induction that the statement is true for the transformation of the other columns. We assume that the first $r$ columns have the desired form. To eliminate the elements $a_{i,r+1}$ of the $(r + 1)$th column with $i \geqslant r + 3$ we apply $(r + 2, i)$ successive rotations. By the induction hypotheses, the row operations result in a linear combination of zero elements of the first $r$ columns, so that these elements are in fact unchanged. As $i > r + 2$, the column operations only change columns with indices greater than $(r + 1)$.

Hence the similarity transformation of $\mathbf{A}$ into an upper Hessenberg matrix $\mathbf{H}$ is achieved by a total of $N^* = \frac{1}{2}(n-1)(n-2)$ Givens rotations. In order to derive the amount of work required we study the transformation of the element $a_{j+1,j}$ more closely. From (6.15) and (6.61) we have

$$a'_{j+1,j} = \frac{a_{j+1,j}|a_{j+1,j}| + \operatorname{sgn}(a_{j+1,j})a_{ij}^2}{\sqrt{a_{j+1,j}^2 + a_{ij}^2}} = \operatorname{sgn}(a_{j+1,j})\sqrt{a_{j+1,j}^2 + a_{ij}^2}.$$

The computation of this new element requires no extra work, if we modify the formulae (6.61) as follows

$$\text{if } a_{j+1,j} \neq 0 \quad w := \operatorname{sgn}(a_{j+1,j})\sqrt{a_{j+1,j}^2 + a_{ij}^2}$$
$$\cos\varphi = a_{j+1,j}/w \quad \sin\varphi = -a_{ij}/w$$
$$\text{if } a_{j+1,j} = 0 \quad w := -a_{ij} \quad \cos\varphi = 0 \quad \sin\varphi = 1 \qquad (6.62)$$

Then we have $a'_{j+1,j} = w$. A Givens rotation for the elimination of the element $a_{ij}$ of the $j$th column requires four essential operations and a square root because of (6.62), then $4(n-j)$ multiplications for the row operations (6.15) and finally $4n$ multiplications for the column operations (6.16). The treatment of the $j$th column requires $4(n-j-1)(2n-j+1)$ multiplications and $(n-j-1)$ square roots. Summing over $j$, from 1 to $(n-2)$, yields the total amount of computational effort to be

$$Z_{\text{HessG}} = \tfrac{10}{3}n^3 - 8n^2 + \tfrac{2}{3}n + 4 \qquad (6.63)$$

multiplications and $N^* = \tfrac{1}{2}(n-1)(n-2)$ square roots.

In the following we shall solve the eigenvalue problem for the Hessenberg matrix **H**. Therefore we have to consider the problem of how the eigenvectors of **A** can be obtained from those of **H**. The matrices **H** and **A** are related by

$$\mathbf{H} = \mathbf{U}_{N^*}^T \cdots \mathbf{U}_2^T \mathbf{U}_1^T \mathbf{A} \mathbf{U}_1 \mathbf{U}_2 \cdots \mathbf{U}_{N^*} = \mathbf{Q}^T \mathbf{A} \mathbf{Q} \quad \mathbf{Q} = \mathbf{U}_1 \mathbf{U}_2 \cdots \mathbf{U}_{N^*}.$$

$\mathbf{U}_k$ denotes the $k$th Jacobi rotation matrix, and **Q** is orthogonal as a product of the $N^*$ orthogonal matrices $\mathbf{U}_k$. The eigenvalue problem $\mathbf{A}\mathbf{x} = \lambda \mathbf{x}$ is transformed by **Q** into

$$\mathbf{Q}^T \mathbf{A} \mathbf{Q} \mathbf{Q}^T \mathbf{x} = \mathbf{H}(\mathbf{Q}^T \mathbf{x}) = \lambda(\mathbf{Q}^T \mathbf{x}),$$

that is, the eigenvectors $\mathbf{x}_j$, $\mathbf{y}_j$ of **A** and **H** respectively, are related by

$$\mathbf{y}_j = \mathbf{Q}^T \mathbf{x}_j \quad \text{or} \quad \mathbf{x}_j = \mathbf{Q}\mathbf{y}_j = \mathbf{U}_1 \mathbf{U}_2 \cdots \mathbf{U}_{N^*} \mathbf{y}_j. \qquad (6.64)$$

An eigenvector $\mathbf{x}_j$ is obtained from $\mathbf{y}_j$ by successively multiplying the latter by the Jacobi rotation matrices, $\mathbf{U}_k$, in *reverse* order of their occurrence in the transformation of **A** into Hessenberg form. Thus we need all the information about the rotations. It would be natural to store the values of $\cos\varphi$ and $\sin\varphi$, but this would require $(n-1)(n-2) \approx n^2$ storage places. However, following a clever idea of Stewart (1976) only a single value is required from which $\cos\varphi$ and $\sin\varphi$ can be retrieved in a numerically stable way with a small effort. He defines the quantity

$$\rho := \begin{cases} 1 & \text{if } \sin\varphi = 1 \\ \sin\varphi & \text{if } |\sin\varphi| < \cos\varphi \\ \operatorname{sgn}(\sin\varphi)/\cos\varphi & \text{if } |\sin\varphi| \geqslant \cos\varphi \text{ and } \sin\varphi \neq 1 \end{cases} \qquad (6.65)$$

Here we essentially use the restriction that the angle $\varphi$ assures us that $\cos\varphi \geqslant 0$. The computation of $\cos\varphi$ and $\sin\varphi$ from $\rho$ requires analogous distinctions.

The value $\rho$ completely characterizes the rotation, it can be stored in place of $a_{ij}$, the eliminated matrix element. When the transformation of **A** into Hessenberg form **H** is completed, the information required for the reverse transformation of the eigenvectors is found below the subdiagonal of the matrix array.

The transformation of a nonsymmetric matrix **A** of order $n$ into Hessenberg form possesses a simple algorithmic description. The preparation of the values $\rho$ is, however, not included in the following algorithm. $\delta$ denotes the smallest positive number such that $1 + \delta \neq 1$ holds on the computer.

## 6.3 Transformation Methods

for $j = 1, 2, \ldots, n-2$:
  for $i = j+2, j+3, \ldots, n$:
    if $a_{ij} \neq 0$ then
      if $|a_{j+1,j}| < \delta \times |a_{ij}|$ then
        $w = -a_{ij}$; $c = 0$; $s = 1$
      else
        $w = \text{sgn}(a_{j+1,j})\sqrt{a_{j+1,j}^2 + a_{ij}^2}$
        $c = a_{j+1,j}/w$; $s = -a_{ij}/w$ 
        (6.66)
      $a_{j+1,j} = w$; $a_{ij} = 0$
      for $k = j+1, j+2, \ldots, n$:
        $h = c \times a_{j+1,k} - s \times a_{ik}$
        $a_{ik} = s \times a_{j+1,k} + c \times a_{ik}$; $a_{j+1,k} = h$
      for $k = 1, 2, \ldots, n$:
        $h = c \times a_{k,j+1} - s \times a_{ki}$
        $a_{ki} = s \times a_{k,j+1} + c \times a_{ki}$; $a_{k,j+1} = h$

*Example* 6.3 The matrix

$$\mathbf{A} = \begin{pmatrix} 7 & 3 & 4 & -11 & -9 & -2 \\ -6 & 4 & -5 & 7 & 1 & 12 \\ -1 & -9 & 2 & 2 & 9 & 1 \\ -8 & 0 & -1 & 5 & 0 & 8 \\ -4 & 3 & -5 & 7 & 2 & 10 \\ 6 & 1 & 4 & -11 & -7 & -1 \end{pmatrix} \quad (6.67)$$

is reduced to the Hessenberg form by means of an orthogonal similarity transformation. Four Givens transformations eliminating the elements of the first column yield the matrix $\mathbf{A}^{(4)}$, where the elements are rounded to six decimal places:

$$\mathbf{A}^{(4)} \doteq \begin{pmatrix} 7.000\,000 & -7.276\,069 & 3.452\,379 & -9.536\,895 & -5.933\,434 & -6.323\,124 \\ -12.369\,317 & 4.130\,719 & -6.658\,726 & 8.223\,249 & 0.509\,438 & 19.209\,667 \\ 0 & -0.571\,507 & 4.324\,324 & 7.618\,758 & 9.855\,831 & -1.445\,000 \\ 0 & -0.821\,267 & 3.340\,098 & -0.512\,443 & -0.698\,839 & -5.843\,013 \\ 0 & 1.384\,035 & -3.643\,194 & 0.745\,037 & -0.162\,309 & 5.627\,779 \\ 0 & -6.953\,693 & 0.630\,344 & -4.548\,600 & -3.872\,656 & 4.219\,708 \end{pmatrix}$$

A continuation of the transformation leads to the Hessenberg matrix $\mathbf{H}$ in which the values $\rho$, (6.65), are substituted for the zero elements below the subdiagonal

$$H \doteq \begin{pmatrix} 7.000\,000 & -7.276\,069 & -5.812\,049 & 0.139\,701 & 9.015\,201 & -7.936\,343 \\ -12.369\,317 & 4.130\,719 & 18.968\,509 & -1.207\,073 & -10.683\,309 & 2.415\,951 \\ -0.164\,399 & -7.160\,342 & 2.447\,765 & -0.565\,594 & 4.181\,396 & -3.250\,955 \\ -1.652\,189 & -1.750\,721 & -8.598\,771 & 2.915\,100 & 3.416\,858 & 5.722\,969 \\ -0.369\,800 & 1.706\,882 & -1.459\,149 & -1.046\,436 & -2.835\,101 & 10.979\,178 \\ 0.485\,071 & -4.192\,677 & 4.429\,714 & 1.459\,546 & -1.414\,293 & 5.341\,517 \end{pmatrix}$$

(6.68)

### 6.3.2 Transformation into tridiagonal form

If the same sequence of similarity transformations of the preceding section is applied to a *symmetric* matrix $A$, the resulting matrix $J = Q^T A Q$ is *tridiagonal* because the symmetry is maintained.

$$J = \begin{pmatrix} \alpha_1 & \beta_1 & & & & \\ \beta_1 & \alpha_2 & \beta_2 & & & \\ & \beta_2 & \alpha_3 & \beta_3 & & \\ & & \ddots & \ddots & \ddots & \\ & & & \beta_{n-2} & \alpha_{n-1} & \beta_{n-1} \\ & & & & \beta_{n-1} & \alpha_n \end{pmatrix}$$

(6.69)

This process is called the *Givens method* (Givens, 1954, 1958). If the symmetry of the transformed matrices is taken into account the above algorithm is simplified, because it can operate with the matrix elements in and below the diagonal. A typical situation is shown in Figure 6.2 for the elimination of the element $a_{ij} (i \geq j+2)$.

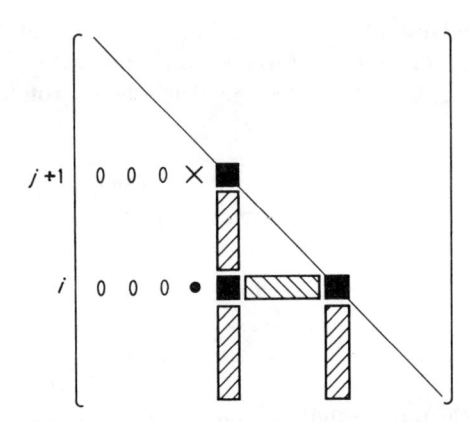

Figure 6.2 Givens rotation for a symmetric matrix

## 6.3 Transformation Methods

From Figure 6.2 we can see the following parts of a Givens rotation: (1) In the $j$th column only the element $a_{j+1,j}$ must be replaced. (2) The three elements of the cross points are transformed. (3) Treatment of the matrix elements between the cross points of the $(j+1)$th column and the $i$th row. (4) Transformation of the elements below the $i$th row that lie in the $(j+1)$th and the $i$th column.

To reduce the number of multiplications, we write the formulae (6.17) and (6.18) as follows

$$a''_{pp} = a_{pp} - [(a_{pp} - a_{qq})\sin\varphi + 2a_{qp}\cos\varphi]\sin\varphi =: a_{pp} - z,$$
$$a''_{qq} = a_{qq} + [(a_{pp} - a_{qq})\sin\varphi + 2a_{qp}\cos\varphi]\sin\varphi =: a_{qq} + z.$$

With this modification the elimination of $a_{ij}$ requires $4 + 9 + 4(n-j-2) = 4(n-j) + 5$ multiplications and a square root. The reduction of the $j$th column therefore requires $(n-j-1)(4n-4j+5)$ multiplications and $(n-j-1)$ square roots. We sum over $j$, from 1 to $(n-2)$, to find that the amount of computational effort required for the Givens method consists of

$$Z_{\text{Givens}} = \tfrac{4}{3}n^3 - \tfrac{3}{2}n^2 - \tfrac{29}{6}n + 5 \tag{6.70}$$

multiplications and $N^* = \tfrac{1}{2}(n-1)(n-2)$ square roots.

The algorithmic description of the Givens method results from (6.66) by just replacing the last two loops, that is the last six lines, by

```
d = a_{j+1,j+1} - a_{ii}; z = (d × s + 2 × c × a_{i,j+1}) × s
a_{i,j+1} = d × c × s + a_{i,j+1} × (c × c − s × s)
a_{j+1,j+1} = a_{j+1,j+1} − z; a_{ii} = a_{ii} + z
for k = j + 2, j + 3, …, i − 1:
    h = c × a_{k,j+1} − s × a_{ik}
    a_{ik} = s × a_{k,j+1} + c × a_{ik}; a_{k,j+1} = h
for k = i + 1, i + 2, …, n:
    h = c × a_{k,j+1} − s × a_{ki}
    a_{ki} = s × a_{k,j+1} + c × a_{ki}; a_{k,j+1} = h
```

*Example 6.4* The transformation of the symmetric matrix (Gregory and Karney, 1969)

$$\mathbf{A} = \begin{pmatrix} 5 & 4 & 3 & 2 & 1 \\ 4 & 6 & 0 & 4 & 3 \\ 3 & 0 & 7 & 6 & 5 \\ 2 & 4 & 6 & 8 & 7 \\ 1 & 3 & 5 & 7 & 9 \end{pmatrix}$$

into tridiagonal form by means of the Givens method yields

$$\mathbf{J} \doteq \begin{pmatrix} 5.000\,000 & 5.477\,226 & 0 & 0 & 0 \\ 5.477\,226 & 13.933\,333 & 9.298\,506 & 0 & 0 \\ -0.600\,000 & 9.298\,506 & 9.202\,474 & -2.664\,957 & 0 \\ -0.371\,391 & -3.758\,508 & -2.664\,957 & 4.207\,706 & 2.154\,826 \\ -0.182\,574 & -1.441\,553 & 0.312\,935 & 2.154\,826 & 2.656\,486 \end{pmatrix} \tag{6.71}$$

Below the subdiagonal of $\mathbf{J}$, (6.71), the values $\rho$, (6.65), of the Givens rotations are stored so that a back transformation of the eigenvectors of $\mathbf{J}$ into those of $\mathbf{A}$ is possible.

### 6.3.3 Fast Givens transformation

We consider the multiplication of a matrix $\mathbf{A}$, of order $n$, by a $(p, q)$ rotation matrix $\mathbf{U}(p, q; \varphi)$, (6.14), resulting in $\mathbf{A}' = \mathbf{U}^T\mathbf{A}$. As only the matrix elements of the $p$th and $q$th rows are changed, we direct our attention to these elements and write, for simplicity, $c = \cos \varphi$, $s = \sin \varphi$. From formulae (6.15) the computation of a pair of matrix elements

$$\begin{aligned} a'_{pj} &= c\,a_{pj} - s\,a_{qj} \\ a'_{qj} &= s\,a_{pj} + c\,a_{qj} \end{aligned} \tag{6.72}$$

requires four multiplications. Following an idea of Gentleman (Gentleman, 1973; Hammarling, 1974; Rath, 1982) the number of operations can be essentially reduced by half if the matrix $\mathbf{A}$, as well as the transformed matrix $\mathbf{A}'$, are factored with suitable nonsingular diagonal matrices $\mathbf{D}$ and $\mathbf{D}'$

$$\mathbf{A} = \mathbf{D}\tilde{\mathbf{A}} \quad \mathbf{A}' = \mathbf{D}'\tilde{\mathbf{A}}' \quad \mathbf{D} = \mathrm{diag}\,(d_1, d_2, \ldots, d_n). \tag{6.73}$$

Hence from (6.72) we get the new relationships

$$\begin{aligned} d'_p \tilde{a}'_{pj} &= c d_p \tilde{a}_{pj} - s d_q \tilde{a}_{qj} \\ d'_q \tilde{a}'_{qj} &= s d_p \tilde{a}_{pj} + c d_q \tilde{a}_{qj} \end{aligned} \tag{6.74}$$

In order to compute the two elements $\tilde{a}'_{pj}$ and $\tilde{a}'_{qj}$ using only two multiplications, there are essentially four possible definitions of the diagonal elements $d'_p$ and $d'_q$

$$\begin{aligned} &\text{(a)} \quad d'_p = c d_p \quad \text{and} \quad d'_q = c d_q \\ &\text{(b)} \quad d'_p = s d_q \quad \text{and} \quad d'_q = s d_p \\ &\text{(c)} \quad d'_p = c d_p \quad \text{and} \quad d'_q = s d_p \\ &\text{(d)} \quad d'_p = s d_q \quad \text{and} \quad d'_q = c d_q \end{aligned} \tag{6.75}$$

## 6.3 Transformation Methods

In the following we shall use case (a) of (6.75), if $|c| \geq |s|$, and case (b) otherwise. We do this because later we will always apply a sequence of multiplications, of rotation matrices, so that the corresponding diagonal elements of $\mathbf{D}$ will be multiplied by $c$ or $s$. With this choice, the chance of underflow for the diagonal elements can be reduced. Hence we get

$$\text{case (a)} \quad \tilde{a}'_{pj} = \tilde{a}_{pj} - \left(\frac{sd_q}{cd_p}\right)\tilde{a}_{qj}$$

$$\tilde{a}'_{qj} = \left(\frac{sd_p}{cd_q}\right)\tilde{a}_{pj} + \tilde{a}_{qj}$$

(6.76)

$$\text{case (b)} \quad \tilde{a}'_{pj} = \left(\frac{cd_p}{sd_q}\right)\tilde{a}_{pj} - \tilde{a}_{qj}$$

$$\tilde{a}'_{qj} = \tilde{a}_{pj} + \left(\frac{cd_q}{sd_p}\right)\tilde{a}_{qj}$$

(6.77)

With formulae (6.76) or (6.77) the decisive step has been made to compute the new elements $\tilde{a}'_{pj}$ and $\tilde{a}'_{qj}$ of the diagonally scaled matrix $\tilde{\mathbf{A}}'$, with two multiplications, after preparing the appropriate factors. We even go a step further and take into account that a multiplication of $\mathbf{A}$ by $\mathbf{U}^T$ aims to annihilate the matrix element $a_{qk} \neq 0$ with a given index $k$. Due to (6.74)

$$d'_q \tilde{a}'_{qk} = sd_p \tilde{a}_{pk} + cd_q \tilde{a}_{qk} = 0$$

should hold or equivalently

$$T := \cot \varphi = \frac{c}{s} = -\frac{d_p \tilde{a}_{pk}}{d_q \tilde{a}_{qk}}.$$

(6.78)

Using (6.78), the two factors in (6.77) are given by the expressions

$$-\frac{cd_p}{sd_q} = \frac{d_p^2 \tilde{a}_{pk}}{d_q^2 \tilde{a}_{qk}} =: f_1 \qquad -\frac{cd_q}{sd_p} = \frac{\tilde{a}_{pk}}{\tilde{a}_{qk}} =: f_2$$

(6.79)

whereas the two multipliers in (6.76) are the reciprocals of (6.79). Now from (6.79) we see that we do not need the diagonal elements $d_p$ and $d_q$ to compute the multipliers $f_1$ and $f_2$ but their squares, $d_p^2$ and $d_q^2$. Therefore we work with the squares and replace (6.75) by

$$\text{(a) } (d'_p)^2 = c^2(d_p)^2 \quad \text{and} \quad (d'_q)^2 = c^2(d_q)^2$$
$$\text{(b) } (d'_p)^2 = s^2(d_q)^2 \quad \text{and} \quad (d'_q)^2 = s^2(d_p)^2 \tag{6.80}$$

The necessary values $c^2$ and $s^2$ of (6.80) can be computed either from $T^2 = \cot^2 \varphi$ or from its reciprocal $t^2 = \tan^2 \varphi$ on the basis of trigonometric identities. The decision whether case (a) or (b) occurs can be made by means of the value of $T^2 = f_1 f_2$ using formulae (6.79). On the assumption $\tilde{a}_{qk} \neq 0$, it can be computed almost problem-free, bar the possibility of overflow. If $T^2 \geqslant 1$, we have to choose case (a) and we have to take the reciprocal values for the multipliers

$$\text{(a) } t^2 = 1/T^2 \quad c^2 = \cos^2 \varphi = \frac{1}{1+t^2}$$
$$\text{(b) } T^2 = f_1 f_2 \quad s^2 = \sin^2 \varphi = \frac{1}{1+T^2} \tag{6.81}$$

With this modification the square root in formula (6.62) could be eliminated as well. The *fast Givens transformation* is based on the two fundamental ideas of updating the *squares* of the diagonal elements of $\mathbf{D}$ of the factored matrix $\mathbf{A} = \mathbf{D}\tilde{\mathbf{A}}$ and of computing the matrix $\tilde{\mathbf{A}}$ such that $\mathbf{D}'\tilde{\mathbf{A}}'$ equals the transformed matrix $\mathbf{A}'$. If after a sequence of Givens transformations the transformed matrix is needed explicitly, the factored representation must be multiplied out. Now $n$ square roots and a number of multiplications are required, depending on the structure of the transformed matrix.

Now we switch to Givens rotations in connection with similarity transformations. The matrix $\mathbf{A}$ is multiplied from both sides thus we have to use the matrices in the following factored form

$$\mathbf{A} = \mathbf{D}\tilde{\mathbf{A}}\mathbf{D} \quad \mathbf{A}'' = \mathbf{D}'\tilde{\mathbf{A}}''\mathbf{D}' \quad \mathbf{D} = \mathrm{diag}(d_1, d_2, \ldots, d_n). \tag{6.82}$$

With the representation $\mathbf{A}' = \mathbf{D}'\tilde{\mathbf{A}}'\mathbf{D}$, where the diagonal matrix $\mathbf{D}$ on the right originates from the factorization of $\mathbf{A}$, the first step of the transformation $\mathbf{A}' = \mathbf{U}^T\mathbf{A}$ can be taken from above with no changes, because the diagonal elements of $\mathbf{D}$ that appear on both sides in the formulae similar to (6.74) cancel. For the second step we get, using (6.16) and (6.82),

$$a''_{ip} = d'_i \tilde{a}''_{ip} d'_p = ca'_{ip} - sa'_{iq} = cd'_i \tilde{a}'_{ip} d_p - sd'_i \tilde{a}'_{iq} d_q$$
$$a''_{iq} = d'_i \tilde{a}''_{iq} d'_q = sa'_{ip} + ca'_{iq} = sd'_i \tilde{a}'_{ip} d_p + cd'_i \tilde{a}'_{iq} d_q$$

Here the diagonal elements $d'_i$ cancel for each index $i$ so that for the two cases (a) and (b) we get formulae that are analogous to those of (6.76) and (6.77).

## 6.3 Transformation Methods

$$\text{case (a)} \quad \tilde{a}''_{ip} = \tilde{a}'_{ip} - \left(\frac{sd_q}{cd_p}\right)\tilde{a}'_{iq} = \tilde{a}'_{ip} + f_1\tilde{a}'_{iq}$$

$$\tilde{a}''_{iq} = \left(\frac{sd_p}{cd_q}\right)\tilde{a}'_{ip} + \tilde{a}'_{iq} = -f_2\tilde{a}'_{ip} + \tilde{a}'_{iq}$$

$$\text{case (b)} \quad \tilde{a}''_{ip} = \left(\frac{cd_p}{sd_q}\right)\tilde{a}'_{ip} - \tilde{a}'_{iq} = -f_1\tilde{a}'_{ip} - \tilde{a}'_{iq}$$

$$\tilde{a}''_{iq} = \tilde{a}'_{ip} + \left(\frac{cd_q}{sd_p}\right)\tilde{a}'_{iq} = \tilde{a}'_{ip} - f_2\tilde{a}'_{iq}$$

(6.83)

We now apply the fast version of Givens rotations in order to transform a nonsymmetric matrix **A** into the Hessenberg form. The elimination of the element $a_{ij}$ with $i \geq j+2$ of the $j$th column requires the computation of the values $f_1$ and $f_2$ by means of the formulae (6.79) and then of $T^2 = f_1 f_2$, needing four essential operations. Depending on the situation, the reciprocals of these three values have to be computed by three more essential operations. The computation of $c^2$ or $s^2$ and of the two changing diagonal elements, according to (6.80), requires three operations. Hence, up to this point the number of essential operations is either 7 or 10. On the assumption that the two cases are likely to occur with the same probability, we assume that the number of operations is 9. An additional multiplication occurs while computing $\tilde{a}'_{j+1,j}$, according to (6.76) or (6.77), and the row and column operations require a total of $2(n-j+n)$ multiplications. Hence the amount of work required to eliminate the elements $a_{ij}$ of the $j$th column is $(n-j-1)(4n-2j+10)$, hence summing over $j$, from 1 to $(n-2)$, yields the number of $(5n^3 - 35n + 30)/3$ operations. We obtain the desired Hessenberg matrix **H**, which has $\frac{1}{2}(n^2 + 3n - 2)$ nonzero matrix elements, from the factored representation by additional $(n^2 + 3n - 2)$ multiplications and $n$ square roots. Therefore the total amount of work required to transform **A** into Hessenberg form, by means of the fast Givens rotations, is

$$Z_{\text{HessFG}} = \tfrac{5}{3}n^3 + n^2 - \tfrac{26}{3}n + 8. \tag{6.84}$$

In comparison to (6.63), the number of multiplications is halved for large orders $n$, and the number of square roots decreases from $N^* = \frac{1}{2}(n-1)(n-2)$ to $n$.

The fast Givens transformation has the minor disadvantage that it is not possible to condense the information on the rotations into a single number $\rho$, (6.65), now two values are required.

Although the transformations that are applied to the scaled matrices **A** are no longer orthogonal, an analysis of the errors shows that the fast version of the Givens rotations has essentially the same good properties as the usual Givens transformation (Parlett, 1980; Rath, 1982).

The algorithmic implementation of the fast Givens transformation is similar to (6.66). Obviously the determination of the factors $f_1$ and $f_2$ and the proper choice and execution of the cases (a) or (b) require corresponding extensions.

*Example* 6.5 The matrix **A** (6.67), of order $n = 6$ will be reduced to Hessenberg form by means of the fast Givens transformation. To eliminate $a_{31}$, because $\mathbf{D} = \mathbf{I}$, according to (6.79) we have $f_1 = f_2 = 6$, and due to $T^2 = f_1 f_2 = 36 > 1$ case (a) must be chosen, that is we have $f_1 = f_2 = 1/6$ and $c^2 \doteq 0.972\,972\,97$. The corresponding transformation yields

$$\tilde{\mathbf{A}}_1 \doteq \begin{pmatrix} 7.000\,00 & 3.666\,67 & 3.500\,00 & -11.000\,00 & -9.000\,00 & -2.000\,00 \\ -6.166\,67 & 1.722\,22 & -5.083\,33 & 7.333\,33 & 2.500\,00 & 12.166\,67 \\ 0 & -9.194\,44 & 4.444\,44 & 0.833\,33 & 8.833\,33 & -1.000\,00 \\ -8.000\,00 & -0.166\,67 & -1.000\,00 & 5.000\,00 & 0 & 8.000\,00 \\ -4.000\,00 & 2.166\,67 & -5.500\,00 & 7.000\,00 & 2.000\,00 & 10.000\,00 \\ 6.000\,00 & 1.666\,67 & 3.833\,33 & -11.000\,00 & -7.000\,00 & -1.000\,00 \end{pmatrix}$$

and $\mathbf{D}_1^2 \doteq \mathrm{diag}(1, 0.972\,972\,97, 0.972\,972\,97, 1, 1, 1)$. The elimination of $\tilde{a}_{41}^{(1)}$ is performed with multipliers of case (b), that is $f_1 = 0.75$ and $f_2 \doteq 0.770\,833$, and hence we have $T^2 = 0.578\,125$ and $s^2 \doteq 0.633\,663\,37$. The transformed matrix is

$$\tilde{\mathbf{A}}_2 \doteq \begin{pmatrix} 7.000\,00 & 8.250\,00 & 3.500\,00 & 12.145\,83 & -9.000\,00 & -2.000\,00 \\ 12.625\,00 & 11.343\,75 & 4.812\,50 & 6.968\,75 & -1.875\,00 & -17.125\,00 \\ 0 & 6.062\,50 & 4.444\,44 & -9.836\,81 & 8.833\,33 & -1.000\,00 \\ 0 & -4.867\,19 & -4.312\,50 & -0.831\,16 & 2.500\,00 & 6.000\,00 \\ -4.000\,00 & -8.625\,00 & -5.500\,00 & -3.229\,17 & 2.000\,00 & 10.000\,00 \\ 6.000\,00 & 9.750\,00 & 3.833\,33 & 10.145\,83 & -7.000\,00 & -1.000\,00 \end{pmatrix}$$

with $\mathbf{D}_2^2 \doteq \mathrm{diag}(1, 0.633\,663, 0.972\,973, 0.616\,537, 1, 1)$. After ten steps we obtain the desired matrix in factored form

$$\tilde{\mathbf{A}}_{10} \doteq \begin{pmatrix} 7.000\,00 & 11.250\,00 & 6.843\,85 & -0.196\,11 & -33.401\,05 & -18.479\,11 \\ 19.125\,00 & 9.875\,00 & 34.535\,04 & -2.619\,95 & -61.199\,28 & -8.697\,70 \\ 0 & -13.036\,49 & 3.394\,00 & -0.934\,93 & 18.242\,20 & 8.913\,38 \\ 0 & 0 & -14.213\,84 & 5.744\,63 & 17.771\,16 & -18.706\,22 \\ 0 & 0 & 0 & -5.442\,54 & -38.916\,88 & -94.714\,19 \\ 0 & 0 & 0 & 0 & 12.200\,70 & 28.959\,13 \end{pmatrix}$$

and $\mathbf{D}_{10}^2 \doteq \mathrm{diag}(1, 0.418\,301, 0.721\,204, 0.507\,448, 0.072\,850, 0.184\,450)$. From this the Hessenberg matrix $\mathbf{H} = \mathbf{D}_{10} \tilde{\mathbf{A}}_{10} \mathbf{D}_{10}$ results which essentially coincides with that of (6.68). The two matrices differ insofar as the signs of the elements of a row and the corresponding column may be different.

### 6.3.4 Hyman's method

The first practical application of the transformation of an nonsymmetric matrix **A** into a Hessenberg matrix **H** will be the derivation of a simple and

## 6.3 Transformation Methods

useful procedure for computing the eigenvalues and eigenvectors of **H**. The basic idea of Hyman's method (Hyman, 1957) consists for a given $\lambda$ computing the *value* of the characteristic polynomial $P(\lambda):=|\mathbf{H}-\lambda\mathbf{I}|$ and the value of its first derivative $P'(\lambda)$, so that the eigenvalues, being the zeros of $P(\lambda)$, may be found iteratively by means of Newton's method.

The characteristic polynomial of a Hessenberg matrix **H** is given by

$$P(\lambda) = \begin{vmatrix} h_{11}-\lambda & h_{12} & h_{13} & \cdots & h_{1,n-1} & h_{1n} \\ h_{21} & h_{22}-\lambda & h_{23} & \cdots & h_{2,n-1} & h_{2n} \\ 0 & h_{32} & h_{33}-\lambda & \cdots & h_{3,n-1} & h_{3n} \\ \vdots & \vdots & \vdots & \ddots & \vdots & \vdots \\ 0 & 0 & 0 & \cdots & h_{n-1,n-1}-\lambda & h_{n-1,n} \\ 0 & 0 & 0 & \cdots & h_{n,n-1} & h_{nn}-\lambda \end{vmatrix}. \quad (6.85)$$

We assume that the elements along the lower off-diagonal $h_{21}, h_{32}, \ldots, h_{n,n-1}$ are nonzero. Otherwise the characteristic polynomial is the product of determinants of lower order, and the problem of computing the eigenvalues of **H** is reduced to a problem of smaller order. It is known that the value of a determinant is not changed by adding an arbitrary linear combination of the other columns to a column. Hence we can add multiples of the first $(n-1)$ columns to the last one without changing the value of the determinant. The multipliers $x_j$ of the $j$th column are to be determined in such a way that the last column becomes a multiple of the first unit vector $\mathbf{e}_1$. This requirement leads to the following $(n-1)$ equations for $x_1, x_2, \ldots, x_{n-1}$.

$$h_{21}x_1 + (h_{22}-\lambda)x_2 + h_{23}x_3 + \cdots + h_{2,n-2}x_{n-2} + h_{2,n-1}x_{n-1} + h_{2n} = 0$$
$$h_{32}x_2 + (h_{33}-\lambda)x_3 + \cdots + h_{3,n-2}x_{n-2} + h_{3,n-1}x_{n-1} + h_{3n} = 0$$
$$\cdots\cdots\cdots\cdots\cdots\cdots\cdots\cdots\cdots\cdots\cdots\cdots\cdots\cdots\cdots$$
$$h_{n-1,n-2}x_{n-2} + (h_{n-1,n-1}-\lambda)x_{n-1} + h_{n-1,n} = 0$$
$$h_{n,n-1}x_{n-1} + (h_{nn}-\lambda) = 0$$
$$(6.86)$$

From our hypothesis $h_{i+1,i} \neq 0$ for $i=1,2,\ldots,n-1$, the system (6.86) of linear equations has a nonsingular upper triangular matrix so that the unknowns can be computed by means of the process of back-substitution. If we put $x_n = 1$ for convenience we obtain the explicit formulae

$$x_i = \left((\lambda - h_{i+1,i+1})x_{i+1} - \sum_{j=i+2}^{n} h_{i+1,j}x_j\right)\Big/h_{i+1,i}, \quad (i=n-1, n-2, \ldots, 1)$$

(6.87)

With these values the first component of the last column is given by

$$(h_{11} - \lambda)x_1 + h_{12}x_2 + \cdots + h_{1,n-1}x_{n-1} + h_{1n}x_n =: p(\lambda). \tag{6.88}$$

The determinant (6.85) is thus changed into

$$P(\lambda) = \begin{vmatrix} h_{11} - \lambda & h_{12} & h_{13} & \cdots & h_{1,n-1} & p(\lambda) \\ h_{21} & h_{22} - \lambda & h_{23} & \cdots & h_{2,n-1} & 0 \\ 0 & h_{32} & h_{33} - \lambda & \cdots & h_{3,n-1} & 0 \\ \vdots & \vdots & \vdots & \ddots & \vdots & \vdots \\ 0 & 0 & 0 & \cdots & h_{n-1,n-1} - \lambda & 0 \\ 0 & 0 & 0 & \cdots & h_{n,n-1} & 0 \end{vmatrix}, \tag{6.89}$$

so that the characteristic polynomial $P(\lambda)$ has the representation

$$P(\lambda) = (-1)^{n+1} h_{21} h_{32} \cdots h_{n,n-1} p(\lambda). \tag{6.90}$$

The value of the characteristic polynomial of a Hessenberg matrix **H** for a given value of $\lambda$ is, up to a constant factor, equal to $p(\lambda)$ defined by (6.88).

From (6.87), the coefficients $x_i$ are polynomials in $\lambda$ of exact degree $(n - i)$. If we differentiate (6.87) with respect to $\lambda$ we get with $x'_n = 0$

$$x'_i = \left( x_{i+1} + (\lambda - h_{i+1,i+1})x'_{i+1} - \sum_{j=i+2}^{n-1} h_{i+1,j} x'_j \right) \bigg/ h_{i+1,i} \quad (i = n-1, n-2, \ldots, 1) \tag{6.91}$$

a similar recursion formula for the derivatives $x'_i$. Furthermore, from (6.88) we obtain

$$p'(\lambda) = -x_1 + (h_{11} - \lambda)x'_1 + h_{12}x'_2 + \cdots + h_{1,n-1}x'_{n-1}. \tag{6.92}$$

From (6.90) we have $P'(\lambda) = (-1)^{n+1} h_{21} h_{32} \cdots h_{n,n-1} p'(\lambda)$, and because Newton's method only requires the quotient $P(\lambda)/P'(\lambda)$, it is sufficient to determine the values $p(\lambda)$ and $p'(\lambda)$ to perform an iteration step.

Hyman's method consists of two steps: firstly the coefficients $x_i$ and their derivatives $x'_i =: y_i$ are simultaneously computed, and secondly the desired values of $p(\lambda)$ and $p'(\lambda)$ are determined. In the following summary of the algorithm we put $y_n = x'_n = 0$ for unifying purposes, and we consider $\lambda$ to be given

## 6.3 Transformation Methods

$$\begin{aligned}
&x_n = 1; \, y_n = 0 \\
&\text{for } i = n-1, n-2, \ldots, 1: \\
&\quad s = (\lambda - h_{i+1,i+1}) \times x_{i+1}; \, t = x_{i+1} + (\lambda - h_{i+1,i+1}) \times y_{i+1} \\
&\quad \text{for } j = i+2, i+3, \ldots, n: \\
&\qquad s = s - h_{i+1,j} \times x_j; \, t = t - h_{i+1,j} \times y_j \\
&\quad x_i = s/h_{i+1,i}; \, y_i = t/h_{i+1,i} \\
&s = (h_{11} - \lambda) \times x_1; \, t = (h_{11} - \lambda) \times y_1 - x_1 \\
&\text{for } i = 2, 3, \ldots, n: \\
&\quad s = s + h_{1i} \times x_i; \, t = t + h_{1i} \times y_i \\
&p(\lambda) = s; \, p'(\lambda) = t
\end{aligned}$$

(6.93)

The computation of the pair of values $p(\lambda)$ and $p'(\lambda)$ for a given real value $\lambda$, by means of the algorithm (6.93), requires

$$Z_{\text{Hyman}} = n^2 + 3n - 2 \tag{6.94}$$

multiplicative operations. The eigenvalues of real, nonsymmetric matrices may be complex, therefore, a computer program should also allow $\lambda$ to be complex. The amount of work that is required is increased, but it should be taken into account that the matrix elements $h_{ij}$ are real, and that the frequent multiplications of real matrix elements with a complex number requires only two operations. It follows that the number of real multiplications is essentially doubled in comparison with (6.94) if $\lambda$ is complex. However, the eigenvalues are pairwise complex conjugate, and hence as soon as one complex eigenvalue $\lambda_k$ has been computed the corresponding complex conjugate eigenvalue $\lambda_{k+1} = \bar{\lambda}_k$ has been found.

If several, or all the eigenvalues of the Hessenberg matrix **H** are to be computed successively the technique of implicit deflation of known zeros of the characteristic polynomial must be applied (see Section 5.5).

Instead of Newton's method producing a quadratically convergent sequence, *Laguerre's method* with cubic convergence can be applied (Parlett, 1964). For this method the second derivative $p''(\lambda)$ must be known. It can be obtained by deriving first a recursion formula for the $x_i''$ from (6.91) and then by differentiating (6.92) with respect to $\lambda$ to get an expression for $p''(\lambda)$.

As soon as an eigenvalue $\lambda_k$ of the Hessenberg matrix **H** is determined as a zero of the polynomial $p(\lambda)$ it follows from (6.88), (6.86) and because of $p(\lambda_k) = 0$, that the vector $\mathbf{x}_k := (x_1, x_2, \ldots, x_n)^T$, with the components $x_i$ computed according to (6.87), is an eigenvector of **H** corresponding to $\lambda_k$ due to the relation $\mathbf{H}\mathbf{x}_k = \lambda_k \mathbf{x}_k$. Theoretically, the Hyman algorithm also yields the corresponding eigenvector. An error analysis shows that the computation of the value $p(\lambda)$ is numerically stable apart from the known problem of evaluating a polynomial (Wilkinson, 1965). For the components $x_i$, however, no equivalent statement holds. The numerical instability that may be observed in the computation of the $x_i$ can be so severe, especially when the order $n$ of the matrix $\mathbf{H}$ becomes large, that the components of the resulting vector $\mathbf{x}_k$ are completely wrong!

The eigenvector $\mathbf{x}_k$, corresponding to $\lambda_k$, can be computed by means of the *inverse vector iteration*. Let $\bar{\lambda}$ be the computed approximation of the eigenvalue $\lambda_k$. Then we construct the sequence of iterated vectors $\mathbf{z}^{(v)}$ from

$$(\mathbf{H} - \bar{\lambda}\mathbf{I})\mathbf{z}^{(v)} = \mathbf{z}^{(v-1)} \quad (v = 1, 2, \ldots) \tag{6.95}$$

where $\mathbf{z}^{(0)}$ denotes a suitable initial vector. On the simplifying assumption that the Hessenberg matrix $\mathbf{H}$ has a system of $n$ linearly independent eigenvectors $\mathbf{x}_1, \mathbf{x}_2, \ldots, \mathbf{x}_n$ the initial vector $\mathbf{z}^{(0)}$ has the unique representation

$$\mathbf{z}^{(0)} = \sum_{i=1}^{n} c_i \mathbf{x}_i. \tag{6.96}$$

As $(\mathbf{H} - \bar{\lambda}\mathbf{I})\mathbf{x}_i = (\lambda_i - \bar{\lambda})\mathbf{x}_i$, it follows for the iterated vectors $\mathbf{z}^{(v)}$ that

$$\mathbf{z}^{(v)} = \sum_{i=1}^{n} \frac{c_i}{(\lambda_i - \bar{\lambda})^v} \mathbf{x}_i. \tag{6.97}$$

On the additional assumptions

$$c_k \neq 0 \quad \text{and} \quad 0 < |\lambda_k - \bar{\lambda}| = \varepsilon \ll \delta := \min_{i \neq k} |\lambda_i - \bar{\lambda}| \tag{6.98}$$

from (6.97) we obtain the representation for $\mathbf{z}^{(v)}$

$$\mathbf{z}^{(v)} = \frac{1}{(\lambda_k - \bar{\lambda})^v} \left[ c_k \mathbf{x}_k + \sum_{\substack{i=1 \\ i \neq k}}^{n} c_i \left( \frac{\lambda_k - \bar{\lambda}}{\lambda_i - \bar{\lambda}} \right)^v \mathbf{x}_i \right]. \tag{6.99}$$

Hence the sequence $\mathbf{z}^{(v)}$ quickly converges towards the direction of the eigenvector $\mathbf{x}_k$, provided that after each iteration step the resulting vector $\mathbf{z}^{(v)}$ is normalized. In spite of the aforementioned instability, the Hyman's method provides a good initial vector $\mathbf{z}^{(0)}$, with $c_k \neq 0$, for which a single iteration step, (6.95), is usually sufficient to produce an iterate that coincides with $\mathbf{x}_k$ to a reasonable degree of accuracy. The convergence rate depends, according to (6.98) and (6.99), on the mutual position of the eigenvalues and on the approximate value $\bar{\lambda}$.

The system of linear equations (6.95) is solved for $\mathbf{z}^{(v)}$ by means of the Gaussian algorithm with the relative maximal column pivoting strategy. $\mathbf{H} - \bar{\lambda}\mathbf{I}$ is a Hessenberg matrix, thus it is advantageous to apply the technique for tridiagonal systems of equations that was developed in Section 1.3.3. If the row permutations are performed before the elimination step, then, analogously to (1.121), a final scheme results with a Hessenberg structure. The resulting right triangular matrix is, in general, filled with nonzero elements. The procedure simplifies both the elimination steps of the decomposition and the forward substitution.

The matrix $\mathbf{H} - \bar{\lambda}\mathbf{I}$ is almost singular because $\bar{\lambda}$ represents a good approxi-

## 6.4 QR Algorithm

mation of an eigenvalue. Consequently the system of equations (6.95), is ill-conditioned. Therefore it should be expected that the computed solution vector $\tilde{z}^{(v)}$ has a large relative error in comparison with the exact vector $z^{(v)}$. However, it is observed that the error $\tilde{z}^{(v)} - z^{(v)}$ has a dominant component in the direction of the eigenvector, corresponding to the eigenvalue of the smallest absolute size of the matrix $H - \bar{\lambda} I$. Consequently, the error of the computed solution is essentially proportional to $x_k$, that is just in the desired direction. This observation makes the inverse vector iteration a useful method for the computation of eigenvectors.

*Example* 6.6 The eigenvalues and eigenvectors of the Hessenberg matrix $H$ (6.68) of Example 6.3 are computed by means of Hyman's method. The algorithm, (6.93), combined with Newton's method yields the rounded values of the following table, where the real initial value $\lambda^{(0)} = 2.5$ is used.

| $\mu$ | $\lambda^{(\mu)}$ | $p(\lambda^{(\mu)})$ | $p'(\lambda^{(\mu)})$ | $\Delta\lambda$ |
|---|---|---|---|---|
| 0 | 2.500 000 000 | $1.7571 \times 10^{-1}$ | $-0.405\,02$ | $4.3384 \times 10^{-1}$ |
| 1 | 2.933 837 624 | $1.9506 \times 10^{-2}$ | $-0.305\,37$ | $6.3877 \times 10^{-2}$ |
| 2 | 2.997 714 268 | $6.4983 \times 10^{-4}$ | $-0.284\,69$ | $2.2826 \times 10^{-3}$ |
| 3 | 2.999 996 881 | $8.8553 \times 10^{-7}$ | $-0.283\,91$ | $3.1190 \times 10^{-6}$ |
| 4 | 3.000 000 000 | $-6 \times 10^{-11}$ | | |

Hyman's method directly provides the (normed) eigenvector corresponding to the eigenvalue $\lambda_1 = 3$

$$y_1 \doteq (0.247\,016, 0.113\,164, 0.475\,263, 0.456\,783, 0.600\,192, 0.362\,520)^T$$

of the Hessenberg matrix $H$. Application of the inverse vector iteration gives no further improvement. The back transformation of $y_1$ into the eigenvector $x_1$ of $A$ by means of the technique (6.64), that uses the $\rho$ values given in (6.68), yields

$$x_1 \doteq (0.247\,016, -0.123\,508, 0.823\,387, 0.411\,693, -0.123\,508, 0.247\,016)^T.$$

The remaining eigenvalues of $H$ are $\lambda_2 = 4$, $\lambda_{3,4} = 1 \pm 2i$, $\lambda_{5,6} = 5 \pm 6i$.

## 6.4 QR Algorithm

The most reliable method for computing the eigenvalues of a Hessenberg matrix $H$ or a symmetric, tridiagonal matrix $J$ consists of the construction of a sequence of orthogonally similar matrices which converges towards a limit matrix that yields the desired eigenvalues.

### 6.4.1 Fundamentals of the QR transformation

We collect together some theoretical facts that are useful to establish the method and its subsequent implementation.

*Theorem 6.4* Each square matrix $\mathbf{A}$ can be decomposed into the product of an orthogonal matrix $\mathbf{Q}$ and a right triangular matrix $\mathbf{R}$ so that we have

$$\boxed{\mathbf{A} = \mathbf{QR}} \tag{6.100}$$

The factorization (6.100) is called the *QR decomposition* of the matrix $\mathbf{A}$.

*Proof* We show the existence of the QR decomposition by construction. For this reason we multiply the given matrix $\mathbf{A}$ successively by suitably chosen rotation matrices $\mathbf{U}^T(p, q; \varphi)$ from the left such that the matrix elements below the diagonal become zero in a particular order. The elimination can be achieved, for instance, columnwise in the following order

$$a_{21}, a_{31}, \ldots, a_{n1}, a_{32}, a_{42}, \ldots, a_{n2}, a_{43}, \ldots, a_{n,n-1} \tag{6.101}$$

by means of $(p, q)$ rotations with the corresponding rotation index pairs

$$(1, 2), (1, 3), \ldots, (1, n), (2, 3), (2, 4), \ldots, (2, n), (3, 4), \ldots, (n-1, n). \tag{6.102}$$

Thus a $(j, i)$ rotation matrix $\mathbf{U}_k^T = \mathbf{U}^T(j, i; \varphi_k)$ is applied to eliminate the matrix element $a_{ij}^{(k-1)}$ of $\mathbf{A}^{(k-1)}$

$$\mathbf{A}^{(k)} = \mathbf{U}_k^T \mathbf{A}^{(k-1)} \quad \mathbf{A}^{(0)} = \mathbf{A} \quad (k = 1, 2, \ldots, \tfrac{1}{2}n(n-1)). \tag{6.103}$$

The only effect of this multiplication is that the $j$th and the $i$th rows of $\mathbf{A}^{(k-1)}$ are linearly combined. According to (6.15) the values of $\cos \varphi_k$ and $\sin \varphi_k$ are determined from the condition

$$a_{ij}^{(k)} = a_{jj}^{(k-1)} \sin \varphi_k + a_{ij}^{(k-1)} \cos \varphi_k = 0 \tag{6.104}$$

analogously to (6.62). If $a_{ij}^{(k-1)} = 0$ already holds, then no rotation is applied and we put $\mathbf{U}_k = \mathbf{I}$.

It remains to verify that the sequence of matrices $\mathbf{A}^{(k)}$ defined by (6.102) and (6.103) generates a right triangular matrix $\mathbf{A}^{(N)} = \mathbf{R}$ for $N = \tfrac{1}{2}n(n-1)$. The first $(n-1)$ matrix multiplications obviously eliminate the elements $a_{i1}$ with $2 \leq i \leq n$. Now we assume that the first $(j-1)$ columns already have the desired form. The elimination of an arbitrary element $a_{ij}$ of the $j$th column with $i > j$ is achieved by a multiplication from the left by a $(j, i)$ rotation matrix. In the first $(j-1)$ columns matrix elements are linearly combined that are zero from our assumption. The vanishing elements that have already been generated are not destroyed and thus we have

$$\mathbf{U}_N^T \mathbf{U}_{N-1}^T \cdots \mathbf{U}_2^T \mathbf{U}_1^T \mathbf{A}^{(0)} = \mathbf{R}.$$

## 6.4 QR Algorithm

The matrix $\mathbf{Q}^T := \mathbf{U}_N^T \mathbf{U}_{N-1}^T \cdots \mathbf{U}_2^T \mathbf{U}_1^T$ is orthogonal since it is a product of orthogonal matrices, and hence the existence of the QR decomposition $\mathbf{A}^{(0)} = \mathbf{A} = \mathbf{QR}$ (6.100) is proved.

An immediate consequence of the proof of Theorem 6.4 is the following theorem.

*Theorem 6.5 The QR decomposition of a Hessenberg matrix $\mathbf{H}$ or a tridiagonal matrix $\mathbf{J}$ of order $n$ is feasible with $(n-1)$ rotation matrices.*

*Proof.* If the given matrix $\mathbf{A}$ is either a Hessenberg or a tridiagonal matrix, then the elimination of $a_{21}$, which is the only element of the first column below the diagonal that is nonzero, is achieved by a left multiplication of the matrix $\mathbf{U}^T(1, 2; \varphi_1) = \mathbf{U}_1^T$. Thereby the structure of the submatrix of order $(n-1)$ that is obtained from $\mathbf{A}^{(1)} = \mathbf{U}_1^T \mathbf{A}$ by omitting the first row and the first column is the same as that of $\mathbf{A}$. This observation also holds for the following elimination steps. The multiplication of $\mathbf{A}$ from the left by the $(n-1)$ rotation matrices with the index pairs $(1, 2), (2, 3), \ldots, (n-1, n)$ results in the QR decomposition $\mathbf{A} = \mathbf{QR}$ with $\mathbf{Q} := \mathbf{U}_1 \mathbf{U}_2 \cdots \mathbf{U}_{n-1}$. In the case of a tridiagonal matrix $\mathbf{J}$ the right triangular matrix $\mathbf{R}$ usually contains nonzero elements in the two adjacent upper diagonals, because each multiplication from the left by $\mathbf{U}^T(i, i+1; \varphi_i)$ usually generates a non-vanishing element at the place $(i, i+2)$ for $i = 1, 2, \ldots, n-2$.

The statement of Theorem 6.5 is vital for the algorithm to be developed, because the QR decomposition of a Hessenberg matrix is much less expensive than that of a full matrix.

On the basis of the QR decomposition of a square matrix $\mathbf{A}$ we define the *QR transformation* of $\mathbf{A}$ as follows.

$$\boxed{\mathbf{A} = \mathbf{QR} \quad \mathbf{A}' = \mathbf{RQ} \quad \text{QR transformation}} \tag{6.105}$$

*Theorem 6.6 The matrix $\mathbf{A}'$ defined by the QR transformation (6.105) is orthogonally similar to the matrix $\mathbf{A}$.*

*Proof* The matrix $\mathbf{Q}$ of the QR decomposition of $\mathbf{A}$ is orthogonal and hence nonsingular therefore the inverse $\mathbf{Q}^{-1} = \mathbf{Q}^T$ exists. From (6.105) the orthogonal similarity of the two matrices $\mathbf{A}'$ and $\mathbf{A}$ follows according to

$$\mathbf{R} = \mathbf{Q}^{-1}\mathbf{A} \quad \text{and} \quad \mathbf{A}' = \mathbf{RQ} = \mathbf{Q}^{-1}\mathbf{AQ}.$$

*Theorem 6.7 The QR transformed matrix of a Hessenberg matrix $\mathbf{H}$ of order $n$ is again of Hessenberg form.*

*Proof* From Theorem 6.5 the matrix $\mathbf{Q}$ of the QR decomposition of a Hessenberg matrix $\mathbf{H}$ of order $n$ is given as the product of the $(n-1)$ rotation matrices

$$\mathbf{Q} = \mathbf{U}(1,2;\varphi_1)\mathbf{U}(2,3;\varphi_2)\cdots\mathbf{U}(n-1,n;\varphi_{n-1}). \tag{6.106}$$

The QR transformed matrix $\mathbf{H}' = \mathbf{RQ}$ is obtained, according to (6.106), by successive multiplication of the right triangular matrix $\mathbf{R}$ from the right with the rotation matrices. The matrix $\mathbf{RU}(1,2;\varphi_1)$ is no longer an upper triangular matrix because the linear combinations of the first two columns generates a non-vanishing matrix element at the place $(2,1)$. The next multiplication by $\mathbf{U}(2,3;\varphi_2)$ produces, because of the corresponding column operations, exactly one nonvanishing element at the position $(3,2)$. The general multiplication from the right by $\mathbf{U}(k,k+1;\varphi_k)$ usually generates a nonzero matrix element at the place $(k+1,k)$ without changing the previous columns. Hence, it follows that $\mathbf{H}'$ is a Hessenberg matrix.

The amount of computational work required for a QR transformation of a Hessenberg matrix $\mathbf{H}$ of order $n$ consists of computing the upper triangular matrix $\mathbf{R}$ and calculating $\mathbf{H}'$. The general $j$th step of the first part requires $4 + 4(n - j)$ multiplications and a square root, if the formulae (6.62) and (6.15) are used. The matrix multiplication by $\mathbf{U}(j, j+1; \varphi_j)$ in the second part requires $4j + 2$ multiplications, if we take into account that the diagonal element and the new off-diagonal element are each obtained by only one operation. After summing over $j$, from 1 to $(n-1)$, we get a total of

$$Z_{\text{QR Hess}} = (4n+6)(n-1) = 4n^2 + 2n - 6 \tag{6.107}$$

multiplications and $(n-1)$ square roots.

*Theorem 6.8 The QR transformed matrix of a symmetric, tridiagonal matrix is again symmetric and tridiagonal.*

*Proof* From Theorem 6.6 the matrix $\mathbf{A}'$ is orthogonally similar to $\mathbf{A}$, and the symmetry is maintained due to $\mathbf{A}'^{\text{T}} = (\mathbf{Q}^{-1}\mathbf{A}\mathbf{Q})^{\text{T}} = \mathbf{Q}^{\text{T}}\mathbf{A}^{\text{T}}\mathbf{Q}^{-1\text{T}} = \mathbf{Q}^{-1}\mathbf{A}\mathbf{Q} = \mathbf{A}'$. Moreover, a tridiagonal matrix is a special Hessenberg matrix, and consequently, $\mathbf{A}'$ is a Hessenberg matrix from Theorem 6.7. Taking the symmetry into account, the matrix $\mathbf{A}'$ must be tridiagonal.

The QR-algorithm is motivated by the following *theorem of Schur* (Schur, 1909). Which, as we only wish to compute the eigenvalues of real matrices, will be stated in its real form.

*Theorem 6.9 For each real matrix $\mathbf{A}$ of order n, there exists an orthogonal matrix $\mathbf{U}$ of order n such that the similar matrix $\mathbf{R} := \mathbf{U}^{-1}\mathbf{A}\mathbf{U}$ has the quasi-triangular form*

## 6.4 QR Algorithm

$$\mathbf{R} := \mathbf{U}^{-1}\mathbf{A}\mathbf{U} = \begin{pmatrix} \mathbf{R}_{11} & \mathbf{R}_{12} & \mathbf{R}_{13} & \cdots & \mathbf{R}_{1m} \\ 0 & \mathbf{R}_{22} & \mathbf{R}_{23} & \cdots & \mathbf{R}_{2m} \\ 0 & 0 & \mathbf{R}_{33} & \cdots & \mathbf{R}_{3m} \\ \vdots & \vdots & \vdots & \ddots & \vdots \\ 0 & 0 & 0 & \cdots & \mathbf{R}_{mm} \end{pmatrix}. \tag{6.108}$$

*The matrices $\mathbf{R}_{ii}$ ($i = 1, 2, \ldots, m$) are either of order one or of order two and have in the last case a pair of complex conjugate eigenvalues.*

*Proof* We show the existence of such an orthogonal matrix $\mathbf{U}$ in three steps.

(1) Let $\lambda$ be a real eigenvalue of $\mathbf{A}$, and let $\mathbf{x}$ be the corresponding normed eigenvector with $\|\mathbf{x}\|_2 = 1$. The vector $\mathbf{x}$ can be complemented by a further $(n-1)$ vectors to form an orthonormal basis of $\mathbf{R}^n$. We define the orthogonal matrix $\mathbf{U}_1 := (\mathbf{x}, \tilde{\mathbf{U}}_1)$, where $\tilde{\mathbf{U}}_1 \in \mathbf{R}^{n \times (n-1)}$ has as its columns the remaining $(n-1)$ basis vectors. Then we have $\mathbf{A}\mathbf{U}_1 = (\lambda\mathbf{x}, \mathbf{A}\tilde{\mathbf{U}}_1)$ and, moreover,

$$\mathbf{U}_1^{-1}\mathbf{A}\mathbf{U}_1 = \mathbf{U}_1^T\mathbf{A}\mathbf{U}_1 = \begin{pmatrix} \mathbf{x}^T \\ \tilde{\mathbf{U}}_1^T \end{pmatrix}(\lambda\mathbf{x}, \mathbf{A}\tilde{\mathbf{U}}_1) = \left(\begin{array}{c|c} \lambda & \mathbf{x}^T\mathbf{A}\tilde{\mathbf{U}}_1 \\ \hline 0 & \tilde{\mathbf{U}}_1^T\mathbf{A}\tilde{\mathbf{U}}_1 \end{array}\right) =: \mathbf{A}_1. \tag{6.109}$$

The matrix $\mathbf{A}_1$, (6.109), is orthogonally similar to $\mathbf{A}$, the eigenvalue $\lambda$ appears in the left upper corner. Below it the first column contains a zero vector, from the orthogonality of $\mathbf{x}$ with respect to the columns of $\tilde{\mathbf{U}}_1$. The submatrix $\tilde{\mathbf{U}}_1^T\mathbf{A}\tilde{\mathbf{U}}_1$ is real and of order $(n-1)$.

(2) If $\lambda = \alpha + i\beta$ with $\beta \neq 0$ is a complex eigenvalue of $\mathbf{A}$ and $\mathbf{x} = \mathbf{u} + i\mathbf{v}$ is the corresponding eigenvector, then $\bar{\lambda} = \alpha - i\beta$ is necessarily a complex conjugate eigenvalue of $\mathbf{A}$ with the corresponding complex conjugate eigenvector $\bar{\mathbf{x}} = \mathbf{u} - i\mathbf{v}$. Thus we have

$$\mathbf{A}(\mathbf{u} \pm i\mathbf{v}) = (\alpha \pm i\beta)(\mathbf{u} \pm i\mathbf{v})$$

and hence for the real and imaginary parts

$$\mathbf{A}\mathbf{u} = \alpha\mathbf{u} - \beta\mathbf{v} \quad \mathbf{A}\mathbf{v} = \beta\mathbf{u} + \alpha\mathbf{v}. \tag{6.110}$$

As eigenvectors that correspond to different eigenvalues are linearly independent, this being true for $\mathbf{x}$ and $\bar{\mathbf{x}}$, the two real vectors $\mathbf{u}$ and $\mathbf{v}$ are linearly independent. We define the matrix $\mathbf{Y} := (\mathbf{u}, \mathbf{v}) \in \mathbf{R}^{n \times 2}$ for which, from (6.110), the following relationship holds

$$\mathbf{A}\mathbf{Y} = \mathbf{Y}\begin{pmatrix} \alpha & \beta \\ -\beta & \alpha \end{pmatrix} =: \mathbf{Y}\boldsymbol{\Gamma}. \tag{6.111}$$

In the subspace spanned by $\mathbf{u}$ and $\mathbf{v}$ there exist two orthonormal vectors $\mathbf{x}_1$ and $\mathbf{x}_2$ which we combine in the matrix $\mathbf{X} \in \mathbf{R}^{n \times 2}$. The following relationship holds between $\mathbf{Y}$ and $\mathbf{X}$

$$\mathbf{Y} = \mathbf{X}\mathbf{C} \quad \text{where } \mathbf{C} \in \mathbf{R}^{2 \times 2} \text{ is nonsingular.}$$

Hence the matrix equation (6.111) is equivalent to

$$AXC = XC\Gamma \quad \text{or} \quad AX = XC\Gamma C^{-1} =: XS. \tag{6.112}$$

The matrix $S = C\Gamma C^{-1} \in \mathbf{R}^{2 \times 2}$ is similar to $\Gamma$ and so has $\lambda = \alpha + i\beta$ and $\bar{\lambda} = \alpha - i\beta$ as pair of complex conjugate eigenvalues.

We complement $x_1$ and $x_2$ by $(n-2)$ further vectors to give an orthonormal basis of $\mathbf{R}^n$. We then define the orthogonal matrix $U_2 := (x_1, x_2, \tilde{U}_2) = (X, \tilde{U}_2)$, where $\tilde{U}_2 \in \mathbf{R}^{n \times (n-2)}$ has as its columns the remaining $(n-2)$ vectors of the above basis. From (6.112) we have

$$U_2^{-1} A U_2 = U_2^T A U_2 = \begin{pmatrix} X^T \\ \tilde{U}_2^T \end{pmatrix} A(X, \tilde{U}_2) = \left( \begin{array}{c|c} S & X^T A \tilde{U}_2 \\ \hline 0 & \tilde{U}_2^T A \tilde{U}_2 \end{array} \right) =: A_2. \tag{6.113}$$

The matrix $A_2$, (6.113), that is orthogonally similar to $A$, contains in the left-hand upper corner the matrix $S$. Below this matrix of order two $\tilde{U}_2^T A X = \tilde{U}_2^T X S = 0 \in \mathbf{R}^{(n-2) \times 2}$ is a zero matrix because the vectors $x_1$ and $x_2$ are orthogonal with respect to the column vectors of $\tilde{U}_2$. Finally, $\tilde{U}_2^T A \tilde{U}_2$ represents a real matrix of order $(n-2)$.

(3) The structure of the matrices $A_1$, (6.109), and $A_2$, (6.113), shows that the set of eigenvalues of $A$ consists in the first case of $\lambda$ and the eigenvalues of $\tilde{U}_1^T A \tilde{U}_1$ and in the second case of the pair of complex conjugate eigenvalues of $S$ and the eigenvalues of $\tilde{U}_2^T A \tilde{U}_2$. The orthogonal similarity transformation can be analogously applied to the submatrix $\tilde{A}_i := \tilde{U}_i^T A \tilde{U}_i$ of order $(n-i)$ with $i = 1$ or $i = 2$ for a further real eigenvalue or pair of complex conjugate eigenvalues. Let $\tilde{V}_i \in \mathbf{R}^{(n-i) \times (n-i)}$ be the orthogonal matrix that transforms $\tilde{A}_i$ into the form (6.109) or (6.113), respectively, then

$$V := \left( \begin{array}{c|c} I_i & 0 \\ \hline 0 & \tilde{V}_i \end{array} \right) \in \mathbf{R}^{n \times n}$$

represents the orthogonal matrix which transforms $A_i$, $V^T A_i V$, with a further step into the quasi-triangular form. The consequent continuation of the transformation yields the statement of the theorem, where $U$ is given as the product of the orthogonal matrices.

Schur's theorem is independent of the multiplicities of the eigenvalues of the matrix $A$. The eigenvalues of $A$ can either be read directly or can be easily computed from the submatrices $R_{ii}$ of the quasi-triangular matrix $R$. However, Theorem 6.9 contains a statement of existence, and the proof is unfortunately not constructive.

In order to anticipate wrong interpretations of the theorem, we point out that the columns of the orthogonal matrix $U$ in (6.108) have in general nothing to do with the eigenvectors of $A$. This is quite obvious in the case of a complex eigenvalue of $A$ because the matrix equation $AU = UR$ which is equivalent to (6.108) contains real matrices.

## 6.4.2 Practical implementation, real eigenvalues

The QR-transformation, (6.105), is now applied repeatedly to define a sequence of orthogonally similar matrices with the aim of implementing the statement of Schur's theorem in a constructive way. In order to reduce the amount of work and because of Theorem 6.7, the QR transformation is applied to a Hessenberg matrix $\mathbf{H} = \mathbf{H}_1$ and a sequence of similar Hessenberg matrices is constructed according to the following rule.

$$\boxed{\mathbf{H}_k = \mathbf{Q}_k \mathbf{R}_k \quad \mathbf{H}_{k+1} = \mathbf{R}_k \mathbf{Q}_k \quad (k = 1, 2, \ldots)} \tag{6.114}$$

The formulae (6.114) is the original *QR algorithm* of Francis (Francis, 1961). The following convergence property can be shown for the sequence of matrices $\mathbf{H}_k$ (Francis, 1961; Parlett, 1980; Wilkinson, 1965).

*Theorem* 6.10 Let $\lambda_i$ be the eigenvalues of $\mathbf{H}$ satisfying the condition $|\lambda_1| > |\lambda_2| > \cdots > |\lambda_n|$. Let $\mathbf{X} \in \mathbf{R}^{n \times n}$ be the nonsingular matrix whose columns are the eigenvectors $\mathbf{x}_i$ of $\mathbf{H}$. If the LR decomposition of $\mathbf{X}^{-1}$ exists, then the matrices $\mathbf{H}_k$ converge as $k \to \infty$ towards a right triangular matrix such that $\lim_{k \to \infty} h_{ii}^{(k)} = \lambda_i$, $(i = 1, 2, \ldots, n)$ holds. If the matrix $\mathbf{H}$ has pairs of complex conjugate eigenvalues such that their absolute values and the absolute values of the real eigenvalues are pairwise different, and the (complex) LR decomposition of $\mathbf{X}^{-1}$ exists, then the matrices $\mathbf{H}_k$ converge towards a quasitriangular matrix (6.108).

Theorem 6.10 guarantees by restrictive assumptions the convergence of the sequence of matrices $\mathbf{H}_k$ (6.114) towards a quasitriangular matrix. However, the convergence of those matrix elements $h_{i+1,i}^{(k)}$ that do converge towards zero can be quite slow. Therefore the original and simple QR algorithm is usually not efficient enough due to the large number of iterations. However, the QR algorithm has a close relationship to the vector iteration method (Golub, 1983, Stewart, 1976), so that a corresponding analysis of the convergence behaviour of the subdiagonal elements shows that for large enough $k$ asymptotically we have

$$|h_{i+1,i}^{(k)}| \approx \left|\frac{\lambda_{i+1}}{\lambda_i}\right|^k \quad (i = 1, 2, \ldots, n-1). \tag{6.115}$$

If $|\lambda_{i+1}/\lambda_i| < 1$ then the absolute values of the subdiagonal elements $h_{i+1,i}^{(k)}$ converge towards zero like geometric sequences. The linear convergence is determined by the quotient of the moduli of successive eigenvalues which can be arbitrarily close to one. For a pair of complex conjugate eigenvalues $\lambda_i$, $\lambda_{i+1} = \bar{\lambda}_i$ the sequence $h_{i+1,i}^{(k)}$ cannot tend to zero because of (6.115).

However, the convergence result (6.115) gives us the hint that the linear convergence of certain subdiagonal elements can be improved considerably by an appropriate *spectral shift*. First of all we consider the case when $\mathbf{H}$ has only real eigenvalues and postpone the treatment of the case of complex eigenvalues. The matrix $\mathbf{H} - \sigma\mathbf{I}$ with $\sigma \in \mathbf{R}$ has eigenvalues $\lambda_i - \sigma$, $(i = 1, 2, \ldots, n)$. We assume that they are numbered so that $|\lambda_1 - \sigma| > |\lambda_2 - \sigma| > \cdots > |\lambda_n - \sigma|$ holds. For the sequence of matrices $\tilde{\mathbf{H}}_k$ (6.114) with $\tilde{\mathbf{H}}_1 = \mathbf{H} - \sigma\mathbf{I}$ we now have, because of (6.115),

$$|\tilde{h}_{i+1,i}^{(k)}| \approx \left|\frac{\lambda_{i+1} - \sigma}{\lambda_i - \sigma}\right|^k \quad (i = 1, 2, \ldots, n-1). \tag{6.116}$$

If $\sigma$ is a good approximation of $\lambda_n$ such that $|\lambda_n - \sigma| \ll |\lambda_i - \sigma|$, $i = 1, 2, \ldots, n-1$, then it is clear that $\tilde{h}_{n,n-1}^{(k)}$ tends to zero quite fast, and the matrix $\tilde{\mathbf{H}}_k$ becomes reducible after a few iteration steps.

The technique of the spectral shift is not only applied once but in each step of the QR algorithm, and the shift $\sigma_k$ is chosen in an appropriate way on the basis of the information available from $\mathbf{H}_k$. Thus instead of (6.114) we define

$$\boxed{\mathbf{H}_k - \sigma_k\mathbf{I} = \mathbf{Q}_k\mathbf{R}_k \quad \mathbf{H}_{k+1} = \mathbf{R}_k\mathbf{Q}_k + \sigma_k\mathbf{I} \quad (k = 1, 2, \ldots)} \tag{6.117}$$

to be the *QR algorithm with explicit shift*. The shift that is applied to $\mathbf{H}_k$ cancels out when $\mathbf{H}_{k+1}$ is computed, thus the matrices $\mathbf{H}_{k+1}$ and $\mathbf{H}_k$ are again orthogonally similar. On the basis of the relationship mentioned above of the QR algorithm with the vector iteration, the choice of the shift, $\sigma_k$, according to

$$\boxed{\sigma_k = h_{nn}^{(k)} \quad (k = 1, 2, \ldots)} \tag{6.118}$$

is appropriate (Stewart, 1973). Finally with this choice of $\sigma_k$ the sequence of the subdiagonal elements $h_{n,n-1}^{(k)}$ converges *quadratically* to zero (Stewart, 1973).

The obvious implementation of the QR algorithm (6.117), indicated by the proofs of Theorems 6.5 and 6.7, consists of first performing the QR decomposition of $\mathbf{H}_k - \sigma_k\mathbf{I}$ by means of $(n-1)$ rotation matrices, and then computing the matrix $\mathbf{H}_{k+1}$. If we proceed in this way the $(n-1)$ pairs of values $c_i := \cos\varphi_i$ and $s_i := \sin\varphi_i$ of the matrices $\mathbf{U}(i, i+1; \varphi_i) =: \mathbf{U}_i$ have to be stored. However, this is not necessary as will become clear from the following consideration. For a QR transformation the following relationships hold

$$\begin{aligned}\mathbf{U}_{n-1}^T \cdots \mathbf{U}_3^T\mathbf{U}_2^T\mathbf{U}_1^T(\mathbf{H}_k - \sigma_k\mathbf{I}) &= \mathbf{Q}_k^T(\mathbf{H}_k - \sigma_k\mathbf{I}) = \mathbf{R}_k \\ \mathbf{H}_{k+1} = \mathbf{R}_k\mathbf{Q}_k + \sigma_k\mathbf{I} &= \mathbf{R}_k\mathbf{U}_1\mathbf{U}_2\mathbf{U}_3 \cdots \mathbf{U}_{n-1} + \sigma_k\mathbf{I} \\ &= \mathbf{U}_{n-1}^T \cdots \mathbf{U}_3^T\mathbf{U}_2^T\mathbf{U}_1^T(\mathbf{H}_k - \sigma_k\mathbf{I})\mathbf{U}_1\mathbf{U}_2\mathbf{U}_3 \cdots \mathbf{U}_{n-1} + \sigma_k\mathbf{I}\end{aligned} \tag{6.119}$$

The associativity of the matrix multiplications means that we can choose the order of the operations, to compute $\mathbf{H}_{k+1}$ in (6.119), in a convenient manner. As soon as the two matrix multiplications $\mathbf{U}_2^T\mathbf{U}_1^T(\mathbf{H}_k - \sigma_k\mathbf{I})$ have been performed

## 6.4 QR Algorithm

in the course of the QR decomposition, the resulting matrix has the structure that is given in Figure 6.3 for $n = 6$.

At this stage we take into account that the subsequent multiplications from the left by $U_3^T$, $U_4^T$,... only act on the third and subsequent rows. This means that the first two rows and also the first two columns of the current matrix are not changed by the QR decomposition again. Consequently, the multiplication from the right by $U_1$, which causes linear combinations of the first and second columns, operates on the existing definite elements of the matrix $R_k$ of the QR decomposition. As soon as the multiplication by $U_3^T$ is executed the multiplication from the right by $U_2$ can be performed due to an analogous reasoning.

Hence the QR transformation (6.119) can be performed in an alternating manner such that in the general $i$th step ($2 \leq i \leq n-1$) the QR decomposition is carried out with respect to the $i$th and $(i+1)$th row while in the $(i-1)$th and $i$th columns the Hessenberg matrix $H_{k+1}$ is essentially formed. The subtraction of $\sigma_k$ from the diagonal elements of $H_k$ and the later addition of $\sigma_k$ to the diagonal elements can easily be included in the process. If we build the matrix $H_{k+1}$ at the place of $H_k$ and if we denote the continuously changing matrix by $H$, the QR step, (6.119), for a given value $\sigma$ can be summarized algorithmically in (6.120). $\delta$ denotes the smallest positive number of the computer such that $1 + \delta \neq 1$.

$$
\begin{aligned}
&h_{11} = h_{11} - \sigma \\
&\text{for } i = 1, 2, \ldots, n: \\
&\quad \text{if } i < n \text{ then} \\
&\quad\quad \text{if } |h_{ii}| < \delta \times |h_{i+1,i}| \text{ then} \\
&\quad\quad\quad w = |h_{i+1,i}|;\ c = 0;\ s = \operatorname{sgn}(h_{i+1,i}) \\
&\quad\quad \text{else} \\
&\quad\quad\quad w = \sqrt{h_{ii}^2 + h_{i+1,i}^2};\ c = h_{ii}/w;\ s = -h_{i+1,i}/w \\
&\quad\quad h_{ii} = w;\ h_{i+1,i} = 0;\ h_{i+1,i+1} = h_{i+1,i+1} - \sigma \\
&\quad\quad \text{for } j = i+1, i+2, \ldots, n: \\
&\quad\quad\quad g = c \times h_{ij} - s \times h_{i+1,j} \\
&\quad\quad\quad h_{i+1,j} = s \times h_{ij} + c \times h_{i+1,j};\ h_{ij} = g \\
&\quad \text{if } i > 1 \text{ then} \\
&\quad\quad \text{for } j = 1, 2, \ldots, i: \\
&\quad\quad\quad g = \tilde{c} \times h_{j,i-1} - \tilde{s} \times h_{ji} \\
&\quad\quad\quad h_{ji} = \tilde{s} \times h_{j,i-1} + \tilde{c} \times h_{ji};\ h_{j,i-1} = g \\
&\quad\quad h_{i-1,i-1} = h_{i-1,i-1} + \sigma \\
&\quad \tilde{c} = c;\ \tilde{s} = s \\
&h_{nn} = h_{nn} + \sigma
\end{aligned} \qquad (6.120)
$$

The subdiagonal elements of $H_k$ tend to zero, so we need a criterion for deciding when an element $|h_{i+1,i}^{(k)}|$ can be considered to be sufficiently small to

$$\begin{pmatrix} \times & \times & \times & \times & \times & \times & \times \\ 0 & \times & \times & \times & \times & \times & \times \\ 0 & 0 & \times & \times & \times & \times & \times \\ 0 & 0 & \times & \times & \times & \times & \times \\ 0 & 0 & 0 & \times & \times & \times & \times \\ 0 & 0 & 0 & 0 & \times & \times & \times \\ 0 & 0 & 0 & 0 & 0 & \times & \times \end{pmatrix}$$

Figure 6.3  Structure of $\mathbf{U}_2^T \mathbf{U}_1^T (\mathbf{H}_k - \sigma_k \mathbf{I})$

be set equal to zero. A safe criterion that also takes account of eigenvalues of $\mathbf{H}_1$ that are small in absolute value is

$$|h_{i+1,i}^{(k)}| < \delta \cdot \min\{|h_{ii}^{(k)}|, |h_{i+1,i+1}^{(k)}|\}. \tag{6.121}$$

As soon as a subdiagonal element $h_{i+1,i}^{(k)}$ satisfies the condition (6.121) the Hessenberg matrix $\mathbf{H}_k$ is reducible. The computation of the eigenvalues of $\mathbf{H}_k$ is reduced to the problem of determining the eigenvalues of the two submatrices of Hessenberg form. In the case considered of real eigenvalues of $\mathbf{H}_1$, and because of the applied strategy of choosing the spectral shifts (6.118), it is most probable that the subdiagonal element $h_{n,n-1}^{(k)}$ first satisfies the condition (6.121). The matrix $\mathbf{H}_k$ is reduced to a Hessenberg matrix $\hat{\mathbf{H}}$ of order $(n-1)$ and a matrix of the order one, which must necessarily contain an eigenvalue $\lambda$.

$$\mathbf{H}_k = \begin{pmatrix} X & X & X & X & X & | & X \\ X & X & X & X & X & | & X \\ 0 & X & X & X & X & | & X \\ 0 & 0 & X & X & X & | & X \\ 0 & 0 & 0 & X & X & | & X \\ \hline 0 & 0 & 0 & 0 & 0 & | & \lambda \end{pmatrix} = \left( \begin{array}{c|c} \hat{\mathbf{H}} & \hat{\mathbf{h}} \\ \hline \mathbf{0}^T & \lambda \end{array} \right) \tag{6.122}$$

As soon as the situation (6.122) occurs the QR algorithm is continued with the submatrix $\hat{\mathbf{H}} \in \mathbf{R}^{(n-1) \times (n-1)}$ which has the remaining eigenvalues. Here an essential advantage of the QR algorithm becomes apparent. The order of the matrix is reduced with each computed eigenvalue. This is not the case for Hyman's method. There we had to work with the same given Hessenberg matrix, and the deflation of a computed eigenvalue was only performed implicitly.

*Example 6.7* The eigenvalues of the Hessenberg matrix $\mathbf{H}$ (6.68) are computed by means of the QR algorithm. The spectral shifts $\sigma_k$ for the QR steps are determined according to the rule (6.124) given in the following section. The first six QR steps yield the following interesting values.

## 6.4 QR Algorithm

| k | $\sigma_k$ | $h_{66}^{(k+1)}$ | $h_{65}^{(k+1)}$ | $h_{55}^{(k+1)}$ |
|---|---|---|---|---|
| 1 | 2.342 469 183 | 2.358 475 293 | $-1.656\,19 \times 10^{-1}$ | 5.351 616 732 |
| 2 | 2.357 470 399 | 2.894 439 187 | $-4.692\,53 \times 10^{-2}$ | 0.304 289 831 |
| 3 | 2.780 738 412 | 2.984 399 174 | $-7.645\,77 \times 10^{-3}$ | 1.764 372 360 |
| 4 | 2.969 415 975 | 2.999 399 624 | $-1.041\,38 \times 10^{-4}$ | 5.056 039 373 |
| 5 | 2.999 618 708 | 2.999 999 797 | $-3.908\,52 \times 10^{-8}$ | 3.998 363 523 |
| 6 | 3.000 000 026 | 3.000 000 000 | $1.152\,98 \times 10^{-15}$ | 3.871 195 470 |

We observe that the values $\sigma_k$ and $h_{66}^{(k+1)}$ converge towards an eigenvalue $\lambda_1 = 3$ and that $h_{65}^{(k+1)}$ converges quadratically to zero. After the sixth QR step the condition (6.121) is satisfied with $\delta = 10^{-12}$. The submatrix $\hat{\mathbf{H}}$ of order five in $\mathbf{H}_7$ is approximately

$$\hat{\mathbf{H}} \doteq \begin{pmatrix} 4.991\,623 & 5.988\,209 & -3.490\,458 & -5.233\,181 & 1.236\,387 \\ -5.982\,847 & 5.021\,476 & 6.488\,094 & 7.053\,759 & -16.317\,615 \\ 0 & -0.033\,423 & 0.160\,555 & 5.243\,584 & 11.907\,163 \\ 0 & 0 & -0.870\,774 & 1.955\,150 & -1.175\,364 \\ 0 & 0 & 0 & -0.107\,481 & 3.871\,195 \end{pmatrix}$$

We observe that the subdiagonal elements of $\hat{\mathbf{H}}$ have decreased in absolute value in comparison with those of $\mathbf{H}$ (6.68). Therefore the shift $\sigma_7$ of the following QR step is in fact a good approximation of another real eigenvalue. The most interesting values of the three subsequent QR steps for $\hat{\mathbf{H}}$ are

| k | $\sigma_k$ | $h_{55}^{(k+1)}$ | $h_{54}^{(k+1)}$ | $h_{44}^{(k+1)}$ |
|---|---|---|---|---|
| 7 | 3.935 002 781 | 4.003 282 759 | $-2.367\,60 \times 10^{-3}$ | 0.782 074 |
| 8 | 4.000 557 588 | 4.000 000 470 | $2.911\,08 \times 10^{-7}$ | $-0.391\,982$ |
| 9 | 4.000 000 053 | 4.000 000 000 | $-2.299\,20 \times 10^{-15}$ | $-1.506\,178$ |

The condition (6.121) is satisfied, and we obtain the eigenvalue $\lambda_2 = 4$ and the submatrix of order four

$$\hat{\hat{\mathbf{H}}} \doteq \begin{pmatrix} 4.993\,897 & 5.997\,724 & -10.776\,453 & 2.629\,846 \\ -6.000\,388 & 4.999\,678 & -1.126\,468 & -0.830\,809 \\ 0 & -0.003\,201 & 3.512\,604 & 2.722\,802 \\ 0 & 0 & -3.777\,837 & -1.506\,178 \end{pmatrix}$$

The eigenvalues of $\hat{\hat{\mathbf{H}}}$ are pairwise conjugate. They will be computed in the following section in Example 6.8.

### 6.4.3 QR double step, complex eigenvalues

Let the real Hessenberg matrix $\mathbf{H} = \mathbf{H}_1$ also possess pairs of complex conjugate eigenvalues. The spectral shifts (6.118) $\sigma_k = h_{nn}^{(k)}$ are obviously not useful approximations of complex eigenvalues. Hence their choice must be adapted, and the submatrix of order two in the bottom right-hand corner of $\mathbf{H}_k$ is considered

$$\mathbf{C}_k := \begin{pmatrix} h_{n-1,n-1}^{(k)} & h_{n-1,n}^{(k)} \\ h_{n,n-1}^{(k)} & h_{n,n}^{(k)} \end{pmatrix}. \tag{6.123}$$

If the eigenvalues $\mu_1^{(k)}$ and $\mu_2^{(k)}$ of $\mathbf{C}_k$ are real, instead of using (6.118), the spectral shift $\sigma_k$ is fixed by that eigenvalue $\mu_1^{(k)}$ which is nearest to $h_{nn}^{(k)}$, that is we define

$$\boxed{\sigma_k = \mu_1^{(k)} \in \mathbf{R} \quad \text{where} \quad |\mu_1^{(k)} - h_{nn}^{(k)}| \leqslant |\mu_2^{(k)} - h_{nn}^{(k)}| \quad (k = 1, 2, \ldots)} \tag{6.124}$$

However, if the eigenvalues of $\mathbf{C}_k$ are complex conjugate, the spectral shifts for the two following QR transformations are fixed by the complex conjugate values

$$\boxed{\sigma_k = \mu_1^{(k)} \quad \sigma_{k+1} = \mu_2^{(k)} = \bar{\sigma}_k \quad \mu_1^{(k)} \in \mathbf{C}} \tag{6.125}$$

The matrix $\mathbf{H}_k - \sigma_k \mathbf{I}$ has, because of (6.125), complex diagonal elements. Theorem 6.4 can be generalized to complex matrices $\mathbf{A}$. Therefore we have the factorization

$$\mathbf{A} = \mathbf{U}\mathbf{R} \tag{6.126}$$

where $\mathbf{U}$ denotes a *unitary matrix* with $\mathbf{U}^H \mathbf{U} = \mathbf{I}$, $\mathbf{U}^H = \bar{\mathbf{U}}^T$, and $\mathbf{R}$ is a complex right triangular matrix with real diagonal elements. The unitary QR decomposition (6.126) will only be needed for a short theoretical consideration but not for practical use.

The two steps of the QR algorithm using the explicit spectral shifts $\sigma_k$ and $\sigma_{k+1}$, according to (6.125), read with the unitary decompositions (6.126), as follows

$$\mathbf{H}_k - \sigma_k \mathbf{I} = \mathbf{U}_k \mathbf{R}_k \qquad \mathbf{H}_{k+1} = \mathbf{R}_k \mathbf{U}_k + \sigma_k \mathbf{I}, \tag{6.127}$$

$$\mathbf{H}_{k+1} - \sigma_{k+1} \mathbf{I} = \mathbf{U}_{k+1} \mathbf{R}_{k+1} \qquad \mathbf{H}_{k+2} = \mathbf{R}_{k+1} \mathbf{U}_{k+1} + \sigma_{k+1} \mathbf{I}. \tag{6.128}$$

In the generalization of Theorem 6.6, the matrix $\mathbf{H}_{k+2}$ is unitarily similar to $\mathbf{H}_k$, and therefore we have

$$\mathbf{H}_{k+2} = \mathbf{U}_{k+1}^H \mathbf{U}_k^H \mathbf{H}_k \mathbf{U}_k \mathbf{U}_{k+1} = (\mathbf{U}_k \mathbf{U}_{k+1})^H \mathbf{H}_k (\mathbf{U}_k \mathbf{U}_{k+1}). \tag{6.129}$$

From (6.128) and (6.127) for the product of the matrices we obtain

## 6.4 QR Algorithm

$$\begin{aligned}
\mathbf{U}_k \mathbf{U}_{k+1} \mathbf{R}_{k+1} \mathbf{R}_k &= \mathbf{U}_k (\mathbf{H}_{k+1} - \sigma_{k+1} \mathbf{I}) \mathbf{R}_k \\
&= \mathbf{U}_k (\mathbf{H}_{k+1} - \sigma_{k+1} \mathbf{I}) \mathbf{U}_k^H (\mathbf{H}_k - \sigma_k \mathbf{I}) \\
&= \mathbf{U}_k (\mathbf{R}_k \mathbf{U}_k + \sigma_k \mathbf{I} - \sigma_{k+1} \mathbf{I}) \mathbf{U}_k^H (\mathbf{H}_k - \sigma_k \mathbf{I}) \\
&= (\mathbf{H}_k - \sigma_{k+1} \mathbf{I})(\mathbf{H}_k - \sigma_k \mathbf{I}) \\
&= \mathbf{H}_k^2 - (\sigma_k + \sigma_{k+1}) \mathbf{H}_k + \sigma_k \sigma_{k+1} \mathbf{I}.
\end{aligned} \quad (6.130)$$

The two quantities

$$\boxed{\begin{aligned}
s &:= \sigma_k + \sigma_{k+1} = h^{(k)}_{n-1,n-1} + h^{(k)}_{n,n} \\
t &:= \sigma_k \cdot \sigma_{k+1} = h^{(k)}_{n-1,n-1} h^{(k)}_{n,n} - h^{(k)}_{n-1,n} h^{(k)}_{n,n-1}
\end{aligned}} \quad (6.131)$$

are real, from (6.125), and therefore the matrix (6.130)

$$\mathbf{X} := \mathbf{H}_k^2 - s \mathbf{H}_k + t \mathbf{I} \quad (6.132)$$

is also real. As a consequence of (6.130) $(\mathbf{U}_k \mathbf{U}_{k+1})(\mathbf{R}_{k+1} \mathbf{R}_k) = \mathbf{X}$ represents a unitary QR decomposition of the real matrix $\mathbf{X}$. However, each real matrix $\mathbf{X}$ possesses a real QR decomposition (6.100). Therefore it must be possible to choose the unitary matrices $\mathbf{U}_k$ and $\mathbf{U}_{k+1}$ in such a way that their product $\mathbf{U}_k \mathbf{U}_{k+1} =: \mathbf{Q}$ is an orthogonal matrix. According to (6.129) the matrix $\mathbf{H}_{k+2}$ defined by the *QR double step* (6.127) and (6.128) is again a real Hessenberg matrix.

In order to avoid complex arithmetic, a possible procedure for computing $\mathbf{H}_{k+2}$ from $\mathbf{H}_k$ seems to consist of the following three steps: (1) Compute the matrix $\mathbf{X}$ following (6.132); (2) Determine the QR decomposition $\mathbf{X} = \mathbf{QR}$; (3) Compute $\mathbf{H}_{k+2} = \mathbf{Q}^T \mathbf{H}_k \mathbf{Q}$. However, this procedure is too expensive because the computation of $\mathbf{X}$ requires about $n^3/6$ multiplications. An efficient method for computing $\mathbf{H}_{k+2}$ from $\mathbf{H}_k$ is based on the following theorem.

*Theorem 6.11* The orthogonal similarity transformation of a matrix $\mathbf{A}$ into a Hessenberg matrix $\mathbf{H} = \mathbf{Q}^T \mathbf{A} \mathbf{Q}$ is uniquely determined by the first column of $\mathbf{Q}$, if the subdiagonal elements $h_{i+1,i}$ $(i = 1, 2, \ldots, n-1)$ of the irreducible matrix $\mathbf{H}$ are positive.

*Proof* Let $\mathbf{q}_k$ denote the $k$th column of the orthogonal matrix $\mathbf{Q}$. From the hypothesis $\mathbf{q}_1$ is known. We want to show by induction, with respect to the index $k$, that the $(k+1)$th column $\mathbf{q}_{k+1}$ and the matrix elements $h_{ik}$, $(i = 1, 2, \ldots, k+1)$ of the $k$th column of $\mathbf{H}$ are uniquely determined on the assumption that the columns $\mathbf{q}_1, \mathbf{q}_2, \ldots, \mathbf{q}_k$ are known. The orthogonal similarity of $\mathbf{A}$ and $\mathbf{H}$ is equivalent to the matrix equation $\mathbf{QH} = \mathbf{AQ}$. Equating the $k$th column of both sides gives the equation

$$h_{1k} \mathbf{q}_1 + h_{2k} \mathbf{q}_2 + \cdots + h_{kk} \mathbf{q}_k + h_{k+1,k} \mathbf{q}_{k+1} = \mathbf{A} \mathbf{q}_k. \quad (6.133)$$

The vectors $q_i$ are orthonormal, that is $q_i^T q_j = \delta_{ij}$, thus from (6.133), after left multiplication by $q_i^T$, for $i = 1, 2, \ldots, k$, we obtain the matrix elements $h_{ik}$ uniquely defined by

$$h_{ik} = q_i^T A q_k \quad (i = 1, 2, \ldots, k).$$

The vector $q_{k+1}$ is given, according to (6.133) by

$$q_{k+1} = \frac{1}{h_{k+1,k}} \left( A q_k - \sum_{i=1}^{k} h_{ik} q_i \right).$$

The value of $h_{k+1,k}$, which is assumed to be strictly positive, is uniquely determined together with $q_{k+1}$ by the condition $q_{k+1}^T q_{k+1} = 1$. For $k = n$ we have $q_{n+1} = 0$ in (6.133), and in this case the uniqueness of the matrix elements $h_{in}$ of the last column of $H$ follows from the previous argument.

The requirement of positive subdiagonal elements is necessary for uniqueness because $q_{k+1}$ and $h_{k+1,k}$ are distinguished only by the sign. In the following we shall apply Theorem 6.11 only in the sense that the Hessenberg matrix, $H$, and the orthogonal matrix, $Q$, are essentially determined by the first column of $Q$, that is apart from certain signs.

To apply the statement of Theorem 6.11 to the direct computation $H_{k+2} = Q^T H_k Q$ we observe the following essential facts. First of all, the orthogonal matrix $Q$ is defined by the QR decomposition of $X$. However, the first column of $Q$ is essentially fixed by the first column of $X$. To see this let $Q_0$ be that orthogonal matrix which only reduces the first column of $X$ to the desired form in the course of the QR decomposition, so that we have

$$Q_0^T X = \left( \begin{array}{c|c} r_{11} & g^T \\ \hline 0 & \tilde{X} \end{array} \right). \tag{6.134}$$

The subsequent Givens rotations, (6.102), of the QR decomposition no longer change the first column of $Q_0$. The second point is that the orthogonal matrix $\tilde{Q}$ which transforms an arbitrary matrix $A$ into Hessenberg form has the property that its first column, $\tilde{q}_1$, is equal to the unit vector $e_1$. In fact, this is an immediate consequence of the representation of $\tilde{Q} = U_1 \cdot U_2 \cdots U_{N^*}$ as the product of the rotation matrices, $U_k$, with the rotation index pairs (6.59) whereby the index one does not occur. Therefore the first column $q_1$ of the product matrix

$$Q := Q_0 \tilde{Q} \tag{6.135}$$

is equal to the first column of $Q_0$. This implies the following procedure: the given Hessenberg matrix $H_k$ is subjected to an orthogonal similarity transformation by $Q_0$ in a first step which yields

$$B = Q_0^T H_k Q_0. \tag{6.136}$$

In the next step the matrix $B$ is transformed into Hessenberg form again by

## 6.4 QR Algorithm

means of $\tilde{\mathbf{Q}}$ so that we have

$$\mathbf{H}_{k+2} = \tilde{\mathbf{Q}}^T \mathbf{B} \tilde{\mathbf{Q}} = \tilde{\mathbf{Q}}^T \mathbf{Q}_0^T \mathbf{H}_k \mathbf{Q}_0 \tilde{\mathbf{Q}} = \mathbf{Q}^T \mathbf{H}_k \mathbf{Q}. \tag{6.137}$$

The essential uniqueness means that the desired transformation has been achieved.

To perform the sketched procedure we first have to fix the matrix $\mathbf{Q}_0$ by the elements of the first column of $\mathbf{X}$ defined by (6.132). $\mathbf{H}_k$ is a Hessenberg matrix so only the first three elements, denoted by $x_1, x_2$ and $x_3$, of the first column of $\mathbf{X}$ are nonzero. From (6.132) they are defined as follows

$$\boxed{\begin{aligned} x_1 &= h_{11}^2 + h_{12} h_{21} - s h_{11} + t \\ x_2 &= h_{21}(h_{11} + h_{22} - s) \\ x_3 &= h_{21} h_{32} \end{aligned}} \tag{6.138}$$

This special structure of $\mathbf{X}$ means that the matrix $\mathbf{Q}_0$ can be defined as the product of two rotation matrices

$$\mathbf{Q}_0 = \mathbf{U}(1,2;\varphi_1) \mathbf{U}(1,3;\varphi_2) \tag{6.139}$$

with angles $\varphi_1$ and $\varphi_2$ that are determined from the pairs of values $(x_1, x_2)$ and $(x'_1, x_3)$ according to (6.104), where $x'_1$ denotes the new value of $x_1$ after the first transformation step.

The similarity transformation (6.136) of $\mathbf{H}_k$ by $\mathbf{Q}_0$ only changes the first three rows and columns of $\mathbf{H}_k$ so that the matrix $\mathbf{B}$ has the following special structure, for instance for $n = 7$:

$$\mathbf{B} = \mathbf{Q}_0^T \mathbf{H}_k \mathbf{Q}_0 = \begin{pmatrix} X & X & X & X & X & X & X \\ X & X & X & X & X & X & X \\ \boxed{X} & X & X & X & X & X & X \\ \boxed{X} & 0 & X & X & X & X & X \\ 0 & 0 & 0 & X & X & X & X \\ 0 & 0 & 0 & 0 & X & X & X \\ 0 & 0 & 0 & 0 & 0 & X & X \end{pmatrix} \tag{6.140}$$

The Hessenberg form has been destroyed only with respect to the first column of $\mathbf{B}$ where in general, two nonzero matrix elements have been produced below the subdiagonal. The matrix $\mathbf{B}$ must be transformed similarly into Hessenberg form again. This can be achieved by the appropriate Givens rotations of Section 6.3.1, by taking into account the special structure of $\mathbf{B}$. The treatment of the first column of $\mathbf{B}$ requires only two Givens rotations with matrices $\mathbf{U}_{23} = \mathbf{U}(2,3;\varphi_3)$ and $\mathbf{U}_{24} = \mathbf{U}(2,4;\varphi_4)$. The first and second rotations cause linear combinations of the second and third rows and columns, and second and fourth rows and columns, respectively, thus the transformed matrix will have the following structure for $n = 7$:

$$\mathbf{B}_1 := \mathbf{U}_{24}^T \mathbf{U}_{23}^T \mathbf{B} \mathbf{U}_{23} \mathbf{U}_{24} = \begin{pmatrix} X & X & X & X & X & X & X \\ X & X & X & X & X & X & X \\ 0 & X & X & X & X & X & X \\ 0 & \boxed{X} & X & X & X & X & X \\ 0 & \boxed{X} & 0 & X & X & X & X \\ 0 & 0 & 0 & 0 & X & X & X \\ 0 & 0 & 0 & 0 & 0 & X & X \end{pmatrix} \qquad (6.141)$$

The two nonvanishing matrix elements of the first column of $\mathbf{B}$ below the subdiagonal are changed by the transformation (6.141) into two other nonvanishing matrix elements of the second column of $\mathbf{B}_1$ below the subdiagonal. Apart from the second column $\mathbf{B}_1$ has Hessenberg structure. Pairs of similar Givens transformations will shift the two elements below the subdiagonal successively downwards to the right until the Hessenberg form is set up once again. The treatment of the $(n-2)$th column requires only one Givens rotation, of course.

This completes the description of the computational details for computing $\mathbf{H}_{k+2}$ from $\mathbf{H}_k$ by a so-called QR double step with the complex conjugate spectral shifts $\sigma_k$ and $\sigma_{k+1} = \bar{\sigma}_k$ following (6.125). The two shifts $\sigma_k$ and $\sigma_{k+1}$ are nowhere explicitly performed. They are only used by the values $s$ and $t$, given by (6.131), and the values $x_1$, $x_2$ and $x_3$ (6.138) to fix the first column of $\mathbf{Q}$. This means that the shifts are contained implicitly in the orthogonal matrix $\mathbf{Q}_0$. Therefore the transition from $\mathbf{H}_k$ to $\mathbf{H}_{k+2}$ is called a *QR double step with implicit spectral shifts* (Francis, 1961).

The amount of computational effort required for a QR double step is composed of computing $\mathbf{B}$, (6.140), and of transforming $\mathbf{B}$ into $\mathbf{H}_{k+2}$. The preparation of the values $c = \cos \varphi$ and $s = \sin \varphi$ of the two rotation matrices for $\mathbf{Q}_0$ requires eight multiplicative operations and two square roots. The execution of the two Givens transformations requires $8n + 24$ multiplications. The same holds for the two transformations that are needed for the treatment of the $j$th column if the structure is taken into account. The last step requires only half of the operations thus, the amount of computational effort required for a QR double step consists of about

$$Z_{\text{QR double}} \simeq 8n^2 + 20n - 44 \qquad (6.142)$$

essential operations and $(2n-3)$ square roots. It is doubled in comparison to the amount of effort (6.107) of a single QR step with explicit spectral shift if the order $n$ is large. The amount of computational effort of a double step can be reduced if either the fast Givens transformation (Rath, 1982) or the *Householder transformation* (Golub, 1983; Stewart, 1973; Wilkinson and Reinsch, 1971) is used. At best the amount of work is $Z_{\text{QR double}} \approx 5n^2$.

## 6.4 QR Algorithm

The algorithmic implementation of the QR double step in the form described must take into account the structure of the given Hessenberg matrix, because we have to consider the fact that below the subdiagonal values are stored that are needed for the back transformation of eigenvectors. In the program (6.143) we denote the two matrix elements below the subdiagonal by $x_2$ and $x_3$. Moreover, we can take into account that when the element $h_{j+2,j} = x_2$ of the $j$th column is eliminated the new nonvanishing matrix element $h_{j+3,j+1} =: x_2$ is produced. The same holds for $h_{j+3,j} = x_3$. The matrix $\mathbf{H}_{k+2}$ is generated in place of $\mathbf{H}_k$ and is denoted by $\mathbf{H}$. The transformations by $\mathbf{Q}_0$ and by the subsequent matrices $\mathbf{U}(p, q; \varphi)$ are combined. However, this requires some appropriate assignments of values and distinction of cases. Let $\delta$ denote the smallest positive number of the computer such that $1 + \delta \neq 1$.

$s = h_{n-1,n-1} + h_{n,n}; \; t = h_{n-1,n-1} \times h_{n,n} - h_{n-1,n} \times h_{n,n-1}$
$x_1 = (h_{11} - s) \times h_{11} + h_{12} \times h_{21} + t$
$x_2 = h_{21} \times (h_{11} + h_{22} - s); \; x_3 = h_{21} \times h_{32}$
for $p = 1, 2, \ldots, n - 1$:
  for $i = 2, 3$:
    if $x_i \neq 0$ then
      if $|x_1| < \delta \times |x_i|$ then
        $w = -x_i; \; c = 0; \; s = 1$
      else
        $w = \sqrt{x_1^2 + x_i^2}; \; c = x_1/w; \; s = -x_i/w$
      $x_1 = w; \; q = p + i - 1$
      for $k = q - 1, q, \ldots, n$:
        $g = c \times h_{pk} - s \times h_{qk}$
        $h_{qk} = s \times h_{pk} + c \times h_{qk}; \; h_{pk} = g$
      if $i = 3$ then
        $g = c \times h_{pp} - s \times x_2; \; x_2 = s \times h_{pp} + c \times x_2; \; h_{pp} = g$
      for $j = 1, 2, \ldots, p + 1$:
        $g = c \times h_{jp} - s \times h_{jq}$
        $h_{jq} = s \times h_{jp} + c \times h_{jq}; \; h_{jp} = g$
      if $i = 2 \wedge p < n - 1$ then
        $x_2 = -s \times h_{p+2,q}; \; h_{p+2,q} = c \times h_{p+2,q}$
      if $i = 3$ then
        $g = c \times x_2 - s \times h_{qq}; \; h_{qq} = s \times x_2 + c \times h_{qq}; \; x_2 = g.$
        if $p < n - 2$ then
          $x_3 = -s \times h_{p+3,q}; \; h_{p+3,q} = c \times h_{p+3,q}$
        else
          $x_3 = 0$
  if $p > 1$ then $h_{p,p-1} = x_1$
  $x_1 = h_{p+1,p}$

(6.143)

*Example 6.8* The two eigenvalues of the submatrix $\mathbf{C}_k$ (6.123) of the Hessenberg matrix $\hat{\mathbf{H}}$ of Example 6.7 are complex conjugate. With $s_1 \doteq 2.006\,425$ and $t_1 \doteq 4.995\,697$ the QR double step with implicit spectral shift yields the matrix

$$\hat{\mathbf{H}}_2 \doteq \begin{pmatrix} 4.999\,999 & -5.999\,999 & -3.368\,568 & 2.979\,044 \\ 5.999\,997 & 5.000\,002 & -9.159\,566 & 4.566\,268 \\ 0 & 2.292 \times 10^{-6} & 1.775\,362 & -5.690\,920 \\ 0 & 0 & 0.808\,514 & 0.224\,637 \end{pmatrix}.$$

After another QR double step with $s_2 \doteq 1.999\,999$ and $t_2 \doteq 5.000\,001$ the resulting Hessenberg matrix is reducible:

$$\hat{\mathbf{H}}_3 \doteq \begin{pmatrix} 5.000\,000 & -6.000\,000 & -6.320\,708 & 5.447\,002 \\ 6.000\,000 & 5.000\,000 & 4.599\,706 & -5.847\,384 \\ 0 & 0 & 0.296\,328 & -5.712\,541 \\ 0 & 0 & 0.786\,892 & 1.703\,672 \end{pmatrix}.$$

From this the two pairs of complex conjugate eigenvalues $\lambda_{3,4} = 1 \pm 2i$ and $\lambda_{5,6} = 5 \pm 6i$ are computed.

### 6.4.4 QR algorithm for tridiagonal matrices

In Section 6.3.2 it is shown how a symmetric matrix $\mathbf{A}$ can be reduced by an orthogonal similarity transformation to a symmetric and tridiagonal matrix

$$\mathbf{J} = \begin{pmatrix} \alpha_1 & \beta_1 & & & & \\ \beta_1 & \alpha_2 & \beta_2 & & & \\ & \beta_2 & \alpha_3 & \beta_3 & & \\ & & \ddots & \ddots & \ddots & \\ & & & \beta_{n-2} & \alpha_{n-1} & \beta_{n-1} \\ & & & & \beta_{n-1} & \alpha_n \end{pmatrix} \qquad (6.144)$$

We assume that the matrix $\mathbf{J}$ is irreducible so that $\beta_i \neq 0$ for $i = 1, 2, \ldots, (n-1)$. Following Theorem 6.8 the QR algorithm generates a sequence of orthogonally similar tridiagonal matrices $\mathbf{J}_{k+1} = \mathbf{Q}_k^T \mathbf{J}_k \mathbf{Q}_k$, $\mathbf{J}_1 = \mathbf{J}$, for $k = 1, 2, \ldots$. Hence, a very efficient procedure will result for computing all eigenvalues of $\mathbf{J}$. The spectral shift $\sigma_k$ could be chosen, according to (6.118), as $\sigma_k = \alpha_n^{(k)}$. However, choosing the shift $\sigma_k$ as the one of the two real eigenvalues of the submatrix

$$\mathbf{C}_k := \begin{pmatrix} \alpha_{n-1}^{(k)} & \beta_{n-1}^{(k)} \\ \beta_{n-1}^{(k)} & \alpha_n^{(k)} \end{pmatrix} \qquad (6.145)$$

which is nearer to $\alpha_n^{(k)}$ has some advantages concerning the convergence

## 6.4 QR Algorithm

properties of the algorithm. The eigenvalues of $C_k$ are given by

$$\mu_{1,2}^{(k)} = \frac{\alpha_{n-1}^{(k)} + \alpha_n^{(k)}}{2} \pm \sqrt{\left(\frac{\alpha_{n-1}^{(k)} - \alpha_n^{(k)}}{2}\right)^2 + \beta_{n-1}^{(k)2}}$$

$$= \alpha_n^{(k)} + d \pm \sqrt{d^2 + \beta_{n-1}^{(k)2}} \quad d = \frac{\alpha_{n-1}^{(k)} - \alpha_n^{(k)}}{2}.$$

We obtain the desired eigenvalue that is nearer to $\alpha_n^{(k)}$ if the opposite sign of the square root is taken to that of $d$. If by chance $d = 0$, both eigenvalues $\mu_{1,2}^{(k)}$ are the same distance from $\alpha_n^{(k)}$, and we may choose $\sigma_k = \alpha_n^{(k)}$. Thus the spectral shift is given by the formula

$$\boxed{\sigma_k = \alpha_n^{(k)} + d - \mathrm{sgn}\,(d)\sqrt{d^2 + \beta_{n-1}^{(k)2}} \quad d = \tfrac{1}{2}(\alpha_{n-1}^{(k)} - \alpha_n^{(k)})} \qquad (6.146)$$

However, if $\alpha_n^{(k)} = \alpha_{n-1}^{(k)} = 0$ then $d = 0$, and the shift must be taken as $\sigma_k = \beta_{n-1}^{(k)}$ in this exceptional case.

The QR step $J_k - \sigma_k I = Q_k R_k$, $J_{k+1} = R_k Q_k + \sigma_k I$ with $J_{k+1} = Q_k^T J_k Q_k$ is performed by applying the technique of the implicit spectral shift. The considerations of Section 6.4.3 are simplified and applied in a similar way. From Theorem 6.11 we obtain, by specializing, the following theorem.

*Theorem 6.12 The orthogonal similarity transformation of a symmetric matrix $A$ into a tridiagonal matrix $J = Q^T A Q$ is uniquely determined by the first column of $Q$ if the subdiagonal elements $\beta_i$, $(i = 1, 2, \ldots, n - 1)$ of the irreducible matrix $J$ are positive.*

The matrix $Q_0$, which is used to transform the first column of $J_k - \sigma_k I$ in the course of the QR decomposition, is in the present situation a Jacobi matrix $U(1, 2; \varphi_0)$ whose values $c = \cos \varphi_0$ and $s = \sin \varphi_0$ are determined by the two elements $\alpha_1 - \sigma_k$ and $\beta_1$. As before we form the orthogonally similar matrix $B = Q_0^T J_k Q_0$ which must be subsequently reduced to a tridiagonal matrix by an appropriate sequence of orthogonal similarity transformations. The matrix $B$ has quite a simple structure because only the first two rows and columns of $J_k$ are changed. Suppressing the upper index $k$, the matrix $B$ for $n = 6$ is

$$B = Q_0^T J_k Q_0 = \begin{pmatrix} \alpha_1' & \beta_1' & y & & & \\ \beta_1' & \alpha_2' & \beta_2' & & & \\ y & \beta_2' & \alpha_3 & \beta_3 & & \\ & & \beta_3 & \alpha_4 & \beta_4 & \\ & & & \beta_4 & \alpha_5 & \beta_5 \\ & & & & \beta_5 & \alpha_6 \end{pmatrix} \qquad (6.147)$$

The tridiagonal structure is only destroyed with respect to the first row and column. A Givens rotation with a matrix $\mathbf{U}_1 = \mathbf{U}(2, 3; \varphi_1)$ eliminates the element $y$ in (6.147) if the angle $\varphi_1$ is chosen properly, but generates a new pair of nonvanishing elements outside the three diagonals. The result is

$$\mathbf{U}_1^T \mathbf{B} \mathbf{U}_1 = \begin{pmatrix} \alpha_1' & \beta_1'' & & & & \\ \beta_1'' & \alpha_2'' & \beta_2'' & y' & & \\ & \beta_2'' & \alpha_3' & \beta_3' & & \\ & y' & \beta_3' & \alpha_4 & \beta_4 & \\ & & & \beta_4 & \alpha_5 & \beta_5 \\ & & & & \beta_5 & \alpha_6 \end{pmatrix} \qquad (6.148)$$

Each subsequent Givens rotation shifts the disturbing element downwards to the right. After $(n-2)$ such transformation steps the matrix $\mathbf{B}$ is reduced to a tridiagonal form and hence represents the desired matrix $\mathbf{J}_{k+1}$.

A QR step with implicit spectral shift $\sigma$ for a tridiagonal matrix $\mathbf{J}$ is suitably implemented in such a manner that the elements of the two diagonals are stored as two vectors $\boldsymbol{\alpha} = (\alpha_1, \alpha_2, \ldots, \alpha_n)^T$ and $\boldsymbol{\beta} = (\beta_1, \beta_2, \ldots, \beta_{n-1})^T$ with which the transformation from $\mathbf{J}_k$ into $\mathbf{J}_{k+1}$ is performed. Furthermore we see that all $(n-1)$ Givens rotations change two consecutive rows and columns. The diagonal elements and the corresponding off-diagonal elements can be obtained according to the formulae (6.17), (6.18) and (6.19). The representation (6.26) is not applicable because the angle $\varphi$ is fixed differently. The mentioned formulae can be changed as follows to reduce the number of operations. We set $c = \cos \varphi$, $s = \sin \varphi$ and $q = p + 1$ to obtain

$$\alpha_p'' = \alpha_p - 2\beta_p cs - (\alpha_p - \alpha_{p+1})s^2 = \alpha_p - z$$
$$\alpha_{p+1}'' = \alpha_{p+1} + 2\beta_p cs + (\alpha_p - \alpha_{p+1})s^2 = \alpha_{p+1} + z$$
$$\beta_p'' = (\alpha_p - \alpha_{p+1})cs + \beta_p(c^2 - s^2)$$

with

$$z = [2\beta_p c + (\alpha_p - \alpha_{p+1})s]s$$

Finally, the rotation angle of the transformation matrix $\mathbf{Q}_0$ is determined differently than that of the subsequent Givens rotations $\mathbf{U}_k$. To unify the situation in the following algorithmic description of a QR step, the variables $x$ and $y$ are adequately defined. The spectral shift $\sigma$ is assumed to be given according to (6.146).

## 6.4 QR Algorithm

```
x = α₁ - σ; y = β₁
for p = 1, 2, ..., n - 1:
    if |x| ≤ δ × |y| then
        w = -y; c = 0; s = 1
    else
        w = √(x² + y²); c = x/w; s = -y/w
    d = αₚ - αₚ₊₁; z = (2 × c × βₚ + d × s) × s
    αₚ = αₚ - z; αₚ₊₁ = αₚ₊₁ + z
    βₚ = d × c × s + (c² - s²) × βₚ; x = βₚ
    if p > 1 then βₚ₋₁ = w
    if p < n - 1 then y = -s × βₚ₊₁; βₚ₊₁ = c × βₚ₊₁
```
(6.149)

A QR step for a symmetric, tridiagonal matrix $\mathbf{J}_k$ of order $n$, with the algorithm (6.149), requires about

$$Z_{QR,\text{trid}} \simeq 15(n-1) \qquad (6.150)$$

essential operations and $(n-1)$ square roots. The high efficiency of the procedure is based on the fact that the last off-diagonal element $\beta_{n-1}^{(k)}$ converges *cubically* to zero (Gourlay, 1973; Wilkinson, 1968). This fact has the consequence that the computation of all eigenvalues of a tridiagonal matrix of large order $n$ requires, on average, only two or three QR steps per eigenvalue because the spectral shifts (6.146) become excellent approximations of the next eigenvalue as the computation proceeds. Moreover, the order of the tridiagonal matrices to be treated decreases.

*Example 6.9* The QR algorithm is indeed efficient for computing all eigenvalues of the tridiagonal matrix $\mathbf{J}$, (6.71), of Example 6.4. With the spectral shifts $\sigma_k$, (6.146), three QR steps yield the following essential values:

| $k$ | $\sigma_k$ | $\alpha_5^{(k+1)}$ | $\beta_4^{(k+1)}$ |
|---|---|---|---|
| 1 | 1.141 933 723 | 1.330 722 500 | $-0.148\,259\,790$ |
| 2 | 1.323 643 137 | 1.327 045 601 | $1.469\,39 \times 10^{-4}$ |
| 3 | 1.327 045 590 | 1.327 045 600 | $-4.388 \times 10^{-13}$ |

The cubic convergence of $\beta_4^{(k)}$ to zero is obvious. The tridiagonal matrix is reducible, and we have found with $\alpha_5^{(4)}$ the first eigenvalue $\lambda_1 \doteq 1.327\,045\,600$. The reduced matrix of order four is

$$\hat{\mathbf{J}} \doteq \begin{pmatrix} 22.354\,976 & 0.881\,561 & & \\ 0.881\,561 & 7.399\,454 & 0.819\,413 & \\ & 0.819\,413 & 1.807\,160 & 3.010\,085 \\ & & 3.010\,085 & 2.111\,365 \end{pmatrix}.$$

The most important elements of the subsequent QR steps for $\hat{\mathbf{J}}$ are summarized in the following table.

| $k$ | $\sigma_k$ | $\alpha_4^{(k+1)}$ | $\beta_3^{(k+1)}$ |
|---|---|---|---|
| 4 | 4.973 188 459 | 4.846 875 984 | 0.116 631 770 |
| 5 | 4.849 265 094 | 4.848 950 120 | $6.527\,73 \times 10^{-6}$ |
| 6 | 4.848 950 120 | 4.848 950 120 | $-1.2 \times 10^{-16}$ |

Hence, the second eigenvalue is $\lambda_2 \doteq 4.848\,950\,120$, and the remaining submatrix of order three is

$$\hat{\hat{\mathbf{J}}} \doteq \begin{pmatrix} 22.406\,874 & 0.003\,815 & \\ 0.003\,815 & 3.850\,210 & 4.257\,076 \\ & 4.257\,076 & 2.566\,920 \end{pmatrix},$$

from which the other eigenvalues are computed by two and one QR step, respectively, in the order $\lambda_3 \doteq -1.096\,595\,182$, $\lambda_4 \doteq 7.513\,724\,154$ and $\lambda_5 \doteq 22.406\,875\,308$.

### 6.4.5 Computation of eigenvectors

The above description of the QR algorithm primarily yields the eigenvalues of a Hessenberg matrix $\mathbf{H}$ or a tridiagonal, symmetric matrix $\mathbf{J}$. We now treat the problem of determining the corresponding eigenvectors.

The product of the orthogonal matrices $\mathbf{Q}_k$, $k = 1, 2, \ldots, M$, that are used in the course of the QR algorithm is an orthogonal matrix

$$\mathbf{Q} := \mathbf{Q}_1 \mathbf{Q}_2 \cdots \mathbf{Q}_M, \tag{6.151}$$

which transforms either $\mathbf{H}$ into a quasitriangular matrix, $\mathbf{R} = \mathbf{Q}^T \mathbf{H} \mathbf{Q}$ (6.108), or $\mathbf{J}$ into a diagonal matrix, $\mathbf{D} = \mathbf{Q}^T \mathbf{J} \mathbf{Q}$. This matrix $\mathbf{Q}$ reduces the given eigenvalue problem $\mathbf{H}\mathbf{x} = \lambda \mathbf{x}$ or $\mathbf{J}\mathbf{x} = \lambda \mathbf{x}$ into

$$\mathbf{R}\mathbf{y} = \lambda \mathbf{y} \quad \text{or} \quad \mathbf{D}\mathbf{y} = \lambda \mathbf{y} \quad \text{with} \quad \mathbf{y} = \mathbf{Q}^T \mathbf{x}. \tag{6.152}$$

For a known eigenvalue $\lambda_k$ the corresponding eigenvector $\mathbf{y}_k$ of $\mathbf{R}$ can be easily computed. For the case of the matrix $\mathbf{D}$ we have $\mathbf{y}_k = \mathbf{e}_k$ if $\lambda_k$ is the $k$th diagonal element of $\mathbf{D}$. The eigenvector $\mathbf{x}_k$ of $\mathbf{H}$ is given, according to (6.152), by $\mathbf{x}_k = \mathbf{Q}\mathbf{y}_k$, and that of $\mathbf{J}$ is equal to the $k$th column of $\mathbf{Q}$.

A method for computing the eigenvectors consists of forming the matrix $\mathbf{Q}$, (6.151), explicitly as the product of all $\mathbf{Q}_k$. Each of the matrices $\mathbf{Q}_k$ is the product of simple rotation matrices, so that $\mathbf{Q}$ can be computed recursively in a similar way to (6.48). The amount of computational effort required for a single QR step increases a lot because the multiplication from the right by a rotation matrix has to be executed for all columns of $\mathbf{Q}$ even when the order of the matrix $\mathbf{H}$, or $\mathbf{J}$, to be treated is smaller than $n$ after some eigenvalues have been determined. The sketched procedure has the advantage of yielding with $\mathbf{Q}$ $n$ orthonormal eigenvectors for the tridiagonal, symmetric matrix $\mathbf{J}$.

In order to avoid the time consuming computation of the matrix $\mathbf{Q}$, a second method for computing the eigenvectors $\mathbf{x}_k$ uses the *inverse vector iteration* of Section 6.3.4. The initial vector $\mathbf{z}^{(0)}$ can be found either by Hyman's method or by following a suggestion of Wilkinson as the solution of the system of equations

$$\tilde{\mathbf{R}}\mathbf{z}^{(0)} = \mathbf{e} \quad \mathbf{e} = (1, 1, \ldots, 1)^{\mathrm{T}} \tag{6.153}$$

where $\tilde{\mathbf{R}}$ denotes the right triangular matrix that results from the Gaussian algorithm for decomposing the matrix $(\mathbf{H} - \bar{\lambda}\mathbf{I})$ or $(\mathbf{J} - \bar{\lambda}\mathbf{I})$, respectively, by applying the relative maximal column pivoting strategy, and where $\bar{\lambda}$ represents the known approximation of $\lambda_k$. The vector $\mathbf{z}^{(0)}$ is computed from (6.153) by the process of backward substitution. This is equivalent to half an iteration step of the inverse vector iteration.

The second method for computing the eigenvectors essentially requires a decomposition of $(\mathbf{H} - \bar{\lambda}\mathbf{I})$ for each eigenvalue. A decomposition of a Hessenberg matrix $\mathbf{H}$ of order $n$ requires about $\frac{1}{2}n^2$, the process of the forward substitution about $n$, and the back substitution about $\frac{1}{2}n^2$ multiplicative operations. Usually two iteration steps of the inverse vector iteration are sufficient, thus we can say that the amount of computational effort required to determine all $n$ eigenvectors consists of about $2n^3$ operations. For a tridiagonal matrix $\mathbf{J}$ it is only proportional to $n^2$.

If the matrix $\mathbf{H}$, or $\mathbf{J}$, is the result of a previous similarity transformation of a matrix $\mathbf{A}$, the eigenvectors must be transformed back into those of $\mathbf{A}$ according to the procedure of Section 6.3.1.

## 6.5 Exercises

**6.1.** Analyze the sensitivity of the zeros $z_i$ of the two polynomials when a single coefficient $p_k$ is perturbed by $\varepsilon p_k$.

(a) $P_8(x) = x^8 - 37.3x^7 + 592x^6 - 5207.8x^5 + 27\,661x^4$
$\qquad - 90\,255.7x^3 + 174\,786x^2 - 180\,115.2x + 72\,576$
with $z_1 = 1$, $z_2 = 3$, $z_3 = 4$, $z_4 = 4.5$, $z_5 = 4.8$, $z_6 = 5$, $z_7 = 7$, $z_8 = 8$.

(b) $P_{12}(x) = x^{12} - 78x^{11} + 1001x^{10} - 5005x^9 + 12\,870x^8 - 19\,448x^7 + 18\,564x^6$
$\qquad - 11\,628x^5 + 4845x^4 - 1330x^3 + 231x^2 - 23x + 1$
with $z_i = [1 - \cos((2i-1)\pi/25)]/2 \quad (i = 1, 2, \ldots, 12)$.

What are the consequences for the result if a computer arithmetic with eight- or twelve-decimal digits is used, and the relative errors, $\varepsilon$, of the corresponding size are inevitable in the representation of the polynomial coefficients?

**6.2.** Transform the matrix

$$\mathbf{A} = \begin{pmatrix} 1 & -6 \\ -6 & -4 \end{pmatrix}$$

into diagonal form by means of a Jacobi rotation. What are the eigenvalues and eigenvectors?

**6.3.** Apply a step of the classical Jacobi method to the symmetric matrix

$$\mathbf{A} = \begin{pmatrix} 2 & 3 & -4 \\ 3 & 6 & 2 \\ -4 & 2 & 10 \end{pmatrix}.$$

How much does $S(\mathbf{A})$ decrease? Then reduce the matrix $\mathbf{A}$ to tridiagonal form by Givens' method. Explain why the two resulting tridiagonal matrices differ. How large is the decrease in $S(\mathbf{A})$ in the latter case? Also apply the fast Givens transformation to reduce $\mathbf{A}$ to tridiagonal form.

**6.4.** Use computer programs to determine the eigenvalues and eigenvectors of the following symmetric matrices by means of the classical and the cyclic Jacobi method, and also by means of the reduction to tridiagonal form, QR algorithm and inverse vector iteration.

$$\mathbf{A}_1 = \begin{pmatrix} 3 & -2 & 4 & 5 \\ -2 & 7 & 3 & 8 \\ 4 & 3 & 10 & 1 \\ 5 & 8 & 1 & 6 \end{pmatrix} \qquad \mathbf{A}_2 = \begin{pmatrix} 5 & -5 & 5 & 0 \\ -5 & 16 & -8 & 7 \\ 5 & -8 & 16 & 7 \\ 0 & 7 & 7 & 21 \end{pmatrix}$$

$$\mathbf{A}_3 = \begin{pmatrix} 1 & 1 & 1 & 1 & 1 & 1 \\ 1 & 2 & 3 & 4 & 5 & 6 \\ 1 & 3 & 6 & 10 & 15 & 21 \\ 1 & 4 & 10 & 20 & 35 & 56 \\ 1 & 5 & 15 & 35 & 70 & 126 \\ 1 & 6 & 21 & 56 & 126 & 252 \end{pmatrix} \qquad \mathbf{A}_4 = \begin{pmatrix} 19 & 5 & -12 & 6 & 7 & 16 \\ 5 & 13 & 9 & -18 & 12 & 4 \\ -12 & 9 & 18 & 4 & 6 & 14 \\ 6 & -18 & 4 & 19 & 2 & -16 \\ 7 & 12 & 6 & 2 & 5 & 15 \\ 16 & 4 & 14 & -16 & 15 & 13 \end{pmatrix}$$

**6.5.** Perform two iteration steps with Hyman's method for the Hessenberg matrix

$$\mathbf{H} = \begin{pmatrix} 2 & -5 & -13 & -25 \\ 4 & 13 & 29 & 60 \\ 0 & -2 & -15 & -48 \\ 0 & 0 & 5 & 18 \end{pmatrix}$$

using the initial value $\lambda^{(0)} = 2.5$ and subsequently compute with the resulting approximate eigenvalue the corresponding eigenvector by applying the inverse vector iteration. Do the same for the initial value $\lambda^{(0)} = 7.5$.

## 6.5 Exercises

**6.6.** Write computer programs for the transformation into Hessenberg form, of Hyman's method, QR transformation and vector iteration and with them compute the eigenvalues and eigenvectors of the following nonsymmetric matrices.

$$A_1 = \begin{pmatrix} -3 & 9 & 0 & 1 \\ 1 & 6 & 0 & 0 \\ -23 & 23 & 4 & 3 \\ -12 & 15 & 1 & 3 \end{pmatrix} \quad A_2 = \begin{pmatrix} 28 & 17 & -16 & 11 & 9 & -2 & -27 \\ -1 & 29 & 7 & -6 & -2 & 28 & 1 \\ -11 & -1 & 12 & 3 & -8 & 10 & 11 \\ -6 & -11 & 12 & 8 & -12 & -5 & 6 \\ -3 & 1 & -4 & 3 & 8 & 4 & 3 \\ 14 & 16 & -7 & 6 & 2 & 4 & -14 \\ -37 & -18 & -9 & 5 & 7 & 26 & 38 \end{pmatrix}$$

$$A_3 = \begin{pmatrix} 1 & 2 & 3 & 4 & 5 & 6 \\ 6 & 1 & 2 & 3 & 4 & 5 \\ 5 & 6 & 1 & 2 & 3 & 4 \\ 4 & 5 & 6 & 1 & 2 & 3 \\ 3 & 4 & 5 & 6 & 1 & 2 \\ 2 & 3 & 4 & 5 & 6 & 1 \end{pmatrix} \quad A_4 = \begin{pmatrix} 3 & -5 & 4 & -2 & 0 & 8 & 1 \\ 4 & 2 & -1 & 7 & 6 & 2 & 9 \\ -5 & 8 & -2 & 3 & 1 & 4 & 2 \\ -6 & -4 & 2 & 5 & -8 & 1 & -3 \\ 1 & -2 & 7 & 5 & 2 & 8 & 4 \\ 8 & 1 & -7 & 6 & 4 & 0 & -1 \\ -1 & -8 & -9 & -1 & 3 & -3 & 2 \end{pmatrix}$$

Hint: The matrices $A_1$ and $A_2$ have real eigenvalues.

**6.7.** Let $H$ be an irreducible Hessenberg matrix of order $n$. Show by means of the transformation formulae of the individual steps that the diagonal elements of $R$ of the QR decomposition, $H = QR$, satisfy the inequalities

$$|r_{ii}| \geq |h_{i+1,i}| \quad \text{for } i = 1, 2, \ldots, n-1.$$

If $\bar{\lambda}$ is a good approximation of an eigenvalue of $H$, deduce from this result that in the QR decomposition of $(H - \bar{\lambda}I) = QR$ only the last diagonal element $r_{nn}$ can be arbitrarily small in absolute value.

**6.8.** Investigate experimentally the influence of the real spectral shifts $\sigma_k$ given by (6.118) or (6.124) on the number of iteration steps of the QR algorithm applied to the matrices of problems 6.4 and 6.6.

**6.9.** Develop an algorithm for a QR step for a Hessenberg matrix using the technique of the implicit spectral shift in the case of a real shift $\sigma$.

# 7

# Method of Least Squares

In several branches of science, for instance in experimental physics or biology, we are often faced with the problem of determining unknown parameters of a function from a series of measurements or observations. The function is given either on the basis of a known natural law or of model assumptions. In order to take into account the unavoidable observational errors, the number of measurements is usually much larger than the number of parameters. The resulting overdetermined system of linear or nonlinear equations for the unknown parameters have, in general, no solution. However, we can only claim that the errors, or residuals, of the equations, which have to be admitted, are minimal in a sense that has to be defined more precisely. If the observational errors are assumed to be normally distributed, then the *method of least squares*, introduced by Gauss, is most appropriate for theoretical probability reasons (Ludwig, 1969). The Gaussian principle of curve fitting leads to simpler computational techniques than the Chebyshev principle, which is more suitable with regards to the approximation of functions.

## 7.1 Linear Problems, Normal Equations

We consider an overdetermined system of $N$ linear equations in $n$ unknowns $x_1, x_2, \ldots, x_n$

$$\sum_{k=1}^{n} c_{ik} x_k + d_i = r_i \quad (i=1,2,\ldots,N) \quad n < N \tag{7.1}$$

where we have introduced the *residuals* $r_i$. We write the so called *error equations* (7.1) in matrix form as follows

$$\boxed{\mathbf{Cx} + \mathbf{d} = \mathbf{r} \quad \mathbf{C} \in \mathbf{R}^{N \times n} \quad \mathbf{x} \in \mathbf{R}^n \quad \mathbf{d}, \mathbf{r} \in \mathbf{R}^N.} \tag{7.2}$$

## 7.1 Linear Problems, Normal Equations

For the following we assume that the matrix $\mathbf{C}$ has the maximal rank $n$, that is, its column vectors are linearly independent. The unknowns $x_k$ of the error equations are to be determined according to the Gaussian principle, such that the *sum of squares of the residuals* $r_i$ is minimal. This requirement is equivalent to minimizing the square of the Euclidean norm of the residual vector. From (7.2) we obtain

$$\begin{aligned}\mathbf{r}^T\mathbf{r} &= (\mathbf{Cx}+\mathbf{d})^T(\mathbf{Cx}+\mathbf{d}) = \mathbf{x}^T\mathbf{C}^T\mathbf{Cx}+\mathbf{x}^T\mathbf{C}^T\mathbf{d}+\mathbf{d}^T\mathbf{Cx}+\mathbf{d}^T\mathbf{d}\\ &= \mathbf{x}^T\mathbf{C}^T\mathbf{Cx}+2(\mathbf{C}^T\mathbf{d})^T\mathbf{x}+\mathbf{d}^T\mathbf{d}.\end{aligned} \qquad (7.3)$$

Hence, the square of the Euclidean length of $\mathbf{r}$ is represented by a quadratic function $F(\mathbf{x})$ of the $n$ unknowns $x_k$. To simplify the notation we define

$$\boxed{\mathbf{A}:=\mathbf{C}^T\mathbf{C} \quad \mathbf{b}:=\mathbf{C}^T\mathbf{d} \quad \mathbf{A}\in\mathbf{R}^{n\times n} \quad \mathbf{b}\in\mathbf{R}^n.} \qquad (7.4)$$

As $\mathbf{C}$ has maximal rank, the symmetric matrix $\mathbf{A}$ is *positive definite*. Indeed the corresponding quadratic form satisfies

$$Q(\mathbf{x}) = \mathbf{x}^T\mathbf{A}\mathbf{x} = \mathbf{x}^T\mathbf{C}^T\mathbf{C}\mathbf{x} = (\mathbf{Cx})^T(\mathbf{Cx}) \geq 0 \quad \text{for all } \mathbf{x}\in\mathbf{R}^n$$
$$Q(\mathbf{x}) = 0 \Leftrightarrow (\mathbf{Cx}) = \mathbf{0} \Leftrightarrow \mathbf{x} = \mathbf{0}.$$

From (7.4) it follows that the quadratic function $F(\mathbf{x})$ to be minimized is

$$\boxed{F(\mathbf{x}):=\mathbf{r}^T\mathbf{r} = \mathbf{x}^T\mathbf{A}\mathbf{x}+2\mathbf{b}^T\mathbf{x}+\mathbf{d}^T\mathbf{d} = \text{Min}!} \qquad (7.5)$$

A necessary condition for a minimum of $F(\mathbf{x})$ at the point $\mathbf{x}$, is the vanishing of its gradient $\nabla F(\mathbf{x})$. The $i$th component of the gradient $\nabla F(\mathbf{x})$ is obtained from the explicit representation of (7.5)

$$\frac{\partial F(\mathbf{x})}{\partial x_i} = 2\sum_{k=1}^{n} a_{ik}x_k + 2b_i \quad (i=1,2,\ldots,n). \qquad (7.6)$$

After division by 2 from (7.6) we obtain the linear system of equations

$$\boxed{\mathbf{A}\mathbf{x}+\mathbf{b}=\mathbf{0}} \qquad (7.7)$$

for the unknowns $x_1, x_2, \ldots, x_n$. We call (7.7) the *normal equations* of the error equations (7.2). The matrix $\mathbf{A}$ is positive definite, thus from the assumption on $\mathbf{C}$, the unknowns $x_k$ are uniquely determined by the normal equations (7.7), and they can be computed by Cholesky's method. The function $F(\mathbf{x})$ is indeed minimized by these values, because the Hessian matrix of $F(\mathbf{x})$, the matrix of the second partial derivatives, is equal to the positive definite matrix $\mathbf{A}$.

The classical treatment of the error equations (7.2) according to the Gaussian principle, consists of the following simple solution steps.

> 1. $\mathbf{A} = \mathbf{C}^T\mathbf{C}$  $\mathbf{b} = \mathbf{C}^T\mathbf{d}$  (normal equations $\mathbf{Ax} + \mathbf{b} = \mathbf{0}$)
> 2. $\mathbf{A} = \mathbf{LL}^T$  (Cholesky's decomposition)
> 3. $\mathbf{Ly} - \mathbf{b} = \mathbf{0}$  $\mathbf{L}^T\mathbf{x} + \mathbf{y} = \mathbf{0}$  (forward/backward substitution)
> [4. $\mathbf{r} = \mathbf{Cx} + \mathbf{d}$  (computation of residuals)]

(7.8)

We obtain an easily remembered procedure for the computation of the matrix elements $a_{ik}$ and the components $b_i$ of the normal equations, if we introduce the column vectors $\mathbf{c}_i$ of the matrix $\mathbf{C}$. Then the following representations hold

$$a_{ik} = \mathbf{c}_i^T \mathbf{c}_k \quad b_i = \mathbf{c}_i^T \mathbf{d} \quad (i, k = 1, 2, \ldots, n), \tag{7.9}$$

so that $a_{ik}$ is equal to the scalar product of the $i$th and $k$th column vectors of $\mathbf{C}$, and $b_i$ is the scalar product of the $i$th column vector $\mathbf{c}_i$ and the constant vector $\mathbf{d}$ of the error equations. For symmetry reasons, only the elements of $\mathbf{A}$ in and below the diagonal must be computed. The amount of work required to obtain the normal equations consists of $Z_{\text{normequ}} = nN(n+3)/2$ multiplications. The solution of $N$ error equations (7.1) in $n$ unknowns by means of the algorithm (7.8), including the computation of the residuals, due to (1.102), requires

$$Z_{\text{erroreq}} = \tfrac{1}{2}nN(n+5) + \tfrac{1}{6}n^3 + \tfrac{3}{2}n^2 + \tfrac{1}{3}n \tag{7.10}$$

multiplicative operations and $n$ square roots.

*Example* 7.1  A physical quantity $z$ is observed at some definite, not equidistant, times $t_i$ according to (7.11).

| $i$ | 1 | 2 | 3 | 4 | 5 | 6 | 7 |
|---|---|---|---|---|---|---|---|
| $t_i$ | 0.04 | 0.32 | 0.51 | 0.73 | 1.03 | 1.42 | 1.60 |
| $z_i$ | 2.63 | 1.18 | 1.16 | 1.54 | 2.65 | 5.41 | 7.67 |

(7.11)

It is known that $z$ is a quadratic function in time $t$, whose parameters are to be estimated by applying the method of least squares. We set

$$z(t) = \alpha_0 + \alpha_1 t + \alpha_2 t^2 \tag{7.12}$$

## 7.1 Linear Problems, Normal Equations

and obtain the $i$th error equation

$$\alpha_0 + \alpha_1 t_i + \alpha_2 t_i^2 - z_i = r_i \quad (i = 1, 2, \ldots, 7)$$

and hence the system of error equations without residuals

| $\alpha_0$ | $\alpha_1$ | $\alpha_2$ | 1 |
|---|---|---|---|
| 1 | 0.04 | 0.0016 | $-2.63$ |
| 1 | 0.32 | 0.1024 | $-1.18$ |
| 1 | 0.51 | 0.2601 | $-1.16$ |
| 1 | 0.73 | 0.5329 | $-1.54$ |
| 1 | 1.03 | 1.0609 | $-2.65$ |
| 1 | 1.42 | 2.0164 | $-5.41$ |
| 1 | 1.60 | 2.5600 | $-7.67$ |

(7.13)

The first three columns of (7.13) correspond to the matrix $\mathbf{C}$, and the fourth column equals the vector $\mathbf{d}$ of the error equations (7.2). The normal equations are given by (computed with a precision of six decimal digits)

| $\alpha_0$ | $\alpha_1$ | $\alpha_2$ | 1 |
|---|---|---|---|
| 7.000 00 | 5.650 00 | 6.534 30 | $-22.2400$ |
| 5.650 00 | 6.534 30 | 8.606 52 | $-24.8823$ |
| 6.534 30 | 8.606 52 | 12.107 1 | $-34.6027$ |

(7.14)

The Cholesky decomposition and the forward and backward substitutions yield

$$\mathbf{L} = \begin{pmatrix} 2.645\,75 & & \\ 2.135\,50 & 1.404\,97 & \\ 2.469\,73 & 2.371\,87 & 0.617\,867 \end{pmatrix}$$

$$\mathbf{y} = \begin{pmatrix} -8.405\,93 \\ -4.933\,49 \\ -3.464\,66 \end{pmatrix} \quad \boldsymbol{\alpha} = \begin{pmatrix} 2.749\,28 \\ -5.955\,01 \\ 5.607\,45 \end{pmatrix}. \tag{7.15}$$

The resulting quadratic function

$$z(t) = 2.749\,28 - 5.955\,01 t + 5.607\,45 t^2 \tag{7.16}$$

has the residual vector $\mathbf{r} \doteq (-0.1099, 0.2379, 0.0107, -0.1497, -0.0854, 0.1901, -0.0936)^T$. The measured points and the graph of the quadratic function are

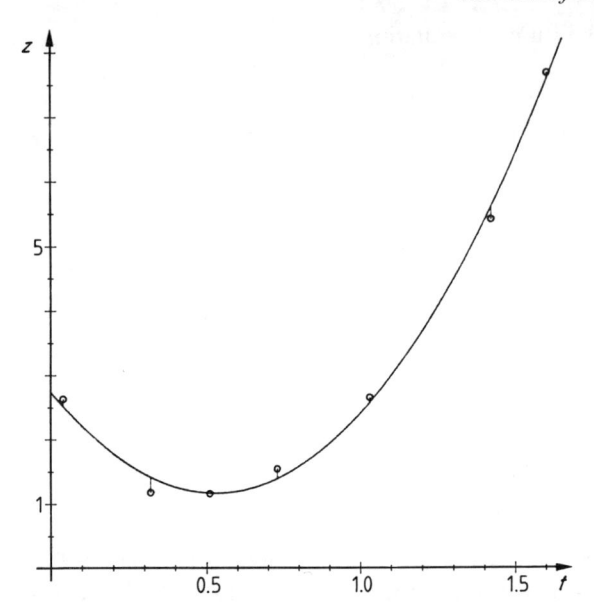

Figure 7.1 Method of least squares for a quadratic function

represented in Figure 7.1. The residuals $r_i$ are the differences between the curve $z(t)$ at $t_i$, and the measured values $z_i$. They can be interpreted as corrections of the measurements, such that the corrected measured points lie on the curve.

The method of the normal equations may suffer when the condition number of the matrix $\mathbf{A}$ of the normal equations is quite large. As a consequence, the computed solution $\tilde{\mathbf{x}}$ may have a correspondingly large relative error (see Section 1.2.1). The matrix elements $a_{ik}$ and the components $b_i$ of the constant vector are computed as scalar products, (7.9), thus rounding errors are usually inevitable.

The matrix $\mathbf{A}$ of the normal equations (7.14) has the condition number $\kappa_2(\mathbf{A})$ $= \lambda_{max}/\lambda_{min} \doteq 23.00/0.090\,00 \doteq 256$. If we calculate to a precision of six decimal digits, only the first three digits of the computed solution $\tilde{a}$ are guaranteed to be correct, see Section 1.2.1. A calculation with a precision of twelve decimal digits yields for $z(t)$, the following coefficients rounded to seven digits

$$z(t) \doteq 2.749\,198 - 5.954\,657t + 5.607\,247t^2. \tag{7.17}$$

*Example 7.2* To illustrate the possible ill-condition of the normal equations, we consider a typical problem of data fitting. In order to get an analytic description $y = f(x)$ of the characteristic of a nonlinear transfer element, the following output quantities $y_i$ have been observed for exact input quantities $x_i$.

## 7.2 Methods of Orthogonal Transformation

| $x$ | 0.2 | 0.5 | 1.0 | 1.5 | 2.0 | 3.0 |
|---|---|---|---|---|---|---|
| $y$ | 0.3 | 0.5 | 0.8 | 1.0 | 1.2 | 1.3 |

(7.18)

The transfer element behaves linearly for small input $x$, and the characteristic has a horizontal asymptote for large $x$. In order to take this behaviour into account we use the representation for $f(x)$

$$f(x) = \alpha_1 \frac{x}{1+x} + \alpha_2(1 - e^{-x}) \tag{7.19}$$

with the two unknown parameters $\alpha_1$ and $\alpha_2$. A calculation with six decimal digits yields the following system of error equations, normal equations and the left triangular matrix $\mathbf{L}$ of the Cholesky decomposition

| $\alpha_1$ | $\alpha_2$ | 1 |
|---|---|---|
| 0.166 667 | 0.181 269 | −0.3 |
| 0.333 333 | 0.393 469 | −0.5 |
| 0.500 000 | 0.632 121 | −0.8 |
| 0.600 000 | 0.776 870 | −1.0 |
| 0.666 667 | 0.864 665 | −1.2 |
| 0.750 000 | 0.950 213 | −1.3 |

| $\alpha_1$ | $\alpha_2$ | 1 |
|---|---|---|
| 1.755 83 | 2.232 66 | −2.991 67 |
| 2.232 66 | 2.841 34 | −3.806 56 |

$$\mathbf{L} = \begin{pmatrix} 1.325\,08 & \\ 1.684\,92 & 0.048\,7852 \end{pmatrix} \tag{7.20}$$

The forward and the backward substitutions yield $\alpha_1 = 0.384\,196$ and $\alpha_2 = 1.037\,82$, the desired representation of the characteristic being

$$f(x) = 0.384\,196 \frac{x}{1+x} + 1.037\,82(1 - e^{-x}) \tag{7.21}$$

and the residual vector $\mathbf{r} \doteq (-0.0478,\ 0.0364,\ 0.0481,\ 0.0368,\ -0.0465,\ -0.0257)^T$. From the two eigenvalues $\lambda_1 \doteq 4.596\,27$ and $\lambda_2 \doteq 0.000\,9006$ of the matrix of the normal equations, the condition number $\kappa_2(\mathbf{A}) \doteq 5104$ follows. The computed values of the parameters $\alpha_1$ and $\alpha_2$ may have a correspondingly large relative error. Therefore, the same problem has been treated to twelve digit accuracy. The different values $\alpha_1 \doteq 0.382\,495$ and $\alpha_2 \doteq 1.039\,15$ have been obtained, for which the residuals coincide with the values given above. This result indicates the high sensitivity of the parameters $\alpha_1$ and $\alpha_2$.

### 7.2 Methods of Orthogonal Transformation

The mentioned problem of a possible ill-condition of the normal equations demands that safer numerical procedures should be used to solve the error equations, according to the method of least squares. The computation of the normal equations must be avoided, and the desired solution should be determined by a direct treatment of the error equations. In the following, two variants will be described.

### 7.2.1 Givens transformation

The essential point for the procedures to be developed is the fact that the length of a vector remains invariant under orthogonal transformations. To solve the error equations $\mathbf{Cx} + \mathbf{d} = \mathbf{r}$ according to the Gaussian principle, they can be transformed by means of an orthogonal matrix $\mathbf{Q} \in \mathbf{R}^{N \times N}$ without changing the sum of the squares of the residuals. Therefore the system of error equations (7.2) is replaced by the equivalent system

$$\mathbf{Q}^T\mathbf{Cx} + \mathbf{Q}^T\mathbf{d} = \mathbf{Q}^T\mathbf{r} = \hat{\mathbf{r}}. \tag{7.22}$$

In (7.22) the orthogonal matrix $\mathbf{Q}$ will subsequently be chosen in such a way that $\mathbf{Q}^T\mathbf{C}$ has a special structure. Theorem 6.4 is generalized and we have the following theorem.

*Theorem 7.1* For every matrix $\mathbf{C} \in \mathbf{R}^{N \times n}$ of maximal rank $n < N$, there exists an orthogonal matrix $\mathbf{Q} \in \mathbf{R}^{N \times N}$ such that

$$\mathbf{C} = \mathbf{Q}\hat{\mathbf{R}} \quad \text{with} \quad \hat{\mathbf{R}} = \begin{pmatrix} \mathbf{R} \\ \hline \mathbf{0} \end{pmatrix} \quad \mathbf{R} \in \mathbb{R}^{n \times n}, \mathbf{0} \in \mathbf{R}^{(N-n) \times n} \tag{7.23}$$

holds, where $\mathbf{R}$ is a nonsingular right triangular matrix, and $\mathbf{0}$ represents a zero matrix.

*Proof* Analogous to the proof of Theorem 6.4 we realize that the successive multiplication of the matrix $\mathbf{C}$ from the left by rotation matrices $\mathbf{U}^T(p, q; \varphi)$ with the rotation indices

$$(1, 2), (1, 3), \ldots, (1, N), (2, 3), (2, 4), \ldots, (2, N), (3, 4), \ldots, (n, N) \tag{7.24}$$

and with rotation angles chosen according to (6.104), eliminates the current matrix elements in the order

$$c_{21}, c_{31}, \ldots, c_{N1}, c_{32}, c_{42}, \ldots, c_{N2}, c_{43}, \ldots, c_{Nn} \tag{7.25}$$

After $N^* = n(2N - n - 1)/2$ steps of transformation, (7.23) holds with

$$\mathbf{U}_{N^*}^T \cdots \mathbf{U}_2^T \mathbf{U}_1^T \mathbf{C} = \mathbf{Q}^T\mathbf{C} = \hat{\mathbf{R}} \quad \text{or} \quad \mathbf{C} = \mathbf{Q}\hat{\mathbf{R}}. \tag{7.26}$$

As the orthogonal matrix $\mathbf{Q}$ is nonsingular the ranks of $\mathbf{C}$ and $\hat{\mathbf{R}}$ are equal, and hence the right triangular matrix $\mathbf{R}$ is nonsingular.

## 7.2 Methods of Orthogonal Transformation

With the matrix $\mathbf{Q}$ chosen according to Theorem 7.1 the system (7.22) is now

$$\hat{\mathbf{R}}\mathbf{x} + \hat{\mathbf{d}} = \hat{\mathbf{r}} \quad \hat{\mathbf{d}} = \mathbf{Q}^T\mathbf{d}. \tag{7.27}$$

From (7.23) the orthogonally transformed system of error equations (7.27) has the form

$$\begin{aligned} r_{11}x_1 + r_{12}x_2 + \cdots + r_{1n}x_n + \hat{d}_1 &= \hat{r}_1 \\ r_{22}x_2 + \cdots + r_{2n}x_n + \hat{d}_2 &= \hat{r}_2 \\ &\vdots \\ r_{nn}x_n + \hat{d}_n &= \hat{r}_n \\ \hat{d}_{n+1} &= \hat{r}_{n+1} \\ &\vdots \\ \hat{d}_N &= \hat{r}_N \end{aligned} \tag{7.28}$$

The method of least squares requires the sum of squares of the transformed residuals $\hat{r}_i$ to be minimal. The values of the last $(N - n)$ residuals are fixed by the corresponding $\hat{d}_j$, independently of the unknowns $x_k$. The sum of squares of the residuals is minimal if and only if $\hat{r}_1 = \hat{r}_2 = \cdots = \hat{r}_n = 0$ holds, and it is equal to the sum of squares of the last $(N - n)$ residuals $\hat{r}_j$. Consequently the unknowns $x_k$ are given by the system of linear equations

$$\mathbf{R}\mathbf{x} + \hat{\mathbf{d}}_1 = 0 \tag{7.29}$$

where $\hat{\mathbf{d}}_1 \in \mathbf{R}^n$ denotes the vector consisting of the $n$ first components of $\hat{\mathbf{d}} = \mathbf{Q}^T\mathbf{d}$. The solution $\mathbf{x}$ is obtained from (7.29) by the process of backward substitution.

If we only want to know the unknowns $x_k$ of the system of error equations $\mathbf{C}\mathbf{x} + \mathbf{d} = \mathbf{r}$ the algorithm has already been completely described. Moreover, if the residuals $r_i$ are required, then they can be obtained, in principle, by substitution of the $x_k$ into the given error equations. This procedure requires the matrix $\mathbf{C}$ and the vector $\mathbf{d}$ to still be available. With respect to storage requirements, it is more economical to compute the residual vector $\mathbf{r}$, from (7.22) and (7.26), from $\hat{\mathbf{r}}$ according to

$$\mathbf{r} = \mathbf{Q}\hat{\mathbf{r}} = \mathbf{U}_1 \mathbf{U}_2 \cdots \mathbf{U}_{N*} \hat{\mathbf{r}}. \tag{7.30}$$

The information on the rotation matrices $\mathbf{U}_k$ can be stored by means of the values $\rho$ (6.65) at the places of the eliminated matrix elements $c_{ij}$. The first $n$ components of the residual vector $\hat{\mathbf{r}}$ are equal to zero, and the last $(N - n)$ components are defined by the corresponding $\hat{d}_j$. The desired residual vector $\mathbf{r}$ is obtained from $\hat{\mathbf{r}}$ by successive multiplication with the rotation matrices $\mathbf{U}_k$ in the opposite order of their application during the transformation of $\mathbf{C}$ into $\hat{\mathbf{R}}$.

The treatment of the error equations (7.2) according to the Gaussian principle by means of the orthogonal transformation, with Givens rotations, can be summarized as follows:

$$
\begin{aligned}
&1.\ \mathbf{C} = \mathbf{Q}\hat{\mathbf{R}} && \text{(QR decomposition, Givens rotations)} \\
&2.\ \hat{\mathbf{d}} = \mathbf{Q}^T \mathbf{d} && \text{(transformation of } \mathbf{d}\text{)} \\
&3.\ \mathbf{R}\mathbf{x} + \hat{\mathbf{d}}_1 = \mathbf{0} && \text{(backward substitution)} \\
&[4.\ \mathbf{r} = \mathbf{Q}\hat{\mathbf{r}} && \text{(backtransformation of } \hat{\mathbf{r}}\text{)}]
\end{aligned}
$$
(7.31)

The first and the second step of (7.31) are usually performed simultaneously if a single system of error equations has to be solved. As soon as several systems (7.2) must be successively treated with the same matrix $\mathbf{C}$ but with different vectors $\mathbf{d}$, it is appropriate to separate the two steps. In this case the information on the rotations must be available for the second step, as indicated above.

The amount of work required for the procedure (7.31) consists of

$$Z_{eeq} = 2nN(n+6) - \tfrac{2}{3}n^3 - \tfrac{13}{2}n^2 - \tfrac{35}{6}n \tag{7.32}$$

multiplicative operations and $n(2N - n - 1)$ square roots. In comparison to (7.10) it is larger by a factor between 2 and 4, depending on the ratio between $N$ and $n$. The additional effort is, however, justified by the fact that the computed solution $\tilde{\mathbf{x}}$ of the error equations has a much smaller relative error if the same precision is used.

The computation of the solution $\mathbf{x}$ of a system of error equations $\mathbf{Cx} + \mathbf{d} = \mathbf{r}$ by means of the procedure (7.31) has the algorithmic description (7.33). It is assumed that the matrix $\hat{\mathbf{R}}$ is built at the place of $\mathbf{C}$ and that the values $\rho$ (6.65) are stored at the places of the eliminated matrix elements $c_{ij}$. The two steps 1 and 2 of (7.31) are performed simultaneously. To avoid a conflict in the name of variables we denote $\gamma = \cos\varphi$ and $\sigma = \sin\varphi$. Finally, $\delta$ is the smallest positive number of the computer with $1 + \delta \neq 1$.

```
for j = 1, 2, ..., n:
    for i = j + 1, j + 2, ..., N:
        if c_ij ≠ 0 then
            if |c_jj| < δ × |c_ij| then
                w = −c_ij; γ = 0; σ = 1; ρ = 1
            else
                w = sgn(c_jj) × √(c_jj² + c_ij²)
                γ = c_jj/w; σ = −c_ij/w
                if |σ| < γ then ρ = σ else ρ = sgn(σ)/γ
            c_jj = w; c_ij = ρ
            for k = j + 1, j + 2, ..., n:
                h = γ × c_jk − σ × c_ik
```

## 7.2 Methods of Orthogonal Transformation

$$c_{ik} = \sigma \times c_{jk} + \gamma \times c_{ik}; \; c_{jk} = h$$
$$h = \gamma \times d_j - \sigma \times d_i; \; d_i = \sigma \times d_j + \gamma \times d_i; \; d_j = h$$
for $i = n, n-1, \ldots, 1$: (7.33)
$\quad s = d_i; \; r_i = 0$
$\quad$ for $k = i+1, i+2, \ldots, n$:
$\quad\quad s = s + c_{ik} \times x_k$
$\quad x_i = -s/c_{ii}$
for $i = n+1, n+2, \ldots, N$:
$\quad r_i = d_i$
for $j = n, n-1, \ldots, 1$:
$\quad$ for $i = N, N-1, \ldots, j+1$:
$\quad\quad \rho = c_{ij}$
$\quad\quad$ if $\rho = 1$ then $\gamma = 0; \; \sigma = 1$
$\quad\quad$ else
$\quad\quad\quad$ if $|\rho| < 1$ then $\sigma = \rho; \; \gamma = \sqrt{1-\sigma^2}$
$\quad\quad\quad$ else $\gamma = 1/|\rho|; \; \sigma = \text{sgn}(\rho) \times \sqrt{1-\gamma^2}$
$\quad\quad h = \gamma \times r_j + \sigma \times r_i; \; r_i = -\sigma \times r_j + \gamma \times r_i; \; r_j = h$

*Example 7.3* The overdetermined system of Equations (7.13) of the Example 7.1 is solved by means of the method of orthogonal transformation (7.31). A calculation using six decimal digits yields the matrix $\hat{\mathbf{R}}$ of the QR decomposition of **C** with the values $\rho$ instead of the zeros, the transformed vector $\hat{\mathbf{d}}$ and the residual vector **r**

$$\hat{\mathbf{R}} = \begin{pmatrix} 2.645\,75 & 2.135\,49 & 2.469\,73 \\ -1.414\,21 & 1.404\,97 & 2.371\,87 \\ -0.577\,351 & -1.688\,78 & 0.617\,881 \\ -0.500\,000 & -1.516\,16 & -2.531\,86 \\ -0.447\,213 & -1.495\,16 & -2.194\,33 \\ -0.408\,248 & -1.469\,45 & -1.989\,63 \\ -0.377\,965 & -0.609\,538 & -0.635\,138 \end{pmatrix}$$

$$\hat{\mathbf{d}} = \begin{pmatrix} -8.405\,94 \\ -4.933\,53 \\ -3.464\,60 \\ 0.128\,686 \\ 0.234\,145 \\ 0.211\,350 \\ -0.165\,290 \end{pmatrix} \quad \mathbf{r} = \begin{pmatrix} -0.110\,017 \\ 0.237\,881 \\ 0.010\,8260 \\ -0.149\,594 \\ -0.085\,3911 \\ 0.190\,043 \\ -0.093\,7032 \end{pmatrix} \quad (7.34)$$

The solution vector $\boldsymbol{\alpha} = (2.749\,20, -5.954\,63, 5.607\,23)^T$ is obtained by the process of backsubstitution applied to the matrix $\mathbf{R}$ and the first three components of $\hat{\mathbf{d}}$. It differs from the 'exact' solution (7.17) at most by three units in the fifth decimal place and has an essentially smaller relative error in comparison with (7.16).

*Example* 7.4  The method of the orthogonal transformation applied to the system of error equations (7.20) of Example 7.2 yields an approximate solution with a smaller error. The essential results of a calculation to six decimal digits are

$$\hat{\mathbf{R}} = \begin{pmatrix} 1.325\,08 & 1.684\,92 \\ -2.236\,07 & 0.048\,6849 \\ -1.673\,32 & -2.472\,37 \\ -0.693\,334 & -0.653\,865 \\ -0.610\,277 & -0.403\,536 \\ -0.566\,004 & 0.086\,2074 \end{pmatrix}$$

$$\hat{\mathbf{d}} = \begin{pmatrix} -2.257\,73 \\ -0.050\,5901 \\ -0.045\,6739 \\ -0.031\,4087 \\ -0.077\,8043 \\ -0.031\,3078 \end{pmatrix} \quad \mathbf{r} = \begin{pmatrix} -0.047\,8834 \\ 0.036\,3745 \\ 0.048\,1185 \\ 0.036\,7843 \\ -0.046\,4831 \\ -0.025\,7141 \end{pmatrix} \quad (7.35)$$

The resulting parameter values $\alpha_1 = 0.382\,528$ and $\alpha_2 = 1.039\,13$ are now correct up to four decimal places.

The numerical examples suggest that there is a relation between the classical method of the normal equations and the method of orthogonal transformation. This is indeed the case, as can be seen from the fact that the matrix $\mathbf{C}$ of the error equations is decomposed into $\mathbf{C} = \mathbf{Q}\hat{\mathbf{R}}$ (7.23), so that for the matrix $\mathbf{A}$ of the normal equations, taking the orthogonality of $\mathbf{Q}$ and the structure of $\hat{\mathbf{R}}$ into account, we have

$$\mathbf{A} = \mathbf{C}^T\mathbf{C} = \hat{\mathbf{R}}^T\mathbf{Q}^T\mathbf{Q}\hat{\mathbf{R}} = \hat{\mathbf{R}}^T\hat{\mathbf{R}} = \mathbf{R}^T\mathbf{R}. \quad (7.36)$$

The Cholesky decomposition of a symmetric, positive definite matrix $\mathbf{A} = \mathbf{L}\mathbf{L}^T$ is unique if the diagonal elements of $\mathbf{L}$ are positive. Thus from (7.36) it follows that the matrix $\mathbf{R}$ must essentially coincide with $\mathbf{L}^T$, apart from possibly having different signs of rows. Although theoretically the two procedures yield essentially the same triangular matrices, there exists an essential numerical difference.

We first consider the generation of the diagonal elements of $\mathbf{R}$. The sequence of orthogonal Givens transformations keeps the Euclidean norms of the column

## 7.2 Methods of Orthogonal Transformation

vectors $\mathbf{c}_j$ of $\mathbf{C}$ invariant. Therefore, after the elimination of the elements in the first column we have $|r_{11}| = \|\mathbf{c}_1\|_2$. Since the changed element $c'_{12}$ remains unchanged during the elimination of the elements of the second column, it follows that for the second diagonal element $|r_{22}| = \|\mathbf{c}'_2 - c'_{12}\mathbf{e}_1\|_2 \leq \|\mathbf{c}_2\|_2$. In general we have $|r_{jj}| \leq \|\mathbf{c}_j\|_2$, $j = 1, 2, \ldots, n$. It is now important that the diagonal elements $r_{jj}$ arise from the given vectors $\mathbf{c}_j$ as Euclidean norms of partial vectors after orthogonal transformations.

The diagonal elements $l_{jj}$ of $\mathbf{L}$ of the Cholesky decomposition of $\mathbf{A} = \mathbf{L}\mathbf{L}^T$ result from the diagonal elements $a_{jj}$ after subtracting from them squares of matrix elements $l_{jk}$. According to (7.9), $a_{jj} = \mathbf{c}_j^T \mathbf{c}_j = \|\mathbf{c}_j\|_2^2$ equals the square of the Euclidean norm of the column vector $\mathbf{c}_j$. In the course of the Cholesky decomposition of $\mathbf{A}$, the calculation is performed with the squares of norms, up to the point when $l_{jj}$ is produced as the square root of the norm of a vector. From the fact that the square of the norm is reduced in the Cholesky procedure, the relative error caused by cancellation can be much larger in the case of a large reduction of the value compared to that produced by the orthogonal transformation which operates on the vectors $\mathbf{c}_j$. Therefore the matrix $\mathbf{R}$ is numerically more exact than the matrix $\mathbf{L}$ of the Cholesky decomposition.

Finally, the solution $\mathbf{x}$ of the error equations is computed from the system of equations (7.29) with the more accurate right triangular matrix $\mathbf{R}$ and the vector $\hat{\mathbf{d}}_1$. This is produced from $\mathbf{d}$ by a sequence of orthogonal transformations in a numerically stable way. Theoretically, the vector $\hat{\mathbf{d}}_1$ is essentially equal to the vector $\mathbf{y}$, that is obtained from $\mathbf{L}\mathbf{y} - \mathbf{b} = 0$ by the process of the forward substitution, but this is already less accurate than $\hat{\mathbf{d}}_1$ because of the lower accuracy of $\mathbf{L}$.

The amount of computational work of the procedure (7.31) can be reduced by applying the fast Givens transformation of Section 6.3.3. With this method from (7.27) we obtain the transformed matrix $\hat{\mathbf{R}}$ and the transformed vectors $\hat{\mathbf{d}}$ and $\hat{\mathbf{r}}$ in the following factored form

$$\hat{\mathbf{R}} = \mathbf{D}\tilde{\mathbf{R}} \quad \hat{\mathbf{d}} = \mathbf{D}\tilde{\mathbf{d}} \quad \hat{\mathbf{r}} = \mathbf{D}\tilde{\mathbf{r}} \quad \mathbf{D} \in \mathbf{R}^{N \times N} \tag{7.37}$$

The matrix $\mathbf{D}$ is nonsingular, thus it is irrelevant to the determination of the solution $\mathbf{x}$ from the system of error equations that is analogous to (7.28). If $\tilde{\mathbf{R}}$ denotes the right triangular matrix in $\hat{\mathbf{R}}$ and $\hat{\tilde{\mathbf{d}}}_1$ the vector defined by the first $n$ components of $\tilde{\mathbf{d}}$ we obtain $\mathbf{x}$ from

$$\tilde{\mathbf{R}}\mathbf{x} + \hat{\tilde{\mathbf{d}}}_1 = 0 \tag{7.38}$$

The computation of the residual vector $\mathbf{r}$ from $\hat{\mathbf{r}}$, analogous to (7.30), requires the information about the transformations. As two numbers are necessary for each step, it is more appropriate to compute $\mathbf{r}$ from the given error equations $\mathbf{C}\mathbf{x} + \mathbf{d} = \mathbf{r}$. If only the sum of squares of the residuals are of interest, it can be directly obtained from the last $(N - n)$ components of $\mathbf{d}$ and the corresponding diagonal elements of $\mathbf{D}$.

The amount of computational effort required to eliminate the current matrix elements $c_{ij}$, $i = j+1, j+2, \ldots, N$ of the $j$th column including the transformation of the components of the vector **d** consists of $(N-j)(2n-2j+12)$ operations. After summation over $j$, from 1 to $n$, the total amount of work for the determination of the solution **x** and the residual vector **r** is

$$Z_{\text{eeqFG}} = nN(n+12) - \tfrac{1}{3}n^3 - \tfrac{11}{2}n^2 - \tfrac{31}{6}n. \qquad (7.39)$$

In comparison with (7.32) the number of multiplicative operations is halved for large $n$ and $N$, and, moreover, all square roots are eliminated. For small $n$ and $N$ the term $12nN$ dominates, so that the fast version is less efficient. This is true for Examples 7.3 and 7.4.

### 7.2.2 Special computational techniques

Systems of error equations originating from problems of surveying have the property that the matrix **C** is *sparse*, because each error equation contains only a few unknowns compared to $n$. The method of orthogonal transformation, as described in the last section, can take the sparsity of **C** into account by omitting rotations corresponding to vanishing matrix elements $c_{ij}$. Moreover, it is obvious from the sequence (7.25), that the matrix elements $c_{ij}$ of the $i$th row with $i > j$, which are equal to zero in the given matrix **C**, and situated to the left of the first nonvanishing element, remain unchanged during the course of the elimination. In order to make use of this fact, with respect to storage requirements, we denote

$$f_i(\mathbf{C}) := \min\{j | c_{ij} \neq 0, j = 1, 2, \ldots, n\} \quad (i = 1, 2, \ldots, N) \qquad (7.40)$$

to be the index of the first matrix element of **C** of the $i$th row that is nonzero. The nonsingular right triangular matrix **R** will be produced in the first $n$ rows of **C** thus we assume that

$$f_i(\mathbf{C}) \leq i \quad (i = 1, 2, \ldots, n) \qquad (7.41)$$

The condition (7.41) can always be satisfied by a proper permutation of the error equations, or if need be by a different numbering of the unknowns. On this assumption, the orthogonal transformation of **C** can be performed with those matrix elements $c_{ij}$ whose index pairs $(i, j)$ belong to the *envelope* of **C**, defined by

$$\text{Env}(\mathbf{C}) := \{(i,j) | f_i(\mathbf{C}) \leq j \leq n; i = 1, 2, \ldots, N\}. \qquad (7.42)$$

To store the relevant matrix elements in an economical way, they are arranged rowwise in a one-dimensional array, such that each row starts with the first nonvanishing element. For the matrix $\mathbf{C} \in \mathbf{R}^{6 \times 4}$

## 7.2 Methods of Orthogonal Transformation

$$C = \begin{pmatrix} c_{11} & 0 & c_{13} & c_{14} \\ 0 & c_{22} & 0 & c_{24} \\ 0 & c_{32} & c_{33} & 0 \\ 0 & 0 & c_{43} & c_{44} \\ 0 & 0 & c_{53} & c_{54} \\ c_{61} & 0 & 0 & c_{64} \end{pmatrix} \quad \text{with} \quad \begin{array}{l} f_1 = 1 \\ f_2 = 2 \\ f_3 = 2 \\ f_4 = 3 \\ f_5 = 3 \\ f_6 = 1 \end{array} \tag{7.43}$$

the arrangement is as follows:

C: | $c_{11}c_{12}c_{13}c_{14}$ | $c_{22}c_{23}c_{24}$ | $c_{32}c_{33}c_{34}$ | $c_{43}c_{44}$ | $c_{53}c_{54}$ | $c_{61}c_{62}c_{63}c_{64}$ |

$$\tag{7.44}$$

For those vanishing matrix elements of $C$, whose index pairs belong to the envelope, storage places must be provided because they may become nonzero during the process. In order to have access to the matrix elements $c_{ij}$ in the arrangement (7.44) an auxiliary vector $z \in R^N$ is needed whose $i$th component indicates the position of the last matrix element $c_{in}$ of the $i$th row. For (7.44) it is given by

$$z = (4, 7, 10, 12, 14, 18)^T.$$

The matrix element $c_{ij}$ with $(i,j) \in \text{Env}(C)$ is then the $k$th element of a general arrangement (7.44) where $k$ is given by

$$(i,j) \in \text{Env}(C) \rightarrow k = z_i + j - n. \tag{7.45}$$

The transformation of $C$ into $\hat{R}$ is possible without test if the matrix elements $c_{ij}$ with $(i,j) \in \text{Env}(C)$ are eliminated rowwise instead of columnwise. Analogous to the proof of Theorem 6.4 it can be verified that the corresponding sequence of rotations for the rowwise elimination achieves the desired transformation. The index of the first nonvanishing matrix element of the $i$th row is defined by

$$f_i(C) = n - z_i + z_{i-1} + 1 \quad (i = 2, 3, \ldots, N). \tag{7.46}$$

With these hints it is easy to adapt the algorithm (7.33). The loop statements for $i$ and $j$ have to be interchanged, the initial value for $j$ has to be replaced by $f_i(C)$, (7.46), and the indices of the matrix elements $c_{ij}$ have to be replaced according to (7.45).

So far we have assumed that the matrix $C$ and the vector $d$ are completely stored for an implementation of the method. Now we consider the case where the error equations either are built up successively, or can be read from an external storage device. We discuss a variant to compute the solution $x$ of the error equations (7.2) with a minimum amount of storage (George and Heath, 1980).

This is of some importance to the treatment of large systems of error equations on small computers.

In detail the transformation is assumed to be performed rowwise, and the vector $\hat{\mathbf{d}}$ is computed simultaneously. We investigate the treatment of the $i$th error equation and its contribution to the resulting system (7.29). The first error equation obviously contributes the initial values of the first row of $\mathbf{R}$ and of the first component of $\hat{\mathbf{d}}_1$. For $2 \leqslant i \leqslant n$ at most $(i-1)$ rotations must be performed to eliminate the $(i-1)$ elements $c_{ij}, j = 1, 2, \ldots, i-1$. The remaining transformed equation yields the initial values of the $i$th row of $\mathbf{R}$ and the $i$th component of $\hat{\mathbf{d}}_1$. In the subsequent error equations $(i > n)$ all $n$ coefficients $c_{ij}$ have to be eliminated by corresponding rotations by means of the $n$ rows of the arising matrix $\mathbf{R}$. As such an $i$th error equation remains unchanged after completing its treatment it is no longer required. According to (7.28), the transformed constant term $\tilde{d}_i$ at most contributes itself to the sum of treated squares of the residuals. Hence, the error equations can be treated independently, and the system of equations (7.29) is built gradually. The implementation only requires storage space for the right triangular matrix $\mathbf{R}$, for the constant vector $\hat{\mathbf{d}}_1$ and for one error equation. Therefore, the amount of required storage space is only about $S \approx n(n+1)/2 + 2n$, if $\mathbf{R}$ is stored accordingly.

In the following algorithmic description of the above process, we assume the usual indexing is used for the matrix $\mathbf{R} = (r_{ij})$, for clarity the components of the vector $\hat{\mathbf{d}}_1$ are denoted by $d_i$, and the $i$th error equation is written without the index $i$ in the form

$$c_1 x_1 + c_2 x_2 + \cdots + c_n x_n + \tilde{d} = r$$

whose coefficients have to be defined as indicated. The process of the backward substitution for solving $\mathbf{R}\mathbf{x} + \mathbf{d}_1 = \mathbf{0}$ can be taken from (7.33).

```
for i = 1, 2, ..., N:
    define c₁, c₂, ..., cₙ, d̃
    for j = 1, 2, ..., min(i − 1, n):
        if cⱼ ≠ 0 then
            if |rⱼⱼ| < δ × |cⱼ| then
                w = −cⱼ; γ = 0; σ = 1
            else
                w = sgn(rⱼⱼ) × √(rⱼⱼ² + cⱼ²)
                γ = rⱼⱼ/w; σ = −cⱼ/w
            rⱼⱼ = w
            for k = j + 1, j + 2, ..., n:
                h = γ × rⱼₖ − σ × cₖ
                cₖ = σ × rⱼₖ + γ × cₖ; rⱼₖ = h
            h = γ × dⱼ − σ × d̃; d̃ = σ × dⱼ + γ × d̃; dⱼ = h
```

## 7.2 Methods of Orthogonal Transformation

> if $i \leqslant n$ then
>   for $k = i, i+1, \ldots, n$:
>   $r_{ik} = c_k$
> $d_i = \tilde{d}$

### 7.2.3 Householder transformation

A different approach to the problem of transforming the system of error equations $\mathbf{Cx} + \mathbf{d} = \mathbf{r}$ orthogonally into the equivalent system (7.27) $\hat{\mathbf{R}}\mathbf{x} + \hat{\mathbf{d}} = \hat{\mathbf{r}}$ consists of the use of so called *Householder matrices* (Householder, 1958)

$$\mathbf{U} := \mathbf{I} - 2\mathbf{w}\mathbf{w}^T \quad \text{with} \quad \mathbf{w}^T\mathbf{w} = 1 \quad \mathbf{w} \in \mathbf{R}^N \quad \mathbf{U} \in \mathbf{R}^{N \times N}. \tag{7.47}$$

The matrix $\mathbf{U}$ (7.47) is symmetric and *involutory* as, $\|\mathbf{w}\|_2$ equal to one, we have

$$\mathbf{UU} = (\mathbf{I} - 2\mathbf{w}\mathbf{w}^T)(\mathbf{I} - 2\mathbf{w}\mathbf{w}^T) = \mathbf{I} - 2\mathbf{w}\mathbf{w}^T - 2\mathbf{w}\mathbf{w}^T + 4\mathbf{w}\mathbf{w}^T\mathbf{w}\mathbf{w}^T = \mathbf{I}.$$

From this it follows, noting $\mathbf{U}^T = \mathbf{U}$, that $\mathbf{U}$ is orthogonal. The Householder matrix $\mathbf{U}$, interpreted as the matrix representation of a linear map from $\mathbf{R}^N$ onto $\mathbf{R}^N$, corresponds to a reflection with respect to a certain hyperplane. To see this let $\mathbf{s} \in \mathbf{R}^N$ be an arbitrary vector that is orthogonal to $\mathbf{w}$. Its image vector

$$\mathbf{s}' := \mathbf{U}\mathbf{s} = (\mathbf{I} - 2\mathbf{w}\mathbf{w}^T)\mathbf{s} = \mathbf{s} - 2\mathbf{w}(\mathbf{w}^T\mathbf{s}) = \mathbf{s}$$

is identical to $\mathbf{s}$. The image vector of a vector $\mathbf{z} \in \mathbf{R}^N$ that is proportional to $\mathbf{w}$, that is $\mathbf{z} = c\mathbf{w}$,

$$\mathbf{z}' := \mathbf{U}\mathbf{z} = c\mathbf{U}\mathbf{w} = c(\mathbf{I} - 2\mathbf{w}\mathbf{w}^T)\mathbf{w} = c(\mathbf{w} - 2\mathbf{w}(\mathbf{w}^T\mathbf{w})) = -c\mathbf{w} = -\mathbf{z}$$

is opposite to $\mathbf{z}$. An arbitrary $\mathbf{x} \in \mathbf{R}^N$ is the unique sum $\mathbf{x} = \mathbf{s} + \mathbf{z}$ of two vectors $\mathbf{s}$ and $\mathbf{z}$ with the mentioned properties. Therefore for its image we have

$$\mathbf{x}' := \mathbf{U}\mathbf{x} = \mathbf{U}(\mathbf{s} + \mathbf{z}) = \mathbf{s} - \mathbf{z}$$

so that the vector $\mathbf{x}$ is reflected with respect to the hyperplane orthogonal to $\mathbf{w}$ and passing through the origin.

The practical significance of Householder matrices (7.47) is that by an appropriate choice of the vector $\mathbf{w}$, specific components of the image vector $\mathbf{x}' = \mathbf{U}\mathbf{x}$ of an arbitrary vector $\mathbf{x} \in \mathbf{R}^N$ can be made to vanish. Of course, the squares of the Euclidean norm of the two vectors $\mathbf{x}$ and $\mathbf{x}'$ are equal. We exploit this fact in the following to transform a given matrix $\mathbf{C} \in \mathbf{R}^{N \times n}$ into the desired form $\hat{\mathbf{R}}$ (7.23) by using a sequence of $n$ Householder matrices, each treating an individual column.

In the first transformation step, an appropriate Householder matrix $\mathbf{U}_1 = \mathbf{I} - 2\mathbf{w}_1\mathbf{w}_1^T$ is chosen in such a way that the first column of the matrix $\mathbf{C}' = \mathbf{U}_1\mathbf{C}$ equals a multiple of the first unit vector $\mathbf{e}_1$. Together with the condition $\|\mathbf{w}_1\| = 1$, $N$ conditions have to be satisfied. Therefore we set $\mathbf{w}_1 = (w_1, w_2, \ldots, w_N)^T$ whose $N$ components are to be determined from these requirements. Let us denote by $\mathbf{c}_1$

the first column of **C**. Then for the first column of **C'**

$$\mathbf{c}'_1 = \mathbf{U}_1\mathbf{c}_1 = (\mathbf{I} - 2\mathbf{w}_1\mathbf{w}_1^T)\mathbf{c}_1 = \mathbf{c}_1 - 2\mathbf{w}_1(\mathbf{w}_1^T\mathbf{c}_1) = \mathbf{c}_1 - 2(\mathbf{w}_1^T\mathbf{c}_1)\mathbf{w}_1. \tag{7.48}$$

Consequently for the components of $\mathbf{c}'_1$ we get the following representations and conditions

$$c'_{11} = c_{11} - 2w_1 \sum_{i=1}^{N} w_i c_{i1}, \tag{7.49}$$

$$c'_{k1} = c_{k1} - 2w_k \sum_{i=1}^{N} w_i c_{i1} = 0 \quad (k = 2, 3, \ldots, N). \tag{7.50}$$

As the Euclidean norms of the vectors $\mathbf{c}_1$ and $\mathbf{c}'_1$ are equal we must have, from (7.50),

$$(c'_{11})^2 = \sum_{i=1}^{N} c_{i1}^2 =: s^2 \quad s > 0 \tag{7.51}$$

where $s$ is a known quantity which can be computed from the elements of the first column of **C**. We introduce the auxiliary quantity

$$h := 2 \sum_{i=1}^{N} w_i c_{i1} \tag{7.52}$$

and rewriting (7.49) to (7.51) we get the set of equations

$$c'_{11} = c_{11} - w_1 h = \pm s \tag{7.53}$$

$$c_{k1} - w_k h = 0 \quad (k = 2, 3, \ldots, N). \tag{7.54}$$

Multiplying (7.53) by $w_1$ and the $k$th equation of (7.54) by $w_k$ and then adding all equations we obtain

$$\sum_{k=1}^{N} w_k c_{k1} - h \sum_{k=1}^{N} w_k^2 = \tfrac{1}{2}h - h = -\tfrac{1}{2}h = \pm sw_1. \tag{7.55}$$

It follows from (7.53) and (7.55) that

$$c_{11} \pm 2sw_1^2 = \pm s \quad \text{or} \quad 2sw_1^2 = s \mp c_{11}.$$

In order to determine the value $w_1$ from the last equation in a numerically safe way, avoiding cancellation of digits, we choose the sign to be equal to that of $c_{11}$. We summarize the determination of the components of $\mathbf{w}_1$:

$$\boxed{\begin{aligned} s &= + \sqrt{\sum_{i=1}^{N} c_{i1}^2} \\ w_1 &= + \sqrt{\tfrac{1}{2}\left(1 + \frac{|c_{11}|}{s}\right)} \end{aligned}}$$

## 7.2 Methods of Orthogonal Transformation

$$h = \begin{cases} 2sw_1 & \text{if } c_{11} \geq 0 \\ -2sw_1 & \text{if } c_{11} < 0 \end{cases}$$

$$w_k = \frac{c_{k1}}{h} \quad (k = 2, 3, \ldots, N)$$

(7.56)

With the Householder matrix $\mathbf{U}_1 = \mathbf{I} - 2\mathbf{w}_1\mathbf{w}_1^T$ thus defined, the transformation of the matrix $\mathbf{C}$ is performed most efficiently as follows. The general element of $\mathbf{C}'$ is given by

$$c'_{ij} = \sum_{k=1}^{N} (\delta_{ik} - 2w_i w_k) c_{kj} = c_{ij} - 2w_i \sum_{k=1}^{N} w_k c_{kj} =: c_{ij} - w_i p_j$$

(7.57)

where we have introduced the auxiliary sums

$$p_j := 2 \sum_{k=1}^{N} w_k c_{kj} \quad (j = 2, 3, \ldots, n).$$

(7.58)

If we take (7.53) into account, the essential elements of $\mathbf{C}'$ are determined according to

$$\left. \begin{aligned} c'_{11} &= \begin{cases} -s & \text{if } c_{11} \geq 0 \\ s & \text{if } c_{11} < 0 \end{cases} \\ p_j &= 2 \sum_{k=1}^{N} w_k c_{kj} \\ c'_{ij} &= c_{ij} - w_i p_j \quad (i = 1, 2, \ldots, N) \end{aligned} \right\} \quad (j = 2, 3, \ldots, n)$$

(7.59)

The second column of $\mathbf{C}'$ is treated completely analogously by means of a Householder matrix $\mathbf{U}_2 = \mathbf{I} - 2\mathbf{w}_2\mathbf{w}_2^T$ with $\mathbf{w}_2 = (0, w_2, w_3, \ldots, w_N)^T$ leading to $\mathbf{C}'' = \mathbf{U}_2\mathbf{C}'$. The first component of $\mathbf{w}_2$ is set equal to zero for two reasons. We only want to eliminate the $(N-2)$ last elements of the second column of $\mathbf{C}'$. thus together with $\mathbf{w}_2^T\mathbf{w}_2 = 1$ we have only $(N-1)$ conditions to satisfy. On the other hand the first row and the first column of $\mathbf{C}'$ will not be changed by the multiplication by $\mathbf{U}_2$. The last $(N-1)$ components of $\mathbf{w}_2$ are obtained analogously to (7.56), and the elements of $\mathbf{C}''$ are computed, after some minor modifications in a similar manner to (7.59).

After the first $(l-1)$ columns of $\mathbf{C}$ have been transformed to the desired form by means of Householder matrices $\mathbf{U}_1, \mathbf{U}_2, \ldots, \mathbf{U}_{l-1}$, the $l$th column is treated with $\mathbf{U}_l = \mathbf{I} - 2\mathbf{w}_l\mathbf{w}_l^T$ where the first $(l-1)$ components of $\mathbf{w}_l = (0, 0, \ldots, 0, w_l, w_{l+1}, \ldots, w_N)^T$ vanish. Therefore the first $(l-1)$ columns and rows of the current matrix $\mathbf{C}$ remain unchanged by the corresponding step of reduction.

After $n$ transformation steps, the aim (7.27) is achieved with

$$\mathbf{U}_n\mathbf{U}_{n-1}\cdots\mathbf{U}_2\mathbf{U}_1\mathbf{C} = \hat{\mathbf{R}} \quad \mathbf{U}_n\mathbf{U}_{n-1}\cdots\mathbf{U}_2\mathbf{U}_1\mathbf{d} = \hat{\mathbf{d}}. \tag{7.60}$$

The desired solution $\mathbf{x}$ is finally obtained from the system (7.29) by the process of backsubstitution.

If the residual vector $\mathbf{r}$ is also desired we have

$$\mathbf{U}_n\mathbf{U}_{n-1}\cdots\mathbf{U}_2\mathbf{U}_1\mathbf{r} = \hat{\mathbf{r}} = (0,0,\ldots,0,\hat{d}_{n+1},\hat{d}_{n+2},\ldots,\hat{d}_N)^\mathrm{T}. \tag{7.61}$$

From the symmetry of the Householder matrices $\mathbf{U}_l$ from (7.61) it follows that

$$\mathbf{r} = \mathbf{U}_1\mathbf{U}_2\cdots\mathbf{U}_{n-1}\mathbf{U}_n\hat{\mathbf{r}}. \tag{7.62}$$

The known residual vector $\hat{\mathbf{r}}$ must be multiplied successively by the matrices $\mathbf{U}_n, \mathbf{U}_{n-1},\ldots, \mathbf{U}_1$. The general multiplication of a vector $\mathbf{y}\in\mathbf{R}^N$ by $\mathbf{U}_l$ requires, noting that,

$$\mathbf{y}' = \mathbf{U}_l\mathbf{y} = (\mathbf{I} - 2\mathbf{w}_l\mathbf{w}_l^\mathrm{T})\mathbf{y} = \mathbf{y} - 2(\mathbf{w}_l^\mathrm{T}\mathbf{y})\mathbf{w}_l \tag{7.63}$$

the calculation of the scalar product $\mathbf{w}_l^\mathrm{T}\mathbf{y} =: a$ and then the subtraction of $(2a)\cdot\mathbf{w}_l$ from $\mathbf{y}$. As the first $(l-1)$ components of $\mathbf{w}_l$ are zero, such a step requires $2(N-l+1)$ multiplications. Therefore the back transformation of $\hat{\mathbf{r}}$ into $\mathbf{r}$ following (7.62) requires a total of $Z_{\text{back}} = n(2N - n + 1)$ operations.

To compute $\mathbf{r}$, we need the vectors $\mathbf{w}_1, \mathbf{w}_2,\ldots, \mathbf{w}_n$ that define the Householder matrices. The components of $\mathbf{w}_l$ can be stored in the matrix $\mathbf{C}$ in the $l$th column at the place of $c_{ll}$ and the eliminated elements $c_{il}$. This is reasonable because the final diagonal elements $r_{ii} = c_{ii}$ of the right triangular matrix $\mathbf{R}$ play a special role during the back substitution. Therefore they are stored as the elements of a vector $\mathbf{t}\in\mathbf{R}^n$.

In the following algorithmic description of the Householder transformation for solving a system of error equations $\mathbf{Cx} + \mathbf{d} = \mathbf{r}$, the orthogonal transformation of $\mathbf{C}$ and then the process of computing $\hat{\mathbf{d}}$, the solution $\mathbf{x}$ and the residual vector $\mathbf{r}$ are given separately. This allows the successive treatment of several systems of error equations, with the same matrix $\mathbf{C}$ but with different vectors $\mathbf{d}$. The nonvanishing components of the vectors $\mathbf{w}_l$ are stored in the corresponding columns of $\mathbf{C}$. Therefore no vector $\mathbf{w}$ is needed, but the formulae (7.59) have to be modified. Moreover, it is not necessary to use suffixed values $p_j$ (7.59). The computation of the vector $\hat{\mathbf{d}}$ from $\mathbf{d}$ according to (7.60) follows the formulae (7.63), as does the residual vector $\mathbf{r}$.

---

for $l = 1, 2, \ldots, n$:
　$s = 0$
　for $i = l, l+1, \ldots, N$:
　　$s = s + c_{il}^2$
　$s = \sqrt{s}$;　$w = \sqrt{(1 + |c_{ll}|/s)/2}$

---

## 7.2 Methods of Orthogonal Transformation

$$
\begin{aligned}
&\text{if } c_{ll} \geq 0 \text{ then } h = 2 \times s \times w;\ t_l = -s \\
&\qquad\text{else } h = -2 \times s \times w;\ t_l = s \\
&c_{ll} = w \\
&\text{for } k = l+1, l+2, \ldots, N: \\
&\quad c_{kl} = c_{kl}/h \\
&\text{for } j = l+1, l+2, \ldots, n: \\
&\quad p = 0 \\
&\quad \text{for } k = l, l+1, \ldots, N: \\
&\qquad p = p + c_{kl} \times c_{kj} \\
&\quad p = p + p \\
&\quad \text{for } i = l, l+1, \ldots, N: \\
&\qquad c_{ij} = c_{ij} - p \times c_{il}
\end{aligned}
\tag{7.64}
$$

$$
\begin{aligned}
&\text{for } l = 1, 2, \ldots, n: \\
&\quad s = 0 \\
&\quad \text{for } k = l, l+1, \ldots, N: \\
&\qquad s = s + c_{kl} \times d_k \\
&\quad s = s + s \\
&\quad \text{for } k = l, l+1, \ldots, N: \\
&\qquad d_k = d_k - s \times c_{kl} \\
&\text{for } i = n, n-1, \ldots, 1: \\
&\quad s = d_i;\ r_i = 0 \\
&\quad \text{for } k = i+1, i+2, \ldots, n: \\
&\qquad s = s + c_{ik} \times x_k \\
&\quad x_i = -s/t_i \\
&\text{for } i = n+1, n+2, \ldots, N: \\
&\quad r_i = d_i \\
&\text{for } l = n, n-1, \ldots, 1: \\
&\quad s = 0 \\
&\quad \text{for } k = l, l+1, \ldots, N: \\
&\qquad s = s + c_{kl} \times r_k \\
&\quad s = s + s \\
&\quad \text{for } k = l, l+1, \ldots, N: \\
&\qquad r_k = r_k - s \times c_{kl}
\end{aligned}
\tag{7.65}
$$

The Householder transformation method solves the problem with the smallest computational effort. The $l$th step of the algorithm (7.64) requires $(N - l + 1) + 4 + (N - l) + 2(n - l)(N - l + 1) = 2(N - l + 1)(n - l + 1) + 3$ multiplications and 2 square roots. After summation over $l$, from 1 to $n$, the total amount of computational effort required for the orthogonal transformation of $\mathbf{C}$ is

$$Z_{\text{Householder}} = Nn(n+1) - \tfrac{1}{3}n(n^2 - 10)$$

multiplicative operations and $2n$ square roots. The amount of work needed to compute $\hat{\mathbf{d}}$ is the same as for the back transformation of the residual vector. Therefore the algorithm (7.65) requires

$$2n(2N - n + 1) + \tfrac{1}{2}n(n + 1) = 4Nn - \tfrac{3}{2}n^2 + \tfrac{5}{2}n$$

multiplications, and the total amount of effort consists of

$$\boxed{Z_{\text{eeq Householder}} = nN(n + 5) - \tfrac{1}{3}n^3 - \tfrac{3}{2}n^2 + \tfrac{35}{6}n} \qquad (7.66)$$

essential operations and $2n$ square roots. In comparison with (7.32), the number of essential operations is, in fact, only about half as many, with the number of square roots being much smaller. Even in comparison with the method of fast Givens transformations, the Householder method is slightly more efficient.

The implementation of the Householder transformation assumes that the matrix $\mathbf{C}$ is stored completely. The treatment of sparse systems of error equations requires additional considerations, for example, keeping the fill-in as small as possible (Duff and Reid, 1976; Gill and Murray, 1976; Golub and Plemmons, 1980; Golub and Van Loan, 1983). Finally, it is impossible to compute the matrix $\mathbf{R}$ successively by treating the error equations separately.

*Example* 7.5 For the system of error equations (7.20), on the basis of a calculation to six decimal digits, the Householder transformation yields the matrix $\tilde{\mathbf{C}}$, which contains the vectors $\mathbf{w}_i$ and the off-diagonal element of $\hat{\mathbf{R}}$, the transformed vector $\hat{\mathbf{d}}$ and the vector $\mathbf{t}$ with the diagonal elements of $\mathbf{R}$

$$\tilde{\mathbf{C}} = \begin{pmatrix} 0.750\,260 & -1.684\,91 \\ 0.167\,646 & 0.861\,177 \\ 0.251\,470 & -0.078\,9376 \\ 0.301\,764 & -0.313\,223 \\ 0.335\,293 & -0.365\,641 \\ 0.377\,205 & -0.142\,624 \end{pmatrix} \qquad \hat{\mathbf{d}} = \begin{pmatrix} 2.257\,73 \\ -0.050\,584\,0 \\ 0.068\,483\,8 \\ 0.073\,162\,8 \\ -0.005\,104\,70 \\ 0.006\,163\,10 \end{pmatrix}$$

$$\mathbf{t} = \begin{pmatrix} -1.325\,08 \\ 0.048\,6913 \end{pmatrix}$$

The backward substitution provides the parameter values $\alpha_1 = 0.382\,867$ and $\alpha_2 = 1.038\,87$, which are correct only up to three decimal places. The size of the errors must be expected according to an error analysis (Kaufman, 1979; Lawson and Hanson, 1974).

## 7.3 Singular Value Decomposition

In Theorem 7.1 a decomposition (7.23) of the matrix $C \in \mathbf{R}^{N \times n}$ is introduced on the assumption that $C$ is of maximal rank $n < N$ in order to apply it to the solution of error equations. We will now get acquainted with a more general orthogonal decomposition of a matrix, where the assumption on the maximal rank can be omitted. This decomposition allows us to describe the set of solutions of the system of error equations in this general situation and to characterize a certain solution with a special property. In a preliminary step we give a generalization of Theorem 7.1. In view of its application we shall formulate it only in the following situation.

*Theorem 7.2* For each matrix $\mathbf{C} \in \mathbf{R}^{N \times n}$ of rank $r < n < N$ there exist two orthogonal matrices $\mathbf{Q} \in \mathbf{R}^{N \times N}$ and $\mathbf{W} \in \mathbf{R}^{n \times n}$ such that

$$\mathbf{Q}^T \mathbf{C} \mathbf{W} = \hat{\mathbf{R}} \quad \text{with} \quad \hat{\mathbf{R}} = \left(\begin{array}{c|c} \mathbf{R} & \mathbf{0}_1 \\ \hline \mathbf{0}_2 & \mathbf{0}_3 \end{array}\right) \quad \hat{\mathbf{R}} \in \mathbf{R}^{N \times n} \quad \mathbf{R} \in \mathbf{R}^{r \times r} \quad (7.67)$$

where $\mathbf{R}$ denotes a nonsingular right triangular matrix of order $r$ and the $\mathbf{0}_i$, $i = 1, 2, 3$, are zero matrices.

*Proof* Let $\mathbf{P} \in \mathbf{R}^{N \times N}$ be a permutation matrix such that in $\mathbf{C}' = \mathbf{P}\mathbf{C}$ the first $r$ row vectors are linearly independent. When the matrix $\mathbf{C}'$ is successively multiplied from the right by rotation matrices $\mathbf{U}(p, q; \varphi) \in \mathbf{R}^{n \times n}$ with the index pairs

$$(1, 2), (1, 3), \ldots, (1, n), (2, 3), (2, 4), \ldots, (2, n), \ldots, (r, r+1), \ldots, (r, n)$$

and with appropriate rotation angles, determined analogously to (6.104), the current matrix elements of the first $r$ rows are eliminated in the order

$$c'_{12}, c'_{13}, \ldots, c'_{1n}, c'_{23}, c'_{24}, \ldots, c'_{2n}, \ldots, c'_{r,r+1}, \ldots, c'_{rn}.$$

We denote the product of the mentioned rotation matrices by $\mathbf{W} \in \mathbf{R}^{n \times n}$. The transformed matrix has the structure

$$\mathbf{C}'\mathbf{W} = \mathbf{P}\mathbf{C}\mathbf{W} = \mathbf{C}'' = \left(\begin{array}{c|c} \mathbf{L} & \mathbf{0}_1 \\ \hline \mathbf{X} & \mathbf{0}_2 \end{array}\right) \quad \mathbf{0}_1 \in \mathbf{R}^{r \times (n-r)} \quad \mathbf{0}_2 \in \mathbf{R}^{(N-r) \times (n-r)} \quad (7.68)$$

where $\mathbf{L} \in \mathbf{R}^{r \times r}$ is a nonsingular left triangular matrix, $\mathbf{X} \in \mathbf{R}^{(N-r) \times r}$ is in general a nonzero matrix and $\mathbf{0}_1$, $\mathbf{0}_2$ denote zero matrices. As a consequence of the assumption on $\mathbf{C}'$, the first $r$ rows of $\mathbf{C}''$ are linearly independent, and hence $\mathbf{L}$ is nonsingular. As the rank of $\mathbf{C}''$ is equal to $r$, $\mathbf{0}_2$ is necessarily a zero matrix.

Following Theorem 7.1 there exists an orthogonal matrix $\mathbf{Q}_1 \in \mathbf{R}^{N \times N}$ such that $\mathbf{Q}_1^T \mathbf{C}'' = \hat{\mathbf{R}}$ has the property (7.67). On the basis of the constructive proof there, it is obviously sufficient to treat the first $r$ columns, because the vanishing elements of the last $(n - r)$ columns of $\mathbf{C}''$ are unchanged. The orthogonal matrix $\mathbf{Q}$ of the statement (7.67) is given by $\mathbf{Q}^T = \mathbf{Q}_1^T \mathbf{P}$.

*Theorem 7.3* For each matrix $C \in \mathbf{R}^{N \times n}$ of rank $r \leq n < N$ there exist two orthogonal matrices $U \in \mathbf{R}^{N \times N}$ and $V \in \mathbf{R}^{n \times n}$ such that the singular value decomposition

$$C = U\hat{S}V^T \quad \text{with} \quad \hat{S} = \begin{pmatrix} S \\ \hline 0 \end{pmatrix} \quad \hat{S} \in \mathbf{R}^{N \times n} \quad S \in \mathbf{R}^{n \times n} \tag{7.69}$$

holds, where $S$ is a diagonal matrix with nonnegative diagonal elements $s_i$ forming a nonincreasing sequence $s_1 \geq s_2 \geq \cdots \geq s_r > s_{r+1} = \cdots = s_n = 0$, and $\mathbf{0}$ represents a zero matrix.

*Proof* We first consider the case $r = n$ and will subsequently treat the general situation $r < n$. For $r = n$ the matrix $A := C^T C \in \mathbf{R}^{n \times n}$ is symmetric and positive definite. Its real and positive eigenvalues $s_i^2$ are assumed to be arranged in the nonincreasing order $s_1^2 \geq s_2^2 \geq \cdots \geq s_n^2 > 0$. From the theorem of principal axes there is an orthogonal matrix $V \in \mathbf{R}^{n \times n}$ such that

$$V^T A V = V^T C^T C V = D \quad \text{with} \quad D = \mathrm{diag}(s_1^2, s_2^2, \ldots, s_n^2) \tag{7.70}$$

holds. Moreover, let $S$ be the nonsingular diagonal matrix with the positive values $s_i$ along the diagonal. Then we define the matrix

$$\hat{U} := CVS^{-1} \in \mathbf{R}^{N \times n} \tag{7.71}$$

which consists of $n$ orthonormal column vectors from $\hat{U}^T \hat{U} = S^{-1} V^T C^T C V S^{-1} = I_n$, where (7.70) has been used. These can be completed into an orthonormal basis of $\mathbf{R}^N$, and therefore we can extend $\hat{U}$ to an orthogonal matrix $U := (\hat{U}, Y) \in \mathbf{R}^{N \times N}$ where $Y^T \hat{U} = Y^T CVS^{-1} = 0$. As a consequence we obtain, in this case with $r = n$,

$$U^T C V = \begin{pmatrix} S^{-1} V^T C^T \\ \hline Y^T \end{pmatrix} CV = \begin{pmatrix} S^{-1} V^T C^T C V \\ \hline Y^T C V \end{pmatrix} = \begin{pmatrix} S \\ \hline 0 \end{pmatrix} = \hat{S}$$

the statement (7.69).

We now apply this result to show (7.69) in the case $r < n$. Following Theorem 7.2 there are orthogonal matrices $Q$ and $W$ such that $Q^T C W = \hat{R}$ with the matrix $\hat{R}$ (7.67). The submatrix of $\hat{R}$, built with the first $r$ columns, has maximal rank $r$ and as a consequence there exist two orthogonal matrices $\tilde{U}$ and $\tilde{V}$, such that

$$\begin{pmatrix} R \\ \hline 0 \end{pmatrix} = \tilde{U}\tilde{S}\tilde{V}^T = \tilde{U}\begin{pmatrix} S_1 \\ \hline 0 \end{pmatrix}\tilde{V}^T \quad \tilde{U} \in \mathbf{R}^{N \times N}$$

$$\cdot \tilde{V} \in \mathbf{R}^{r \times r} \quad \tilde{S} \in \mathbf{R}^{N \times r} \quad S_1 \in \mathbf{R}^{r \times r}$$

holds. We extend the matrix $\tilde{S}$ by $(n - r)$ zero vectors and the matrix $\tilde{V}$ to an

## 7.3 Singular Value Decomposition

orthogonal matrix as follows

$$\hat{S} := \left(\begin{array}{c|c} S_1 & 0 \\ \hline 0 & 0 \end{array}\right) \in \mathbf{R}^{N \times n} \quad \hat{V} := \left(\begin{array}{c|c} \tilde{V} & 0 \\ \hline 0 & I_{n-r} \end{array}\right) \in \mathbf{R}^{n \times n}$$

where $I_{n-r}$ denotes the identity matrix of order $(n-r)$. With the orthogonal matrices

$$U := Q\tilde{U} \in \mathbf{R}^{N \times N} \quad \text{and} \quad V := W\hat{V} \in \mathbf{R}^{n \times n}$$

we obtain

$$U^T C V = \tilde{U}^T Q^T C W \hat{V} = \tilde{U}^T \left(\begin{array}{c|c} R & 0 \\ \hline 0 & 0 \end{array}\right) \left(\begin{array}{c|c} \tilde{V} & 0 \\ \hline 0 & I \end{array}\right) = \left(\begin{array}{c|c} S_1 & 0 \\ \hline 0 & 0 \end{array}\right) = \hat{S}. \quad (7.72)$$

Statement (7.69) now follows from (7.72) by partitioning $\hat{S}$ appropriately.

The $s_i$ are called the *singular values* of the matrix $C$. If we denote the column vectors of $U$ and $V$ by $u_i \in \mathbf{R}^N$ and $v_i \in \mathbf{R}^n$, respectively, then we get the following relationships from (7.69)

$$Cv_i = s_i u_i \quad \text{and} \quad C^T u_i = s_i v_i \quad (i = 1, 2, \ldots, n). \quad (7.73)$$

We call $v_i$ the *right singular vectors* and $u_i$ the *left singular vectors* of the matrix $C$.

The singular value decomposition (7.69) offers another opportunity to replace a system of error equations $Cx + d = r$ by an equivalent system that is orthogonally transformed. Since $VV^T = I$ we can write

$$U^T C V V^T x + U^T d = U^T r = \hat{r}. \quad (7.74)$$

Now we introduce the vectors

$$y := V^T x \in \mathbf{R}^n \quad b := U^T d \in \mathbf{R}^N \quad \text{with} \quad b_i = u_i^T d. \quad (7.75)$$

The singular value decomposition means that the system (7.74) has the special structure

$$\left.\begin{array}{l} s_i y_i + b_i = \hat{r}_i \quad (i = 1, 2, \ldots, r) \\ b_i = \hat{r}_i \quad (i = r+1, r+2, \ldots, N) \end{array}\right\} \quad (7.76)$$

We see that the last $(N-r)$ residuals $\hat{r}_i$ do not depend on the (new) unknowns. Hence the sum of the squares of the residuals is minimal if and only if we have $\hat{r}_i = 0$, for $i = 1, 2, \ldots, r$, and it has the uniquely determined value

$$\rho_{\min} := r^T r = \sum_{i=r+1}^{N} \hat{r}_i^2 = \sum_{i=r+1}^{N} b_i^2 = \sum_{i=r+1}^{N} (u_i^T d)^2. \quad (7.77)$$

The first $r$ unknowns $y_i$ are given by

$$y_i = -b_i/s_i \quad (i = 1, 2, \ldots, r), \quad (7.78)$$

whereas the remaining $(n-r)$ unknowns can be chosen arbitrarily. Therefore, the desired solution vector $\mathbf{x}$ has, according to (7.75), the representation

$$\mathbf{x} = -\sum_{i=1}^{r} \frac{\mathbf{u}_i^T \mathbf{d}}{s_i} \mathbf{v}_i + \sum_{i=r+1}^{n} y_i \mathbf{v}_i \tag{7.79}$$

with the $(n-r)$ free parameters $y_{r+1}, y_{r+2}, \ldots, y_n$. If the matrix $\mathbf{C}$ does not have maximal rank, then the *general solution* is given by the sum of a fixed particular solution in the subspace of the $r$ right singular vectors $\mathbf{v}_i$, corresponding to the *positive* singular values, and of an arbitrary vector from the null space of the matrix $\mathbf{C}$. In fact, from (7.73), for the vanishing singular values we have $\mathbf{C}\mathbf{v}_i = \mathbf{0}$, $(i = r+1, r+2, \ldots, n)$.

The solution set of the system of error equations contains a special solution $\mathbf{x}^*$ of minimal Euclidean norm. The orthogonality of the right singular vectors $\mathbf{v}_i$ means that it is characterized by $y_{r+1} = y_{r+2} = \cdots = y_n = 0$,

$$\boxed{\mathbf{x}^* = -\sum_{i=1}^{r} \frac{\mathbf{u}_i^T \mathbf{d}}{s_i} \mathbf{v}_i, \quad \|\mathbf{x}^*\|_2 = \min_{\mathbf{C}\mathbf{x}+\mathbf{d}=\mathbf{r}} \|\mathbf{x}\|_2.} \tag{7.80}$$

The singular value decomposition of the matrix $\mathbf{C}$ reveals the structure of the general solution $\mathbf{x}$, (7.79), and of the special solution $\mathbf{x}^*$, (7.80), which can be quite helpful for an adequate treatment of delicate error equations whose normal equations are extremely ill-conditioned. In the case $r = n$ some singular values $s_i$, which are the square roots of the eigenvalues of $\mathbf{A} = \mathbf{C}^T\mathbf{C}$, can be very small. In the case $r < n$ rather small positive singular values frequently occur. Because they appear in the denominator of formula (7.80), they may cause large contributions to the solution vector $\mathbf{x}^*$, which are possibly undesirable. Therefore it can be more reasonable to consider, instead of (7.80), the sequence of vectors

$$\boxed{\mathbf{x}^{(k)} := -\sum_{i=1}^{k} \frac{\mathbf{u}_i^T \mathbf{d}}{s_i} \mathbf{v}_i \quad (k = 1, 2, \ldots, r).} \tag{7.81}$$

We obtain $\mathbf{x}^{(k)}$ by setting $y_{k+1} = \cdots = y_r = 0$ so that because of (7.76) the corresponding sums of the squares of the residuals

$$\rho^{(k)} := \sum_{i=k+1}^{N} \hat{r}_i^2 = \rho_{\min} + \sum_{i=k+1}^{r} b_i^2 = \rho_{\min} + \sum_{i=k+1}^{r} (\mathbf{u}_i^T \mathbf{d})^2 \tag{7.82}$$

form a nonincreasing sequence, for increasing $k$, with $\rho^{(r)} = \rho_{\min}$. However, the Euclidean norm $\|\mathbf{x}^{(k)}\|_2$ is nondecreasing, therefore the vectors $\mathbf{x}^{(k)}$ can be considered as approximations of $\mathbf{x}^*$. Depending on the problem or on the specific

## 7.4 Nonlinear Problems

aim that approximation $\mathbf{x}^{(k)}$ is adequate for either $\rho^{(k)} - \rho_{\min}$, which is smaller than a given bound, or in which all contributions are omitted that correspond to singular values, which in turn are smaller than a given bound (Golub and Van Loan, 1983; Lawson and Hanson, 1974).

If the singular value decomposition of $\mathbf{C} = \mathbf{U}\hat{\mathbf{S}}\mathbf{V}^T$ (7.69) is known, the rest of the procedure is prescribed by (7.80) or (7.81). A detailed description of the algorithm for the singular value decomposition is beyond the scope of the book. It basically consists of two steps. The matrix $\mathbf{C}$ is first transformed by two orthogonal matrices $\mathbf{Q}$ and $\mathbf{W}$ analogously to (7.67) in such a way that $\mathbf{R}$ is a *bidiagonal* matrix that contains nonvanishing elements only along the diagonal and the adjacent superdiagonal. The matrix $\hat{\mathbf{R}}$ is then treated by a variant of the QR-algorithm leading to $\hat{\mathbf{S}}$ (Chan, 1982; Golub and Kahan, 1965; Golub and Van Loan, 1983; Lawson and Hanson, 1974). Computer programs are given by (Chan, 1982; Forsythe *et al.*, 1977; Golub and Reinsch, 1970; Lawson and Hanson, 1974).

*Example 7.6* The matrix $\mathbf{C} \in \mathbf{R}^{7 \times 3}$ in (7.13) possesses a singular value decomposition with the matrices

$$\mathbf{U} \doteq \begin{pmatrix} 0.103\,519 & -0.528\,021 & -0.705\,006 & 0.089\,676 & 0.307\,578 & 0.171\,692 & -0.285\,166 \\ 0.149\,237 & -0.485\,300 & -0.074\,075 & -0.259\,334 & -0.694\,518 & -0.430\,083 & 0.046\,303 \\ 0.193\,350 & -0.426\,606 & 0.213\,222 & -0.003\,650 & 0.430\,267 & -0.101\,781 & 0.734\,614 \\ 0.257\,649 & -0.328\,640 & 0.403\,629 & 0.513\,837 & -0.295\,124 & 0.544\,000 & -0.125\,034 \\ 0.368\,194 & -0.143\,158 & 0.417\,248 & -0.597\,155 & 0.297\,387 & 0.046\,476 & -0.471\,858 \\ 0.551\,347 & 0.187\,483 & 0.010\,556 & 0.496\,511 & 0.152\,325 & -0.589\,797 & -0.207\,770 \\ 0.650\,918 & 0.374\,216 & -0.338\,957 & -0.239\,885 & -0.197\,914 & 0.359\,493 & 0.308\,912 \end{pmatrix}$$

and

$$\mathbf{V} \doteq \begin{pmatrix} 0.474\,170 & -0.845\,773 & -0.244\,605 \\ 0.530\,047 & 0.052\,392 & 0.846\,348 \\ 0.703\,003 & 0.530\,965 & -0.473\,142 \end{pmatrix}$$

and the singular values are $s_1 \doteq 4.796\,200$, $s_2 \doteq 1.596\,202$, $s_3 \doteq 0.300\,009$. The vector $\mathbf{b}$ as in (7.75) is $\mathbf{b} \doteq (-10.020\,46, -0.542\,841, 2.509\,634, 0.019\,345, -0.128\,170, -0.353\,416, 0.040\,857)^T$. With $y_1 \doteq 2.089\,251$, $y_2 \doteq 0.340\,083$, $y_3 \doteq -8.365\,203$, the desired components of the solution vector are computed from the columns of $\mathbf{V}$, they are $\alpha_0 \doteq 2.749\,198$, $\alpha_1 \doteq -5.954\,657$, $\alpha_2 \doteq 5.607\,247$.

## 7.4 Nonlinear Problems

Systems of overdetermined nonlinear equations, that also have to be treated by means of the method of least squares, can be solved by two different procedures.

We describe the fundamental ideas from which several variants have been developed for special problems.

### 7.4.1 The Gauss–Newton method

We consider the overdetermined system of $N$ nonlinear equations in the $n$ unknowns $x_1, x_2, \ldots, x_n$ resulting from $N$ observed measurements $l_1, l_2, \ldots, l_N$

$$f_i(x_1, x_2, \ldots, x_n) - l_i = r_i \quad (i = 1, 2, \ldots, N). \tag{7.83}$$

In (7.83) we have admitted that the nonlinear functions $f_i(x_1, x_2, \ldots, x_n)$ depend on the index $i$ of the error equation, although this is rarely the case. Moreover, the fundamental principle of the method of least squares has been applied to formulate (7.83) according to which the theoretical value has to be compared with the observed value, their difference yielding the residual. Only such a formulation is mathematically adequate to apply the Gaussian principle.

The necessary conditions for a minimum of the function

$$F(\mathbf{x}) := \mathbf{r}^T \mathbf{r} = \sum_{i=1}^{N} [f_i(x_1, x_2, \ldots, x_n) - l_i]^2 \tag{7.84}$$

are

$$\frac{1}{2} \frac{\partial F(\mathbf{x})}{\partial x_j} = \sum_{i=1}^{N} [f_i(x_1, x_2, \ldots, x_n) - l_i] \frac{\partial f_i(x_1, x_2, \ldots, x_n)}{\partial x_j} = 0$$

$$(j = 1, 2, \ldots, n) \tag{7.85}$$

representing a system of $n$ nonlinear equations in the unknowns $x_1, x_2, \ldots, x_n$. It is usually quite troublesome to solve.

For this reason the nonlinear error equations (7.83) are first *linearized*. We assume that initial approximations $x_1^{(0)}, x_2^{(0)}, \ldots, x_n^{(0)}$ of the desired unknowns are known. Then we use the substitutions

$$x_j = x_j^{(0)} + \xi_j \quad (j = 1, 2, \ldots, n) \tag{7.86}$$

with the (small) *corrections* $\xi_i$, so that the $i$th error equation of (7.83) can approximately be replaced by

$$\sum_{j=1}^{n} \frac{\partial f_i(x_1^{(0)}, x_2^{(0)}, \ldots, x_n^{(0)})}{\partial x_j} \xi_j + f_i(x_1^{(0)}, x_2^{(0)}, \ldots, x_n^{(0)}) - l_i = \rho_i^{(0)}. \tag{7.87}$$

The residual values of the linearized error equations are different from those of the original equations, and hence they are denoted by $\rho_i$. We define the quantities

$$c_{ij}^{(0)} := \frac{\partial f_i(x_1^{(0)}, x_2^{(0)}, \ldots, x_n^{(0)})}{\partial x_j} \quad d_i^{(0)} := f_i(x_1^{(0)}, x_2^{(0)}, \ldots, x_n^{(0)}) - l_i$$

$$(i = 1, 2, \ldots, N; j = 1, 2, \ldots, n) \tag{7.88}$$

## 7.4 Nonlinear Problems

and see that (7.87) represents a linear system of error equations $\mathbf{C}^{(0)}\boldsymbol{\xi} + \mathbf{d}^{(0)} = \boldsymbol{\rho}^{(0)}$ with $\mathbf{C}^{(0)} = (c_{ij}^{(0)}) \in \mathbf{R}^{N \times n}$ and $\mathbf{d}^{(0)} = (d_1^{(0)}, d_2^{(0)}, \ldots, d_N^{(0)})^T$ for the correction vector $\boldsymbol{\xi} = (\xi_1, \xi_2, \ldots, \xi_n)^T$. This system can be solved by means of the procedures of Section 7.2 or 7.3. The correction vector $\boldsymbol{\xi}^{(1)}$ that is obtained from the linearized system of error equations (7.87) will in general not yield the solution of the system of nonlinear equations (7.83). However, the values

$$x_j^{(1)} := x_j^{(0)} + \xi_j^{(1)} \quad (j = 1, 2, \ldots, n) \tag{7.89}$$

are, in favourable circumstances, better approximations of the unknowns $x_j$ that can be improved iteratively. The procedure is called the *Gauss–Newton method* because the correction $\boldsymbol{\xi}^{(1)}$ is determined from (7.87) according to the Gaussian principle and because the error equations (7.87) coincide in the special case when $N = n$ with the linear equations occurring in Newton's method for solving systems of nonlinear equations. The matrix $\mathbf{C}^{(0)}$, with the elements (7.88), is the Jacobian matrix of the functions $f_i(x_1, x_2, \ldots, x_n)$.

**Example 7.7** To determine the dimensions of a pyramid with a square base, the side of the base, its diagonal, the height, the edge of the pyramid and the height of a side-face are measured (see Figure 7.2). The measured values are (in appropriate length units) $a = 2.8$, $d = 4.0$, $H = 4.5$, $s = 5.0$ and $h = 4.7$. The unknowns of the problem are the length $x_1$ of the base edge and the height $x_2$ of the pyramid. The system of five, partially nonlinear error equations with the five different functions $f_i(x_1, x_2)$, is

$$\begin{aligned} x_1 - a &= r_1 & f_1(x_1, x_2) &:= x_1 \\ \sqrt{2}x_1 - d &= r_2 & f_2(x_1, x_2) &:= \sqrt{2}x_1 \\ x_2 - H &= r_3 & f_3(x_1, x_2) &:= x_2 \\ \sqrt{\tfrac{1}{2}x_1^2 + x_2^2} - s &= r_4 & f_4(x_1, x_2) &:= \sqrt{\tfrac{1}{2}x_1^2 + x_2^2} \\ \sqrt{\tfrac{1}{4}x_1^2 + x_2^2} - h &= r_5 & f_5(x_1, x_2) &:= \sqrt{\tfrac{1}{4}x_1^2 + x_2^2} \end{aligned} \tag{7.90}$$

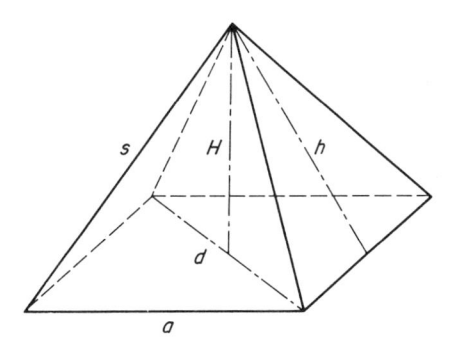

Figure 7.2 Pyramid

The measurements $a$ and $H$ are certainly useful approximations of the unknowns, and we set $x_1^{(0)} = 2.8$, $x_2^{(0)} = 4.5$. With these initial values to six decimal places, we obtain

$$\mathbf{C}^{(0)} \doteq \begin{pmatrix} 1.00000 & 0 \\ 1.41421 & 0 \\ 0 & 1.00000 \\ 0.284767 & 0.915322 \\ 0.148533 & 0.954857 \end{pmatrix} \quad \mathbf{d}^{(0)} \doteq \begin{pmatrix} 0 \\ -0.04021 \\ 0 \\ -0.08370 \\ 0.01275 \end{pmatrix} = \mathbf{r}^{(0)}.$$

The Householder method yields the correction $\boldsymbol{\xi}^{(1)} \doteq (0.0227890, 0.0201000)^\mathrm{T}$, and we get the new approximations $x_1^{(1)} \doteq 2.82279$, $x_2^{(1)} \doteq 4.52010$. For these values we obtain the matrix $\mathbf{C}^{(1)}$ and the constant vector $\mathbf{d}^{(1)}$

$$\mathbf{C}^{(1)} \doteq \begin{pmatrix} 1.00000 & 0 \\ 1.41421 & 0 \\ 0 & 1.00000 \\ 0.285640 & 0.914780 \\ 0.149028 & 0.954548 \end{pmatrix} \quad \mathbf{d}^{(1)} \doteq \begin{pmatrix} 0.02279 \\ -0.00798 \\ 0.02010 \\ -0.05881 \\ 0.03533 \end{pmatrix} = \mathbf{r}^{(1)}.$$

The resulting correction vector is $\boldsymbol{\xi}^{(2)} \doteq (0.00001073, -0.00001090)^\mathrm{T}$ such that $x_1^{(2)} \doteq 2.82280$, $x_2^{(2)} \doteq 4.52009$. The iteration is continued until a certain norm of the correction $\boldsymbol{\xi}^{(k)}$ is sufficiently small. From the fast convergence, the next iteration does not change the approximate solution within the precision used. The vectors $\mathbf{d}^{(k)}$ are equal to the residual vectors of the system of nonlinear error equations (7.90). Their Euclidean norms $\|\mathbf{r}^{(0)}\| \doteq 9.373 \times 10^{-2}$, $\|\mathbf{r}^{(1)}\| \doteq 7.546 \times 10^{-2}$ and $\|\mathbf{r}^{(2)}\| \doteq 7.546 \times 10^{-2}$ decrease monotonically only within the given digits.

**Example 7.8** The position of a ship can be determined, for instance, by radio bearing, measuring the directions to known broadcasting stations. To simplify the problem we assume that the earth curvature need not be taken into account and that a fixed direction is known. Therefore we want to determine the unknown coordinates $x$ and $y$ of a point P by measuring several angles $\alpha_i$ under which radio stations $S_i$ with known coordinates $(x_i, y_i)$ are observed (see Figure 7.3).

The $i$th error equation for the unknowns $x$ and $y$ is

$$\arctan\left(\frac{y - y_i}{x - x_i}\right) - \alpha_i = r_i.$$

For the process of linearization we have to take into account that the angles are in radians and that they must be reduced to the interval of the principal value of the

## 7.4 Nonlinear Problems

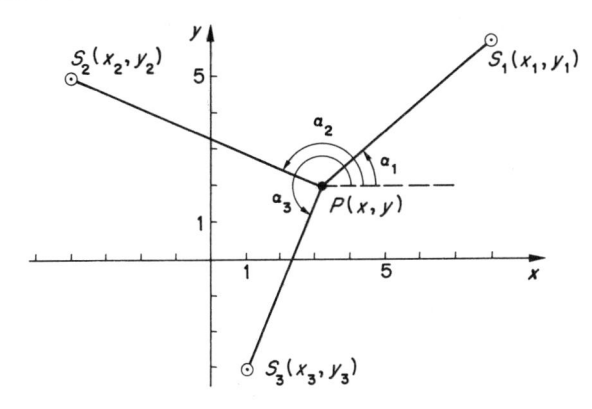

Figure 7.3 Localization by means of radio bearing

arctan-function. With the approximate values $x^{(k)}$ and $y^{(k)}$ the $i$th error equation for the corrections $\xi$ and $\eta$ is

$$\frac{-(y^{(k)} - y_i)}{(x^{(k)} - x_i)^2 + (y^{(k)} - y_i)^2}\xi + \frac{x^{(k)} - x_i}{(x^{(k)} - x_i)^2 + (y^{(k)} - y_i)^2}\eta + \arctan\left(\frac{y^{(k)} - y_i}{x^{(k)} - x_i}\right) - \alpha_i = \rho_i$$

(7.91)

The data that has been used to draw Figure 7.3 is

| $i$ | $x_i$ | $y_i$ | $\alpha_i$ | Principal value |
|---|---|---|---|---|
| 1 | 8 | 6 | 42° | 0.733 038 |
| 2 | −4 | 5 | 158° | −0.383 972 |
| 3 | 1 | −3 | 248° | 1.186 82 |

For the estimated coordinates of the position using $x^{(0)} = 3$, $y^{(0)} = 2$ and to six decimal places, from (7.91) we get the matrix $\mathbf{C}^{(0)}$ and the constant vector $\mathbf{d}^{(0)}$

$$\mathbf{C}^{(0)} \doteq \begin{pmatrix} 0.097\,5610 & -0.121\,951 \\ 0.051\,7241 & 0.120\,690 \\ -0.172\,414 & 0.068\,9655 \end{pmatrix} \quad \mathbf{d}^{(0)} \doteq \begin{pmatrix} -0.058\,297 \\ -0.020\,919 \\ 0.003\,470 \end{pmatrix}$$

The progress of the iteration is summarized in the following table.

| k | $x^{(k)}$ | $y^{(k)}$ | $\mathbf{r}^{(k)T}\mathbf{r}^{(k)}$ | $\zeta^{(k+1)}$ | $\eta^{(k+1)}$ |
|---|---|---|---|---|---|
| 0 | 3.000 00 | 2.000 00 | $3.848\,19 \times 10^{-3}$ | 0.148 633 | $-0.064\,809\,2$ |
| 1 | 3.148 63 | 1.935 19 | $2.422\,27 \times 10^{-3}$ | 0.007 211 19 | 0.008 342 56 |
| 2 | 3.155 84 | 1.943 53 | $2.420\,07 \times 10^{-3}$ | 0.000 214 94 | $-0.000\,146\,32$ |
| 3 | 3.156 05 | 1.943 38 | $2.420\,29 \times 10^{-3}$ | 0.000 003 67 | 0.000 015 63 |
| 4 | 3.156 05 | 1.943 40 | | | |

### 7.4.2 Minimization methods

The sequence of vectors $\mathbf{x}^{(k)}$ that is obtained by applying the Gauss–Newton method need not converge towards the desired solution $\mathbf{x}$ of the system of nonlinear error equations, because of an inadequate choice of the initial approximation $\mathbf{x}^{(0)}$ or the problem may be too delicate. In order to obtain a sequence of vectors $\mathbf{x}^{(k)}$ converging towards $\mathbf{x}$ for any case, it should have the property, motivated by the Gaussian principle, that the sum of squares of the residuals $F(\mathbf{x}) = \mathbf{r}^T\mathbf{r}$, (7.84), satisfies the condition

$$F(\mathbf{x}^{(k)}) < F(\mathbf{x}^{(k-1)}) \quad (k = 1, 2, \ldots). \tag{7.92}$$

This is the usual requirement of a method for finding the minimum of a function $F(\mathbf{x})$. In order to find an $\mathbf{x}^{(k)}$, a so called *direction of descent* $\mathbf{v}^{(k)}$ must be known for which positive values $t$ exist such that for

$$\mathbf{x}^{(k)} = \mathbf{x}^{(k-1)} + t\mathbf{v}^{(k)} \quad t > 0 \tag{7.93}$$

the condition (7.92) is satisfied. A direction of descent is given by the negative gradient of the function $F(\mathbf{x})$ at the point $\mathbf{x}^{(k-1)}$. According to (7.85), (7.83) and (7.88) this direction is given by

$$\mathbf{v}^{(k)} = -\mathbf{C}^{(k-1)T}\mathbf{r}^{(k-1)}, \tag{7.94}$$

where $\mathbf{C}^{(k-1)}$ denotes the Jacobian matrix and $\mathbf{r}^{(k-1)}$ the residual vector corresponding to $\mathbf{x}^{(k-1)}$. If the parameter $t$ is determined from the condition

$$F(\mathbf{x}^{(k)}) = \min_t F(\mathbf{x}^{(k-1)} + t\mathbf{v}^{(k)}) \tag{7.95}$$

we speak of the *method of steepest descent*. The function (7.95) is nonlinear in the unknown $t$. Due to the computational effort, the value of $t$ is usually determined only approximately by means of a search method (Brent, 1973; Jacoby et al., 1972). The locally optimal direction (7.94) produces a sequence $\mathbf{x}^{(k)}$ which often converges slowly towards the solution $\mathbf{x}$. For this reason the method of steepest descent is often quite inefficient.

## 7.4 Nonlinear Problems

**Theorem 7.4** *The correction vector $\xi^{(k+1)}$ resulting from the Gauss–Newton method corresponding to the approximation $\mathbf{x}^{(k)}$ is a direction of descent if $\nabla F(\mathbf{x}^{(k)}) \neq \mathbf{0}$ holds.*

*Proof* We have to show that the angle between the solution vector $\xi^{(k+1)}$ of the linearized error equations $\mathbf{C}^{(k)}\xi^{(k+1)} + \mathbf{d}^{(k)} = \mathbf{\rho}^{(k)}$ and the gradient $\nabla F(\mathbf{x}^{(k)})$ is obtuse. To simplify the notation we drop the upper indices in the following. Hence we have to show that $(\nabla F)^T \xi < 0$. We use the singular value decomposition (7.69) of the matrix $\mathbf{C} = \mathbf{U}\hat{\mathbf{S}}\mathbf{V}^T$ in order to treat the general case. As $\mathbf{d} = \mathbf{r}$ the gradient $\nabla F$ can be written as

$$\nabla F = 2\mathbf{C}^T\mathbf{d} = 2\mathbf{V}(\hat{\mathbf{S}}^T\mathbf{U}^T\mathbf{d}) = 2\sum_{j=1}^{r} s_j(\mathbf{u}_j^T\mathbf{d})\mathbf{v}_j. \tag{7.96}$$

The correction vector $\xi^*$ of minimal Euclidean norm is given by (7.80)

$$\xi^* = -\sum_{i=1}^{r} \frac{(\mathbf{u}_i^T\mathbf{d})}{s_i}\mathbf{v}_i$$

and from the orthonormality of the vectors $\mathbf{v}_i$ we have

$$(\nabla F)^T \xi^* = -2\sum_{j=1}^{r}(\mathbf{u}_j^T\mathbf{d})^2 < 0$$

because in (7.96) not all scalars $\mathbf{u}_i^T\mathbf{d}$ can vanish if $\nabla F \neq \mathbf{0}$ holds.

From Theorem 7.4 a minimization algorithm can be defined by means of the Gauss–Newton method. With a chosen initial vector $\mathbf{x}^{(0)}$ we perform the following steps for $k = 0, 1, 2, \ldots,$: The linearized error equations $\mathbf{C}^{(k)}\xi^{(k+1)} + \mathbf{d}^{(k)} = \mathbf{\rho}^{(k)}$ are solved to obtain the correction vector $\xi^{(k+1)}$ as the direction of descent. In order to achieve a decrease of the function value $F(\mathbf{x}^{(k+1)})$, in comparison to $F(\mathbf{x}^{(k)})$, we test whether the condition $F(\mathbf{y}) < F(\mathbf{x}^{(k)})$ holds for the vectors $\mathbf{y} := \mathbf{x}^{(k)} + t\xi^{(k+1)}$, where the sequence of parameters $t = 1, \frac{1}{2}, \frac{1}{4}, \ldots$ is taken. If the condition is satisfied then we set $\mathbf{x}^{(k+1)} = \mathbf{y}$, and a test on convergence follows. It should be noted that the evaluation of $F(\mathbf{y})$ requires the computation of the residual vector $\mathbf{r}$ corresponding to $\mathbf{y}$. As soon as $\mathbf{y}$ is an acceptable vector, with $\mathbf{r}^{(k+1)} = \mathbf{d}^{(k+1)}$, we have the constant vector of the error equations of the subsequent iteration step available.

The sum of squares of residuals decreases in each step and has a lower bound therefore the convergence of the sequence $\mathbf{x}^{(k)}$ is guaranteed. In the case of an inadequate initial vector $\mathbf{x}^{(0)}$ the convergence can be quite slow at the beginning of the iteration. In a sufficiently small neighbourhood of the solution $\mathbf{x}$ the convergence is almost quadratic (Fletcher, 1980).

A more efficient method has been proposed by Marquardt (1963). In order to define a more advantageous direction of descent he considers the problem of finding the vector $\mathbf{v}$ that solves

$$\|\mathbf{C}\mathbf{v} + \mathbf{d}\|_2^2 + \lambda^2\|\mathbf{v}\|_2^2 = \text{Min!} \quad \lambda > 0 \tag{7.97}$$

where we have set $\mathbf{C} = \mathbf{C}^{(k)}$ and $\mathbf{d} = \mathbf{d}^{(k)}$. For a given value of the parameter $\lambda$ the vector $\mathbf{v}$ is the least square solution of the system of error equations

$$\tilde{\mathbf{C}}\mathbf{v} + \tilde{\mathbf{d}} = \tilde{\boldsymbol{\rho}} \quad \text{with} \quad \tilde{\mathbf{C}} := \begin{pmatrix} \mathbf{C} \\ \hline \lambda \mathbf{I} \end{pmatrix} \in \mathbf{R}^{(N+n) \times n} \quad \tilde{\mathbf{d}} := \begin{pmatrix} \mathbf{d} \\ \hline \mathbf{0} \end{pmatrix} \in \mathbf{R}^{N+n} \quad \tilde{\boldsymbol{\rho}} \in \mathbf{R}^{N+n} \tag{7.98}$$

For each $\lambda > 0$ the matrix $\tilde{\mathbf{C}}$ has maximal rank $n$ independent of the rank of $\mathbf{C}$. In comparison with the Gauss–Newton method, the system of error equations (7.87) is enlarged by $n$ equations and is thus regularized. The solution $\mathbf{y}$ has the following properties.

**Theorem 7.5** *The solution vector* $\mathbf{v} = \mathbf{v}^{(k+1)}$ *of* (7.97) *is a direction of descent if* $\nabla F(\mathbf{x}^{(k)}) \neq \mathbf{0}$.

*Proof* $\tilde{\mathbf{C}}$ has maximal rank, therefore, we can represent the solution $\mathbf{v}$ of (7.98) formally by means of the corresponding normal equations as follows

$$\mathbf{v} = -(\tilde{\mathbf{C}}^T\tilde{\mathbf{C}})^{-1}(\tilde{\mathbf{C}}^T\tilde{\mathbf{d}}) = -(\tilde{\mathbf{C}}^T\tilde{\mathbf{C}})^{-1}(\mathbf{C}^T\mathbf{d}) = -\tfrac{1}{2}(\tilde{\mathbf{C}}^T\tilde{\mathbf{C}})^{-1}(\nabla F). \tag{7.99}$$

Hence we have

$$(\nabla F)^T \mathbf{v} = -\tfrac{1}{2}(\nabla F)^T (\tilde{\mathbf{C}}^T\tilde{\mathbf{C}})^{-1}(\nabla F) < 0 \quad \text{if} \quad \nabla F \neq \mathbf{0},$$

because the matrix $(\tilde{\mathbf{C}}^T\tilde{\mathbf{C}})^{-1}$ is symmetric and positive definite, the angle between $\nabla F$ and $\mathbf{v}$ is obtuse.

**Theorem 7.6** *The Euclidean norm* $\|\mathbf{v}\|_2$ *of the solution vector* $\mathbf{v}$ *of* (7.97) *is a monotonically decreasing function of* $\lambda$.

*Proof* From the special structure of $\tilde{\mathbf{C}}$, the matrix $\mathbf{A}$ of the normal equations corresponding to (7.98) is given by

$$\mathbf{A} = \tilde{\mathbf{C}}^T\tilde{\mathbf{C}} = \mathbf{C}^T\mathbf{C} + \lambda^2 \mathbf{I}.$$

There is an orthogonal matrix $\mathbf{U} \in \mathbf{R}^{n \times n}$ such that for the symmetric and positive semidefinite matrix $\mathbf{C}^T\mathbf{C}$ we have

$$\mathbf{U}^T\mathbf{C}^T\mathbf{C}\mathbf{U} = \mathbf{D} \quad \text{and} \quad \mathbf{U}^T\mathbf{A}\mathbf{U} = \mathbf{D} + \lambda^2 \mathbf{I} \quad \mathbf{D} = \text{diag}(d_1, d_2, \ldots, d_n) \quad d_i \geq 0. \tag{7.100}$$

From (7.99) and (7.100) the square of the Euclidean norm is given by

$$\|\mathbf{v}\|_2^2 = \mathbf{v}^T\mathbf{v} = \mathbf{d}^T\mathbf{C}\mathbf{U}(\mathbf{D} + \lambda^2 \mathbf{I})^{-1}\mathbf{U}^T\mathbf{U}(\mathbf{D} + \lambda^2 \mathbf{I})^{-1}\mathbf{U}^T\mathbf{C}^T\mathbf{d}$$

$$= \sum_{j=1}^{n} \frac{h_j^2}{(d_j + \lambda^2)^2} \quad \text{with} \quad \mathbf{h} := \mathbf{U}^T\mathbf{C}^T\mathbf{d} = (h_1, h_2, \ldots, h_n)^T$$

To minimize the sum of squares of the residuals the Euclidean norm of the

## 7.4 Nonlinear Problems

vector $\mathbf{v}^{(k+1)}$ is controlled in the *Marquardt's method* by the parameter $\lambda$ in such a way that

$$\mathbf{x}^{(k+1)} = \mathbf{x}^{(k)} + \mathbf{v}^{(k+1)} \quad F(\mathbf{x}^{(k+1)}) < F(\mathbf{x}^{(k)}) \quad (k=0,1,2,\ldots). \tag{7.101}$$

The value $\lambda$ is appropriately chosen in the course of the process. If the condition (7.101) is satisfied for a current value of $\lambda$ then it will be decreased for the subsequent step, for instance halved. However, if (7.101) is not satisfied for a current value of $\lambda$ then it is increased, for instance doubled, until a vector $\mathbf{v}^{(k+1)}$ results for which the condition holds. It is clear that the initial guess $\lambda^{(0)}$ must be given together with the initial vector $\mathbf{x}^{(0)}$. It has been proposed that the problem dependent value should be taken

$$\lambda^{(0)} = \|\mathbf{C}^{(0)}\|_F / \sqrt{nN} = \sqrt{\frac{1}{nN} \sum_{i,j} (c_{ij}^{(0)})^2}$$

with the Frobenius norm of the matrix $\mathbf{C}^{(0)}$ corresponding to the initial vector $\mathbf{x}^{(0)}$.

An iteration step of Marquardt's method requires the solution of the system of error equations (7.98) for possibly several values of the parameter $\lambda$. The most efficient procedure consists of two separate steps (Golub and Pereyra, 1973). In a preliminary step the first $N$ error equations, that are independent of $\lambda$, are transformed by means of an orthogonal matrix $\mathbf{Q}_1$ in such a way that we have

$$\mathbf{Q}_1^T \tilde{\mathbf{C}} = \begin{pmatrix} \mathbf{R}_1 \\ \hline \mathbf{0}_1 \\ \hline \lambda \mathbf{I} \end{pmatrix} \quad \mathbf{Q}_1^T \tilde{\mathbf{d}} = \begin{pmatrix} \hat{\mathbf{d}}_1 \\ \hat{\mathbf{d}}_2 \\ \mathbf{0} \end{pmatrix} \quad \mathbf{R}_1 \in \mathbf{R}^{n \times n} \quad \mathbf{0}_1 \in \mathbf{R}^{(N-n) \times n}. \tag{7.102}$$

On the assumption that $\mathbf{C}$ has maximal rank $\mathbf{R}_1$ is a nonsingular right triangular matrix, and the transformation can be performed either by the method of Givens or Householder. In any case, the last $n$ error equations remain unchanged. The transformation is now finished for the given $\lambda$ by means of an orthogonal matrix $\mathbf{Q}_2$ to obtain the following matrices and vectors from the result (7.102)

$$\mathbf{Q}_2^T \mathbf{Q}_1^T \tilde{\mathbf{C}} = \begin{pmatrix} \mathbf{R}_2 \\ \hline \mathbf{0}_1 \\ \hline \mathbf{0}_2 \end{pmatrix} \quad \mathbf{Q}_2^T \mathbf{Q}_1^T \tilde{\mathbf{d}} = \begin{pmatrix} \hat{\hat{\mathbf{d}}}_1 \\ \hat{\mathbf{d}}_2 \\ \hat{\mathbf{d}}_3 \end{pmatrix} \quad \mathbf{R}_2, \mathbf{0}_2 \in \mathbf{R}^{n \times n}. \tag{7.103}$$

We get the desired vector $\mathbf{v}^{(k+1)}$ from $\mathbf{R}_2 \mathbf{v}^{(k+1)} + \hat{\hat{\mathbf{d}}}_1 = \mathbf{0}$ with the nonsingular right triangular matrix $\mathbf{R}_2$ by the process of back substitution. If $\lambda$ has to be increased, then only the second step must be repeated. For this purpose the resulting $\mathbf{R}_1$ and $\hat{\mathbf{d}}_1$ of the first part must be saved. The zero matrix $\mathbf{O}_1$ and the vector $\hat{\mathbf{d}}_2$ of (7.102) are not needed for the second step and therefore it can be performed quite efficiently with low storage requirements. The rather special structure of the error equations of the second step suggests the use of the Givens transformation. The computational technique described in Section 7.2.2 requires the smallest amount of storage space. However, if the Householder transformation is used the second step needs an auxiliary matrix $C_H \in R^{(2n) \times n}$ and an auxiliary vector $d_H \in R^{2n}$, whereby the structure of $C_H$ can be taken into account, to increase the efficiency.

In several least square problems of natural sciences a certain function

$$f(x) = \sum_{j=1}^{\mu} a_j \varphi_j(x; \alpha_1, \alpha_2, \ldots, \alpha_\nu)$$

must be determined with the unknown parameters $a_1, a_2, \ldots, a_\mu; \alpha_1, \alpha_2, \ldots, \alpha_\nu$ by means of $N$ observations of the function $f(x)$ for $N$ different arguments $x_i$. It is assumed that the functions $\varphi_j(x; \alpha_1, \alpha_2, \ldots, \alpha_\nu)$ are nonlinear with respect to the $\alpha_k$. The resulting error equations are linear with respect to the $a_j$ for fixed $\alpha_k$ and nonlinear with respect to the $\alpha_k$ if the $a_j$ are fixed. For an efficient treatment of such problems a special algorithm has been designed (Golub and Pereyra, 1973).

## 7.5 Exercises

**7.1.** The lengths of the edges of a prism and the circumferences perpendicular to the first and second edge are measured as follows:

Edge 1: 26 mm; edge 2: 38 mm; edge 3: 55 mm; circumference $\perp$ edge 1: 188 mm; circumference $\perp$ edge 2: 163 mm.

What are the lengths of the edges resulting from the method of least squares?

**7.2.** In order to determine the amplitude $A$ and the phase angle $\phi$ of an oscillation $x = A \sin(2t + \phi)$ the elongations $x_k$ have been measured at four times, $t_k$.

| $t_k =$ | 0   | $\pi/4$ | $\pi/2$ | $3\pi/4$ |
|---------|-----|---------|---------|----------|
| $x_k =$ | 1.6 | 1.1     | $-1.8$  | $-0.9$   |

To obtain a system of linear equations new unknowns have to be introduced by using a trigonometric formula.

**7.3.** The function $y = \sin(x)$ is to be approximated in the interval $[0, \pi/4]$ by a polynomial $P(x) = a_1 x + a_3 x^3$ that is odd like $\sin(x)$. The coefficients $a_1$ and $a_3$ are to be determined by the method of least squares using the support abscissae $x_k = k\pi/24$, $(k = 1, 2, \ldots, 6)$. Draw the graph of the error function $r(x) := P(x) - \sin(x)$ and compare the result with that obtained in Exercise 2.7.

**7.4.** By means of computer programs for the solution of systems of linear error equations based on the normal equations and on the two variants of orthogonal transformation, the functions

(a) $f(x) = \cos(x)$, $x \in [0, \pi/2]$  (b) $f(x) = e^x$, $x \in [0, 1]$

are to be approximated by polynomials $P_n(x) = a_0 + a_1 x + a_2 x^2 + \cdots + a_n x^n$ of degree $n = 2, 3, \ldots, 8$ such that the sum of squares of the residuals is minimal for $N = 10$ and $N = 20$ equidistant support abscissae. To explain the partially different results compute estimates of the condition numbers of the matrix $A$ of the normal equations, by means

## 7.5 Exercises

of the inverse of the right triangular matrix $\mathbf{R}$ using the relationship $\|\mathbf{A}^{-1}\|_F \leqslant \|\mathbf{R}^{-1}\|_F \|\mathbf{R}^{-T}\|_F = \|\mathbf{R}^{-1}\|_F^2$, where $\|\mathbf{A}\|_F$ denotes the Frobenius norm.

**7.5.** The edges of the base of a prism are measured to be $a = 21$ cm and $b = 28$ cm and its height $c = 12$ cm. Moreover, we get the lengths of the diagonal of the base $d = 34$ cm, of the diagonal of the side face $e = 24$ cm and of the diagonal of the prism $f = 38$ cm. Determine the lengths of the edges of the prism by the method of least squares using the Gauss–Newton method and procedures of minimization.

**7.6.** In order to detect the position of an illegal sender, five direction finding stations are put into action. The positions of the direction finding stations are given in a $(x, y)$ coordinate system, and the direction angles $\alpha$ are measured counterclockwise from the positive $x$-axis.

| Direction finding station | 1 | 2 | 3 | 4 | 5 |
|---|---|---|---|---|---|
| $x$ coordinate | 4 | 18 | 26 | 13 | 0 |
| $y$ coordinate | 1 | 0 | 15 | 16 | 14 |
| direction angle $\alpha$ | 45° | 120° | 210° | 270° | 330° |

Make a large scale drawing of the situation. What are the most likely coordinates of the sender according to the method of least squares? To start the Gauss–Newton method, or one of the methods of minimization, choose the initial guess $P_0(12.6, 8.0)$.

**7.7.** The concentration $z(t)$ of a substance in a chemical process follows the law

$$z(t) = a_1 + a_2 e^{\alpha_1 t} + a_3 e^{\alpha_2 t} \quad \alpha_1, \alpha_2 \in \mathbb{R} \quad \alpha_1, \alpha_2 < 0.$$

To determine the parameters $a_1, a_2, a_3, \alpha_1, \alpha_2$ the following observations for $z(t)$ are available.

| $t_k$ | 0 | 0.5 | 1.0 | 1.5 | 2.0 | 3.0 | 5.0 | 8.0 | 10.0 |
|---|---|---|---|---|---|---|---|---|---|
| $z_k$ | 3.85 | 2.95 | 2.63 | 2.33 | 2.24 | 2.05 | 1.82 | 1.80 | 1.75 |

Use the initial values $a_1^{(0)} = 1.75$, $a_2^{(0)} = 1.20$, $a_3^{(0)} = 0.8$, $\alpha_1^{(0)} = -0.5$, $\alpha_2^{(0)} = -2$ to treat the nonlinear error equations by means of the Gauss–Newton method and of minimization.

# 8

# Numerical Quadrature

Some problems of applied mathematics require the calculation of integrals that cannot be determined analytically. Therefore the numerical computation of integrals, known as quadrature, is of some importance. In the following we consider the approximate calculation of definite integrals. Indefinite integrals are suitably treated as initial value problems of ordinary differential equations (see Chapter 9). Numerical quadrature may be applied, in many instances, for example the calculation of surfaces and volumes of bodies, probabilities and effective profiles; the evaluation of integral transforms and integrals in the complex domain; the construction of conformal mappings of polygonal domains by means of the Schwarz–Christoffel formulae (Henrici, 1985); the treatment of integral equations in connection with the boundary element method and finally in the finite element method (Schwarz, 1984).

There exist several methods for the computation of definite integrals, from which the most suitable is chosen according to the following criteria: (a) smoothness of the integrand and the presence of singularities; (b) available information on the integrand, that is whether a table of values is given, or whether it can be evaluated for arbitrary values; (c) the desired accuracy of the approximate value; (d) the number of cases to be treated. In general, numerical quadrature is quite a stable process that can be performed, even with simple algorithms, in an accurate and efficient manner. We shall describe some of the most important methods together with their field of application. The requirement of simplicity and efficiency is fulfilled well by the transformation methods in combination with the trapezoidal rule. Moreover, we also treat the classical methods of Romberg and Gauss and describe an adaptive procedure. For more extensive treatments of numerical quadrature see Davis and Rabinowitz (1984), Engels (1980), Krylov (1962) and Stroud (1971, 1974).

## 8.1 The Trapezoidal Method

Although the trapezoidal method represents the simplest procedure of quadrature, it is most accurate and most efficient for some important special cases. As other methods are based on the trapezoidal method we start with a detailed discussion about it.

### 8.1.1 Problem and notation

We consider the problem of calculating for the definite integral

$$I := \int_a^b f(x)\,dx \tag{8.1}$$

an approximate value $\tilde{I}$ such that the error satisfies the condition $|\tilde{I} - I| < \varepsilon$ for a prescribed tolerance $\varepsilon > 0$. We assume that $f(x)$ is a real-valued, integrable function defined on $[a, b]$. A finite procedure for computing $\tilde{I}$ is called a *quadrature formula*. It must be a *linear functional* with respect to $f$, as is the case for the right-hand side of (8.1). If only a finite number of function values, $f(x_j)$, $x_j \in [a, b]$, are allowed by the quadrature formula, it must have the form

$$\tilde{I} = \sum_{j=1}^{n} w_j f(x_j) \quad x_j \in [a,b] \quad (j = 1, 2, \ldots, n). \tag{8.2}$$

We call (8.2) an *n-point-formula*. The values $x_j$ are called the *nodes* or support abscissae of integration and the $w_j$ are the corresponding *weights* of the quadrature formula. Sometimes quadrature formulae are considered in which, as well as the function values, $f(x_j)$, the derivatives $f'(x_j)$, $f''(x_j)$, ... also occur. The abscissa $x_j$ is called a multiple node in this case (Golub and Kautsky, 1983).

### 8.1.2 Definition of the trapezoidal method and improvements

To approximate the integral (8.1), the finite interval $[a, b]$ is divided into $n > 0$ subintervals of equal length, $h := (b - a)/n$ such that the support abscissae are obtained by

$$x_j := a + jh \quad (j = 0, 1, \ldots, n). \tag{8.3}$$

We define the *trapezoidal approximation* of the integral (8.1) by the sum

$$T(h) := h\left(\tfrac{1}{2}f(x_0) + \sum_{j=1}^{n-1} f(x_j) + \tfrac{1}{2}f(x_n)\right) = h \sum_{j=0}^{n} {}'' f(x_j). \tag{8.4}$$

Which represents the area under the polygonal curve shown in Figure 8.1. The

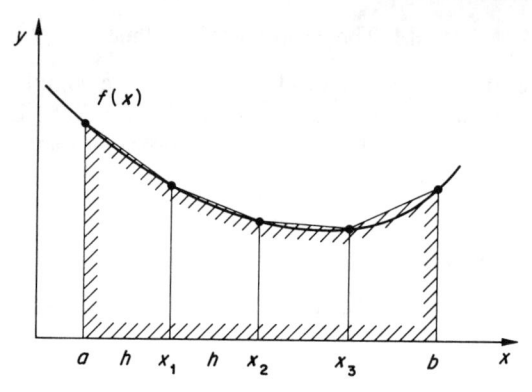

Figure 8.1 Trapezoidal rule

symbol $\sum''$ indicates that the first and the last term of the sum have half the weight.

A similar and obvious approximation of the integral (8.1), corresponding to the Riemannian sum is given by the *midpoint sum*

$$M(h) := h \sum_{j=0}^{n-1} f\left(x_{j+\frac{1}{2}}\right) \quad x_{j+\frac{1}{2}} := a + \left(j + \tfrac{1}{2}\right)h \tag{8.5}$$

where $M(h)$ represents the area under the staircase curve of Figure 8.2.

We immediately obtain the following relationship from (8.4) and (8.5)

$$T\left(\frac{h}{2}\right) = \frac{1}{2}[T(h) + M(h)]. \tag{8.6}$$

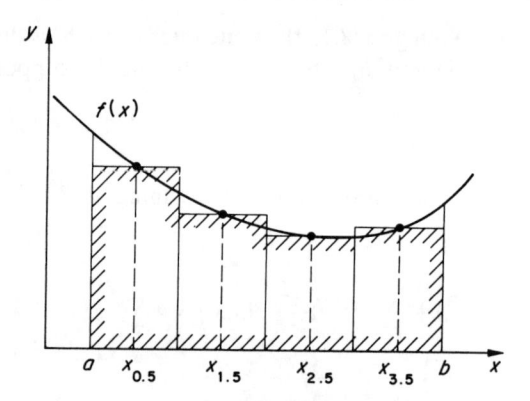

Figure 8.2 Mid-point rule

## 8.1 The Trapezoidal Method

The relationship (8.6) allows the improvement of the trapezoidal approximations by successively halving the step size $h$ in such a manner that, in addition to the computed approximation $T(h)$, also the value $M(h)$ is computed. Each halving of the step size approximately doubles the computational effort, which is measured by the number of function evaluations. However, the previously computed function values are reused in an economical manner. The successive halving of the step size can be stopped, for instance, if $T(h)$ and $M(h)$ differ by less than a given tolerance $\varepsilon > 0$. In general the error $|T(h/2) - I|$ is at most equal to $\varepsilon$.

To summarize, the trapezoidal sums $T(h)$ are calculated by successive halving of the step size $h$ by the following algorithm, where $\varepsilon$ denotes the prescribed tolerance, $f(x)$ is the integrand and $a, b$ are the limits of integration.

$$
\begin{aligned}
&h = b - a;\ n = 1;\ T = h \times (f(a) + f(b))/2 \\
&\text{for } k = 1, 2, \ldots, 10: \\
&\quad M = 0 \\
&\quad \text{for } j = 0, 1, \ldots, n-1: \\
&\quad\quad M = M + f(a + (j + 0.5) \times h) \\
&\quad M = h \times M;\ T = (T + M)/2;\ h = h/2;\ n = 2 \times n \\
&\quad \text{if } |T - M| < \varepsilon:\ \text{STOP}
\end{aligned}
\tag{8.7}
$$

In general, the trapezoidal sums converge quite slowly towards the value $I$, if no additional measures are taken. However, if $f(x)$ is periodic and analytic on $\mathbf{R}$, and if $(b - a)$ is equal to its period, then the algorithm (8.7) requires no improvements (see Section 8.2.1 below).

The trapezoidal method is also suitable for the approximate evaluation of integrals over $\mathbf{R}$ of functions $f(x)$ that decrease sufficiently fast. The definitions (8.4) and (8.5) are generalized for the two-sided unbounded interval, and moreover, we introduce a shift $s$, that can be arbitrarily chosen. Thus we define the trapezoidal and the midpoint sums as follows

$$T(h, s) := h \sum_{j=-\infty}^{\infty} f(s + jh)$$

$$M(h, s) := h \sum_{j=-\infty}^{\infty} f\left[s + \left(j + \frac{1}{2}\right)h\right] = T\left(h, s + \frac{h}{2}\right). \tag{8.8}$$

Analogously to (8.6) we have

$$T\left(\frac{h}{2}, s\right) = \frac{1}{2}[T(h, s) + M(h, s)].$$

Due to the infinite sums, the application of (8.8) is only practicable for sufficiently fast decreasing integrands, for instance, exponentially. A suitable shift $s$ has to be chosen, that is appropriate to the integrand $f(x)$, as well as an initial step size $h_0$.

Then the values $T$ and $M$ are calculated by starting the summation with $j=0$ and then increasing $|j|$. The (infinite) summation over $j$ is stopped as soon as the absolute values of the function values become smaller than a tolerance $\delta$. Thus we get the following modified algorithm for computing the trapezoidal sums for the improper integral

$$I = \int_{-\infty}^{\infty} f(x)\,dx$$

of a sufficiently fast decreasing function $f(x)$.

$$
\begin{aligned}
&h = h_0;\ T = f(s);\ j = 1;\ z = 0 \\
\text{ST:}\ &f1 = f(s + j \times h);\ f2 = f(s - j \times h);\ T = T + f1 + f2;\ j = j + 1 \\
&\text{if } |f1| + |f2| > \delta\text{: } z = 0;\ \text{go to ST} \\
&z = z + 1;\ \text{if } z = 1\text{: go to ST} \\
&T = h \times T \\
&\text{for } k = 1, 2, \ldots, 10\text{:} \\
&\quad M = f(s + 0.5 \times h);\ j = 1;\ z = 0 \\
\text{SM:}\ &\quad f1 = f(s + (j + 0.5) \times h);\ f2 = f(s - (j - 0.5) \times h) \\
&\quad M = M + f1 + f2;\ j = j + 1 \\
&\quad \text{if } |f1| + |f2| > \delta\text{: } z = 0;\ \text{go to SM} \\
&\quad z = z + 1;\ \text{if } z = 1\text{: go to SM} \\
&\quad M = h \times M;\ T = (T + M)/2;\ h = h/2 \\
&\quad \text{if } |T - M| < \varepsilon\text{: STOP}
\end{aligned}
\qquad (8.9)
$$

In order to avoid the unnecessary evaluation of functions in the case of an asymmetrically decreasing integrand, the algorithm (8.9) can be improved in such a way that the sums for increasing and decreasing $j$ are performed separately.

Table 8.1 Improper integral with oscillating integrand

|   | $s = 0$ | Z | $s = 0.12$ | Z | $s = 0.3456$ | Z |
|---|---|---|---|---|---|---|
| T | 1.027 908 2242 | 15 | 0.960 284 3084 | 15 | 0.513 929 7286 | 15 |
| M | −0.898 053 6479 | 17 | −0.830 429 7538 | 17 | −0.384 075 2730 | 17 |
| T | 0.064 927 2881 |  | 0.064 927 2773 |  | 0.064 927 2278 |  |
| M | 0.064 927 2112 | 27 | 0.064 927 2215 | 27 | 0.064 927 2710 | 29 |
| T | 0.064 927 2496 |  | 0.064 927 2494 |  | 0.064 927 2494 |  |
| M | 0.064 927 2494 | 51 | 0.064 927 2494 | 51 | 0.064 927 2494 | 51 |
| T | 0.064 927 2495 |  | 0.064 927 2494 |  | 0.064 927 2494 |  |

## 8.1 The Trapezoidal Method

*Example* 8.1 The approximate evaluation of

$$I = \int_{-\infty}^{\infty} e^{-0.25x^2} \cos(2x)\,dx$$

with an oscillating, fast decreasing integrand, by means of the algorithm (8.9), yields values of trapezoidal sums according to Table 8.1. For the initial step size $h_0 = 2$, the number $Z$ of function evaluations are given for three different shifts $s$. The prescribed tolerances are $\delta = 10^{-14}$ and $\varepsilon = 10^{-10}$. The behaviour of the algorithm is almost independent of the choice of $s$.

### 8.1.3 The Euler–MacLaurin formula

The formula, that was first found by Euler and independently by MacLaurin soon afterwards, allows the asymptotic investigation of the error between the integral and the trapezoidal sum as $h \to 0$. In the following we restrict ourselves to giving a heuristic motivation of the formal structure of the important relationship, and we refer you to Henrici (1977), Schmeisser and Schirmeier (1976) and Stoer (1983) for a more rigorous treatment.

Let $f(x)$ be a function which can be represented as a Taylor series for all $x \in [a, b]$. We write the Taylor series by means of the differential operator D in the form

$$f(x+h) - f(x) = (e^{hD} - 1)f(x) \quad \text{with} \quad D := \frac{d}{dx}.$$

If we solve formally for $f(x)$ we obtain

$$f(x) = E(f(x+h) - f(x)), \tag{8.10}$$

where the inverse operator is defined by the series

$$E := (e^{hD} - 1)^{-1} = \sum_{k=0}^{\infty} \frac{B_k}{k!}(hD)^{k-1}.$$

The coefficients $B_k$ are the Bernoulli numbers

$$B_0 = 1 \quad B_1 = -\tfrac{1}{2} \quad B_2 = \tfrac{1}{6} \quad B_3 = B_5 = B_7 = \cdots = 0,$$
$$B_4 = -\tfrac{1}{30} \quad B_6 = \tfrac{1}{42} \quad B_8 = -\tfrac{1}{30} \quad B_{10} = \tfrac{5}{66}, \ldots,$$

for which we have the following asymptotic relationship (Erdélyi et al., 1953)

$$|B_k| \sim 2 \cdot k!(2\pi)^{-k} \quad k \text{ even } k \to \infty.$$

From (8.10), by a formal application of the inverse operator we obtain the following representation

$$f(x) = \left( (hD)^{-1} - \tfrac{1}{2} + \sum_{k=2}^{\infty} \frac{B_k}{k!} (hD)^{k-1} \right) (f(x+h) - f(x))$$

$$= h^{-1} D^{-1}(f(x+h) - f(x)) - \tfrac{1}{2}(f(x+h) - f(x))$$

$$+ \sum_{k=2}^{\infty} \frac{B_k}{k!} h^{k-1} [f^{(k-1)}(x+h) - f^{(k-1)}(x)].$$

If we take into account that the Bernoulli numbers with odd subscript greater than one vanish, we obtain

$$\frac{h}{2}[f(x) + f(x+h)] = \int_x^{x+h} f(t)\,dt$$

$$+ \sum_{k=1}^{\infty} \frac{B_{2k}}{(2k)!} h^{2k} [f^{(2k-1)}(x+h) - f^{(2k-1)}(x)].$$

Now we sum this formal relationship, for the $n$ subintervals of the trapezoidal integration (8.4), and obtain

$$T(h) = \int_a^b f(x)\,dx + \sum_{k=1}^{\infty} \frac{B_{2k}}{(2k)!} h^{2k} [f^{(2k-1)}(b) - f^{(2k-1)}(a)]. \qquad (8.11)$$

The derivatives $f^{(2k-1)}(x)$ usually increase in absolute value quite fast for increasing $k$ thus the series (8.11) does not converge in general, so that (8.11) does not hold in this form. However, in most cases it can be shown that each finite partial sum of (8.11) is asymptotically equal to $T(h)$ as $h \to 0$. In these cases the Euler–MacLaurin formula (8.11) holds in the form

$$T(h) = \int_a^b f(x)\,dx + \sum_{k=1}^N c_k h^{2k} + R_{N+1}(h) \qquad (8.12)$$

with well determined coefficients $c_k$ independent of $h$ and with a remainder term $R_{N+1}(h)$ that satisfies $R_{N+1}(h) = O(h^{2N+2})$ for each fixed $N$ if $h \to 0$.

In the general case we have $c_1 \neq 0$, and hence the error of the trapezoidal formula is $O(h^2)$, i.e. of second order. This means that by halving the step size $h$, the error is only reduced by a factor of four, but the computational effort is doubled. Improvements of the trapezoidal method, based on (8.12), are discussed in the next section. However, if $c_k = 0$ for $k = 1, 2, \ldots, N$, then the error of the trapezoidal rule is given by the remainder term. It has to be expected that the trapezoidal method yields especially good approximations in this case, but the Euler–MacLaurin formula is not able to make a precise statement about the error. These important special cases are considered in Section 8.2.

## 8.1 The Trapezoidal Method

### 8.1.4 The Romberg procedure

The trapezoidal rule will now be improved on the assumption that an asymptotic expansion (8.12) holds with $c_k \neq 0$. To achieve an improvement, trapezoidal sums $T(h_i)$ are computed for some step sizes $h_i$, and then an extrapolation to the value $T(0) = I$ is applied, by means of the *Neville algorithm* of Section 3.5.2. The algorithm (8.7) for halving the step size, yields the function values $T(h_i)$ for values $h_0 = b - a$, $h_i = h_{i-1}/2$, $i = 1, 2, \ldots$ that form a geometric sequence with the quotient $q = 1/2$. The expansion (8.12) contains only even powers of $h$, therefore the extrapolation is given by the recursion (3.82). The arrangement of the auxiliary values $p_i^{(k)}$ is called the *Romberg scheme* (Romberg, 1955).

*Example* 8.2 The Romberg scheme for the integral

$$\int_1^2 \frac{dx}{x} = \log(2) \doteq 0.693\,147\,1806 \tag{8.13}$$

reads to ten decimal places,

| $h_i$  | $T(h_i) = p_i^{(0)}$ | $p_i^{(1)}$    | $p_i^{(2)}$    | $p_i^{(3)}$    | $p_i^{(4)}$    |
|--------|----------------------|----------------|----------------|----------------|----------------|
| 1      | 0.750 000 0000       |                |                |                |                |
| 0.5    | 0.708 333 3333       | 0.694 444 4444 |                |                |                |
| 0.25   | 0.697 023 8095       | 0.693 253 9683 | 0.693 174 6032 |                |                |
| 0.125  | 0.694 121 8504       | 0.693 154 5307 | 0.693 147 9015 | 0.693 147 4776 |                |
| 0.0625 | 0.693 391 2022       | 0.693 147 6528 | 0.693 147 1843 | 0.693 147 1831 | 0.693 147 1819 |

Usually the uppermost diagonal of the Romberg scheme, consisting of the values $p_k^{(k)}$ is used to detect convergence. If the step size halving algorithm (8.7) is stopped in combination with the Romberg scheme as soon as, for instance, $|p_k^{(k)} - p_{k-1}^{(k-1)}| < \varepsilon$ is satisfied, quite an efficient procedure is obtained that yields accurate results for *smooth* integrands. To avoid wasting computer time, for example for too small a prescribed tolerance $\varepsilon$, the number of halvings of the step size should be bounded.

It can be shown (Bauer et al., 1963), that in general, the diagonal sequence $p_k^{(k)}$ converges *superlinearly*, that is that the quotients

$$q_k := \frac{p_k^{(k)} - I}{p_{k-1}^{(k-1)} - I} \qquad I = \int_a^b f(x)\,dx \tag{8.14}$$

tend to zero with increasing $k$. The quotients obtained from Example 8.2 are $q_1 \doteq 0.002\,28$, $q_2 \doteq 0.002\,11$, $q_3 \doteq 0.001\,08$, $q_4 \doteq 0.000\,46$ confirming the superlinear convergence. However, they show that the convergence is not much better than linear.

*Example 8.3.* The integrand of

$$I = \int_0^1 x^{3/2}\,dx = 0.4$$

seems to be harmless, but it will show that some precautions must be taken when Romberg's procedure is applied.

| $h_i$ | $T(h_i) = p_i^{(0)}$ | $p_i^{(1)}$ | $p_i^{(2)}$ | $p_i^{(3)}$ | $p_i^{(4)}$ |
|---|---|---|---|---|---|
| 1      | 0.500 000 0000 |               |               |               |               |
| 0.5    | 0.426 776 6953 | 0.402 368 9271 |               |               |               |
| 0.25   | 0.407 018 1109 | 0.400 431 9161 | 0.400 302 7820 |               |               |
| 0.125  | 0.401 812 4648 | 0.400 077 2494 | 0.400 053 6050 | 0.400 049 6498 |               |
| 0.0625 | 0.400 463 4013 | 0.400 013 7135 | 0.400 009 4777 | 0.400 008 7773 | 0.400 008 6170 |

The quotients (8.14) are $q_1 \doteq 0.0237$, $q_2 \doteq 0.1278$, $q_3 \doteq 0.1640$, $q_4 \doteq 0.1736$. From these values and the Romberg scheme we recognize that only the column of the $p_i^{(1)}$ yields an essential improvement in comparison with $T(h_i)$. This behaviour can be explained by the singularity of the second derivative of the integrand $f(x)$ at $x = 0$, so that the error law (8.12) is only valid for $N = 1$. For the integral under consideration an additional term of the form $O(h^{5/2})$ exists in the Euler–MacLaurin formula (Rutishauser 1963). This term can be eliminated by means of the Neville algorithm if an additional column $\tilde{p}_i^{(2)}$ is formed from the $p_i^{(1)}$ values, with the divisor $(2^{5/2} - 1)$ instead of $(2^4 - 1)$. Afterwards, the normal Romberg scheme can be used and hence we obtain

| $h_i$ | $\tilde{p}_i^{(2)}$ | $\tilde{p}_i^{(3)}$ | $\tilde{p}_i^{(4)}$ |
|---|---|---|---|
| 0.25   | 0.400 015 9678 |               |               |
| 0.125  | 0.400 001 0892 | 0.400 000 0973 |               |
| 0.0625 | 0.400 000 0701 | 0.400 000 0022 | 0.400 000 0007 |

The corresponding quotients of (8.14) $\tilde{q}_2 \doteq 0.006\,74$, $\tilde{q}_3 \doteq 0.006\,10$, $\tilde{q}_4 \doteq 0.004\,66$ are now decreasing.

We do not pursue the technique of taking the singularity into account, since the more general methods of transformation of Section 8.2 are much more efficient.

The Romberg procedure has an obvious interpretation. The recursion step (3.82) of the Neville algorithm

$$p_i^{(k)} = p_i^{(k-1)} + (p_i^{(k-1)} - p_{i-1}^{(k-1)})/(4^k - 1) \tag{8.15}$$

for $k = 1$ and $i = 1, 2, \ldots$ eliminates the first term $c_1 h^2$ of (8.12), so that asymptotically we have

$$p_i^{(1)} = \int_a^b f(x)\,dx + \sum_{k=2}^N c_k^* h^{2k} + R_{N+1}^*(h). \tag{8.16}$$

## 8.1 The Trapezoidal Method

Correspondingly, the subsequent recursion steps ($k = 2, 3, \ldots$) eliminate the error terms $h^4, h^6, \ldots$ respectively. Hence we have

$$p_i^{(k)} = \int_a^b f(x)\,dx + O(h_i^{2k+2}) \quad (i \geqslant k = 1, 2, \ldots).$$

The special values $p_i^{(1)}$ of the first column of the Romberg scheme represent approximations of the integral $I$ with an error $O(h_i^4)$. If we set $h_{i-1} = h$ and $h_i = h/2$, then according to (8.15) and (8.16) we have

$$p_i^{(1)} = \frac{1}{3}\left[4T\left(\frac{h}{2}\right) - T(h)\right] = \tfrac{1}{3}[T(h) + 2M(h)].$$

From this relationship we obtain Simpson's quadrature formula that is valid for an arbitrary step length $h = (b-a)/n$, $n \in \mathbf{N}$

$$S(h) := \frac{h}{3}\left(\sum_{j=0}^{n}{}'' f(x_j) + 2\sum_{j=0}^{n-1} f(x_{j+1/2})\right) \quad x_\mu := a + \mu h. \tag{8.17}$$

The Romberg procedure has the advantage of being simple, but it is often too expensive if high accuracy is required. If the integrand $f(x)$ is available for arbitrary $x \in [a, b]$, the methods of the Sections 8.2 and 8.4 are more efficient. If an integrand is not too simple the amount of computational effort of a quadrature algorithm is essentially proportional to the number of function evaluations $f(x_j)$. The amount of work required is doubled for each step of Romberg's method. The moderate improvement attained in the accuracy of at most some decimal digits is the worst disadvantage of the Romberg procedure.

The situation is somewhat improved by the sequence of step lengths that has been proposed by *Bulirsch* (1964)

$$h_0 = b - a \quad h_1 = h_0/2 \quad h_2 = h_0/3 \quad h_3 = h_0/4 \quad h_4 = h_0/6 \quad h_5 = h_0/8$$
$$h_6 = h_0/12 \quad h_7 = h_0/16, \ldots \tag{8.18}$$

The sequence (8.18) has the advantage that the amount of computational effort required to compute the trapezoidal sums increases more slowly. Instead of the extrapolation by means of polynomials the rational extrapolation, that is often more accurate, can be applied. A computer program is given in Bulirsch and Stoer (1967).

### 8.1.5 Adaptive quadrature

The high amount of work of the Romberg quadrature is caused by the uniform distribution of the nodes. A reduction of the effort can be attained by choosing the nodes depending on the individual integrand. In *adaptive quadrature methods* certain criteria control the choice automatically.

We first state that an integral can generally be computed faster and more accurately over a short interval, than over a long interval. Therefore it is always advantageous to subdivide the interval of integration $[a, b]$, in a suitable way, prior to the application of a quadrature rule, and then to add the contributions. Adaptive methods subdivide $[a, b]$ successively until the prescribed accuracy is attained for each subinterval by means of the quadrature formula used. The subdivision is finer where $f(x)$ varies greatly and coarser in regions of small variation (see Figure 8.3). The decision whether an interval has to be divided further, is made by comparing two different approximations $\tilde{I}_1$ and $\tilde{I}_2$ of the same subintegral.

In order to explain the principle, we use the value of the trapezoidal rule as an approximation $\tilde{I}_1$ on the subinterval $[a_j, b_j]$

$$\tilde{I}_1 = \tfrac{1}{2} h_j [f(a_j) + f(b_j)] \quad h_j := b_j - a_j.$$

For $\tilde{I}_2$ we choose the value of the Simpson's rule, (8.17), according to Lyness (1969)

$$\tilde{I}_2 = \tfrac{1}{3} [\tilde{I}_1 + 2 h_j f(m_j)] \quad m_j := \tfrac{1}{2} (a_j + b_j).$$

The computed function values can be reused in a most economical way.

As a stopping criterion for the local halving of the interval, we follow a proposal by *Gander* (1985). The halving of an interval is stopped as soon as

$$\tilde{I}_1 + I_S = \tilde{I}_2 + I_S$$

holds in computer arithmetic, where $I_S \neq 0$ is an estimated value of the desired integral $I$. It does not need to be a good approximation of $I \neq 0$, but it must be of the same order of magnitude. This criterion has the effect that contributions of the integral $I$ that are small, in absolute value, do not need to be computed with a small relative error, and that an approximation of $I$ is obtained with almost full accuracy of the computer. If, on the other hand, we only wish to

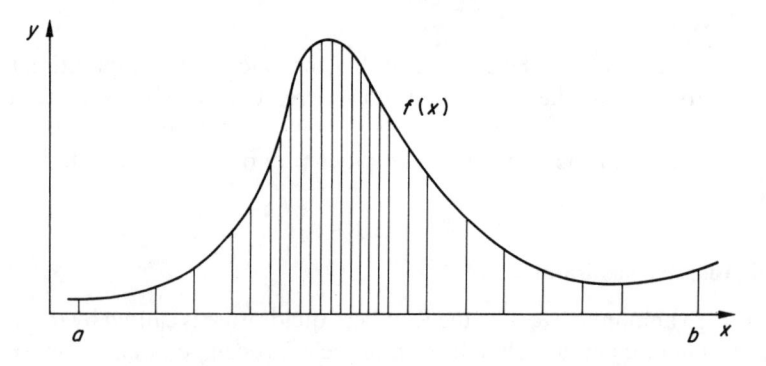

Figure 8.3 Adaptive quadrature

## 8.1 The Trapezoidal Method

approximate $I \neq 0$ with a *relative accuracy* $\varepsilon > 0$, this can be achieved by defining the estimated value, $I_S$, to be about $\varepsilon I/\delta$, where $\delta$ denotes the smallest positive number of the computer such that $1 + \delta \neq 1$.

The most elegant and compact algorithmic description of the adaptive quadrature is achieved by the recursive use of subroutines (Gander, 1985). Disregarding this possibility the algorithmic implementation requires the storage of the abscissae $a_j$ and the corresponding function values, $f_j = f(a_j)$, as vectors so that they can be reused. Moreover, the information on the subintervals is needed for which the integrals still have to be computed. In the following algorithm an index vector $\mathbf{u}$ is used which contains the indices $p$ of the generated abscissae $a_p$ with which the intervals of integration can be defined. The number of components of the vector $\mathbf{u}$ varies in the course of the algorithm (8.19), whereas those of the vectors $\mathbf{a}$ and $\mathbf{f}$ increase monotonically.

$$
\begin{aligned}
&\text{START:} \quad a_0 = a;\ a_1 = b;\ f_0 = f(a);\ f_1 = f(b);\ I = 0 \\
&\qquad\qquad j = 0,\ k = 1;\ p = 1;\ l = 1;\ u_1 = 1 \\
&\text{HALF:} \quad h = a_k - a_j;\ m = (a_j + a_k)/2;\ fm = f(m) \\
&\qquad\qquad I1 = h \times (f_j + f_k)/2;\ I2 = (I1 + 2 \times h \times fm)/3 \\
&\qquad\qquad \text{if } IS + I1 \neq IS + I2 \text{ then} \\
&\qquad\qquad\quad p = p + 1;\ a_p = m;\ f_p = fm;\ k = p \\
&\qquad\qquad\quad l = l + 1;\ u_l = p;\ \text{go to HALF} \\
&\qquad\qquad \text{else} \\
&\qquad\qquad\quad I = I + I2;\ j = u_l;\ l = l - 1;\ k = u_l \\
&\qquad\qquad \text{if } l > 0:\ \text{go to HALF}
\end{aligned}
\qquad (8.19)
$$

The adaptive quadrature is, of course, more efficient if simple quadrature formulae with higher orders of error are combined. Such formulae are considered in Section 8.3.

*Example 8.4* The singular integral of Example 8.3 has been approximately computed by means of (8.19). A computation with an accuracy of fourteen digits, that is $\delta = 10^{-14}$, using the estimated value 0.5 for $I$ yields the following results, where $N$ denotes the number of function evaluations.

| $\varepsilon$ | $10^{-4}$ | $10^{-5}$ | $10^{-6}$ | $10^{-7}$ | $10^{-8}$ |
|---|---|---|---|---|---|
| $\tilde{I}$ | 0.400 000 4636 | 0.400 000 0214 | 0.400 000 0029 | 0.400 000 0005 | 0.400 000 0001 |
| $N$ | 43 | 85 | 207 | 387 | 905 |

*Example 8.5* The integral

$$I = \int_0^1 (e^{-50(x-0.5)^2} + e^{-2x})\, dx \doteq 0.682\,995\,042\,14$$

Figure 8.4 Subintervals of the adaptive quadrature

has been approximately computed by means of the adaptive quadrature (8.19). The integrand has a large variation in the vicinity of $x = 0.5$. Therefore the quadrature requires a fine subdivision there. The resulting subdivision is shown in Figure 8.4 for $\varepsilon = 10^{-4}$ and the estimated value 0.5. The following results are obtained for different $\varepsilon$, with $N$ function evaluations.

| $\varepsilon$ | $10^{-2}$ | $10^{-3}$ | $10^{-4}$ | $10^{-5}$ | $10^{-6}$ |
|---|---|---|---|---|---|
| $\tilde{I}$ | 0.682 994 9919 | 0.682 995 6160 | 0.682 995 0665 | 0.682 995 0432 | 0.682 995 0421 |
| $N$ | 47 | 95 | 215 | 465 | 953 |

Adaptive methods distinguish themselves by the fact that they produce acceptable approximations of integrals, for almost arbitrary integrands that are, for instance, only continuous on subintervals or that are even unbounded. However, the number of function evaluations will be quite high. Therefore it is advisable to use this method only in cases where a preliminary analysis of the integrand is not possible. A collection of computer programs for the adaptive quadrature may be found, for instance, in Piessens et al. (1983).

Finally, we refer to the fact that no absolutely safe automatic quadrature algorithm, working in a deterministic way, can exist. If we applied such a hypothetic procedure to the function $f(x) = 0$ for $[a,b]$ with $a < b$, we would obtain after $N$ function evaluations, $f(x_j)$, at the abscissae, $x_j$, the correct answer $\tilde{I} = 0$. For the function

$$g(x) := \prod_{j=1}^{N} (x - x_j)^2$$

which satisfies $g(x_j) = 0$, $j = 1, 2, \ldots, N$, the algorithm uses exactly the same support abscissae $x_j$, and we also get the result $\tilde{I} = 0$, although we have $I > 0$.

Likewise, it will not be easy to find a practicable and universal quadrature algorithm that integrates, for instance,

$$f(x) = \begin{cases} 1000 & 0.334 < x < 0.335 \\ 0 & \text{elsewhere} \end{cases}$$

from $a = -1$ to $b = 1$ in a correct way.

## 8.2 Transformation Methods

In this section we consider integrals for which all terms of finite order in the error law, (8.12), of the trapezoidal rule vanish. The remainder term is then

## 8.2 Transformation Methods

exponentially small. Integrals with this property arise frequently and have important applications. Moreover, integrals with analytic integrands, $f(x)$, can be reduced to the mentioned case by appropriate transformations.

### 8.2.1 Periodic integrands

A first case where all $c_k$ of the Euler–MacLaurin formula (8.12) vanish, is encountered if the integrand $f(x)$ is analytic on $\mathbf{R}$ and $\tau$ periodic, that is we have

$$f(x+\tau) = f(x) \quad \text{for all } x \in \mathbf{R}$$

and if the integration extends over a full period. Without loss of generality we can set $a = 0$ and $b = \tau$. Then we have

$$f^{(2k-1)}(b) - f^{(2k-1)}(a) = 0 \quad (k = 1, 2, \ldots). \tag{8.20}$$

For each $N$ only the remainder term is present in (8.12). Instead of investigating the remainder term of (8.12) in the case of periodic functions, it is simpler to express the error of the trapezoidal sums by means of the Fourier series of $f(x)$. In the complex notation with $i^2 = -1$ we have

$$f(x) = \sum_{k=-\infty}^{\infty} f_k \exp\left(ikx\frac{2\pi}{\tau}\right) \tag{8.21}$$

with the complex Fourier coefficients

$$f_k := \frac{1}{\tau} \int_0^\tau f(x) \exp\left(-ikx\frac{2\pi}{\tau}\right) dx. \tag{8.22}$$

Since $f(\tau) = f(0)$ the trapezoidal sum, (8.4), for $n$ subintervals is given by

$$T(h) = h \sum_{j=0}^{n-1} f(jh) \quad h = \frac{\tau}{n} \quad n \in \mathbf{N}. \tag{8.23}$$

After substituting the Fourier series (8.21) into (8.23) and interchanging the summations we obtain

$$T(h) = \frac{\tau}{n} \sum_{k=-\infty}^{\infty} f_k \sum_{j=0}^{n-1} \exp\left(ijk\frac{2\pi}{n}\right).$$

Since

$$\sum_{j=0}^{n-1} \exp\left(ijk\frac{2\pi}{n}\right) = \begin{cases} n & \text{for } k \equiv 0 \pmod{n} \\ 0 & \text{else} \end{cases}$$

the sum over $k$ reduces to terms with $k = nl, l \in \mathbf{Z}$, and we obtain

$$T(h) = \tau \sum_{l=-\infty}^{\infty} f_{nl}. \tag{8.24}$$

From (8.22) we have

$$\tau f_0 = \int_0^\tau f(x)\,dx = I$$

and hence the following relation follows from (8.24)

$$\boxed{T(h) - I = \tau(f_n + f_{-n} + f_{2n} + f_{-2n} + \cdots).} \qquad (8.25)$$

For a further discussion we use a result of complex analysis (Henrici, 1977).

**Theorem 8.1** *Let $f(z)$ be $\tau$ periodic and analytic within the strip $|\text{Im}(z)| < \omega, \omega > 0$. Then the Fourier coefficients $f_k$ (8.22) of $f(z)$ decrease at least as fast as a geometric sequence according to*

$$|f_k| = O\left(\exp\left(-|k|\omega\frac{2\pi}{\tau}\right)\right) \quad k \to \infty.$$

For functions $f(x)$ that satisfy the hypotheses of Theorem 8.1 it follows from (8.25) that because $h = \tau/n$

$$\boxed{|T(h) - I| = O\left(\exp\left(-\omega\frac{2\pi}{h}\right)\right).} \qquad (8.26)$$

Therefore the error of the trapezoidal sum $T(h)$ *exponentially* decreases with decreasing step length $h$. For sufficiently small $h$ halving of the step size approximately results in squaring the error. Hence, the number of correct digits increases proportionally to the amount of computational effort.

The algorithm (8.7) is quite suitable for the approximate computation of such integrals. In order to take into account the behaviour of the convergence the stopping criterion $|T - M| < \sqrt{\varepsilon}$ can be used in (8.7), if the last computed trapezoidal sum is allowed to have an error of about $\varepsilon$ compared with $I$.

Integrals of the type under consideration occur frequently in practice. For instance, we mention, the evaluation of the surface of an ellipsoid, the integral representation of the Bessel functions, the evaluation of the real Fourier coefficients and the determination of mean values and of periods (Davis, 1959).

**Example 8.6** The circumference $U$ of an ellipse with the semi-axes $A$ and $B$ with $0 < B < A$ is given by

$$U = \int_0^{2\pi} \sqrt{A^2 \sin^2 \varphi + B^2 \cos^2 \varphi}\,d\varphi = 4A \int_0^{\pi/2} \sqrt{1 - e^2 \cos^2 \varphi}\,d\varphi \qquad (8.27)$$

## 8.2 Transformation Methods

Table 8.2 Trapezoidal sums for periodic integrand

| $n$ | $T\left(\dfrac{2\pi}{n}\right)$ | $q_n$ |
|---|---|---|
| 8  | 4.253 304 8630 |        |
| 16 | 4.287 758 3000 | 0.0405 |
| 24 | 4.289 111 9296 | 0.0681 |
| 32 | 4.289 202 6897 | 0.0828 |
| 40 | 4.289 210 1345 | 0.0919 |
| 48 | 4.289 210 8138 | 0.0980 |
| 56 | 4.289 210 8800 | 0.1024 |
| 64 | 4.289 210 8868 | 0.1057 |

where $e := \sqrt{A^2 - B^2}/A$ is its eccentricity. We calculate the trapezoidal sums with the step lengths $h = 2\pi/n$, $n = 8, 16, 24, \ldots, 64$, for $A = 1$, $B = 0.25$ and hence $e \doteq 0.968\,246$. Table 8.2 shows the resulting values.

The $q$ values are the quotients of consecutive errors. To illustrate the convergence behaviour more clearly, the number $n$ of subintervals is increased as an arithmetic sequence with the difference $d = 8$. According to (8.26) the errors behave like a geometric sequence with the quotient $e^{-d\omega} = e^{-8\omega}$. The integrand (8.27) has branching points at $\varphi = \pm i\omega$ where $\cosh(\omega) = 4/\sqrt{15}$, and is therefore only analytic within the strip $|\mathrm{Im}(\varphi)| < 0.255\,4128$. From Theorem 8.1 and (8.26) it follows that $\lim_{n \to \infty} |q_n| = e^{-8\omega} \doteq 0.1296$, which is in good agreement with the observed quotients in Table 8.2. Since the value of $\omega$ is small for the chosen proportion of the axes, the convergence of the trapezoidal sums is relatively slow. For a proportion $A/B = 2$, even $n = 32$ subintervals, that is eight subintervals per quarter period, are sufficient to produce an approximation accurate to ten digits.

### 8.2.2 Integrals over R

The second special case of the error law, (8.12), is obtained for improper integrals

$$I = \int_{-\infty}^{\infty} f(x)\,dx. \tag{8.28}$$

The integrand is supposed to be absolutely integrable and analytic on the whole real axis. Moreover, we assume that $f^{(k)}(a) \to 0$ as $a \to \pm\infty$, for $k = 0, 1, 2, \ldots$. The formal limiting process $a \to -\infty$ and $b \to \infty$ in (8.11) leads us to believe that the trapezoidal method (8.9) for the computation of the integral (8.28) yields quite good approximations, in this case too. This is indeed the case, because now the error law is obtained by means of Poisson's formula. We restrict ourselves to a more formal discussion and for a rigorous treatment refer you to Henrici (1977). The trapezoidal sum (8.8) defined for (8.28)

$$T(h, s) := h \sum_{j=-\infty}^{\infty} f(jh + s) \tag{8.29}$$

is periodic as a function of $s$, with period $h$. It can be written as a Fourier series

$$T(h, s) = \sum_{k=-\infty}^{\infty} t_k \exp\left(iks\frac{2\pi}{h}\right) \tag{8.30}$$

with the Fourier coefficients

$$t_k = \frac{1}{h} \int_0^h T(h, s) \exp\left(-iks\frac{2\pi}{h}\right) ds. \tag{8.31}$$

We substitute (8.29) into (8.31) and interchange the order of integration and summation and thus obtain

$$t_k = \sum_{j=-\infty}^{\infty} \int_0^h f(jh + s) \exp\left(-iks\frac{2\pi}{h}\right) ds = \int_{-\infty}^{\infty} f(s) \exp\left(-iks\frac{2\pi}{h}\right) ds.$$

Now we introduce the Fourier integral

$$g(t) := \int_{-\infty}^{\infty} f(s) e^{-ist} ds \tag{8.32}$$

of the integrand $f(s)$ and from (8.30) obtain the *Poisson formula*

$$T(h, s) = \text{PV}\left[\sum_{k=-\infty}^{\infty} g\left(k\frac{2\pi}{h}\right) \exp\left(isk\frac{2\pi}{h}\right)\right] \tag{8.33}$$

where PV denotes the principal value which results when the infinite sum is formed symmetrically. Since $g(0) = I$, it follows from (8.33) that

$$T(h, s) - I = \text{PV}\left[\sum_{k \neq 0} g\left(k\frac{2\pi}{h}\right) \exp\left(isk\frac{2\pi}{h}\right)\right]. \tag{8.34}$$

The behaviour of the error as $h \to 0$ is given by the behaviour of the Fourier integral (8.32) as $t \to \infty$ (see Henrici, 1977).

**Theorem 8.2** *Let $f(z)$ be an analytic function within the strip $|\text{Im}(z)| < \omega$, $\omega > 0$. Then the Fourier integral (8.32) asymptotically satisfies $|g(t)| = O(e^{-|t|\omega})$ as $|t| \to \infty$.*

From Theorem 8.2 and (8.34) it follows that

## 8.2 Transformation Methods

$$\boxed{|T(h,s) - I| = O\left(\exp\left(-\omega\frac{2\pi}{h}\right)\right) \quad h > 0.} \qquad (8.35)$$

In formal agreement with (8.26) the error of the trapezoidal sum is again exponentially small as $h \to 0$.

**Example 8.7** For $f(x) = e^{-x^2/2}$ we have

$$I = \int_{-\infty}^{\infty} e^{-x^2/2}\, dx = \sqrt{2\pi} \doteq 2.506\,628\,274\,63 \quad g(t) = \sqrt{2\pi}\, e^{-t^2/2}. \qquad (8.36)$$

By means of (8.9) we obtain $T(2,0) = 2.542\,683\,044$, $T(1,0) = 2.506\,628\,288$, $T(0.5,0) = 2.506\,628\,275$. The integrand decreases very rapidly, therefore the summation on the interval $[-7,7]$ already yields an accuracy of ten digits. If a sufficiently high computer precision was used, then the error of $T(0.5,0)$ would be less than $3 \times 10^{-34}$ in absolute value, from (8.34) and (8.36).

The fast convergence of $T(h, s)$, with respect to decreasing $h$ holds, due to (8.35), as is the case for slowly decreasing $f(x)$. However, the evaluation of $T(h, s)$ by means of (8.9) becomes very time consuming.

**Example 8.8** The function $f(x) = 1/(1 + x^2)$ has the Fourier integral (8.32) $g(t) = \pi e^{-|t|}$. Therefore from (8.33), the trapezoidal sums for the improper integral have the explicit representation

$$T(h,s) = \pi + 2\pi \sum_{k=1}^{\infty} \exp\left(-k\frac{2\pi}{h}\right) \cos\left(sk\frac{2\pi}{h}\right).$$

From this formula the values $T(2,0) = 3.425\,377\,150$, $T(1,0) = 3.153\,348\,095$, $T(0.5,0) = 3.141\,614\,566$, $T(0.25,0) = 3.141\,592\,654$ can be calculated. The fast convergence of the trapezoidal sums towards the limit $I = \pi$ also takes place here, but the evaluation of $T(h, 0)$ by means of (8.9) with the same accuracy would require about $10^{10}$ terms.

### 8.2.3 Transformation methods

In the following, the classical technique of change of variable will be applied to transform the integral (8.1) in such a manner that it can be evaluated by means of a fast convergent quadrature method. We define the transformation by

$$x = \varphi(t) \quad \varphi'(t) > 0 \qquad (8.37)$$

where $\varphi(t)$ is a suitably chosen, simply computable, strictly monotonic and analytic function. Its inverse maps the interval of integration $[a, b]$ injectively onto the interval $[\alpha, \beta]$ with $\varphi(\alpha) = a$, $\varphi(\beta) = b$, thus we obtain

$$I = \int_a^b f(x)\,dx = \int_\alpha^\beta F(t)\,dt \quad \text{with} \quad F(t) := f(\varphi(t))\varphi'(t). \tag{8.38}$$

Main applications of the transformation methods are the treatment of integrals with singular integrands and of integrals over unbounded intervals with slowly decreasing integrands. The first proposals of the transformation methods were made by Goodwin (1949), Sag and Szekeres (1964), Schwartz (1969), Stenger (1973), Takahasi and Mori (1973). In the following we present some of the most important transformations.

(a) *Algebraic substitution* A representative model of an integral with an algebraic boundary singularity is

$$I = \int_0^1 x^{p/q} f(x)\,dx \quad q = 2, 3, \ldots;\ p > -q \quad p \in \mathbb{Z} \tag{8.39}$$

where $f(x)$ is analytic on $[0, 1]$. The condition for $p$ and $q$ guarantees the existence of $I$. The change of the variable

$$x = \varphi(t) = t^q \quad \varphi'(t) = qt^{q-1} > 0 \text{ in } (0, 1)$$

reduces (8.39) to the integral

$$I = q \int_0^1 t^{p+q-1} f(t^q)\,dt$$

which exists because $p + q - 1 \geq 0$ and the integrand has no singularity. It can be efficiently evaluated by means of the Romberg procedure or the Gaussian quadrature (see Section 8.4).

The integral $I = \int_0^1 x^{3/2}\,dx$, of Example 8.3, is reduced by the change of the variable $x = t^2$ to the integral $I = 2\int_0^1 t^4\,dt$ which can be numerically evaluated without any problems.

(b) *tanh-transformation* If integrable singularities, probably of a logarithmic nature, exist at both ends of the interval, then $\varphi(t)$ in (8.37) should be chosen in such a way that the interval of integration is mapped onto the complete real axis, that is $\alpha = -\infty$, $\beta = \infty$. In order to favour the exponential decrease of the transformed integrand, it is required that $\varphi(t)$ tends exponentially towards the limits $a$ and $b$. For the integral

$$I = \int_{-1}^1 f(x)\,dx$$

for instance, the change of variable

$$x = \varphi(t) = \tanh(t) \quad \varphi'(t) = \frac{1}{\cosh^2(t)} \tag{8.40}$$

## 8.2 Transformation Methods

satisfies the required conditions. The transformed integral is

$$I = \int_{-\infty}^{\infty} F(t)\,dt \quad \text{with} \quad F(t) = \frac{f(\tanh(t))}{\cosh^2(t)}. \tag{8.41}$$

Due to the fast increasing denominator, the integrand $F(t)$ is likely to decrease exponentially as $t \to \pm\infty$, so that the integral, (8.41), can be efficiently computed by means of the trapezoidal method (8.9). However, the numerical evaluation of $F(t)$ must be done with some care, because for large values of $|t|$ it may happen that $\tanh(t) = \pm 1$ results, from the limited precision of the computer. If $f(x)$ had singularities at the boundaries it would be impossible to compute $F(t)$ for large $|t|$ for just this reason. In order to eliminate this numerical difficulty, for instance, the relationships

$$\tanh(t) = -1 + e^t/\cosh(t) = -1 + \xi \quad t \leq 0$$
$$\tanh(t) = 1 - e^{-t}/\cosh(t) = 1 - \eta \quad t \geq 0$$

can be used together with locally valid expansions for $f(-1+\xi)$ and $f(1-\eta)$.

Another difficulty of the tanh-transformation is the fact that quite large numbers are produced, so that it may fail on computers with a small range of the exponent.

The transformation (8.40) does not guarantee that the transformed integrand $F(t)$ decreases exponentially on both sides. An example is given by

$$f(x) = \frac{x^2}{(1-x^2)\operatorname{artanh}^2(x)} \quad \text{with} \quad F(t) = \frac{\tanh^2(t)}{t^2}.$$

(c) *sinh-transformation* We consider integrals on an unbounded interval of integration which require too many terms for the trapezoidal method, because $f(x)$ decreases too slowly. In these cases the change of the variable

$$x = \varphi(t) = \sinh(t) \tag{8.42}$$

is suitable so that we obtain

$$I = \int_{-\infty}^{\infty} f(x)\,dx = \int_{-\infty}^{\infty} F(t)\,dt \quad \text{with} \quad F(t) = f(\sinh(t))\cosh(t). \tag{8.43}$$

After a finite number of applications of the sinh-transformation, (8.42), an integrand results that is exponentially decreasing on both sides so that the trapezoidal method, (8.9), can be efficiently applied. Usually a single transformation step is sufficient. There exist quite exceptional examples of integrable functions for which the desired property cannot be achieved by means of a finite number of sinh-transformations (Szekeres, 1961/62)

*Example* 8.9 The sinh-transformation reduces the integral of Example 8.8 to

$$I = \int_{-\infty}^{\infty} \frac{dx}{1+x^2} = \int_{-\infty}^{\infty} \frac{dt}{\cosh(t)} = \pi.$$

The trapezoidal sums are $T(2,0) = 3.232\,618\,532$, $T(1,0) = 3.142\,242\,660$, $T(0.5,0) = 3.141\,592\,687$, $T(0.25,0) = 3.141\,592\,654$. In order to obtain the trapezoidal sums to an accuracy of ten digits, only values for $|t| \leqslant 23$ must be taken into account. The trapezoidal sum $T(0.5,0)$ is already an approximation of $I$ to eight digits, and $T(0.25,0)$ would be accurate to sixteen decimal places if a sufficiently high precision was used.

(d) *exp-transformation*  Integrals with half infinite intervals $(a, \infty)$ can be reduced by means of the simple transformation

$$x = \varphi(t) = a + e^t$$

to

$$I = \int_a^\infty f(x)\,dx = \int_{-\infty}^\infty f(a + e^t)e^t\,dt.$$

A disadvantage of the method of change of variable may be that a quadrature procedure can only be applied when some transformations have been performed by hand. In the following we show that for a given mapping $x = \varphi(t)$ the transformation can be performed in a purely numerical manner, that is by means of values of $\varphi(t)$ and $\varphi'(t)$. To see this, we assume that the approximate evaluation of the transformed integral (8.38) is done by means of an $n$-point quadrature formula (8.2) with nodes $t_j$, weights $v_j$ and the remainder term $R_{n+1}$.

$$I = \int_\alpha^\beta F(t)\,dt = \sum_{j=1}^n v_j F(t_j) + R_{n+1} \tag{8.44}$$

On the basis of the definition of $F(t)$, (8.38), we can interpret (8.44) as a quadrature formula for the original integral with nodes $x_j$ and weights $w_j$, namely

$$\tilde{I} = \sum_{j=1}^n w_j f(x_j) \quad x_j := \varphi(t_j) \quad w_j := v_j \varphi'(t_j) \quad (j = 1, 2, \ldots, n). \tag{8.45}$$

It is easy to compute the values $x_j$ and $w_j$ when they are needed. This is possible because of the choice of $\varphi(t)$ as a simple function. To get an even higher efficiency, a table of $x_j$ and $w_j$ can be prepared which is used when the quadrature formula, (8.45), is applied. More, and in particular special implementations of this idea, can be found in Iri *et al.* (1970), Mori (1978) and Takahasi and Mori (1974).

## 8.3 Interpolation Quadrature Formulae

To compute definite integrals approximately, we now consider methods that are based on approximations of the integrand $f(x)$ by interpolating functions. We discuss the two cases where $f(x)$ is approximated either by an interpolation polynomial or by a spline function. The definite integral of $f(x)$ is then approximated by the integral of the interpolation function.

## 8.3 Interpolation Quadrature Formulae

### 8.3.1 Newton–Cotes quadrature formulae

To approximately compute the definite integral

$$I := \int_a^b f(x)\,dx \tag{8.46}$$

where $f(x)$ is a continuous, real-valued function on $[a,b]$, let $x_0, x_1, \ldots, x_n$ be given pairwise different support abscissae. Without loss of generality we assume that

$$a \leqslant x_0 < x_1 < x_2 < \cdots < x_{n-1} < x_n \leqslant b. \tag{8.47}$$

For these support abscissae, and the corresponding support ordinates $f_k := f(x_k)$, $k = 0, 1, \ldots, n$, there exists a unique interpolation polynomial $P_n(x)$ of degree at most $n$, from Theorem 3.1. Using the Lagrange polynomials $L_k(x)$, (3.3), it has the representation

$$P_n(x) = \sum_{k=0}^{n} f(x_k) L_k(x).$$

To approximate $I$, (8.46), we define the value

$$Q_n := \int_a^b P_n(x)\,dx = \sum_{k=0}^{n} f(x_k) \int_a^b L_k(x)\,dx =: (b-a) \sum_{k=0}^{n} w_k f(x_k). \tag{8.48}$$

The quantities

$$w_k := \frac{1}{b-a} \int_a^b L_k(x)\,dx \quad (k = 0, 1, \ldots, n), \tag{8.49}$$

which only depend on the support abscissae $x_0, x_1, \ldots, x_n$ and on $(b-a)$, are the *integration weights* of the *quadrature formula* (8.48) to the *nodes* $x_k$.

It is clear from the definition (8.48) of the interpolation quadrature formula, that $Q_n$ yields the exact value of the definite integral if $f(x)$ is a polynomial of degree not exceeding $n$. Otherwise $Q_n$ is an approximation of $I$ with an error that depends on the integrand

$$E_n[f] := I - Q_n = \int_a^b f(x)\,dx - (b-a) \sum_{k=0}^{n} w_k f(x_k). \tag{8.50}$$

In certain cases the quadrature formula (8.48) is also exact for polynomials of higher degree. The following notion is introduced to qualify the accuracy of a quadrature formula.

**Definition 8.1** *A quadrature formula* (8.48) *has the degree of precision* $m \in \mathbf{N}$, *if it integrates all polynomials up to the degree* $m$ *exactly, and if* $m$ *is the maximum integer with this property.*

As $E_n[f]$ is a linear functional on $f$, a quadrature formula (8.48) has the degree of precision $m$ if and only if the following statement holds

$$E[x^j] = 0 \quad \text{for } j = 0, 1, \ldots, m \quad \text{and} \quad E[x^{m+1}] \neq 0 \tag{8.51}$$

From the previous considerations a theorem follows.

**Theorem 8.3** *For arbitrarily given $(n+1)$ pairwise different nodes $x_k$ with the property* (8.47), *a uniquely determined interpolation quadrature formula*

$$Q_n = (b-a) \sum_{k=0}^{n} w_k f(x_k) \quad \text{with} \quad w_k = \frac{1}{b-a} \int_a^b L_k(x)\,dx \quad (k = 0, 1, \ldots, n) \tag{8.52}$$

*exists whose degree of precision is at least equal to $n$.*

After these general considerations we present a first class of quadrature formulae with equidistant nodes in the interval $[a, b]$, so that we have

$$x_i := a + ih \quad (i = 0, 1, \ldots, n) \quad h := \frac{b-a}{n}. \tag{8.53}$$

Such quadrature formulae are useful to approximate integrals of functions $f(x)$ that are defined by tables. In order to determine the weights $w_k$ of (8.52) the change of the variable $x = a + (b-a)t$, $dx = (b-a)dt$ is helpful. From the definition of the Lagrange polynomials $L_k(x)$ we obtain

$$w_k = \frac{1}{b-a} \int_a^b \prod_{\substack{i=0 \\ i \neq k}}^{n} \left( \frac{x - x_i}{x_k - x_i} \right) dx = \int_0^1 \prod_{\substack{i=0 \\ i \neq k}}^{n} \left( \frac{nt - i}{k - i} \right) dt. \tag{8.54}$$

For $n = 2$ the weights are obtained from (8.54)

$$w_0 = \int_0^1 \frac{(2t-1)(2t-2)}{(-1)(-2)}\,dt = \frac{1}{2}\int_0^1 (4t^2 - 6t + 2)\,dt = \frac{1}{6},$$

$$w_1 = \int_0^1 \frac{2t(2t-2)}{(1)(-1)}\,dt = -\int_0^1 (4t^2 - 4t)\,dt = \frac{2}{3},$$

$$w_2 = \int_0^1 \frac{2t(2t-1)}{(2)(1)}\,dt = \frac{1}{2}\int_0^1 (4t^2 - 2t)\,dt = \frac{1}{6}.$$

The resulting quadrature formula

$$\boxed{Q_2 = \frac{(b-a)}{6}(f_0 + 4f_1 + f_2) = \frac{h}{3}(f_0 + 4f_1 + f_2) \quad f_i = f(x_i)} \tag{8.55}$$

## 8.3 Interpolation Quadrature Formulae

is *Simpson's rule* or *Kepler's rule*. Its degree of precision is $m = 3$, although only an interpolation polynomial of degree two has been used for its derivation. In fact, we have

$$E_2[x^3] = \int_a^b x^3\,dx - \frac{b-a}{6}\left[a^3 + 4\left(\frac{a+b}{2}\right)^3 + b^3\right]$$
$$= \tfrac{1}{4}(b^4 - a^4) - \tfrac{1}{6}(b-a)[a^3 + \tfrac{1}{2}(a^3 + 3a^2b + 3ab^2 + b^3) + b^3]$$
$$= (b-a)[\tfrac{1}{4}(b^3 + ab^2 + a^2b + a^3) - \tfrac{1}{4}(a^3 + a^2b + ab^2 + b^3)] = 0,$$

but $E_2[x^4] \neq 0$. To see this, it is sufficient to evaluate $I$ and $Q_2$ for $a = -1$ and $b = 1$.

For $n = 3$, the quadrature formula is

$$Q_3 = \frac{3h}{8}(f_0 + 3f_1 + 3f_2 + f_3) \quad h = \frac{b-a}{3} \tag{8.56}$$

It is called the *Newton's (3/8)-rule*. Its degree of precision is $m = 3$, because $E_3[x^4] \neq 0$.

The number of nodes and hence the degree of the interpolation polynomial can be arbitrarily increased. The quadrature formulae for $n = 4$ and $n = 5$ are

$$Q_4 = \frac{2h}{45}(7f_0 + 32f_1 + 12f_2 + 32f_3 + 7f_4) \tag{8.57}$$

$$Q_5 = \frac{5h}{288}(19f_0 + 75f_1 + 50f_2 + 50f_3 + 75f_4 + 19f_5) \tag{8.58}$$

The degree of precision of the formulae (8.57) and (8.58) is $m = 5$.

The formulae (8.55) to (8.58) are called *closed Newton–Cotes quadrature formulae*, because both end points of the interval $[a, b]$ are nodes. It can be shown that they have the degree of precision $m = n + 1$ for $n$ even and $m = n$ for $n$ odd (Isaacson and Keller, 1966). For this reason it is more advantageous to use Newton–Cotes quadrature formulae for even $n$, that is with an odd number of nodes. In order to increase the degree of precision in this class of quadrature formulae it is necessary to add pairs of additional nodes. In principle the degree $n$ can be increased arbitrarily. However, for $n = 8$ and $n \geq 10$ some of the integration weights $w_k$ of the Newton–Cotes quadrature formulae become negative (Stroud, 1974). Moreover, it is known that the interpolation polynomials of high degree oscillate a lot near the end points of the interpolation interval. Therefore the Newton–Cotes formulae should only be used for $n < 8$.

The quadrature error $E_n[f] = I - Q_n$ of a Newton–Cotes formula for $(n+1)$ nodes, for a function $f(x)$ that is $(n+1)$ times continuously differentiable on $[a,b]$, from (3.35) has the representation

$$E_n[f] = \frac{1}{(n+1)!} \int_a^b f^{(n+1)}(\xi(x))\varphi_n(x)\,dx \quad \text{with} \quad \varphi_n(x) := \prod_{i=0}^n (x - x_i). \tag{8.59}$$

As $\varphi_n(x)$ changes its sign at each support abscissa, further treatment of the integral (8.59) requires some extensive consideration in order to obtain a simple description of the quadrature error. Without proof we state the result in the following theorems (Isaacson and Keller, 1966).

**Theorem 8.4** *If $n$ is even, and if $f(x)$ is $(n+2)$ times continuously differentiable on $[a,b]$, then the quadrature error of the closed Newton–Cotes formula (8.52) is given by*

$$E_n[f] = \frac{K_n}{(n+2)!} f^{(n+2)}(\zeta) \quad a < \zeta < b \quad K_n := \int_a^b x\varphi_n(x)\,dx. \tag{8.60}$$

**Theorem 8.5** *For $n$ odd and a function $f(x)$ that is $(n+1)$ times continuously differentiable on $[a,b]$, the quadrature error of the closed Newton–Cotes formula (8.52) is given by*

$$E_n[f] = \frac{K_n}{(n+1)!} f^{(n+1)}(\zeta) \quad a < \zeta < b \quad K_n := \int_a^b \varphi_n(x)\,dx. \tag{8.61}$$

For Simpson's rule (8.55), $n = 2$, from (8.60) we obtain

$$K_2 = \int_a^b x(x - x_0)(x - x_1)(x - x_2)\,dx$$

$$= (b - a) \int_0^1 (a + 2ht)(2ht)(2ht - h)(2ht - 2h)\,dt$$

$$= (b - a)\left( ah^3 \int_0^1 2t(2t - 1)(2t - 2)\,dt + 4h^4 \int_0^1 t^2(2t - 1)(2t - 2)\,dt \right)$$

$$= 16h^5 \int_0^1 (2t^4 - 3t^3 + t^2)\,dt = -\tfrac{4}{15}h^5.$$

The first integral in the large parentheses is equal to zero, because the integrand is an odd function with respect to the middle point of the interval. Hence the quadrature error of Simpson's rule is

$$E_2[f] = -\tfrac{1}{90}h^5 f^{(4)}(\zeta) = -\tfrac{1}{2880}(b-a)^5 f^{(4)}(\zeta) \quad a < \zeta < b. \tag{8.62}$$

Analogously, we determine the quadrature error of Newton's (3/8)-rule, (8.56),

## 8.3 Interpolation Quadrature Formulae

since $n = 3$, from (8.61)

$$K_3 = \int_a^b (x-x_0)(x-x_1)(x-x_2)(x-x_3)\,dx$$

$$= (b-a)\int_0^1 (3ht)(3ht-h)(3ht-2h)(3ht-3h)\,dt$$

$$= 27h^5 \int_0^1 t(3t-1)(3t-2)(t-1)\,dt = -\tfrac{9}{10}h^5.$$

Hence the quadrature error is

$$E_3[f] = -\tfrac{3}{80}h^5 f^{(4)}(\zeta) = -\tfrac{1}{6480}(b-a)^5 f^{(4)}(\zeta) \quad a<\zeta<b. \tag{8.63}$$

The coefficient of the quadrature error of the (3/8)-rule is 2.25 times smaller than that of the Simpson's rule. If a node is added, the error is approximately halved.

The next two Newton–Cotes formulae, (8.57) and (8.58), have the quadrature errors

$$E_4[f] = -\frac{8h^7}{945}f^{(6)}(\zeta) = -\frac{(b-a)^7}{1\,935\,360}f^{(6)}(\zeta) \quad a<\zeta<b$$

$$E_5[f] = -\frac{275h^7}{12\,096}f^{(6)}(\zeta) = -\frac{11(b-a)^7}{37\,800\,000}f^{(6)}(\zeta) \quad a<\zeta<b. \tag{8.64}$$

The coefficient of the quadrature error of $Q_5$ is about 1.78 times smaller than that of $Q_4$, so that the error of $Q_5$ is not even halved compared with that of $Q_4$ with an additional function evaluation.

*Example 8.10* We compute the value of

$$I = \int_0^1 e^{-x}\,dx = 1 - \frac{1}{e} \doteq 0.632\,120\,559$$

approximately by means of the Newton–Cotes formulae (8.55) to (8.58).

$h = \tfrac{1}{2}$   $Q_2 \doteq 0.632\,334$     $E_2 \doteq -0.000\,213$
$h = \tfrac{1}{3}$   $Q_3 \doteq 0.632\,216$     $E_3 \doteq -0.000\,095$
$h = \tfrac{1}{4}$   $Q_4 \doteq 0.632\,120\,875$     $E_4 \doteq -0.000\,000\,317$
$h = \tfrac{1}{5}$   $Q_5 \doteq 0.632\,120\,737$     $E_5 \doteq -0.000\,000\,179.$

The errors of the quadrature rules satisfy the laws mentioned above, hence it is confirmed that it is better to use the Newton–Cotes quadrature formulae only for even $n$.

*Example* 8.11 For the integral

$$I = \int_0^1 \frac{4}{1+x^2} dx = \pi \tag{8.65}$$

the Newton–Cotes formulae (8.55) to (8.58) produce the approximations

$h = \frac{1}{2}$  $Q_2 \doteq 3.133\,333$  $E_2 \doteq 0.008\,259$
$h = \frac{1}{3}$  $Q_3 \doteq 3.138\,462$  $E_3 \doteq 0.003\,131$
$h = \frac{1}{4}$  $Q_4 \doteq 3.142\,118$  $E_4 \doteq -0.000\,525$
$h = \frac{1}{5}$  $Q_5 \doteq 3.141\,878$  $E_5 \doteq -0.000\,286$.

In comparison with the results of Example 8.10 the quadrature errors are much larger.

The value $I$ of a definite integral can be better approximated by means of Newton–Cotes quadrature formulae if the interval of integration $[a,b]$ is subdivided into $N$ subintervals each of equal length. The quadrature formula is then applied to each subinterval, and the contributions are added. In this way we obtain the *summed Newton–Cotes quadrature formulae*. Thus Simpson's rule, (8.55), delivers the *summed Simpson rule*

$$S_2 = \frac{h}{3}\left(f(a) + 4f(x_1) + f(b) + 2\sum_{k=1}^{N-1}[f(x_{2k}) + 2f(x_{2k+1})]\right)$$

$$\text{with } h := \frac{b-a}{2N} \quad x_j := a + jh \quad (j = 1, 2, \ldots, 2N-1). \tag{8.66}$$

The formulation (8.66) allows a direct implementation on a computer. From (8.62) we get the error of the summed Simpson's rule, by adding up the quadrature errors of the subintervals, and then by applying the intermediate value theorem, to yield

$$E_{S_2}[f] = -\frac{(b-a)}{180} h^4 f^{(4)}(\zeta) \quad a < \zeta < b. \tag{8.67}$$

The summed Newton–Cotes quadrature formula, following from $Q_4$, is

$$S_4 = \frac{2h}{45}\Big(7(f(a) + f(b)) + 32(f(x_1) + f(x_3)) + 12f(x_2)$$

$$+ \sum_{k=1}^{N-1}[14f(x_{4k}) + 32(f(x_{4k+1}) + f(x_{4k+3})) + 12f(x_{4k+2})]\Big)$$

$$\text{with } h := \frac{b-a}{4N} \quad x_j := a + jh \quad (j = 1, 2, \ldots, 4N-1).$$

(8.68)

## 8.3 Interpolation Quadrature Formulae

Using (8.64), its quadrature error is

$$E_{S_4}[f] = -\frac{2(b-a)}{945} h^6 f^{(6)}(\zeta) \quad a < \zeta < b. \tag{8.69}$$

*Example* 8.12 In order to get a better approximation of the integral (8.65), we apply the summed Newton–Cotes formulae (8.66) and (8.68) for $N = 2, 3, 4, 6, 8$. In Table 8.3 the numerical results together with the quadrature errors are given. In the case of the summed Simpson's rule the error $E_{S_4}$ decreases faster than $h^4$. Due to the sign change of the fourth derivative of $f(x)$ the individual quadrature errors of the subintervals almost cancel.

*Example* 8.13 To approximately compute

$$I = \int_0^{\pi/2} x \cos(x)\,dx = \frac{\pi}{2} - 1 \doteq 0.570\,796\,3268 \tag{8.70}$$

the summed quadrature formulae $S_2$ and $S_4$ have been used for $N = 2, 3, 4, 6, 8, 12$. The resulting approximations and quadrature errors are given in Table 8.4, with the errors decreasing as $h^4$ and $h^6$, respectively.

The *open Newton–Cotes quadrature formulae* use nodes that are equidistantly distributed in the interior of the interval $[a, b]$. If a single support abscissa $x_1$ in the middle of the interval is chosen, the corresponding interpolation polynomial $P_0(x)$ is constant and equal to $f(x_1)$. In this way we obtain the

Table 8.3 Summed Newton–Cotes formulae, Example 8.12

| N | $S_2$ | $E_{S_2}$ | $S_4$ | $E_{S_4}$ |
|---|---|---|---|---|
| 2 | 3.141 568 6275 | 0.000 024 0261 | 3.141 594 0941 | −0.000 001 4405 |
| 3 | 3.141 591 7809 | 0.000 000 8727 | 3.141 592 6976 | −0.000 000 0440 |
| 4 | 3.141 592 5025 | 0.000 000 1511 | 3.141 592 6611 | −0.000 000 0076 |
| 6 | 3.141 592 6403 | 0.000 000 0133 | 3.141 592 6543 | −0.000 000 0007 |
| 8 | 3.141 592 6512 | 0.000 000 0024 | 3.141 592 6537 | −0.000 000 0001 |

Table 8.4 Summed Newton–Cotes formulae, Example 8.13

| N | $S_2$ | $E_{S_2}$ | $S_4$ | $E_{S_4}$ |
|---|---|---|---|---|
| 2 | 0.571 416 4993 | −0.000 620 1725 | 0.570 795 5084 | 0.000 000 8184 |
| 3 | 0.570 917 0264 | −0.000 120 6997 | 0.570 796 2560 | 0.000 000 0708 |
| 4 | 0.570 834 3204 | −0.000 037 9936 | 0.570 796 3143 | 0.000 000 0125 |
| 6 | 0.570 803 8042 | −0.000 007 4774 | 0.570 796 3257 | 0.000 000 0011 |
| 8 | 0.570 798 6896 | −0.000 002 3628 | 0.570 796 3266 | 0.000 000 0002 |
| 12 | 0.570 796 7931 | −0.000 000 4663 | 0.570 796 3268 | 0.000 000 0000 |

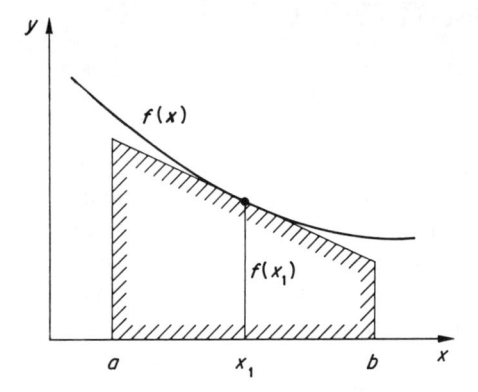

Figure 8.5 Tangent trapezoidal rule

*midpoint rule* or *tangent trapezoidal rule*

$$Q_0^0 = (b-a)f(x_1) \quad x_1 = (a+b)/2. \tag{8.71}$$

The approximation $Q_0^0$ of the integral $I$, can also be interpreted as the area of the trapezium, which is defined by the tangent to the graph of the function at the midpoint (see Figure 8.5). The degree of precision of the midpoint rule is obviously $m = 1$, and its quadrature error is

$$E_0^0[f] = \tfrac{1}{24}(b-a)^3 f''(\zeta) \quad a < \zeta < b.$$

Compared with the trapezoidal rule the coefficient of $f''(\zeta)$ of the error is only half as large and has the opposite sign. In the case of a convex or concave function, $f(x)$, the trapezoidal rule and the tangent trapezoidal rule both yield bounds for the value $I$.

If we subdivide $[a,b]$ into three intervals of equal length, $h = (b-a)/3$, the linear interpolation polynomial, corresponding to the two interior support abscissae, $x_1 = a + h$ and $x_2 = a + 2h$, yields the quadrature formula (see Figure 8.6)

$$Q_1^0 = \frac{3h}{2}(f_1 + f_2) \quad h = (b-a)/3. \tag{8.72}$$

Again it only has the degree of precision $m = 1$ and a quadrature error of

$$E_1^0[f] = \tfrac{1}{108}(b-a)^3 f''(\zeta) \quad a < \zeta < b.$$

In the case of four subintervals, with the three interior nodes $x_1 = a + h$, $x_2 = a + 2h$ and $x_3 = a + 3h$, we obtain the interpolation quadrature formula

## 8.3 Interpolation Quadrature Formulae

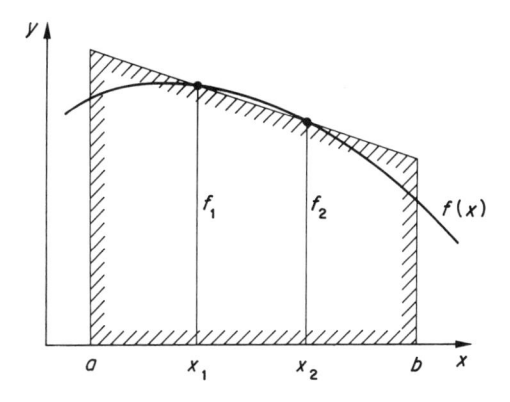

Figure 8.6 Open Newton–Cotes quadrature formula

$$Q_2^0 = \frac{4h}{3}(2f_1 - f_2 + 2f_3) \quad h = (b-a)/4 \quad x_i = a + ih. \tag{8.73}$$

This has the degree of precision $m = 3$ and a quadrature error of

$$E_2^0[f] = \frac{7}{23\,040}(b-a)^5 f^{(4)}(\zeta) \quad a < \zeta < b.$$

The coefficient of $f^{(4)}(\zeta)$, in comparison with that of Simpson's rule, (8.62), is somewhat smaller.

### 8.3.2 Spline quadrature formulae

Another class of interpolation quadrature formulae results from spline interpolation functions. In the following we only consider the cubic spline interpolation functions of Section 3.7 for *equidistant* support abscissae, $x_i = a + ih$, $i = 0, 1, \ldots, n$. In each subinterval $[x_i, x_{i+1}]$, the uniquely defined spline function, $s(x)$, is equal to a cubic polynomial $s_i(x)$, $x \in [x_i, x_{i+1}]$. We first compute the integral of $s_i(x)$, over $[x_i, x_{i+1}]$. From (3.117) and with (3.124), we obtain

$$\begin{aligned}
\tilde{I}_i &:= \int_{x_i}^{x_{i+1}} s_i(x)\,dx = \int_{x_i}^{x_{i+1}} [a_i(x-x_i)^3 + b_i(x-x_i)^2 + c_i(x-x_i) + d_i]\,dx \\
&= \tfrac{1}{4}a_i h^4 + \tfrac{1}{3}b_i h^3 + \tfrac{1}{2}c_i h^2 + d_i h \\
&= \tfrac{1}{24}h^3(y''_{i+1} - y''_i) + \tfrac{1}{6}h^3 y''_i + \tfrac{1}{2}h(y_{i+1} - y_i) - \tfrac{1}{12}h^3(y''_{i+1} + 2y''_i) + hy_i \\
&= \tfrac{1}{2}h(y_i + y_{i+1}) - \tfrac{1}{24}h^3(y''_i + y''_{i+1}) \quad (i = 0, 1, \ldots, n-1).
\end{aligned} \tag{8.74}$$

Summing over $i$, we obtain the *spline quadrature formula*

$$\tilde{I}_S = \int_a^b s(x)\,dx = \frac{h}{2}\left(y_0 + 2\sum_{i=1}^{n-1} y_i + y_n\right) - \frac{h^3}{24}\left(y_0'' + 2\sum_{i=1}^{n-1} y_i'' + y_n''\right).$$

(8.75)

The first part of the formula (8.75) is equal to the trapezoidal sum (8.4). Given function values $y_i := f(x_i)$, $(i = 0, 1, \ldots, n)$, the values $y_i''$, $(i = 1, 2, \ldots, n-1)$ must be determined from the system of equations (3.128) for chosen boundary conditions $y_0''$ and $y_n''$. A natural spline function ($y_0'' = y_n'' = 0$) will, in general, not be adequate enough to express the integrand $f(x)$ in the first and the last subinterval. Therefore a general cubic spline function should be used, for instance, by imposing the conditions $y_0'' = \alpha y_1''$ and $y_n'' = \beta y_{n-1}''$. From (3.128), the following system of equations results for $n = 5$

| $y_1''$ | $y_2''$ | $y_3''$ | $y_4''$ | 1 |
|---|---|---|---|---|
| $4+\alpha$ | 1 | | | $-6(y_2 - 2y_1 + y_0)/h^2$ |
| 1 | 4 | 1 | | $-6(y_3 - 2y_2 + y_1)/h^2$ |
| | 1 | 4 | 1 | $-6(y_4 - 2y_3 + y_2)/h^2$ |
| | | 1 | $4+\beta$ | $-6(y_5 - 2y_4 + y_3)/h^2$ |

(8.76)

By adding the equations (8.76) and by generalizing the result in an obvious manner, we obtain

$$(\alpha - 1)y_1'' + 6\sum_{i=1}^{n-1} y_i'' + (\beta - 1)y_{n-1}'' - 6(y_0 - y_1 - y_{n-1} + y_n)/h^2 = 0$$

and hence

$$\sum_{i=1}^{n-1} y_i'' = (y_0 - y_1 - y_{n-1} + y_n)/h^2 + \tfrac{1}{6}[(1-\alpha)y_1'' + (1-\beta)y_{n-1}''].$$

The expression for this sum is now substituted into (8.75), and so a formula results that is simpler to evaluate.

$$\tilde{I}_S = h\left(\frac{5}{12}y_0 + \frac{13}{12}y_1 + \sum_{i=2}^{n-2} y_i + \frac{13}{12}y_{n-1} + \frac{5}{12}y_n\right)$$

$$-\frac{h^3}{72}[(1+2\alpha)y_1'' + (1+2\beta)y_{n-1}'']$$

(8.77)

$$h := (b-a)/n \quad n \geq 3.$$

## 8.4 Gaussian Quadrature Formulae

Table 8.5 Spline quadrature

| n | $\alpha = \beta = 1$ $\tilde{I}_s$ | E | $\alpha = \beta = 0.5$ $\tilde{I}_s$ | E | $\alpha = \beta = 0$ $\tilde{I}_s$ | E |
|---|---|---|---|---|---|---|
| 4  | 3.140 882 3529 | $7.10 \times 10^{-4}$  | 3.140 147 06 | $-1.45 \times 10^{-3}$ | 3.139 201 68 | $-2.39 \times 10^{-3}$ |
| 8  | 3.141 568 8362 | $2.38 \times 10^{-5}$  | 3.141 452 79 | $-1.40 \times 10^{-4}$ | 3.141 305 65 | $-2.87 \times 10^{-4}$ |
| 12 | 3.141 589 4935 | $3.16 \times 10^{-6}$  | 3.141 553 76 | $-3.89 \times 10^{-5}$ | 3.141 508 45 | $-8.42 \times 10^{-5}$ |
| 16 | 3.141 591 9011 | $7.53 \times 10^{-7}$  | 3.141 576 62 | $-1.60 \times 10^{-5}$ | 3.141 557 26 | $-3.54 \times 10^{-5}$ |
| 24 | 3.141 592 5542 | $9.94 \times 10^{-8}$  | 3.141 587 99 | $-4.67 \times 10^{-6}$ | 3.141 582 19 | $-1.05 \times 10^{-5}$ |
| 32 | 3.141 592 6300 | $2.36 \times 10^{-8}$  | 3.141 590 70 | $-1.96 \times 10^{-6}$ | 3.141 588 24 | $-4.41 \times 10^{-6}$ |
| 48 | 3.141 592 6505 | $3.12 \times 10^{-9}$  | 3.141 592 08 | $-5.78 \times 10^{-7}$ | 3.141 591 35 | $-1.31 \times 10^{-6}$ |
| 64 | 3.141 592 6528 | $7.40 \times 10^{-10}$ | 3.141 592 41 | $-2.43 \times 10^{-7}$ | 3.141 592 10 | $-5.51 \times 10^{-7}$ |

It does not have the form of (8.2) yet, but in principle it is possible to express the two values $y_1''$ and $y_{n-1}''$, for given $\alpha$ and $\beta$, by using the values of $y_i$ and (8.76). Some quadrature formulae of this type can be found in Ahlberg et al. (1967). An equally convenient way is to solve the tridiagonal system of equations (8.76) and substitute the resulting values of $y_1''$ and $y_{n-1}''$ into (8.77). These values depend, of course, on $\alpha$ and $\beta$.

*Example* 8.14 The integral (8.65) is approximated by means of the spline quadrature formula (8.77) for $n = 4, 8, 12, 16, 24, 32, 48, 64$ and with three different values $\alpha = \beta$. The results are summarized in Table 8.5, together with the quadrature errors $E$. We recognize that $\alpha = \beta = 1$ yields the best approximations. This is clear because with decreasing step size $h$ the pairs $y_0''$ and $y_1''$, and $y_{n-1}''$ and $y_n''$, agree better. However, a comparison with the results of Table 8.3 shows that the spline quadrature yields less accurate approximations of $I$ with the same number of function evaluations. The quadrature error is $O(h^5)$ for this integrand, for the same reason as was mentioned in Example 8.12.

The problem of suitably fixing $y_0''$ and $y_n''$ can be tackled in several ways. If the first, or the second, derivative of the integrand is available the corresponding values can be taken into account in the systems (3.134) or (3.128). An approximation of the first derivative can also be obtained by means of numerical differentiation, for instance, by the formula (3.27). Finally it has been proposed (Volk, 1980) to use the function value $y_{1/2} := f(x_{1/2})$, $x_{1/2} = (x_0 + x_1)/2$ and to request that $s_0(x_{1/2}) = y_{1/2}$. This requirement allows us to determine $y_0''$ in such a way that the quadrature error can be estimated optimally.

## 8.4 Gaussian Quadrature Formulae

The quadrature formulae considered so far have been derived on the assumption that the support abscissae of integration are given so that only the corresponding weights have to be determined. In the following we pose the problem of how to choose the nodes $x_k$ and the weights $w_k$ such that the resulting interpolation quadrature formula, (8.52), has a maximal degree of precision.

We wish to find such a quadrature formula $Q_n$, for the special interval $[-1, 1]$

$$\int_{-1}^{1} f(x)\,dx = \sum_{k=1}^{n} w_k f(x_k) + E_n[f] = Q_n + E_n[f] \quad x_k \in [-1, 1] \qquad (8.78)$$

with $n$ nodes, $x_k$, and $n$ weights, $w_k$. Defining the interval of integration to be $[-1, 1]$ is no restriction, because every bounded interval $[a, b]$ can be mapped onto the unit interval by means of a linear transformation. This special interval is considered because certain properties of the Legendre polynomials are used to derive the method.

**Theorem 8.6** *The degree of precision of a quadrature formula (8.78) with n nodes cannot exceed* $(2n - 1)$.

*Proof* We consider the polynomial of degree $2n$

$$q(x) := \prod_{k=1}^{n} (x - x_k)^2$$

which is defined by means of the $n$ pairwise different support abscissae of the quadrature formula. This nonzero polynomial satisfies $q(x) \geq 0$ for all $x \in [-1, 1]$, and hence we have

$$I = \int_{-1}^{1} q(x)\,dx > 0.$$

However, the quadrature formula (8.78) yields, because $q(x_k) = 0$, $k = 1, 2, \ldots, n$, the value $Q = 0$, that is $E[q] \neq 0$, hence the degree of precision must be less than $2n$.

**Theorem 8.7** *There exists a unique quadrature formula*

$$Q = \sum_{k=1}^{n} w_k f(x_k) \quad x_k \in [-1, 1] \qquad (8.79)$$

*with n nodes $x_k$, which has the maximal degree of precision* $(2n - 1)$. *The nodes $x_k$ are the zeros of the nth Legendre polynomial $P_n(x)$ (4.110), and the weights are defined by*

$$w_k = \int_{-1}^{1} \prod_{\substack{j=1 \\ j \neq k}}^{n} \left( \frac{x - x_j}{x_k - x_j} \right)^2 dx > 0 \quad (k = 1, 2, \ldots, n). \qquad (8.80)$$

*Proof* We prove the theorem in three parts: existence, the weights formula and uniqueness.

## 8.4 Gaussian Quadrature Formulae

(a) First we show the existence of a quadrature formula of the degree of precision $(2n-1)$. For this purpose we use the fact that the Legendre polynomial $P_n(x)$ has $n$ simple zeros, $x_1, x_2, \ldots, x_n$, in the interior of $[-1, 1]$, as was proved in Theorem 4.15. For these $n$ pairwise different support abscissae, a uniquely determined interpolation quadrature formula exists, from Theorem 8.3, whose degree of precision is at least $(n-1)$.

Now let $p(x)$ be an arbitrary real polynomial whose degree is at most equal to $(2n-1)$. If we divide $p(x)$ by the $n$th Legendre polynomial $P_n(x)$ the following relationship holds

$$p(x) = q(x)P_n(x) + r(x) \tag{8.81}$$

with degree $(q(x)) \leq n-1$ and degree $(r(x)) \leq n-1$. From this follows

$$\int_{-1}^{1} p(x)\,dx = \int_{-1}^{1} q(x)P_n(x)\,dx + \int_{-1}^{1} r(x)\,dx = \int_{-1}^{1} r(x)\,dx. \tag{8.82}$$

Here we used the fact that, from (4.111), the Legendre polynomial $P_n(x)$ is orthogonal to all Legendre polynomials $P_0(x), P_1(x), \ldots, P_{n-1}(x)$ of lower degree, and hence also to $q(x)$.

For the interpolation quadrature formula corresponding to the nodes $x_k$ and the weights $w_k$, the following relationship is valid for the polynomial $p(x)$, from (8.81), $P_n(x_k) = 0$ and (8.82)

$$\sum_{k=1}^{n} w_k p(x_k) = \sum_{k=1}^{n} w_k q(x_k) P_n(x_k) + \sum_{k=1}^{n} w_k r(x_k) = \sum_{k=1}^{n} w_k r(x_k)$$

$$= \int_{-1}^{1} r(x)\,dx = \int_{-1}^{1} p(x)\,dx. \tag{8.83}$$

The second to last equality in (8.83) is based on the fact that the degree of precision of the interpolation quadrature formula is at least $(n-1)$. Consequently, we have shown that the quadrature formula (8.79) is exact for each polynomial of degree less than $2n$. We deduce from Theorem 8.6 that the degree of precision is maximal.

(b) The weights $w_k$ of the interpolation quadrature formula are now given by

$$w_k = \int_{-1}^{1} L_k(x)\,dx = \int_{-1}^{1} \prod_{\substack{j=1 \\ j \neq k}}^{n} \left(\frac{x - x_j}{x_k - x_j}\right) dx \quad (k = 1, 2, \ldots, n) \tag{8.84}$$

where $L_k(x)$ is the Lagrange polynomial of degree $(n-1)$ for the support abscissae $x_1, x_2, \ldots, x_n$ satisfying $L_k(x_j) = \delta_{kj}$. This representation is not very helpful. Therefore we use the proven fact that the interpolation quadrature formula, corresponding to the nodes $x_k$, has the degree of precision $(2n-1)$ and consequently for the polynomial $L_k^2(x)$ of degree $(2n-2)$ yields the exact value.

Thus we have

$$0 < \int_{-1}^{1} \prod_{\substack{j=1 \\ j \neq k}}^{n} \left(\frac{x - x_j}{x_k - x_j}\right)^2 dx = \sum_{\mu=1}^{n} w_\mu L_k^2(x_\mu) = w_k \quad (k = 1, 2, \ldots, n). \quad (8.85)$$

which proves (8.80). In particular from this representation we conclude that the weights $w_k$ are strictly positive for all $n \in \mathbf{N}$.

(c) To show the uniqueness of the quadrature formula, we assume that there exists another formula

$$Q^* := \sum_{k=1}^{n} w_k^* f(x_k^*) \quad x_k^* \neq x_j^* \quad \text{for all } k \neq j \quad (8.86)$$

which has a degree of precision $(2n - 1)$. From (b) the weights $w_k^*$ of (8.86) satisfy the property $w_k^* > 0$, $k = 1, 2, \ldots, n$. We want to show that the nodes $x_k^*$ coincide with those of (8.79), apart from a permutation. For this purpose we consider the auxiliary polynomial

$$h(x) := L_k^*(x) P_n(x) \quad L_k^*(x) := \prod_{\substack{j=1 \\ j \neq k}}^{n} \left(\frac{x - x_j^*}{x_k^* - x_j^*}\right) \quad \text{degree}(h(x)) = 2n - 1.$$

From our assumption, the quadrature formula (8.86) integrates $h(x)$ exactly, and therefore for $k = 1, 2, \ldots, n$ we have

$$0 = \int_{-1}^{1} h(x) dx = \int_{-1}^{1} L_k^*(x) P_n(x) dx = \sum_{\mu=1}^{n} w_\mu^* L_k^*(x_\mu^*) P_n(x_\mu^*)$$

$$= w_k^* P_n(x_k^*). \quad (8.87)$$

Using the orthogonality of $P_n(x)$ with respect to all polynomials of degree less than $n$ the second integral in (8.87) vanishes. However we know that $w_k^* > 0$, hence $P_n(x_k^*) = 0$, $k = 1, 2, \ldots, n$, that is, the nodes $x_k^*$ of (8.86) are the zeros of the $n$th Legendre polynomial $P_n(x)$. Therefore the nodes are uniquely determined as those of (8.79). Moreover, the weights of the corresponding interpolation quadrature formula are uniquely determined by (8.84).

The quadrature methods of maximal degree of precision that are characterized by Theorem 8.7 are called *Gaussian quadrature formulae*. The nonzero nodes $x_k$ are pairwise symmetric with respect to the origin. Due to (8.80) the weights are equal for these pairs. Therefore it is sufficient to tabulate the nonnegative nodes $x_k$ together with the corresponding weights $w_k$. Usually the nodes are given in decreasing order $1 > x_1 > x_2 > \cdots$ (Abramowitz and Stegun, 1971; Schmeisser and Schirmeier, 1976; Stroud, 1974).

In order to be able to use the Gaussian quadrature formulae, the nodes $x_k$ and the weights $w_k$ must be known numerically. A simple way consists of using the values taken from tables as data sets in a computer program. In place of

## 8.4 Gaussian Quadrature Formulae

that we want to present a stable method for computing the zeros, $x_k$, of the $n$th Legendre polynomial and the weights, $w_k$ (Gautschi, 1970; Golub and Welsch, 1969). The basis of the method is given by the following theorem.

**Theorem 8.8** *The nth Legendre polynomial $P_n(x)$ of order $n$, $n \geqslant 1$ is equal to the determinant of*

$$P_n(x) = \begin{vmatrix} a_1 x & b_1 & & & & \\ b_1 & a_2 x & b_2 & & & \\ & b_2 & a_3 x & b_3 & & \\ & & \ddots & \ddots & \ddots & \\ & & & b_{n-2} & a_{n-1} x & b_{n-1} \\ & & & & b_{n-1} & a_n x \end{vmatrix} \quad \begin{aligned} a_k &= \frac{2k-1}{k} \\ b_k &= \sqrt{\frac{k}{k+1}} \\ (n &= 1,2,3,\ldots). \end{aligned}$$

(8.88)

*Proof* We will show that the determinants for three successive integers $n$ satisfy the recursion formula (4.116) of the Legendre polynomials. We expand the determinant (8.88) with respect to the last row to obtain

$$P_n(x) = a_n x P_{n-1}(x) - b_{n-1}^2 P_{n-2}(x) \quad n \geqslant 3. \tag{8.89}$$

In (8.89) if we replace $n$ by $(n+1)$ and take into account the definitions of $a_k$ and $b_k$ we obtain the recursion formula

$$P_{n+1}(x) = \frac{2n+1}{n+1} x P_n(x) - \frac{n}{n+1} P_{n-1}(x) \quad n \geqslant 2.$$

If we define $P_0(x) = 1$, then it is also valid for $n = 1$, since $P_1(x) = x$.

Following Theorem 8.7, the desired $x_k$ are the zeros of the determinant (8.88). They can be obtained as the eigenvalues of a symmetric and tridiagonal matrix $J_n$ as can be seen from the following. We eliminate the coefficients $a_i$ along the diagonal in such a way that the symmetry of the determinant is maintained. Dividing the $k$th row and the $k$th column by $\sqrt{a_k} = \sqrt{(2k-1)/k}$ from (8.88) we get

$$P_n(x) = \begin{vmatrix} x & \beta_1 & & & & \\ \beta_1 & x & \beta_2 & & & \\ & \beta_2 & x & \beta_3 & & \\ & & \ddots & \ddots & \ddots & \\ & & & \beta_{n-2} & x & \beta_{n-1} \\ & & & & \beta_{n-1} & x \end{vmatrix} \cdot \prod_{k=1}^{n} a_k. \tag{8.90}$$

The off-diagonal elements of the determinant (8.90) are

$$\beta_k = \frac{b_k}{\sqrt{a_k a_{k+1}}} = \sqrt{\frac{k \cdot k \cdot (k+1)}{(k+1)(2k-1)(2k+1)}} = \frac{k}{\sqrt{4k^2-1}} \quad (k=1,2,\ldots,n-1). \tag{8.91}$$

The non-vanishing zeros of $P_n(x)$ are pairwise of opposite value, therefore the $x_k$ are the eigenvalues of the symmetric, tridiagonal matrix

$$\mathbf{J}_n := \begin{pmatrix} 0 & \beta_1 & & & & \\ \beta_1 & 0 & \beta_2 & & & \\ & \beta_2 & 0 & \beta_3 & & \\ & & \ddots & \ddots & \ddots & \\ & & & \beta_{n-2} & 0 & \beta_{n-1} \\ & & & & \beta_{n-1} & 0 \end{pmatrix} \in \mathbf{R}^{n \times n} \tag{8.92}$$

and can therefore be computed by means of the QR algorithm in a stable and efficient way. However, the weights $w_k$ can also be obtained using the eigenvalue problem for $\mathbf{J}_n$. In order to show this connection, we verify that

$$\mathbf{z}^{(k)} := (\alpha_0 \sqrt{a_1} P_0(x_k), \alpha_1 \sqrt{a_2} P_1(x_k), \alpha_2 \sqrt{a_3} P_2(x_k), \ldots, \alpha_{n-1} \sqrt{a_n} P_{n-1}(x_k))^T \in \mathbf{R}^n$$

$$\text{with} \quad \alpha_0 := 1, \alpha_j := 1 \bigg/ \prod_{l=1}^{j} b_l, \quad (j=1,2,\ldots,n-1) \tag{8.93}$$

is an eigenvector of $\mathbf{J}_n$ corresponding to the eigenvalue $x_k$ for $k=1,2,\ldots,n$. For the first component of $\mathbf{J}_n \mathbf{z}^{(k)}$, using (8.88), (8.91) and (8.93) we have

$$\alpha_1 \beta_1 \sqrt{a_2} P_1(x_k) = x_k = x_k [\alpha_0 \sqrt{a_1} P_0(x_k)].$$

For the $i$th component ($1 < i < n$) after several substitutions and due to (8.89), we obtain

$$\alpha_{i-2} \beta_{i-1} \sqrt{a_{i-1}} P_{i-2}(x_k) + \alpha_i \beta_i \sqrt{a_{i+1}} P_i(x_k)$$

$$= \frac{\alpha_{i-1}}{\sqrt{a_i}} [P_i(x_k) + b_{i-1}^2 P_{i-2}(x_k)] = x_k (\alpha_{i-1} \sqrt{a_i} P_{i-1}(x_k)).$$

This relationship also remains valid for the last component $i = n$, because $P_n(x_k) = 0$.

Furthermore, since the Legendre polynomials $P_0(x), P_1(x), \ldots, P_{n-1}(x)$ are integrated exactly by the Gaussian quadrature formula, and the orthogonality property (4.111) holds, this leads to

$$\int_{-1}^{1} P_0(x) P_i(x)\, dx = \int_{-1}^{1} P_i(x)\, dx = \sum_{k=1}^{n} w_k P_i(x_k)$$

$$= \begin{cases} 2 & \text{for } i = 0 \\ 0 & \text{for } i = 1, 2, \ldots, n-1 \end{cases} \tag{8.94}$$

## 8.4 Gaussian Quadrature Formulae

Consequently the weights $w_k$ satisfy the system of equations (8.94). If we multiply the first equation by $\alpha_0\sqrt{a_1} = 1$ and the $j$th equation ($j = 2, 3, \ldots, n$) by $\alpha_{j-1}\sqrt{a_j} \neq 0$, then the matrix of the system resulting from (8.94)

$$\mathbf{Cw} = 2\mathbf{e}_1 \quad \mathbf{w} := (w_1, w_2, \ldots, w_n)^T \quad \mathbf{e}_1 = (1, 0, 0, \ldots, 0)^T \tag{8.95}$$

contains as columns the eigenvectors $\mathbf{z}^{(k)}$ (8.93). However, the eigenvectors of the symmetric matrix $\mathbf{J}_n$ corresponding to pairwise different eigenvalues are mutually orthogonal. Hence by multiplying (8.95) from the left by $\mathbf{z}^{(k)T}$ we obtain

$$(\mathbf{z}^{(k)T}\mathbf{z}^{(k)})w_k = 2\mathbf{z}^{(k)T}\mathbf{e}_1 = 2z_1^{(k)} = 2 \tag{8.96}$$

where $z_1^{(k)} = 1$ is the first component of the non-normalized eigenvector $\mathbf{z}^{(k)}$ (8.93). If $\tilde{\mathbf{z}}^{(k)}$ denotes the normalized eigenvector corresponding to $\mathbf{z}^{(k)}$ from (8.96) we deduce

$$\boxed{w_k = 2(\tilde{z}_1^{(k)})^2 \quad (k = 1, 2, \ldots, n).} \tag{8.97}$$

Again the relationship (8.96) or (8.97) shows that the weights of the Gaussian quadrature formulae are positive for all $n \in \mathbf{N}$.

In Table 8.6, the nodes and weights are given for some values of $n$. The approximate computation of an integral

$$I = \int_a^b f(t)\,dt$$

by means of a Gaussian quadrature formula requires a change of variable

$$t = \frac{b-a}{2}x + \frac{a+b}{2} \tag{8.98}$$

so that we have

$$I = \int_a^b f(t)\,dt = \frac{b-a}{2}\int_{-1}^1 f\left(\frac{b-a}{2}x + \frac{a+b}{2}\right)dx.$$

Table 8.6 Nodes and weights of Gaussian quadrature.

| | | |
|---|---|---|
| $n = 2$ | $x_1 = \frac{\sqrt{3}}{3} \doteq 0.577\,350\,2692$ | $w_1 = 1$ |
| $n = 3$ | $x_1 = \sqrt{\frac{3}{5}} \doteq 0.774\,596\,6692$ <br> $x_2 = 0,$ | $w_1 = \frac{5}{9} \doteq 0.555\,555\,5556$ <br> $w_2 = \frac{8}{9} \doteq 0.888\,888\,8889$ |
| $n = 4$ | $x_1 \doteq 0.861\,136\,3116$ <br> $x_2 \doteq 0.339\,981\,0436$ | $w_1 \doteq 0.347\,854\,8451$ <br> $w_2 \doteq 0.652\,145\,1549$ |
| $n = 5$ | $x_1 \doteq 0.906\,179\,8459$ <br> $x_2 \doteq 0.538\,469\,3101$ <br> $x_3 = 0,$ | $w_1 \doteq 0.236\,926\,8851$ <br> $w_2 \doteq 0.478\,628\,6705$ <br> $w_3 \doteq 0.568\,888\,8889$ |

Table 8.7 Gaussian quadrature

| n | $I_1$ $Q_n$ | $E_n$ | $I_2$ $Q_n$ | $E_n$ |
|---|---|---|---|---|
| 2 | 3.147 540 9836 | −0.005 948 3300 | 0.563 562 244 208 | 0.007 234 082 587 |
| 3 | 3.141 068 1400 | 0.000 524 5136 | 0.570 851 127 976 | −0.000 054 801 181 |
| 4 | 3.141 611 9052 | −0.000 019 2517 | 0.570 796 127 158 | 0.000 000 199 637 |
| 5 | 3.141 592 6399 | 0.000 000 0138 | 0.570 796 327 221 | −0.000 000 000 426 |
| 6 | 3.141 592 6112 | 0.000 000 0424 | 0.570 796 326 794 | 0.000 000 000 001 |
| 7 | 3.141 592 6563 | −0.000 000 0027 | 0.570 796 326 796 | −0.000 000 000 001 |

The quadrature formula is of the form

$$Q = \frac{b-a}{2} \sum_{k=1}^{n} w_k f\left(\frac{b-a}{2} x_k + \frac{a+b}{2}\right) = \frac{b-a}{2} \sum_{k=1}^{n} w_k f(t_k) \tag{8.99}$$

where $t_k$ are the transformed Gaussian nodes using (8.98).

*Example* 8.15 To illustrate the high degree of precision of the Gaussian quadrature formulae, we compute

$$I_1 = \int_0^1 \frac{4}{1+x^2} dx = \pi \quad I_2 = \int_0^{\pi/2} x \cos(x) dx = \frac{\pi}{2} - 1$$

for different values of $n$. Table 8.7 contains the results together with the quadrature errors. The calculation was performed to an accuracy of fourteen digits, using the values of the nodes $x_k$ and the weights $w_k$ from Schmeisser and Schirmeier (1976).

## 8.5 Exercises

**8.1.** Let $a \leq b < c$ be the semi-axes of an ellipsoid. Its surface area $A$ is given by the integral of a periodic function

$$A = ab \int_0^{2\pi} \left[1 + \left(\frac{1}{w} + w\right) \arctan(w)\right] d\varphi \quad w = \sqrt{\frac{c^2 - a^2}{a^2} \cos^2 \varphi + \frac{c^2 - b^2}{b^2} \sin^2 \varphi}.$$

Compute $A$ by means of the algorithm (8.7). From the symmetry, it is sufficient to integrate over one quarter of the period. If $a \geq b > c$, then in real form we have

$$A = ab \int_0^{2\pi} \left[1 + \left(\frac{1}{v} - v\right) \operatorname{arctanh}(v)\right] d\varphi \quad v = iw.$$

**8.2.** Apply appropriate transformations to the following integrals so that subsequently they can be computed in an efficient manner by means of the algorithm (8.9) for integrals over **R**.

## 8.5 Exercises

(a) $\int_0^1 x^{0.21}\sqrt{\log\left(\dfrac{1}{x}\right)}\,dx = \sqrt{\pi/2.662}$

(b) $\int_0^1 \dfrac{dx}{(3-2x)x^{3/4}(1-x)^{1/4}} = \pi\sqrt{2}\cdot 3^{-3/4}$

(c) $\int_0^\infty x^{-0.7} e^{-0.4x} \cos(2x)\,dx = \Gamma(0.3)\,\text{Re}\,(0.4+2i)^{-0.3}$

(d) $\int_0^\infty \dfrac{dx}{x^{1-\alpha}+x^{1+\beta}} = \dfrac{\omega}{\sin(\alpha\omega)} \quad \omega = \dfrac{\pi}{\alpha+\beta} \quad \alpha>0,\ \beta>0$

(e) $\int_{-\infty}^\infty \dfrac{dx}{(1+x^2)^{5/4}} = \left(\dfrac{\pi}{2}\right)^{3/2}\Gamma(1.25)^{-2}$

(f) $\int_{-\infty}^\infty \dfrac{dx}{x^2+e^{4x}} \doteq 3.160\,322\,869\,7485$

**8.3.** Apply more sinh-transformations $t = t_1 = \sinh(t_2),\ldots$ to the integral of Example 8.9 and study the efficiency of the algorithm (8.9) for the corresponding integrals.

**8.4.** The adaptive quadrature algorithm, (8.19), operates in a satisfactory way even for many discontinuous integrands, for instance

$$I = \int_1^x \dfrac{[\xi]}{\xi}\,d\xi = [x]\log(x) - \log([x]!) \quad x>1,$$

where $[x]$ denotes the integer part of $x$. However, the algorithm (8.19) fails for the integral resulting from the change of variable $\xi = e^t$

$$I = \int_0^{\log x} [e^t]\,dt$$

because it stops in general too early. What is the reason for this?

**8.5.** Develop an analogous algorithm to (8.19) for the adaptive quadrature by using the interpolation quadrature formulae $Q_2$, (8.55), and $Q_4$, (8.57). The evaluated function values are to be stored so that they can be reused later. What is the increase in efficiency that results for the computation of the integrals of the Examples 8.3 and 8.5?

**8.6.** Compute the following integrals by means of the Romberg procedure, an adaptive quadrature method, the summed Newton–Cotes quadrature formulae $S_2$, (8.66), and $S_4$, (8.68), for $N = 2, 3, 4, 6, 8, 12$, the spline quadrature, (8.77), for $n = 4, 8, 12, 16, 24, 32, 48$, and the Gaussian formulae for $n = 4, 5, 6, 7, 8$ nodes. Compare the number of function evaluations required to achieve the same accuracy of the approximate values

(a) $\int_0^3 \dfrac{x}{1+x^2}\,dx = \tfrac{1}{2}\log(10)$ (b) $\int_0^{0.95} \dfrac{dx}{1-x} = \log(20)$

(c) $\dfrac{1}{\pi}\int_0^\pi \cos(x\sin\varphi)\,d\varphi = J_0(x)$ (Bessel function)

$J_0(1) \doteq 0.765\,197\,6866 \quad J_0(3) \doteq -0.260\,051\,9549 \quad J_0(5) \doteq -0.177\,596\,7713$

(d) $\displaystyle\int_0^{\pi/2} \frac{d\varphi}{\sqrt{1-m\sin^2\varphi}} = K(m)$  (complete elliptic integral of the first kind)

$K(0.5) \doteq 1.854\,074\,6773 \quad K(0.8) \doteq 2.257\,205\,3268 \quad K(0.96) \doteq 3.016\,112\,4925.$

**8.7.** Compute the nodes $x_k$, and the weights $w_k$ of the Gaussian quadrature formulae for $n = 5, 6, 8, 10, 12, 16, 20$ by means of the QR algorithm, and the inverse vector iteration applied to the tridiagonal matrices $\mathbf{J}_n$ (8.92).

# 9

# Ordinary Differential Equations

Many problems of applied mathematics lead to ordinary differential equations or systems of ordinary differential equations. The solution can only be explicitly given in relatively few cases, therefore numerical methods are needed to produce a sufficiently accurate approximation to the desired solution. In the following we shall present two classes of methods and analyse certain properties that must be taken into account when the procedures are applied in practice. For systems of ordinary differential equations arising in physics, chemistry, biology or engineering sciences we often have quite specific properties that influence the choice of the method.

The procedures will be presented for simplicity and clarity on the basis of the scalar differential equation of first order $y' = f(x, y)$ for a single unknown function $y(x)$. The procedure for systems of differential equations will be discussed later on. An initial condition $y(x_0) = y_0$, prescribing the value $y_0$ at a given point $x_0$, is necessary in order to fix a certain solution among the one-parametric family of solutions of an ordinary differential equation of first order. We do not concern ourselves about questions of existence and uniqueness of a solution, but assume in advance that the corresponding hypotheses are satisfied (Amann, 1983; Collatz, 1981; Walter, 1972). For more detailed and advanced treatments of numerical methods for the solution of ordinary differential equations we refer the reader to Gear (1971), Gekeler (1984), Grigorieff (1972, 1977), Henrici (1962), Jain (1984), Lambert (1973), Lapidus and Seinfeld (1971), Shampine and Gordon (1975), Aiken (1985), Butcher (1987) and Hairer *et al.* (1987).

## 9.1 Single Step Methods

### 9.1.1 Euler and the Taylor series method

We consider the first-order scalar differential equation

$$y'(x) = f(x, y(x)) \tag{9.1}$$

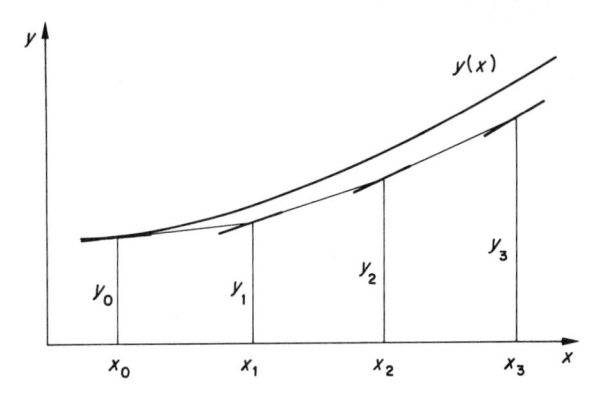

Figure 9.1 Euler's method

for the desired solution $y(x)$ of the independant variable $x$ satisfying the initial condition

$$y(x_0) = y_0, \tag{9.2}$$

where $x_0$ and $y_0$ are prescribed values. At the point $(x_0, y_0)$ the differential equation, (9.1), with the computable value $y'(x_0) = f(x_0, y_0)$, the slope of the tangent of the desired function is fixed. The simplest numerical method for treating the initial value problem (9.1) and (9.2) consists of approximating the solution curve $y(x)$, in the sense of a linearization, by its tangent (see Figure 9.1). With the step size $h$ and the corresponding equidistant support abscissae

$$x_k = x_0 + kh \quad (k = 1, 2, \ldots) \tag{9.3}$$

we can successively obtain the approximations $y_k$ of the exact solution values $y(x_k)$ by means of the general formula

$$y_{k+1} = y_k + hf(x_k, y_k) \quad (k = 0, 1, 2, \ldots). \tag{9.4}$$

At each point $(x_k, y_k)$ *Euler's method* uses the slope of the directional field that is defined by the given differential equation to determine the next approximation $y_{k+1}$. Due to the geometric construction of the approximate values, the procedure is often called the *polygonal method*. It is obviously quite coarse and will only produce good approximations for small step sizes $h$. It represents the simplest member of a *one step method* which uses only the known approximate value $y_k$ at the support abscissa $x_k$ to compute the approximation $y_{k+1}$ at $x_{k+1} = x_k + h$.

*Example* 9.1 We consider the initial value problem

$$y' = -2xy^2 \quad y(0) = 1 \tag{9.5}$$

## 9.1 Single Step Methods

Table 9.1 Euler's method, different step sizes

| $x_k$ | $y(x_k)$ | $h=0.1$ | | $h=0.01$ | | $h=0.001$ | |
|---|---|---|---|---|---|---|---|
| | | $y_k$ | $e_k$ | $y_k$ | $e_k$ | $y_k$ | $e_k$ |
| 0   | 1.000 00 | 1.000 00 | —        | 1.000 00 | —         | 1.000 00 | —         |
| 0.1 | 0.990 10 | 1.000 00 | −0.009 90 | 0.991 07 | −0.000 97 | 0.990 20 | −0.000 10 |
| 0.2 | 0.961 54 | 0.980 00 | −0.018 46 | 0.963 30 | −0.001 76 | 0.961 71 | −0.000 18 |
| 0.3 | 0.917 43 | 0.941 58 | −0.024 15 | 0.919 69 | −0.002 26 | 0.917 66 | −0.000 22 |
| 0.4 | 0.862 07 | 0.888 39 | −0.026 32 | 0.864 48 | −0.002 42 | 0.862 31 | −0.000 24 |
| 0.5 | 0.800 00 | 0.825 25 | −0.025 25 | 0.802 29 | −0.002 29 | 0.800 23 | −0.000 23 |
| 0.6 | 0.735 29 | 0.757 15 | −0.021 85 | 0.737 27 | −0.001 98 | 0.735 49 | −0.000 20 |

with the exact solution $y(x) = 1/(x^2 + 1)$. From Euler's method, we obtain the approximate values $y_k$ for different step sizes $h$ at the same discrete abscissae $x_k$ and the corresponding errors $e_k := y(x_k) - y_k$ as shown in Table 9.1. The observed errors decrease in proportion to the step size $h$.

In a small neighbourhood of the initial point $(x_0, y_0)$ a better approximation of the desired solution $y(x)$ can be attained by means of the Taylor series with remainder term

$$y(x) = y(x_0) + \frac{(x-x_0)}{1!}y'(x_0) + \frac{(x-x_0)^2}{2!}y''(x_0) + \cdots + \frac{(x-x_0)^p}{p!}y^{(p)}(x_0) + R_{p+1}. \tag{9.6}$$

If the remainder term $R_{p+1}$ is neglected in (9.6) we obtain the approximate value $y_{k+1}$ with the step size $h := x_{k+1} - x_k$ by the general formula

$$y_{k+1} = y_k + \frac{h}{1!}y'_k + \frac{h^2}{2!}y''_k + \cdots + \frac{h^p}{p!}y_k^{(p)}. \tag{9.7}$$

Here $y_k^{(m)}$ denotes the value of the $m$th derivative at the point $(x_k, y_k)$. This procedure requires the knowledge of the derivatives of the function $y(x)$ up to a certain order at the abscissa $x_k$. The second and the higher order derivatives can in principle be obtained by successively differentiating the given differential equation (9.1) with respect to $x$ and systematically replacing $y'$. Quite soon, the resulting expressions become so complicated, with respect to the partial derivatives of the given function $f(x, y)$, that the procedure is only practicable in very simple cases.

However, the Taylor series method can be used quite successfully if it is performed differently. The desired Taylor series is written in the following way with unknown coefficients $c_i$

$$y(x) = y(x_0) + c_1(x-x_0) + c_2(x-x_0)^2 + c_3(x-x_0)^3 + \cdots \tag{9.8}$$

which should yield a solution of the differential equation and thus is substituted into (9.1). Then comparing coefficients results in a set of equations from which the unknown coefficients $c_i$ can be determined recursively.

*Example 9.2* To show the features of the procedure, we consider the initial value problem $y' = -2xy^2$, $y(0) = 1$. At the general approximate point $(x_k, y_k)$ we use the Taylor series

$$y(x) = y_k + c_1(x - x_k) + c_2(x - x_k)^2 + c_3(x - x_k)^3 + c_4(x - x_k)^4 + \cdots$$

This expansion is substituted together with its first derivative into the differential equation. Then a comparison of the coefficients of the powers of $h := (x - x_k)$ is applied. Therefore the variable $x$ appearing in the differential equation is written as $x = x_k + h$, and we get the following identity to be satisfied

$$\begin{aligned} c_1 &+ 2c_2 h + 3c_3 h^2 + 4c_4 h^3 + \cdots \\ &= -2(x_k + h)(y_k + c_1 h + c_2 h^2 + c_3 h^3 + \cdots)^2 \\ &= -2(x_k + h)(y_k^2 + 2c_1 y_k h + (c_1^2 + 2c_2 y_k)h^2 + (2c_1 c_2 + 2c_3 y_k)h^3 + \cdots] \\ &= -2x_k y_k^2 + (-2y_k^2 - 4c_1 x_k y_k)h + [-4c_1 y_k - 2x_k(c_1^2 + 2c_2 y_k)]h^2 \\ &\quad + [-2(c_1^2 + 2c_2 y_k) - 4x_k(c_1 c_2 + c_3 y_k)]h^3 + \cdots \end{aligned}$$

We confine ourselves to comparing the coefficients of corresponding powers of $h$ up to the third power, and thus after some minor reformulations we obtain the set of equations

$$\begin{aligned} c_1 &= -2x_k y_k^2 \\ c_2 &= -(y_k + 2c_1 x_k)y_k \\ c_3 &= -[4c_1 y_k + 2x_k(c_1^2 + 2c_2 y_k)]/3 \\ c_4 &= -[\tfrac{1}{2}c_1^2 + c_2 y_k + x_k(c_1 c_2 + c_3 y_k)] \end{aligned}$$

The coefficients $c_1$, $c_2$, $c_3$ and $c_4$ can, in fact, be computed recursively. The approximate value $y_{k+1}$ at the abscissa $x_{k+1} = x_k + h$ is defined for the given step size $h$ by the expression

$$y_{k+1} = y_k + c_1 h + c_2 h^2 + c_3 h^3 + c_4 h^4. \tag{9.9}$$

Table 9.2 contains the computed values $y_k$ obtained with the step size $h = 0.1$, the corresponding coefficients $c_i$ and the errors $e_k := y(x_k) - y_k$. As the Taylor series takes into account the derivatives up to the fourth order, the resulting values $y_k$ are, in comparison with the results of the Euler method, much better approximations.

The number of terms of the Taylor series and hence the quality of the approximation can in principle be increased arbitrarily. In certain applications the Taylor algorithm gives excellent approximations (Halin, 1983; Waldvogel, 1984). It has the disadvantage that the set of recursion formulae must be found for

## 9.1 Single Step Methods

Table 9.2 Taylor algorithm, $h = 0.1$

| $x_k$ | $y_k$ | $c_1$ | $c_2$ | $c_3$ | $c_4$ | $e_k$ |
|---|---|---|---|---|---|---|
| 0   | 1.000 0000 | 0.000 0000  | −1.000 0000 | 0.000 0000 | 1.000 0000  | —          |
| 0.1 | 0.990 1000 | −0.196 0596 | −0.941 4743 | 0.380 5494 | 0.856 7973  | −0.000 0010 |
| 0.2 | 0.961 5455 | −0.369 8279 | −0.782 3272 | 0.656 5037 | 0.499 7400  | −0.000 0071 |
| 0.3 | 0.917 4459 | −0.505 0242 | −0.563 7076 | 0.773 6352 | 0.091 3102  | −0.000 0147 |
| 0.4 | 0.862 0892 | −0.594 5582 | −0.333 1480 | 0.742 3249 | −0.224 7569 | −0.000 0202 |
| 0.5 | 0.800 0218 | −0.640 0348 | −0.127 9930 | 0.614 4390 | −0.389 1673 | −0.000 0218 |
| 0.6 | 0.735 3139 | −0.648 8238 | 0.031 8205  | 0.449 0115 | −0.419 5953 | −0.000 0197 |
| 0.7 | 0.671 1567 | −0.630 6319 | 0.142 1026  | 0.289 7305 | −0.367 6095 | −0.000 0158 |
| 0.8 | 0.609 7675 | −0.594 9063 | 0.208 5909  | 0.159 2475 | −0.282 5582 | −0.000 0114 |
| 0.9 | 0.552 4938 | −0.549 4489 | 0.241 1714  | 0.063 7248 | −0.196 6194 | −0.000 0076 |
| 1.0 | 0.500 0047 |             |             |            |             | −0.000 0047 |

each new differential equation. However, this task can be done either by the computer or by means of an analyzing program that does not only determine the recursion formulae for a given differential equation but also simultaneously generates a computer program for their evaluation (Corliss and Chang, 1982; Halin, 1983). Moreover, the Taylor series method allows a relatively simple control of the step size and the number of terms which must be considered in order to keep the approximation errors within prescribed bounds.

To avoid the aforementioned effort, we shall present different methods that can be directly applied to the given differential equation.

### 9.1.2 Discretization errors, order of convergence

Prior to a first judgement of single step methods we state some facts about error estimates. For this purpose we consider a general computational scheme of the form

$$y_{k+1} = y_k + h\Phi(x_k, y_k, h) \quad (k = 0, 1, 2, \ldots). \tag{9.10}$$

The function $\Phi(x_k, y_k, h)$ describes how the new approximate value $y_{k+1}$ is computed from the information $(x_k, y_k)$ and the step size $h$. In the case of Euler's method, (9.4), the function

$$\Phi(x_k, y_k, h) = f(x_k, y_k) \tag{9.11}$$

does not depend on $h$. For the Taylor algorithm we have the generalization of (9.9)

$$\Phi(x_k, y_k, h) = c_1 + c_2 h + c_3 h^2 + \cdots + c_p h^{p-1} \tag{9.12}$$

where the coefficients $c_i$ depend on the differential equation to be solved, that is on $f(x, y)$, and on the point $(x_k, y_k)$.

A computational rule (9.10) can only be a useful method if the function $\Phi(x, y, h)$ is related to $f(x, y)$. From (9.10) it is found by means of the limiting process

$$\lim_{h \to 0} \frac{y_{k+1} - y_k}{h} = y'(x_k) = \Phi(x_k, y_k, 0).$$

**Definition 9.1** *A one step method (9.10) is called consistent with the differential equation (9.1) if*

$$\Phi(x, y, 0) = f(x, y). \tag{9.13}$$

The consistency condition (9.13) is satisfied by the Euler method, because of (9.11), and by the Taylor series method, according to (9.12) because $c_1 = y'(x_k) = f(x_k, y_k)$.

In the following we assume the step size $h$ to be a fixed constant so that the support abscissae $x_k = x_0 + kh$ are equidistant. Moreover, $y(x)$ will always denote the exact solution of the differential equation $y'(x) = f(x, y(x))$ satisfying the initial condition $y(x_0) = y_0$, and $y(x_k)$ is the value of the exact solution at $x_k$. Finally, $y_k$ will denote the approximate value at the abscissa $x_k$ obtained by means of the integration method (9.10). We assume the arithmetic to be exact and do not take into account rounding errors, because they are usually negligible. With these hypotheses we define the following quantity.

**Definition 9.2** *The local discretization error $d_{k+1}$ at $x_{k+1}$ is defined by the expression*

$$d_{k+1} := y(x_{k+1}) - y(x_k) - h\Phi(x_k, y(x_k), h). \tag{9.14}$$

The local discretization error $d_{k+1}$ indicates how well the exact solution $y(x)$ fulfils the formula (9.10). In the case of Euler's method and of the Taylor series

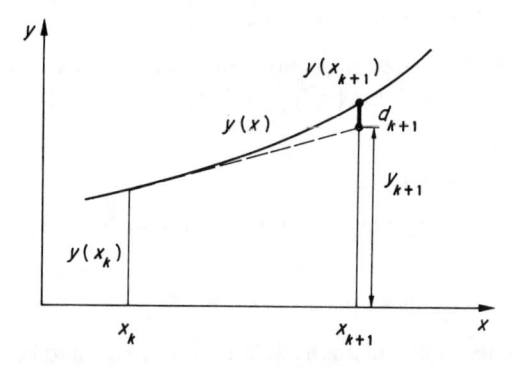

Figure 9.2 Local discretization error

## 9.1 Single Step Methods

method the error $d_{k+1}$ is equal to the difference between the exact value $y(x_{k+1})$ and the computed approximation $y_{k+1}$ if at the abscissa $x_k$ the exact value $y(x_k)$ is used, (see Figure 9.2).

For practical purposes the total error between the computed approximation and the exact solution, after many integration steps, is of interest.

**Definition 9.3** *The global discretization error $g_k$ at $x_k$ is given by the difference*

$$g_k := y(x_k) - y_k. \tag{9.15}$$

In order to be able to estimate the global error $g_k$ the function $\Phi(x, y, h)$ must satisfy a Lipschitz condition, with respect to the variable $y$, on a suitably chosen domain $B$

$$|\Phi(x, y, h) - \Phi(x, y^*, h)| \leqslant L|y - y^*| \quad \text{for} \quad x, y, y^*, h \in B \quad 0 < L < \infty. \tag{9.16}$$

In the case of Euler's method this is the condition for the function $f(x, y)$ to be Lipschitz continuous, which has to be fulfilled to guarantee the existence and uniqueness of a solution of (9.1). For more general one step methods a close relationship exists between (9.16) and the Lipschitz condition for $f(x, y)$.

From the definition of the local discretization error, (9.14), it follows that

$$y(x_{k+1}) = y(x_k) + h\Phi(x_k, y(x_k), h) + d_{k+1},$$

and by subtracting (9.10) we obtain

$$g_{k+1} = g_k + h[\Phi(x_k, y(x_k), h) - \Phi(x_k, y_k, h)] + d_{k+1}.$$

Due to (9.16) we can estimate the following

$$|g_{k+1}| \leqslant |g_k| + h|\Phi(x_k, y(x_k), h) - \Phi(x_k, y_k, h)| + |d_{k+1}| \tag{9.17}$$
$$\leqslant |g_k| + hL|y(x_k) - y_k| + |d_{k+1}| = (1 + hL)|g_k| + |d_{k+1}|.$$

For further treatment we assume that the absolute value of the local discretization error is bounded, that is

$$\max_k |d_k| \leqslant D. \tag{9.18}$$

Consequently the absolute value of the global discretization error satisfies the inequality

$$|g_{k+1}| \leqslant (1 + hL)|g_k| + D, \quad (k = 0, 1, 2, \ldots). \tag{9.19}$$

**Lemma 9.1** *If the values $g_k$ satisfy (9.19), then*

$$|g_n| \leqslant \frac{(1 + hL)^n - 1}{hL} D + (1 + hL)^n |g_0| \leqslant \frac{D}{hL}(e^{nhL} - 1) + e^{nhL}|g_0|. \tag{9.20}$$

*Proof* Repeated application of (9.19) yields the first inequality

$$|g_n| \leq (1+hL)|g_{n-1}| + D$$
$$\leq (1+hL)^2 |g_{n-2}| + [(1+hL) + 1]D$$
$$\leq (1+hL)^3 |g_{n-3}| + [(1+hL)^2 + (1+hL) + 1]D$$
$$\vdots$$
$$\leq (1+hL)^n |g_0| + [(1+hL)^{n-1} + \cdots + (1+hL) + 1]D$$
$$= \frac{(1+hL)^n - 1}{hL} D + (1+hL)^n |g_0|.$$

To show the second inequality we use the fact that the function $e^t$ is convex, so that the tangent at $t=0$ is below the curve. Hence we have $1+t \leq e^t$ for all $t \in \mathbf{R}$. From this the inequalities $(1+hL) \leq e^{hL}$ and $(1+hL)^n \leq e^{nhL}$ follow and hence the second part of the lemma.

As the global discretization error satisfies $g_0 = y(x_0) - y_0 = 0$ an immediate consequence of Lemma 9.1 is the following theorem.

**Theorem 9.2** *The global discretization error $g_n$, at the fixed abscissa $x_n = x_0 + nh$ is bounded by*

$$|g_n| \leq \frac{D}{hL}(e^{nhL} - 1) \leq \frac{D}{hL} e^{nhL}. \tag{9.21}$$

The bound (9.21) of the global discretization error shows that besides the Lipschitz constant, $L$, of the function $\Phi(x, y, h)$, the bound $D$ of the absolute value of the local discretization errors, $d_k$, is the decisive quantity. Therefore the local discretization error $d_{k+1}$ plays the central role in the qualitative judgement of a one-step method. It will be determined by means of Taylor series where we implicitly assume that the function $f(x, y)$ and the solution $y(x)$ are sufficiently many times continuously differentiable.

The local discretization error of Euler's method is given by

$$d_{k+1} = y(x_{k+1}) - y(x_k) - hf(x_k, y(x_k)).$$

We replace $y(x_{k+1})$ by the Taylor series expansion with remainder term at the point $x_k$, noting that $f(x_k, y(x_k)) = y'(x_k)$ and with $0 < \theta < 1$ we obtain

$$d_{k+1} = y(x_k) + hy'(x_k) + \tfrac{1}{2}h^2 y''(x_k + \theta h) - y(x_k) - hy'(x_k) = \tfrac{1}{2}h^2 y''(x_k + \theta h).$$

Let

$$D := \max_{k=0,\ldots,n-1} |d_{k+1}| \leq \tfrac{1}{2}h^2 \max_{x_0 \leq \xi \leq x_n} |y''(\xi)| =: \tfrac{1}{2} h^2 M_2.$$

Thus from (9.21) we obtain the following estimate for the global discretization

## 9.1 Single Step Methods

error of Euler's method

$$|g_n| \leq \frac{hM_2}{2L} \exp[L(x_n - x_0)] \tag{9.22}$$

If we fix the abscissa $x_n$ and make the step size $h = (x_n - x_0)/n$ decrease by increasing $n$, then (9.22) shows that the error bound decreases proportionally to the step size $h$. Consequently, the value $y_n$ converges to the exact value $y(x_n)$ at the fixed abscissa $x_n$ as $h \to 0$, at least if rounding errors do not occur. The convergence is linear, with respect to the step size $h$, and we say that Euler's method is of *order* one (see the numerical results of Example 9.1).

**Definition 9.4** *A one step method, (9.10), is of order p, if its local discretization error $d_k$ satisfies the relationship*

$$\max_{1 \leq k \leq n} |d_k| \leq D = \text{const} \cdot h^{p+1} = O(h^{p+1}) \tag{9.23}$$

*so that the global discretization error $g_n$ is bounded by*

$$|g_n| \leq \frac{\text{const}}{L} e^{nhL} h^p = O(h^p). \tag{9.24}$$

From this definition, the Taylor algorithm (9.7) is of order $p$ because its local discretization error is given by

$$d_{k+1} = y(x_{k+1}) - y(x_k) - hy'(x_k) - \frac{h^2}{2!} y''(x_k) - \cdots - \frac{h^p}{p!} y^{(p)}(x_k)$$

$$= \frac{h^{p+1}}{(p+1)!} y^{(p+1)}(x_k + \theta h) \quad 0 < \theta < 1.$$

The integration method that is used in Example 9.2 has order 4. This explains the high degree of accuracy of the computed approximations there.

### 9.1.3 Improved polygonal method, trapezoidal method, Heun's method

Since Euler's method is of order one an extrapolation can be applied. So we assume that two integrations have been performed first with the step size $h_1 = h$ and the second with the step size $h_2 = h/2$ by means of the polygonal method, (9.4), up to a given abscissa $x$. The resulting values $y_n$ and $y_{2n}$ after $n$ and $2n$ steps, respectively, satisfy approximately

$$y_n \simeq y(x) + c_1 h + O(h^2),$$
$$y_{2n} \simeq y(x) + c_1 \tfrac{1}{2} h + O(h^2).$$

By applying the so-called *Richardson extrapolation*, we define the extrapolated quantity

$$\tilde{y} = 2y_{2n} - y_n \simeq y(x) + O(h^2), \tag{9.25}$$

whose error, relative to $y(x)$, is of second order.

Instead of integrating a differential equation by means of Euler's method with two different step sizes, it is better to directly apply the extrapolation to the values that result from an integration step, with the step size $h$, and from a double step, with half the step size. In both cases we start from the approximate point $(x_k, y_k)$. An ordinary step of the Euler method, with the step size $h$, yields the value

$$y^{(1)}_{k+1} = y_k + hf(x_k, y_k). \tag{9.26}$$

A double step with the step size $h/2$ produces the values

$$y^{(2)}_{k+\frac{1}{2}} = y_k + \frac{h}{2}f(x_k, y_k),$$

$$y^{(2)}_{k+1} = y^{(2)}_{k+\frac{1}{2}} + \frac{h}{2}f\left(x_k + \frac{h}{2}, y^{(2)}_{k+\frac{1}{2}}\right). \tag{9.27}$$

The Richardson extrapolation applied to $y^{(2)}_{k+1}$ and $y^{(1)}_{k+1}$ defines the extrapolated approximation

$$y_{k+1} = 2y^{(2)}_{k+1} - y^{(1)}_{k+1} = 2y^{(2)}_{k+\frac{1}{2}} + hf\left(x_k + \frac{h}{2}, y^{(2)}_{k+\frac{1}{2}}\right) - y_k - hf(x_k, y_k)$$

$$= 2y_k + hf(x_k, y_k) + hf\left(x_k + \frac{h}{2}, y^{(2)}_{k+\frac{1}{2}}\right) - y_k - hf(x_k, y_k)$$

$$= y_k + hf\left(x_k + \frac{h}{2}, y_k + \frac{h}{2}f(x_k, y_k)\right). \tag{9.28}$$

We formulate the result (9.28) as the following algorithm:

$$\begin{array}{l} k_1 = f(x_k, y_k) \\ k_2 = f\left(x_k + \dfrac{h}{2}, y_k + \dfrac{1}{2}hk_1\right) \\ y_{k+1} = y_k + hk_2 \end{array} \tag{9.29}$$

The procedure (9.29) is called the *improved polygonal Euler method*. A single step requires the evaluation of the function $f(x, y)$ for two different pairs of values. The quantity $k_1$ is equal to the slope of the directional field at the point $(x_k, y_k)$. It serves to determine the auxiliary point $(x_k + h/2, y^{(2)}_{k+\frac{1}{2}})$ and the corresponding slope $k_2$. The approximation $y_{k+1}$ is then computed by means of this slope, so

## 9.1 Single Step Methods

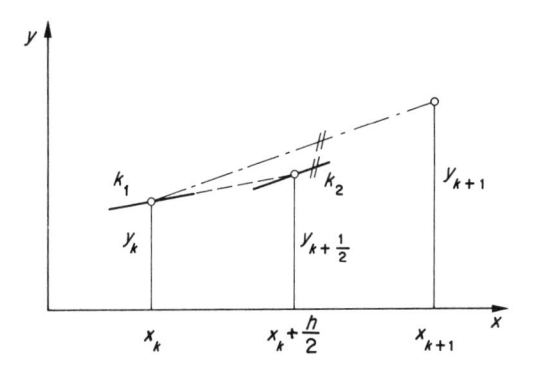

Figure 9.3 Improved polygonal method

that the change in the directional field is taken into account. The geometrical interpretation of the method is given in Figure 9.3.

Before determining the order of the improved polygonal method, we give some formulae that will be useful later on. Differentiation of the given differential equation yields

$$y' = f(x, y)$$
$$y'' = f_x + f_y y' = f_x + f f_y =: F \tag{9.30}$$
$$y''' = f_{xx} + f_{xy} y' + (f_x + f_y y') f_y + f(f_{xy} + f_{yy} y')$$
$$= (f_{xx} + 2 f f_{xy} + f^2 f_{yy}) + (f_x + f f_y) f_y =: G + F f_y \tag{9.31}$$

Thus, by using appropriate Taylor series expansions and the relationships (9.30) and (9.31), for the local discretization error of the improved polygonal method we obtain

$$d_{k+1} = y(x_{k+1}) - y(x_k) - h f\left(x_k + \frac{h}{2}, y(x_k) + \frac{h}{2} f(x_k, y(x_k))\right)$$
$$= h y'(x_k) + \tfrac{1}{2} h^2 y''(x_k) + \tfrac{1}{6} h^3 y'''(x_k) + O(h^4)$$
$$- h\Bigg[ f(x_k, y(x_k)) + \frac{h}{2} f_x(x_k, y(x_k)) + \frac{h}{2} f(x_k, y(x_k)) f_y(x_k, y(x_k))$$
$$+ \frac{1}{2}\left(\frac{h}{2}\right)^2 f_{xx} + \left(\frac{h}{2}\right)^2 f f_{xy} + \frac{1}{2}\left(\frac{h}{2}\right)^2 f^2 f_{yy} + O(h^3) \Bigg]$$
$$= \tfrac{1}{6} h^3 y'''(x_k) - \tfrac{1}{8} h^3 G(x_k, y(x_k)) + O(h^4) = \tfrac{1}{6}(\tfrac{1}{4} G + F f_y) h^3 + O(h^4).$$

In general the principal part of $d_{k+1}$ is proportional to $h^3$, and thus the improved polygonal method is of second order.

Further one step methods for the approximate solution of the initial value problem (9.1) and (9.2) are obtained with the aid of a definite integration of the differential equation $y'(x) = f(x, y(x))$ with respect to the independent variable $x$ over the interval $[x_k, x_{k+1}]$, of the length $h = x_{k+1} - x_k$. On the left-hand side, the integration can be performed explicitly, and so with

$$y(x_{k+1}) - y(x_k) = \int_{x_k}^{x_{k+1}} f(x, y(x)) \, dx \tag{9.32}$$

we obtain an integral equation which is equivalent to the given differential equation. In general no primitive for the right-hand side of (9.32) can be found, because $y(x)$ is an unknown function, therefore the value of the integral is approximated by means of a quadrature formula. We use the simple trapezoidal rule. Consequently the equation, (9.32), will only hold approximately, and we have to replace $y(x_k)$ by its approximate value $y_k$ to obtain the *trapezoidal method*

$$\boxed{y_{k+1} = y_k + \frac{h}{2}(f(x_k, y_k) + f(x_{k+1}, y_{k+1})).} \tag{9.33}$$

For a general nonlinear differential equation the formula (9.33) is an implicit equation for the unknown approximation $y_{k+1}$. Therefore we call (9.33) an *implicit method of integration*. Each step requires the solution of a nonlinear equation.

In the special case of a first-order linear differential equation

$$y'(x) = a(x)y(x) + b(x)$$

where $a(x)$ and $b(x)$ are given functions, the trapezoidal method (9.33) yields the linear equation for $y_{k+1}$

$$y_{k+1} = y_k + \tfrac{1}{2}h[a(x_k)y_k + b(x_k) + a(x_{k+1})y_{k+1} + b(x_{k+1})].$$

From this we obtain an explicit recursion formula

$$y_{k+1} = \frac{[2 + ha(x_k)]y_k + h[b(x_k) + b(x_{k+1})]}{2 - ha(x_{k+1})} \quad (k = 0, 1, 2, \ldots)$$

that allows the successive computation of the values $y_1, y_2, \ldots$.

For a nonlinear differential equation the implicit equation already has the convenient fixed point form, so that a solution, $y_{k+1}$, may be found by means of the method of successive approximation. As the value of $f(x_k, y_k)$ must be computed a suitable initial value $y_{k+1}^{(0)}$ of the fixed point iteration is

$$y_{k+1}^{(0)} = y_k + hf(x_k, y_k) \tag{9.34}$$

defined by Euler's method. The sequence of iterates

$$y_{k+1}^{(n+1)} = y_k + \tfrac{1}{2}h[f(x_k, y_k) + f(x_{k+1}, y_{k+1}^{(n)})] \quad (n = 0, 1, 2, \ldots) \tag{9.35}$$

## 9.1 Single Step Methods

converges towards the fixed point $y_{k+1}$ if the function $f(x, y)$ satisfies the usual Lipschitz condition, where the Lipschitz constant $L$ satisfies $hL/2 < 1$, because then the hypotheses of the Banach fixed point theorem are satisfied locally.

The local discretization error of the trapezoidal method is

$$\begin{aligned} d_{k+1} &= y(x_{k+1}) - y(x_k) - \tfrac{1}{2}h[f(x_k, y(x_k)) + f(x_{k+1}, y(x_{k+1}))] \\ &= y(x_{k+1}) - y(x_k) - \tfrac{1}{2}h[y'(x_k) + y'(x_{k+1})] \\ &= hy'(x_k) + \tfrac{1}{2}h^2 y''(x_k) + \tfrac{1}{6}h^3 y'''(x_k) + O(h^4) \\ &\quad - \tfrac{1}{2}h[y'(x_k) + y'(x_k) + hy''(x_k) + \tfrac{1}{2}h^2 y'''(x_k) + O(h^3)] \\ &= -\tfrac{1}{12}h^3 y'''(x_k) + O(h^4). \end{aligned}$$

According to the principal part of $d_{k+1}$, the trapezoidal method is of order 2. It is the same as that of the improved polygonal method. However, the trapezoidal method has specific stability properties, as will be seen in Section 9.3.

Since the value $y_{k+1}$, defined by (9.33), is only an approximation of $y(x_{k+1})$ we may restrict the fixed point iteration (9.35) to a single step. Thus, by slightly changing the notation, we obtain *Heun's method*:

$$\begin{aligned} y_{k+1}^{(P)} &= y_k + hf(x_k, y_k) \\ y_{k+1} &= y_k + \tfrac{1}{2}h[f(x_k, y_k) + f(x_{k+1}, y_{k+1}^{(P)})] \end{aligned} \quad (9.36)$$

The explicit first order Euler method is used to determine a so-called *predictor value* $y_{k+1}^{(P)}$, which is subsequently corrected by means of the implicit trapezoidal method to obtain $y_{k+1}$. The explicit method of Heun (9.36) is therefore called a *predictor corrector method*. It is of second order, which can be derived in a similar manner to the improved polygonal method. We formulate Heun's method in the following algorithmic way:

$$\begin{aligned} k_1 &= f(x_k, y_k) \\ k_2 &= f(x_k + h, y_k + hk_1) \\ y_{k+1} &= y_k + \tfrac{1}{2}h(k_1 + k_2) \end{aligned} \quad (9.37)$$

To determine $y_{k+1}$ the average of the two slopes $k_1$ and $k_2$ of the directional field are used at the points $(x_k, y_k)$ and $(x_{k+1}, y_{k+1}^{(P)})$.

The improved polygonal method, (9.29), and Heun's method (9.37), are examples of explicit two-stage second order Runge–Kutta methods, which will be generalized in the next section.

**Example 9.3** The initial value problem $y' = -2xy^2$, $y(0) = 1$ has been treated by means of the improved polygonal method and Heun's method each with step

Table 9.3 Improved polygonal method and Heun's method

| $x_k$ | Improved polygonal method | | | | Method of Heun | | | |
|---|---|---|---|---|---|---|---|---|
| | $h=0.1$ | | $h=0.05$ | | $h=0.1$ | | $h=0.05$ | |
| | $y_k$ | $g_k$ | $y_k$ | $g_k$ | $y_k$ | $g_k$ | $y_k$ | $g_k$ |
| 0   | 1.000 00 | —         | 1.000 00 | —         | 1.000 00 | —         | 1.000 00 | —         |
| 0.1 | 0.990 00 | 0.000 10  | 0.990 07 | 0.000 02  | 0.990 00 | 0.000 10  | 0.990 09 | 0.000 01  |
| 0.2 | 0.961 18 | 0.000 36  | 0.961 45 | 0.000 09  | 0.961 37 | 0.000 17  | 0.961 52 | 0.000 02  |
| 0.3 | 0.916 74 | 0.000 69  | 0.917 27 | 0.000 16  | 0.917 25 | 0.000 19  | 0.917 42 | 0.000 01  |
| 0.4 | 0.861 10 | 0.000 96  | 0.861 84 | 0.000 23  | 0.861 95 | 0.000 11  | 0.862 08 | −0.000 01 |
| 0.5 | 0.798 89 | 0.001 11  | 0.799 74 | 0.000 26  | 0.800 03 | −0.000 03 | 0.800 04 | −0.000 04 |
| 0.6 | 0.734 18 | 0.001 11  | 0.735 03 | 0.000 26  | 0.735 53 | −0.000 23 | 0.735 38 | −0.000 09 |
| 0.7 | 0.670 14 | 0.001 00  | 0.670 91 | 0.000 23  | 0.671 59 | −0.000 45 | 0.671 28 | −0.000 14 |
| 0.8 | 0.608 95 | 0.000 80  | 0.609 57 | 0.000 18  | 0.610 40 | −0.000 64 | 0.609 93 | −0.000 18 |
| 0.9 | 0.551 91 | 0.000 58  | 0.552 36 | 0.000 13  | 0.553 29 | −0.000 80 | 0.552 70 | −0.000 21 |
| 1.0 | 0.499 64 | 0.000 36  | 0.499 92 | 0.000 08  | 0.500 92 | −0.000 92 | 0.500 24 | −0.000 24 |

sizes $h = 0.1$ and $h = 0.05$. The approximations $y_k$ and the corresponding global discretization errors $g_k = y(x_k) - y_k$ are given in Table 9.3. The numerical results show the improved accuracy of the methods and that their order of convergence is 2.

### 9.1.4 Runge–Kutta methods

We shall describe the principle for deriving one step methods of higher order. The corresponding calculations quickly become complicated and involved, and therefore we confine ourselves to a detailed treatment of three-stage Runge–Kutta methods. Subsequently we shall cite some selected methods of higher order that have special properties.

Similar to the derivation of the trapezoidal method, our starting point is the integral equation (9.32)

$$y(x_{k+1}) - y(x_k) = \int_{x_k}^{x_{k+1}} f(x, y(x))\,dx.$$

The value of the integral is now approximated by a quadrature formula for three support abscissae $\xi_1, \xi_2, \xi_3$ of the interval $[x_k, x_{k+1}]$ and corresponding weights $c_1, c_2, c_3$, where the nodes, $\xi_i$, and the weights, $c_i$, are arbitrary for the time being. Later they will be determined in such a way that the resulting procedure has maximal order. Thus we obtain the following formula for the approximation $y_{k+1}$

$$y_{k+1} = y_k + h[c_1 f(\xi_1, y(\xi_1)) + c_2 f(\xi_2, y(\xi_2)) + c_3 f(\xi_3, y(\xi_3))]. \quad (9.38)$$

In (9.38) the nodes, $\xi_i$, and the unknown values $y(\xi_i)$ must be fixed. We will use the idea of the predictor method and ensure that the method is explicit. This

## 9.1 Single Step Methods

aim implies the following fixing of the support abscissae

$$\xi_1 = x_k \quad \xi_2 = x_k + a_2 h \quad \xi_3 = x_k + a_3 h \quad 0 < a_2, a_3 \leq 1. \tag{9.39}$$

The first support abscissa is set equal to $x_k$ so that we have $y(\xi_1) = y_k$. The two remaining values $y(\xi_i)$ are defined by the predictor formulae

$$y(\xi_2): \quad y_2^* = y_k + h b_{21} f(x_k, y_k)$$
$$y(\xi_3): \quad y_3^* = y_k + h b_{31} f(x_k, y_k) + h b_{32} f(x_k + a_2 h, y_2^*) \tag{9.40}$$

with the three additional parameters $b_{21}, b_{31}$ and $b_{32}$ to be chosen later. The first predictor value $y_2^*$ depends on the slope at the point $(x_k, y_k)$, whereas the second one, $y_3^*$, also depends on the slope at the auxiliary point $(\xi_2, y_2^*)$.

After substitution of (9.39) and (9.40) into (9.38) we obtain an explicit one step method, (9.10), which we define as follows:

$$\begin{aligned} k_1 &= f(x_k, y_k) \\ k_2 &= f(x_k + a_2 h, y_k + h b_{21} k_1) \\ k_3 &= f[x_k + a_3 h, y_k + h(b_{31} k_1 + b_{32} k_2)] \\ y_{k+1} &= y_k + h(c_1 k_1 + c_2 k_2 + c_3 k_3) \end{aligned} \tag{9.41}$$

As the function $f(x, y)$ must be evaluated three times per integration step, the method (9.41) is called a *three-stage Runge–Kutta method*. We wish to determine the eight parameters $a_2, a_3, b_{21}, b_{31}, b_{32}, c_1, c_2, c_3$ in such a way that the one step method, (9.41), has an order as high as possible. Before the representation of the local discretization error is determined, we demand that the parameters must satisfy the following conditions

$$a_2 = b_{21} \quad a_3 = b_{31} + b_{32}. \tag{9.42}$$

The conditions of (9.42) are motivated by the requirement that the predictors $y_2^*$ and $y_3^*$ should be exact for the special differential equation $y' = 1$.

The local discretization error of the method (9.41) is given by

$$d_{k+1} = y(x_{k+1}) - y(x_k) - h(c_1 \bar{k}_1 + c_2 \bar{k}_2 + c_3 \bar{k}_3), \tag{9.43}$$

where $\bar{k}_i$ denote the expressions that arise from $k_i$ by replacing $y_k$ by $y(x_k)$. The expressions of $\bar{k}_1, \bar{k}_2$ and $\bar{k}_3$ are successively expanded into Taylor series at $x_k$ and substituted into (9.43). We use the identities (9.30) and (9.31) and omitting the values $x_k, y(x_k)$, to simplify the notation, we obtain

$$\begin{aligned} \bar{k}_1 &= f(x_k, y(x_k)) = f \\ \bar{k}_2 &= f[x_k + a_2 h, y(x_k) + a_2 h f(x_k, y(x_k))] \\ &= f + a_2 h f_x + a_2 h f f_y + \tfrac{1}{2} a_2^2 h^2 f_{xx} + a_2^2 h^2 f f_{xy} + \tfrac{1}{2} a_2^2 h^2 f^2 f_{yy} + O(h^3) \\ &= f + a_2 h F + \tfrac{1}{2} a_2^2 h^2 G + O(h^3) \end{aligned}$$

$$\bar{k}_3 = f(x_k + a_3 h, y(x_k) + h(b_{31}\bar{k}_1 + b_{32}\bar{k}_2))$$
$$= f + a_3 h f_x + h(b_{31}\bar{k}_1 + b_{32}\bar{k}_2)f_y$$
$$+ \tfrac{1}{2}a_3^2 h^2 f_{xx} + a_3(b_{31}\bar{k}_1 + b_{32}\bar{k}_2)h^2 f_{xy}$$
$$+ \tfrac{1}{2}(b_{31}\bar{k}_1 + b_{32}\bar{k}_2)^2 h^2 f_{yy} + O(h^3)$$
$$= f + h[a_3 f_x + (b_{31} + b_{32})ff_y]$$
$$+ h^2(a_2 b_{32} F f_y + \tfrac{1}{2}a_3^2 f_{xx} + a_3(b_{31} + b_{32})ff_{xy}$$
$$+ \tfrac{1}{2}(b_{31} + b_{32})^2 f^2 f_{yy}) + O(h^3)$$
$$= f + a_3 h F + h^2(a_2 b_{32} F f_y + \tfrac{1}{2}a_3^2 G) + O(h^3)$$
$$d_{k+1} = hf\{1 - c_1 - c_2 - c_3\} + h^2 F\{\tfrac{1}{2} - a_2 c_2 - a_3 c_3\}$$
$$+ h^3[Ff_y(\tfrac{1}{6} - a_2 c_3 b_{32}) + G(\tfrac{1}{6} - \tfrac{1}{2}a_2^2 c_2 - \tfrac{1}{2}a_3^2 c_3)] + O(h^4)$$

If the method is to be of the order 3, at least, the six parameters $c_1, c_2, c_3, a_2, a_3$ and $b_{32}$ must satisfy the system of four nonlinear equations:

$$\boxed{\begin{aligned} c_1 + c_2 + c_3 &= 1 \\ a_2 c_2 + a_3 c_3 &= \tfrac{1}{2} \\ a_2 c_3 b_{32} &= \tfrac{1}{6} \\ a_2^2 c_2 + a_3^2 c_3 &= \tfrac{1}{3} \end{aligned}} \qquad (9.44)$$

The number of equations (9.44) is smaller than the number of unknowns, thus the obvious question arises of whether a method of order 4 is possible. However, this is not the case because the coefficient of $h^4$ of the Taylor series for $d_{k+1}$ contains a term that is independent of the six parameters. Thus the maximal attainable order for explicit, three-stage Runge–Kutta methods (9.41) is 3.

The system of equations (9.44) has a solution set with two parameters. Let us choose, for instance, $a_2$ and $a_3$ as the parameters. If $a_2 \neq a_3$ and $a_2 \neq \tfrac{2}{3}$ then the values of $c_2$ and $c_3$ are determined by the second and fourth equation, and the remaining unknowns by the first and third equation

$$\boxed{\begin{aligned} c_2 &= \frac{3a_3 - 2}{6a_2(a_3 - a_2)} \qquad c_3 = \frac{2 - 3a_2}{6a_3(a_3 - a_2)} \\ c_1 &= \frac{6a_2 a_3 + 2 - 3(a_2 + a_3)}{6a_2 a_3} \qquad b_{32} = \frac{a_3(a_3 - a_2)}{a_2(2 - 3a_2)} \end{aligned}} \qquad (9.45)$$

It is left to the reader to find the parameters if either $a_2 = a_3$ or $a_2 = \tfrac{2}{3}$. The parameters are chosen amongst the two-parametric set of three-stage Runge–

## 9.1 Single Step Methods

Kutta methods of order 3 in order to satisfy different criteria. When calculations had to be performed by hand simple and easily remembered numbers were of most interest. For modern computers this is no longer essential. Therefore we may, for instance, require that the principal part of $d_{k+1}$ is a minimum in a certain sense, especially for a certain class of differential equations. Moreover, the problem of the step size control can be decisive for the choice of the parameters.

With $a_2 = \frac{1}{3}$ and $a_3 = \frac{2}{3}$ we obtain a first Runge–Kutta method of third order. From (9.45) and (9.42) we have $c_2 = 0, c_3 = \frac{3}{4}, c_1 = \frac{1}{4}, b_{32} = \frac{2}{3}, b_{31} = a_3 - b_{32} = 0$, and thus *Heun's third-order method* is defined by

$$\begin{aligned} k_1 &= f(x_k, y_k) \\ k_2 &= f(x_k + \tfrac{1}{3}h, y_k + \tfrac{1}{3}hk_1) \\ k_3 &= f(x_k + \tfrac{2}{3}h, y_k + \tfrac{2}{3}hk_2) \\ y_{k+1} &= y_k + \tfrac{1}{4}h(k_1 + 3k_3) \end{aligned} \qquad (9.46)$$

This method, with easily remembered numbers, has the property that the slope $k_3$ only depends on the previous slope $k_2$, and that the approximation $y_{k+1}$ is determined by using the slopes $k_1$ and $k_3$ only. The slope $k_2$ is needed as an auxiliary quantity for $k_3$. In Figure 9.4 the quantities are shown to give a geometrical illustration of the method (9.46).

A different third-order Runge–Kutta method is given for $a_2 = \frac{1}{2}$ and $a_3 = 1$. From (9.45) we get $c_2 = \frac{2}{3}, c_3 = \frac{1}{6}, c_1 = \frac{1}{6}, b_{32} = 2$ and from (9.42) $b_{31} = a_3 - b_{32} = -1$. Thus *Kutta's third-order method* is defined by (Kutta, 1901)

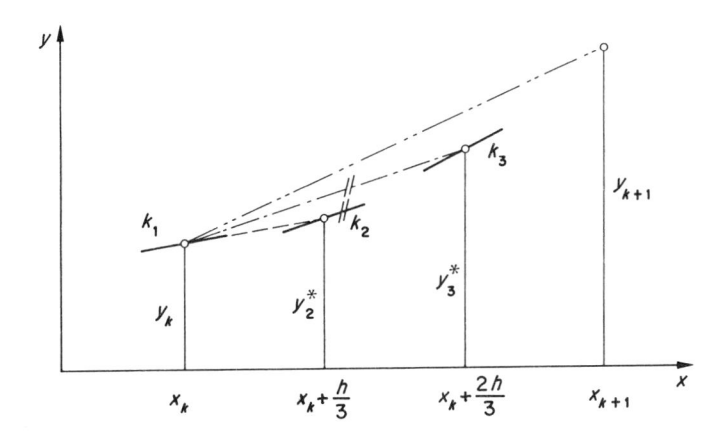

Figure 9.4 Heun's method of third order

$$\begin{aligned}
k_1 &= f(x_k, y_k) \\
k_2 &= f(x_k + \tfrac{1}{2}h, y_k + \tfrac{1}{2}hk_1) \\
k_3 &= f(x_k + h, y_k - hk_1 + 2hk_2) \\
y_{k+1} &= y_k + \tfrac{1}{6}h(k_1 + 4k_2 + k_3)
\end{aligned} \qquad (9.47)$$

It uses Simpson's quadrature formula (8.55).

The suitable choice of the step size $h$ plays a central role. Therefore we describe a principle of *step size control* on the basis of the methods developed thus far. For this purpose we aim to estimate the local discretization error of the method used by means of a method of higher order. The amount of work required for calculating the desired estimate should be small. Thus to fulfil this requirement the Runge–Kutta method of higher order should use the same $k_i$ values as the method of integration.

The improved polygonal method, (9.29), and Kutta's method (9.47), satisfy this requirement. The local discretization error $d_{k+1}^{(IP)}$ of the improved polygonal method is given by

$$d_{k+1}^{(IP)} = y(x_{k+1}) - y(x_k) - h\bar{k}_2$$

and that of Kutta's method by

$$d_{k+1}^{(K)} = y(x_{k+1}) - y(x_k) - \tfrac{1}{6}h(\bar{k}_1 + 4\bar{k}_2 + \bar{k}_3).$$

From these two representations we deduce

$$d_{k+1}^{(IP)} = \tfrac{1}{6}h(\bar{k}_1 + 4\bar{k}_2 + \bar{k}_3) - h\bar{k}_2 + d_{k+1}^{(K)}.$$

The unknown values $\bar{k}_i$ are now replaced by the known approximations $k_i$, and so because $d_{k+1}^{(K)} = O(h^4)$ we get

$$d_{k+1}^{(IP)} \approx \tfrac{1}{6}h(k_1 + 4k_2 + k_3) - hk_2 + O(h^4) = \tfrac{1}{6}h(k_1 - 2k_2 + k_3) + O(h^4). \qquad (9.48)$$

With the additional function evaluation for $k_3$, the expression $h(k_1 - 2k_2 + k_3)/6$ yields an estimate of the local discretization error, $d_{k+1}^{(IP)}$, of the improved polygonal method. By considering its absolute value one can decide whether the step size has to be decreased in order to satisfy a prescribed accuracy condition, or whether it can be increased for the following integration step.

Accidentally we have found two Runge–Kutta methods that can be combined for a step size control. We now wish to find a convenient method in order to supplement Heun's method (9.37). This method prescribes $a_2 = 1$. The remaining five parameters of the three-stage Runge–Kutta method must be determined from the system of equations, (9.44), reduced to

## 9.1 Single Step Methods

$$c_1 + c_2 + c_3 = 1$$
$$c_2 + a_3 c_3 = \tfrac{1}{2}$$
$$c_3 b_{32} = \tfrac{1}{6}$$
$$c_2 + a_3^2 c_3 = \tfrac{1}{3}.$$

From the second and fourth equations we obtain $a_3 c_3 (1 - a_3) = \tfrac{1}{6}$. By arbitrarily choosing $a_3 = \tfrac{1}{2}$, the other parameters are $c_3 = \tfrac{2}{3}$, $c_2 = \tfrac{1}{6}$, $c_1 = \tfrac{1}{6}$, $b_{32} = \tfrac{1}{4}$ and $b_{31} = a_3 - b_{32} = \tfrac{1}{4}$. The corresponding third-order Runge–Kutta method is

$$\boxed{\begin{aligned}
k_1 &= f(x_k, y_k) \\
k_2 &= f(x_k + h, y_k + h k_1) \\
k_3 &= f(x_k + \tfrac{1}{2}h, y_k + \tfrac{1}{4}h(k_1 + k_2)) \\
y_{k+1} &= y_k + \tfrac{1}{6}h(k_1 + k_2 + 4 k_3)
\end{aligned}} \qquad (9.49)$$

Analogous to (9.48) the difference of the approximations of the two methods yields

$$d_{k+1}^{(H)} \approx \tfrac{1}{6}h(k_1 + k_2 + 4 k_3) - \tfrac{1}{2}h(k_1 + k_2) + O(h^4) = \tfrac{1}{3}h(-k_1 - k_2 + 2 k_3) + O(h^4) \qquad (9.50)$$

an estimate of the local discretization error $d_{k+1}^{(H)}$ of Heun's second-order method.

*Example 9.4* The step size control is illustrated on the basis of the second order Heun method (9.37) with the estimate (9.50) for the initial value problem $y' = -2xy^2$, $y(0) = 1$. The absolute value of the discretization error should not be larger than $\varepsilon = 10^{-6}$ for each step. Whenever the absolute value of the estimate $d_{k+1}^{(H)}$ is too large, the step size is halved. As soon as it drops below the value $10^{-7}$, the step size is doubled. The results are summarized in Table 9.4. The step size initially given is $h = 0.05$, it must be changed immediately. With the step size $h = 0.025$ six integration steps can be successfully performed. For $x_k = 0.125$, the estimate $|d_{k+1}^{(H)}|$ is smaller than $10^{-7}$, so the step size is doubled on trial. However, the subsequent estimate is too large, so that the step size $h$ is reduced to the previous value. The seeming failure of the strategy can be explained by the sign change of the local discretization error. After that integration step, the choice $h = 0.05$ is successful, and a repeated increase of the step size would be possible due to the estimate $d_{k+1}^{(H)}$, but the computer program allows an increase of $h$ only after two consecutive successful integration steps. At $x_k = 0.300$ the step size must be halved. The global discretization errors $g_k$ are determined by means of the exact solution.

When generalizing (9.41) to an explicit four-stage Runge–Kutta method, we must admit thirteen parameters $a_2, b_{21}, a_3, b_{31}, b_{32}, a_4, b_{41}, b_{42}, b_{43}, c_1, c_2, c_3, c_4$

Table 9.4 Heun's second-order method with step size control

| $x_k$ | $y_k$ | $10^8 g_k$ | $h$ | $k_1$ | $k_2$ | $k_3$ | $10^8 d_{k+1}^{(H)}$ |
|---|---|---|---|---|---|---|---|
| 0 | 1.000 000 00 | 0 | 0.050 | 0 | −0.100 0000 | −0.049 8751 | 416(!) |
|  |  |  | 0.025 | 0 | −0.050 0000 | −0.024 9844 | 26 |
| 0.025 | 0.999 375 00 | 39 | 0.025 | −0.049 9375 | −0.099 6257 | −0.074 7662 | 26 |
| 0.050 | 0.997 505 46 | 77 | 0.025 | −0.099 5017 | −0.148 5091 | −0.123 9909 | 24 |
| 0.075 | 0.994 405 33 | 114 | 0.025 | −0.148 3263 | −0.196 2962 | −0.172 2985 | 21 |
| 0.100 | 0.990 097 54 | 147 | 0.025 | −0.196 0586 | −0.242 6528 | −0.219 3460 | 16 |
| 0.125 | 0.984 613 65 | 173 | 0.025 | −0.242 3660 | −0.287 2707 | −0.264 8130 | 9 |
| 0.150 | 0.977 993 19 | 192 | 0.050 | −0.286 9412 | −0.371 4456 | −0.329 1543 | 130(!) |
|  |  |  | 0.025 | −0.286 9412 | −0.329 8718 | −0.308 4071 | 1 |
| 0.175 | 0.970 283 03 | 199 | 0.050 | −0.329 5072 | −0.409 3871 | −0.369 4444 | 9 |
| 0.225 | 0.951 810 67 | 372 | 0.050 | −0.407 6746 | −0.477 1559 | −0.442 5056 | −301(!) |
|  |  |  | 0.025 | −0.407 6746 | −0.443 3230 | −0.425 5273 | −48 |
| 0.250 | 0.941 173 20 | 327 | 0.025 | −0.442 9035 | −0.475 7979 | −0.459 3917 | −68 |
| 0.275 | 0.929 689 44 | 260 | 0.025 | −0.475 3774 | −0.505 4196 | −0.490 4532 | −91 |
| 0.300 | 0.917 429 47 | 172 | 0.025 | −0.505 0061 | −0.532 1361 | −0.518 6407 | −116(!) |

for which we again require that

$$a_k = \sum_{j=1}^{k-1} b_{kj} \quad (k = 2, 3, 4). \tag{9.51}$$

Analogous to the three-stage Runge–Kutta method the local discretization error is expanded into a Taylor series with respect to the step size $h$. After eliminating the three parameters of (9.51), the requirement for a fourth order method leads to a system of eight nonlinear equations for the ten remaining parameters (for instance see Gear (1971) and Grigorieff (1972)). Once again the system has a two-parametric set of solutions that may be described by using $a_2$ and $a_3$ as parameters (Gear, 1971; Ralston and Rabinowitz, 1978). In the following we cite some of the more important cases.

Historically the *classical fourth-order Runge–Kutta method* is the oldest method and is given by

$$\begin{aligned}
k_1 &= f(x_k, y_k) \\
k_2 &= f(x_k + \tfrac{1}{2}h, y_k + \tfrac{1}{2}hk_1) \\
k_3 &= f(x_k + \tfrac{1}{2}h, y_k + \tfrac{1}{2}hk_2) \\
k_4 &= f(x_k + h, y_k + hk_3) \\
y_{k+1} &= y_k + \tfrac{1}{6}h(k_1 + 2k_2 + 2k_3 + k_4)
\end{aligned} \tag{9.52}$$

The method (9.52) is surprisingly simple with respect to the parameter values

## 9.1 Single Step Methods

and has the property that the successive computation of the slopes $k_i$, $(i \geq 3)$ only requires the preceding value $k_{i-1}$.

Another fourth-order Runge–Kutta method is directly related to the so-called Newton's (3/8)-quadrature rule, (8.56), namely

$$
\begin{aligned}
k_1 &= f(x_k, y_k) \\
k_2 &= f(x_k + \tfrac{1}{3}h, y_k + \tfrac{1}{3}hk_1) \\
k_3 &= f(x_k + \tfrac{2}{3}h, y_k - \tfrac{1}{3}hk_1 + hk_2) \\
k_4 &= f(x_k + h, y_k + hk_1 - hk_2 + hk_3) \\
y_{k+1} &= y_k + \tfrac{1}{8}h(k_1 + 3k_2 + 3k_3 + k_4)
\end{aligned}
\tag{9.53}
$$

Apart from these fourth-order Runge–Kutta methods many other variants exist with different objectives (Grigorieff, 1972; Lambert, 1973). On the assumption that the absolute values of the partial derivatives of $f(x, y)$ are bounded, methods have been derived for which the absolute value of the principal part of the local discretization error is a minimum (Ceschino and Kuntzmann, 1966; Grigorieff, 1972; Lambert, 1973; Ralston and Rabinowitz, 1978).

A step size control is not feasible for the combination of three- and four-stage Runge–Kutta methods as above, because the corresponding systems of equations do not admit solutions so that $k_1, k_2$ and $k_3$ are the same for both methods.

However, if the number of function evaluations per step is increased to five, Fehlberg (1970) skilfully found a combination of a third- and fourth-order Runge–Kutta method so that the fifth function evaluation can be used as the first one of the subsequent integration step. Therefore only four evaluations of $f(x, y)$ are required if the integration step is successful.

An estimate of $d_{k+1}$ of a fourth-order Runge–Kutta method can be obtained with the aid of an appropriate method of order five. To accomplish this one must consider a six-stage method, because Butcher (1965) has shown that the maximal attainable order $p$ of an explicit $m$-stage Runge–Kutta method is given according to Table 9.5.

With this consideration England (1969) has found the following fourth-order Runge–Kutta method

$$
\begin{aligned}
k_1 &= f(x_k, y_k) \\
k_2 &= f(x_k + \tfrac{1}{2}h, y_k + \tfrac{1}{2}hk_1) \\
k_3 &= f(x_k + \tfrac{1}{2}h, y_k + \tfrac{1}{4}hk_1 + \tfrac{1}{4}hk_2) \\
k_4 &= f(x_k + h, y_k - hk_2 + 2hk_3) \\
y_{k+1} &= y_k + \tfrac{1}{6}h(k_1 + 4k_3 + k_4)
\end{aligned}
\tag{9.54}
$$

Table 9.5 Maximal order of explicit Runge–Kutta methods

| m | 1 | 2 | 3 | 4 | 5 | 6 | 7 | 8 | 9 |
|---|---|---|---|---|---|---|---|---|---|
| p | 1 | 2 | 3 | 4 | 4 | 5 | 6 | 6 | 7 |

This method (9.54) can be extended to a six-stage Runge–Kutta method of order five by the additional formulae

$$k_5 = f(x_k + \tfrac{2}{3}h, y_k + \tfrac{1}{27}h(7k_1 + 10k_2 + k_4))$$
$$k_6 = f(x_k + \tfrac{1}{5}h, y_k + \tfrac{1}{625}h(28k_1 - 125k_2 + 546k_3 + 54k_4 - 378k_5))$$
$$y_{k+1} = y_k + \tfrac{1}{336}h(14k_1 + 35k_4 + 162k_5 + 125k_6)$$

(9.55)

If the expression (9.54) of the approximation $y_{k+1}$ is subtracted from the more accurate approximation (9.55) we obtain, analogous to (9.48), the estimate of the local discretization error $d_{k+1}$ of the fourth-order method (9.54)

$$d_{k+1} \approx \tfrac{1}{336}h(-42k_1 - 224k_3 - 21k_4 + 162k_5 + 125k_6) + O(h^6). \tag{9.56}$$

An improved variant with a much smaller local discretization error has been proposed by Fehlberg (1970), where the fourth-order Runge–Kutta method uses five of the six necessary function evaluations. He has developed similar combinations of Runge–Kutta methods of higher order for the step size control (Fehlberg, 1964, 1969).

A different approach for an automatic step size control of Runge–Kutta methods is used by *Zonneveld* (Stroud, 1974; Zonneveld, 1964). He estimates the local discretization error by means of an additional, suitably defined function evaluation in such a manner that a linear combination of $k_i$ values is computed that are also used for the integration step. With this idea it is possible to define a four-stage Runge–Kutta method of order three with a step size control. In a similar way the classical method, (9.52), of order four can be extended by an additional function evaluation to obtain an estimate of its local discretization error.

### 9.1.5 Implicit Runge–Kutta methods

The numerical solution of differential equations with certain properties, which will be discussed in Section 9.3, requires special methods. The implicit Runge–Kutta methods belong to this class, that is characterized by the fact that the slopes $k_1, k_2, \ldots$ are defined by an implicit system of equations. Again we describe the principle in detail only for a one-stage method because the derivation of multi-stage methods is quite elaborate. Such a method can be defined by the two

## 9.1 Single Step Methods

formulae
$$k_1 = f(x_k + a_1 h, y_k + b_{11} h k_1)$$
$$y_{k+1} = y_k + h c_1 k_1 \tag{9.57}$$

in which the free parameters $a_1, b_{11}$ and $c_1$ are to be determined in such a way that the order of the method (9.57) is maximal. We again require that $a_1 = b_{11}$ thus reducing the number of unknown parameters to two. The expansion of the corresponding local discretization error into a Taylor series in $h$ is now somewhat more complicated because the equation is implicit in $k_1$. $\bar{k}_1$ denotes the $\bar{k}_1$ by the value resulting from (9.57) if $y_k$ is replaced by $y(x_k)$. We expand $\bar{k}_1$ into a Taylor series at the point $(x_k, y(x_k))$, but shall omit the variables to simplify the notation.

$$\bar{k}_1 = f(x_k + a_1 h, y(x_k) + a_1 h \bar{k}_1) \tag{9.58}$$
$$= f + a_1 h (f_x + \bar{k}_1 f_y) + \tfrac{1}{2} a_1^2 h^2 (f_{xx} + 2\bar{k}_1 f_{xy} + \bar{k}_1^2 f_{yy}) + O(h^3)$$

To solve this implicit equation for $\bar{k}_1$ we use a series expansion

$$\bar{k}_1 = \alpha_0 + \alpha_1 h + \alpha_2 h^2 + \cdots \tag{9.59}$$

We determine the coefficients $\alpha_i$ by substituting (9.59) into (9.58)

$$\alpha_0 + \alpha_1 h + \alpha_2 h^2 + O(h^3)$$
$$= f + a_1 h [f_x + (\alpha_0 + \alpha_1 h) f_y] + \tfrac{1}{2} a_1^2 h^2 (f_{xx} + 2\alpha_0 f_{xy} + \alpha_0^2 f_{yy}) + O(h^3).$$

By comparing coefficients, with (9.30) and (9.31), we obtain

$$\alpha_0 = f$$
$$\alpha_1 = a_1(f_x + \alpha_0 f_y) = a_1(f_x + f f_y) = a_1 F$$
$$\alpha_2 = a_1 \alpha_1 f_y + \tfrac{1}{2} a_1^2 (f_{xx} + 2 f f_{xy} + f^2 f_{yy}) = a_1^2 (F f_y + \tfrac{1}{2} G)$$

With these values the local discretization error of the method (9.57) is

$$d_{k+1} = y(x_{k+1}) - y(x_k) - h c_1 \bar{k}_1$$
$$= h f + \tfrac{1}{2} h^2 F + \tfrac{1}{6} h^3 (G + F f_y) - h c_1 f - h^2 c_1 a_1 F$$
$$\quad - h^3 a_1^2 c_1 (F f_y + \tfrac{1}{2} G) + O(h^4)$$
$$= (1 - c_1) h f + (\tfrac{1}{2} - a_1 c_1) h^2 F + [(\tfrac{1}{6} - a_1^2 c_1) F f_y$$
$$\quad + (\tfrac{1}{6} - \tfrac{1}{2} a_1^2 c_1) G] h^3 + O(h^4).$$

If we define $c_1 = 1$ and $a_1 = \tfrac{1}{2}$, the coefficients of $h$ and $h^2$ vanish, whereas the coefficient of $h^3$ is equal to $-F f_y/12 + G/24$. Thus the order of the implicit Runge–Kutta method

$$k_1 = f(x_k + \tfrac{1}{2} h, y_k + \tfrac{1}{2} h k_1)$$
$$y_{k+1} = y_k + h k_1 \tag{9.60}$$

is 2. Each integration step requires the solution of the implicit equation for $k_1$. The fixed point iteration can obviously by applied. In comparison to the implicit trapezoidal method, (9.33), the procedure (9.60) has no specific advantages because the principal part of the local discretization error is approximately the same size.

The general formulae of a two-stage, implicit Runge–Kutta method

$$\begin{aligned}
k_1 &= f(x_k + a_1 h, y_k + h b_{11} k_1 + h b_{12} k_2) \\
k_2 &= f(x_k + a_2 h, y_k + h b_{21} k_1 + h b_{22} k_2) \\
y_{k+1} &= y_k + h(c_1 k_1 + c_2 k_2)
\end{aligned} \qquad (9.61)$$

contain eight parameters that should satisfy the relationships $a_1 = b_{11} + b_{12}$, $a_2 = b_{21} + b_{22}$. The remaining six free parameters allow the construction of a fourth-order method. The correspondingly elaborate analysis of the local discretization error yields, apart from a permutation of the $k$ values, the unique implicit fourth-order Runge–Kutta method

$$\begin{aligned}
k_1 &= f\left(x_k + \frac{3-\sqrt{3}}{6} h, y_k + \frac{1}{4} h k_1 + \frac{3-2\sqrt{3}}{12} h k_2\right) \\
k_2 &= f\left(x_k + \frac{3+\sqrt{3}}{6} h, y_k + \frac{3+2\sqrt{3}}{12} h k_1 + \frac{1}{4} h k_2\right) \\
y_{k+1} &= y_k + \frac{h}{2}(k_1 + k_2)
\end{aligned} \qquad (9.62)$$

It can be shown, (Butcher, 1963), that an $m$-stage implicit Runge–Kutta method, after a suitable choice of the $m(m+1)$ free parameters, has the maximal attainable order $2m$ for all $m$. Examples are given in Butcher (1964). In spite of the considerable increase of the order the implicit Runge–Kutta methods are not very attractive for general usage because each integration step requires the solution of a system of equations, that is in general nonlinear, for the $m$ unknowns $k_i$. If this system is solved by means of the fixed point iteration, each step of integration involves much more than $m$ function evaluations. Moreover, the fixed point iteration is convergent for step sizes $h$ that satisfy the sufficient condition

$$hBL < 1 \quad \text{with} \quad B := \max_i \sum_j |b_{ij}| \qquad (9.63)$$

where $L$ denotes the Lipschitz constant of the function $f(x, y)$. However, the implicit Runge–Kutta methods have a stability property that can be decisive for the integration of systems of so-called stiff differential equations (see Section 9.3).

## 9.1 Single Step Methods

### 9.1.6 Differential equations of higher order and systems

In the following we describe how to proceed to numerically solve differential equations of higher order or systems of differential equations. Such problems can be reduced to systems of first-order differential equations by introducing new unknowns.

We explain the procedure for an initial value problem of a fourth-order differential equation that is assumed to be explicit, that is

$$y^{(4)}(x) = f(x, y(x), y'(x), y''(x), y^{(3)}(x)) \tag{9.64}$$

where $f$ denotes a function of five variables that is in general nonlinear. The initial conditions

$$y(x_0) = y_0 \quad y'(x_0) = y'_0 \quad y''(x_0) = y''_0 \quad y^{(3)}(x_0) = y_0^{(3)} \tag{9.65}$$

prescribe the value of the function $y(x)$ and of its first three derivatives at a given abscissa $x_0$. We also consider the first, second and third derivatives of $y(x)$ as unknown functions, and define

$$y_1(x) := y(x) \quad y_2(x) := y'(x) \quad y_3(x) := y''(x) \quad y_4(x) := y^{(3)}(x). \tag{9.66}$$

Then the given initial value problem (9.64), (9.65) can equivalently be formulated as a system of four first-order differential equations for the four functions $y_1(x)$, $y_2(x)$, $y_3(x)$, $y_4(x)$

$$\begin{aligned} y'_1(x) &= y_2(x) \\ y'_2(x) &= y_3(x) \\ y'_3(x) &= y_4(x) \\ y'_4(x) &= f(x, y_1(x), y_2(x), y_3(x), y_4(x)) \end{aligned} \tag{9.67}$$

subject to the initial conditions

$$y_1(x_0) = y_0 \quad y_2(x_0) = y'_0 \quad y_3(x_0) = y''_0 \quad y_4(x_0) = y_0^{(3)}. \tag{9.68}$$

In the next step we introduce the following vector functions and the vector

$$\mathbf{y}(x) = \begin{pmatrix} y_1(x) \\ y_2(x) \\ y_3(x) \\ y_4(x) \end{pmatrix} \quad \mathbf{f}(x, \mathbf{y}) = \begin{pmatrix} y_2(x) \\ y_3(x) \\ y_4(x) \\ f(x, y_1(x), y_2(x), y_3(x), y_4(x)) \end{pmatrix} \quad \mathbf{y}_0 = \begin{pmatrix} y_0 \\ y'_0 \\ y''_0 \\ y_0^{(3)} \end{pmatrix}. \tag{9.69}$$

With this notation the initial value problem (9.67), (9.68) can be written in the following abbreviated form

$$\boxed{\mathbf{y}'(x) = \mathbf{f}(x, \mathbf{y}(x)) \quad \mathbf{y}(x_0) = \mathbf{y}_0.} \tag{9.70}$$

From a formal point of view (9.70) is the same initial value problem as the scalar differential equation (9.1), (9.2) with the only difference being that $y(x)$ is replaced by the vector function $\mathbf{y}(x)$ and $f(x, y(x))$ by $\mathbf{f}(x, \mathbf{y}(x))$. Therefore the treated one step methods above can be directly applied to treat (9.70).

From natural laws the equations of motion in mechanics lead us to systems of second-order differential equations. For instance, the differential equations describing the motion of two coupled pendula are

$$u''(x) = f(x, u(x), u'(x), v(x), v'(x))$$
$$v''(x) = g(x, u(x), u'(x), v(x), v'(x))$$
(9.71)

for the angles of elongation $u(x)$ and $v(x)$ subject to initial conditions

$$u(x_0) = u_0 \quad u'(x_0) = u'_0 \quad v(x_0) = v_0 \quad v'(x_0) = v'_0.$$
(9.72)

Apart from $u(x)$ and $v(x)$ we also consider their first derivatives $u'(x)$ and $v'(x)$ as unknown functions and define

$$y_1(x) := u(x) \quad y_2(x) := u'(x) \quad y_3(x) := v(x) \quad y_4(x) := v'(x).$$
(9.73)

The initial value problem (9.71), (9.72) can be written, by means of the notation of (9.73), as the equivalent system of four differential equations

$$y'_1(x) = y_2(x)$$
$$y'_2(x) = f(x, y_1(x), y_2(x), y_3(x), y_4(x))$$
$$y'_3(x) = y_4(x)$$
$$y'_4(x) = g(x, y_1(x), y_2(x), y_3(x), y_4(x))$$
(9.74)

subject to the initial conditions

$$y_1(x_0) = u_0 \quad y_2(x_0) = u'_0 \quad y_3(x_0) = v_0 \quad y_4(x_0) = v'_0.$$
(9.75)

With definitions analogous to (9.69) we obviously obtain, instead of (9.74), (9.75), a problem of type (9.70).

Problems in chemistry and biology usually lead to systems of ordinary differential equations, (9.70), for the state variables $y_1(x), y_2(x), \ldots, y_n(x)$ describing the change per unit time of the state variables on the basis of known laws or of model assumptions as functions of the state of the system. In these cases each component of the vector function $\mathbf{f}(x, \mathbf{y}(x))$ will in general depend on all unknown functions $y_i(x)$.

All methods treated so far can be easily generalized to systems of first-order differential equations, this includes those methods to be developed in the following for solving a single first-order differential equation. We just have to replace certain scalar quantities by vectors. For illustration we give the *classical Runge–Kutta method*, (9.52), for a system of $n$ differential equations.

## 9.1 Single Step Methods

$$\begin{aligned}
\mathbf{k}_1 &= \mathbf{f}(x_k, \mathbf{y}_k) \\
\mathbf{k}_2 &= \mathbf{f}(x_k + \tfrac{1}{2}h, \mathbf{y}_k + \tfrac{1}{2}h\mathbf{k}_1) \\
\mathbf{k}_3 &= \mathbf{f}(x_k + \tfrac{1}{2}h, \mathbf{y}_k + \tfrac{1}{2}h\mathbf{k}_2) \\
\mathbf{k}_4 &= \mathbf{f}(x_k + h, \mathbf{y}_k + h\mathbf{k}_3) \\
\mathbf{y}_{k+1} &= \mathbf{y}_k + \tfrac{1}{6}h(\mathbf{k}_1 + 2\mathbf{k}_2 + 2\mathbf{k}_3 + \mathbf{k}_4)
\end{aligned} \qquad (9.76)$$

In (9.76) $\mathbf{y}_k$, $\mathbf{y}_{k+1}$, $\mathbf{k}_1$, $\mathbf{k}_2$, $\mathbf{k}_3$ and $\mathbf{k}_4$ represent vectors in $\mathbf{R}^n$.

If we reformulate the implicit, fourth-order Runge–Kutta method (9.62) for a system of $n$ differential equations (9.70), each integration step requires the solution of a system of $2n$ coupled, and in general, nonlinear equations for the $n$ components of $\mathbf{k}_1$ and $\mathbf{k}_2$, respectively.

When the Runge–Kutta method (9.54), (9.55), proposed by England is applied to a system (9.70) in order to have a step size control, an estimate of the size of the local discretization error, which is a vector $\mathbf{d}_{k+1} \in \mathbf{R}^n$, must be measured by a vector norm.

*Example 9.5* We show the numerical integration of a system of differential equations representing a simple biological model. We consider the classical prey–predator model of Lotka–Volterra that is described by the nonlinear, autonomous system of differential equations

$$\begin{aligned}
\dot{y}_1(t) &= a y_1(t)(1 - y_2(t)) \\
\dot{y}_2(t) &= y_2(t)(y_1(t) - 1)
\end{aligned} \qquad (9.77)$$

where $y_1(t)$ and $y_2(t)$ measure the size of the populations of the prey and predator animals, respectively, as a function of time $t$, and where $a > 0$ denotes a parameter. We wish to solve (9.77) approximately for $a = 10$ subject to the initial condition $y_1(0) = 3$, $y_2(0) = 1$ by means of the fourth-order Runge–Kutta method (9.54), using a step size control on the basis of (9.55) and (9.56). We require the Euclidean vector norm of the estimate of the local discretization error to be less than $\varepsilon = 10^{-6}$. As we have $\|\mathbf{d}_{k+1}\| = O(h^5)$ the step size $h$ is controlled in such a way that it is decreased by the factor 0.9 if the condition $\|\mathbf{d}_{k+1}\| \leq \varepsilon$ is not satisfied, and increased by the factor $1/0.9 \doteq 1.111$ if $\|\mathbf{d}_{k+1}\| < 0.5\varepsilon$. In Figure 9.5 the solutions $y_1(t)$ and $y_2(t)$, the norm of the estimate for $\mathbf{d}_{k+1}$ and the used step size $h$ are shown as functions of $t$. With the initial step size $h = 0.025$ the integration up to $t = 5$, that is more than two periods, required 103 integration steps with a minimal step size $h_{\min} = 0.025$ and a maximal step size $h_{\max} \doteq 0.0885$. With the constant step size $h = 0.025$ we would need 200 integration steps. The vector functions for (9.77) are

$$\mathbf{y}(t) = \begin{pmatrix} y_1(t) \\ y_2(t) \end{pmatrix} \quad \mathbf{f}(t, \mathbf{y}(t)) = \begin{pmatrix} a y_1(t)(1 - y_2(t)) \\ y_2(t)(y_1(t) - 1) \end{pmatrix} \quad \mathbf{y}_0 = \begin{pmatrix} 3 \\ 1 \end{pmatrix}.$$

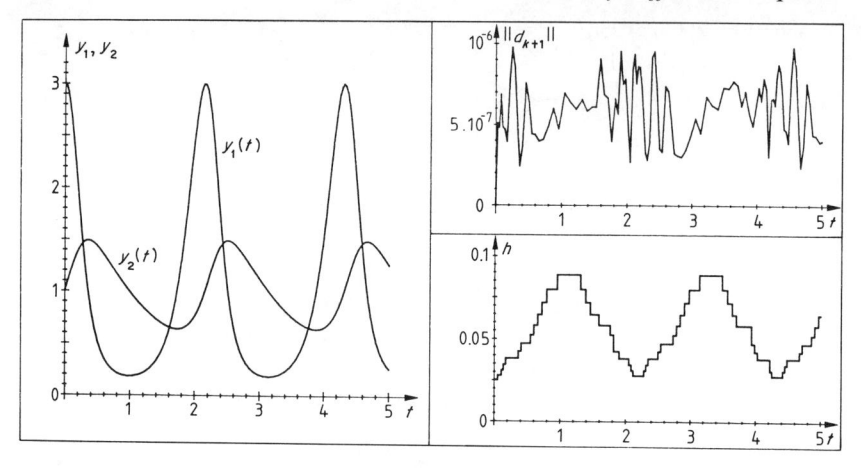

Figure 9.5 Solutions and step size control

## 9.2 Multistep Methods

A one step method determines the approximation $y_{k+1}$ at the abscissa $x_{k+1} = x_k + h$ solely on the basis of the approximation point $(x_k, y_k)$. In contrast *multistep methods* use the information at the previous support abscissae $x_{k-1}, x_{k-2}, \ldots, x_{k-m}$, that are now assumed to be *equidistant*, to compute $y_{k+1}$. First we develop the historically oldest variants and then consider general linear multistep methods.

### 9.2.1 Adams–Bashforth methods

Again the integral equation

$$y(x_{k+1}) = y(x_k) + \int_{x_k}^{x_{k+1}} f(x, y(x)) \, dx \qquad (9.78)$$

is the starting point. To explain the principle we want to approximate the integral by means of the known values $f_k := f(x_k, y_k)$, $f_{k-1} := f(x_{k-1}, y_{k-1})$, $f_{k-2} := f(x_{k-2}, y_{k-2})$ and $f_{k-3} := f(x_{k-3}, y_{k-3})$ and an interpolation quadrature formula. Thus we use the uniquely defined interpolation polynomial $P_3(x)$ of degree three for the four support points $(x_{k-3}, f_{k-3})$, $(x_{k-2}, f_{k-2})$, $(x_{k-1}, f_{k-1})$, $(x_k, f_k)$ with the equidistant support abscissae $x_{k-j} = x_k - jh$, $j = 0, 1, 2, 3$, (see Figure 9.6). It can be represented by means of the corresponding Lagrange polynomials $L_{k-j}(x)$, (3.3), as follows

$$P_3(x) = \sum_{j=0}^{3} f_{k-j} L_{k-j}(x). \qquad (9.79)$$

The integral in (9.78) is approximated by the integral of $P_3(x)$ over the interval

## 9.2 Multistep Methods

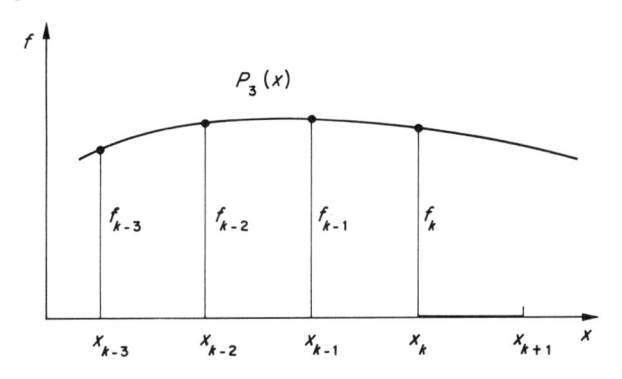

Figure 9.6 Approximation of the integral

$[x_k, x_{k+1}]$, and so we obtain

$$y_{k+1} = y_k + \sum_{j=0}^{3} f_{k-j} \int_{x_k}^{x_{k+1}} L_{k-j}(x) \, dx. \tag{9.80}$$

The elementary evaluation of the four integrals is only shown for $j = 0$. The change of variable $x = x_k + h\xi$, $dx = h \, d\xi$ yields

$$I_0 = \int_{x_k}^{x_{k+1}} L_k(x) \, dx = \int_{x_k}^{x_{k+1}} \frac{(x - x_{k-3})(x - x_{k-2})(x - x_{k-1})}{(x_k - x_{k-3})(x_k - x_{k-2})(x_k - x_{k-1})} \, dx$$

$$= h \int_0^1 \frac{(3 + \xi)(2 + \xi)(1 + \xi)}{3 \cdot 2 \cdot 1} \, d\xi = \frac{h}{6} \int_0^1 (6 + 11\xi + 6\xi^2 + \xi^3) \, d\xi = \frac{55}{24} h.$$

From (9.80) we deduce the *Adams–Bashforth method*

$$\boxed{y_{k+1} = y_k + \tfrac{1}{24} h (55 f_k - 59 f_{k-1} + 37 f_{k-2} - 9 f_{k-3}).} \tag{9.81}$$

The information at the four consecutive support abscissae $x_{k-3}, x_{k-2}, x_{k-1}, x_k$ is linearly combined, resulting in an explicit formula for $y_{k+1}$, therefore we call (9.81) an *explicit, linear four-step method*. An obvious property of this method is that each integration step requires only one function evaluation $f(x_k, y_k)$ because the previous values $f_{k-1}, f_{k-2}, f_{k-3}$ have already been computed. However, a change of the step size is no longer simple, because new equidistant support abscissae $x_{k-j}^*$ with usually unknown approximate values $y_{k-j}^*$ are required. These can be obtained, for instance, by means of a suitable Hermite interpolation.

The local discretization error of a multistep method is defined by Definition 9.2. For method (9.81) it is given by

$$d_{k+1} := y(x_{k+1}) - y(x_k) - \tfrac{1}{24}h[55f(x_k,y(x_k)) - 59f(x_{k-1},y(x_{k-1}))$$
$$+ 37f(x_{k-2},y(x_{k-2})) - 9f(x_{k-3},y(x_{k-3}))]$$
$$= y(x_{k+1}) - y(x_k) - \tfrac{1}{24}h[55y'(x_k) - 59y'(x_{k-1})$$
$$+ 37y'(x_{k-2}) - 9y'(x_{k-3})]$$
$$= hy' + \tfrac{1}{2}h^2y'' + \tfrac{1}{6}h^3y^{(3)} + \tfrac{1}{24}h^4y^{(4)} + \tfrac{1}{120}h^5y^{(5)} + O(h^6)$$
$$- \tfrac{1}{24}h[55y' - 59(y' - hy'' + \tfrac{1}{2}h^2y^{(3)} - \tfrac{1}{6}h^3y^{(4)} + \tfrac{1}{24}h^4y^{(5)} + O(h^5))$$
$$+ 37(y' - 2hy'' + 2h^2y^{(3)} - \tfrac{4}{3}h^3y^{(4)} + \tfrac{2}{3}h^4y^{(5)} + O(h^5))$$
$$- 9(y' - 3hy'' + \tfrac{9}{2}h^2y^{(3)} - \tfrac{9}{2}h^3y^{(4)} + \tfrac{27}{8}h^4y^{(5)} + O(h^5))]$$
$$= \tfrac{251}{720}h^5y^{(5)} + O(h^6). \tag{9.82}$$

The relationship between the local discretization errors $d_k$ of a multistep method, and the global discretization error $g_n = y(x_n) - y_n$ at an abscissa $x_n = x_0 + nh$ obeys more complicated rules than those of one-step methods. We shall resume this point after the discussion of general multistep methods in Section 9.2.3 and content ourselves for the time being with the following definition.

**Definition 9.5** *An m-step method is of order p if its local discretization error $d_k$ satisfies the estimate*

$$\max_{m \leq k \leq n} |d_k| \leq D = \text{const} \cdot h^{p+1} = O(h^{p+1}). \tag{9.83}$$

The numerical constant of the principal part of the local discretization error will be essential for comparison. In the following we call it the coefficient of the principal part. The Adams–Bashforth method (9.81) therefore has order 4, and its coefficient of the principal part is 251/720. On additional assumptions, the global discretization error can be estimated by a bound that is similar to (9.24).

By varying the number of the backward support abscissae similar methods can obviously be derived. The corresponding five- and six-step methods are

$$\boxed{\begin{aligned} y_{k+1} = y_k + \tfrac{1}{720}h(1901f_k - 2774f_{k-1} + 2616f_{k-2} - 1274f_{k-3} \\ + 251f_{k-4}) \end{aligned}} \tag{9.84}$$

$$\boxed{\begin{aligned} y_{k+1} = y_k + \tfrac{1}{1440}h(4277f_k - 7923f_{k-1} + 9982f_{k-2} \\ - 7298f_{k-3} + 2877f_{k-4} - 475f_{k-5}) \end{aligned}} \tag{9.85}$$

It is a simple calculation to see that each type of explicit $m$-step Adams–Bashforth method has the order $p = m$. Hence the order can simply be increased whereby the property that each integration step requires only one function evaluation is maintained. This minimal amount of effort, explains why Adams–

## 9.2 Multistep Methods

Bashforth methods are often used. Moreover, the combination of two methods of different orders allows us to estimate the local discretization error $d_{k+1}$ with practically no computational effort. For instance, the estimate of the local discretization error of the four-step method, (9.81), is obtained by means of a combination with (9.84)

$$d_{k+1} \approx \tfrac{1}{720}h(251f_k - 1004f_{k-1} + 1506f_{k-2} - 1004f_{k-3} + 251f_{k-4}) + O(h^6)$$

that can be used for a step size control. However, the local discretization errors of an Adams–Bashforth method is much larger than that of an explicit Runge–Kutta method of the same order. Therefore in comparison a smaller step size is necessary for the multistep methods so that their discretization errors are similar in absolute value.

In order to be able to use an $m$-step method, besides the initial condition $y(x_0) = y_0$, $(m-1)$ more initial values $y_1, y_2, \ldots, y_{m-1}$ are necessary. For reasons that will become clear later they must be determined in such a way that their errors correspond to the order of the multistep method used. To obtain the desired approximations $y_j$ at the abscissae $x_j = x_0 + jh$, $j = 1, 2, \ldots, m-1$, the Runge–Kutta methods are quite suitable and possibly a smaller step size $h^* \leqslant h$ is adequate. There also exist starting procedures based on multistep methods (Grigorieff, 1977).

### 9.2.2 Adams–Moulton methods

To obtain a better approximation of the integral in (9.78), the unknown value $f_{k+1} := f(x_{k+1}, y_{k+1})$ at the new abscissa $x_{k+1}$ is also used in addition to the known values of the function $f(x, y)$ at the abscissae $x_{k-3}, x_{k-2}, x_{k-1}, x_k$. As five support ordinates $f_{k-3}, f_{k-2}, f_{k-1}, f_k, f_{k+1}$ are available, we can construct an interpolation polynomial $P_4(x)$ of degree four. It can be defined by means of the Lagrange polynomials in the following way

$$P_4(x) = \sum_{j=-1}^{3} f_{k-j} L_{k-j}(x). \tag{9.86}$$

The integrand in (9.78) is replaced by $P_4(x)$, and so we obtain

$$y_{k+1} = y_k + \sum_{j=-1}^{3} f_{k-j} \int_{x_k}^{x_{k+1}} L_{k-j}(x)\,dx. \tag{9.87}$$

After an elementary evaluation of the integrals, from (9.87) we obtain the *Adams–Moulton method*

$$y_{k+1} = y_k + \frac{h}{720}[251 f(x_{k+1}, y_{k+1}) + 646 f_k - 264 f_{k-1} + 106 f_{k-2} - 19 f_{k-3}]. \tag{9.88}$$

The procedure of integration (9.88) is implicit, as is seen by the values on the right-hand side. The implicit equation (9.88) for $y_{k+1}$ can be solved by means of the fixed point iteration. The convergence is guaranteed for all step sizes $h$ satisfying $251hL < 720$ where $L$ denotes the Lipschitz constant of $f(x, y)$. As the information at four support abscissae is used, (9.88) is called an *implicit four-step method*. The order of the Adams–Moulton method (9.88) is five because its local discretization error is

$$d_{k+1} = -\frac{3}{160} h^6 y^{(6)}(x_k) + O(h^7).$$

The implicit Adams–Moulton method, (9.88), has an order that is one larger than the explicit Adams–Bashforth method, (9.81). Moreover, the coefficient of the principal part of the local discretization error of the implicit four-step method is much smaller than that of the explicit five-step method of the same order.

Varying the number of backward abscissae leads to similar implicit m-step Adams–Moulton methods whose order is $p = m + 1$. We simply cite the examples of the implicit three- and five-step methods,

$$y_{k+1} = y_k + \tfrac{1}{24} h [9 f(x_{k+1}, y_{k+1}) + 19 f_k - 5 f_{k-1} + f_{k-2}] \tag{9.89}$$

$$y_{k+1} = y_k + \tfrac{1}{1440} h [475 f(x_{k+1}, y_{k+1}) + 1427 f_k - 798 f_{k-1} + 482 f_{k-2} - 173 f_{k-3} + 27 f_{k-4}] \tag{9.90}$$

The determination of $y_{k+1}$ from the implicit equation, required in each integration step, is once again avoided by means of the predictor–corrector technique. The implicit formula is only used to improve a good initial approximation $y_{k+1}^{(0)}$ by means of a single step of the fixed point iteration. It is clear that the initial value $y_{k+1}^{(0)}$ for an implicit m-step method of Adams–Moulton should be found on the basis of the explicit m-step method of Adams–Bashforth. Such a combination of two three-step methods yielding a *predictor–corrector method* is

$$\begin{aligned} y_{k+1}^{(P)} &= y_k + \tfrac{1}{12} h (23 f_k - 16 f_{k-1} + 5 f_{k-2}) \\ y_{k+1} &= y_k + \tfrac{1}{24} h [9 f(x_{k+1}, y_{k+1}^{(P)}) + 19 f_k - 5 f_{k-1} + f_{k-2}] \end{aligned} \tag{9.91}$$

The *Adams–Bashforth–Moulton (ABM) method*, (9.91), has a local discretization error $d_{k+1} = O(h^5)$ so that its order is 4. It can generally be shown that a predictor–corrector method has the order $p = m + 1$, where it is defined by combining an explicit m-step predictor of order $m$ with an implicit m-step

## 9.2 Multistep Methods

Table 9.6 Coefficients of the principal parts of local discretization errors

| $m$ | 3 | 4 | 5 | 6 |
|---|---|---|---|---|
| Adams–Bashforth $C_{m+1}^{(AB)}$ | $\dfrac{3}{8}$ | $\dfrac{251}{720}$ | $\dfrac{95}{288}$ | $\dfrac{19087}{60480}$ |
| Adams–Moulton $C_{m+2}^{(AM)}$ | $-\dfrac{19}{720}$ | $-\dfrac{3}{160}$ | $-\dfrac{863}{60480}$ | $-\dfrac{275}{24192}$ |

corrector of order $m + 1$ (Grigorieff, 1977; Lambert, 1973). This analysis reveals that the coefficient $C_{m+2}^{(PC)}$ of the principal part of the local discretization error is given as a linear combination of the constants $C_{m+1}^{(P)}$ and $C_{m+2}^{(C)}$ of the two underlying methods, whereby different derivatives of $y(x)$ occur. However, we have already mentioned that the coefficients $C_{m+1}^{(AB)}$ of the Adams–Bashforth methods are much larger in absolute value than the constants $C_{m+2}^{(AM)}$ of the Adams–Moulton methods (see Table 9.6). Therefore the size of the local discretization error of the predictor–corrector method is essentially determined by the principal part of the explicit predictor formula.

The situation is improved if the predictor formula has the same order as the corrector formula, that is we have to combine two methods with different step numbers. As an example we give the predictor–corrector method consisting of the explicit four-step Adams–Bashforth method, (9.81), as predictor and the implicit three-step Adams–Moulton method, (9.89), as corrector

$$\begin{aligned}y_{k+1}^{(P)} &= y_k + \tfrac{1}{24}h(55f_k - 59f_{k-1} + 37f_{k-2} - 9f_{k-3}) \\ y_{k+1} &= y_k + \tfrac{1}{24}h[9f(x_{k+1}, y_{k+1}^{(P)}) + 19f_k - 5f_{k-1} + f_{k-2}]\end{aligned} \qquad (9.92)$$

The integration scheme (9.92) has the same order 4 as (9.91). However, the coefficient of the principal part of the local discretization error is $C_5^{(PC)} = -19/720 = C_5^{(AM)}$. Such combinations of predictor–corrector methods always have a principal part of the local discretization error equal to that of the corrector formula (Grigorieff, 1977; Lambert, 1973). Therefore, the local discretization error of the method (9.92) is, in absolute value, smaller than that of (9.91). The practical use of the predictor–corrector method (9.92) requires four initial values $y_0$, $y_1, y_2, y_3$, from the fact that the predictor is a four-step method.

*Example 9.6* We solve the initial value problem $y' = xe^{x-y}$, $y(1) = 0$ by means of the ABM methods, (9.91) and (9.92), and choose the step size $h = 0.1$. In order to compare the two methods under the same conditions, the three initial values $y_1, y_2, y_3$ are computed by means of the classical Runge–Kutta method (9.52).

Table 9.7 Runge–Kutta and ABM methods

| $x_k$ | $y(x_k)$ | Runge–Kutta $y_k$ | ABM $y_k^{(P)}$ | (9.91) $y_k$ | ABM $y_k^{(P)}$ | (9.92) $y_k$ |
|---|---|---|---|---|---|---|
| 1.0 | 0 | 0 | | | | |
| 1.1 | 0.262 6847 | 0.262 6829 | | | | |
| 1.2 | 0.509 2384 | 0.509 2357 | | | | |
| 1.3 | 0.742 3130 | 0.742 3100 | | | | |
| 1.4 | 0.963 9679 | 0.963 9648 | 0.964 1567 | 0.963 9538 | 0.963 8945 | 0.963 9751 |
| 1.5 | 1.175 8340 | 1.175 8309 | 1.175 9651 | 1.175 8147 | 1.175 7894 | 1.175 8460 |
| 1.6 | 1.379 2243 | 1.379 2213 | 1.379 3138 | 1.379 2027 | 1.379 2028 | 1.379 2382 |
| 1.7 | 1.575 2114 | 1.575 2087 | 1.575 2742 | 1.575 1894 | 1.575 1995 | 1.575 2258 |
| 1.8 | 1.764 6825 | 1.764 6799 | 1.764 7268 | 1.764 6610 | 1.764 6775 | 1.764 6965 |
| 1.9 | 1.948 3792 | 1.948 3769 | 1.948 4106 | 1.948 3588 | 1.948 3785 | 1.948 3924 |
| 2.0 | 2.126 9280 | 2.126 9258 | 2.126 9503 | 2.126 9089 | 2.126 9298 | 2.126 9401 |
| 2.5 | 2.958 7436 | 2.958 7422 | 2.958 7472 | 2.958 7315 | 2.958 7478 | 2.958 7508 |
| 3.0 | 3.717 7359 | 3.717 7350 | 3.717 7359 | 3.717 7287 | 3.717 7388 | 3.717 7399 |

Table 9.7 contains the approximations that are obtained by means of the fourth-order Runge–Kutta method, (9.52), as well as the two predictor–corrector methods. For these methods the predictor values $y_{k+1}^{(P)}$ are also given. The multistep methods are slightly less accurate than the Runge–Kutta method.

### 9.2.3 General linear multistep methods

The two classes of multistep methods treated so far are based on the integration of the differential equation over the interval $[x_k, x_{k+1}]$. Now we consider a more general approach by defining a linear $m$-step method in such a way that the approximations $y_k, y_{k-1}, \ldots, y_{k-m+1}$ are also included in the form of a linear combination to determine $y_{k+1}$. For the equidistant support abscissae $x_{k-j} = x_k - jh$, and with the abbreviation $s := k - m + 1$, we seek an $m$-step-method in the form

$$\sum_{j=0}^{m} a_j y_{s+j} = h \sum_{j=0}^{m} b_j f(x_{s+j}, y_{s+j}) \quad (m \geqslant 2). \tag{9.93}$$

In (9.93) we must have $a_m \neq 0$, and in the following, without loss of generality we set $a_m = 1$. In order that (9.93) defines a real $m$-step method the coefficients $a_0$ and $b_0$ are not allowed to vanish simultaneously. If $b_m = 0$, then (9.93) is an *explicit*, otherwise (9.93) is an *implicit* $m$-step method. The remaining $2m$, or $(2m + 1)$, parameters of (9.93) are to be determined in such a way that the resulting method has properties that are specified in the following.

**Definition 9.6** *The linear multistep method (9.93) is of order $p$ if in the expansion*

## 9.2 Multistep Methods

of the local discretization error $d_{k+1}$ into a power series in $h$ at an arbitrary abscissa $\bar{x}$ the following holds

$$d_{k+1} = \sum_{j=0}^{m} [a_j y(x_{s+j}) - hb_j f(x_{s+j}, y(x_{s+j}))]$$
$$= c_0 y(\bar{x}) + c_1 hy'(\bar{x}) + \cdots + c_p h^p y^{(p)}(\bar{x}) + c_{p+1} h^{p+1} y^{(p+1)}(\bar{x}) + \cdots$$
$$c_0 = c_1 = \cdots = c_p = 0 \quad \text{and} \quad c_{p+1} \neq 0. \tag{9.94}$$

The definition of the order does not specify $\bar{x}$ precisely. It can be proved by a simple argument that $p$ does not depend on $\bar{x}$ from which the power series is formed and, moreover, that the coefficient, $c_{p+1}$, of the principal part of the local discretization error is independent of $\bar{x}$ (Lambert, 1973). A suitable choice of $\bar{x}$ allows us to represent the coefficients $c_i$ of the expansion as simple functions of the parameters $a_j$ and $b_j$. These can be obtained by using Taylor series expansions for $y(x_{s+j})$ and $y'(x_{s+j})$ on the implicit assumption that $y(x)$ is sufficiently many times continuously differentiable. For the suitable choice $\bar{x} = x_s$ we have

$$y(x_{s+j}) = y(\bar{x} + jh) = \sum_{l=0}^{q} \frac{(jh)^l}{l!} y^{(l)}(\bar{x}) + R_{q+1},$$
$$y'(x_{s+j}) = y'(\bar{x} + jh) = \sum_{l=0}^{q-1} \frac{(jh)^l}{l!} y^{(l+1)}(\bar{x}) + \bar{R}_q. \tag{9.95}$$

From (9.94) and (9.95) we deduce the equations

$$c_0 = a_0 + a_1 + a_2 + \cdots + a_m$$
$$c_1 = a_1 + 2a_2 + \cdots + ma_m - (b_0 + b_1 + b_2 + \cdots + b_m)$$
$$c_2 = \frac{1}{2!}(a_1 + 2^2 a_2 + \cdots + m^2 a_m) - \frac{1}{1!}(b_1 + 2b_2 + \cdots + mb_m) \tag{9.96}$$
$$\vdots$$
$$c_l = \frac{1}{l!}(a_1 + 2^l a_2 + \cdots + m^l a_m) - \frac{1}{(l-1)!}(b_1 + 2^{l-1} b_2 + \cdots + m^{l-1} b_m)$$
$$(l = 2, 3, \ldots, q).$$

For a given $m$ the parameters $a_j$ and $b_j$ can be determined in such a way that the order is maximal. By doing this we prescribe conditions concerning the structure of the desired multistep method.

*Example 9.7* We seek an explicit three-step method of maximal order

$$a_0 y_{k-2} + a_1 y_{k-1} + a_2 y_k + a_3 y_{k+1} = h(b_0 f_{k-2} + b_1 f_{k-1} + b_2 f_k)$$

such that $a_0 = a_2 = 0$. Since $a_3 = 1$, the four remaining parameters can be found as the unique solution of the four linear equations

$$c_0 = a_1 + a_3 = 0$$
$$c_1 = a_1 + 3a_3 - b_0 - b_1 - b_2 = 0$$
$$2c_2 = a_1 + 9a_3 - 2b_1 - 4b_2 = 0$$
$$6c_3 = a_1 + 27a_3 - 3b_1 - 12b_2 = 0$$

namely $a_1 = -1$, $b_0 = 1/3$, $b_1 = -2/3$, $b_2 = 7/3$. The resulting multistep method of order $p = 3$ is

$$y_{k+1} = y_{k-1} + \tfrac{1}{3}h(7f_k - 2f_{k-1} + f_{k-2}). \tag{9.97}$$

This is a *Nyström method* which is a representative of the class of explicit $m$-step methods with the property $a_0 = a_1 = \cdots = a_{m-3} = a_{m-1} = 0$ and $a_m = 1$. The $(m+1)$ free parameters $a_{m-2}, b_0, b_1, \ldots, b_{m-1}$ can be chosen in such a way that the first $(m+1)$ coefficients $c_0, c_1, \ldots, c_m$ of the expansion (9.94) vanish, so that the order of the $m$-step Nyström methods is $p = m$.

*Example* 9.8 What is the implicit three-step method of maximal order

$$a_0 y_{k-2} + a_1 y_{k-1} + a_2 y_k + a_3 y_{k+1} = h(b_0 f_{k-2} + b_1 f_{k-1} + b_2 f_k + b_3 f_{k+1})$$

with $a_0 = a_2 = 0$? Since $a_3 = 1$, the five free parameters can be determined from the first five equations of (9.96).

$$c_0 = a_1 + a_3 = 0$$
$$c_1 = a_1 + 3a_3 - b_0 - b_1 - b_2 - b_3 = 0$$
$$2c_2 = a_1 + 9a_3 - 2b_1 - 4b_2 - 6b_3 = 0$$
$$6c_3 = a_1 + 27a_3 - 3b_1 - 12b_2 - 27b_3 = 0$$
$$24c_4 = a_1 + 81a_3 - 4b_1 - 32b_2 - 108b_3 = 0.$$

From this it follows that $a_1 = -1$, $b_0 = 0$, $b_1 = \tfrac{1}{3}$, $b_2 = \tfrac{4}{3}$, $b_3 = \tfrac{1}{3}$. As $b_0 = 0$, we obtain an implicit two-step method that is of order 4.

$$y_{k+1} = y_{k-1} + \tfrac{1}{3}h(f(x_{k+1}, y_{k+1}) + 4f_k + f_{k-1}) \tag{9.98}$$

We recognize in (9.98) Simpson's quadrature rule (8.55). (9.98) is a *Milne–Simpson* method belonging to the class of implicit $m$-step methods that are characterized by $a_0 = a_1 = \cdots = a_{m-3} = a_{m-1} = 0$ and $a_m = 1$. The $(m+2)$ free parameters $a_{m-2}, b_0, b_1, \ldots, b_m$ allow us to construct Milne–Simpson methods of order $p = (m+1)$.

*Example* 9.9 Finally, we look for an explicit two-step method of maximal order

$$a_0 y_{k-1} + a_1 y_k + a_2 y_{k+1} = h(b_0 f_{k-1} + b_1 f_k)$$

## 9.2 Multistep Methods

with no restrictions on the coefficients. With four free parameters we can attain the order $p = 3$, if they satisfy the system of linear equations

$$c_0 = a_0 + a_1 + a_2 = 0$$
$$c_1 = a_1 + 2a_2 - b_0 - b_1 = 0$$
$$2c_2 = a_1 + 4a_2 - 2b_1 = 0$$
$$6c_3 = a_1 + 8a_2 - 3b_1 = 0$$

Its solution $a_0 = -5$, $a_1 = 4$, $a_2 = 1$, $b_0 = 2$, $b_1 = 4$ yields the two-step method

$$y_{k+1} = 5y_{k-1} - 4y_k + h(4f_k + 2f_{k-1}). \tag{9.99}$$

Although its order is $p = 3$, it will be shown that the method defined by (9.99) is not useful because it violates a fundamental condition which we will find out about later on.

As an auxiliary means for investigation we define the *first* and the *second characteristic polynomials* of the multistep method, (9.93), as

$$\rho(z) := \sum_{j=0}^{m} a_j z^j, \quad \sigma(z) := \sum_{j=0}^{m} b_j z^j. \tag{9.100}$$

Moreover, we introduce the following notion, similar to Definition 9.1.

**Definition 9.7** *A multistep method, (9.93), is consistent if its order, p, is at least one.*

From (9.96) a linear multistep method (9.93) is consistent if and only if the two characteristic polynomials satisfy the equations

$$c_0 = \rho(1) = 0 \quad c_1 = \rho'(1) - \sigma(1) = 0. \tag{9.101}$$

These equations will be important later on.

Now we want to understand, by means of a plausible consideration, that the consistency of a multistep method is a necessary condition for the convergence, as $h \to 0$, of the generated approximations $y_k$ to $y(x)$, the value of the solution of the differential equation. Let $\bar{x}$ be a fixed abscissa. We choose it equal to $x_s = x_{k-m+1}$. Then we consider (9.93) subject to the initial condition $y(\bar{x}) = y_s$ and furthermore we assume that the values $y_{s+j}$, $j = 1, 2, \ldots, m$, converge in the following manner towards $y(\bar{x})$ as $h \to 0$

$$y_{s+j} = y(\bar{x}) + O(h^r) \quad r \geq 1 \quad (j = 1, 2, \ldots, m). \tag{9.102}$$

After substituting (9.102) into (9.93) it follows that as $h \to 0$

$$\sum_{j=0}^{m} a_j y(\bar{x}) = 0$$

and because $y(\bar{x})$ is an arbitrary value, in particular $y(\bar{x}) \neq 0$, the first condition of (9.101) must necessarily be satisfied. This condition follows from the sole assumption that the approximate values $y_k$, at a fixed point $\bar{x}$, converge towards the function value $y(\bar{x})$. Now we also use the assumption that $y(x)$ is a solution of the differential equation. We have for the difference quotients

$$\lim_{h \to 0} \frac{y_{s+j} - y_s}{jh} = y'(\bar{x}) \quad (j = 1, 2, \ldots, m)$$

and we assume representations of the kind

$$y_{s+j} = y(\bar{x}) + jhy'(\bar{x}) + O(h^q) \quad q \geq 2 \quad (j = 1, 2, \ldots, m). \tag{9.103}$$

Substitution of (9.103) into (9.93) yields

$$\sum_{j=0}^{m} a_j y_{s+j} = y(\bar{x}) \sum_{j=0}^{m} a_j + hy'(\bar{x}) \sum_{j=1}^{m} j a_j + O(h^q) = h \sum_{j=0}^{m} b_j f(x_{s+j}, y_{s+j}).$$

The coefficient of $y(\bar{x})$ vanishes, and we can divide the last equation by $h > 0$, and then let $h \to 0$. As $\lim_{h \to 0} f(x_{s+j}, y_{s+j}) = f(\bar{x}, y(\bar{x}))$, for $j = 1, 2, \ldots m$, it follows that

$$y'(\bar{x}) \sum_{j=1}^{m} j a_j = f(\bar{x}, y(\bar{x})) \sum_{j=0}^{m} b_j.$$

From $y'(\bar{x}) = f(\bar{x}, y(\bar{x}))$ the second condition of consistency $\rho'(1) = \sigma(1)$ must necessarily be satisfied.

All multistep methods considered so far are consistent, because their orders are at least equal to one. However, the property of consistency is not sufficient for a multistep method to have the convergence property that was assumed above. We investigate the situation with the aid of the linear *test initial value problem*

$$y'(x) = \lambda y(x) \quad y(0) = 1 \quad \lambda \in \mathbf{R} \quad \lambda < 0 \tag{9.104}$$

with the solution $y(x) = e^{\lambda x}$. We apply the method (9.93) to the test differential equation and after replacing $f_l = \lambda y_l$ obtain

$$\sum_{j=0}^{m} (a_j - h\lambda b_j) y_{s+j} = 0. \tag{9.105}$$

For $m = 3$ the equation (9.105) reads explicitly

$$(a_0 - h\lambda b_0) y_{k-2} + (a_1 - h\lambda b_1) y_{k-1} + (a_2 - h\lambda b_2) y_k + (a_3 - h\lambda b_3) y_{k+1} = 0. \tag{9.106}$$

Hence the approximations $y_k$ that are computed by means of the $m$-step method satisfy the $m$th-order linear, homogeneous *difference equation* (9.105). Their coefficients are constant for a fixed step size $h$. The following consideration is based upon the third-order difference equation (9.106). Its general solution is

## 9.2 Multistep Methods

determined with the aid of trial functions of the form $y_k = z^k$, $z \neq 0$. By substituting this into (9.106) and subsequently dividing by $z^{k-2} \neq 0$, for $z$ we obtain the following algebraic equation of degree three

$$(a_0 - h\lambda b_0) + (a_1 - h\lambda b_1)z + (a_2 - h\lambda b_2)z^2 + (a_3 - h\lambda b_3)z^3 = 0 \tag{9.107}$$

which can be written by means of the two characteristic polynomials of the multistep method

$$\varphi(z) := \rho(z) - h\lambda\sigma(z) = 0. \tag{9.108}$$

(9.107) and (9.108) are called the *characteristic equation* of the corresponding difference equation. The equation (9.107) has three solutions $z_1, z_2, z_3$ which we assume to be pairwise different for the time being. Then $z_1^k, z_2^k$ and $z_3^k$ form a fundamental system of independent solutions of the difference equation (9.106). Due to the linearity of (9.106), the general solution is

$$y_k = c_1 z_1^k + c_2 z_2^k + c_3 z_3^k, \quad \text{where } c_1, c_2, c_3 \text{ are arbitrary.} \tag{9.109}$$

The constants $c_1, c_2, c_3$ are determined from the initial values $y_0, y_1, y_2$ that must be known if a 3-step method is used. The equations following from these conditions are

$$\begin{aligned} c_1 + c_2 + c_3 &= y_0 \\ z_1 c_1 + z_2 c_2 + z_3 c_3 &= y_1 \\ z_1^2 c_1 + z_2^2 c_2 + z_3^2 c_3 &= y_2 \end{aligned} \tag{9.110}$$

On the assumption $z_i \neq z_j$ for $i \neq j$ the system of linear equations (9.110) has a unique solution, because its *Vandermonde determinant* is nonzero. With the coefficients $c_i$ determined from (9.110) we have an explicit expression for the approximations $y_k$ ($k \geq 3$) of the solution of the test differential equation, and we can discuss their qualitative behaviour as $h \to 0$ and $k \to \infty$. We have assumed $\lambda < 0$, therefore the exact solution $y(x)$ decreases exponentially. Of course, this should be the case for the computed values $y_k$, especially for arbitrarily small step sizes $h$. From (9.109) this is true if and only if all zeros $z_i$ of the characteristic equation (9.108) are in absolute value less than one. However as $h \to 0$ the solutions $z_i$ of $\varphi(z) = 0$ become the zeros of the first characteristic polynomial $\rho(z)$, due to (9.108), and consequently these zero are not allowed to lie outside the closed unit disc of the complex plane. This necessary condition for the usefulness of a multistep method for integrating the test initial value problem has been shown on the simplifying assumption of pairwise different solutions $z_i$ of (9.108).

For instance in the case of a six-step method if the solutions are $z_1, z_2 = z_3$, $z_4 = z_5 = z_6$, but otherwise distinct, the general solution of the corresponding difference equation has the form (Henrici 1962, 1972)

$$y_k = c_1 z_1^k + c_2 z_2^k + c_3 k z_2^k + c_4 z_4^k + c_5 k z_4^k + c_6 k(k-1) z_4^k. \tag{9.111}$$

The representation (9.111) again implies that $y_k$ decreases for increasing $k$ if and only if the solutions $z_i$ have moduli smaller than one. As $h \to 0$ no multiple zeros are allowed to have moduli equal to one. These considerations motivate the following definition.

**Definition 9.8** *A multistep method* (9.93) *is zero-stable if the zeros $z_i$ of the first characteristic polynomial $\rho(z)$*
(a) *are in absolute value at most equal to one, and*
(b) *multiple zeros are located in the interior of the unit disc.*

All $m$-step Adams–Bashforth and Adams–Moulton methods satisfy the zero-stability condition because their first characteristic polynomials $\rho(z) = z^m - z^{m-1} = z^{m-1}(z-1)$ have the simple zero $z_1 = 1$ and the zero $z_2 = z_3 = \cdots = z_m = 0$ of multiplicity $(m-1)$. The same is true for the Nyström and the Milne–Simpson methods because their first characteristic polynomials $\rho(z) = z^m - z^{m-2} = z^{m-2}(z^2 - 1)$ have two simple zeros $z_1 = 1$ and $z_2 = -1$ on the unit circle and the zero $z_3 = \cdots = z_m = 0$ of multiplicity $(m-2)$ at the origin.

On the other hand the multistep method (9.99) with the first characteristic polynomial $\rho(z) = z^2 + 4z - 5$, has the zeros $z_1 = 1$ and $z_2 = -5$, and is not zero-stable. To illustrate the situation further in Table 9.8, for $\lambda = -1$ and for several values of the step size $h$, the zeros $z_1$ and $z_2$ of the characteristic equation $\varphi(z) = z^2 + 4(1+h)z - (5-2h) = 0$ together with the coefficients $c_1$ and $c_2$ of the special solution $y_k = c_1 z_1^k + c_2 z_2^k$ of the difference equation are given, whereby the exact initial values $y_0 = 1$ and $y_1 = e^{-h}$ have been used. The numbers show that the second term of $y_k$ increases in an oscillating way as $k \to \infty$, whereas the first term would represent the desired solution qualitatively because $z_1$ is a good approximation of $e^{-h}$.

It can be shown that consistent and zero-stable multistep methods are convergent (Henrici, 1962), that is the computed approximations $y_k$ at a fixed abscissa $\bar{x} = x_0 + kh$ converge towards the value $y(\bar{x})$ of the solution as $h \to 0$ if $kh = \bar{x} - x_0$.

We finish the treatment of multistep methods by relating the local and the

Table 9.8 Instability of the method (9.99)

| $h$ | $e^{-h}$ | $z_1$ | $z_2$ | $c_1$ | $c_2$ |
|---|---|---|---|---|---|
| 0.4 | 0.670 320 046 | 0.669 870 315 | −6.269 870 | 1.000 064 805 | $-6.481 \times 10^{-5}$ |
| 0.2 | 0.818 730 753 | 0.818 695 388 | −5.618 695 | 1.000 005 494 | $-5.494 \times 10^{-6}$ |
| 0.1 | 0.904 837 418 | 0.904 834 939 | −5.304 835 | 1.000 000 399 | $-3.992 \times 10^{-7}$ |
| 0.05 | 0.951 229 425 | 0.951 229 261 | −5.151 229 | 1.000 000 027 | $-2.688 \times 10^{-8}$ |
| 0.02 | 0.980 198 673 | 0.980 198 669 | −5.060 199 | 1.000 000 001 | $-7.195 \times 10^{-10}$ |
| 0.01 | 0.990 049 834 | 0.990 049 834 | −5.030 050 | 1.000 000 000 | $-4.634 \times 10^{-11}$ |
| ↓ | ↓ | ↓ | ↓ | | |
| 0 | 1 | 1 | −5 | | |

## 9.2 Multistep Methods

global discretization errors. In order to simplify the analysis that is quite involved for the general case, we restrict ourselves to explicit $m$-step methods with the additional property

$$a_j \leq 0 \quad (j = 0, 1, 2, \ldots, m-1) \quad a_m = 1. \tag{9.112}$$

The class of zero-stable multistep methods considered above have this property, with either $a_{m-1} = -1$ or $a_{m-2} = -1$. Let us define the maximal absolute error between the $m$ initial values $y_0, y_1, \ldots, y_m$ to be prescribed and the exact values of the solution $y(x_0), y(x_1), \ldots, y(x_m)$ to be

$$\max_{0 \leq i \leq m-1} |y(x_i) - y_i| =: G < \infty. \tag{9.113}$$

**Theorem 9.3** *Let $g_n = y(x_n) - y_n$ denote the global discretization error at the abscissa $x_n = x_0 + nh$ of an explicit $m$-step method (9.93). If the property (9.112) holds, then the estimate*

$$|g_n| \leq \left(G + \frac{D}{hLB}\right) e^{nhLB} \quad n \geq m, \tag{9.114}$$

*holds, where $G$ is the maximal error (9.113) of the initial values, $D = \max_{m \leq k \leq n} |d_k|$ denotes the maximal absolute value of the local discretization errors $d_k$, $L$ is the Lipschitz constant of the function $f(x, y)$, and $B$ is defined by*

$$B := \sum_{j=0}^{m-1} |b_j|. \tag{9.115}$$

*Proof* The explicit multistep methods under consideration are

$$y_{k+1} = \sum_{j=0}^{m-1} [-a_j y_{s+j} + h b_j f(x_{s+j}, y_{s+j})]. \tag{9.116}$$

The definition (9.94) of the local discretization error $d_{k+1}$ yields

$$y(x_{k+1}) = \sum_{j=0}^{m-1} [-a_j y(x_{s+j}) + h b_j f(x_{s+j}, y(x_{s+j}))] + d_{k+1}. \tag{9.117}$$

We subtract (9.116) from (9.117), take the absolute value and use the Lipschitz continuity of $f(x, y)$ with the Lipschitz constant $L$, to obtain the estimate for the global discretization error

$$|g_{k+1}| \leq \sum_{j=0}^{m-1} \{|a_j| + hL|b_j|\} |g_{s+j}| + |d_{k+1}| \quad (k \geq m-1). \tag{9.118}$$

The bound of $|g_{k+1}|$ depends on the $m$ previous global discretization errors. To treat the difference inequality (9.118) further we define

$$C_j := |a_j| + hL|b_j| \quad (j = 0, 1, \ldots, m-1) \quad C := \sum_{j=0}^{m-1} C_j. \tag{9.119}$$

Due to (9.112) and the consistency condition

$$\sum_{j=0}^{m-1} a_j = -a_m = -1 \quad \text{we have } C = 1 + hLB \geq 1.$$

Hence from (9.118) we get

$$|g_{k+1}| \leq \sum_{j=0}^{m-1} C_j |g_{s+j}| + D \quad (k = m-1, m, \ldots). \tag{9.120}$$

We now apply (9.120) successively for $k = m-1, m, m+1$ in order to find the general formula

$$|g_m| \leq \sum_{j=0}^{m-1} C_j |g_j| + D \leq G \sum_{j=0}^{m-1} C_j + D = GC + D$$

$$|g_{m+1}| \leq \sum_{j=0}^{m-1} C_j |g_{j+1}| + D \leq G \sum_{j=0}^{m-2} C_j + C_{m-1}(GC + D) + D$$

$$\leq (GC + D) \sum_{j=0}^{m-1} C_j + D = C^2 G + (C + 1)D$$

$$|g_{m+2}| \leq \sum_{j=0}^{m-1} C_j |g_{j+2}| + D \leq [C^2 G + (C+1)D]C + D = C^3 G + (C^2 + C + 1)D$$

By induction we can show the following estimate:

$$|g_{m+l}| \leq C^{l+1} G + D \sum_{\mu=0}^{l} C^\mu \leq C^{m+l} G + D \sum_{\mu=0}^{m+l-1} C^\mu \quad (l \geq 0). \tag{9.121}$$

The second, larger bound in (9.121) is obtained by increasing the exponent, $C \geq 1$, and by including some additional terms in the sum. If we set $m + l = n$ in (9.121)

$$|g_n| \leq C^n G + D \sum_{\mu=0}^{n-1} C^\mu = C^n G + D \frac{C^n - 1}{C - 1}$$

$$= (1 + hLB)^n G + \frac{D}{hLB} [(1 + hLB)^n - 1]$$

$$\leq G e^{nhLB} + \frac{D}{hLB} (e^{nhLB} - 1) \leq \left( G + \frac{D}{hLB} \right) e^{nhLB}.$$

An analogous result to Theorem 9.3 can be shown for general and also implicit multistep methods (Henrici, 1962). The estimate (9.114) tells us two things. If we assume that the $m$ initial values are exact, so that $G = 0$, and that the multistep method is of order $p$, so that $D = \text{const} \cdot h^{p+1}$, then from (9.114) we have

$$|g_n| \leq \frac{\text{const}}{LB} h^p e^{nhLB} = O(h^p). \tag{9.122}$$

## 9.3 Stability

In this case the global discretization error $g_n$ is proportional to $h^p$, and in fact the method has the *order of convergence p*. However, the initial values $y_0, y_1, \ldots, y_{m-1}$ are usually only approximations and in this case the result (9.114) then reveals the relevant fact that these initial errors have a decisive influence on the global discretization error. The estimate (9.122) remains true for $G \neq 0$ only if

$$\max_{0 \leq i \leq m-1} |y(x_i) - y_i| = \max_{0 \leq i \leq m-1} |g_i| = G = O(h^p). \tag{9.123}$$

Therefore the $m$ initial values $y_0, y_1, \ldots, y_{m-1}$ must be given with the same order of accuracy, $p$, as is the order of the $m$-step method used to integrate a differential equation numerically. If a one step method is used to produce these values, it must be of the same order. This rule has been respected in Example 9.6.

## 9.3 Stability

The choice of a suitable procedure to solve a system of first-order differential equations approximately requires some care, because the properties of the given differential equations and of the resulting solution functions must be taken into account. If we do not proceed carefully it may well happen that the computed approximations have very little to do with the desired solution functions, or may even be meaningless. In the following we shall analyze instabilities that occur in the case of an inadequate use of the methods that must be avoided (Dahlquist, 1985; Rutishauser, 1952).

### 9.3.1 Inherent instability

We investigate how the solution $y(x)$ depends upon the initial value $y(x_0)$ by means of a class of differential equations whose set of solutions can be given explicitly. The initial value problem is

$$y'(x) = \lambda(y(x) - F(x)) + F'(x) \quad y(x_0) = y_0, \tag{9.124}$$

where $F(x)$ is a function that is at least continuously differentiable on an interval containing $x_0$. As $y_{\text{hom}}(x) = Ce^{\lambda x}$ is the general solution of the homogeneous differential equation and $y_{\text{part}}(x) = F(x)$ is a particular solution of the non-homogeneous differential equation, the solution of (9.124) is

$$y(x) = (y_0 - F(x_0)) \exp[\lambda(x - x_0)] + F(x). \tag{9.125}$$

For the special initial condition $y(x_0) = F(x_0)$ we have $y(x) = F(x)$, and the exponential term does not occur. For the slightly perturbed initial value $\hat{y}_0 = F(x_0) + \varepsilon$, where $\varepsilon$ is small in absolute value, the solution is

$$\hat{y}(x) = \varepsilon \exp[\lambda(x - x_0)] + F(x). \tag{9.126}$$

Let $\lambda \in \mathbf{R}$ be positive. Then the first term of $\hat{y}(x)$ increases exponentially with increasing $x$, so that the neighbouring solution $\hat{y}(x)$ deviates from $y(x)$ more and more. Therefore we can say that the solution $y(x)$ is highly sensitive to small changes, $\varepsilon$, of the initial condition. The problem posed is said to be called ill-conditioned. The sensitivity of the solution completely depends upon the given initial value problem, we call this phenomenon *inherent instability*. It manifests itself in such a way that the computed approximations $y_n$ will depart from the exact values $y(x_n) = F(x_n)$ according to (9.126) in an exponential manner. The inherent instability can only be tackled by means of high order methods combined with a high precision in order to keep the discretization and the rounding errors small.

*Example* 9.10 We consider the initial value problem

$$y'(x) = 10\left(y(x) - \frac{x^2}{1+x^2}\right) + \frac{2x}{(1+x^2)^2} \quad y(0) = y_0 = 0$$

of the type (9.124) with the solution $y(x) = x^2/(1 + x^2)$. The classical fourth-order Runge–Kutta method (9.52) yields, with the step size $h = 0.01$, the approximation $\hat{y}(x)$ which is represented in Figure 9.7 together with $y(x)$ on the interval $[0, 2.2]$. The inherent instability is obvious.

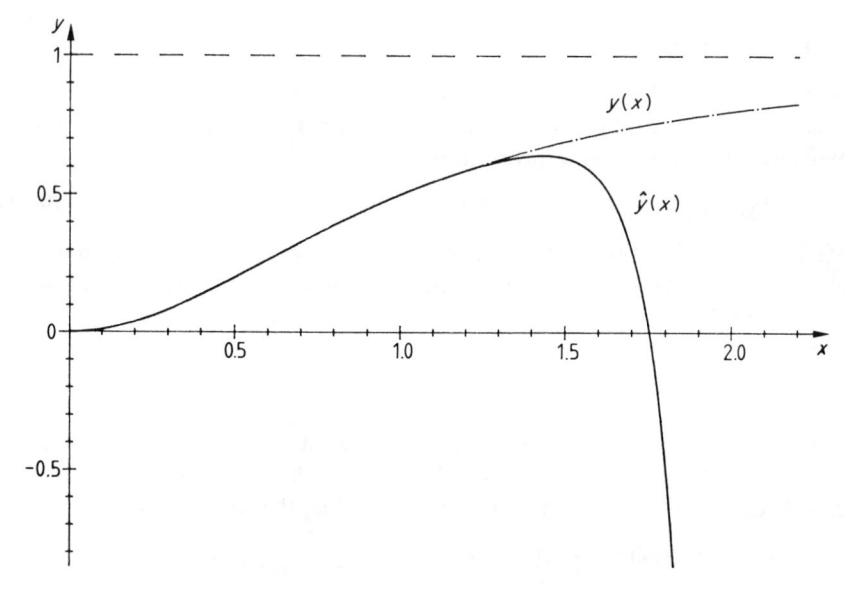

Figure 9.7 Inherent instability

## 9.3 Stability

### 9.3.2 Absolute stability

Once more we consider the linear test initial value problem

$$y'(x) = \lambda y(x) \quad y(0) = 1 \quad \lambda \in \mathbf{R} \quad \text{or} \quad \lambda \in \mathbf{C} \tag{9.127}$$

with the known solution $y(x) = e^{\lambda x}$. The following results remain valid for nonlinear differential equations because they can be locally approximated by a linear differential equation after a linearization with respect to $y$. For an integration step with the small step size $h$, the approximations qualitatively have the same behaviour.

A typical representative of the one-step methods is the classical fourth-order Runge–Kutta method, (9.52). We investigate how it works by applying it to the initial value problem (9.127). We get successively

$$k_1 = \lambda y_k$$
$$k_2 = \lambda(y_k + \tfrac{1}{2}hk_1) = (\lambda + \tfrac{1}{2}h\lambda^2)y_k$$
$$k_3 = \lambda(y_k + \tfrac{1}{2}hk_2) = (\lambda + \tfrac{1}{2}h\lambda^2 + \tfrac{1}{4}h^2\lambda^3)y_k$$
$$k_4 = \lambda(y_k + hk_3) = (\lambda + h\lambda^2 + \tfrac{1}{2}h^2\lambda^3 + \tfrac{1}{4}h^3\lambda^4)y_k$$
$$y_{k+1} = y_k + \tfrac{1}{6}h(k_1 + 2k_2 + 2k_3 + k_4) = (1 + h\lambda + \tfrac{1}{2}h^2\lambda^2 + \tfrac{1}{6}h^3\lambda^3$$
$$\qquad + \tfrac{1}{24}h^4\lambda^4)y_k \tag{9.128}$$

According to (9.128) $y_{k+1}$ is produced from $y_k$ by a multiplication with the factor

$$F(h\lambda) := 1 + h\lambda + \tfrac{1}{2}h^2\lambda^2 + \tfrac{1}{6}h^3\lambda^3 + \tfrac{1}{24}h^4\lambda^4 \tag{9.129}$$

which depends on the product $h\lambda$ and obviously approximates $e^{h\lambda}$, with an error $O(h^5)$. For the solution, we have $y(x_{k+1}) = e^{h\lambda} y(x_k)$, and hence the previous statement agrees with the fact that the local discretization error of the classical Runge–Kutta method is $d_{k+1} = O(h^5)$. The multiplier $F(h\lambda)$ in (9.129) is, in fact, a good approximation of $e^{h\lambda}$ for small absolute values of $h\lambda$.

Let us first consider the simpler case of a real $\lambda$. For $\lambda > 0$ we have $h\lambda > 0$ and $F(h\lambda) > 1$. This means that the approximations, $y_n$, are computed qualitatively in a correct way. This case is not very interesting because the processes of natural sciences that are described by differential equations usually contain components that decrease exponentially. This is the case for $\lambda < 0$. The computed approximate solution $y_n$ decreases like $y(x_n)$ if and only if the condition $|F(h\lambda)| < 1$ is satisfied. As $F(h\lambda)$ is a polynomial of degree four in $h\lambda$ this condition cannot be satisfied for all negative values of $h\lambda$. Systems of differential equations often have solutions consisting of components that decay exponentially but in an oscillating way. Such components correspond to complex values $\lambda$. Now the solution $y(x)$ is complex, and again we have the relationship $y(x_{k+1}) = e^{h\lambda} y(x_k)$. The complex multiplier $e^{h\lambda}$ has, in the only case of real interest $\text{Re}(\lambda) < 0$, a modulus less than one. The necessary and sufficient condition for the computed approximations $y_n$ to decrease in absolute value like $y(x_n)$ is therefore given by $|F(h\lambda)| < 1$.

Similar conditions must be satisfied for all explicit Runge–Kutta methods. A simple calculation shows that for all explicit $p$-stage Runge–Kutta methods of order $p \leqslant 4$ applied to the test initial value problem, (9.127), the multiplier $F(h\lambda)$ is always equal to the first $(p+1)$ terms of the Taylor series of $e^{h\lambda}$. Runge–Kutta methods of higher order $p$ require $m > p$ stages so that $F(h\lambda)$ is a polynomial of degree $m$ which agrees with the first $(p+1)$ terms of the Taylor series of $e^{h\lambda}$. The coefficients of the polynomial, up to the exponent $m$, depend on the specific method.

The necessary and sufficient condition is expressed by means of following definition.

**Definition 9.9** *For a one-step method, which for the test initial value problem* (9.127) *leads to* $y_{k+1} = F(h\lambda)y_k$, *the set*

$$B := \{\mu \in \mathbf{C} \mid |F(\mu)| < 1\} \tag{9.130}$$

*is called the region of absolute stability.*

Consequently the step size $h > 0$ must be chosen in such a way that for $\text{Re}(\lambda) < 0$ we have $h\lambda \in B$. If this condition is violated the procedure produces meaningless results, that is, the procedure is unstable. This *stability condition* must be taken into consideration especially in the case where systems of differential equations have to be solved, because then the step size $h$ must be chosen in such a way that for all decay constants $\lambda_j$ with $\text{Re}(\lambda_j) < 0$ the conditions $h\lambda_j \in B$ are all simultaneously satisfied.

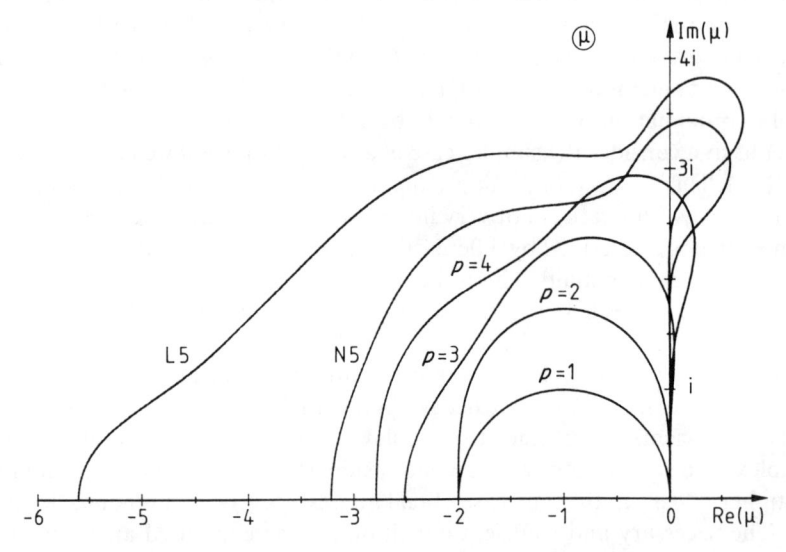

Figure 9.8 Stability regions for explicit Runge–Kutta methods

## 9.3 Stability

Figure 9.8 shows the boundaries of the regions of absolute stability for explicit Runge–Kutta methods of the orders $p = 1, 2, 3, 4, 5$. The regions are symmetric with respect to the real axis therefore the boundaries are only given for the upper half of the complex plane. In the case $p = 5$ the boundary N5 is given for the special Nyström Runge–Kutta method (Grigorieff, 1972) which has the property that $F(h\lambda)$ agrees with the first six terms of the Taylor series of $e^{h\lambda}$.

The regions of absolute stability become larger with increasing order. The region of stability of the Euler method is the interior of a unit disc whose centre is $\mu = -1$. A measure of the size of the stability region is the *stability interval* comprising the real negative values $\mu$. Table 9.9 contains the information on the stability intervals of the methods of the orders $p = 1, 2, 3, 4, 5$ whereby in the last case Nyström's method is assumed.

Lawson (1966) has derived a fifth-order six-stage Runge–Kutta method with the extremely large stability interval $(-5.60, 0)$. The corresponding boundary of the stability region is drawn in Figure 9.8 and is labelled by L5.

We now turn to the implicit one-step methods to which the trapezoidal method and the implicit Runge–Kutta methods belong. The trapezoidal method (9.33) applied to the test initial value problem (9.127) leads us to the explicit computational scheme

$$y_{k+1} = y_k + \tfrac{1}{2}h(\lambda y_k + \lambda y_{k+1}) \quad \text{hence} \quad y_{k+1} = \frac{1 + \tfrac{1}{2}h\lambda}{1 - \tfrac{1}{2}h\lambda} y_k =: F(h\lambda) y_k.$$

(9.131)

The decisive function $F(h\lambda)$ is rational with the property

$$|F(\mu)| = \left|\frac{2 + \mu}{2 - \mu}\right| < 1 \quad \text{for all } \mu \text{ with } \mathrm{Re}(\mu) < 0.$$

(9.132)

This is due to the fact that in absolute value the real part of the numerator is always less than the real part of the denominator for all $\mathrm{Re}(\mu) < 0$ and the imaginary parts have opposite signs and are equal in size. Therefore the region of absolute stability comprises the whole left half-plane. The trapezoidal method is said to be *absolutely stable* because no bounds on the step size $h$ have to be taken into account in order to obtain a stable integration. The problem-oriented choice of $h$ will be treated in connection with systems of stiff differential equations.

Table 9.9 Stability intervals of Runge–Kutta methods

| $p$ | 1 | 2 | 3 | 4 | 5 |
|---|---|---|---|---|---|
| Interval | $(-2.0, 0)$ | $(-2.0, 0)$ | $(-2.51, 0)$ | $(-2.78, 0)$ | $(-3.21, 0)$ |

For the one-stage implicit Runge–Kutta method, (9.60), we obtain

$$k_1 = \lambda(y_k + \tfrac{1}{2}hk_1) \quad \text{hence} \quad k_1 = \frac{\lambda}{1 - \tfrac{1}{2}h\lambda} y_k$$

$$y_{k+1} = y_k + hk_1 = \frac{1 + \tfrac{1}{2}h\lambda}{1 - \tfrac{1}{2}h\lambda} y_k =: F(h\lambda) y_k.$$

$F(h\lambda)$ is identical to that of the trapezoidal method, thus the implicit Runge–Kutta method (9.60) is also absolutely stable.

When we apply the two-stage implicit Runge–Kutta method, (9.62), to the test differential equation we have to solve the two linear equations for $k_1$ and $k_2$ and then substitute the expressions into the formula for $y_{k+1}$. A simple calculation yields the result

$$y_{k+1} = \frac{1 + \tfrac{1}{2}h\lambda + \tfrac{1}{12}h^2\lambda^2}{1 - \tfrac{1}{2}h\lambda + \tfrac{1}{12}h^2\lambda^2} y_k =: F(h\lambda) y_k. \tag{9.133}$$

The rational function $F(h\lambda) = F(\mu)$ defined by (9.133), similar to (9.131), is a so-called *Padé approximation* of $e^\mu$. It has the property $|F(\mu)| < 1$ for all $\mu$ with $\text{Re}(\mu) < 0$. Therefore (9.62) is also an absolutely stable method. Without using the properties of a Padé approximation we can verify the statement of the absolute stability by determining the boundary of the region of absolute stability. The boundary is defined by the points $\mu \in \mathbb{C}$ satisfying $|F(\mu)| = 1$, and hence the equation

$$1 + \tfrac{1}{2}\mu + \tfrac{1}{12}\mu^2 = e^{i\theta}(1 - \tfrac{1}{2}\mu + \tfrac{1}{12}\mu^2) \quad 0 \leq \theta \leq 2\pi$$

must hold. Consequently, $\mu$ has to be a solution of the following quadratic equation

$$\mu^2 + 6\frac{1 + e^{i\theta}}{1 - e^{i\theta}}\mu + 12 = 0 \quad \text{or} \quad \mu^2 + \frac{6i \sin\theta}{1 - \cos\theta}\mu + 12 = 0. \tag{9.134}$$

It is readily seen that (9.134) has only purely imaginary solutions, so that the imaginary axis is the boundary of the stability region. For real negative values of $\mu$ we obviously have $|F(\mu)| < 1$.

The same stability problem exists for linear multistep methods. The considerations of Section 9.2.3 that ended with the notion of the zero-stability show that the approximations $y_k$ of a general $m$-step method, (9.93), for the test initial value problem, (9.127), satisfy the linear difference equation (9.105) of order $m$. Its general solution is given by

$$y_k = c_1 z_1^k + c_2 z_2^k + \cdots + c_m z_m^k, \tag{9.135}$$

where the $m$ zeros $z_1, z_2, \ldots, z_m$ of the characteristic equation $\varphi(z) = \rho(z) - h\lambda\sigma(z) = 0$ are assumed to be pairwise different for simplicity. The general solution, (9.135), only decreases in the interesting case $\text{Re}(\lambda) < 0$ if and only if

## 9.3 Stability

all $z_j$ have a modulus less than one. The same condition must be satisfied if multiple zeros $z_j$ occur (see (9.111)).

**Definition 9.10** *For a linear multistep method, (9.93), the set of complex values $\mu = h\lambda$, for which the characteristic equation $\varphi(z) = \rho(z) - h\lambda\sigma(z) = 0$ only has zeros $z_j \in \mathbf{C}$ in the interior of the unit disc, is called the region of absolute stability.*

The explicit four-step *Adams–Bashforth* method, (9.81), has the characteristic equation, after multiplication by 24,

$$24\varphi(z) = 24z^4 - (24 + 55\mu)z^3 + 59\mu z^2 - 37\mu z + 9\mu = 0. \tag{9.136}$$

The boundary of the stability region of multistep methods can easily be determined as the geometric locus of the values $\mu \in \mathbf{C}$ for which $|z| = 1$ holds. Hence it is sufficient to calculate for $z = e^{i\theta}$, $0 \leq \theta \leq 2\pi$ the corresponding values $\mu$. In the case of the characteristic equation (9.136) we obtain the explicit expression

$$\mu = \frac{24z^4 - 24z^3}{55z^3 - 59z^2 + 37z - 9} = \frac{\rho(z)}{\sigma(z)} \quad z = e^{i\theta} \quad 0 \leq \theta \leq 2\pi.$$

Again the boundary of the stability region is obviously symmetric with respect to the real axis. Therefore it is drawn for the four-step Adams–Bashforth method (9.81) only in the upper half of the plane and is labelled by AB4. The interval of

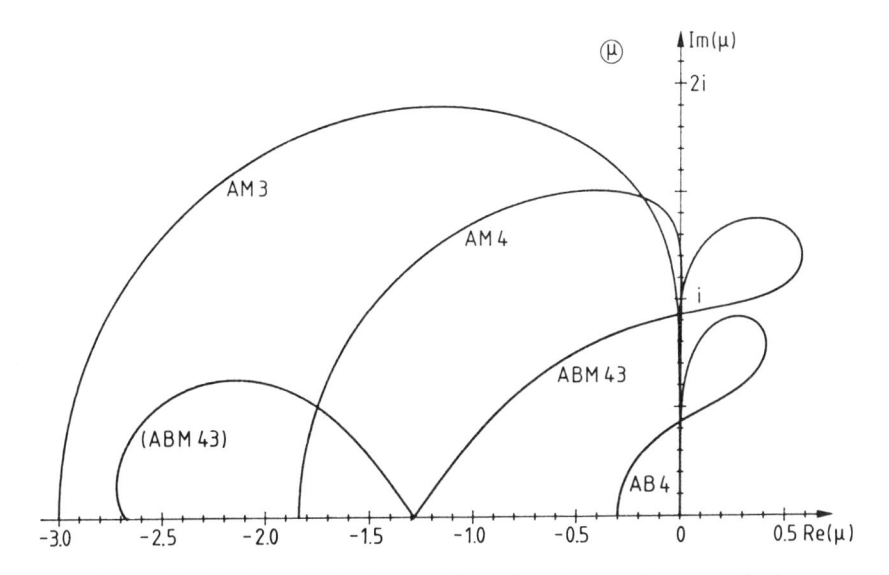

Figure 9.9 Regions of absolute stability of various multistep methods

stability $(-0.3, 0)$ is about nine times smaller than that of the fourth-order Runge–Kutta methods. The explicit Adams–Bashforth methods have in general small stability regions.

The characteristic equation of the implicit four-step *Adams–Moulton* method, (9.88), after multiplication by 720, is

$$(720 - 251\mu)z^4 - (720 + 646\mu)z^3 + 264\mu z^2 - 106\mu z + 19\mu = 0. \qquad (9.137)$$

The corresponding boundary of the region of absolute stability is drawn in Figure 9.9, too and labelled by AM4. In comparison with the explicit four-step method the stability region is much larger, and the interval of stability is $(-1.836, 0)$. Although the method is implicit, the stability region is finite, and the method is not absolutely stable.

The region of absolute stability of the implicit three-step fourth-order Adams–Moulton method (9.89), whose boundary is labelled by AM3 in Figure 9.9, is even larger. The interval of stability $(-3.0, 0)$ is even larger than that of the classical Runge–Kutta method of the same order.

Usually the Adams–Moulton method is used in connection with the Adams–Bashforth method as a predictor–corrector procedure. The method (9.92) yields, for the test initial value problem, the following predictor and corrector value

$$y_{k+1}^{(P)} = y_k + \tfrac{1}{24}h\lambda(55y_k - 59y_{k-1} + 37y_{k-2} - 9y_{k-3})$$
$$y_{k+1} = y_k + \tfrac{1}{24}h\lambda\{9[y_k + \tfrac{1}{24}h\lambda(55y_k - 59y_{k-1} + 37y_{k-2} - 9y_{k-3})]$$
$$+ 19y_k - 5y_{k-1} + y_{k-2}\}.$$

If we add the term $9y_{k+1}$ and compensate for it we obtain the difference equation for the approximation $y_k$

$$y_{k+1} - y_k - \tfrac{1}{24}h\lambda(9y_{k+1} + 19y_k - 5y_{k-1} + y_{k-2})$$
$$+ \tfrac{9}{24}h\lambda[y_{k+1} - y_k - \tfrac{1}{24}h\lambda(55y_k - 59y_{k-1} + 37y_{k-2} - 9y_{k-3})] = 0. \qquad (9.138)$$

In (9.138) we recognize the coefficients of the first and second characteristic polynomials $\rho_{AM}(z)$, $\sigma_{AM}(z)$, $\rho_{AB}(z)$ and $\sigma_{AB}(z)$ of the two methods involved. Moreover, the fraction $\tfrac{9}{24}$ can be identified to be the coefficient $b_3^{(AM)}$ of the implicit three-step Adams–Moulton method, and therefore the characteristic equation of (9.138) is

$$\varphi_{ABM}(z) = z[\rho_{AM}(z) - \mu\sigma_{AM}(z)] + b_3^{(AM)}\mu[\rho_{AB}(z) - \mu\sigma_{AB}(z)] = 0. \qquad (9.139)$$

The characteristic equation (9.139) is typical of all predictor–corrector methods. It can be used in a similar manner as above to determine the boundary of the region of absolute stability with the minor difference that for a given value of $z = e^{i\theta}$ a quadratic equation for $\mu$ must be solved. In Figure 9.9 the boundary of the stability region of the ABM method (9.92) is shown. It is labelled by ABM43 to indicate that the explicit four-step Adams–Bashforth method is combined with

## 9.3 Stability

the implicit three-step Adams–Moulton method. The stability region is smaller than that of Adams–Moulton (AM3) due to the fact that the predictor–corrector value is used instead of the exact solution $y_{k+1}$ of the implicit equation. The interval of stability of the method (9.92) is $(-1.28, 0)$.

The Nyström and Milne–Simpson methods including their combinations, to make predictor–corrector methods, are useless with respect to the absolute stability. This is due to the fact that for $h > 0$ the zero $z_2 = -1$ of the first characteristic polynomial $\rho(z)$ of all these methods becomes a solution of the characteristic equation $\varphi(z) = \rho(z) - h\lambda\sigma(z) = 0$ that is outside the unit disc whenever $\text{Re}(h\lambda) < 0$. The region of absolute stability does not contain a neighbourhood of the origin in the left half-plane. To illustrate this situation we consider the three-step Nyström method (9.97). Its characteristic equation is

$$\varphi(z) = z^3 - \tfrac{7}{3}\mu z^2 - (1 - \tfrac{2}{3}\mu)z - \tfrac{1}{3}\mu = 0.$$

Its zeros are given in Table 9.10 for some real, negative values of $\mu$.

The characteristic equation of the Milne–Simpson method (9.98) is

$$\varphi(z) = (1 - \tfrac{1}{3}\mu)z^2 - \tfrac{4}{3}\mu z - (1 + \tfrac{1}{3}\mu) = 0$$

it has the solutions $z_1 = (2\mu + \sqrt{9 + 3\mu^2})/(3 - \mu)$, $z_2 = (2\mu - \sqrt{9 + 3\mu^2})/(3 - \mu)$. We have $z_2 < -1$ for all $\mu < 0$.

Even though the Nyström or Milne–Simpson methods, considered as a closed class of integration schemes that are consistent and zero-stable, are almost useless, they may, however, be useful in certain combinations with other methods. For instance, the combination of the method (9.99) that is not even zero-stable as the predictor with the Milne–Simpson method (9.98) as the corrector formula has a surprising property (Stetter, 1965). The methods have the orders $p = 3$ and $p = 4$, respectively

$$\boxed{\begin{aligned} y_{k+1}^{(P)} &= -4y_k + 5y_{k-1} + h(4f_k + 2f_{k-1}) \\ y_{k+1} &= y_{k-1} + \tfrac{1}{3}h(f(x_{k+1}, y_{k+1}^{(P)}) + 4f_k + f_{k-1}) \end{aligned}} \quad (9.140)$$

The characteristic equation of the predictor–corrector method (9.140) is

$$\begin{aligned} \varphi(z) &= (z^2 - 1) - \tfrac{1}{3}\mu(z^2 + 4z + 1) + \tfrac{1}{3}\mu[(z^2 + 4z - 5) - \mu(4z + 2)] \\ &= z^2 - \tfrac{4}{3}\mu^2 z - (1 + 2\mu + \tfrac{2}{3}\mu^2) = 0. \end{aligned}$$

Table 9.10 Absolute instability of the Nyström method

| $\mu$ | $-0.01$ | $-0.02$ | $-0.05$ | $-0.10$ | $-0.20$ | $-0.30$ |
|---|---|---|---|---|---|---|
| $z_1$ | 0.9900 | 0.9802 | 0.9512 | 0.9048 | 0.8185 | 0.7396 |
| $z_2$ | $-1.0167$ | $-1.0334$ | $-1.0841$ | $-1.1697$ | $-1.3457$ | $-1.5281$ |
| $z_3$ | 0.0033 | 0.0066 | 0.0162 | 0.0315 | 0.0605 | 0.0885 |

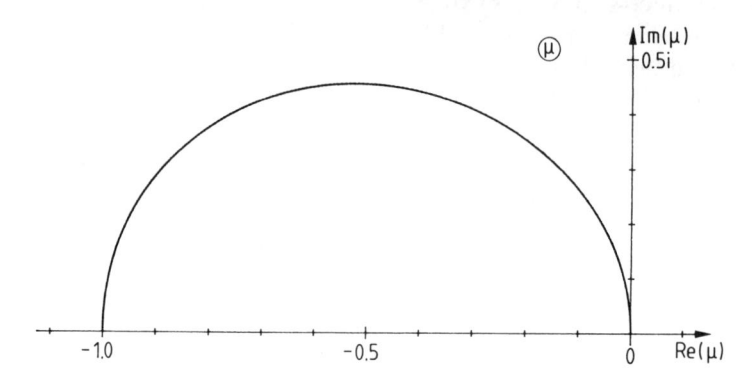

Figure 9.10 Region of absolute stability of the method (9.140)

The boundary of the region of absolute stability that is relevant for $\operatorname{Re}(\mu) < 0$ is shown in Figure 9.10 for the upper half-plane. The stability interval is $(-1.0, 0)$ and is only slightly smaller than that of the ABM43 method. Therefore the predictor–corrector method (9.140) of order 4 based on two useless two-step methods has a useful region of absolute stability.

### 9.3.3 Stiff differential equations

The solutions of systems of differential equations describing physical, chemical or biological processes often have the property that they have components that decay exponentially with quite different speed. If a method is applied whose region of absolute stability does not comprise the whole left part of the complex plane, the step size $h$ must be chosen in such a way that the complex values $\mu = h\lambda_j$, where $\lambda_j$ are the so called decay constants, belong to the region of absolute stability so that a stable integration can be guaranteed.

*Example* 9.11  We show the situation and the problem by means of a system of three linear, homogeneous differential equations

$$\begin{aligned} y_1' &= -0.5y_1 + 32.6y_2 + 35.7y_3 \\ y_2' &= \phantom{-0.5y_1 +} - 48y_2 + \phantom{3}9y_3 \\ y_3' &= \phantom{-0.5y_1 + 3}9y_2 - 72y_3 \end{aligned} \qquad (9.141)$$

subject to the initial conditions $y_1(0) = 4$, $y_2(0) = 13$, $y_3(0) = 1$. The solution is found by means of

$$y_1(x) = a_1 e^{\lambda x} \quad y_2(x) = a_2 e^{\lambda x} \quad y_3(x) = a_3 e^{\lambda x}.$$

## 9.3 Stability

Substitution into (9.141) yields the eigenvalue problem

$$(-0.5 - \lambda)a_1 + 32.6a_2 + \quad 35.7a_3 = 0$$
$$(-48 - \lambda)a_2 + \quad 9a_3 = 0 \qquad (9.142)$$
$$9a_2 + (-72 - \lambda)a_3 = 0.$$

From this a nontrivial triplet of values $(a_1, a_2, a_3)^T =: \mathbf{a}$ is to be determined as an eigenvector of the coefficient matrix $\mathbf{A}$ of the system of differential equations (9.141). The three eigenvalues of (9.142) are $\lambda_1 = -0.5$, $\lambda_2 = -45$ and $\lambda_3 = -75$. For each eigenvalue a solution of (9.141) exists, and the general solution is obtained as a linear combination of these fundamental solutions. From the initial conditions we obtain the solutions

$$y_1(x) = 15e^{-0.5x} - 12e^{-45x} + e^{-75x}$$
$$y_2(x) = \qquad\qquad 12e^{-45x} + e^{-75x} \qquad (9.143)$$
$$y_3(x) = \qquad\qquad 4e^{-45x} - 3e^{-75x}.$$

The different decay constants of the solution components are given by the eigenvalues $\lambda_j$. For a numerical integration of (9.141) we want to use the classical fourth-order Runge–Kutta method (9.52). In order to integrate the fastest decaying component $e^{-75x}$ with an accuracy of at least four digits a step size $h_1 = 0.0025$ is necessary. This step size is determined by the requirement that $e^{-75h_1}$ agrees with $F(-75h_1)$ given by (9.129) to five decimal places. If we integrate (9.141) over 60 steps up to the abscissa $x_1 = 0.150$, then $e^{-75 \cdot 0.150} = e^{-11.25} \doteq 0.000\,013$ is already much smaller than $e^{-45 \cdot 0.150} = e^{-6.75} \doteq 0.001\,171$. This fast decay, and at this point small component, no longer needs such an accurate integration. Hence, we can increase the step size. In order that the component $e^{-45x}$ is now treated with comparable accuracy, we must choose the step size $h_2 = 0.005$. After 30 more integration steps we reach $x_2 = x_1 + 30h_2 = 0.300$. At this stage $e^{-45 \cdot 0.300} = e^{-13.5} \doteq 0.000\,0014$ is small so that we can increase the step size once again. If we considered the slowly decaying component $e^{-0.5x}$ alone, a step size of $\tilde{h} = 0.4$ would be adequate to attain the desired accuracy. However, this step size violates the condition of stability since $\tilde{\mu} = -75\tilde{h} = -30$ does not belong to the interval of stability $(-2.78, 0)$. The maximal step size $h^*$ has to satisfy the inequality $h^* \leq 2.78/75 \doteq 0.037$. With $h_3 = 0.035$ an integration of (9.141) until $x = 24$, where we have $|y_1(x)| \leq 0.0001$, requires a further 678 integration steps. In Figure 9.11 the Euclidean norm of the global error $\|\mathbf{g}_k\| = \|\mathbf{y}(x_k) - \mathbf{y}_k\|$ is shown in a logarithmic scale as a function of $x$ for the interval $[0, 1.2]$ if the step sizes are chosen according to the previous considerations. Each increase of the step size generates a temporary unsteady increase of the norm of the global discretization error, because the corresponding component of the solution is integrated less accurately. The slow, approximately linear increase of $\|\mathbf{g}_k\|$ in the logarithmic scale that is observed for $x \geq 0.6$ corresponds to the

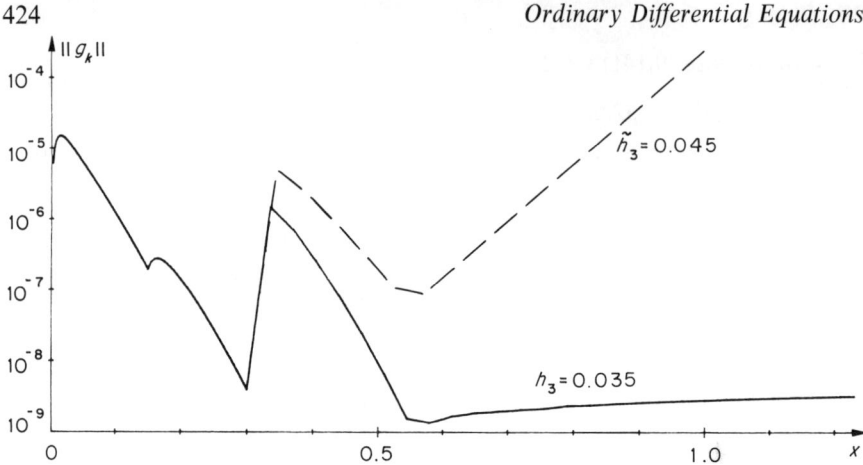

Figure 9.11 Stable and unstable integration of a system of differential equations

predicted behaviour according to the estimate (9.21). If in place of $h_3$ the step size $\tilde{h}_3 = 0.045$ is used to continue the integration at $x_2 = 0.3$, the norm of the global error clearly shows the instability of the Runge–Kutta method, because the condition of absolute stability is violated since $-75 \cdot 0.045 = -3.35 < -2.78$.

A linear, nonhomogeneous system of differential equations

$$\mathbf{y}'(x) = \mathbf{A}\mathbf{y}(x) + \mathbf{b}(x) \quad \mathbf{A} \in \mathbf{R}^{n \times n} \quad \mathbf{y}, \mathbf{b} \in \mathbf{R}^n \tag{9.144}$$

is called *stiff* if the eigenvalues $\lambda_j$, $j = 1, 2, \ldots, n$, of the matrix $\mathbf{A}$ have quite different negative real parts. A measure of the *stiffness* $S$ of the system of differential equations (9.144) is given by the quotient of the moduli of the largest and smallest real parts of the eigenvalues

$$S := \max_j |\mathrm{Re}(\lambda_j)| \Big/ \min_j |\mathrm{Re}(\lambda_j)|. \tag{9.145}$$

For the system of differential equations (9.141) we have $S = 150$. This is not very stiff, because the stiffness $S$ often reaches values between $10^3$ and $10^6$. In order to be able to treat such systems by means of a step size $h$ that is not too small only methods whose region of absolute stability comprises the whole left half-plane $\mathrm{Re}(\mu) < 0$ can be considered. This condition is satisfied by the trapezoidal method (9.33) and by the implicit Runge–Kutta methods (9.60) and (9.62). There also exist some absolutely stable, implicit multistep methods. However, all mentioned integration schemes require the solution of a system of equations for the unknowns $k_i$ in each step.

The problem of stiffness especially exists for nonlinear systems of differential equations

$$\mathbf{y}'(x) = \mathbf{f}(x, \mathbf{y}(x)) \quad \mathbf{y}(x) \in \mathbf{R}^n. \tag{9.146}$$

## 9.3 Stability

The stiffness of (9.146) can be defined by means of a linearization. The *local* behaviour of the *exact* solution $\mathbf{y}(x)$ is analyzed in the neighbourhood of $x_k$ subject to the initial condition $\mathbf{y}(x_k) = \mathbf{y}_k$ where $\mathbf{y}_k$ denotes the computed approximation at $x_k$. Let $\mathbf{y}(x)$ be of the form

$$\mathbf{y}(x) = \mathbf{y}_k + \mathbf{z}(x) \quad \text{for} \quad x_k \leqslant x \leqslant x_k + h \tag{9.147}$$

and we assume that $h$, as well as the norm of the vector $\mathbf{z}(x) = (z_1(x), z_2(x), \ldots, z_n(x))^T$, are small. After substitution of (9.147) into (9.146) we linearize the right-hand side of the differential equation and for the $i$th equation, $i = 1, 2, \ldots, n$, obtain

$$\begin{aligned} z_i'(x) &= f_i(x_k + (x - x_k), y_{1k} + z_1(x), y_{2k} + z_2(x), \ldots, y_{nk} + z_n(x)) \\ &\approx f_i(x_k, \mathbf{y}_k) + (x - x_k)\frac{\partial f_i(x_k, \mathbf{y}_k)}{\partial x} + \sum_{j=1}^{n} \frac{\partial f_i(x_k, \mathbf{y}_k)}{\partial y_j} z_j(x). \end{aligned} \tag{9.148}$$

The resulting $n$ linear, nonhomogeneous differential equations for the functions $z_1(x), z_2(x), \ldots, z_n(x)$ can be combined by means of the *Jacobian matrix*

$$\mathbf{J}(x_k) := \begin{pmatrix} \dfrac{\partial f_1}{\partial y_1} & \dfrac{\partial f_1}{\partial y_2} & \cdots & \dfrac{\partial f_1}{\partial y_n} \\ \dfrac{\partial f_2}{\partial y_1} & \dfrac{\partial f_2}{\partial y_2} & \cdots & \dfrac{\partial f_2}{\partial y_n} \\ \vdots & \vdots & & \vdots \\ \dfrac{\partial f_n}{\partial y_1} & \dfrac{\partial f_n}{\partial y_2} & \cdots & \dfrac{\partial f_n}{\partial y_n} \end{pmatrix}_{(x_k, \mathbf{y}_k)} \in \mathbf{R}^{n \times n} \tag{9.149}$$

and of the vectors

$$\mathbf{z}(x) := \begin{pmatrix} z_1(x) \\ z_2(x) \\ \vdots \\ z_n(x) \end{pmatrix} \quad \mathbf{f}_k := \begin{pmatrix} f_1(x_k, \mathbf{y}_k) \\ f_2(x_k, \mathbf{y}_k) \\ \vdots \\ f_n(x_k, \mathbf{y}_k) \end{pmatrix} \quad \mathbf{g}_k := \begin{pmatrix} \dfrac{\partial f_1}{\partial x} \\ \dfrac{\partial f_2}{\partial x} \\ \vdots \\ \dfrac{\partial f_n}{\partial x} \end{pmatrix}_{(x_k, \mathbf{y}_k)} \in \mathbf{R}^n$$

to the following initial value problem

$$\mathbf{z}'(x) \approx \mathbf{J}(x_k)\mathbf{z}(x) + \mathbf{f}_k + (x - x_k)\mathbf{g}_k \quad \mathbf{z}(x_k) = \mathbf{0}. \tag{9.150}$$

Therefore the qualitative behaviour of $\mathbf{y}(x)$ in the neighbourhood of $x_k$ is described by $\mathbf{z}(x)$ solving (9.150). The nonlinear system of differential equations (9.146) is said to be *stiff* if the eigenvalues $\lambda_j$ of the Jacobian matrix $\mathbf{J}(x_k)$ defined by (9.149) have quite different negative real parts, so that the value $S$ (9.145) is large. The measure of the stiffness of (9.146) now depends on $x_k$ and on the approximate solution $\mathbf{y}_k$ so that $S$ may change by a great extent during the course of the numerical integration.

*Example* 9.12 We consider the nonlinear system of first-order differential equations for three unknown functions

$$\begin{aligned} \dot{y}_1 &= -0.1y_1 + 100y_2y_3 \\ \dot{y}_2 &= \phantom{-}0.1y_1 - 100y_2y_3 - 500y_2^2 \\ \dot{y}_3 &= \phantom{-}500y_2^2 - 0.5y_3 \end{aligned} \tag{9.151}$$

subject to the initial condition $y_1(0) = 4$, $y_2(0) = 2$, $y_3(0) = 0.5$. The system (9.151) describes the kinetic reaction of three chemical substances $Y_1$, $Y_2$, $Y_3$ according to the law of mass reaction. The functions $y_1(t)$, $y_2(t)$, $y_3(t)$ represent the corresponding concentrations of the substances at the time $t$. The reactions run with very different time constants expressed by the coefficients in (9.151) that differ greatly in size. The Jacobian matrix $\mathbf{J}(t)$ of (9.151) is

$$\mathbf{J}(t) = \begin{pmatrix} -0.1 & 100y_3 & 100y_2 \\ 0.1 & -100y_3 - 1000y_2 & -100y_2 \\ 0 & 1000y_2 & -0.5 \end{pmatrix}. \tag{9.152}$$

The matrix elements of $\mathbf{J}$ depend on the course of the chemical reaction, and the eigenvalues of $\mathbf{J}(t)$ depend on $t$. For the initial time $t = 0$ the eigenvalues of

$$\mathbf{J}(0) = \begin{pmatrix} -0.1 & 50 & 200 \\ 0.1 & -2050 & -200 \\ 0 & 2000 & -0.5 \end{pmatrix}$$

are $\lambda_1 \doteq -0.000\,249$, $\lambda_2 \doteq -219.0646$, $\lambda_3 \doteq -1831.535$. Consequently we have $S \doteq 7.35 \times 10^6$, and the system (9.151) is quite stiff at $t = 0$. To illustrate the problem of stiffness we integrate (9.151) by means of the classical Runge–Kutta method (9.52). The absolutely largest, negative eigenvalue $\lambda_3$ requires a small initial step size $h = 0.0002$ so that the corresponding fast decaying component of the solution is integrated with a local, relative discretization error of about $10^{-4}$. After 25 steps of integration the stiffness of the system has decreased (see Table 9.11). Now because $\lambda_3 \doteq -372.48$ the step size can be increased to $h = 0.001$, since the fast decreasing component that is already small can be treated with a lower relative accuracy. The same rules hold for further integration; the results are summarized in Table 9.11. For selected times the approximations of the three functions, the eigenvalues $\lambda_1$, $\lambda_2$ and $\lambda_3$ of the Jacobian matrix $\mathbf{J}$, the measure $S$ of the stiffness and the step sizes used are given. At the beginning $S$ decreases fast, increases temporarily and finally decreases monotonically. From

## 9.4 Exercises

Table 9.11 Integration of a stiff system of differential equations

| $t$ | $y_{1,k}$ | $y_{2,k}$ | $y_{3,k}$ | $\lambda_1$ | $\lambda_2$ | $\lambda_3$ | $S$ | $h$ |
|---|---|---|---|---|---|---|---|---|
| 0     | 4.0000 | 2.0000 | 0.5000 | −0.000 25 | −219.06  | −1831.5  | $7.35 \times 10^6$ | 0.0002 |
| 0.001 | 4.1379 | 0.9177 | 1.4438 | −0.000 54 | −86.950  | −975.74  | $1.80 \times 10^6$ |        |
| 0.002 | 4.2496 | 0.5494 | 1.6996 | −0.000 90 | −45.359  | −674.62  | $7.52 \times 10^5$ |        |
| 0.003 | 4.3281 | 0.3684 | 1.8013 | −0.001 33 | −26.566  | −522.56  | $3.94 \times 10^5$ |        |
| 0.004 | 4.3846 | 0.2630 | 1.8493 | −0.001 84 | −16.588  | −431.91  | $2.35 \times 10^5$ |        |
| 0.005 | 4.4264 | 0.1952 | 1.8743 | −0.002 43 | −10.798  | −372.48  | $1.54 \times 10^5$ | 0.0010 |
| 0.006 | 4.4581 | 0.1489 | 1.8880 | −0.003 10 | −7.2503  | −331.06  | $1.07 \times 10^5$ |        |
| 0.008 | 4.5016 | 0.0914 | 1.9001 | −0.004 65 | −3.5318  | −278.45  | $5.99 \times 10^4$ |        |
| 0.010 | 4.5287 | 0.0588 | 1.9038 | −0.006 20 | −1.9132  | −247.81  | $4.00 \times 10^4$ | 0.0025 |
| 0.020 | 4.5735 | 0.0097 | 1.8985 | −0.004 44 | −0.5477  | −199.60  | $4.50 \times 10^4$ |        |
| 0.030 | 4.5795 | 0.0035 | 1.8892 | −0.001 78 | −0.5063  | −192.48  | $1.08 \times 10^5$ |        |
| 0.040 | 4.5804 | 0.0026 | 1.8799 | −0.001 34 | −0.5035  | −190.65  | $1.42 \times 10^5$ |        |
| 0.050 | 4.5805 | 0.0025 | 1.8705 | −0.001 28 | −0.5032  | −189.60  | $1.48 \times 10^5$ | 0.0050 |
| 0.10  | 4.5803 | 0.0025 | 1.8245 | −0.001 34 | −0.5034  | −185.03  | $1.39 \times 10^5$ |        |
| 0.15  | 4.5800 | 0.0025 | 1.7796 | −0.001 40 | −0.5036  | −180.60  | $1.29 \times 10^5$ |        |
| 0.20  | 4.5798 | 0.0026 | 1.7358 | −0.001 47 | −0.5039  | −176.29  | $1.20 \times 10^5$ |        |
| 0.25  | 4.5796 | 0.0027 | 1.6931 | −0.001 54 | −0.5042  | −172.08  | $1.12 \times 10^5$ | 0.010  |
| 0.50  | 4.5782 | 0.0030 | 1.4951 | −0.001 96 | −0.5060  | −152.63  | $7.80 \times 10^4$ |        |
| 0.75  | 4.5765 | 0.0034 | 1.3207 | −0.002 47 | −0.5086  | −135.56  | $5.48 \times 10^4$ |        |
| 1.00  | 4.5745 | 0.0038 | 1.1670 | −0.003 11 | −0.5124  | −120.63  | $3.88 \times 10^4$ | 0.020  |
| 2.00  | 4.5601 | 0.0061 | 0.7177 | −0.007 10 | −0.5482  | −77.88   | $1.10 \times 10^4$ |        |
| 3.00  | 4.5290 | 0.0089 | 0.4583 | −0.012 55 | −0.6514  | −54.71   | $4.36 \times 10^3$ |        |
| 4.00  | 4.4720 | 0.0117 | 0.3216 | −0.016 20 | −0.8284  | −43.62   | $2.69 \times 10^3$ |        |
| 5.00  | 4.3899 | 0.0134 | 0.2590 | −0.017 49 | −0.9863  | −38.92   | $2.23 \times 10^3$ | 0.050  |
| 10.0  | 3.8881 | 0.0141 | 0.2060 | −0.018 54 | −1.1115  | −34.14   | $1.84 \times 10^3$ |        |

$t = 0.25$ the step size $h$ is bounded by the interval of absolute stability of the Runge–Kutta method used.

## 9.4. Exercises

**9.1.** Determine the exact solution of the initial value problem

$$y' = \frac{2x}{y^2} \quad y(0) = 1$$

and compute the approximate solutions in $[0, 3]$
  (a) by means of Euler's method with the step sizes $h = 0.1, 0.01, 0.001$;
  (b) by means of the improved polygonal method and Heun's method with the step sizes $h = 0.1, 0.05, 0.025, 0.01$;
  (c) by means of a Runge–Kutta method of order three and four with the step sizes $h = 0.2, 0.1, 0.05, 0.025$;
  (d) by means of the Taylor series method for which the recursion formulae for the coefficients up to the fourth order should be derived. Choose the step sizes $h = 0.2, 0.1, 0.05$.

Verify the orders of convergence of the methods by means of the computed global discretization errors at $x_k = 0.2k$, $k = 1, 2, \ldots, 15$.

**9.2.** What are the conditions for the three parameters $a$, $c_1$ and $c_2$ of the two-stage Runge–Kutta method

$$k_1 = f(x_k, y_k) \quad k_2 = f(x_k + ah, y_k + ahk_1)$$
$$y_{k+1} = y_k + h(c_1 k_1 + c_2 k_2),$$

so that it has maximal order? Show that the order cannot exceed two. Then derive Heun's method and the improved polygonal method.

**9.3.** Solve the following initial value problems

(a) $y' = \dfrac{1}{1+4x^2} - 8y^2 \quad y(0) = 0 \quad x \in [0, 4]$

(b) $y' = \dfrac{1}{1+4x^2} + 0.4y^2 \quad y(0) = 0 \quad x \in [0, 4]$

(c) $y' = \dfrac{1 - x^2 - y^2}{1 + x^2 + xy} \quad y(0) = 0 \quad x \in [0, 10]$

by means of a Runge–Kutta method of second and fourth order in the given intervals. Apply a step size control in such a way that the estimate of the absolute value of the local discretization error is at most equal to $\varepsilon$, ($\varepsilon = 10^{-4}, 10^{-6}, 10^{-8}$). To control the step size invent different strategies and experiment with them.

**9.4.** The small elongation $x(t)$ of an oscillating pendulum with friction satisfies the second-order differential equation

$$\ddot{x}(t) + 0.12\dot{x}(t) + 2x(t) = 0.$$

with the initial condition $x(0) = 1$, $\dot{x}(0) = 0$. Solve the corresponding system of first-order differential equations approximately by means of the classical fourth-order Runge–Kutta method with three different step sizes. Represent the approximate solution $x_k$ graphically. Moreover, compute the global discretization error with the aid of the exact solution and investigate its behaviour. Is the numerical integration of the two complex components of the solution accurate with respect to the amplitude and the phase?

**9.5.** The system of differential equations

$$\dot{x} = 1.2x - x^2 - \dfrac{xy}{x+0.2}$$

$$\dot{y} = \dfrac{1.5xy}{x+0.2} - y$$

describes a biological prey–predator model, where $x(t)$ measures the number of preys and $y(t)$ that of the predators. Solve the system by means of the classical fourth-order Runge–Kutta method on the interval $0 \leq t \leq 30$ with the step size $h = 0.1$. Compute the solutions for the two different initial conditions $x(0) = 1$, $y(0) = 0.75$ and $\bar{x}(0) = 0.75$, $\bar{y}(0) = 0.25$ and represent the solutions in the $(x, y)$ phase plane. What does the result tell you?

As a different approach, use England's method with automatic step size control.

## 9.4 Exercises

**9.6.** Derive the explicit three-step Adams–Bashforth method and the implicit three-step Adams–Moulton method on the basis of the integral equation and as special cases of a general multistep method. Then show that the orders of the methods are three and four, respectively, and verify the coefficients of their principal parts of the local discretization errors given in Table 9.6.

**9.7.** Solve the differential equations of the Exercises 9.1 and 9.3 approximately by means of the Adams–Bashforth–Moulton methods ABM33 (9.91) and ABM43 (9.92) with the step size $h = 0.1$. In order to be able to compare the two methods on a fair basis in both cases three initial values $y_1, y_2, y_3$ should be computed by means of the classical Runge–Kutta method.

**9.8.** What are the regions of absolute stability of the following multistep methods?

(a) AB3: $y_{k+1} = y_k + \frac{1}{12}h(23f_k - 16f_{k-1} + 5f_{k-2})$
(b) AM2: $y_{k+1} = y_k + \frac{1}{12}h(5f(x_{k+1}, y_{k+1}) + 8f_k - f_{k-1})$
(c) Predictor–corrector method ABM32.

Show for the ABM32 method that the coefficient of the principal part of the local discretization error is in fact equal to that of the AM2 method.

**9.9.** The problem of a stable integration is to be analyzed for the linear, homogeneous system of differential equations (Lambert, 1973)

$$\begin{aligned} y_1' &= -21y_1 + 19y_2 - 20y_3 & y_1(0) &= 1 \\ y_2' &= 19y_1 - 21y_2 + 20y_3 & y_2(0) &= 0 \\ y_3' &= 40y_1 - 40y_2 - 40y_3 & y_3(0) &= -1. \end{aligned}$$

The eigenvalues of the matrix of the system are $\lambda_1 = -2$, $\lambda_{2,3} = -40 \pm 40i$. The system has to be solved by means of the trapezoidal method and by the classical fourth-order Runge–Kutta method. What step sizes $h$ must be chosen in order to guarantee an accuracy of the approximations to four significant digits in the interval $[0, 0.3]$? To answer this question the values of $e^{h\lambda}$ and $F(h\lambda)$ have to be compared. What step sizes are then appropriate to continue the integration for the interval $[0.3, 5]$? Check your statements by means of a numerical integration of the system using the predicted step sizes. What maximal step size is possible in the case of the ABM43 method (9.92)?

**9.10.** Investigate the nonlinear system of differential equations

$$\begin{aligned} \dot{y}_1 &= -0.01y_1 + 0.01y_2 \\ \dot{y}_2 &= y_1 - y_2 - y_1 y_3 \\ \dot{y}_3 &= y_1 y_2 - 100 y_3 \end{aligned}$$

subject to the initial condition $y_1(0) = 0$, $y_2(0) = 1$, $y_3(0) = 1$ with respect to stiffness as a function of $t$.

**9.11.** What are the multistep methods of maximal order subject to the following restrictions:

(a) Explicit $m$-step method with $b_0 = b_1 = \cdots = b_{m-2} = b_m = 0$ for $m = 2$ and $m = 3$.
(b) Implicit $m$-step method with $b_0 = b_1 = \cdots = b_{m-1} = 0$ and $b_m \neq 0$ for $m = 1, 2, 3$.

What are the orders, which methods are zero-stable and what are the regions of absolute stability?

# 10

# Partial Differential Equations

Many processes or stationary states of physics, chemistry or biology can be described by means of functions of several independent variables which from natural laws must satisfy certain partial differential equations. There is a great variety of partial differential equations and systems of partial differential equations that arise in the applications, and their appropriate numerical treatment often requires special procedures. Therefore in the following we must restrict ourselves to solving second-order partial differential equations for an unknown function with two or three independent variables. Moreover, the partial differential equations are either *elliptic* or *parabolic*. In the first case the independent variables represent space coordinates, and the desired function usually describes a *stationary state*. In the other case one variable is the time and the others are space coordinates, and the function required describes a *non-stationary process* as a function of time and more specifically a *diffusion process*. Among these two classes of problems a certain set of relevant problems occur for which we are going to develop some simple methods of solution and discuss their fundamental properties. More extensive representations can be found in Ames (1977), Collatz (1966), Gladwell and Wait (1979), Jain (1984), Marsal (1976), Meis and Marcowitz (1978), Mitchell and Griffiths (1980), Parter (1979), Smith (1978), Törnig et al. (1985), Twizell (1984), Vemuri and Karplus (1981) and Hackbusch (1986), where other types of partial differential equations are also considered.

## 10.1 Elliptic Boundary Value Problems, Finite Differences

### 10.1.1 Formulation of the problem

We look for a function $u(x, y)$ in a given region $G \subset \mathbf{R}^2$ that satisfies a *second-order linear partial differential equation*

## 10.1 Elliptic Boundary Value Problems

$$Au_{xx} + 2Bu_{xy} + Cu_{yy} + Du_x + Eu_y + Fu = H \qquad (10.1)$$

The given coefficients $A, B, C, D, E, F$ and $H$ in (10.1) can be piecewise continuous functions of $x$ and $y$. Analogously to the classification of conic sections

$$Ax^2 + 2Bxy + Cy^2 + Dx + Ey + F = 0$$

the partial differential equations (10.1) are divided into three classes according to the following definition.

**Definition 10.1** *In a region $G$, a second-order partial differential equation (10.1) with $A^2 + B^2 + C^2 \neq 0$ is called*

(a) *elliptic*     *if* $AC - B^2 > 0$     *for all* $(x, y) \in G$,
(b) *hyperbolic*   *if* $AC - B^2 < 0$     *for all* $(x, y) \in G$,
(c) *parabolic*    *if* $AC - B^2 = 0$     *for all* $(x, y) \in G$.

The classical representatives of *elliptic differential equations* are

$$u_{xx} + u_{yy} = 0 \qquad \text{Laplace equation} \qquad (10.2)$$
$$u_{xx} + u_{yy} = f(x, y) \qquad \text{Poisson equation} \qquad (10.3)$$

The Laplace equation occurs in problems of elasticity and hydrodynamics. The solution of the Poisson equation can describe the stationary temperature distribution in a homogeneous medium or the stress in certain torsion problems.

In order to fix the desired solution of an elliptic differential equation in a unique way certain *boundary conditions* must be prescribed on the boundary of the region $G$. For simplicity we assume that the region $G$ is bounded and that its boundary may consist of several curves (see Figure 10.1). We denote the union of all boundary curves by $\Gamma$. The boundary must consist of piecewise continuously differentiable curves on which the normal direction $\mathbf{n}$ pointing out

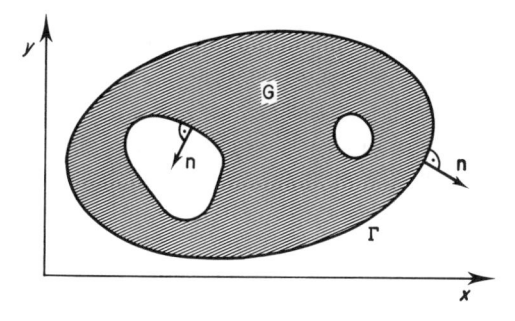

Figure 10.1 Region $G$ and boundary $\Gamma$

of the region $G$ can be defined. The boundary $\Gamma$ is assumed to consist of three disjoint parts $\Gamma_1, \Gamma_2$ and $\Gamma_3$ in such a way that

$$\Gamma_1 \cup \Gamma_2 \cup \Gamma_3 = \Gamma \tag{10.4}$$

We admit empty boundary parts. The appropriate boundary conditions for (10.1) or more specifically (10.2) or (10.3) are

$$u = \varphi \quad \text{on } \Gamma_1 \quad (\textit{Dirichlet} \text{ condition}) \tag{10.5}$$

$$\frac{\partial u}{\partial n} = \gamma \quad \text{on } \Gamma_2 \quad (\textit{Neumann} \text{ condition}) \tag{10.6}$$

$$\frac{\partial u}{\partial n} + \alpha u = \beta \quad \text{on } \Gamma_3 \quad (\textit{Cauchy} \text{ condition}) \tag{10.7}$$

where $\varphi$, $\gamma$, $\alpha$ and $\beta$ are given functions on the corresponding boundary parts. Usually they are defined as functions of the arc length $s$ on the boundary. The conditions (10.5), (10.6) and (10.7) are often called first, second and third boundary conditions, respectively. If an elliptic differential equation is only subject to Dirichlet boundary conditions, that is $\Gamma_1 = \Gamma$, then the problem is called a *Dirichlet boundary value problem*. If we have $\Gamma_2 = \Gamma$ then we have to solve a *Neumann boundary value problem*. This problem has a solution if and only if the function $\gamma(s)$ satisfies the integral condition $\int_\Gamma \gamma(s)\,ds = \iint_G f(x, y)\,dx\,dy$ (Williams, 1980).

### 10.1.2 Discretization of the problem

In the following we wish to approximately solve the Laplace or Poisson equation in a given region $G$ subject to boundary conditions (10.5) to (10.7). We start with simple problems and then build up to more complicated situations. The *finite difference method* consists of the following steps that are formulated in quite a general way.

*First step* The desired function $u(x, y)$ is replaced by its values at discrete points of the region $G$ and of the boundary $\Gamma$. For this *discretization* of $u(x, y)$ it is quite natural to use a regular square net with *mesh size h* in the region $G$, see Figure 10.2. Of course, the values of $u$ at the *grid points* are only required if they are not already known from Dirichlet boundary conditions. In the case of curved boundary parts it will be necessary to introduce points as the intersection of grid lines with the boundary. In Figure 10.2 the grid points are marked by dots.

We denote the value of the exact solution function $u(x, y)$ at a grid point P with the coordinates $x_i$ and $y_j$ by $u(x_i, y_j)$ and the corresponding approximate value by $u_{i,j}$.

## 10.1 Elliptic Boundary Value Problems

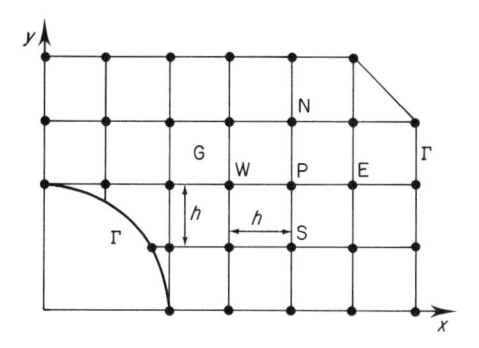

Figure 10.2 Region with mesh and grid points

The regular square net used to generate the grid points has some simplifying properties which will be important in the following discussions. In certain problems it may be more appropriate, or may even be necessary to use a net with variable mesh sizes in the $x$ and $y$ direction in order to take into account the shape of the region or the behaviour of the desired solution (Marsal, 1976). Moreover, regular triangular or hexagonal nets may be extremely useful (Collatz, 1966; Marsal, 1976) because a regular hexagonal net allows a locally finer discretization with no problems.

*Second step* For the chosen discretization of the function the partial differential equation must be suitably approximated at the grid points by means of the discrete function values $u_{i,j}$. In the case of a regular square net the first and the second partial derivatives can be approximated by means of difference quotients as in Section 3.2.2. For the first partial derivatives it is advantageous to use the central difference quotient (3.28). For a *regular interior point* $P(x_i, y_j)$ which has four neighbouring grid points at a distance $h$ away, we use the approximations

$$u_x(x_i, y_j) \approx \frac{u_{i+1,j} - u_{i-1,j}}{2h} \quad u_y(x_i, y_j) \approx \frac{u_{i,j+1} - u_{i,j-1}}{2h} \tag{10.8}$$

$$u_{xx}(x_i, y_j) \approx \frac{u_{i+1,j} - 2u_{i,j} + u_{i-1,j}}{h^2}$$

$$u_{yy}(x_i, y_j) \approx \frac{u_{i,j+1} - 2u_{i,j} + u_{i,j-1}}{h^2} \tag{10.9}$$

where we have already defined the difference quotients by means of the approximate values at the grid points. In order to obtain an easily remembered notation, avoiding double indices, we denote the four neighbours of P by the compass points N, W, S and E (see Figure 10.2) and thus define

$$u_P := u_{i,j} \quad u_N := u_{i,j+1} \quad u_W := u_{i-1,j} \quad u_S := u_{i,j-1} \quad u_E := u_{i+1,j}. \tag{10.10}$$

The Poisson equation (10.3) is now approximated at the grid point P by the *difference equation*

$$\frac{u_E - 2u_P + u_W}{h^2} + \frac{u_N - 2u_P + u_S}{h^2} = f_P \quad f_P := f(x_i, y_j)$$

which, after multiplication by $-h^2$, is equivalent to

$$\boxed{4u_P - u_N - u_W - u_S - u_E + h^2 f_P = 0.} \tag{10.11}$$

It is often written in the *operator form*

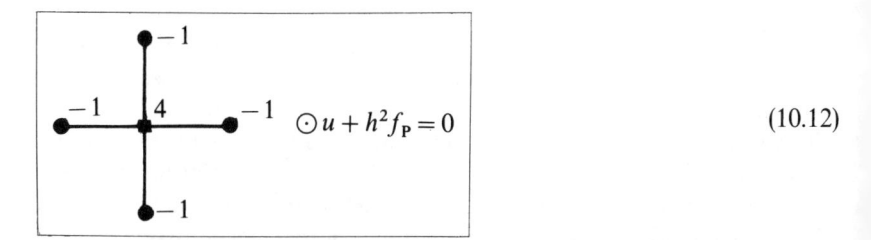

$$\begin{array}{c}\phantom{-1}\bullet-1\\ -1\;\boxed{4}\;-1\\ \phantom{-1}\bullet-1\end{array} \odot u + h^2 f_P = 0 \tag{10.12}$$

in which the multiplication coefficients of $u$ are given by the corresponding values. The grid point P, for which the operator form is valid, is marked by a square.

*Third step* The prescribed boundary conditions of the boundary value problem must be taken into account, and the difference approximation of the differential equation may also have to be adapted to the boundary conditions.

The simplest situation occurs if only Dirichlet boundary conditions are to be satisfied, and the net can be chosen in such a way that without exception regular interior grid points are generated. In this case the difference equation (10.11) can be applied to all interior grid points with unknown function value, and the known values on the boundary can simply be substituted. If there exist irregular grid points as in Figure 10.2, appropriate difference equations must be derived for them. We shall consider the treatment of such irregular grid points near the boundary in Section 10.1.3.

In general Neumann and Cauchy boundary conditions (10.6) and (10.7) require somewhat more elaborate procedures which will be systematically treated in Section 10.1.3. At this stage we only consider a simple situation that can be handled easily. We assume that the boundary coincides with a net line parallel to the $y$ axis, and that the Neumann boundary condition requires the vanishing of the normal derivative (see Figure 10.3). The outer normal vector **n** points in

## 10.1 Elliptic Boundary Value Problems

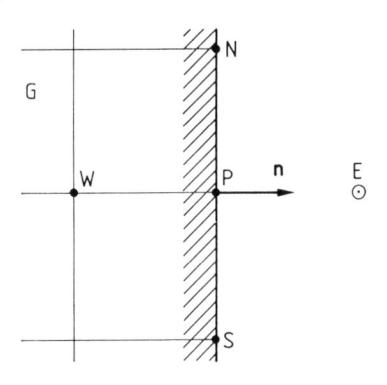

Figure 10.3 Special Neumann boundary condition

the direction of the positive $x$ axis. If we temporarily introduce the auxiliary point E with the value $u_E$, the derivative in the normal direction can be approximated by means of the central difference quotient so that we have

$$\left.\frac{\partial u}{\partial n}\right|_P \approx \frac{u_E - u_W}{2h} = 0 \quad u_E = u_W.$$

The condition for a vanishing normal derivative often means that the function $u(x,y)$ is symmetric with respect to the boundary. From this symmetry the function $u(x,y)$ can be extended beyond the boundary, and it is justified to use the general difference equation (10.11). From (10.11) it follows that

$$4u_P - u_N - 2u_W - u_S + h^2 f_P = 0$$

and we divide this equation by 2 for a reason that will become clear later

$$\boxed{2u_P - \tfrac{1}{2}u_N - u_W - \tfrac{1}{2}u_S + \tfrac{1}{2}h^2 f_P = 0} \tag{10.13}$$

Therefore the corresponding operator form of the difference equation (10.13) for a regular boundary point P with the three neighbouring grid points N, W and S is

$$\odot u + \tfrac{1}{2}h^2 f_P = 0 \quad \left.\frac{\partial u}{\partial n}\right|_P = \left.\frac{\partial u}{\partial x}\right|_P = 0. \tag{10.14}$$

*Fourth step* For a numerical determination of the unknown function values at the grid points we need some equations. According to the two preceding steps we have a linear difference equation for each such grid point at hand, and so we are able to give a system of linear equations for the unknown values. To avoid double indices we now number the grid points of the net whose function values are unknown. The numbering of the grid points should be done following certain rules in order to obtain a suitable structure for the resulting system of equations that allows an efficient solution. The system of linear equations represents the *discrete form* of the given boundary value problem.

*Example* 10.1 In the domain $G$ of Figure 10.4 we want to solve the Poisson equation

$$u_{xx} + u_{yy} = -2 \quad \text{in } G \tag{10.15}$$

subject to the boundary conditions

$$u = 0 \quad \text{along DE and EF} \tag{10.16}$$

$$u = 1 \quad \text{along AB and BC} \tag{10.17}$$

$$\frac{\partial u}{\partial n} = 0 \quad \text{along CD and FA.} \tag{10.18}$$

The solution of the boundary value problem can describe the stress state of a beam under torsion. Its section is ringlike and is obtained from $G$ by reflectioning with respect to the sides CD and FA many times. For reasons of symmetry the

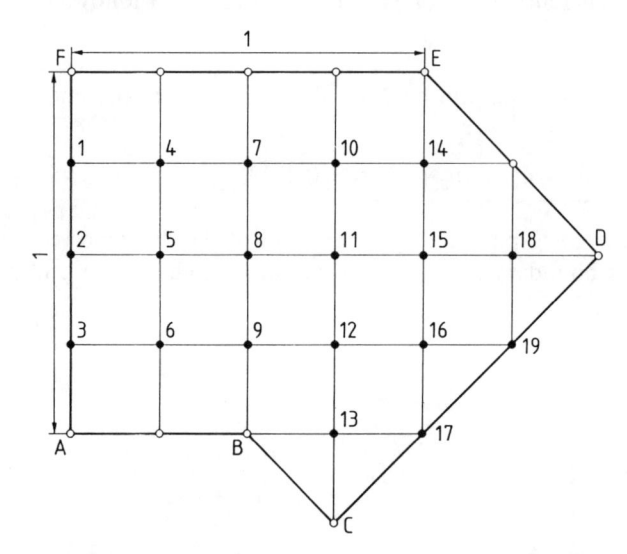

Figure 10.4 Domain G with mesh and grid points, $h = 0.25$

## 10.1 Elliptic Boundary Value Problems

problem can be solved in the domain $G$ of Figure 10.4, whereby the Neumann boundary conditions (10.18) along the boundaries CD and FA follow from this symmetry. The boundary value problem (10.15) to (10.18) can be interpreted differently, for instance, it can also describe the stationary temperature distribution $u(x, y)$ in the ringlike section of a long reservoir if, due to a chemical reaction, a constant heat production goes on. The temperature is kept equal to the normed value one at the interior wall of the reservoir and equal to zero outside.

To discretize the boundary value problem we use the regular net with the mesh size $h = 0.25$ as shown in Figure 10.4. The grid points are either boundary or regular interior points. The grid points with unknown function values are marked by dots, and those with known values, from (10.16) and (10.17), are indicated by empty circles.

For all grid points in the interior of the domain the difference equation (10.11), or its operator form (10.12), is valid with $f_P = -2$. For the grid points on the boundary FA we get the operator form just by reflecting (10.14). For the grid points on CD, for symmetry reasons, with $u_S = u_W$ and $u_E = u_N$ from (10.11) we obtain the difference equation $4u_P - 2u_N - 2u_W + h^2 f_P = 0$ which is divided by 2 for a reason that will soon become clear. We summarize the operator equations for the present boundary value problem as follows

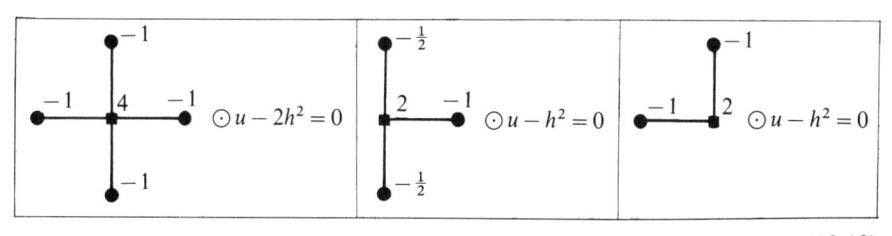

$$\odot u - 2h^2 = 0 \qquad \odot u - h^2 = 0 \qquad \odot u - h^2 = 0$$

(10.19)

Now we number the grid points with unknown function value columnwise, see Figure 10.4. For the corresponding 19 unknowns $u_1, u_2, \ldots, u_{19}$ we can formulate the system of linear equations. It is reasonable to write down the difference equations in the order of the numbered points and to take into account possible Dirichlet boundary conditions. In this way the system of linear equations (10.20) is produced where only the nonzero coefficients are given.

The matrix **A** of the system is *symmetric*. If the difference equations for the boundary grid points with a Neumann condition had not been divided by 2, the matrix **A** would be nonsymmetric. The matrix **A** is diagonally dominant in the weak sense, and irreducible as can easily be verified. The diagonal elements are positive, therefore **A** is *positive definite* (Maess, 1985; Schwarz et al., 1972), and hence the system of linear equations (10.20) has a unique solution that can be computed by means of Cholesky's method according to Section 1.3.1. The

| | $u_1$ | $u_2$ | $u_3$ | $u_4$ | $u_5$ | $u_6$ | $u_7$ | $u_8$ | $u_9$ | $u_{10}$ | $u_{11}$ | $u_{12}$ | $u_{13}$ | $u_{14}$ | $u_{15}$ | $u_{16}$ | $u_{17}$ | $u_{18}$ | $u_{19}$ | 1 |
|---|---|---|---|---|---|---|---|---|---|---|---|---|---|---|---|---|---|---|---|---|
| | 2 | −1/2 | | −1 | | | | | | | | | | | | | | | | −0.0625 |
| | −1/2 | 2 | −1/2 | | −1 | | | | | | | | | | | | | | | −0.0625 |
| | | −1/2 | 2 | | | −1 | | | | | | | | | | | | | | −0.5625 |
| | −1 | | | 4 | −1 | | −1 | | | | | | | | | | | | | −0.125 |
| | | −1 | | −1 | 4 | −1 | | −1 | | | | | | | | | | | | −0.125 |
| | | | −1 | | −1 | 4 | | | −1 | | | | | | | | | | | −1.125 |
| | | | | −1 | | | 4 | −1 | | −1 | | | | | | | | | | −0.125 |
| | | | | | −1 | | −1 | 4 | −1 | | −1 | | | | | | | | | −0.125 |
| | | | | | | −1 | | −1 | 4 | | | −1 | | | | | | | | −1.125 |
| | | | | | | | −1 | | | 4 | −1 | | | −1 | | | | | | −0.125 |
| | | | | | | | | −1 | | −1 | 4 | −1 | | | −1 | | | | | −0.125 |
| | | | | | | | | | −1 | | −1 | 4 | −1 | | | −1 | | | | −0.125 |
| | | | | | | | | | | | | −1 | 4 | | | | −1 | | | −2.125 |
| | | | | | | | | | | −1 | | | | 4 | −1 | | | −1 | | −0.125 |
| | | | | | | | | | | | −1 | | | −1 | 4 | −1 | | | −1 | −0.125 |
| | | | | | | | | | | | | −1 | | | −1 | 4 | −1 | | | −0.125 |
| | | | | | | | | | | | | | −1 | | | −1 | 2 | | | −0.0625 |
| | | | | | | | | | | | | | | −1 | | | | 4 | −1 | −0.125 |
| | | | | | | | | | | | | | | | −1 | | | −1 | 2 | −0.0625 |

(10.20)

## 10.1 Elliptic Boundary Value Problems

numbering used for the grid points and the unknowns has the consequence that the sparse coefficient matrix **A** has a band structure with the band width $m = 4$. If the technique of Section 1.3.2 is used, the system of equations (10.20) can be solved efficiently. Its solution, rounded to five decimal places, together with the given boundary values is presented in (10.21) according to the disposition of the grid points.

$$\begin{array}{llllllll}
0 & 0 & 0 & 0 & 0 & & & \\
0.41686 & 0.41101 & 0.39024 & 0.34300 & 0.24049 & 0 & & \\
0.72044 & 0.71193 & 0.68195 & 0.61628 & 0.49398 & 0.28682 & 0 & \quad (10.21)\\
0.91603 & 0.90933 & 0.88436 & 0.82117 & 0.70731 & 0.52832 & & \\
1 & 1 & 1 & 0.95174 & 0.86077 & & & \\
 & & & 1 & & & & 
\end{array}$$

### 10.1.3 Grid points near the boundary, general boundary conditions

We consider irregular interior grid points and points on the boundary where Neumann or Cauchy conditions have to be satisfied. The systematic treatment of such situations is shown by means of typical examples so that its generalization is obvious. The fundamental problem consists of constructing an appropriate difference approximation of a given differential operator, in our case that will be $u_{xx} + u_{yy}$.

*Example* 10.2 We consider an irregular interior grid point P that is supposed to lie near the boundary $\Gamma$ as shown in Figure 10.5. The boundary curve $\Gamma$ intersects the net lines at the points W' and S' that have the distances $ah$ and $bh$ from P, respectively, with $0 < a, b < 1$.

We aim to find approximations of the second partial derivatives $u_{xx}$ and $u_{yy}$ at

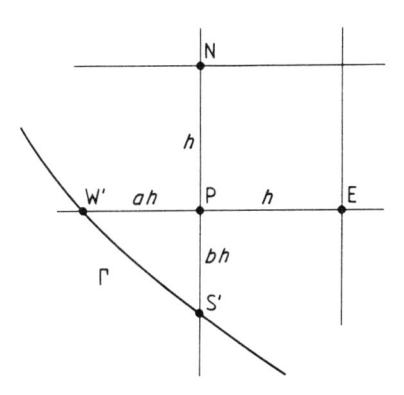

Figure 10.5 Irregular grid point near the boundary

the grid point P(x, y) that are linear combinations of the values $u_P$, $u_E$ and $u_{W'}$ and of $u_P, u_N$ and $u_{S'}$, respectively. Let $u(x, y)$ be sufficiently many times continuously differentiable. With the aid of the Taylor series expansions the following relationships hold for the function values of the corresponding points.

$$u(x+h, y) = u(x, y) + hu_x(x, y) + \tfrac{1}{2}h^2 u_{xx}(x, y) + \tfrac{1}{6}h^3 u_{xxx}(x, y) + \cdots$$
$$u(x-ah, y) = u(x, y) - ahu_x(x, y) + \tfrac{1}{2}a^2 h^2 u_{xx}(x, y) - \tfrac{1}{6}a^3 h^3 u_{xxx}(x, y) + \cdots$$
$$u(x, y) = u(x, y)$$

Now we form the linear combination with coefficients $c_1, c_2, c_3$ and obtain

$$c_1 u(x+h, y) + c_2 u(x-ah, y) + c_3 u(x, y)$$
$$= (c_1 + c_2 + c_3) u(x, y) + (c_1 - ac_2) h u_x(x, y) + (c_1 + a^2 c_2) \tfrac{1}{2} h^2 u_{xx}(x, y) + \cdots$$

We require that this linear combination is an approximation of the second partial derivative $u_{xx}$ at the point P(x, y), and so we obtain the three necessary conditions

$$c_1 + c_2 + c_3 = 0$$
$$(c_1 - ac_2)h = 0$$
$$\tfrac{1}{2}h^2(c_1 + a^2 c_2) = 1.$$

From these the values follow

$$c_1 = \frac{2}{h^2(1+a)} \quad c_2 = \frac{2}{h^2 a(1+a)} \quad c_3 = -\frac{2}{h^2 a}.$$

As the exact function values $u$ at the grid points E, P and W' are not available, we use their approximations $u_E, u_P$ and $u_{W'}$ and thus get the difference approximation

$$u_{xx}(P) \approx \frac{2}{h^2}\left(\frac{u_E}{1+a} + \frac{u_{W'}}{a(1+a)} - \frac{u_P}{a}\right). \tag{10.22}$$

In a similar way we find

$$u_{yy}(P) \approx \frac{2}{h^2}\left(\frac{u_N}{1+b} + \frac{u_{S'}}{b(1+b)} - \frac{u_P}{b}\right). \tag{10.23}$$

From (10.22) and (10.23) at the irregular grid point P of Figure 10.5, after multiplication by $-h^2$, we obtain the difference equation for the Poisson equation (10.3)

$$\boxed{2\left(\frac{1}{a}+\frac{1}{b}\right)u_P - \frac{2}{1+b}u_N - \frac{2}{a(1+a)}u_{W'} - \frac{2}{b(1+b)}u_{S'} - \frac{2}{1+a}u_E + h^2 f_P = 0}$$

$$\tag{10.24}$$

## 10.1 Elliptic Boundary Value Problems

If $a \neq b$ then the coefficients of $u_N$ and $u_E$ in (10.24) are different. This fact will in general have the consequence that the matrix $\mathbf{A}$ of the system of difference equations will be *nonsymmetric*. To see this we assume that the grid points E and N of Figure 10.5 are regular interior points for which the five-point difference equation (10.11) can be applied. If P, E and N have the numbers $i, j$ and $k$, respectively, we have $a_{ij} = -2/(1+a) \neq a_{ji} = -1$, $a_{ik} = -2/(1+b) \neq a_{ki} = -1$. Even by scaling the difference equations it is in general impossible to attain symmetry. In the special case $a = b$ it is advisable to multiply (10.24) by the factor $(1+a)/2$ so that $u_N$ and $u_E$ obtain the desired coefficients $-1$, and the symmetry of $\mathbf{A}$ is maintained, at least with respect to the grid point P. In this case the difference equation (10.24) is modified to

$$\frac{2(1+a)}{a} u_P - u_N - \frac{1}{a} u_{W'} - \frac{1}{a} u_{S'} - u_E + \tfrac{1}{2}(1+a)h^2 f_P = 0 \quad a = b. \qquad (10.25)$$

*Example* 10.3 A Neumann condition is to be satisfied on the boundary part $\Gamma_2$. We first consider the simple situation where the boundary point P is a grid point, and the boundary curve is assumed as shown in Figure 10.6. The angle between the outer normal vector $\mathbf{n}$ at the point P and the positive $x$ axis is $\psi$, measured anticlockwise. We also use the expression of the normal derivative to approximate the differential operator $u_{xx} + u_{yy}$ at the point $P(x, y)$ by an appropriate linear combination.

Obviously we shall use the function values of $u$ at the grid points P, W and S of Figure 10.6. Together with the normal derivative at P these are four quantities. It is easy to see that it is in general impossible to find a linear combination of these four values that approximates the given differential

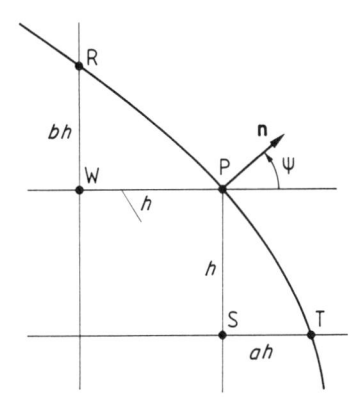

Figure 10.6 Neumann condition at the boundary point P

expression at P. We need at least five quantities if all Taylor series expansions do not contain a term with the mixed second partial derivative. Otherwise six quantities are necessary because we now have to satisfy six conditions. We choose as the two further boundary points $R(x-h, y+bh)$ and $T(x+ah, y-h)$ where $0 < a, b < 1$. Then the following Taylor series expansions hold with respect to the central point $P(x, y)$ whereby we partially omit the obvious values,

$$
\begin{array}{llll}
P: u(x, y) & = u & & c_P \\
W: u(x-h, y) & = u - hu_x & + \tfrac{1}{2} h^2 u_{xx} + \cdots & c_W \\
S: u(x, y-h) & = u \quad\quad - hu_y & + \tfrac{1}{2} h^2 u_{yy} + \cdots & c_S \\
R: u(x-h, y+bh) = u - hu_x + bhu_y + \tfrac{1}{2} h^2 u_{xx} - bh^2 u_{xy} + \tfrac{1}{2} b^2 h^2 u_{yy} + \cdots & c_R \\
T: u(x+ah, y-h) = u + ahu_x - hu_y + \tfrac{1}{2} a^2 h^2 u_{xx} - ah^2 u_{xy} + \tfrac{1}{2} h^2 u_{yy} + \cdots & c_T \\
P: \dfrac{\partial u(x,y)}{\partial n} = u_x \cos\psi + u_y \sin\psi & & & c_n
\end{array}
$$

From the linear combination of the six expansions we get the six equations for the coefficients $c_P, c_W, c_S, c_R, c_T$ and $c_n$

$$
\begin{array}{lrcl}
u: & c_P + c_W + c_S + c_R + c_T & = 0 & \\
u_x: & -hc_W \quad\quad -hc_R + ahc_T + c_n \cos\psi & = 0 & \\
u_y: & -hc_S + bhc_R - hc_T + c_n \sin\psi & = 0 & (10.26)\\
u_{xx}: & \tfrac{1}{2}h^2 c_W + \tfrac{1}{2}h^2 c_R + \tfrac{1}{2}a^2 h^2 c_T & = 1 & \\
u_{xy}: & -bh^2 c_R - ah^2 c_T & = 0 & \\
u_{yy}: & \tfrac{1}{2}h^2 c_S + \tfrac{1}{2}b^2 h^2 c_R + \tfrac{1}{2}h^2 c_T & = 1 &
\end{array}
$$

The system of equations (10.26) has a unique solution. If we define $N := (a+1)\sin\psi + (b+1)\cos\psi$, some selected values are

$$
c_n = \frac{2(a+b+2)}{hN} \quad c_T = \frac{2(\sin\psi - \cos\psi)}{ah^2 N} \quad c_R = \frac{2(\cos\psi - \sin\psi)}{bh^2 N}. \quad (10.27)
$$

The expressions for the other coefficients are more complicated. We see that the difference approximation depends on the geometry of the domain $G$ in the neighbourhood of the boundary point P. For numerically given values of $a, b$ and $\psi$ it is most convenient merely to solve the system of linear equations (10.26). With the resulting values of the coefficients the difference equation for the grid point P reads as

$$
c_P u_P + c_W u_W + c_S u_S + c_R u_R + c_T u_T + c_n \left.\frac{\partial u}{\partial n}\right|_P - f_P = 0, \quad (10.28)
$$

which is suitably multiplied by $-h^2$ so that the coefficient of $u_P$ becomes positive.

## 10.1 Elliptic Boundary Value Problems

The Neumann boundary condition in P is taken into account by substituting the prescribed value of the normal derivative into (10.28).

For the special case $\psi = 45°$ we have $\cos\psi = \sin\psi = \sqrt{2}/2$ and according to (10.27) the coefficients $c_T$ and $c_R$ are equal to zero. Hence the values $u_T$ and $u_R$ of the corresponding boundary points do not occur in the difference equation (10.28). For this simple situation the coefficients can be given by explicit expressions because $N = \frac{1}{2}(a+b+2)\sqrt{2}$

$$c_n = \frac{2\sqrt{2}}{h} \quad c_S = \frac{2}{h^2} \quad c_W = \frac{2}{h^2} \quad c_P = -\frac{4}{h^2}.$$

After a multiplication by $-h^2/2$ the difference equation is

$$2u_P - u_W - u_S - h\sqrt{2}\left.\frac{\partial u}{\partial n}\right|_P + \tfrac{1}{2}h^2 f_P = 0. \tag{10.29}$$

If the boundary condition is $\partial u/\partial n|_P = 0$, then we essentially get the last operator equation of (10.19) that has been derived by a different approach.

*Example* 10.4 The treatment of a Cauchy boundary condition, (10.7), at a general boundary point P follows the procedure for a Neumann condition. We consider the situation shown in Figure 10.7 to explain how to proceed. Let the boundary point $P(x, y)$ not be the intersection point of grid lines. The direction of the outer normal vector is defined by the angle $\psi$.

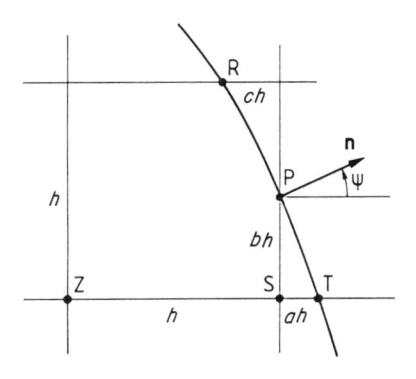

Figure 10.7 Cauchy condition at the boundary point P

Once again we shall need six quantities to approximate $u_{xx} + u_{yy}$. Besides the expression $\partial u/\partial n + \alpha u$ prescribed by the right-hand side of the Cauchy condition at the point P, in a natural way we use the values of $u$ at the points P, S, R and T. The sixth quantity is chosen as the function value $u$ at that grid point nearest to P in the interior of the domain. If $b \leqslant \frac{1}{2}$ then we take the point Z. With the abbreviation $B := 1 - b$ the relevant expansions are

$$
\begin{array}{llll}
\text{P:} & u(x, y) & = u & \quad c_P \\
\text{S:} & u(x, y - bh) & = u \quad - bhu_y \quad\quad\quad\quad\quad\quad\quad + \tfrac{1}{2} b^2 h^2 u_{yy} + \cdots & \quad c_S \\
\text{Z:} & u(x - h, y - bh) & = u - hu_x - bhu_y + \tfrac{1}{2} h^2 u_{xx} + bh^2 u_{xy} + \tfrac{1}{2} b^2 h^2 u_{yy} + \cdots & \quad c_Z \\
\text{R:} & u(x - ch, y + Bh) & = u - chu_x - Bhu_y + \tfrac{1}{2} c^2 h^2 u_{xx} - cBh^2 u_{xy} + \tfrac{1}{2} B^2 h^2 u_{yy} + \cdots & \quad c_R \\
\text{T:} & u(x + ah, y - bh) & = u + ahu_x - bhu_y + \tfrac{1}{2} a^2 h^2 u_{xx} - abh^2 u_{xy} + \tfrac{1}{2} b^2 h^2 u_{yy} + \cdots & \quad c_T \\
\text{P:} & \dfrac{\partial u}{\partial n} + \alpha u & = \alpha u + u_x \cos\psi + u_y \sin\psi & \quad c_n
\end{array}
$$

The coefficients $c_P, c_S, c_Z, c_R, c_T$ and $c_n$ of the linear combination that is to approximate $u_{xx} + u_{yy}$ have to satisfy the following six equations

$$
\begin{array}{lllllll}
u: & c_P + & c_S + & c_Z + & c_R + & c_T + \alpha c_n & = 0 \\
u_x: & & & -hc_Z - & chc_R + & ahc_T + c_n \cos\psi & = 0 \\
u_y: & & -bhc_S & -bhc_Z + & Bhc_R - & bhc_T + c_n \sin\psi & = 0 \\
u_{xx}: & & & \tfrac{1}{2} h^2 c_Z + & \tfrac{1}{2} c^2 h^2 c_R + & \tfrac{1}{2} a^2 h^2 c_T & = 1 \\
u_{xy}: & & & bh^2 c_Z - & cBh^2 c_R - & abh^2 c_T & = 0 \\
u_{yy}: & & \tfrac{1}{2} b^2 h^2 c_S + & \tfrac{1}{2} b^2 h^2 c_Z + & \tfrac{1}{2} B^2 h^2 c_R + & \tfrac{1}{2} b^2 h^2 c_T & = 1
\end{array}
$$

For numerically given values of $a, b, c, h$ and $\psi$ the system of linear equations can be solved. The resulting coefficients yield the following difference approximation of the Poisson equation

$$c_P u_P + c_S u_S + c_Z u_Z + c_R u_R + c_T u_T + c_n \gamma - f_P = 0, \tag{10.30}$$

where we have already substituted the known value $\gamma$ of the Cauchy condition (10.7). The boundary condition at the point P is expressed implicitly by the coefficients of the $u$ values of the difference equation (10.30) and by the explicit constant contribution $c_n \gamma$.

In general $c_Z$ will be nonzero in (10.30). If the grid point Z is a regular interior point, as indicated by the Figure 10.7, then the five-point difference equation (10.11) has to be used for Z. However, this equation does not contain the function value $u_P$, and therefore the matrix $\mathbf{A}$ of the system of equations is in any event *nonsymmetric*. Let us assume that the point P gets the number $i$, and Z the number $j \neq i$. Then the matrix element $a_{ij}$ is in fact nonzero whereas the symmetric element $a_{ji}$ is equal to zero.

*Example* 10.5 We consider the boundary value problem for the region $G$ defined

## 10.1 Elliptic Boundary Value Problems

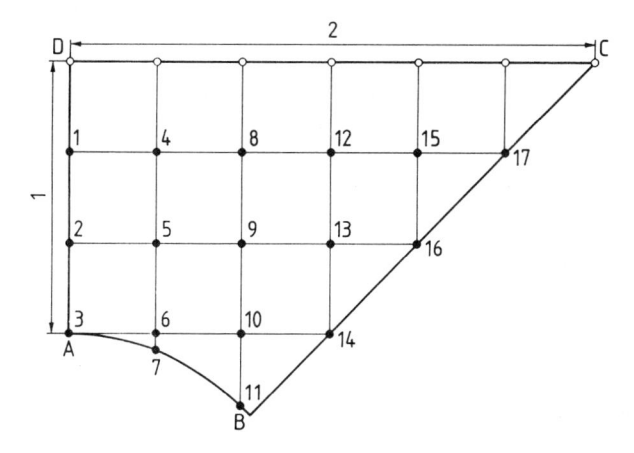

Figure 10.8 Domain G of the boundary value problem with mesh and grid points, $h = 1/3$

in Figure 10.8

$$\begin{aligned}
u_{xx} + u_{yy} &= -2 &&\text{in } G \\
u &= 0 &&\text{on } CD \\
\frac{\partial u}{\partial n} &= 0 &&\text{on } BC \text{ and } DA \\
\frac{\partial u}{\partial n} + 2u &= -1 &&\text{on } AB.
\end{aligned} \qquad (10.31)$$

The boundary curve AB is a segment of a circle with radius $r = 1$, whose centre is the intersection of the two lines DA and CB. The boundary value problem can be interpreted as a heat conduction problem where the stationary temperature distribution $u(x, y)$ in the section of a long container is sought in which there is a constant density of heat source. The container has a pipe inside taking heat away. For reasons of symmetry it is sufficient to determine the solution in the region $G$.

To discretize the problem we use the net shown in Figure 10.8 with the mesh size $h = \frac{1}{3}$. The resulting grid points with unknown function values are marked by circular dots and numbered columnwise.

For most of the grid points the operator equations (10.19) can be applied. A special treatment is necessary for the points 3, 6, 7, 10 and 11. The difference equations for the points 6 and 10 are immediately obtained from (10.24). The relative distances between points 6 and 7 and 10 and 11 are easily determined to be

$$b_6 = \overline{P_6 P_7}/h = 3 - 2\sqrt{2} \doteq 0.171\,573$$
$$b_{10} = \overline{P_{10} P_{11}}/h = 3 - \sqrt{5} \doteq 0.763\,932$$

and from (10.24), because of $a = 1$, we obtain the operator equations

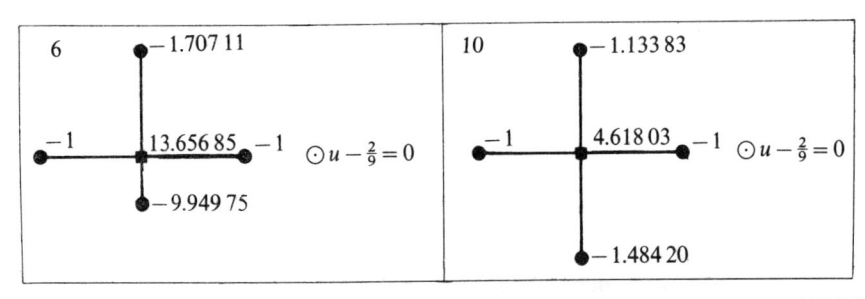

(10.32)

At point 3 two boundary parts join for which a Neumann and a Cauchy condition must be satisfied, respectively. We can treat this situation according to the procedure outlined in Examples 10.3 and 10.4. We may make use of the essential simplification that at point 3 we have relative to the boundary DA $\partial u/\partial n = -u_x$ and relative to AB $\partial u/\partial n = -u_y$. In this situation it is sufficient to have five quantities to obtain an approximation of the differential expression namely

| | | |
|---|---|---|
| 3: | $u(x, y) = u$ | $c_3$ |
| 2: | $u(x, y + h) = u + hu_y + \tfrac{1}{2}h^2 u_{yy} + \cdots$ | $c_2$ |
| 6: | $u(x + h, y) = u + hu_x + \tfrac{1}{2}h^2 u_{xx} + \cdots$ | $c_6$ |
| $3_1$: | $\dfrac{\partial u(x, y)}{\partial n_1} = -u_x$ | $c_n$ |
| $3_2$: | $\dfrac{\partial u}{\partial n_2} + 2u = 2u - u_y$ | $c_m$ |

The corresponding system of equations determines the coefficients

$$c_2 = \frac{2}{h^2} \quad c_6 = \frac{2}{h^2} \quad c_n = \frac{2}{h} \quad c_m = \frac{2}{h} \quad c_3 = -\frac{4}{h^2} - \frac{4}{h}$$

and after multiplying by $-h^2/2$ we obtain the difference equation

$$2(1 + h)u_3 - u_2 - u_6 - h\left(\frac{\partial u}{\partial n_1}\right) - h\left(\frac{\partial u}{\partial n_2} + 2u\right) - h^2 = 0.$$

If we take into account the two different boundary conditions at point 3, we

## 10.1 Elliptic Boundary Value Problems

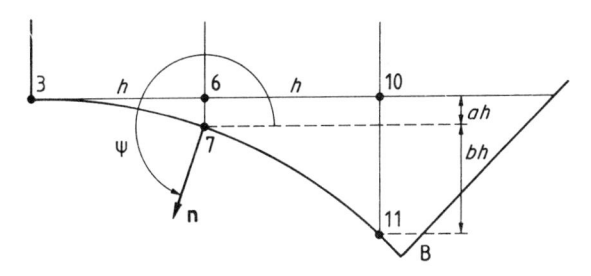

Figure 10.9 Derivation of the difference equation for the boundary point 7

get the operator equation

$$\begin{array}{c} \bullet^{-1} \\ \big|_{2(1+h)\bullet}{-1} \quad \odot u - h^2 + h = 0. \end{array} \tag{10.33}$$

The difference equation (10.33) is valid for the point A of the domain for arbitrary mesh sizes $h$ if the boundary conditions under consideration are to be satisfied.

The derivation of the difference equations for points 7 and 11 follows the procedure outlined in Example 10.4. Figure 10.9 shows the situation for the point 7 with the surrounding grid points used to obtain the difference equation. The angle $\psi$ between the positive $x$ axis and the direction of the normal vector **n** is $\psi \doteq 250.53°$, and the trigonometric function values are $\cos\psi = -\frac{1}{3}$ and $\sin\psi = -\sqrt{8/3} \doteq -0.942\,809$. Other useful values are $a = b_6 = 3 - 2\sqrt{2} \doteq 0.171\,573$ and $b = b_{10} - b_6 = 2\sqrt{2} - \sqrt{5} \doteq 0.592\,359$. With these values the Taylor series expansions of the function at the five points, as well as the expression on the left-hand side of the Cauchy boundary condition for point 7, can be formulated. If the second and third equations are divided by $h$, the fourth and sixth equations are divided by $h^2/2$, and the fifth equation is divided by $h^2$, then the system of equations for the six unknown parameters is

| $c_7$ | $c_3$ | $c_6$ | $c_{10}$ | $c_{11}$ | $c_n$ | 1 |
|---|---|---|---|---|---|---|
| 1 | 1 | 1 | 1 | 1 | 2 | 0 |
| 0 | −1 | 0 | 1 | 1 | −1 | 0 |
| 0 | 0.171 573 | 0.171 573 | 0.171 573 | −0.592 359 | −2.828 427 | 0 |
| 0 | 1 | 0 | 1 | 1 | 0 | −18 |
| 0 | −0.171 573 | 0 | 0.171 573 | −0.592 359 | 0 | 0 |
| 0 | 0.029 4373 | 0.029 4373 | 0.029 4373 | 0.350 889 | 0 | −18 |

448   Partial Differential Equations

We obtain the values $c_7 \doteq 597.59095$, $c_3 \doteq -6.33443$, $c_6 \doteq 518.25323$, $c_{10} \doteq 17.44645$, $c_{11} \doteq 6.88798$, $c_n \doteq 30.66886$. We take into account the value $-1$ of the Cauchy condition, multiply the difference equation by $-h^2 = -\frac{1}{9}$ and thus obtain the operator equation valid for point 7

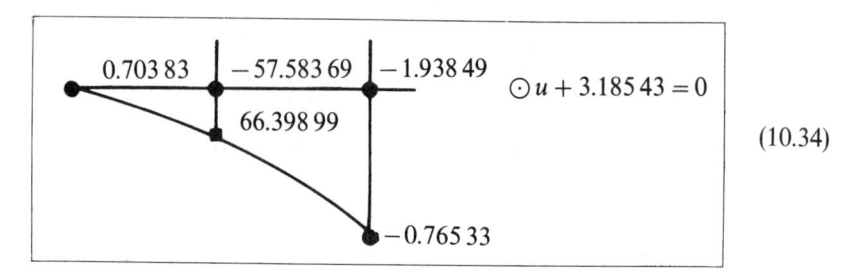

$$\begin{array}{c} 0.70383 \quad -57.58369 \quad -1.93849 \\ 66.39899 \\ -0.76533 \end{array} \odot u + 3.18543 = 0 \qquad (10.34)$$

The large coefficients of the function values $u_6$ and $u_7$ of this difference equation are remarkable. This is due to the small distance between points 6 and 7 and also to the Cauchy condition.

For the remaining grid point 11, Figure 10.10 shows the situation where the marked points are those function values which are used to approximate the differential expression. For the angle $\psi$ we have $\cos\psi = -\frac{2}{3}$ and $\sin\psi = -\sqrt{5/3} \doteq -0.745356$. Analogously to before the derived system of linear equations for the desired coefficients $c_{11}$, $c_7$, $c_6$, $c_{10}$, $c_{14}$ and $c_n$

| $c_{11}$ | $c_7$ | $c_6$ | $c_{10}$ | $c_{14}$ | $c_n$ | 1 |
|---|---|---|---|---|---|---|
| 1 | 1 | 1 | 1 | 1 | 2 | 0 |
| 0 | $-1$ | $-1$ | 0 | 1 | $-2$ | 0 |
| 0 | 0.592359 | 0.763932 | 0.763932 | 0.763932 | $-2.236068$ | 0 |
| 0 | 1 | 1 | 0 | 1 | 0 | $-18$ |
| 0 | $-0.592359$ | $-0.763932$ | 0 | 0.763932 | 0 | 0 |
| 0 | 0.350889 | 0.583592 | 0.583592 | 0.583592 | 0 | $-18$ |

yields in a similar way the operator equation for point 11

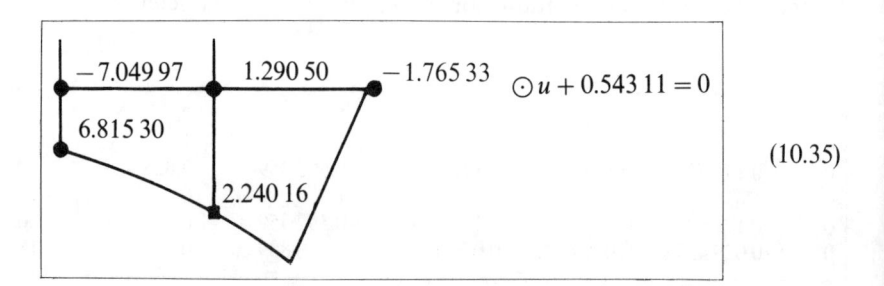

$$\begin{array}{c} -7.04997 \quad 1.29050 \quad -1.76533 \\ 6.81530 \\ 2.24016 \end{array} \odot u + 0.54311 = 0 \qquad (10.35)$$

| | $u_1$ | $u_2$ | $u_3$ | $u_4$ | $u_5$ | $u_6$ | $u_7$ | $u_8$ | $u_9$ | $u_{10}$ | $u_{11}$ | $u_{12}$ | $u_{13}$ | $u_{14}$ | $u_{15}$ | $u_{16}$ | $u_{17}$ | 1 |
|---|---|---|---|---|---|---|---|---|---|---|---|---|---|---|---|---|---|---|
| | 2 | −0.5 | | | | | | | | | | | | | | | | −0.111 |
| | −0.5 | 2 | −0.5 | | | | | | | | | | | | | | | −0.111 |
| | | 1 | 2.67 | | | | | | | | | | | | | | | 0.222 |
| | −1 | | | 4 | −1 | | | −1 | | | | | | | | | | −0.222 |
| | | −1 | | −1 | 4 | −1 | | | −1 | | | | | | | | | −0.222 |
| | | | −1 | | −1.71 | 13.7 | −9.95 | | | −1 | −1.48 | | | | | | | −0.222 |
| | | | 0.704 | | | −57.6 | 66.4 | | | −1.94 | −0.765 | | | | | | | 3.19 |
| | | | | −1 | | | | 4 | −1 | | | −1 | | | | | | −0.222 |
| | | | | | −1 | | | −1 | 4 | −1 | | | −1 | | | | | −0.222 |
| | | | | | | −1 | | | −1.13 | 4.62 | −1.48 | | | −1 | | | | −0.222 |
| | | | | | | | 6.82 | | | 1.29 | 2.24 | | | −1.77 | | | | 0.543 |
| | | | | | | | | −1 | | | | 4 | −1 | | −1 | | | −0.222 |
| | | | | | | | | | −1 | | | −1 | 4 | −1 | | −1 | | −0.222 |
| | | | | | | | | | | −1 | | | −1 | 2 | | | −1 | −0.111 |
| | | | | | | | | | | | | −1 | | | 4 | −1 | | −0.222 |
| | | | | | | | | | | | | | | | −1 | 2 | −1 | −0.111 |
| | | | | | | | | | | | | | | | −1 | | 2 | −0.111 |

(10.36)

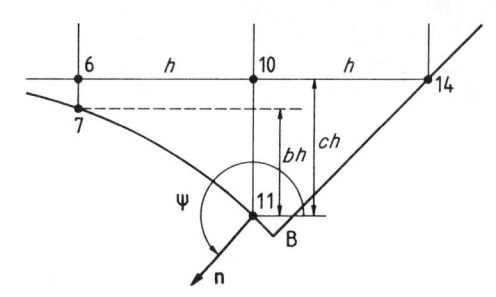

Figure 10.10 Derivation of the difference equation for the boundary point 11

After all these preliminaries for the irregular points near the boundary and for the points on the curved boundary we are able to write down the system of difference equations corresponding to the boundary value problem (10.31). Due to lack of space the noninteger valued coefficients are given to three decimal places in (10.36). The system of linear equations has been solved with the values of the operator equations (10.32) to (10.35). Other quantities not explicitly given have been rounded to six digits.

The matrix of the system of equations is not symmetric and does not even have a symmetric structure because we have $a_{73} \neq 0$ but $a_{37} = 0$. The matrix has a banded structure whereby the leftside bandwidth is $m_1 = 5$, due to the eleventh equation, and the rightside bandwidth is $m_2 = 4$. Although the matrix is not diagonally dominant (see 11th equation), the system can be solved by means of the Gaussian algorithm with diagonal pivot strategy with no problems. The process only involves matrix elements within the prescribed band so that a technique should be applied that is economical with respect to computational effort and storage requirements, similar to that of Section 1.3.2.

The solution of the system (10.36), rounded to four decimal places, is given in the approximate arrangement of the grid points together with the given boundary values in (10.37). The results provide an intuitive picture of the temperature distribution attaining its maximum in the interior of the domain and decreasing towards the inner boundary due to heat loss.

$$
\begin{array}{ccccccc}
0 & 0 & 0 & 0 & 0 & 0 & 0 \\
0.2912 & 0.3006 & 0.3197 & 0.3272 & 0.2983 & 0.2047 & \\
0.3414 & 0.3692 & 0.4288 & 0.4687 & 0.4391 & & \\
0.1137 & 0.1840 & 0.3353 & 0.4576 & & & \\
 & 0.1217 & 0.1337 & & & & \\
\end{array}
\qquad (10.37)
$$

The derivation of difference equations for points on the boundary with Neumann or Cauchy condition is troublesome and prone to errors. However, the determination of the coefficients can be performed by means of a relatively simple computer program. It only needs to use the information about neigh-

## 10.1 Elliptic Boundary Value Problems

bouring points (relative coordinates with respect to the boundary point) the type of the boundary condition including the angle $\psi$ and the corresponding values $\gamma$ or $\alpha$ and $\beta$, respectively, as well as the elliptic differential equation to be approximated. If we restrict ourselves to the Poisson equation, then the value of $f(x, y)$ at the boundary point is sufficient. It is possible to go a step further and to use the computer to generate the complete system of difference equations corresponding to a given boundary value problem, for instance, in the form of operator equations. The information on the differential equation, the domain, the boundary curves and boundary conditions and the desired net must be given in a certain way (Engeli, 1962).

### 10.1.4 Discretization errors

The computed function values at the grid points, as a solution of the system of linear difference equations, are only approximations of the corresponding values of the solution $u(x, y)$ of the given boundary value problem. In order to get at least qualitative estimates, we first determine the *local discretization error* of the difference approximation. In the following we restrict ourselves to considering the Poisson equation and the difference equations used so far. The local discretization error of a difference equation is defined as the value that results when the exact solution of the differential equation is substituted into the difference equation. Therefore in the case of the five-point difference equation (10.11) valid for a regular interior point $P(x, y)$ we obtain

$$d_P := \frac{1}{h^2}[u(x, y+h) + u(x-h, y) + u(x, y-h) + u(x+h, y)$$
$$- 4u(x, y)] - f(x, y). \qquad (10.38)$$

For the function $u(x, y)$ we use the Taylor series expansions

$$u(x \pm h, y) = u \pm h u_x + \tfrac{1}{2}h^2 u_{xx} \pm \tfrac{1}{6}h^3 u_{xxx} + \tfrac{1}{24}h^4 u_{xxxx}$$
$$\pm \tfrac{1}{120}h^5 u_{xxxxx} + \tfrac{1}{720}h^6 u_{xxxxxx} \pm \cdots \qquad (10.39)$$
$$u(x, y \pm h) = u \pm h u_y + \tfrac{1}{2}h^2 u_{yy} \pm \tfrac{1}{6}h^3 u_{yyy} + \tfrac{1}{24}h^4 u_{yyyy}$$
$$\pm \tfrac{1}{120}h^5 u_{yyyyy} + \tfrac{1}{720}h^6 u_{yyyyyy} \pm \cdots$$

In (10.39) the values of $u$ and the partial derivatives are taken for the values $(x, y)$. Their substitution into (10.38) yields

$$d_P = (u_{xx} + u_{yy} - f(x, y))_P + \tfrac{1}{12}h^2(u_{xxxx} + u_{yyyy})_P$$
$$+ \tfrac{1}{360}h^4(u_{xxxxxx} + u_{yyyyyy})_P + \cdots$$

We assumed that $u(x, y)$ is a solution of the Poisson equation thus the expression in the first brackets is equal to zero. Hence the local discretization error of the five-point difference equation at a regular interior grid point is given by

$$d_P = \tfrac{1}{12}h^2(u_{xxxx} + u_{yyyy})_P + \tfrac{1}{360}h^4(u_{xxxxxx} + u_{yyyyyy})_P + \cdots. \qquad (10.40)$$

From (10.40) we deduce that $d_P = O(h^2)$. This result remains valid for a boundary point with the Neumann condition $\partial u/\partial n = 0$ whenever the boundary coincides either with a net line or a diagonal of it.

For an irregular grid point P near the boundary according to Figure 10.5, from the Taylor expansions, similar to those of (10.39), and the derivation in Example 10.2, it is obvious that the local discretization error is proportional to the mesh size $h$ that is multiplied by some third partial derivatives. Consequently, for the local discretization error we have $d_P = O(h)$. The same is true for the difference equations for boundary points, which we derived in the Examples 10.3 to 10.5, since in all these cases the terms with third partial derivatives do not cancel when the local discretization error is analyzed.

In the next step we want to find the relationship between the local discretization error $d_P$ and the error $e_P := u(x, y) - u_P$ of the computed approximation at the point $P(x, y)$. To do this we assume that at all grid points a difference equation can be used whose local discretization error has the form (10.40). For a typical regular interior grid point P, according to (10.38), we have

$$\frac{1}{h^2}[u(x, y+h) + u(x-h, y) + u(x, y-h) + u(x+h, y)$$
$$- 4u(x, y)] - f(x, y) - d_P = 0,$$

also the approximations satisfy the difference equation

$$\frac{1}{h^2}(u_N + u_W + u_S + u_E - 4u_P) - f_P = 0.$$

By subtracting the two equations we have the following equation for the error

$$\frac{1}{h^2}(e_N + e_W + e_S + e_E - 4e_P) - d_P = 0. \qquad (10.41)$$

When (10.40) is taken into account, the usual multiplication of (10.41) by $-h^2$ yields, for each regular point P,

$$4e_P - e_N - e_W - e_S - e_E + C_P h^4 + D_P h^6 + \cdots = 0 \qquad (10.42)$$

where $C_P, D_P, \ldots$ are constants that depend on P and on the solution $u(x, y)$. The discrete errors satisfy a system of linear equations whose matrix $\mathbf{A}$ is identical to that of the system of difference equations. The components of the constant vector of (10.42) are $O(h^4)$. We combine the errors at the grid points in the error vector $\boldsymbol{\varepsilon}$ and the constants $C_P$ and $D_P$ in the vectors $\boldsymbol{\gamma}$ and $\boldsymbol{\delta}$, respectively. So from (10.42) we obtain the system

$$\mathbf{A}\boldsymbol{\varepsilon} + h^4\boldsymbol{\gamma} + h^6\boldsymbol{\delta} + \cdots = \mathbf{0}.$$

## 10.1 Elliptic Boundary Value Problems

Due to the nonsingularity of **A**, we can solve the last equation for $\varepsilon$

$$\varepsilon = -\mathbf{A}^{-1}(h^4\gamma + h^6\delta + \cdots)$$

From this, if we use the Euclidean vector norm and the subordinate spectral norm, we obtain the error estimate

$$\|\varepsilon\|_e \leq \|\mathbf{A}^{-1}\|_e \{h^4 \|\gamma\|_e + h^6 \|\delta\|_e + \cdots\}. \tag{10.43}$$

It can be shown that the spectral norm of the inverse of **A** satisfies $\|\mathbf{A}^{-1}\|_e \leq Kh^{-2}$, see Example 10.6. As a consequence of (10.43) we obtain the following error estimate

$$\|\varepsilon\|_e \leq K\{h^2 \|\gamma\|_e + h^4 \|\delta\|_e + \cdots\}. \tag{10.44}$$

The Euclidean norm of the error decreases as $h^2$. The *order of convergence* of the five-point formula (10.11) is two. At the same time it follows that the approximate solutions at the grid points converge, as $h \to 0$, to the exact values of the solution $u(x, y)$ of the boundary value problem on the implicit assumption that the solution $u(x, y)$ is sufficiently many times continuously differentiable on the closed domain. The inequality (10.44) can also be obtained by different methods (Collatz, 1966; Finkenstein, 1977).

An analogous statement of convergence is true if difference equations with a local discretization error $O(h)$ are present. In (10.43) we then have a $h^3$ term, and consequently the bound for the norm of the error is only proportional to $h$.

Once more we stress that the above convergence behaviour is only true if the solution $u(x, y)$ is at least four times continuously differentiable on $\bar{G} = G \cup \Gamma$. For instance, this assumption is not fulfilled if Dirichlet boundary conditions are discontinuous, or if the domain has obtuse corners. Then the low order partial derivatives of the solution are singular at the mentioned points. Such cases require special analytic techniques to describe the convergence behaviour correctly (Bramble *et al.*, 1968). It is often quite advantageous for the numerical solution of such problems to treat the singularity by means of adequate analytic tools (Gladwell and Wait, 1979; Mitchell and Griffiths, 1980).

*Example* 10.6 To show the increase of the spectral norm of $\mathbf{A}^{-1}$ mentioned above, we consider the boundary value problem

$$u_{xx} + u_{yy} = f(x, y) \quad \text{in } G = \{(x, y) | 0 < x < a, 0 < y < b\}, \quad u = \varphi(s) \quad \text{on } \Gamma.$$

Let the mesh size $h$ be chosen in such a way that we have $h = a/(N + 1) = b/(M + 1)$ with $N, M \in \mathbf{N}$. Figure 10.11 shows the case $N = 5$ and $M = 3$. The discretized problem has in general $N \cdot M$ interior grid points for which the five-point formula (10.11) can be applied.

If the grid points are numbered columnwise, the matrix **A** of order $N \cdot M$ has a regular block structure. For $N = 5$ and $M = 3$ it looks like

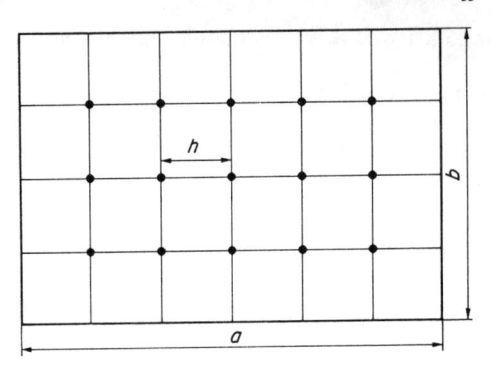

Figure 10.11 Rectangular domain

$$A = \begin{pmatrix} B & -I & & & \\ -I & B & -I & & \\ & -I & B & -I & \\ & & -I & B & -I \\ & & & -I & B \end{pmatrix} \in \mathbb{R}^{15 \times 15}$$

where $B = \begin{pmatrix} 4 & -1 & \\ -1 & 4 & -1 \\ & -1 & 4 \end{pmatrix}$. (10.45)

We can write the matrix $A$ as a sum of two matrices $A_1$ and $A_2$

$$A_1 := \begin{pmatrix} J & & & & \\ & J & & & \\ & & J & & \\ & & & J & \\ & & & & J \end{pmatrix}$$

$$A_2 := \begin{pmatrix} 2I & -I & & & \\ -I & 2I & -I & & \\ & -I & 2I & -I & \\ & & -I & 2I & -I \\ & & & -I & 2I \end{pmatrix} \quad J := \begin{pmatrix} 2 & -1 & \\ -1 & 2 & -1 \\ & -1 & 2 \end{pmatrix}. \quad (10.46)$$

The first matrix $A_1$ is block diagonal whereas $J$ is tridiagonal. The two symmetric matrices $A_1$ and $A_2$ commute, that is we have $A_1 A_2 = A_2 A_1$. Therefore, there exists an orthogonal matrix $U$ such that both matrices $A_1$ and $A_2$ are simultaneously reduced to diagonal matrices by means of the corresponding similarity transformation (Varga, 1961)

$$U^T A_1 U = D_1 \quad U^T A_2 U = D_2 \quad U^T = U^{-1} \quad (10.47)$$

## 10.1 Elliptic Boundary Value Problems

Furthermore, we have

$$U^T A U = U^T(A_1 + A_2)U = D_1 + D_2, \qquad (10.48)$$

so that the eigenvalues of $A$ can be found as the sum of the eigenvalues of $A_1$ and $A_2$. The eigenvalues of $A_1$ each have the multiplicity $N$ and are given by (Zurmühl and Falk, 1984)

$$\lambda_j^{(1)} = 2 - 2\cos\left(\frac{j\pi}{M+1}\right) \quad (j = 1, 2, \ldots, M). \qquad (10.49)$$

If the rows and columns of $A_2$ are simultaneously permuted in such a way that it corresponds to a rowwise numbering of the grid points, then $A_2$ becomes a block structure with tridiagonal matrices $J \in \mathbf{R}^{N \times N}$. Therefore the eigenvalues of $A_2$ each have the multiplicity $M$ and are

$$\lambda_k^{(2)} = 2 - 2\cos\left(\frac{k\pi}{N+1}\right) \quad (k = 1, 2, \ldots, N). \qquad (10.50)$$

If we recall the permutation with respect to the eigenvalues of $A_2$, then the eigenvalues of $A$ have the representation

$$\begin{aligned}
\lambda_{jk} &= 4 - 2\cos\left(\frac{j\pi}{M+1}\right) - 2\cos\left(\frac{k\pi}{N+1}\right) \\
&= 4\left[\sin^2\left(\frac{j\pi}{2M+2}\right) + \sin^2\left(\frac{k\pi}{2N+2}\right)\right]
\end{aligned}$$
$$(j = 1, 2, \ldots, M; k = 1, 2, \ldots, N). \qquad (10.51)$$

The spectral norm of $A^{-1}$ is equal to the reciprocal of the smallest eigenvalue of the symmetric and positive definite matrix $A$. From (10.51), on the assumption that $M \gg 1$ and $N \gg 1$, we obtain the smallest eigenvalue

$$\begin{aligned}
\lambda_{\min} = \lambda_{11} &= 4\left[\sin^2\left(\frac{\pi}{2M+2}\right) + \sin^2\left(\frac{\pi}{2N+2}\right)\right] \approx \pi^2\left(\frac{1}{(M+1)^2} + \frac{1}{(N+1)^2}\right) \\
&= \pi^2\left(\frac{h^2}{a^2} + \frac{h^2}{b^2}\right) = \pi^2\left(\frac{1}{a^2} + \frac{1}{b^2}\right)h^2 =: \frac{1}{C}h^2.
\end{aligned}$$

Therefore we have $\|A^{-1}\|_e \approx C \cdot h^{-2}$ for sufficiently small $h$. Note that it is easy to find the constant $K$ from the representation of $\lambda_{11}$ yielding an upper bound of the spectral norm of $A^{-1}$.

*Example* 10.7 To increase the accuracy of the approximate solution of the difference equations, the mesh size $h$ must be decreased. The error estimate (10.44) reduces the error approximately by a factor of four if the mesh size $h$ is halved. However, the number of grid points, and hence the order of the

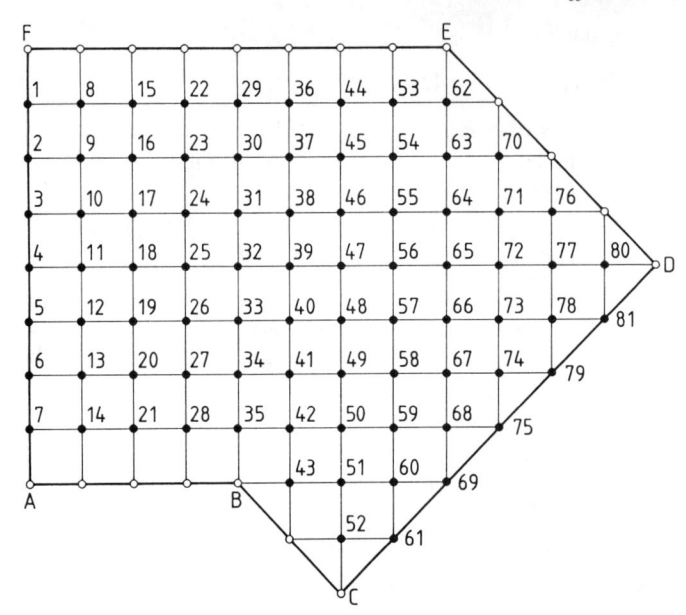

Figure 10.12 Mesh and grid points, $h = 0.125$

system of linear equations, is approximately quadrupled. As an illustration, we solve the boundary value problem (10.15) to (10.18) of Example 10.1 with the mesh size $h = 0.125$ and obtain $n = 81$ grid points with unknown function values, see Figure 10.12. The operator equations (10.19) can be used. If the grid points are numbered columnwise the system of equations is banded and its bandwith is $m = 9$.

For comparison the resulting approximations rounded to five decimal places are given in (10.52) for the desired solution at the grid points of Figure 10.4.

$$\begin{array}{llllllll}
0 & 0 & 0 & 0 & 0 \\
0.417\,71 & 0.411\,78 & 0.390\,70 & 0.342\,27 & 0.232\,86 & 0 \\
0.721\,53 & 0.712\,70 & 0.681\,49 & 0.614\,00 & 0.488\,58 & 0.283\,86 & 0 & \quad(10.52)\\
0.916\,86 & 0.909\,79 & 0.882\,68 & 0.818\,15 & 0.702\,44 & 0.523\,89 \\
1 & 1 & 1 & 0.948\,36 & 0.856\,02 \\
& & & 1
\end{array}$$

A comparison with the results (10.21), for twice the mesh size, shows quite a satisfactory agreement. The largest difference is found to be eight units in the third decimal place. In spite of the corners of the domain, if we assume the order of convergence to be two, an extrapolation shows that the values in (10.52) approximate the desired solution to an accuracy of at least two decimal places.

Each reduction of the mesh size $h$ causes a considerable increase in the number

## 10.1 Elliptic Boundary Value Problems

of unknowns. Moreover, the condition number of the matrix $\mathbf{A}$ of the system of difference equations becomes large, because of the increase of the spectral norm $\|\mathbf{A}^{-1}\|_e$. A different possibility of improving the accuracy of the approximate solution is to increase the order of the difference equations. For the construction of such difference approximations more function values must be used. It would be quite natural to approximate the second partial derivative by means of the difference approximation

$$u_{xx}(x,y) \approx \frac{1}{12h^2}[-u(x-2h,y) + 16u(x-h,y) - 30u(x,y)$$
$$+ 16u(x+h,y) - u(x+2h,y)],$$

whose discretization error is $O(h^4)$. For the Poisson equation, (10.3), after a multiplication by $-12h^2$ we obtain the operator equation for a regular interior grid point

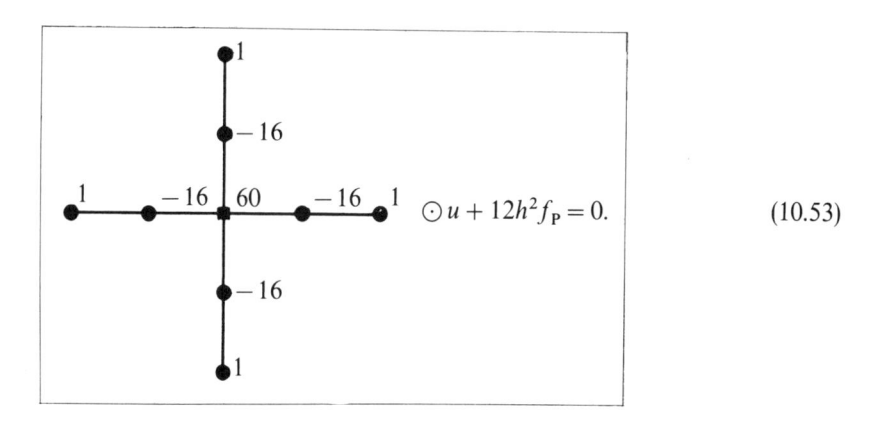

$$\odot u + 12h^2 f_P = 0. \tag{10.53}$$

(stencil: center 60; horizontal and vertical neighbours $-16$; next neighbours $1$)

As this difference equation uses function values of grid points which are in the $x$ and $y$ direction relatively far away from the central point, it is clear that special difference approximations must be derived for numerous grid points near the boundary. Because of this disadvantage we do not pursue (10.53) any further.

Instead of approximating the second partial derivatives separately we want to approximate the differential expression $\Delta u = u_{xx} + u_{yy}$ at the point P as an entity by a linear combination of function values. So we consider the eight neighbouring grid points of $P(x,y)$ according to Figure 10.13 and denote them by the compass directions.

The Taylor series expansions of the solution $u(x,y)$ at the nine grid points will serve to help us. Besides the expansions (10.39) we have analogous ones for

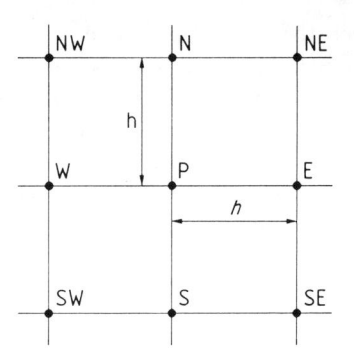

Figure 10.13 Part of the mesh showing the neighbouring points

the additional grid points NE, NW, SW and SE. For instance, we have

$$u(x \pm h, y + h) = u \pm hu_x + hu_y + \tfrac{1}{2}h^2(u_{xx} \pm 2u_{xy} + u_{yy})$$
$$+ \tfrac{1}{6}h^3(\pm u_{xxx} + 3u_{xxy} \pm 3u_{xyy} + u_{yyy}) \qquad (10.54)$$
$$+ \tfrac{1}{24}h^4(u_{xxxx} \pm 4u_{xxxy} + 6u_{xxyy} \pm 4u_{xyyy} + u_{yyyy}) + \cdots$$

Due to symmetry we combine the function values of four grid points to obtain simple expressions and define

$$\Sigma_1 := u(x, y + h) + u(x - h, y) + u(x, y - h) + u(x + h, y)$$
$$= 4u + h^2(u_{xx} + u_{yy}) + \tfrac{1}{12}h^4(u_{xxxx} + u_{yyyy}) + O(h^6) \qquad (10.55)$$

$$\Sigma_2 := u(x + h, y + h) + u(x - h, y + h) + u(x - h, y - h) + u(x + h, y - h)$$
$$= 4u + 2h^2(u_{xx} + u_{yy}) + \tfrac{1}{6}h^4(u_{xxxx} + 6u_{xxyy} + u_{yyyy}) + O(h^6). \qquad (10.56)$$

It is impossible to find a linear combination of $u$, $\Sigma_1$ and $\Sigma_2$ such that the fourth partial derivatives are eliminated and yields a difference approximation of $\Delta u$ with the aimed local discretization error $O(h^4)$. However, we can achieve a favourable combination of the fourth partial derivatives that can be treated further. So, in a first step from (10.55) and (10.56) we obtain

$$4\Sigma_1 + \Sigma_2 - 20u = 6h^2(u_{xx} + u_{yy}) + \tfrac{1}{2}h^4(u_{xxxx} + 2u_{xxyy} + u_{yyyy}) + O(h^6). \qquad (10.57)$$

However the expression in the second parentheses is equal to

$$u_{xxxx} + 2u_{xxyy} + u_{yyyy} = (u_{xx} + u_{yy})_{xx} + (u_{xx} + u_{yy})_{yy} = \Delta(\Delta u).$$

At this stage we use the fact that $u(x, y)$ should be a solution of the Poisson

## 10.1 Elliptic Boundary Value Problems

equation so that we have $\Delta u = f$ and hence $\Delta(\Delta u) = \Delta f = f_{xx} + f_{yy}$. Moreover, we remember that in all previous Taylor series the values are evaluated at the point P. Therefore we can replace the expression within the second parentheses *without error* by the value $\Delta f$ at the point P, and so we obtain

$$(u_{xx} + u_{yy})_P \approx \frac{1}{6h^2}(4\Sigma_1 + \Sigma_2 - 20u) - \frac{h^2}{12}(\Delta f)_P.$$

The exact values of the solution that have been used to define $\Sigma_1$ and $\Sigma_2$ are replaced by the approximations at the corresponding grid points. After multiplying by $-6h^2$ we obtain the following difference equation for a regular interior grid point P

$$\boxed{20u_P - 4(u_N + u_W + u_S + u_E) - u_{NE} - u_{NW} - u_{SW} - u_{SE} + \tfrac{1}{2}h^4(\Delta f)_P + 6h^2 f_P = 0}$$

(10.58)

It is represented in the easy remembered operator form

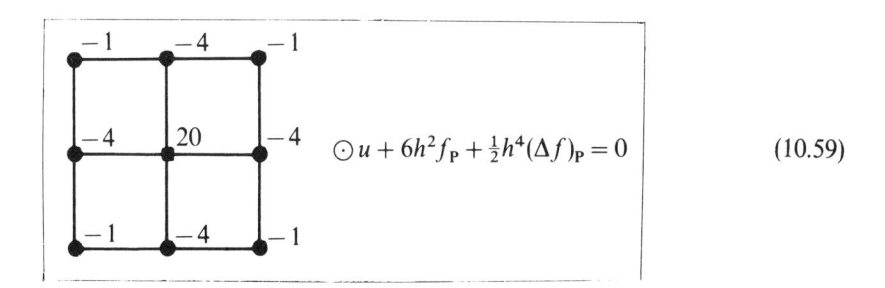

$$\begin{array}{|ccc|} \hline -1 & -4 & -1 \\ -4 & 20 & -4 \\ -1 & -4 & -1 \\ \hline \end{array} \quad \odot u + 6h^2 f_P + \tfrac{1}{2}h^4(\Delta f)_P = 0 \qquad (10.59)$$

The constant term in (10.59) is composed of the function value and the sum $f_{xx} + f_{yy}$ evaluated at the point P. If $f(x, y) =$ constant, then the last term is zero, of course. For the Laplace equation both terms are not present. The local discretization error of the difference approximation is obviously $O(h^4)$.

*Example* 10.8 We want to approximately solve the boundary value problem (10.15) to (10.18) for the region of Figure 10.4 with a net of the mesh size $h = 0.25$ by now using the difference equation (10.58). Since $f = -2$, the operator equation (10.59) simplifies. For the grid points on the boundary FA and CD the corresponding difference equations can easily be derived by symmetry considerations and subsequent division by 2. The same holds for the grid point 16 of Figure 10.4. The four types of difference equations are summarized in (10.60).

[Four operator stencils:]

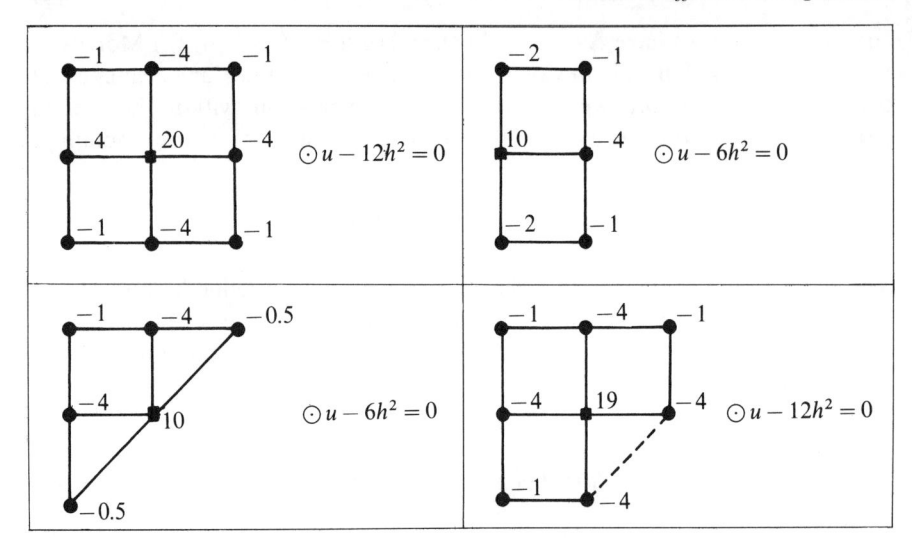

Stencil 1: center 20, edges −4, corners −1; $\odot u - 12h^2 = 0$

Stencil 2: center 10, with values −2, −1, −4, −2, −1; $\odot u - 6h^2 = 0$

Stencil 3 (triangular): center 10, with values −1, −4, −0.5, −4, −0.5; $\odot u - 6h^2 = 0$

Stencil 4: center 19, with values −1, −4, −1, −4, −4, −1, −4 (and dashed diagonal to −4); $\odot u - 12h^2 = 0$

(10.60)

Irregular grid points, with respect to this approximation, are those near the boundary with Dirichlet boundary condition, namely the points numbered 13, 14 and 18. Let us begin with the point 14. The grid point having been denoted by NE is missing. Its function value must be suitably eliminated from the difference equation (10.58). The simplest method to achieve this consists of applying a linear interpolation, which is somewhat crude in view of the high order of the difference equation. Along the straight line from P to NE we eliminate the function value $u_{\text{NE}}$ by means of the boundary point R midway between the two points. Thus we have

$$u_R = \tfrac{1}{2}(u_P + u_{\text{NE}}) \quad \text{i.e.} \quad u_{\text{NE}} = 2u_R - u_P,$$

and this leads to the following operator equation for the point 14

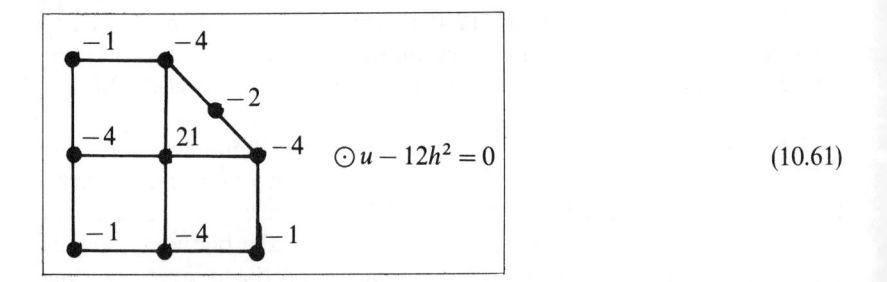

[Stencil: center 21, with values −1, −4, −2, −4, −4, −1, −4, −1]; $\odot u - 12h^2 = 0$ \hfill (10.61)

With the same procedure we can derive from the fourth operator equation (10.60) the corresponding difference equations for the points 13 and 18.

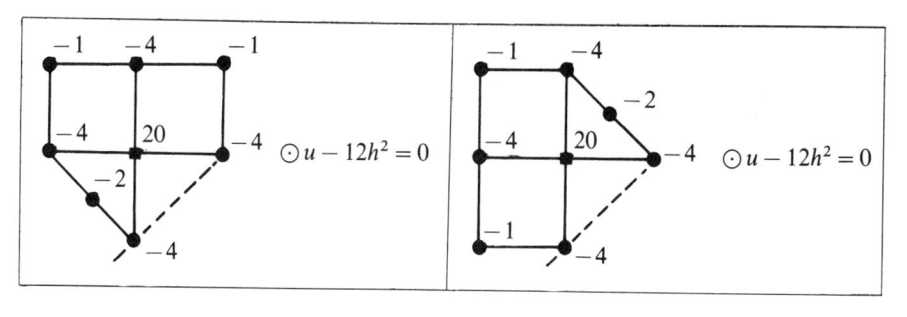

## 10.1 Elliptic Boundary Value Problems

$$\odot u - 12h^2 = 0 \qquad \odot u - 12h^2 = 0 \tag{10.62}$$

The system of linear equations (10.63) (see page 462) that is obtained from the difference equations (10.60) to (10.62) has a symmetric matrix **A**. In comparison with (10.20) the matrix **A** has more nonzero elements, and as a direct consequence of the more complicated operator equations its bandwith is $m = 5$.

The resulting approximations, rounded to five decimal places, are given in (10.64) in the arrangement of the grid points.

$$
\begin{array}{llllllll}
0 & 0 & 0 & 0 & 0 & & & \\
0.418\,88 & 0.413\,15 & 0.392\,82 & 0.346\,14 & 0.238\,62 & 0 & & \\
0.723\,27 & 0.714\,69 & 0.684\,21 & 0.617\,59 & 0.492\,53 & 0.288\,80 & 0 & \\
0.918\,23 & 0.911\,52 & 0.885\,30 & 0.821\,84 & 0.706\,21 & 0.527\,76 & & \\
1 & 1 & 1 & 0.953\,43 & 0.859\,93 & & & \\
& & & 1 & & & & \\
\end{array} \tag{10.64}
$$

Instead of linear interpolation as an auxiliary means of eliminating a function value that does not exist we can apply quadratic interpolation. Again we consider the grid point 14 and now use the quadratic interpolation polynomial along the line from the point SW to NE corresponding to the non-equidistant support abscissae SW, P and R with the function values $u_{SW}$, $u_P$ and $u_R$, respectively. We denote by $H := h\sqrt{2}$ the distance between SW and P. Newton's interpolation polynomial with the parameter $t$, is

$$P_2(t) = u_{SW} + (t+H)\frac{u_P - u_{SW}}{H} + (t+H)t\frac{2u_R - 3u_P + u_{SW}}{\frac{3}{2}H^2}.$$

Its value for $t = H$ is defined to be $u_{NE}$ and is given by

$$u_{NE} := \tfrac{1}{3}u_{SW} - 2u_P + \tfrac{8}{3}u_R.$$

This extrapolated value is substituted into the difference equation. Instead of (10.61) and (10.62) we obtain the following operator equations valid for the points 14, 13 and 18.

$$
\begin{array}{|ccc|ccc|ccc|cccc|cccc|cc|c|}
\hline
u_1 & u_2 & u_3 & u_4 & u_5 & u_6 & u_7 & u_8 & u_9 & u_{10} & u_{11} & u_{12} & u_{13} & u_{14} & u_{15} & u_{16} & u_{17} & u_{18} & u_{19} & 1 \\
\hline
10 & -2 & -2 & & & & & & & & & & & & & & & & & -0.375 \\
-2 & 10 & -2 & & & & & & & & & & & & & & & & & -0.375 \\
-2 & -2 & 10 & & & & & & & & & & & & & & & & & -3.375 \\
\hline
-4 & -1 & -1 & 20 & -4 & -4 & -4 & -1 & -1 & & & & & & & & & & & -0.750 \\
-1 & -4 & -1 & -4 & 20 & -4 & -1 & -4 & -1 & & & & & & & & & & & -0.750 \\
-1 & -1 & -4 & -4 & -4 & 20 & -1 & -1 & -4 & & & & & & & & & & & -6.750 \\
\hline
 & & & -4 & -1 & -1 & 20 & -4 & -4 & -4 & -1 & -1 & & & & & & & & -0.750 \\
 & & & -1 & -4 & -1 & -4 & 20 & -4 & -1 & -4 & -1 & & & & & & & & -0.750 \\
 & & & -1 & -1 & -4 & -4 & -4 & 20 & -1 & -1 & -4 & & & & & & & & -5.750 \\
\hline
 & & & & & & -4 & -1 & -1 & 20 & -4 & -4 & -4 & -4 & -1 & -1 & -1 & & & -0.750 \\
 & & & & & & -1 & -4 & -1 & -4 & 20 & -4 & -1 & -1 & -4 & -1 & -1 & & & -0.750 \\
 & & & & & & -1 & -1 & -4 & -4 & -4 & 20 & -4 & -1 & -1 & -4 & -1 & & & -1.750 \\
 & & & & & & & & & -4 & -4 & -4 & 20 & -1 & -1 & -1 & -4 & & & -10.75 \\
\hline
 & & & & & & & & & -4 & -1 & -1 & & 21 & -4 & -4 & -4 & -1 & -1 & -0.750 \\
 & & & & & & & & & -1 & -4 & -1 & & -4 & 20 & -4 & -1 & -4 & -1 & -0.750 \\
 & & & & & & & & & -1 & -1 & -4 & & -4 & -4 & 19 & -4 & -1 & -4 & -0.750 \\
 & & & & & & & & & & & & & -4 & -1 & -4 & 10 & -1 & -0.5 & -0.875 \\
\hline
 & & & & & & & & & & & & & -1 & -4 & -1 & & 20 & -4 & -0.750 \\
 & & & & & & & & & & & & & -1 & -1 & -4 & -0.5 & -4 & 10 & -0.375 \\
\hline
\end{array}
$$

$$(10.63)$$

## 10.1 Elliptic Boundary Value Problems

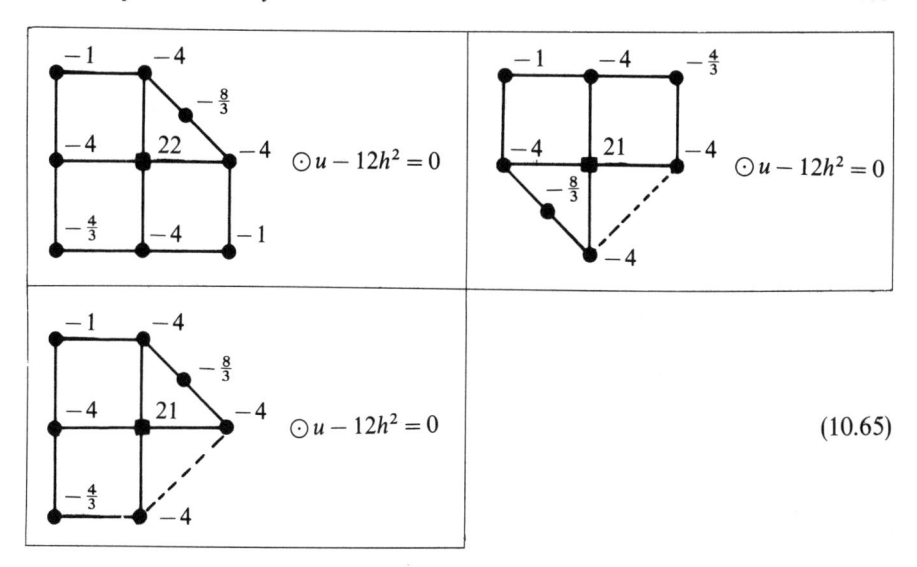

$$\odot u - 12h^2 = 0 \qquad \odot u - 12h^2 = 0$$

$$\odot u - 12h^2 = 0 \tag{10.65}$$

The first twelve equations of the system (10.63) are unchanged by these new equations. Therefore we only give the last seven equations in (10.66).

| $\ldots u_9$ | $u_{10}$ | $u_{11}$ | $u_{12}$ | $u_{13}$ | $u_{14}$ | $u_{15}$ | $u_{16}$ | $u_{17}$ | $u_{18}$ | $u_{19}$ | 1 |
|---|---|---|---|---|---|---|---|---|---|---|---|
| $-1$ |  |  | $-4$ | 21 |  |  | $-1.333$ | $-4$ |  |  | $-11.41667$ |
|  | $-4$ | $-1.333$ |  |  | 22 | $-4$ |  |  | $-1$ |  | $-0.750$ |
|  | $-1$ | $-4$ | $-1$ |  | $-4$ | 20 | $-4$ |  | $-4$ | $-1$ | $-0.750$ |
|  |  | $-1$ | $-4$ | $-1$ |  | $-4$ | 19 | $-4$ | $-1$ | $-4$ | $-0.750$ |
|  |  |  | $-1$ | $-4$ |  |  | $-4$ | 10 |  | $-0.5$ | $-0.875$ |
|  |  |  |  |  | $-1$ | $-4$ | $-1.333$ |  | 21 | $-4$ | $-0.750$ |
|  |  |  |  |  |  | $-1$ | $-4$ | $-0.5$ | $-4$ | 10 | $-0.375$ |

(10.66)

The matrix of the system is not symmetric, as a consequence of the operator equations (10.65), and cannot be made symmetric. Yet, the system of equations can be solved by means of the Gaussian algorithm with diagonal strategy. The solution is given in (10.67) according to the arrangement of the grid points.

$$
\begin{array}{llllllll}
0 & 0 & 0 & 0 & 0 & & & \\
0.41861 & 0.41281 & 0.39223 & 0.34505 & 0.23622 & 0 & & \\
0.72286 & 0.71416 & 0.68326 & 0.61595 & 0.49011 & 0.28494 & 0 & \quad (10.67) \\
0.91792 & 0.91108 & 0.88432 & 0.81975 & 0.70355 & 0.52476 & & \\
1 & 1 & 1 & 0.94982 & 0.85706 & & & \\
& & & 1 & & & &
\end{array}
$$

A comparison with the approximations (10.52) shows that the values differ at most by three digits in the third decimal place. The quadratic interpolation for eliminating function values in difference equations for irregular grid points causes an improvement but at the expense of losing the symmetry of the system of linear equations.

*Example* 10.9 The boundary value problem (10.15) to (10.18) for the region G of Figure 10.4 has been approximately solved by means of the difference equations (10.60) to (10.62) for the net of Figure 10.12 with mesh size $h = 0.125$. For this purpose we had to derive additional operator equations of the type (10.62) for some grid points near the boundary. The resulting system of linear equations is symmetric. Its solution is given for the selected grid points of Figure 10.4 in the corresponding arrangement in (10.68).

|         |         |         |         |         |         |         |   |         |
|---------|---------|---------|---------|---------|---------|---------|---|---------|
| 0       | 0       | 0       | 0       | 0       |         |         |   |         |
| 0.41822 | 0.41232 | 0.39139 | 0.34309 | 0.23216 | 0       |         |   |         |
| 0.72226 | 0.71340 | 0.68209 | 0.61432 | 0.48836 | 0.28417 | 0       |   | (10.68) |
| 0.91744 | 0.91036 | 0.88287 | 0.81830 | 0.70223 | 0.52376 |         |   |         |
| 1       | 1       | 1       | 0.94855 | 0.85588 |         |         |   |         |
|         |         |         | 1       |         |         |         |   |         |

If we assume, in spite of the corners of the domain, that the error of the approximations behaves like $h^4$ it follows from extrapolating the values of (10.68) and (10.64) that the results (10.68) are approximations of $u(x, y)$ at these grid points to four decimal places. The nine-point difference equation (10.58), compared with the five-point difference equation (10.11), yields more accurate approximations as was expected. Moreover, the errors decrease more rapidly with decreasing mesh size $h$.

### 10.1.5 Supplements

We can increase the order of the difference approximation for the Poisson equation without enlarging the number of grid points, in comparison to the nine-point formula (10.58). There we used the differential equation at the point P so that a certain term of the Taylor series expansion of the local discretization error was eliminated. Now the differential expression itself is used in some of the neighbouring grid points beside the function values. The value of the differential expression at the corresponding grid points can be replaced by the known function on the right-hand side of the given differential equation. The Taylor series of $u_{xx} + u_{yy}$ at the four neighbouring grid points N, W, S and E are

$$u_{xx}(x \pm h, y) + u_{yy}(x \pm h, y) = u_{xx} + u_{yy} \pm hu_{xxx} \pm hu_{xyy} + \tfrac{1}{2}h^2 u_{xxxx}$$
$$+ \tfrac{1}{2}h^2 u_{xxyy} + \cdots$$
$$u_{xx}(x, y \pm h) + u_{yy}(x, y \pm h) = u_{xx} + u_{yy} \pm hu_{xxy} \pm hu_{yyy} + \tfrac{1}{2}h^2 u_{xxyy}$$
$$+ \tfrac{1}{2}h^2 u_{yyyy} + \cdots$$

## 10.1 Elliptic Boundary Value Problems

and hence the sum of these four differential expressions is

$$\Sigma_3 := u_{xx}(x, y+h) + u_{yy}(x, y+h) + u_{xx}(x-h, y) + u_{yy}(x-h, y)$$
$$+ u_{xx}(x, y-h) + u_{yy}(x, y-h) + u_{xx}(x+h, y) + u_{yy}(x+h, y)$$
$$= 4(u_{xx} + u_{xy}) + h^2(u_{xxxx} + 2u_{xxyy} + u_{yyyy}) + \cdots. \qquad (10.69)$$

Together with $\Sigma_1$ and $\Sigma_2$, as defined in (10.55) and (10.56), the fourth partial derivatives can be eliminated by means of the following linear combination

$$8\Sigma_1 + 2\Sigma_2 - h^2\Sigma_3 - 40u = 8h^2(u_{xx} + u_{yy})_P + O(h^6) \qquad (10.70)$$

Accordingly from (10.70) we obtain the approximation

$$(u_{xx} + u_{yy})_P \approx \frac{1}{8h^2}(8\Sigma_1 + 2\Sigma_2 - h^2\Sigma_3 - 40u)$$

in which we replace the quantities $u_{xx} + u_{yy}$ occurring in $\Sigma_3$ by the values of $f$ at the corresponding four grid points. In $\Sigma_1$ and $\Sigma_2$ the function values $u$ are again replaced by their approximations, and so for the Poisson equation $u_{xx} + u_{yy} = f(x, y)$ we obtain the difference equation valid for a regular interior grid point **P**

$$\boxed{\begin{array}{c} 20u_P - 4(u_N + u_W + u_S + u_E) - u_{NE} - u_{NW} - u_{SW} - u_{SE} \\ + \tfrac{1}{2}h^2(8f_P + f_N + f_W + f_S + f_E) = 0 \end{array}} \qquad (10.71)$$

Its operator form is

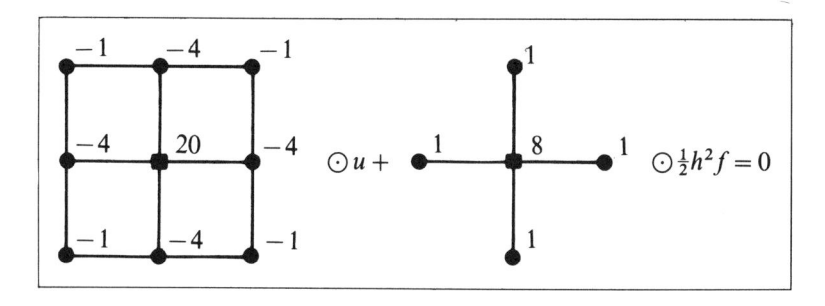

$$\begin{array}{|ccc|}\hline -1 & -4 & -1 \\ -4 & 20 & -4 \\ -1 & -4 & -1 \\ \hline \end{array} \odot u + \begin{array}{ccc} & 1 & \\ 1 & 8 & 1 \\ & 1 & \end{array} \odot \tfrac{1}{2}h^2 f = 0$$

(10.72)

The value of the function $f(x, y)$ appears in (10.71) at several grid points, because the differential equation has been used at several points to derive the difference approximation. For this reason (10.72) could be called a *multiple-point operator*, but it is usually called a *Hermitian operator* (Collatz, 1966), to underline the analogy to the Hermitian interpolation where higher order derivatives are used for a better approximation. It is obvious from the derivation that the local discretization error of (10.71) is at least $O(h^4)$. However, a more careful analysis shows that it is $O(h^6)$, so that the order of the Hermitian formula (10.72) is six. Moreover, it is of interest that for the special case $f(x, y) = $ constant the operator equation (10.72) simplifies and for $f = -2$ reduces to the first equation (10.60). Therefore the difference equation (10.60) is of order six, and this explains the high accuracy of the approximate solutions (10.64), (10.67) and (10.68).

The discretization of a general elliptic partial differential equation, (10.1), is performed according to the procedure previously sketched. In the simplest case the partial derivatives are approximated by difference quotients (10.8) and (10.9). At a regular interior grid point the second mixed partial derivative is usually approximated by

$$u_{xy}(x_i, y_j) \approx \frac{u_{i+1,j+1} - u_{i-1,j+1} - u_{i+1,j-1} - u_{i-1,j-1}}{4h^2}$$

which results from a double application of the central difference quotients (10.8). Whenever the given partial differential equation contains a term $u_{xy}$ then the simplest difference approximation has the structure of the nine-point formula. Complicated boundary conditions can be systematically treated by means of Taylor series expansions. Likewise this technique is adequate to approximate the differential expression by means of a linear combination of function values. In this way we directly obtain information on the local discretization error (Engeli, 1962).

In certain situations it is appropriate to use a net with variable mesh sizes in the $x$ and $y$ direction, so that a locally finer discretization is generated. This can be advisable for subregions where it is known that the solution varies fast or that the derivatives even become singular, for example at an obtuse corner. For the region

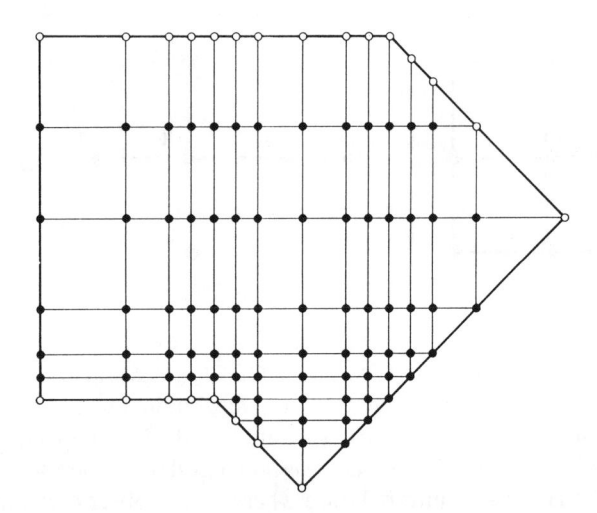

Figure 10.14 Domain with irregular mesh

## 10.1 Elliptic Boundary Value Problems

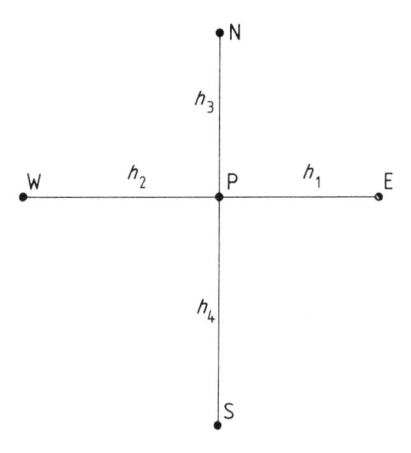

Figure 10.15 Grid points of the irregular mesh

$G$ of the boundary value problem, (10.15) to (10.18), the net shown in Figure 10.14 takes into account the obtuse corner B. The chosen mesh size $h = 1/6$ in the $y$ direction near the point B causes, because of the Neumann condition on CD, a small mesh size in a subregion where it is not necessary.

The approximation of the Poisson equation must now use function values at grid points that are not equidistant (see Figure 10.15). From (10.22) and (10.23) the following approximations of the second partial derivatives can be derived

$$u_{xx}(P) \approx 2\left(\frac{u_E}{h_1(h_1 + h_2)} + \frac{u_W}{h_2(h_1 + h_2)} - \frac{u_P}{h_1 h_2}\right)$$

$$u_{yy}(P) \approx 2\left(\frac{u_N}{h_3(h_3 + h_4)} + \frac{u_S}{h_4(h_3 + h_4)} - \frac{u_P}{h_3 h_4}\right).$$

For a typical grid point P we obtain the difference equation

$$\boxed{\begin{aligned}2\left(\frac{1}{h_1 h_2} + \frac{1}{h_3 h_4}\right)u_P &- \frac{2u_N}{h_3(h_3 + h_4)} - \frac{2u_W}{h_2(h_1 + h_2)} - \frac{2u_S}{h_4(h_3 + h_4)} \\ &- \frac{2u_E}{h_1(h_1 + h_2)} + f_P = 0.\end{aligned}} \qquad (10.73)$$

The local discretization error of (10.73) is only $O(h)$ where $h = \max(h_1, h_2, h_3, h_4)$ and whenever $h_1 \neq h_2$ or $h_3 \neq h_4$. The resulting system of linear difference equations is in general nonsymmetric for obvious reasons.

## 10.2 Parabolic Initial Boundary Value Problems

The mathematical description of time-dependent diffusion or heat conduction problems leads to a parabolic partial differential equation for the desired function that now depends on time and space variables. We first extensively treat the one-dimensional case, describe two methods of discretization with different properties and subsequently consider the two-dimensional case.

### 10.2.1 One-dimensional problems, explicit method

The simplest parabolic differential equation is

$$u_t = u_{xx} \tag{10.74}$$

for a function $u(x, t)$ of space variable $x$ and time $t$. Usually the function $u(x, t)$ is required on a bounded interval of $x$, which, without loss of generality, we assume to be the unit interval $(0, 1)$ and for positive values of $t$. Therefore the region $G$ in which the solution has to be determined consists of an infinite half strip of the $(x, t)$ plane. The solution of the differential equation (10.74) is subject to conditions that can be divided into two classes. An *initial condition*

$$u(x, 0) = f(x) \quad 0 < x < 1 \tag{10.75}$$

prescribes the values of the wanted solution at the time $t = 0$. Moreover, for $x = 0$ and for $x = 1$ *boundary conditions* for all $t > 0$ must be given. Either the value of $u$ is prescribed as a function of the time $t$, Dirichlet condition, or a linear combination of the partial derivative of $u$, with respect to $x$, and the function value must be equal to a given value, which is in general time-dependent, Cauchy condition. An example of such boundary conditions is

$$u(0, t) = \varphi(t) \quad u_x(1, t) + \alpha(t)u(1, t) = \beta(t) \quad t > 0 \tag{10.76}$$

where $\varphi(t)$, $\alpha(t)$ and $\beta(t)$ are given functions of time $t$.

The initial boundary value problem (10.74) to (10.76) is discretized in a similar way as the elliptic boundary value problems. First a net is defined on the domain $[0, 1] \times [0, \infty]$ with, in general, two different mesh sizes $h$ and $k$ in the $x$ and $t$ directions, respectively. Then we want to compute approximations of $u(x, t)$ at the so defined grid points. Next the partial differential equation is replaced by a difference approximation, whereby the boundary conditions are simultaneously taken into account. This will allow us to compute approximate values of the function $u(x, t)$ stepwise for increasing time $t$.

## 10.2 Parabolic Initial Boundary Value Problems

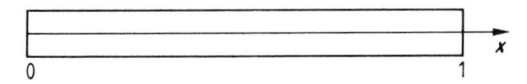

Figure 10.16 Heat conduction in a rod

*Example* 10.10 We consider the heat conduction in a homogenous bar of unit length and of constant section. It is assumed to be insulated on its longitudinal surface so that no heat radiation takes place. At its left end, see Figure 10.16, the temperature changes periodically, and its right end is insulated. If the temperature distribution is known at the time $t = 0$, we look for the temperature distribution in the bar as a function of the space variable $x$ and the time $t$.

The initial boundary value problem for the desired temperature distribution $u(x, y)$ is

$$\begin{aligned} &u_t = u_{xx} \quad \text{for } 0 < x < 1, t > 0; \\ &u(x, 0) = 0 \quad \text{for } 0 < x < 1; \\ &u(0, t) = \sin(\pi t) \quad u_x(1, t) = 0 \quad \text{for } t > 0. \end{aligned} \qquad (10.77)$$

Figure 10.17 shows the net in a part of the half strip for the mesh sizes $h = 1/n$ and $k$. The grid points for which the function values are known either by the initial or by the boundary condition are marked by an empty circle, whereas the grid points with unknown function values are marked by full circles. The grid points have the coordinates $x_i = ih$, $i = 0, 1, \ldots, n$, and $t_j = jk$, $j = 0, 1, 2, \ldots$. We denote the desired approximations of the solution $u(x_i, t_j)$ by $u_{i,j}$. To approximate the partial differential equation at an interior point $P(x_i, t_j)$ we replace the first partial

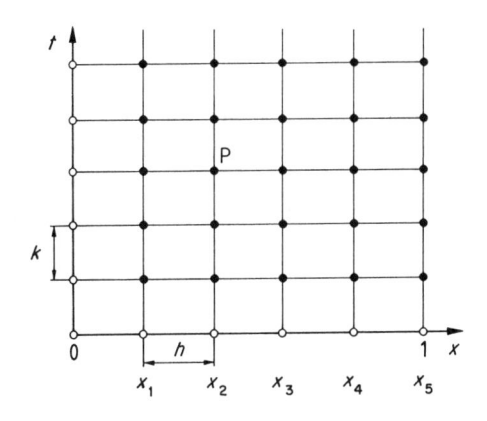

Figure 10.17 Mesh of the half-stripe

derivative, with respect to $t$, by the so-called forward difference quotient

$$u_t(P) \approx \frac{u_{i,j+1} - u_{i,j}}{k}$$

and the second partial derivative, with respect to $x$, by the second difference quotient

$$u_{xx}(P) \approx \frac{u_{i+1,j} - 2u_{i,j} + u_{i-1,j}}{h^2}.$$

Equating the two expressions yields the difference equation

$$u_{i,j+1} - u_{i,j} = \frac{k}{h^2}(u_{i+1,j} - 2u_{i,j} + u_{i-1,j})$$

or equivalently

$$u_{i,j+1} = r u_{i-1,j} + (1 - 2r)u_{i,j} + r u_{i+1,j} \quad r := \frac{k}{h^2}$$

$$(i = 1, 2, \ldots, n-1; \; j = 0, 1, 2, \ldots). \tag{10.78}$$

The boundary condition for the left boundary is easily taken into account because the value $u_{0,j} = \sin(\pi j k)$ can be used in (10.78) for $i = 1$. The Neumann condition on the right boundary is satisfied by means of a symmetry consideration, so that from (10.78) we obtain

$$u_{n,j+1} = 2r u_{n-1,j} + (1 - 2r)u_{n,j} \quad (j = 0, 1, 2, \ldots). \tag{10.79}$$

At time $t = 0$, corresponding to $j = 0$, the function values $u_{i,0}$ are known for $i = 0, 1, \ldots, n$ from the initial condition. The formulae (10.78) and (10.79) allow us to compute the approximations $u_{i,j+1}$, $i = 1, 2, \ldots, n$, for fixed index $j$ from the values $u_{i,j}$ in an explicit manner. Therefore the approximate solution can be successively determined with increasing $j$, that is in the time direction. The discretization applied to the parabolic differential equation is called *Richardson's explicit method*.

We have computed approximate solutions of the initial boundary value problem (10.77) by means of the schemes (10.78) and (10.79) for a fixed mesh size $h = 0.1$ but with different time steps $k = 0.002$, $k = 0.005$ and $k = 0.01$. The results obtained are partly given in Tables 10.1 to 10.3.

## 10.2 Parabolic Initial Boundary Value Problems

Table 10.1 Heat conduction, $h = 0.1$, $k = 0.002$, $r = 0.2$; explicit method

| $t$ | $j$ | $u_{0,j}$ | $u_{1,j}$ | $u_{2,j}$ | $u_{3,j}$ | $u_{4,j}$ | $u_{6,j}$ | $u_{8,j}$ | $u_{10,j}$ |
|---|---|---|---|---|---|---|---|---|---|
| 0   | 0   | 0      | 0      | 0      | 0      | 0      | 0      | 0      | 0      |
| 0.1 | 50  | 0.3090 | 0.2139 | 0.1438 | 0.0936 | 0.0590 | 0.0212 | 0.0069 | 0.0035 |
| 0.2 | 100 | 0.5878 | 0.4580 | 0.3515 | 0.2657 | 0.1980 | 0.1067 | 0.0599 | 0.0456 |
| 0.3 | 150 | 0.8090 | 0.6691 | 0.5476 | 0.4441 | 0.3578 | 0.2320 | 0.1611 | 0.1383 |
| 0.4 | 200 | 0.9511 | 0.8222 | 0.7050 | 0.6009 | 0.5107 | 0.3727 | 0.2909 | 0.2639 |

Table 10.2 Heat conduction, $h = 0.1$, $k = 0.005$, $r = 0.5$; explicit method

| $t$ | $j$ | $u_{0,j}$ | $u_{1,j}$ | $u_{2,j}$ | $u_{3,j}$ | $u_{4,j}$ | $u_{6,j}$ | $u_{8,j}$ | $u_{10,j}$ |
|---|---|---|---|---|---|---|---|---|---|
| 0   | 0   | 0      | 0      | 0      | 0      | 0      | 0      | 0      | 0      |
| 0.1 | 20  | 0.3090 | 0.2136 | 0.1430 | 0.0927 | 0.0579 | 0.0201 | 0.0061 | 0.0027 |
| 0.2 | 40  | 0.5878 | 0.4578 | 0.3510 | 0.2650 | 0.1970 | 0.1053 | 0.0583 | 0.0439 |
| 0.3 | 60  | 0.8090 | 0.6689 | 0.5472 | 0.4435 | 0.3569 | 0.2306 | 0.1594 | 0.1365 |
| 0.4 | 80  | 0.9511 | 0.8222 | 0.7049 | 0.6006 | 0.5101 | 0.3716 | 0.2895 | 0.2624 |
| 0.5 | 100 | 1.0000 | 0.9007 | 0.8049 | 0.7156 | 0.6350 | 0.5060 | 0.4263 | 0.3994 |
| 0.6 | 120 | 0.9511 | 0.8955 | 0.8350 | 0.7736 | 0.7147 | 0.6142 | 0.5487 | 0.5260 |
| 0.7 | 140 | 0.8090 | 0.8063 | 0.7904 | 0.7661 | 0.7376 | 0.6804 | 0.6387 | 0.6235 |
| 0.8 | 160 | 0.5878 | 0.6408 | 0.6737 | 0.6916 | 0.6985 | 0.6941 | 0.6828 | 0.6776 |
| 0.9 | 180 | 0.3090 | 0.4147 | 0.4954 | 0.5555 | 0.5992 | 0.6510 | 0.6731 | 0.6790 |
| 1.0 | 200 | 0      | 0.1497 | 0.2718 | 0.3699 | 0.4474 | 0.5528 | 0.6076 | 0.6245 |

Table 10.3 Heat conduction, $h = 0.1$, $k = 0.01$, $r = 1.0$; explicit method

| $t$ | $j$ | $u_{0,j}$ | $u_{1,j}$ | $u_{2,j}$ | $u_{3,j}$ | $u_{4,j}$ | $u_{6,j}$ | $u_{8,j}$ | $u_{10,j}$ |
|---|---|---|---|---|---|---|---|---|---|
| 0    | 0  | 0      | 0      | 0       | 0       | 0       | 0       | 0       | 0 |
| 0.05 | 5  | 0.1564 | 0.0312 | 0.1256  | −0.0314 | 0.0314  | 0       | 0       | 0 |
| 0.10 | 10 | 0.3090 | 5.4638 | −8.2955 | 8.8274  | −6.7863 | −2.0107 | −0.1885 | 0 |

In the first two cases, $k = 0.002$ and $k = 0.005$, we obtain approximations that are qualitatively correct, whereas we must expect larger errors in the second case because of the coarser time step $k$. In the third case with $k = 0.01$ we only need to perform a few steps to see that the results are meaningless. The explicit method is obviously unstable for this combination of mesh sizes $h$ and $k$ with a value $r = 1.0$.

To investigate the properties of Richardson's explicit method first we determine the local discretization error of the computational scheme (10.78). It is defined by means of the solution $u(x,t)$ of the problem (10.77) by

$$d_{i,j+1} := u(x_i, t_{j+1}) - ru(x_{i-1}, t_j) - (1 - 2r)u(x_i, t_j) - ru(x_{i+1}, t_j)$$
$$= u + ku_t + \tfrac{1}{2}k^2 u_{tt} + \cdots$$
$$- r(u - hu_x + \tfrac{1}{2}h^2 u_{xx} - \tfrac{1}{6}h^3 u_{xxx} + \tfrac{1}{24}h^4 u_{xxxx} - + \cdots)$$
$$- (1 - 2r)u$$
$$- r(u + hu_x + \tfrac{1}{2}h^2 u_{xx} + \tfrac{1}{6}h^3 u_{xxx} + \tfrac{1}{24}h^4 u_{xxxx} + + \cdots)$$
$$= k(u_t - u_{xx}) + \tfrac{1}{2}k^2 u_{tt} - \tfrac{1}{12}kh^2 u_{xxxx} + \cdots$$

where we have used $k = rh^2$. The coefficient of $k$ is equal to zero because $u(x,t)$ is supposed to satisfy the differential equation. Hence for the local discretization error we have

$$d_{i,j+1} = \tfrac{1}{2}k^2 u_{tt}(x_i, t_j) - \tfrac{1}{12}kh^2 u_{xxxx}(x_i, t_j) + \cdots = O(k^2) + O(kh^2). \quad (10.80)$$

In order to get an estimate of the *global discretization error*, $g_{i,j+1}$, of the method we use the fact that the computational scheme essentially performs an integration with respect to time. The formulae (10.78) and (10.79) correspond to *Euler's method* (9.4) for solving a system of ordinary differential equations. This system is obtained if the given partial differential equation is only discretized with respect to the space variable $x$. The second partial derivative is approximated by the second difference quotient, the boundary conditions on the left and right boundary are taken into account, and finally the $n$ functions $y_i(t) := u(x_i, t)$, $i = 1, 2, \ldots, n$, are defined corresponding to the support abscissae $x_i$. Then the following system of ordinary first-order differential equations results

$$\dot{y}_1(t) = \frac{1}{h^2}(-2y_1(t) + y_2(t) + \sin(\pi t))$$

$$\dot{y}_i(t) = \frac{1}{h^2}(y_{i-1}(t) - 2y_i(t) + y_{i+1}(t)) \quad (i = 2, 3, \ldots, n-1) \quad (10.81)$$

$$\dot{y}_n(t) = \frac{1}{h^2}(2y_{n-1}(t) - 2y_n(t)).$$

If Euler's method is applied to (10.81) with the time step $k$ then we obtain (10.78) and (10.79). Due to this relationship it is clear that the global discretization error loses a factor $k$ in comparison to the local error $d_{i,j+1}$. Therefore we have $g_{i,j+1} = O(k) + O(h^2)$, and we can say that Richardson's explicit method is of first order with respect to the time integration and of second order with respect to the discretization in the space variable.

We have to study the central problem of *absolute stability* of the explicit method because of the relationship mentioned. For this reason we write the

## 10.2 Parabolic Initial Boundary Value Problems

computational scheme (10.78) and (10.79) as follows, whereby the boundary condition on the left boundary is included

$$\mathbf{u}_{j+1} = \mathbf{A}\mathbf{u}_j + \mathbf{b}_j \quad (j = 0, 1, 2, \ldots) \tag{19.82}$$

The quantities used in (10.82) are

$$\mathbf{u}_j := \begin{pmatrix} u_{1,j} \\ u_{2,j} \\ u_{3,j} \\ \vdots \\ u_{n-1,j} \\ u_{n,j} \end{pmatrix} \quad \mathbf{A} := \begin{pmatrix} 1-2r & r & & & & \\ r & 1-2r & r & & & \\ & r & 1-2r & r & & \\ & & r & \ddots & \ddots & \\ & & & \ddots & 1-2r & r \\ & & & & 2r & 1-2r \end{pmatrix}$$

$$\mathbf{b}_j := \begin{pmatrix} r \sin(\pi jk) \\ 0 \\ 0 \\ \vdots \\ 0 \\ 0 \end{pmatrix} \tag{10.83}$$

The matrix $\mathbf{A}$ is tridiagonal and depends on $k$ and $h$ via the parameter $r$. The absolute stability holds if and only if the eigenvalues $\lambda_\nu$ of the matrix $\mathbf{A}$ have moduli smaller than one. Therefore we need statements about the eigenvalues as a function of $r$ and write

$$\mathbf{A} = \mathbf{I} - r\mathbf{J} \quad \text{with } \mathbf{J} := \begin{pmatrix} 2 & -1 & & & & \\ -1 & 2 & -1 & & & \\ & -1 & 2 & -1 & & \\ & & \ddots & \ddots & \ddots & \\ & & & -1 & 2 & -1 \\ & & & & -2 & 2 \end{pmatrix} \in \mathbf{R}^{n \times n}.$$

$$\tag{10.84}$$

The eigenvalues $\lambda_\nu$ of $\mathbf{A}$ are related with the eigenvalues $\mu_\nu$ of $\mathbf{J}$ by $\lambda_\nu = 1 - r\mu_\nu$, $\nu = 1, 2, \ldots, n$. The eigenvalues of $\mathbf{J}$ are real because $\mathbf{J}$ is similar to the symmetric matrix $\hat{\mathbf{J}} := \mathbf{D}^{-1}\mathbf{J}\mathbf{D}$ with $\mathbf{D} := \text{diag}(1, 1, \ldots, 1, \sqrt{2})$. The matrix $\hat{\mathbf{J}}$ is positive definite because the Gaussian algorithm for $\hat{\mathbf{J}}$ can be performed with positive pivotal elements if the diagonal strategy is applied. Consequently, the eigenvalues of $\mathbf{J}$ are positive and bounded by four because of the row sum norm (1.67). Moreover, the matrix $\hat{\mathbf{J}} - 4\mathbf{I}$ is negative definite, and hence no eigenvalue of $\hat{\mathbf{J}}$ can be equal to four. Therefore because $r > 0$ the eigenvalues of $\mathbf{A}$ satisfy

$$1 - 4r < \lambda_v < 1$$

and the condition of the absolute stability is satisfied if and only if

$$\boxed{r \leq \tfrac{1}{2} \quad \text{or} \quad k \leq \tfrac{1}{2} h^2} \tag{10.85}$$

In Example 10.10, where $h = 0.1$ and $k = 0.01$, the condition (10.85) is violated because $r = 1$, thus explaining the meaningless values of Table 10.3.

The condition for the time step $k$, (10.85) which guarantees the stability of the explicit method is highly restrictive whenever a small mesh size $h$ must be used. In this case the numerical solution of the initial boundary value problem, until a time $T \gg 1$, requires so many steps that the total amount of work may become prohibitively large. That is why other difference approximations are required that have better stability properties.

The investigation of the absolute stability has been done for the problem (10.77). The condition (10.85) remains valid for the differential equation $u_t = u_{xx}$ with different types of boundary conditions, as can be shown by similar reasoning (Smith, 1978).

### 10.2.2 One-dimensional problems, implicit method

From the point of view of the difference approximation the explicit method has the disadvantage that the two difference quotients approximate the corresponding partial derivatives best at different points of the region $G$. With this in mind, to improve the approximation $u_{xx}$ is now replaced by the arithmetic mean of the two second difference quotients that can be defined for the grid points $P(x_i, t_j)$ and $N(x_i, t_{j+1})$ at two consecutive time levels (see Figure 10.18).

In this way it can be said that the partial differential equation $u_t = u_{xx}$ is

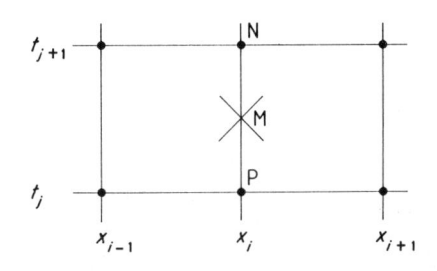

Figure 10.18 Part of the mesh

## 10.2 Parabolic Initial Boundary Value Problems

approximated at the mid point M. With

$$u_{xx} \approx \frac{1}{2h^2}(u_{i+1,j} - 2u_{i,j} + u_{i-1,j} + u_{i+1,j+1} - 2u_{i,j+1} + u_{i-1,j+1})$$

$$u_t \approx \frac{1}{k}(u_{i,j+1} - u_{i,j})$$

by equating the two difference approximations, and multiplying by $2k$ and subsequent ordering, we obtain the following difference equation for an interior point P

$$\begin{aligned} &-ru_{i-1,j+1} + (2+2r)u_{i,j+1} - ru_{i+1,j+1} \\ &= ru_{i-1,j} + (2-2r)u_{i,j} + ru_{i+1,j} \qquad r = \frac{k}{h^2} \end{aligned} \qquad (10.86)$$

The following considerations are based upon the initial boundary value problem (10.77). The boundary conditions lead to the additional difference equations

$$\begin{aligned} &(2+2r)u_{1,j+1} - ru_{2,j+1} \\ &= (2-2r)u_{1,j} + ru_{2,j} + r[\sin(\pi jk) + \sin(\pi(j+1)k)] \end{aligned} \qquad (10.87)$$

$$-2ru_{n-1,j+1} + (2+2r)u_{n,j+1} = 2ru_{n-1,j} + (2-2r)u_{n,j}. \qquad (10.88)$$

When we write down the set of equations (10.86) to (10.88) for a fixed index $j$, a system of linear equations for the $n$ unknowns $u_{1,j+1}, u_{2,j+1}, \ldots, u_{n,j+1}$ is obtained whose coefficient matrix is tridiagonal. For each time step a system of equations has to be solved, therefore, the described *Crank–Nicolson method* is implicit.

The *local discretization error* of the computational scheme (10.86) is given by

$$\begin{aligned} d_{i,j+1} := &-ru(x_{i-1}, t_{j+1}) + (2+2r)u(x_i, t_{j+1}) - ru(x_{i+1}, t_{j+1}) \\ &-ru(x_{i-1}, t_j) - (2-2r)u(x_i, t_j) - ru(x_{i+1}, t_j). \end{aligned}$$

If we substitute the Taylor series expansions, relative to $P(x_i, t_j)$, we obtain the following representation for $d_{i,j+1}$

$$\begin{aligned} d_{i,j+1} = &\, 2k(u_t - u_{xx}) + k^2(u_{tt} - u_{xxt}) \\ &+ \tfrac{1}{3}k^3 u_{ttt} - \tfrac{1}{6}h^2 k u_{xxxx} - \tfrac{1}{2}k^3 u_{xxtt} + \tfrac{1}{12}k^4 u_{tttt} + \cdots. \end{aligned}$$

The expression in the first parentheses is equal to zero because $u(x, t)$ is assumed to be a solution of $u_t = u_{xx}$. For the same reason the expression in the second parentheses is zero because it is equal to the partial derivative with respect to $t$ of $u_t - u_{xx} = 0$. Moreover, $u_{ttt} = u_{xxtt}$, and consequently we have

$$d_{i,j+1} = -\tfrac{1}{6}k^3 u_{xxtt} - \tfrac{1}{6}h^2 k u_{xxxx} + \cdots = O(k^3) + O(h^2 k). \tag{10.89}$$

The connection to the *global discretization error* $g_{i,j+1}$ of the implicit Crank–Nicolson method is provided by the fact that the formulae (10.86) to (10.88) are identical to those obtained when the system of differential equations (10.81) is approximately solved by means of the *trapezoidal method* (9.33) with the time step $k$. Therefore we have $g_{i,j+1} = O(k^2) + O(h^2)$ so that the implicit Crank–Nicolson method is of order two with respect to $h$ and $k$.

Next we show the *absolute stability* of the implicit method when applied to the problem (10.77). With vectors $\mathbf{u}_j$, (10.83), and the matrix $\mathbf{J}$, (10.84), the formulae (10.86) to (10.88) are equivalent to

$$(2\mathbf{I} + r\mathbf{J})\mathbf{u}_{j+1} = (2\mathbf{I} - r\mathbf{J})\mathbf{u}_j + \mathbf{b}_j \tag{10.90}$$

where $\mathbf{b}_j = r[\sin(\pi j k) + \sin(\pi(j+1)k]\mathbf{e}_1$. The matrix $2\mathbf{I} + r\mathbf{J}$ is diagonally dominant since $r > 0$, and hence it is nonsingular. With its inverse we can write (10.90) in the form

$$\mathbf{u}_{j+1} = (2\mathbf{I} + r\mathbf{J})^{-1}(2\mathbf{I} - r\mathbf{J})\mathbf{u}_j + (2\mathbf{I} + r\mathbf{J})^{-1}\mathbf{b}_j. \tag{10.91}$$

The method is absolutely stable if and only if the eigenvalues $\lambda_v$ of the matrix $\mathbf{B} := (2\mathbf{I} + r\mathbf{J})^{-1}(2\mathbf{I} - r\mathbf{J})$ have moduli less than one. As already stated above the eigenvalues $\mu_v$ of $\mathbf{J}$ satisfy $0 < \mu_v < 4$, and therefore the eigenvalues of $\mathbf{B}$ satisfy

$$-1 < \lambda_v = \frac{2 - r\mu_v}{2 + r\mu_v} < 1 \quad \text{for all } v \text{ and all } r > 0.$$

The implicit Crank–Nicolson method is, in fact, absolutely stable because the value $r = k/h^2$ is not subject to any restrictions concerning stability. Of course, we cannot choose $k$ arbitrarily large because otherwise the global discretization errors may become too large. From (10.89) the choice $k = h$, that is $r = 1/h$, is often adequate. The integration in the direction of time is performed with much larger time steps than would be possible with the explicit method.

Although the computation of the vector $\mathbf{u}_{j+1}$, as the solution of the system of equations (10.90), has to be done for each time step, it does not require too much effort, because the matrix $(2\mathbf{I} + r\mathbf{J})$ is tridiagonal, diagonally dominant and constant for all $j$, hence we need only apply the LR-decomposition once and we can use the diagonal strategy. The amount of effort required for this step consists of about $2n$ essential operations, see Section 1.3.3. For each integration step we only need to apply the processes of the forward and backward substitution to the

## 10.2 Parabolic Initial Boundary Value Problems

changing constant vector, about $3n$ multiplicative operations, whose computation requires $2n$ more multiplications if the $i$th component is evaluated according to the formula $\rho u_{i,j} + r(u_{i-1,j} + u_{i+1,j})$ with $\rho := 2 - 2r$. For $r = 1$ this formula is simplified because $\rho = 0$. Therefore the amount of computational effort required for a step of the implicit Crank–Nicolson method is $Z_{\text{CN}} \simeq 5n$ essential operations. According to (10.78) a step of the explicit Richardson method requires about $2n$ multiplications. However, the time step $k$ of the implicit method does not have to satisfy a stability condition, and that is why the implicit scheme is much more efficient, since a much larger $k$ can be chosen. The larger amount of work per step is highly compensated for by a much smaller number of steps.

*Example* 10.11 We treat the initial boundary value problem (10.77) by means of the implicit Crank–Nicolson method and use different combinations of the mesh sizes $h$ and $k$ in order to illustrate the properties previously investigated. In Tables 10.4 and 10.5 selected results are given for $h = 0.1$ and $k = 0.01$ ($r = 1$) and $k = 0.1$ ($r = 10$), respectively. In the first case an approximate solution is obtained

Table 10.4 Heat conduction, $h = 0.1$, $k = 0.01$, $r = 1$; implicit method

| $t$ | $j$ | $u_{0,j}$ | $u_{1,j}$ | $u_{2,j}$ | $u_{3,j}$ | $u_{4,j}$ | $u_{6,j}$ | $u_{8,j}$ | $u_{10,j}$ |
|---|---|---|---|---|---|---|---|---|---|
| 0   | 0   | 0      | 0      | 0      | 0      | 0      | 0      | 0      | 0      |
| 0.1 | 10  | 0.3090 | 0.2141 | 0.1442 | 0.0942 | 0.0597 | 0.0219 | 0.0075 | 0.0039 |
| 0.2 | 20  | 0.5878 | 0.4582 | 0.3518 | 0.2662 | 0.1986 | 0.1076 | 0.0609 | 0.0467 |
| 0.3 | 30  | 0.8090 | 0.6691 | 0.5478 | 0.4445 | 0.3583 | 0.2328 | 0.1622 | 0.1395 |
| 0.4 | 40  | 0.9511 | 0.8222 | 0.7051 | 0.6011 | 0.5110 | 0.3733 | 0.2918 | 0.2649 |
| 0.5 | 50  | 1.0000 | 0.9005 | 0.8048 | 0.7156 | 0.6353 | 0.5069 | 0.4276 | 0.4008 |
| 0.6 | 60  | 0.9511 | 0.8952 | 0.8345 | 0.7730 | 0.7141 | 0.6140 | 0.5487 | 0.5262 |
| 0.7 | 70  | 0.8090 | 0.8057 | 0.7894 | 0.7649 | 0.7363 | 0.6791 | 0.6374 | 0.6224 |
| 0.8 | 80  | 0.5878 | 0.6401 | 0.6725 | 0.6899 | 0.6966 | 0.6918 | 0.6803 | 0.6751 |
| 0.9 | 90  | 0.3090 | 0.4140 | 0.4940 | 0.5535 | 0.5968 | 0.6479 | 0.6697 | 0.6754 |
| 1.0 | 100 | 0      | 0.1490 | 0.2704 | 0.3678 | 0.4448 | 0.5492 | 0.6036 | 0.6203 |

Table 10.5 Heat conduction, $h = 0.1$, $k = 0.1$, $r = 10$; implicit method

| $t$ | $j$ | $u_{0,j}$ | $u_{1,j}$ | $u_{2,j}$ | $u_{3,j}$ | $u_{4,j}$ | $u_{6,j}$ | $u_{8,j}$ | $u_{10,j}$ |
|---|---|---|---|---|---|---|---|---|---|
| 0   | 0  | 0      | 0      | 0      | 0      | 0      | 0      | 0      | 0      |
| 0.1 | 1  | 0.3090 | 0.1983 | 0.1274 | 0.0818 | 0.0527 | 0.0222 | 0.0104 | 0.0073 |
| 0.2 | 2  | 0.5878 | 0.4641 | 0.3540 | 0.2637 | 0.1934 | 0.1025 | 0.0587 | 0.0459 |
| 0.3 | 3  | 0.8090 | 0.6632 | 0.5436 | 0.4422 | 0.3565 | 0.2295 | 0.1575 | 0.1344 |
| 0.4 | 4  | 0.9511 | 0.8246 | 0.7040 | 0.5975 | 0.5064 | 0.3684 | 0.2866 | 0.2594 |
| 0.5 | 5  | 1.0000 | 0.8969 | 0.8022 | 0.7132 | 0.6320 | 0.5017 | 0.4215 | 0.3946 |
| ⋮   | ⋮  |        |        |        |        |        |        |        |        |
| 1.0 | 10 | 0      | 0.1498 | 0.2701 | 0.3672 | 0.4440 | 0.5477 | 0.6014 | 0.6179 |

that agrees well with the results of Tables 10.1 and 10.2, although now a larger step size $k$ is used. This is a consequence of the higher order of the approximation, with respect to $k$. In the second case, with $r = 10$, the implicit method is indeed stable, but the discretization errors are much larger as must be expected for the time step $k = 0.1$. They are quite apparent for the approximations of the first time steps. The time step $k$ should be properly adapted to the variation in time of the boundary condition at $x = 0$, and at the same time to the size of $h$ so that the two principal parts of the global discretization error are comparably large.

For reasons of comparison the problem has also been solved for the mesh sizes $h = 0.05$ and $k = 0.01$, that is with $r = 4.0$. Selected results are given in Table 10.6 for the same support abscissae $x_i$ as in the previous tables. The approximations agree with the exact solution apart from a maximum difference of three digits in

Table 10.6 Heat conduction, $h = 0.05$, $k = 0.01$, $r = 4$; implicit method

| $t$ | $j$ | $u_{0,j}$ | $u_{2,j}$ | $u_{4,j}$ | $u_{6,j}$ | $u_{8,j}$ | $u_{12,j}$ | $u_{16,j}$ | $u_{20,j}$ |
|---|---|---|---|---|---|---|---|---|---|
| 0   | 0   | 0      | 0      | 0      | 0      | 0      | 0      | 0      | 0      |
| 0.1 | 10  | 0.3090 | 0.2140 | 0.1439 | 0.0938 | 0.0592 | 0.0214 | 0.0071 | 0.0037 |
| 0.2 | 20  | 0.5878 | 0.4581 | 0.3516 | 0.2659 | 0.1982 | 0.1070 | 0.0602 | 0.0460 |
| 0.3 | 30  | 0.8090 | 0.6691 | 0.5477 | 0.4442 | 0.3580 | 0.2323 | 0.1615 | 0.1387 |
| 0.4 | 40  | 0.9511 | 0.8222 | 0.7050 | 0.6010 | 0.5108 | 0.3729 | 0.2912 | 0.2642 |
| 0.5 | 50  | 1.0000 | 0.9006 | 0.8048 | 0.7156 | 0.6352 | 0.5067 | 0.4273 | 0.4005 |
| 0.6 | 60  | 0.9511 | 0.8953 | 0.8346 | 0.7732 | 0.7143 | 0.6140 | 0.5487 | 0.5261 |
| 0.7 | 70  | 0.8090 | 0.8058 | 0.7897 | 0.7652 | 0.7366 | 0.6794 | 0.6377 | 0.6226 |
| 0.8 | 80  | 0.5878 | 0.6403 | 0.6728 | 0.6903 | 0.6971 | 0.6924 | 0.6809 | 0.6757 |
| 0.9 | 90  | 0.3090 | 0.4142 | 0.4943 | 0.5540 | 0.5974 | 0.6487 | 0.6705 | 0.6763 |
| 1.0 | 100 | 0      | 0.1492 | 0.2707 | 0.3684 | 0.4454 | 0.5501 | 0.6046 | 0.6214 |

Table 10.7 Stationary state of the temperature, $h = 0.05$, $k = 0.01$, $r = 4$

| $t$ | $j$ | $u_{0,j}$ | $u_{2,j}$ | $u_{4,j}$ | $u_{6,j}$ | $u_{8,j}$ | $u_{12,j}$ | $u_{16,j}$ | $u_{20,j}$ |
|---|---|---|---|---|---|---|---|---|---|
| 2.00 | 200 | 0       | −0.1403 | −0.2532 | −0.3425 | −0.4120 | −0.5040 | −0.5505 | −0.5644 |
| 2.25 | 225 | 0.7071  | 0.5176  | 0.3505  | 0.2057  | 0.0825  | −0.1018 | −0.2089 | −0.2440 |
| 2.50 | 250 | 1.0000  | 0.8726  | 0.7495  | 0.6344  | 0.5301  | 0.3620  | 0.2572  | 0.2217  |
| 2.75 | 275 | 0.7071  | 0.7167  | 0.7099  | 0.6921  | 0.6679  | 0.6148  | 0.5739  | 0.5588  |
| 3.00 | 300 | 0       | 0.1410  | 0.2546  | 0.3447  | 0.4148  | 0.5080  | 0.5551  | 0.5693  |
| 3.25 | 325 | −0.7071 | −0.5171 | −0.3497 | −0.2045 | −0.0810 | 0.1039  | 0.2114  | 0.2467  |
| 3.50 | 350 | −1.0000 | −0.8724 | −0.7491 | −0.6338 | −0.5293 | −0.3609 | −0.2559 | −0.2203 |
| 3.75 | 375 | −0.7071 | −0.7165 | −0.7097 | −0.6918 | −0.6674 | −0.6142 | −0.5732 | −0.5580 |
| 4.00 | 400 | 0       | −0.1410 | −0.2545 | −0.3445 | −0.4146 | −0.5076 | −0.5547 | −0.5689 |
| 4.25 | 425 | 0.7071  | 0.5172  | 0.3497  | 0.2046  | 0.0811  | −0.1037 | −0.2112 | −0.2464 |
| 4.50 | 450 | 1.0000  | 0.8724  | 0.7491  | 0.6338  | 0.5293  | 0.3610  | 0.2560  | 0.2204  |
| 4.75 | 475 | 0.7071  | 0.7165  | 0.7097  | 0.6918  | 0.6675  | 0.6142  | 0.5732  | 0.5581  |
| 5.00 | 500 | 0       | 0.1410  | 0.2545  | 0.3445  | 0.4146  | 0.5076  | 0.5547  | 0.5689  |

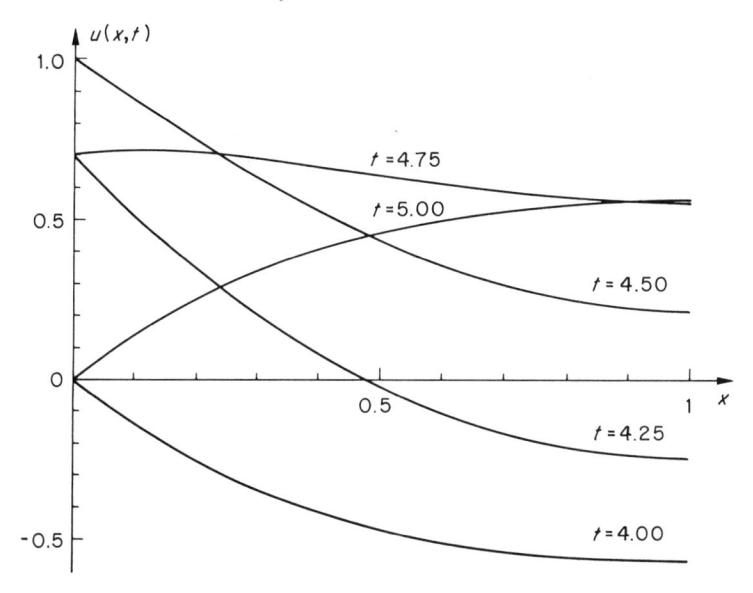

## 10.2 Parabolic Initial Boundary Value Problems

Figure 10.19 Temperature distribution in the stationary state

the fourth decimal place. For this fine space discretization the superiority of the implicit method over the explicit is quite obvious, because for the latter the time step must satisfy $k \leqslant 0.00125$ so that the number of steps would be eight times as large. The amount of work of the implicit method is only 2.5 times larger thus it is more than three times as efficient.

Finally, the state has been determined where the temperature distributions in the beam repeat themselves periodically. This is already the case after two periods ($t = 4$). In Table 10.7 the approximate values of the temperature distribution are given for $t \geqslant 2$, and in Figure 10.19 the temperature distributions are shown for some equidistant times of the half period.

### 10.2.3 Diffusion equation with variable coefficients

Diffusion processes with diffusion coefficients and source densities which are functions of the space coordinates are described by means of parabolic differential equations for the mass concentration $u(x, t)$

$$\frac{\partial u}{\partial t} = \frac{\partial}{\partial x}\left(a(x)\frac{\partial u}{\partial x}\right) + p(x)u + q(x) \quad 0 < x < 1 \quad t > 0 \tag{10.92}$$

where $a(x) > 0$, $p(x)$ and $q(x)$ are given functions of $x$. Of course, the unknown function is subject to initial and boundary conditions. To discretize the problem we use the net in Figure 10.17. We then approximate the differential expression on the right-hand side of (10.92) at the grid point $P(x_i, t_j)$ by applying the first central difference quotient twice. To simplify the notation we define the auxiliary function values

$$a(x_i + \tfrac{1}{2}h) =: a_{i+\frac{1}{2}} \quad a(x_i - \tfrac{1}{2}h) =: a_{i-\frac{1}{2}},$$

and then we have

$$\frac{\partial}{\partial x}\left(a(x)\frac{\partial u}{\partial x}\right)_P \approx \frac{1}{h^2}[a_{i+\frac{1}{2}}(u_{i+1,j} - u_{i,j}) - a_{i-\frac{1}{2}}(u_{i,j} - u_{i-1,j})].$$

Furthermore, we denote $p_i := p(x_i)$ and $q_i := q(x_i)$ to be the known values of the functions. According to the implicit Crank–Nicolson method, the difference approximation of (10.92) is

$$\frac{1}{k}(u_{i,j+1} - u_{i,j})$$

$$= \frac{1}{2}\bigg(\frac{1}{h^2}[a_{i+\frac{1}{2}}(u_{i+1,j+1} - u_{i,j+1}) - a_{i-\frac{1}{2}}(u_{i,j+1} - u_{i-1,j+1})] + p_i u_{i,j+1}$$

$$+ q_i + \frac{1}{h^2}[a_{i+\frac{1}{2}}(u_{i+1,j} - u_{i,j}) - a_{i-\frac{1}{2}}(u_{i,j} - u_{i-1,j})] + p_i u_{i,j} + q_i\bigg).$$

The equation is multiplied by $2k$, the terms are suitably ordered and so for an interior point we get the difference equation

$$\boxed{\begin{aligned}-ra_{i-\frac{1}{2}}u_{i-1,j+1} + [2 + r(a_{i-\frac{1}{2}} + a_{i+\frac{1}{2}}) - h^2 p_i]u_{i,j+1} - ra_{i+\frac{1}{2}}u_{i+1,j+1} \\ = ra_{i-\frac{1}{2}}u_{i-1,j} + [2 - r(a_{i-\frac{1}{2}} + a_{i+\frac{1}{2}}) - h^2 p_i]u_{i,j} + ra_{i+\frac{1}{2}}u_{i+1,j} + 2kq_i \\ (i = 1, 2, \ldots, n-1; \; j = 0, 1, 2, \ldots) \quad r = k/h^2.\end{aligned}}$$

(10.93)

When the boundary conditions are taken into account, for a fixed index $j$, we obtain a tridiagonal system of linear equations for the unknowns $u_{i,j+1}$. Usually the matrix of the system is diagonally dominant.

*Example* 10.12 Let us solve

$$\boxed{\begin{aligned}\frac{\partial u}{\partial t} &= \frac{\partial}{\partial x}\left((1 + 2x^2)\frac{\partial u}{\partial x}\right) + 4x(1-x)u + 5\sin(\pi x) \quad 0 < x < 1 \\ u(x,0) &= 0 \quad 0 < x < 1 \\ u(0,t) &= 0 \quad u_x(1,t) + 0.4u(1,t) = 0 \quad t > 0.\end{aligned}}$$

(10.94)

## 10.2 Parabolic Initial Boundary Value Problems

The Dirichlet condition can be simply taken into account in (10.93) for $i = 1$ by setting $u_{0,j} = u_{0,j+1} = 0$. The Cauchy condition on the right boundary is approximated by means of the central difference quotient on the assumption that the function $u(x, t)$ is also defined outside the interval. We use

$$\frac{u_{n+1,j} - u_{n-1,j}}{2h} + 0.4 u_{n,j} = 0 \quad u_{n+1,j} = u_{n-1,j} - 0.8 h u_{n,j}.$$

The auxiliary values $u_{n+1,j}$ and $u_{n+1,j+1}$ are eliminated in (10.93), and if we assume that $a(x)$ is also defined outside the $x$ interval, then the corresponding difference equation, valid for the grid points on the right boundary, is given by

$$\begin{aligned}
& -r(a_{n-\frac{1}{2}} + a_{n+\frac{1}{2}})u_{n-1,j+1} + \{2 + r[a_{n-\frac{1}{2}} + (1 + 0.8h)a_{n+\frac{1}{2}} - h^2 p_n]\} u_{n,j+1} \\
& = r(a_{n-\frac{1}{2}} + a_{n+\frac{1}{2}})u_{n-1,j} + \{2 - r[a_{n-\frac{1}{2}} + (1 + 0.8h)a_{n+\frac{1}{2}} - h^2 p_n]\} u_{n,j} + 2kq_n.
\end{aligned}$$

The discrete form of the initial boundary value problem (10.94) has been solved for $n = 10, h = 0.1, k = 0.01, r = 1$. We write the matrix $\mathbf{A}$ of the tridiagonal system of equations analogous to (10.90) as $\mathbf{A} = 2\mathbf{I} + r\tilde{\mathbf{J}}$. Then the matrix $\tilde{\mathbf{J}} \in \mathbf{R}^{10 \times 10}$ has the following selected elements

$$\tilde{\mathbf{J}} = \begin{pmatrix}
2.0464 & -1.045 & & & & & \\
-1.045 & 2.1636 & -1.125 & & & & \\
& -1.125 & 2.3616 & -1.245 & & & \\
& & \ddots & \ddots & \ddots & & \\
& & & -2.125 & 4.5636 & -2.445 & \\
& & & & -2.445 & 5.2464 & -2.805 \\
& & & & & -6.01 & 6.2664
\end{pmatrix}$$

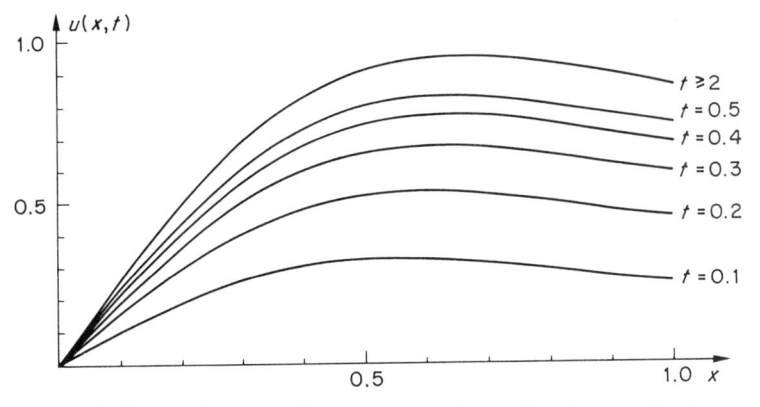

Figure 10.20 Distributions of the concentration as functions of the time $t$

Table 10.8 Diffusion problem, $h = 0.1$, $k = 0.01$, $r = 1$

| t | j | $u_{1,j}$ | $u_{2,j}$ | $u_{3,j}$ | $u_{4,j}$ | $u_{5,j}$ | $u_{6,j}$ | $u_{8,j}$ | $u_{10,j}$ |
|---|---|---|---|---|---|---|---|---|---|
| 0 | 0 | 0 | 0 | 0 | 0 | 0 | 0 | 0 | 0 |
| 0.1 | 10 | 0.1044 | 0.1963 | 0.2660 | 0.3094 | 0.3276 | 0.3255 | 0.2888 | 0.2533 |
| 0.2 | 20 | 0.1591 | 0.3010 | 0.4124 | 0.4872 | 0.5265 | 0.5365 | 0.5037 | 0.4560 |
| 0.3 | 30 | 0.1948 | 0.3695 | 0.5085 | 0.6044 | 0.6581 | 0.6765 | 0.6470 | 0.5915 |
| 0.4 | 40 | 0.2185 | 0.4150 | 0.5722 | 0.6821 | 0.7454 | 0.7695 | 0.7421 | 0.6814 |
| 0.5 | 50 | 0.2342 | 0.4451 | 0.6145 | 0.7336 | 0.8033 | 0.8311 | 0.8052 | 0.7410 |
| ⋮ | | | | | | | | | |
| 1.0 | 100 | 0.2612 | 0.4969 | 0.6871 | 0.8222 | 0.9029 | 0.9371 | 0.9136 | 0.8435 |
| ⋮ | | | | | | | | | |
| 1.5 | 150 | 0.2647 | 0.5035 | 0.6964 | 0.8336 | 0.9157 | 0.9507 | 0.9276 | 0.8567 |
| ⋮ | | | | | | | | | |
| 2.0 | 200 | 0.2651 | 0.5044 | 0.6976 | 0.8351 | 0.9173 | 0.9525 | 0.9294 | 0.8584 |
| ⋮ | | | | | | | | | |
| 2.5 | 250 | 0.2652 | 0.5045 | 0.6978 | 0.8353 | 0.9175 | 0.9527 | 0.9296 | 0.8586 |

A similar consideration to that applied to estimate the eigenvalues of the matrix **J** (10.84) leads to the following result for the eigenvalues $\mu_v$ of $\tilde{\mathbf{J}}$

$$0 < \mu_v < 12.2764.$$

For the corresponding explicit Richardson method the condition of the absolute stability is $r \leq 1/6.1382 \doteq 0.163$. Hence the time step must satisfy the condition $k \leq 0.001\,63$. For the implicit and absolutely stable method we may use a time step that is six times larger, $k = 0.01$. Selected results are given in Table 10.8. The stationary constant state is reached, to four decimal places, for $t = 2.0$. The approximate solution $u(x, t)$ is shown in Figure 10.20 for some values of $t$.

### 10.2.4 Two-dimensional problems

The classical parabolic differential equation for a function $u(x, y, t)$ of the two space variables $x$, $y$ and the time variable $t$ is

$$u_t = u_{xx} + u_{yy}. \tag{10.95}$$

A solution is required in a region $G \subset \mathbf{R}^2$ of the $(x, y)$ plane with the boundary $\Gamma$ for times $t > 0$. The solution of (10.95) is subject to an *initial condition*

$$u(x, y, 0) = f(x, y) \quad \text{in } G \tag{10.96}$$

and to similar *boundary conditions* on the boundary $\Gamma$ as for elliptic boundary value problems. The function $u(x, y, t)$ depends on the time, thus the functions occurring in the Dirichlet, Neumann and Cauchy conditions, (10.5) to (10.7), may also depend on $t$. The set of values $(x, y, t)$ for which the solution is sought consists of the half cylinder in $\mathbf{R}^3$ above the region $G$.

## 10.2 Parabolic Initial Boundary Value Problems

To discretize the initial boundary value problem we use a regular three-dimensional net which is generated from a regular net in the region $G$ with the mesh size $h$, see Figure 10.2, by translating it by multiples of $k$ in the time direction. Then we look for approximations $u_{\mu,\nu,j}$ of the solution $u(x, y, t)$ at the grid points $P(x_\mu, y_\nu, t_j)$.

The approximation of the differential equation (10.95) is done in two stages. The expression $u_{xx} + u_{yy}$ with partial derivatives with respect to the space variables is approximated for a fixed time layer $j$, according to the procedure described in Section 10.1, at each grid point by an appropriate difference expression. For the simplest case of a regular interior point $P(x_\mu, y_\nu, t_j)$ we use

$$(u_{xx} + u_{yy})_P \approx \frac{1}{h^2}(u_{\mu,\nu+1,j} + u_{\mu-1,\nu,j} + u_{\mu,\nu-1,j} + u_{\mu+1,\nu,j} - 4u_{\mu,\nu,j})$$

whereas for a grid point on or near the boundary, approximations of the types of Section 10.1.3 are adequate.

The partial derivative with respect to the time $t$ may, for instance, be approximated by means of the forward difference quotient in P

$$u_t \approx \frac{1}{k}(u_{\mu,\nu,j+1} - u_{\mu,\nu,j}).$$

This defines the *explicit Richardson method* with the computational scheme for a regular interior grid point

$$u_{\mu,\nu,j+1} = u_{\mu,\nu,j} + r(u_{\mu,\nu+1,j} + u_{\mu-1,\nu,j} + u_{\mu,\nu-1,j} + u_{\mu+1,\nu,j} - 4u_{\mu,\nu,j}). \quad (10.97)$$

Again we use the quantity $r = k/h^2$. In order to simplify the index notation we number the grid points with unknown value $u$ in each time layer from 1 to $n$ (see Section 10.1.2). Then we combine the approximate values of the $j$th time layer, with $t_j = jk$, in the vector

$$\mathbf{u}_j := (u_{1,j}, u_{2,j}, \ldots, u_{n,j})^T \in \mathbf{R}^n \quad (10.98)$$

where the first index $i$ of $u_{i,j}$ refers to the number of the grid point. The computational scheme (10.97) can be written as

$$\mathbf{u}_{j+1} = (\mathbf{I} - r\mathbf{A})\mathbf{u}_j + \mathbf{b}_j \quad (j = 0, 1, 2, \ldots). \quad (10.99)$$

The matrix $\mathbf{A} \in \mathbf{R}^{n \times n}$ is the coefficient matrix of the system of difference equations which discretizes the Poisson equation in the region $G$, and $\mathbf{b}_j$ contains the constants that originate from the boundary conditions.

The explicit method satisfies the condition of absolute stability if and only if the absolute values of the eigenvalues of the matrix $(\mathbf{I} - r\mathbf{A})$ are less than one. The resulting condition for $r$ can in general only be given for symmetric and positive definite matrices $\mathbf{A}$. In this case the eigenvalues $\lambda_\nu$ of $(\mathbf{I} - r\mathbf{A})$ are related to the eigenvalues $\mu_\nu$ of $\mathbf{A}$ by

$$\lambda_\nu = 1 - r\mu_\nu \quad (\nu = 1, 2, \ldots, n) \quad \mu_\nu > 0.$$

As $1 - r\mu_v > -1$ must be satisfied for all $v$ we deduce that

$$r < 2/\max_v(\mu_v). \tag{10.100}$$

For a matrix $\mathbf{A}$, completely defined by the five-point formula (10.11), for instance, (10.20) or (10.45), we have on the basis of the row sum norm $\max_v(\mu_v) \leq 8$, so that the condition (10.101)

$$\boxed{r < \tfrac{1}{4} \quad \text{i.e.} \quad k < \tfrac{1}{4}h^2} \tag{10.101}$$

must be satisfied for the explicit method (10.97) to be absolutely stable. In this case the largest eigenvalue of $\mathbf{A}$ is strictly less than 8, so that in (10.101) equality is valid. Anyway, the size of the time step $k$ is highly restricted by $k \leq \tfrac{1}{4}h^2$.

For this reason the *implicit Crank–Nicolson method* must be applied. In (10.97) the expression in parentheses is simply replaced by the arithmetic mean of the expressions of the $j$th and the $(j+1)$th time layer. Instead of (10.99) we obtain a computational scheme of the form

$$(2\mathbf{I} + r\mathbf{A})\mathbf{u}_{j+1} = (2\mathbf{I} - r\mathbf{A})\mathbf{u}_j + \mathbf{b}_j \quad (j = 0, 1, 2, \ldots). \tag{10.102}$$

This method is absolutely stable for symmetric and positive definite matrices $\mathbf{A}$ and for nonsymmetric matrices $\mathbf{A}$ whose eigenvalues $\mu_v$ all have a positive real part, because the eigenvalues $\lambda_v$ of $(2\mathbf{I} + r\mathbf{A})^{-1}(2\mathbf{I} - r\mathbf{A})$ are then in absolute value less than one for all $r > 0$.

The computation of the values $\mathbf{u}_{j+1}$ at the grid points of the $(j+1)$th time layer requires the solution of a system of linear equations, (10.102), with the matrix $(2\mathbf{I} + r\mathbf{A})$ that is constant for all time steps and usually diagonally dominant. Thus the LR decomposition has to be performed just once, and then the processes of the forward and backward substitution have to be applied to the known right-hand side of (10.102). In the case of a small mesh size, $h$, the order of the matrix $(2\mathbf{I} + r\mathbf{A})$ as well as the bandwidth is large, so that the storage requirement and the computational effort per time step is considerable.

In order to substantially reduce the mentioned requirements, Peaceman and Rachford (1955) proposed a discretization so that in each time step a sequence of tridiagonal systems of equations has to be solved. The idea consists of combining two different difference approximations per step. For this reason the time step $k$ is halved, and the auxiliary values $u_{\mu,v,j+\frac{1}{2}} =: u^*_{\mu,v}$ for the time $t_j + k/2 =: t_{j+\frac{1}{2}}$ are defined as the solution of the difference equations

$$\frac{2}{k}(u^*_{\mu,v} - u_{\mu,v,j}) = \frac{1}{h^2}(u^*_{\mu+1,v} - 2u^*_{\mu,v} + u^*_{\mu-1,v})$$
$$+ \frac{1}{h^2}(u_{\mu,v+1,j} - 2u_{\mu,v,j} + u_{\mu,v-1,j}). \tag{10.103}$$

## 10.2 Parabolic Initial Boundary Value Problems

To approximate $u_{xx}$, the second difference quotient is used with the auxiliary values of the time layer $t_{j+\frac{1}{2}}$. The second partial derivative $u_{yy}$, however, is approximated by means of, known, approximations of the time layer $t_j$. The derivative $u_t$ is approximated by the usual forward difference quotient with the half step size $k/2$, of course. In the next step we take the auxiliary values $u^*_{\mu,\nu}$ for fixed $\nu$ together, that is those values that correspond to grid points along a net line parallel to the $x$ axis. For each such group of values the set of equations (10.103) is a tridiagonal system with the typical representative

$$-ru^*_{\mu-1,\nu} + (2+2r)u^*_{\mu,\nu} - ru^*_{\mu+1,\nu}$$
$$= ru_{\mu,\nu-1,j} + (2-2r)u_{\mu,\nu,j} + ru_{\mu,\nu+1,j} \quad r = k/h^2. \tag{10.104}$$

The determination of all the auxiliary values $u^*_{\mu,\nu}$ requires the solution of a set of tridiagonal systems of equations, each corresponding to a line of the net that is parallel to the $x$ axis. With these auxiliary values, the approximations $u_{\mu,\nu,j+1}$ of the time layer $t_{j+1}$ are determined from the difference equations

$$\frac{2}{k}(u_{\mu,\nu,j+1} - u^*_{\mu,\nu}) = \frac{1}{h^2}(u^*_{\mu+1,\nu} - 2u^*_{\mu,\nu} + u^*_{\mu-1,\nu}) \tag{10.105}$$

$$+ \frac{1}{h^2}(u_{\mu,\nu+1,j+1} - 2u_{\mu,\nu,j+1} + u_{\mu,\nu-1,j+1}).$$

Now $u_{xx}$ is approximated by means of the known auxiliary values and $u_{yy}$ by means of the desired approximations of the $(j+1)$th time layer. For this part it is important that we unite the unknown values $u_{\mu,\nu,j+1}$ for fixed $\mu$ in (10.105), that is those values that correspond to grid points on a net line parallel to the $y$ axis. Each of these groups of unknowns, (10.105), is the solution of another tridiagonal system of equations whose typical representative is

$$-ru_{\mu,\nu-1,j+1} + (2+2r)u_{\mu,\nu,j+1} - ru_{\mu,\nu+1,j+1}$$
$$= ru^*_{\mu-1,\nu} + (2-2r)u^*_{\mu,\nu} + ru^*_{\mu+1,\nu} \quad r = k/h^2. \tag{10.106}$$

Once again we have to solve a sequence of tridiagonal systems of linear equations for the unknowns $u_{\mu,\nu,j+1}$, each corresponding to grid points on net lines parallel to the $y$ axis. Due to the change of direction in which the grid points are grouped, the Peaceman and Rachford procedure is usually called the *alternating directions method*.

The tridiagonal systems of equations (10.104) and (10.106) are of the same type as in the one-dimensional problems. The matrices are diagonally dominant. The storage requirement is minimal, and the amount of computational effort required to solve all tridiagonal systems, in order to perform a time step, is only proportional to the number of grid points of the time layer.

*Example* 10.13 A simple and obvious situation is encountered in the case of the rectangular region $G$ of Figure 10.11, if the initial boundary value problem (10.95) is to be solved with Dirichlet boundary conditions. Let the mesh size $h$ be chosen in such a way that $h = a/(N+1) = b/(M+1)$ with $N, M \in \mathbf{N}$. Hence, there are $n = N \cdot M$ interior grid points with unknown function value. The tridiagonal systems of equations for the auxiliary values $u^*$, that have to be solved for each line parallel to the $x$ axis, have the same matrix, that is for $N = 5$

$$\mathbf{H} := \begin{pmatrix} 2+2r & -r & & & \\ -r & 2+2r & -r & & \\ & -r & 2+2r & -r & \\ & & -r & 2+2r & -r \\ & & & -r & 2+2r \end{pmatrix}$$

For a given $r$ it is sufficient to compute the LR decomposition in a preparatory step and only to perform the processes of the forward and the backward substitution in order to determine the $M$ groups of auxiliary values. For the computation of the approximations $u$ of the $(j+1)$th time layer, $N$ tridiagonal systems of equations each with $M$ unknowns and the constant matrix, for example for $M = 3$

$$\mathbf{V} := \begin{pmatrix} 2+2r & -r & \\ -r & 2+2r & -r \\ & -r & 2+2r \end{pmatrix}$$

must be solved in a similar way. If the known right-hand sides of the systems (10.104) and (10.106) are calculated such that a minimal number of multiplications is needed, an integration step requires only $10n$ essential operations.

*Example* 10.14 For a less simple region $G$ and other boundary conditions, the method of alternating directions has a correspondingly more involved implementation. To illustrate the situation we consider the parabolic equation $u_t = u_{xx} + u_{yy}$ in the domain $G$ of Figure 10.4, with the boundary conditions (10.16) to (10.18) and the initial condition $u(x, y, 0) = 0$ in $G$. We use the numbering of the grid points of Figure 10.4, and give the tridiagonal matrices of the systems of equations that have to be solved for the auxiliary values at grid points belonging to the four horizontal lines. If the boundary conditions on FA and CD are taken into account the difference approximations are not divided by two. For each matrix the corresponding $u^*$ values are indicated.

$$\mathbf{H}_1 := \begin{pmatrix} 2+2r & -2r & & & \\ -r & 2+2r & -r & & \\ & -r & 2+2r & -r & \\ & & -r & 2+2r & -r \\ & & & -r & 2+2r \end{pmatrix}$$

$$(u_1^*, u_4^*, u_7^*, u_{10}^*, u_{14}^*)^T$$

## 10.3 Finite Element Method

$$\mathbf{H}_2 := \begin{pmatrix} 2+2r & -2r & & & & \\ -r & 2+2r & -r & & & \\ & -r & 2+2r & -r & & \\ & & -r & 2+2r & -r & \\ & & & -r & 2+2r & -r \\ & & & & -r & 2+2r \end{pmatrix}$$

$$(u_2^*, u_5^*, u_8^*, u_{11}^*, u_{15}^*, u_{18}^*)^T$$

$$\mathbf{H}_3 := \begin{pmatrix} 2+2r & -2r & & & & \\ -r & 2+2r & -r & & & \\ & -r & 2+2r & -r & & \\ & & -r & 2+2r & -r & \\ & & & -r & 2+2r & -r \\ & & & & -2r & 2+2r \end{pmatrix}$$

$$(u_3^*, u_6^*, u_9^*, u_{12}^*, u_{16}^*, u_{19}^*)^T$$

$$\mathbf{H}_4 := \begin{pmatrix} 2+2r & -r \\ -2r & 2+2r \end{pmatrix}$$

$$(u_{13}^*, u_{17}^*)^T$$

The second half step is only described by four different tridiagonal matrices for the six vertical lines because the matrices of order three are identical for the first three lines. Their determination is left to the reader.

### 10.3 Finite Element Method

In the following we consider the energy method which determines the solution of elliptic boundary value problems via an equivalent variational problem. We shall present the basic idea of the finite element method for discretizing the problem and shall describe the procedure in full detail for a particular case. For an extensive treatment of the method we refer you to Schwarz (1988) and the literature cited there.

#### 10.3.1 Fundamentals

Let $G$ be a bounded region of the $(x, y)$ plane with a boundary $\Gamma$ that may consist of several closed curves, and is piecewise continuously differentiable, see Figure 10.1. We consider the integral expression

$$\mathbf{I} := \iint_G [\tfrac{1}{2}(u_x^2 + u_y^2) - \tfrac{1}{2}\rho(x,y)u^2 + f(x,y)u]\,dx\,dy + \oint_\Gamma (\tfrac{1}{2}\alpha(s)u^2 - \beta(s)u)\,ds \tag{10.107}$$

where $\rho(x,y)$ and $f(x,y)$ are functions defined on $G$, $s$ denotes the arc length on $\Gamma$, and $\alpha(s)$ and $\beta(s)$ are given functions of the arc length. In addition to (10.107) the function $u(x,y)$ is subject to boundary conditions on $\Gamma_1$, part of the boundary $\Gamma$, which may comprise the whole boundary or may be empty, that is

$$u = \varphi(s) \quad \text{on} \quad \Gamma_1, \Gamma_1 \subset \Gamma. \tag{10.108}$$

Next we show that the function $u(x,y)$ which makes the integral expression $I$ a stationary point under condition (10.108) solves a certain elliptic boundary value problem on the restrictive assumption that $u(x,y)$ is sufficiently many times continuously differentiable. This will allow us to treat the extremal problem for $I$, (10.107), subject to the condition (10.108) in order to produce the solution of an elliptic boundary value problem. For most of the applications, the integral $I$ represents an energy and not only takes on a stationary value but even a minimum due to extremal principles (Hamilton, Rayleigh or Fermat principle) (Funk, 1970). This is the case when $\rho(x,y) \leq 0$ on $G$ and $\alpha(s) \geq 0$ on $\Gamma$. Due to this relationship we speak of the *energy method*.

The necessary condition for the integral expression $I$ to attain a stationary point consists of the vanishing of the first variation. The rules of variational calculus (Courant and Hilbert, 1968; Funk, 1970) yield

$$\delta I = \iint_G (u_x \delta u_x + u_y \delta u_y - \rho(x,y) u \delta u + f(x,y) \delta u) \, dx \, dy$$

$$+ \oint_\Gamma (\alpha(s) u \delta u - \beta(s) \delta u) \, ds. \tag{10.109}$$

As $u_x \delta u_x + u_y \delta u_y = \operatorname{grad} u \cdot \operatorname{grad} \delta u$ we can apply Green's formula which on the assumption $u(x,y) \in C^2(G \cup \Gamma)$, $v(x,y) \in C^1(G \cup \Gamma)$ states that

$$\iint_G \operatorname{grad} u \operatorname{grad} v \, dx \, dy = -\iint_G (u_{xx} + u_{yy}) v \, dx \, dy + \oint_\Gamma \frac{\partial u}{\partial n} v \, ds$$

where $\partial u/\partial n$ denotes the derivative of $u$ in the direction of the outward normal $\mathbf{n}$ of the boundary $\Gamma$. From (10.109) we obtain

$$\delta I = -\iint_G (u_{xx} + u_{yy} + \rho(x,y) u - f(x,y)) \delta u \, dx \, dy$$

$$+ \oint_\Gamma \left( \frac{\partial u}{\partial n} + \alpha(s) u - \beta(s) \right) \delta u \, ds. \tag{10.110}$$

The first variation $\delta I$ must be zero for every admissible variation $\delta u$ of the function $u$. With the aid of a first restriction on the admissible variations, to those satisfying $\delta u = 0$ on $\Gamma$, from (10.110) we deduce that Euler's differential equation

$$u_{xx} + u_{yy} + \rho(x,y) u = f(x,y) \quad \text{in } G \tag{10.111}$$

must be satisfied. Hence the function $u(x,y)$ that is assumed to be twice

## 10.3 Finite Element Method

continuously differentiable, and to make the integral $I$ (10.107) stationary necessarily satisfies the elliptic differential equation (10.111) in $G$.

If $\Gamma_1$ with prescribed boundary values (10.108), where we must have $\delta u = 0$, does not comprise of the whole boundary $\Gamma$, then the following necessary condition on part $\Gamma_2$ of the boundary must be satisfied

$$\frac{\partial u}{\partial n} + \alpha(s)u = \beta(s) \quad \text{on } \Gamma_2 = \Gamma \setminus \Gamma_1. \tag{10.112}$$

The boundary condition (10.112) is a *natural condition* which the solution $u(x, y)$ of the variational problem (10.107), (10.108) has necessarily to satisfy.

If a function $u(x, y) \in C^2(G \cup \Gamma)$ makes the functional (10.107) stationary, subject to the condition (10.108), then $u(x, y)$ is a solution of the elliptic differential equation (10.111) satisfying the boundary conditions (10.108) and (10.112). Of course, the question arises whether a solution of the variational problem exists, that is whether a function $u(x, y)$ exists that makes $I$ stationary under the condition (10.108). This fundamental question is related to the choice of the space of admissible functions to solve the variational problem. A look at $I$ reveals that $I(u(x, y))$ is defined on much weaker assumptions about $u(x, y)$ than was made above. It is sufficient, for instance, that $u(x, y)$ is piecewise continuously differentiable. By means of an extension of the space of admissible functions we can define a certain *Sobolev space*. For this space the existence of a solution of the variational problem can be shown by means of relatively simple arguments of functional analysis. However, the solution from this Sobolev space does not need to be twice continuously differentiable, and hence does not need to be a solution of the elliptic boundary value problem. The statement given above on the existence of a solution is valid on quite general hypotheses about the functions $\rho(x, y)$, $f(x, y)$, $\alpha(s)$ and $\beta(s)$. Furthermore, if we assume that these functions, as well as the boundary of the domain, are sufficiently smooth, it can be shown, with severe mathematical difficulties, that the solution of the variational problem is, in fact, a solution of the elliptic boundary value problem.

From a practical point of view the problem sketched is of low importance. Anyway in many cases the variational formulation is the most natural approach to describe a physical state, and it is not necessary to go back to the boundary value problem at all. In the following the extremal problem will be approximately solved in such a way that an approximation of $u(x, y)$ will be determined as an element of a finite-dimensional function space of particular functions that are piecewise continuously differentiable.

When we considered the variational problem an essential distinction between the boundary conditions became clear that will be decisive for the practical application of the energy method. The natural boundary condition, (10.112), concerning the normal derivative is contained in the formulation of the extremal problem in an implicit way in the boundary integral of $I$, and we do not have to take care of it in an explicit manner. This will in fact simplify the procedure.

By specifying the functions of the integral expression $I$ we obtain the following special elliptic differential equations

$u_{xx} + u_{yy} = 0$      Laplace equation ($\rho = f = 0$)

$u_{xx} + u_{yy} = f(x, y)$ Poisson equation ($\rho = 0$)

$u_{xx} + u_{yy} + \rho(x, y)u = 0$    Helmholtz equation ($f = 0$)

Special natural boundary conditions are

$$\frac{\partial u}{\partial n} = \beta(s) \qquad \text{Neumann condition } (\alpha = 0)$$

$$\frac{\partial u}{\partial n} + \alpha(s)u = 0 \quad \text{Cauchy condition } (\beta = 0).$$

### 10.3.2 Principle of the finite element method

We first describe the fundamental procedure of the method and subsequently go into details. The integral expression $I$, (10.107), is our starting point. It will first be suitably approximated and then the condition for a stationary value will be formulated whereby the Dirichlet boundary conditions are taken into account.

First the domain $G$ is discretized by subdividing it into subdomains, the so-called *elements*. In the following we only consider *triangulations* of the domain where $G$ is covered by triangular elements in such a way that neighbouring triangles have either a complete side or only a corner in common, see Figure 10.21. The given domain $G$ is replaced by the total area of the triangles. A domain with a curved boundary can be approximated by such a triangulation in quite a flexible manner if near the boundary a locally finer discretization is chosen. The triangles should not have angles that are too obtuse so that we avoid numerical difficulties.

In the second step, the desired function $u(x, y)$ is restricted to a certain class of

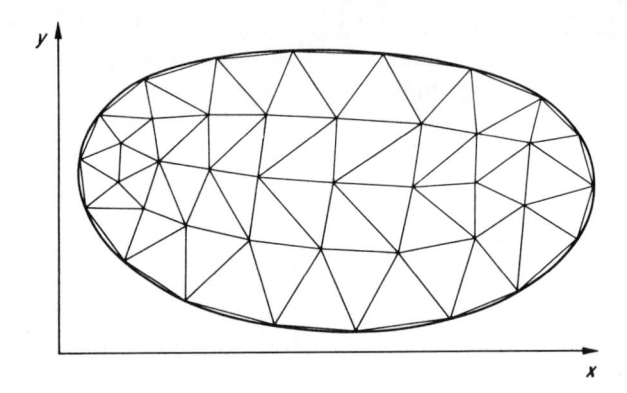

Figure 10.21 Triangulation of the domain

## 10.3 Finite Element Method

functions on each triangle. Linear, quadratic or higher degree polynomials of the two variables $x$ and $y$ are most useful, namely

$$\tilde{u}(x, y) = c_1 + c_2 x + c_3 y \tag{10.113}$$

$$\tilde{u}(x, y) = c_1 + c_2 x + c_3 y + c_4 x^2 + c_5 xy + c_6 y^2. \tag{10.114}$$

These functions defined on each element must be at least continuous along the side of two adjacent triangles so that as a whole the resulting function is admissible for the treatment of the extremal problem, in particular it must be continuous and must have a piecewise continuous first derivative. To satisfy this condition of continuity, the coefficients $c_k$ in (10.113) or (10.114) must be expressed by means of function values at certain *nodal points* of the triangle, or the function $\tilde{u}(x, y)$ is directly represented by means of a linear combination of *basis functions* which are defined analogously to the Lagrange polynomials, with corresponding interpolation properties with respect to the nodal points.

A linear function $\tilde{u}(x, y)$, (10.113), is uniquely determined in the triangle by the three function values at the corners. The continuity of the linear functions along the common side of two adjacent triangles follows from the fact that both functions reduce to linear functions of the arc length, which are uniquely determined by the values at the end points.

On a triangle, the quadratic function, (10.114), is uniquely determined by the six function values at the three corners and the three midpoints of the sides. These functions satisfy the condition of continuity along the common side of two neighbouring triangles because they reduce to quadratic functions of the arc length, which are uniquely fixed by the function values at the end points and the midpoint.

The third step consists of representing the integral $I$ as a function of the function values at the nodal points, the so called *nodal variables*, according to the chosen type of function $\tilde{u}(x, y)$. Therefore we have to compute the contributions of each triangular element and of the sides along the boundary. Then all the contributions are summed. In order to perform the summation in a systematic way the nodal points are numbered, and let $u_j$ denote the nodal variable at the point with the number $j$. The integrands are either quadratic or linear in $u$ and $\tilde{u}(x, y)$ is linear with respect to the coefficients $c_k$ and also with respect to the nodal variables $u_j$. As a direct consequence, the integral $I(\tilde{u}(x, y))$ is a quadratic function of the nodal variables $u_j$. It describes the integral expression over a linear function space, defined by the functions on the elements, whose dimension is equal to the number of nodal points of the chosen discretization.

In the next step, Dirichlet type boundary conditions (10.108) are taken into account, which prescribe the values of nodal variables at certain nodal points on the boundary. These known quantities have to be substituted, in principle, into the previously obtained quadratic function for $I$. For the remaining unknown nodal variables $u_1, u_2, \ldots, u_n$, which we combine in the vector $\mathbf{u} := (u_1, u_2, \ldots, u_n)^T$, we obtain a quadratic function of the form

$$F := \tfrac{1}{2}\mathbf{u}^T \mathbf{A}\mathbf{u} + \mathbf{b}^T \mathbf{u} + d \quad \mathbf{A} \in \mathbf{R}^{n \times n} \quad \mathbf{b} \in \mathbf{R}^n. \tag{10.115}$$

$\mathbf{A}$ is a symmetric matrix which is also positive definite if the integral $I$ (10.107) represents an energy or if $\rho(x, y) \leq 0$, $\alpha(s) \geq 0$, and at least one condition (10.108) is prescribed. The vector $\mathbf{b}$ results from the linear contributions of $I$ and from contributions that are generated when known values for nodal variables are substituted into the quadratic part of the original quadratic function for $I$. A similar remark holds for the constant $d$ in (10.115).

The condition for $F$ to be stationary leads to the system of linear equations

$$\mathbf{A}\mathbf{u} + \mathbf{b} = \mathbf{0} \tag{10.116}$$

with a *symmetric* and *positive definite* matrix $\mathbf{A}$. The solution of (10.116) yields values $u_j$ representing approximations of the function values $u(x, y)$ at the corresponding nodal points. The chosen type of function $\tilde{u}(x, y)$ defines the approximate solution in the interior of the triangular elements so that, for instance, it is possible to construct the level curves of the approximate solution function.

### 10.3.3 Elementwise treatment

For the chosen triangulation and for the chosen type of approximating function $\tilde{u}(x, y)$ we have to determine the contributions by the triangular elements and the boundary pieces to the integral. These contributions are either purely quadratic or linear functions of the nodal variables $u_j$, and we want to derive the matrices of the quadratic forms and the coefficients of the linear forms.

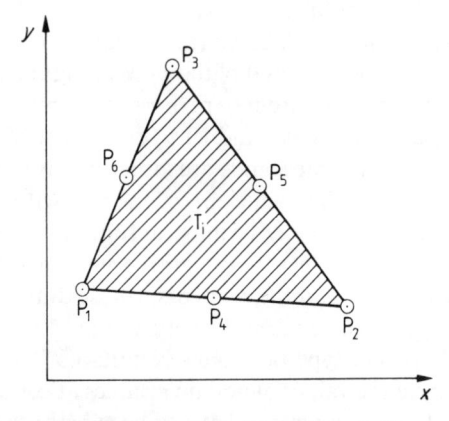

Figure 10.22 Triangular element in general position, nodal points for the quadratic approximation

## 10.3 Finite Element Method

They represent the central part of the construction of the system of linear equations (10.116).

In the following we consider the quadratic function (10.114) and for simplicity assume that the functions $\rho(x, y)$, $f(x, y)$, $\alpha(s)$ and $\beta(s)$ in the integral $I$ (10.107) are constant at least for each element and boundary edge, respectively, so that the corresponding values $\rho$, $f$, $\alpha$ and $\beta$ can be extracted from the integrals.

Let $T_i$ be a triangle in a general position with the six nodal points $P_1$ to $P_6$. We agree that the corners $P_1$, $P_2$ and $P_3$ are ordered counterclockwise, see Figure 10.22. Let the coordinates of the corners $P_i$ be $(x_i, y_i)$. In order to determine the value of the integral over such an element

$$\iint_{T_i} (\tilde{u}_x^2 + \tilde{u}_y^2) \, dx \, dy \tag{10.117}$$

for an arbitrary approximating function in an easy way, $T_i$ is mapped by means of a linear transformation onto an isosceles, right-angled *normal triangle T* according to Figure 10.23. The corresponding transformation is

$$\begin{aligned} x &= x_1 + (x_2 - x_1)\xi + (x_3 - x_1)\eta \\ y &= y_1 + (y_2 - y_1)\xi + (y_3 - y_1)\eta. \end{aligned} \tag{10.118}$$

The integral (10.117) over the region $T_i$ is transformed following the rules of analysis. If the transformed function is denoted by the same symbol, we have

$$\tilde{u}_x = \tilde{u}_\xi \xi_x + \tilde{u}_\eta \eta_x \quad \tilde{u}_y = \tilde{u}_\xi \xi_y + \tilde{u}_\eta \eta_y. \tag{10.119}$$

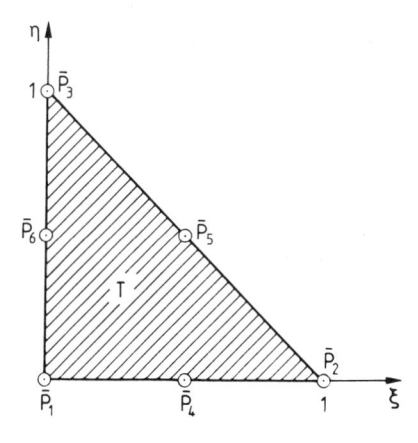

Figure 10.23 Normal triangle with nodal points for the quadratic approximation

Moreover, it follows from (10.118) that

$$\xi_x = \frac{y_3 - y_1}{J} \quad \eta_x = -\frac{y_2 - y_1}{J} \quad \xi_y = -\frac{x_3 - x_1}{J} \quad \eta_y = \frac{x_2 - x_1}{J}, \quad (10.120)$$

where

$$J := (x_2 - x_1)(y_3 - y_1) - (x_3 - x_1)(y_2 - y_1) > 0 \quad (10.121)$$

is the Jacobian of the mapping (10.118), which is equal to twice the area of the triangle $T_i$ from our ordering of the corners. Substitution of (10.119) and (10.120) into (10.117) yields

$$\iint_{T_i} (\tilde{u}_x^2 + \tilde{u}_y^2) \, dx \, dy = \iint_T (a\tilde{u}_\xi^2 + 2b\tilde{u}_\xi \tilde{u}_\eta + c\tilde{u}_\eta^2) \, d\xi \, d\eta \quad (10.122)$$

with the constant coefficients depending on the element $T_i$

$$\boxed{\begin{aligned} a &= [(x_3 - x_1)^2 + (y_3 - y_1)^2]/J \\ b &= -[(x_3 - x_1)(x_2 - x_1) + (y_3 - y_1)(y_2 - y_1)]/J \\ c &= [(x_2 - x_1)^2 + (y_2 - y_1)^2]/J \end{aligned}} \quad (10.123)$$

If the linear transformation (10.118) is applied to the quadratic function (10.114) we obtain a quadratic function of the same type in the variables $\xi$ and $\eta$. We use *basis functions* or *form functions* to represent it which allows us to give an explicit expression of $\tilde{u}(\xi, \eta)$ that directly depends on the function values, $u_j$, at the nodal points, $P_i$, of the normal triangle $T$. Analogously to the Lagrange polynomials (3.4) we define the six basis functions $N_i(\xi, \eta)$ on the normal triangle $T$ subject to the interpolation conditions with respect to the six nodal points $\bar{P}_j(\xi_j, \eta_j)$

$$N_i(\xi_j, \eta_j) = \begin{cases} 1 & \text{if } i = j \\ 0 & \text{if } i \neq j \end{cases} \quad (i, j = 1, 2, \ldots, 6). \quad (10.124)$$

The form functions $N_i(\xi, \eta)$ can easily be found by noticing that such a function is equal to zero along the whole side of the triangle whenever it has to be zero at the three nodal points of that side. Hence, we have

$$\boxed{\begin{aligned} N_1(\xi, \eta) &= (1 - \xi - \eta)(1 - 2\xi - 2\eta) & N_4(\xi, \eta) &= 4\xi(1 - \xi - \eta) \\ N_2(\xi, \eta) &= \xi(2\xi - 1) & N_5(\xi, \eta) &= 4\xi\eta \\ N_3(\xi, \eta) &= \eta(2\eta - 1) & N_6(\xi, \eta) &= 4\eta(1 - \xi - \eta) \end{aligned}} \quad (10.125)$$

Two form functions are shown in Figure 10.24.

With the aid of the basis functions $N_i(\xi, \eta)$, (10.125), we can represent the

## 10.3 Finite Element Method

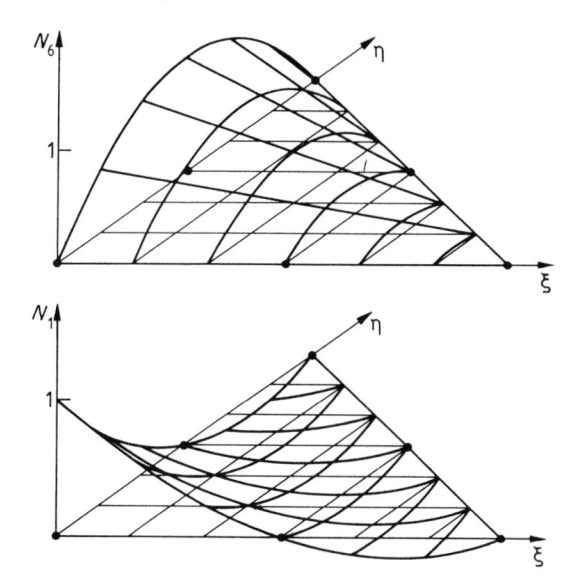

Figure 10.24 Basis functions $N_1(\xi, \eta)$ and $N_6(\xi, \eta)$

quadratic function $\tilde{u}(\xi, \eta)$ in the normal triangle $T$

$$\tilde{u}(\xi, \eta) = \sum_{i=1}^{6} u_i N_i(\xi, \eta) = \mathbf{u}_e^T \mathbf{N}(\xi, \eta), \tag{10.126}$$

where $u_i$ is the function value at the nodal point $\bar{P}_i$. We combine the six function values in the vector $\mathbf{u}_e := (u_1, u_2, \ldots, u_6)^T$, indicating by the subscript $e$ that the vector belongs to an element. Similarly, we introduce the vector of the form functions $\mathbf{N}(\xi, \eta) := (N_1(\xi, \eta), N_2(\xi, \eta), \ldots, N_6(\xi, \eta))^T$. From (10.126) the partial derivatives are given by the expressions

$$\tilde{u}_\xi = \mathbf{u}_e^T \mathbf{N}_\xi(\xi, \eta) \quad \tilde{u}_\eta = \mathbf{u}_e^T \mathbf{N}_\eta(\xi, \eta) \tag{10.127}$$

which will be needed in (10.122). For the three integrals which do not depend on the geometry of the triangular element $T_i$ we obtain, using the identity

$$(\mathbf{u}_e^T \mathbf{N}_\xi)^2 = (\mathbf{u}_e^T \mathbf{N}_\xi)(\mathbf{N}_\xi^T \mathbf{u}_e) = \mathbf{u}_e^T \mathbf{N}_\xi \mathbf{N}_\xi^T \mathbf{u}_e$$

$$I_1 := \iint_T \tilde{u}_\xi^2 \, d\xi \, d\eta = \iint_T [\mathbf{u}_e^T \mathbf{N}_\xi(\xi, \eta)]^2 \, d\xi \, d\eta$$

$$= \mathbf{u}_e^T \left( \iint_T \mathbf{N}_\xi \mathbf{N}_\xi^T \, d\xi \, d\eta \right) \mathbf{u}_e = \mathbf{u}_e^T \mathbf{S}_1 \mathbf{u}_e$$

$$I_2 := 2 \iint_T \tilde{u}_\xi \tilde{u}_\eta \, d\xi \, d\eta = \mathbf{u}_e^T \left( \iint_T (\mathbf{N}_\xi \mathbf{N}_\eta^T + \mathbf{N}_\eta \mathbf{N}_\xi^T) \, d\xi \, d\eta \right) \mathbf{u}_e = \mathbf{u}_e^T \mathbf{S}_2 \mathbf{u}_e$$

$$I_3 := \iint_T \tilde{u}_\eta^2 \, d\xi \, d\eta = \mathbf{u}_e^T \left( \iint_T \mathbf{N}_\eta \mathbf{N}_\eta^T \, d\xi \, d\eta \right) \mathbf{u}_e = \mathbf{u}_e^T \mathbf{S}_3 \mathbf{u}_e. \tag{10.128}$$

In the sense of matrix algebra, the product $\mathbf{N}_\xi \mathbf{N}_\xi^T$ of a column vector and a row vector represents a matrix of the order six, and the integral is understood componentwise. In order to also obtain a representation for $I_2$ with a symmetric matrix, the term $2\tilde{u}_\xi \tilde{u}_\eta$ is split up into two parts. The matrices $\mathbf{S}_1, \mathbf{S}_2$ and $\mathbf{S}_3$ are symmetric and of order six. They must be computed once for the quadratic function $\tilde{u}$ and can then be used to determine the contribution of a general triangular element $T_i$ by means of (10.122) with the coefficients defined by (10.123). The three matrices $\mathbf{S}_i$ in (10.128) are easily obtained by an elementary, somewhat lengthy calculation using the formula for the integrals

$$I_{p,q} := \iint_T \xi^p \eta^q \, d\xi \, d\eta = \frac{p!q!}{(p+q+2)!} \quad p, q \in \mathbf{N}_0.$$

We summarize the result in the so-called *stiffness element matrix* $\mathbf{S}_e \in \mathbf{R}^{6 \times 6}$ of a triangular element $T_i$ in the case of a quadratic function approximation

$$\iint_{T_i} (u_x^2 + u_y^2) \, dx \, dy = \mathbf{u}_e^T \mathbf{S}_e \mathbf{u}_e$$

$$\mathbf{S}_e = \frac{1}{6} \begin{pmatrix} 3(a+2b+c) & a+b & b+c & -4(a+b) & 0 & -4(b+c) \\ a+b & 3a & -b & -4(a+b) & 4b & 0 \\ b+c & -b & 3c & 0 & 4b & -4(b+c) \\ -4(a+b) & -4(a+b) & 0 & 8(a+b+c) & -8(b+c) & 8b \\ 0 & 4b & 4b & -8(b+c) & 8(a+b+c) & -8(a+b) \\ -4(b+c) & 0 & -4(b+c) & 8b & -8(a+b) & 8(a+b+c) \end{pmatrix}$$

(10.129)

The geometry of $T_i$ is contained in the coefficients $a$, $b$ and $c$ defined by (10.123). Similar triangular elements in arbitrary positions have the same stiffness element matrices due to (10.123).

The second integral of $I$ is also transformed to the normal triangle in a similar way and is represented by

$$I_4 := \iint_{T_i} \tilde{u}^2(x, y) \, dx \, dy = J \iint_T \tilde{u}^2(\xi, \eta) \, d\xi \, d\eta = J \iint_T (\mathbf{u}_e^T \mathbf{N})^2 \, d\xi \, d\eta$$

$$= \mathbf{u}_e^T \left( J \iint_T \mathbf{N} \mathbf{N}^T \, d\xi \, d\eta \right) \mathbf{u}_e = \mathbf{u}_e^T \mathbf{M}_e \mathbf{u}_e.$$

The evaluation of the matrix elements of the *mass element matrix* $\mathbf{M}_e \in \mathbf{R}^{6 \times 6}$ of a triangular element is again elementary, and we have

## 10.3 Finite Element Method

$$\iint_{T_i} \tilde{u}^2(x,y)\,dx\,dy = \mathbf{u}_e^T \mathbf{M}_e \mathbf{u}_e$$

$$\mathbf{M}_e = \frac{J}{360}\begin{pmatrix} 6 & -1 & -1 & 0 & -4 & 0 \\ -1 & 6 & -1 & 0 & 0 & -4 \\ -1 & -1 & 6 & -4 & 0 & 0 \\ 0 & 0 & -4 & 32 & 16 & 16 \\ -4 & 0 & 0 & 16 & 32 & 16 \\ 0 & -4 & 0 & 16 & 16 & 32 \end{pmatrix} \qquad (10.130)$$

The geometry of the triangular element appears in the mass element matrix $\mathbf{M}_e$ only as the common factor $J$, which is twice the area of the triangle. The shape of the triangle has no influence on the matrix.

The third integral of $I$, with a linear integrand in $u$, can be written as

$$I_5 := \iint_{T_i} \tilde{u}(x,y)\,dx\,dy = J \iint_T \tilde{u}(\xi,\eta)\,d\xi\,d\eta$$

$$= J \iint_T \mathbf{u}_e^T \mathbf{N}\,d\xi\,d\eta = \mathbf{u}_e^T \mathbf{b}_e = \mathbf{b}_e^T \mathbf{u}_e.$$

The components of the *element vector* $\mathbf{b}_e$ are simply obtained by integrating the basis functions over the normal triangle to obtain

$$\iint_{T_i} \tilde{u}(x,y)\,dx\,dy = \mathbf{b}_e^T \mathbf{u}_e \qquad \mathbf{b}_e = \tfrac{1}{6}J(0,0,0,1,1,1)^T. \qquad (10.131)$$

The result (10.131) represents an interpolating quadrature formula for a triangle on the basis of a quadratic interpolation. We note the remarkable fact that only the function values at the midpoints of the sides enter the formula with the same positive weights, and that the geometry of the triangle appears only in form of the double area.

Finally, we have to treat the two boundary integrals. Our triangulation means that the boundary is approximated by a polygon. Therefore we have to concern ourselves with the computation of the boundary integrals over a side of a triangle. On the side, the approximating function $\tilde{u}(x,y)$ is a quadratic function of the arc length and is uniquely determined by the function values at the three nodal points. Therefore we consider a boundary edge $R_i$ of length $L$ in a general position. Let $P_A$ and $P_B$ denote the two end-points and $P_M$ the middle point,

```
    o─────────────o─────────────o
    P_A           P_M           P_B
```

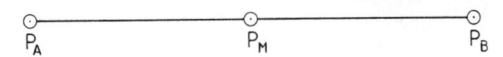

Figure 10.25 Boundary edge with nodal points

see Figure 10.25. The function values at these nodal points are accordingly denoted by $u_A$, $u_B$ and $u_M$, and we define the vector $\mathbf{u}_R := (u_A, u_M, u_B)^T$. To evaluate the boundary integrals we use the change of the variable $s = L\sigma$ which maps the edge onto the unit interval. The quadratic function $\tilde{u}(\sigma)$ is suitably represented by means of the three Lagrange polynomials

$$N_1(\sigma) = (1-\sigma)(1-2\sigma) \quad N_2(\sigma) = 4\sigma(1-\sigma) \quad N_3(\sigma) = -\sigma(1-2\sigma)$$

(10.132)

which are called *basis* or *form functions*. We combine them in the vector $\mathbf{N}(\sigma) := (N_1(\sigma), N_2(\sigma), N_3(\sigma))^T \in \mathbf{R}^3$, and thus for the first boundary integral

$$I_6 := \int_{R_i} \tilde{u}^2(s)\,ds = L\int_0^1 \tilde{u}^2(\sigma)\,d\sigma$$

$$= L\int_0^1 (\mathbf{u}_R^T \mathbf{N}(\sigma))^2\,d\sigma = \mathbf{u}_R^T \left(L\int_0^1 \mathbf{N}\mathbf{N}^T\,d\sigma\right)\mathbf{u}_R$$

$$= \mathbf{u}_R^T \mathbf{M}_R \mathbf{u}_R.$$

The result of the integration of the matrix of order three is the so-called *mass element matrix* $\mathbf{M}_R$ of a boundary edge $R_i$

$$\int_{R_i} \tilde{u}^2(s)\,ds = \mathbf{u}_R^T \mathbf{M}_R \mathbf{u}_R \quad \mathbf{M}_R = \frac{L}{30}\begin{pmatrix} 4 & 2 & -1 \\ 2 & 16 & 2 \\ -1 & 2 & 4 \end{pmatrix}.$$

(10.133)

The second boundary integral is given by Simpson's rule, (8.55), because $\tilde{u}(s)$ is a quadratic function of $s$ and is equal to the interpolation polynomial. Hence, with the boundary element vector $\mathbf{b}_R \in \mathbf{R}^3$ we can write

$$\int_{R_i} \tilde{u}(s)\,ds = \mathbf{b}_R^T \mathbf{u}_R \quad \mathbf{b}_R = \tfrac{1}{6}L(1, 4, 1)^T$$

(10.134)

## 10.3 Finite Element Method

### 10.3.4 Compilation and solution of the linear equations

For a practical implementation of the finite element method on a computer it is best to number all $N$ nodal points of the chosen triangulation, including those at which function values are prescribed by a Dirichlet boundary condition (10.108). The nodal points should be numbered in such a way that the maximal difference between numbers that belong to the same element is as small as possible, so that the band width of the matrix $\mathbf{A}$ of the system (10.116) is minimal. Algorithms for finding an optimal numbering exist (Schwarz, 1988).

Using this numbering the contributions of each triangular element and of each boundary edge are summed yielding a quadratic function of the $N$ nodal variables $u_1, u_2, \ldots, u_N$

$$\tilde{F} = \tfrac{1}{2}\mathbf{u}^T\tilde{\mathbf{A}}\mathbf{u} + \tilde{\mathbf{b}}^T\mathbf{u} \quad \tilde{\mathbf{A}} \in \mathbf{R}^{N \times N} \quad \tilde{\mathbf{b}}, \mathbf{u} \in \mathbf{R}^N \tag{10.135}$$

where $\tilde{\mathbf{A}}$ is built up from the stiffness and mass element matrices and where $\tilde{\mathbf{b}}$ is generated by the element vectors. This step is called the *compilation process*. The *total stiffness matrix* $\tilde{\mathbf{A}}$ is symmetric, but it is singular. The boundary condition (10.108) is usually taken into account at the corresponding boundary nodal points as follows. Let us assume that the condition $u_k = \varphi_k$ is given at the nodal point with number $k$. Therefore the quadratic form $\tfrac{1}{2}\mathbf{u}^T\tilde{\mathbf{A}}\mathbf{u}$ of the quadratic function (10.135) contains the linear terms $\tfrac{1}{2}(\tilde{a}_{ik}u_i\varphi_k + \tilde{a}_{ki}\varphi_k u_i) = \tilde{a}_{ik}u_i\varphi_k$, $i \neq k$ which can be combined with the linear part $\tilde{\mathbf{b}}^T\mathbf{u}$. This only means that $\varphi_k$ times the $k$th column of $\tilde{\mathbf{A}}$ must be added to the vector $\tilde{\mathbf{b}}$. After completing the operation, the $k$th column and the $k$th row of $\tilde{\mathbf{A}}$ and the $k$th component of $\tilde{\mathbf{b}}$ are, in principle, superfluous and could be omitted. In place of that the $k$th column and row of $\tilde{\mathbf{A}}$ are replaced by zero elements, and then we define $\tilde{a}_{kk} = 1$ and $\tilde{b}_k = -\varphi_k$. As soon as all boundary nodal points with Dirichlet boundary conditions are treated as described, a quadratic function of the type (10.115) is obtained but with $N$ nodal variables. In the next step we have to solve the system of linear equations $\mathbf{A}\mathbf{u} + \mathbf{b} = \mathbf{0}$ with the modified matrix $\mathbf{A} \in \mathbf{R}^{N \times N}$ and the modified vector $\mathbf{b} \in \mathbf{R}^N$. The matrix $\mathbf{A}$ is now positive definite, and the system of equations contains some trivial equations which correspond to the Dirichlet type boundary conditions. The solution vector $\mathbf{u} \in \mathbf{R}^N$ also contains those nodal variables whose values are prescribed by boundary conditions. This fact is quite practical for a further evaluation of the result, for instance for the construction of level curves of the solution.

The symmetric and positive definite system of equations $\mathbf{A}\mathbf{u} + \mathbf{b} = \mathbf{0}$ can be solved by means of Cholesky's method taking the band structure into account. However, the band width of $\mathbf{A}$ is quite variable in most applications so that it is advantageous to apply the *envelope oriented* technique of storage and solution. Iterative methods for solving larger systems of linear equations can also be efficient because they take the sparsity of $\mathbf{A}$ into account. For more details see Schwarz (1988).

### 10.3.5 Examples

The following examples have been solved on a desk-top computer HP 85 by means of programs by Schwarz (1988). The figures have been drawn with the plotter.

*Example* 10.15  We solve the boundary problem (10.15) to (10.18) for the region $G$ shown in Figure 10.4. The corresponding variational problem is

$$I = \iint_G [\tfrac{1}{2}(u_x^2 + u_y^2) - 2u]\,dx\,dy = \text{Extr!} \tag{10.136}$$

subject to the boundary conditions

$$\begin{aligned} u &= 0 \quad \text{on DE and EF,} \\ u &= 1 \quad \text{on AB and BC.} \end{aligned} \tag{10.137}$$

We use a coarse triangulation of the domain into 13 triangular elements, see Figure 10.26, so that we can compare the results with those of Example 10.1. Among the $N = 32$ nodal points there exist 12 with prescribed boundary values so that $n = 20$ nodal points have unknown function values. As all triangular elements are isosceles and right-angled, the stiffness element matrices are identical. If we start the numbering of the nodal points of each element with the corner corresponding to the right angle, because $a = c = 1$ and $b = 0$ we have

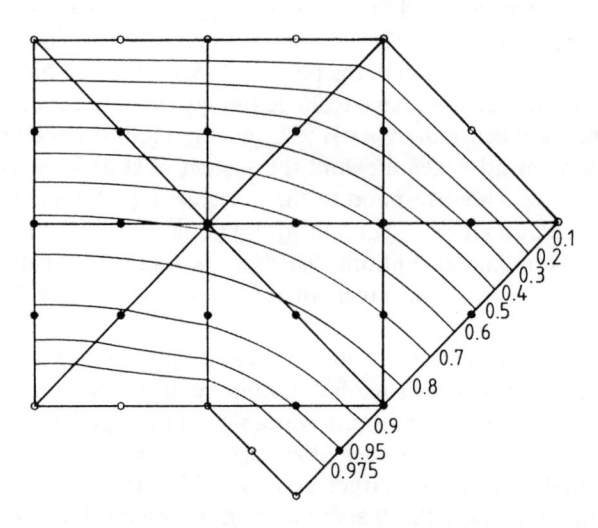

Figure 10.26 Domain with triangulation and level curves of the approximate solution

## 10.3 Finite Element Method

$$S_e = \frac{1}{6}\begin{pmatrix} 6 & 1 & 1 & -4 & 0 & -4 \\ 1 & 3 & 0 & -4 & 0 & 0 \\ 1 & 0 & 3 & 0 & 0 & -4 \\ -4 & -4 & 0 & 16 & -8 & 0 \\ 0 & 0 & 0 & -8 & 16 & -8 \\ -4 & 0 & -4 & 0 & -8 & 16 \end{pmatrix}.$$

The system of equations can be easily compiled by means of $S_e$ and the element vector $b_e$ (Stiefel, 1976). The resulting approximate solution of the boundary value problem is given in (10.138) in the arrangement with the nodal points, rounded to four decimal places. The discretization errors are about as large as those of (10.21).

$$\begin{array}{cccccc} 0 & 0 & 0 & 0 & 0 & \\ 0.4176 & 0.4113 & 0.3899 & 0.3410 & 0.2371 & 0 \\ 0.7229 & 0.7127 & 0.6822 & 0.6119 & 0.4826 & 0.2825 & 0 \\ 0.9167 & 0.9094 & 0.8834 & 0.8169 & 0.7000 & 0.5225 & \\ 1 & 1 & 1 & 0.9474 & 0.8529 & & \\ & & & 1 & 0.9525 & & \\ & & & & 1 & & \end{array} \quad (10.138)$$

In Figure 10.26 some level curves of the approximate solution $\tilde{u}(x, y)$ are shown. The continuity of the approximate solution is obvious, but the discontinuity of

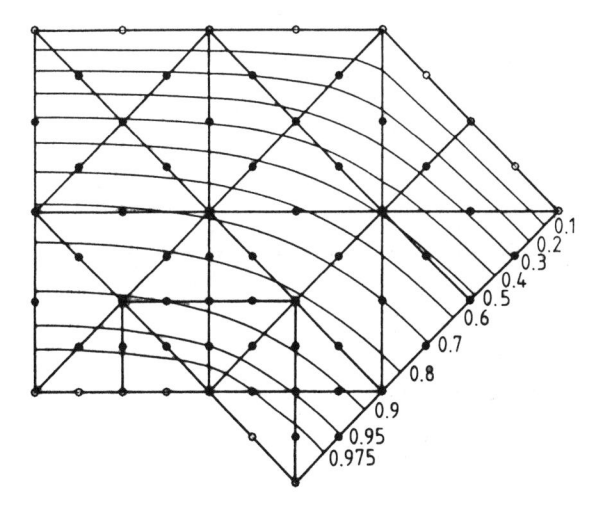

Figure 10.27 Locally finer triangulation and level curves

the first partial derivatives can be seen on the border of two adjacent triangles. The course of the level curves in the vicinity of the obtuse corner is conspicuous. This is caused by the singularity of the partial derivatives of the solution function $u(x, y)$ (Mitchell and Griffiths, 1980). To improve the situation, the boundary value problem has been treated with the finer triangulation of Figure 10.27. In the neighbourhood of the obtuse corner a locally finer triangulation has been chosen. For a total of $N = 65$ nodal points the function values are unknown at $n = 49$ nodal points. The level curves are now smoother, but the influence of the corners is still obvious.

The approximations at those nodal points that correspond to Figure 10.26 are given

$$
\begin{array}{ccccccc}
0 & 0 & 0 & 0 & 0 & & \\
0.4179 & 0.4117 & 0.3908 & 0.3396 & 0.2298 & 0 & \\
0.7214 & 0.7129 & 0.6808 & 0.6128 & 0.4877 & 0.2824 & 0 \\
0.9171 & 0.9101 & 0.8817 & 0.8169 & 0.7006 & 0.5228 & \\
1 & 1 & 1 & 0.9468 & 0.8544 & & \\
& & & 1 & 0.9531 & & \\
& & & & 1 & &
\end{array}
\tag{10.139}
$$

*Example* 10.16  We consider the boundary value problem (10.31) of Example 10.5 for the region $G$ with the curved boundary AB of Figure 10.8. The corresponding

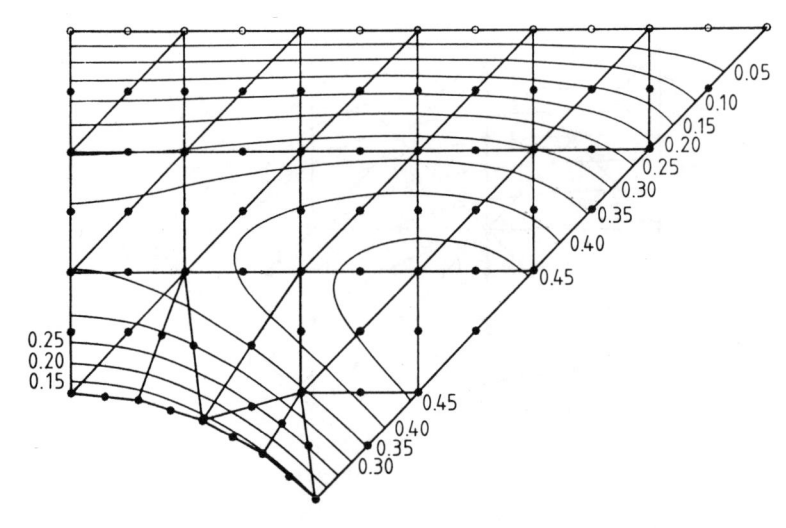

Figure 10.28  Triangulation of the domain with curved boundary, level curves of the approximate solution

variational problem is

$$I = \iint_G [\tfrac{1}{2}(u_x^2 + u_y^2) - 2u]\,dx\,dy + \int_{AB} (u^2 + u)\,ds = \text{Extr!} \tag{10.140}$$

$u = 0$ on CD.

The boundary integral extends only over the circular arc AB with $\alpha(s) = 2$ and $\beta(s) = -1$. Figure 10.28 shows the triangulation used. The circular arc is approximated by four straight-line boundary parts whereby the five nodal corner points are equidistantly distributed over the arc. The area of the approximating domain is slightly larger than that of the given domain. Among the total of $N = 80$ nodal points there are $n = 67$ points with unknown function values. The approximations obtained at selected nodal points, that correspond to those of Figure 10.8 with the exception of the points on the circular arc, are given in (10.141). The level curves shown in Figure 10.28 visualize the solution.

$$\begin{array}{ccccccc}
0 & 0 & 0 & 0 & 0 & 0 & 0 \\
0.2956 & 0.3043 & 0.3223 & 0.3295 & 0.3016 & 0.2094 & \\
0.3483 & 0.3754 & 0.4326 & 0.4715 & 0.4423 & & \\
0.1133 & 0.1312 & 0.3412 & 0.4605 & & & \\
 & & 0.1474 & & & &
\end{array} \tag{10.141}$$

## 10.4 Exercises

**10.1.** We consider the boundary value problem for the region G of Figure 10.29:

$u_{xx} + u_{yy} = -1$ in $G$,

$u = 1$ on AB    $\dfrac{\partial u}{\partial n} = 0$ on BC,

$u = 0$ on CD    $\dfrac{\partial u}{\partial n} + 2u = 1$ on DA.

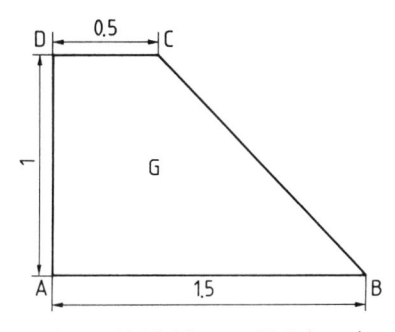

Figure 10.29 Trapezoidal domain

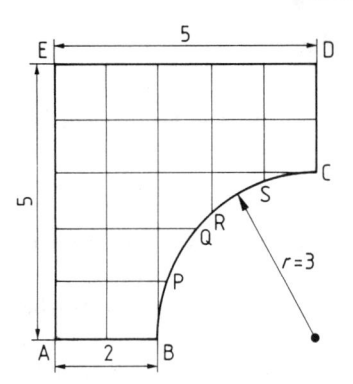

Figure 10.30 Domain with circular boundary

Solve it approximately by means of the difference method on the basis of the five-point approximation using the step sizes $h = \frac{1}{4}$, $h = \frac{1}{6}$ and $h = \frac{1}{8}$. What are the difference equations for mesh points on the side DA? For which numbering of the nodal points is the band width of the system of the difference equations minimal?

**10.2.** Solve the following elliptic boundary value problem with respect to the region $G$ of Figure 10.30, whose boundary BC is a quarter of a circle with radius $r = 3$:

$$u_{xx} + u_{yy} = -10 \quad \text{in } G$$

$$\frac{\partial u}{\partial n} = 0 \quad \text{on AB} \quad u = 4 \text{ on} \quad \text{BC} \quad \frac{\partial u}{\partial n} = 0 \quad \text{on CD},$$

$$\frac{\partial u}{\partial n} + 4u = 1 \quad \text{on DE} \quad u = 0 \quad \text{on EA}.$$

What are the operator equations for the mesh points of Figure 10.30, where the mesh size is $h = 1$? Is the system of difference equations symmetric?

**10.3.** Replace the Dirichlet boundary condition on the curved side BC of the boundary value problem of Exercise 10.2 by the Cauchy condition

$$\frac{\partial u}{\partial n} + 3u = 2.$$

The new problem requires the derivation of further difference equations for the points B, P, Q, R, S, and C on the boundary. What is the structure of the system of difference equations now? Its solution can be determined by means of the Gaussian algorithm with diagonal pivotal strategy.

**10.4.** Solve the elliptic boundary value problem of Exercise 10.1 approximately by applying the nine-point difference equation (10.58). In order to obtain the difference equations that are valid for the boundary points on DA approximate $\partial u/\partial n = -\partial u/\partial x$ by the central difference quotient so that the values $u_{NW}$, $u_W$ and $u_{SW}$ can be eliminated in the formula (10.58). Compare the resulting approximate solution with that of Exercise 10.1.

## 10.4 Exercises

**10.5.** Solve the parabolic initial boundary value problem
$$u_t = u_{xx} + 1 \quad \text{for } 0 < x < 1, t > 0;$$
$$u(x,0) = 0 \quad \text{for } 0 < x < 1,$$
$$u_x(0,t) - 0.5u(0,t) = 0 \quad u_x(1,t) + 0.2u(1,t) = 0 \quad \text{for } t > 0$$
by means of the explicit and the implicit method using $h = 0.1$ and $h = 0.05$ for different time steps $k$. Derive the condition of the absolute stability for the explicit method in the case of the given boundary conditions. Analytically determine the stationary solution $u(x,t)$ as $t \to \infty$, where $u_t = 0$, and compare it with the computed approximation.

**10.6.** A diffusion problem is described as follows
$$\frac{\partial u}{\partial t} = \frac{\partial}{\partial x}\left((x^2 - x + 1)\frac{\partial u}{\partial x}\right) + (2 - x)\sin(t) \quad 0 < x < 1, t > 0$$
$$u(x,0) = 0 \quad 0 < x < 1$$
$$u(0,t) = 0 \quad u_x(1,t) + 0.3u(1,t) = 0 \quad t > 0.$$
The diffusion conductivity depends on the location, and the source density is space and time dependent. Compute an approximate solution using the space step size $h = 0.1$ by means of the explicit and the implicit method. What is the condition of absolute stability in the case of Richardson's method?

**10.7.** Solve the parabolic differential equation
$$u_t = u_{xx} + u_{yy} + 1 \quad \text{in } G \times (0, \infty)$$
subject to the initial condition
$$u(x,y,0) = 0 \quad \text{in } G$$
and the time-independent boundary conditions

$u = 1$ on AB  $\quad \dfrac{\partial u}{\partial n} = 0$ on BC,

$u = 0$ on CD  $\quad \dfrac{\partial u}{\partial n} + 2u = 1$ on DA

where $G$ is the region of the Figure 10.29. Use the explicit Richardson method, the implicit Crank–Nicolson method and the method of alternating directions. Choose the step sizes $h = \frac{1}{4}$ and $h = \frac{1}{6}$. The stationary solution as $t \to \infty$ is equal to the solution of Exercise 10.1.

**10.8.** What are the stiffness element matrices $S_e$ in the case of quadratic approximating functions for
  (a) an equilateral triangle;
  (b) a right-angled triangle with the small sides of length $\overline{P_1P_2} = \alpha h$ and $\overline{P_1P_3} = \beta h$;
  (c) an isosceles triangle of side lengths $h$ and angle $\gamma$? What follows for the matrix elements of $S_e$ in the cases (b) and (c) if $\beta \ll \alpha$ or if $\gamma$ is small, respectively? What are the consequences of triangular elements with an acute angle for the stiffness matrix **A**?

**10.9.** Linear approximating functions in the finite element method.
  (a) Determine and describe the form functions for the linear approximation. With them

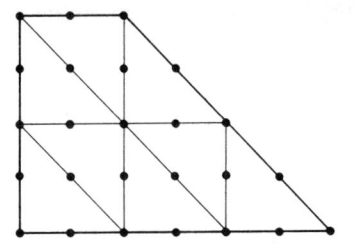

Figure 10.31 Coarse triangulation of the trapezoidal domain

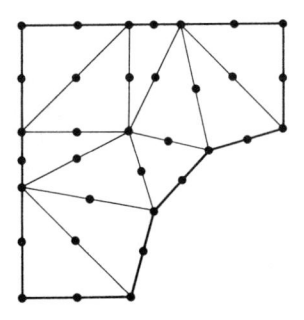

Figure 10.32 Triangulation of the domain with circular boundary

derive the element matrices to approximate the integral expression (10.107) on the assumption that the functions $\rho(x, y)$, $f(x, y)$, $\alpha(s)$ and $\beta(s)$ are constant for each element.

(b) What are the stiffness element matrices $S_e$ for an isosceles right-angled triangle, an equilateral triangle, a right-angled triangle with the small sides of length $\alpha h$ and $\beta h$, and for an isosceles triangle of side lengths $h$ and the angle $\gamma$? What happens in the case of triangles with an acute angle?

(c) Verify that the finite element method, applied to the Poisson equation $u_{xx} + u_{yy} = f(x, y)$, in the case of linear elements for each internal nodal point yields the five-point difference equation (10.11) if a regular triangulation into congruent, isosceles right-angled triangles is used. This is true independently of how many triangular elements adjoin in the nodal point under consideration.

**10.10.** What are the integral expressions to be minimized in the case of the boundary value problems of Exercises 10.1 to 10.3? Solve these boundary value problems by means of the finite element method using quadratic approximating functions. We propose, for instance, to use the triangulations shown in Figures 10.31 and 10.32.

# References

Abramowitz, M., and Stegun, I. A. (1971) *Handbook of mathematical functions*, Dover, New York.
Ahlberg, J., Nilson, E., and Walsh, J. (1967) *The theory of splines and their applications*, Academic Press, New York.
Aiken, R. C. (ed.) (1985) *Stiff computation*, Oxford University Press, Oxford.
Amann, H. (1983). *Gewöhnliche Differentialgleichungen*, de Gruyter., Berlin.
Ames, W. F. (1977). *Numerical methods for partial differential equations*, 2nd edn. Academic Press, New York.
Bathe, K.-J., and Wilson, E. (1976). *Numerical methods in finite element analysis*, Prentice-Hall, Englewood Cliffs, N.J.
Bauer, F. L., Rutishauser, H., and Stiefel, E. (1963). *Proc. of Symposia in Applied Mathematics, Amer. Math. Society*, **15**, 199–218.
Benoit: (1924). *Bull. géodésique* **3**, 67–77.
Blum, E., and Oettli, W. (1975). *Mathematische Optimierung.*, Springer, Berlin.
Böhmer, K. (1974). *Spline-Funktionen*, Teubner, Stuttgart.
Boor, C. de (1978). *A practical guide to splines*, Springer, New York.
Boor, C. de (1980). *SIAM J. Sci. Stat. Comput.*, **1**, 173–178.
Bramble, J. H., Hubbard, B. E., and Zlamal, M. (1968). *SIAM J. Numer. Anal.*, **5**, 1–25.
Brent, R. P. (1973). *Algorithms for minimization without derivatives*, Prentice-Hall, Englewood Cliffs, N.J.
Brigham, E. O. (1974). *The fast Fourier transform*, Prentice-Hall, Englewood Cliffs, N.J.
Bulirsch, R. (1964). *Numer. Math.*, **6**, 6–16.
Bulirsch, R., and Stoer, J. (1967) *Numer. Math.*, **9**, 271–278.
Bunse, W., and Bunse-Gerstner, A. (1984). *Numerische lineare Algebra*, Teubner, Stuttgart.
Butcher, J. C.(1963). *J. Austral. Math Soc.*, **3**, 185–201.
Butcher, J. C. (1964). *Math. Comp.*, **18**, 50–64.
Butcher, J. C. (1965). *Math. Comp.*, **19**, 408–417.
Butcher, J. C. (1987). *The numerical analysis of ordinary differential equations*, John Wiley, Chichester.
Ceschino, F., and Kuntzmann, J. (1966). *Numerical solution of initial value problems*, Prentice-Hall, Englewood Cliffs, N.J.

Chan, T. F. (1982). *ACM Trans. Math. Soft.*, **8**, 72–83.
Chan, T. F. (1982). *ACM Trans. Math. Soft.*, **8**, 84–88.
Clegg, J. C. (1970). *Variationsrechnung*, Teubner, Stuttgart.
Clenshaw, C. W. (1955). *Math. Tab. Wash.*, **9**, 118–120.
Cline, A. K., Moler, C. B., Stewart, G. W. and Wilkinson, J. H. (1979). *SIAM J. Numer. Anal.*, **16**, 368–375.
Collatz, L. (1966). *The numerical treatment of differential equations.*, 3rd edn. Springer, Berlin.
Collatz, L. (1981). *Differentialgleichungen. 6 Aufl.*, Teubner, Stuttgart.
Collatz, L., and Albrecht, J. (1972, 1973). *Aufgaben aus der angewandten Mathematik I + II.*, Akademie-Verlag, Berlin.
Collatz, L., and Wetterling, W. (1971). *Optimierungsaufgaben. 2. Aufl.*, Springer, Berlin.
Cooley, J. W., and Tukey, J. W., (1965). *Math. Comput.*, **19**, 297–301.
Corliss, G., and Chang, Y. F. (1982). *ACM Trans. Math. Soft.*, **8**, 114–144.
Courant, R., and Hilbert, D. (1968). *Methoden der mathematischen Physik I. 3. Aufl.*, Springer, Berlin.
Dahlquist, G. (1985). *BIT*, **25**, 188–204.
Dantzig, G. B. (1966). *Lineare Programmierung und Erweiterungen.*, Springer, Berlin.
Davis, P. J. (1959). In: Langer, R. E. (ed.): *On numerical approximation.*, The Univ. of Wisconsin Press, Madison, 45–60.
Davis, P. J., and Rabinowitz, P. (1984). *Methods of numerical integration*, 2nd edn. Academic Press, New York.
Dowell, M., and Jarrat, P. (1972). *BIT.*, **12**, 503–508.
Duff, I. S., and Reid, J. K. (1976). *J. Inst. Math. Appl.*, **17**, 267–280.
Engeli, M. (1962). *Automatisierte Behandlung elliptischer Randwertprobleme. Dissertation, Nr. 3295* ETH Zürich.
Engels, H. (1980). *Numerical quadrature and cubature*, Academic Press, London.
England, R. (1969). *Comput. J.*, **12**, 166–170.
Epperson, J. F. (1987). *Amer. Math. Monthly*, **94**, 329–341.
Erdélyi, A., Magnus, W., Oberhettinger, F., and Tricomi, F. G. (1953). *Higher transcendental functions*, Vol. I. McGraw-Hill, New York.
Fehlberg, E. (1964). *ZAMM*, **44**, T17–T29.
Fehlberg, E. (1969). *Computing*, **4**, 93–106.
Fehlberg, E. (1970). *Computing*, **6**, 61–71.
Finkenstein, Graf Finck von, K. (1977, 1978) *Einführung in die numerische Mathematik I + II*, Carl Hanser, Munich.
Fletcher, R. (1980). *Practical methods of optimization. Vol. I: Unconstrained optimization*, John Wiley, Chichester.
Forsythe, G. E., Henrici, P. (1960). *Trans. Amer. Math Soc.*, **94**, 1–23.
Forsythe, G. E., Malcolm, M. A., and Moler, C. B. (1977). *Computer methods for mathematical computations*, Prentice-Hall, Englewood Cliffs, N J.
Forsythe, G. E., and Moler, C. B. (1967). *Computer solution of linear algebraic systems*, Prentice-Hall, Englewood Cliffs, N J.
Fox, L., and Parker, I. B. (1968). *Chebyshev polynomials in numerical analysis*, Oxford University Press, London.
Francis, J. G. F. (1961). *Comp. J.*, **4**, 265–272, 332–345.
Funk, P. (1970). *Variationsrechnung und ihre Anwendung in Physik und Technik. 2. Aufl.*, Springer, Berlin.
Gander, W. (1985). *Computermathematik*, Birkhäuser, Basel.
Gautschi, W. (1970). *Math. Comp.*, **24**, 245–260.

Gear C. W. (1971). *Numerical initial value problems in ordinary differential equations*, Prentice-Hall, Englewood Cliffs, NJ.
Gekeler, E. (1984). *Discretization methods for stable initial value problems* (Lecture Notes in Mathematics, 1044) Springer, Berlin.
Gentleman, M. (1973). *J. Inst. Math. Appl.*, **12**, 329–336.
George, A., and Heath, M. T. (1980). *Lin. Alg. Appl.*, **34**, 69–83.
Gill, P. E., and Murray, W. (1976). In Bunch J. R. and Rose D. J. (eds) *Sparse matrix computations*, Academic, New York, 177–200.
Givens, J. W. (1954). *Numerical computation of the characteristic values of a real symmetric matrix Rep. ORNL-1574*, Oak Ridge Nat. Lab., Oak Ridge.
Givens, W. (1958). *SIAM J. Appl. Math.*, **6**, 26–50.
Gladwell, I., and Wait, R. (ed.) (1979). *A survey of numerical methods for partial differential equations*, Clarendon, Oxford.
Glashoff, K., and Gustafson, S. Å. (1978) *Einführung in die lineare Optimierung*. Wissensch, Buchgesellschaft, Darmstadt.
Golub, G. H., and Kahan, W. (1965). *SIAM J. Numer. Anal. Ser B.*, **2**, 205–224.
Golub, G. H., and Kautsky, J. (1983). *Numer. Math.*, **41**, 147–163.
Golub, G. H., and Pereyra, V. (1973). *SIAM J. Numer. Anal.*, **10**, 413–432.
Golub, G. H., and Plemmons, R. J. (1980). *Lin. Alg. Appl.*, **34**, 3–27.
Golub, G. H., and Reinsch, C. (1970). *Numer. Math.*, **14**, 403–420.
Golub, G. H., and Van Loan, Ch. (1983). *Matrix computations*, Oxford Academic, Oxford.
Golub, G. H., and Welsch, J. A. (1969). *Math. Comp.*, **23**, 221–230.
Good, I. J. (1960). *J. Roy. Statist. Soc. Ser. B.*, **20**, (1958) 361–372; Addendum: **22**, 372–375.
Goodwin, E. T. (1949). *Proc. Camb. Phil. Soc.*, **45**, 241–245.
Gourlay, A. R., and Watson, G. A. (1973). *Computational methods for matrix eigenproblems*, Wiley, London.
Gregory, R. T., and Karney, D. L. (1969). *A collection of matrices for testing computational algorithms*, John Wiley, New York.
Grigorieff, R. D. (1972). *Numerik gewöhnlicher Differentialgleichungen. Band 1, Einschrittverfahren*, Teubner, Stuttgart.
Grigorieff, R. D. (1977). *Numerik gewöhnlicher Differentialgleichungen. Band 2, Mehrschrittverfahren*, Teubner, Stuttgart.
G-A E Subcommittee on measurement concept. (1967). *Proc. IEEE*, **55**, 1664–1674.
Hackbusch, W. (1986). *Theorie und Numerik elliptischer Differentialgleichungen*, Teubner, Stuttgart.
Hainer, K. (1983). *Numerik mit BASIC-Tischrechnern*, Teubner, Stuttgart.
Hairer, E., Nørsett, S. P., and Wanner, G. (1987). *Solving ordinary differential equations I. Nonstiff problems*, Springer, Berlin.
Halin, H. J. (1983). In: *1983 summer computer simulation conference, Vancouver B.C.*, Vol. 2, North-Holland, Amsterdam, 1032–1078.
Hall, G., and Watt, J. M. (ed.) (1976). *Modern numerical methods for ordinary differential equations*, Clarendon Press, Oxford.
Hammarling, S. (1974). *J. Inst. Math. Appl.*, **13**, 215–218.
Henrici, P. (1958). *SIAM J. Appl. Math.*, **6**, 144–162.
Henrici, P. (1962). *Discrete variable methods in ordinary differential equations*, John Wiley, New York.
Henrici, P. (1972). *Elemente der numerischen Analysis, Bd. 1 und 2*, Bibliographisches Institut, Mannheim.
Henrici, P. (1977). *Applied and computational complex analysis*, vol. 2, John Wiley, New York.

Henrici, P. (1985). *Applied and computational complex analysis*, vol. 3., John Wiley, New York.
Householder, A. S. (1958). *J. Assoc. Comp. Mach.*, **5**, 339–342.
Hyman, M. A. (1957). *Eigenvalues and eigenvectors of general matrices. Twelfth National Meeting A.C.M., Houston, Texas 1957.*
Iri, M., Moriguti, S., and Takasawa, Y. (1970) *RIMS Kokyuroku Kyoto Univ.*, **91**, 82–118 (in Japanese).
Isaacson, E. and Keller, H. B. (1966). *Analysis of numerical methods*, John Wiley, New York.
Jacobi, C. G. J. (1846). *Crelle's Journal*, **30**, 51–94.
Jacoby, S. L. S. Kowalik, J. S., and Pizzo, J. T. (1972). *Iterative methods for nonlinear optimization problems*, Prentice-Hall, Englewood Cliffs, NJ.
Jain, M. K. (1984). *Numerical solution of differential equations.* 2nd edn. John Wiley, New York.
Jennings, A. (1977) *Matrix computation for engineers and scientists*, John Wiley, New York.
Jordan-Engeln, G., and Reutter, F. (1982). *Numerische Mathematik für Ingenieure. 3. Aufl.* Bibliographisches Institut, Mannheim.
Kall, P. (1976). *Mathematische Methoden des Operations Research*, Teubner, Stuttgart.
Kaufman, L. (1979). *ACM Trans. Math. Soft.*, **5**, 442–450.
Krylov, F. I. (1962). *Approximate calculation of integrals*, (Translation from Russian by A. H. Stroud), Macmillan, New York.
Künzi, H. P., Tzschach, H. G., and Zehnder, C. A. (1967). *Numerische Methoden der mathematischen Optimierung mit ALGOL und FORTRAN- Programmen*, Teubner, Stuttgart.
Kutta, W. (1901). *Z. Math. Phys.*, **46**, 435–453.
Lambert, J. D. (1973). *Computational methods in ordinary differential equations*, John Wiley, New York.
Lapidus, L., and Seinfeld, J. H. (1971). *Numerical solution of ordinary differential equations*, Academic Press, New York.
Lawson, C. L., and Hanson, R. J. (1974). *Solving least squares problems*, Prentice-Hall, Englewood Cliffs, NJ.
Lawson, J. D. (1966). *SIAM J. Numer. Anal.*, **3**, 593–597.
Ludwig, R. (1969). *Methoden der Fehler- und Ausgleichsrechnung*, Friedr. Vieweg, Braunschweig.
Lyness, J. N. (1969). *J. ACM*, **16**, 483–495.
Maess, G. (1985). *Vorlesungen über numerische Mathematik I. Lineare Algebra*, Birkhäuser, Basel.
Marquardt, D. W. (1963). *J. Soc. Indust. Appl. Math.*, **11**, 431–441.
Marsal, D. (1976). *Die numerische Lösung partieller Differentialgleichungen in Wissenschaft und Technik*, Bibliographisches Institut, Mannheim.
Meis, Th., and Marcowitz, U. (1978). *Numerische Behandlung partieller Differentialgleichungen*, Springer, Berlin.
Mitchell, A. R., and Griffiths, D. F. (1980). *The finite difference method in partial differential equations*, John Wiley, Chichester.
Mori, M. (1978), *RIMS Kyoto Univ*, **14**, 713–729.
Muller, D. E. (1956). *Math. Tables Aids Comput.*, **10**, 208–215.
Ortega, J. M., and Rheinboldt, W. C. (1970). *Iterative solution of nonlinear equations in several variables*, Academic Press, New York.
Parlett, B. N. (1964). *Math. Comp.*, **18**, 464–485.
Parlett, B. N. (1980), *The symmetric eigenvalue problem*, Prentice-Hall Englewood Cliffs, NJ.

# References

Parter, S. V. (ed.) (1979). *Numerical methods for partial differential equations*, Academic Press, New York.
Peaceman, D. W., and Rachford, H. H. (1955). *J. Soc. Industr. Appl. Math.*, **3**, 28–41.
Piessens, R., de Doncker-Kapenga, E., Überhuber, C. W., and Kahaner, D. K.: QUADPACK. (1983). *A subroutine package for automatic integration*, Springer, Berlin.
Ralston, A., and Rabinowitz, Ph. (1978). *A first course in numerical analysis.* 2nd edn., McGraw Hill, New York.
Rath, W. (1982). *Numer. Math.*, **40**, 47–56.
Romberg, W. (1955). *Vereinfachte numerische Integration. Det. Kong. Norske Videnskabers Selskab Forhandlinger*, **28**, Nr. 7. Trondheim.
Runge, C. (1901). *Z. Math. Phys.*, **46**, 224–243.
Runge, C. (1903). *Z. Math. Phys.*, **48**, 443–456.
Runge, C. (1905). *Z. Math. Phys.*, **52**, 117–123.
Runge, C., and König, H. (1924). *Vorlesungen über numerisches Rechnen*, Springer, Berlin.
Rutishauser, H. (1952). *ZaMP*, **3**, 65–74.
Rutishauser, H. (1960). *ZaMP*, **11**, 508–513.
Rutishauser, H. (1963). *Numer. Math.*, **5**, 48–54.
Rutishauser, H. (1966). *Numer. Math.*, **9**, 1–10.
Rutishauser, H. (1976). *Vorlesungen über numerische Mathematik. Band 1 und 2*, Birkhäuser, Basel.
Sag, W. T., and Szekeres, G. (1964). *Math. Comp.*, **18**, 245–253.
Sauer, R., and Szabo, I. (1968). *Mathematische Hilfsmittel des Ingenieurs. Band 3*, Springer, Berlin.
Schmeisser, G., and Schirmeier, H. (1976). *Praktische Mathematik*, de Gruyter, Berlin.
Schönhage, A. (1961). *Numer. Math.*, **3**, 374–380.
Schönhage, A. (1964). *Numer. Math.*, **6**, 410–412.
Schröder, G. (1964). *Über die Konvergenz einiger Jacobi-Verfahren zur Bestimmung der Eigenwerte symmetrischer Matrizen*, Westdeutscher Verlag, Köln.
Schur, I. (1909). *Math. Annalen*, **66**, 488–510.
Schwarz, H. R. (1977) *Computing*, **18**, 107–116.
Schwarz, H. R. (1988). *Finite element methods.* Academic Press, London.
Schwarz, H. R. (1988). *FORTRAN-Programme zur Methode der finiten Elemente.* 2. Aufl., Teubner, Stuttgart.
Schwarz, H. R. (1972). Rutishauser, H., Stiefel, E. *Numerik symmetrischer Matrizen.* 2. Aufl., Teubner, Stuttgart.
Schwartz, C. (1969). *J. Comp. Phys.*, **4**, 19–29.
Shampine, L. F., and Gordon, M. K. (1975). *Computer solution of ordinary differential equations. The initial value problem.* Freeman, San Francisco.
Singleton, R. C. (1968). *Comm. ACM*, **11**, 773–776.
Singleton, R. C. (1968). *Comm. ACM*, **11**, 776.
Smirnow, W. I. (1972). *Lehrgang der höheren Mathematik*, Teil II. Deutscher Verlag der Wissenschaften, Berlin.
Smith, G. D. (1978). *Numerical solution of partial differential equations: Finite difference methods.* 2nd edn, Clarendon Press, Oxford.
Späth, H. (1973). *Spline-Algorithmen zur Konstruktion glatter Kurven und Flächen*, Oldenbourg, Munich.
Stenger, F. (1973). *J. Inst. Math. Appl.*, **12**, 103–114.
Stetter, H. J. (1965). *Math. Comp.*, **9**, 84–89.
Stewart, G. W. (1973). *Introduction to matrix computations*, Academic Press, New York.
Stewart, G. W. (1976). *Numer. Math.*, **25**, 137–138.
Stiefel, E. (1976). *Einführung in die numerische Mathematik*, 5. Aufl. Teubner, Stuttgart.

Stoer, J. (1983). *Einführung in die numerische Mathematik I*, 4. Aufl., Springer, Berlin.
Stroud, A. H. (1971). *Approximate calculation of multiple integrals*, Prentice-Hall, Englewood Cliffs, NJ.
Stroud, A. H. (1974). *Numerical quadrature and solution of ordinary differential equations*, Springer, New York.
Stroud, A. H., and Secrest, D. (1966). *Gaussian quadrature formulas*, Prentice-Hall, Englewood Cliffs, NJ.
Stummel, F., and Hainer, K. (1982). *Praktische Mathematik*. 2. Aufl. Teubner Stuttgart.
Szekeres, G. (1961, 1962). *J. Australian Math. Soc.*, **2**, 301–320.
Takahasi, H., and Mori, M. (1973). *Numer. Math.*, **21**, 206–219.
Takahasi, H., and Mori, M. (1974). *Publ. RIMS Kyoto Univ.*, **9**, 721–741.
Terebesi, P. (1930). *Rechenschablonen für die harmonische Analyse und Synthese*, Springer, Berlin.
Törnig, W. (1979). *Numerische Mathematik für Ingenieure und Physiker. Band 1: Numerische Methoden der Algebra*, Springer, Berlin.
Törnig, W. (1979), *Numerische Mathematik für Ingenieure und Physiker Band 2: Eigenwertprobleme und numerische Methoden der Analysis*, Springer, Berlin.
Törnig, W., Gipser, M., and Kaspar, B. (1985). *Numerische Lösung von partiellen Differentialgleichungen der Technik. Differenzenverfahren, finite Elemente und die Behandlung grosser Gleichungssysteme*. Teubner, Stuttgart.
Twizell, E. H. (1984). *Computational methods for partial differential equations*, John Wiley, New York.
Varga, R. S. (1961). *Matrix iterative analysis*. Prentice-Hall, Englewood Cliffs.
Vemuri, V., and Karplus, W. J. (1981). *Digital computer treatment of partial differential equations*, Prentice-Hall, Englewood Cliffs.
Volk, W. (1980). *Error estimates for some interpolation problems with splines*, Report HMI-B328, Hahn-Meitner-Institut für Kernforschung, Berlin.
Waldvogel, J. (1984). *ZaMP*, **35** 780–789.
Walter, W. (1972). *Gewöhnliche Differentialgleichungen*, Springer, Berlin.
Werner, H., and Schaback, R. (1972). *Praktische Mathematik II*, Springer, Berlin.
Werner, W. (1984). *Math. Comput.*, **43**, 205–217.
Wilkinson, J. H. (1962). *Numer. Math.*, **4**, 296–300.
Wilkinson, J. H. (1965). *The algebraic eigenvalue problem*. Clarendon Press, Oxford.
Wilkinson, J. H. (1968). *Lin. Alg. and Its Appl.*, **1**, 409–420.
Wilkinson, J. H. (1969). *Rundungsfehler*, Springer, Berlin.
Wilkinson, J. H., and Reinsch, C. (ed.) (1971). *Handbook for automatic computation. Vol. II, Linear Algebra*, Springer, Berlin.
Willers, Fr. A. (1957). *Methoden der praktischen Analysis*. 3. Aufl. de Gruyter, Berlin.
Williams, W. E. (1980). *Partial differential equations*, Clarendon Press, Oxford.
Winograd, S. (1978). *Math. Comput.*, **32**, 175–199.
Young, D. M. and Gregory, R. T. (1973). *A survey of numerical mathematics*. Vol. I + II, Addison-Wesley, Reding, MA.
Zonneveld, J. A. (1964). *Automatic numerical integration. Math. Centre Tracts*, No. 8. Mathematisch Centrum, Amsterdam.
Zurmühl, R. (1965). *Praktische Mathematik für Ingenieure und Physiker*. 5. Aufl. Springer, Berlin.
Zurmühl, R., and Falk, S. (1984). *Matrizen und ihre Anwendungen. Teil I: Grundlagen. 5. Aufl.*, Springer, Berlin.
Zurmühl, R., and Falk, S. (1986). *Matrizen und ihre Anwendungen. Teil 2: Numerische Methoden. 5. Aufl.*, Springer, Berlin.

# Index

absolutely stable  417, 472, 476, 482, 484
Adams–Bashforth method  399, 410, 419
Adams–Bashforth–Moulton  402
Adams–Moulton method  401, 410, 420
adaptive quadrature  340
Aitken algorithm  109
Aitken's $\delta^2$ process  199
alternante  81
alternating directions  485

backward error analysis  29
backward substitution  4, 15, 33, 39, 138, 222, 265, 296, 301, 312, 327, 476, 484
Bairstow method  233
Banach space  192
band matrix  35, 439, 450, 499
bandwidth  35, 456, 461, 499
barycentric formula  87
basis function  491, 494, 498
basis variable  56
bidiagonal matrix  38, 319
binomial coefficients  90
bisection method  203
bit reversal  166
boundary condition  431, 468, 479, 482, 488

Cauchy condition  432, 439, 443, 468, 481, 482, 490

cellar row  44, 59
central difference quotient  93, 433, 480
characteristic equation  409, 418
characteristic polynomial  238, 265, 407
Chebyshev abscissae  172, 179
Chebyshev approximation  78
Chebyshev polynomials  99, 143, 169, 178
Cholesky decomposition  33, 35, 137, 296, 304
Cholesky method  33, 136, 295, 437, 499
Clenshaw's algorithm  178
companion matrix  233
compatible  22
compilation process  499
condition number  25, 131, 240, 298, 457
consistent  21, 376, 407
continued fraction  120
contraction  192, 195, 210, 217
convergence  195
convergence factor  196, 222
convolution  157, 176
Cooley  164
coordinate shift  69, 72
Crank–Nicolson method  475, 480, 484
cubic convergence  267, 289
cubic interpolation  98
cubic spline function  128, 133
cycle  64, 250
cyclic Jacobi method  250

decomposition  6, 15, 32, 222
deflation  228
degeneration  58, 63
degree of precision  351, 362
dependent variables  42
diagonal dominant  10, 130, 134, 136, 437, 476, 484
diagonal strategy  9, 38, 130, 134, 450, 463, 476
difference equation  408, 434, 459, 470, 480
difference quotient  92, 433
difference scheme  104
differential equation  371
diffusion  468, 479
direction of descent  324
Dirichlet condition  432, 468, 481, 482, 491, 499
discrete Fourier transformation  155, 160
discrete orthogonality  149, 180
discretization  432
divided difference  102

eigenvalue  238, 366, 423, 455, 483
eigenvector  238, 290, 423
element vector  497, 498
elements  490
elimination  3
elliptic equation  431, 489
energy method  488
England  391
envelope  306, 499
equidistant  89, 97, 104, 372
error equations  294, 309, 320
error estimate  194, 203
essential restriction  52
Euler–MacLaurin formula  336
Euler's method  372, 417, 472
exchange step  42
explicit shift  276
extrapolation  94, 111, 123, 337, 379
extremal principle  126, 488
extremal problem  49

fast Fourier transform  159, 164, 176
fast Givens transformation  260, 262, 284, 305
feasible corner  55
feasible domain  51, 53
feasible point  51

feasible solution  70
Fehlberg  391
final equation  3
finite difference method  432
finite element method  487
first order  95
fixed point  191, 216
fixed point iteration  191, 216, 219, 382, 402
fixed point theorem  192, 216
forward substitution  6, 15, 33, 39, 137, 222, 268, 296, 476, 484
Fourier coefficients  146, 174
Fourier polynomial  146, 148, 150
Fourier series  143, 146
free variable  68

Gaussian algorithm  4, 31, 33, 38, 130, 219, 222, 268, 291, 450, 463, 473
Gaussian quadrature  187, 348, 364
Gauss–Newton method  321
Gershgorin's theorem  131
Givens method  258, 327
Givens rotation  254, 262
global discretization error  377, 411, 472, 476
grid point  432, 468

heat conduction  468, 477
Helmholtz equation  490
Hermite interpolation  107, 399
Hermitian operator  465
Hessenberg matrix  254, 271
Hessian matrix  218, 295
Heun's method  383, 387
Horner scheme  226, 233
Householder matrix  309
Householder method  312, 327
Householder transformation  284
Hyman's method  265
hyperbolic equation  431

identity matrix  17
ill-conditioned  269, 414
implicit deflation  229, 267
implicit method  382, 392, 402, 404, 417, 475, 480
implicit spectral shift  284, 287
improved polygonal method  380
independent variables  42
inherent instability  414

*Index* 515

initial boundary value problem 468, 481
initial condition 372, 395, 468, 479, 482
interpolation method 213
interpolation polynomial 85, 150, 178, 213, 351
interpolation quadrature formula 351, 363, 398, 497
inverse divided difference 118
inverse interpolation 114, 216
inverse vector iteration 268, 291
inversion 17, 44
involutory 309
irregular interior point 439, 460
iterative improvement 17, 228
iterative method 191, 206, 215, 499

Jacobi 244
Jacobi method 246, 250
Jacobi rotation 244, 254
Jacobian 217, 219, 321, 425, 494

Kutta's method 387

Lagrange interpolation 86
Lagrange polynomials 86, 351, 363, 398
Laguerre's method 267
Laplace equation 431, 490
least squares method 152, 294
left triangular matrix 5, 32, 38, 137, 315
Legendre polynomials 143, 182, 362, 365
linear convergence 196, 218, 224, 248, 379
linear equations 1, 436, 480, 492
linear interpolation 98, 460
linear multistep method 404, 418
linear program 50
local discretization error 376, 399, 405, 411, 451, 471, 475
Lotka–Volterra 397
lower triangular matrix 5

Marquardt's method 325
mass element matrix 496, 498
matrix inversion 17, 44
matrix norm 21
maximal column pivoting strategy 11
mean square 186
midpoint sum 332, 358
Milne–Simpson method 406, 410, 421
minimax property 178

Muller's method 213
multistep method 398

natural condition 128, 489
natural cubic spline function 128
natural matrix norm 22
natural quintic spline function 128
Neumann condition 432, 439, 470, 482, 490
Neville algorithm 109, 337
Newton–Cotes quadrature 353, 356, 357
Newton–Gregory interpolation 105
Newton interpolation 100, 461
Newton's method 209, 218, 229, 265
Newton's successive relaxation 224
nodal point 491
nodal variable 491
node 331, 351, 362
nonbasis variable 56
nonlinear equations 191, 216, 231
nonlinear error equations 320
nonlinear successive relaxation 223
nonsingular 1
nonsymmetric 441, 444, 450
norm 20
normal equations 295
normal triangle 493
numerical differentiation 90
Nyström method 406, 410, 421

objective function 50
one step method 372, 415
operations research 49
operator form 434, 457, 459
optimization 50
order 379, 400, 404, 466
order of convergence 197, 202, 204, 210, 213, 219, 222, 233, 413, 453
orthogonal basis 143
orthogonal functions 144
orthogonal matrix 241, 242, 270, 300, 309, 315
orthogonal polynomials 143, 169, 172, 183
orthogonal transformation 244, 255, 300
orthonormal vectors 241, 248

parabolic equation 431, 468, 479, 482
partial differential equation 430

Peaceman and Rachford 484
periodic function 143
periodic spline interpolation 136
permutation matrix 8
pivotal column 43
pivotal element 3, 43, 56
pivotal row 43
pivotal strategy 9
Poisson equation 431, 434, 457, 490
Poisson's formula 346
polygonal method 372
positive definite 30, 134, 136, 295, 304, 437, 473, 499
predictor 383
predictor corrector method 383, 402, 420
principal axes 241

QR algorithm 275, 366
QR decomposition 270, 302
QR double step 281
QR transformation 271
quadratic convergence 197, 210, 218, 249, 251, 267, 276
quadratic divisor 230
quadratic form 30
quadratic interpolation 98, 461
quadratic mean 144, 186
quadrature 330
quadrature formula 331, 351, 362
quasi-triangular 272

rational interpolation 116
recursion formula 172, 184
reflection 309
region of absolute stability 416, 419
regula falsi 204
regular interior point 433, 457, 459
relative maximal column pivoting strategy 14, 39, 268
relaxation factor 224
residual 78, 152, 294
residual vector 17
reversibility 58
Richardson extrapolation 379
Richardson method 470, 482, 483
right triangular matrix 4, 38, 270, 291, 300, 315, 327
Romberg scheme 114, 337, 348
rotation matrix 242, 315
Runge 100, 155, 158, 176

Runge–Kutta method 383, 385, 390, 392, 396, 415, 417, 423

scaling 12
Schur's theorem 272
secant method 206
second order 94
sensitivity 26, 239, 299, 414
similar matrices 25, 241, 271
similarity transformation 242, 254, 262
simplex algorithm 58
simplex scheme 59
simplified Newton's method 213, 222
Simpson's quadrature formula 339, 353, 388, 406, 498
singular value 317
singular value decomposition 316
singular vector 317
SOR–Newton method 224
sparse matrix 306, 439, 499
spectral norm 24, 453
spectral radius 24, 217
spectral shift 276
spline function 126, 359
stability interval 417, 420
stationary iterative method 192
stationary step 64
steepest descent 324
Steffensen method 201
step size control 388
stiff differential equation 394, 424
stiffness element matrix 496, 500
stiffness matrix 499
subordinate norm 22
successive relaxation 223
summed trapezoidal rule 176
superlinear convergence 207, 215, 337
support abscissae 85, 150, 179, 372
support coefficients 87
support ordinates 85
symmetric matrix 30, 130, 134, 136, 241, 258, 295, 366, 437, 454, 461, 499

Taylor series method 373
test initial value problem 408, 415
Thiele's continued fraction 120
trapezoidal approximation 331
trapezoidal method 382, 417, 476
trapezoidal rule 148, 382
triangular decomposition 32, 38

## Index

triangulation 490
tridiagonal 37, 130, 134, 258, 271, 286, 366, 454, 473, 475, 476, 484
Tukey 164
two-phase method 73

unitary matrix 280
upper triangular matrix 4, 265

variational problem 126, 487, 500
vector iteration 275
vector norm 20

weights 331, 351, 362

zero-stable 410
zeros of polynomials 224
Zonneveld 392

$A = A$ IDENTITY
$A = B \therefore B = A$ SYMMETRY
$A = B, B = C, \therefore A = C$ TRANSITIVITY

$Ax + B = C$
$Dx + E = F$

$$\begin{bmatrix} A & B & C \\ D & E & F \end{bmatrix} = \begin{bmatrix} D & E & F \\ A & B & C \end{bmatrix}$$